# Introduction to Galaxy Formation and Evolution

Present-day elliptical, spiral and irregular galaxies are large systems made of stars, gas and dark matter. Their properties result from a variety of physical processes that have occurred during the nearly 14 billion years since the Big Bang.

This comprehensive textbook, which bridges the gap between introductory and specialised texts, explains the key physical processes of galaxy formation, from the cosmological recombination of primordial gas to the evolution of the different galaxies that we observe in the Universe today.

In a logical sequence, it introduces cosmology, illustrates the properties of galaxies in the present-day Universe, then explains the physical processes behind galaxy formation in the cosmological context, taking into account the most recent developments in this field. This text ends on how to find distant galaxies with multi-wavelength observations, and how to extract the physical and evolutionary properties of galaxies based on imaging and spectroscopic data.

**Andrea Cimatti** is Professor of Astrophysics at the University of Bologna. He has worked in Germany, in the USA and in Italy at the INAF Arcetri Astrophysical Observatory. His research uses the largest telescopes in space and on Earth to study galaxies and cosmology. He is one of the founders of the ESA's *Euclid* space mission. He is a recipient of the Bessel Prize of the Alexander von Humboldt Foundation, among others. He teaches courses on fundamental astronomy and on galaxy formation and evolution. He has published key papers based on observational studies of distant galaxies.

**Filippo Fraternali** is Associate Professor of Gas Dynamics and Evolution of Galaxies at the Kapteyn Astronomical Institute of the University of Groningen. He obtained his Ph.D. from the University of Bologna, where he was an assistant professor between 2006 and 2017. He did postdoctoral research in The Netherlands and was a Marie-Curie fellow at the University of Oxford. He has published extensively on various topics of galaxy formation and evolution.

**Carlo Nipoti** is Associate Professor of Astrophysics at the University of Bologna, where he received his Ph.D. in Astronomy, and he was a temporary lecturer of Theoretical Physics at the University of Oxford. He teaches courses on the physics of galaxies and on the dynamics of astrophysical systems at undergraduate and graduate levels. His research is in the field of theoretical astrophysics, with special interest in the formation, evolution and dynamics of galaxies, on which he is author of valuable papers.

# Introduction to Galaxy Formation and Evolution

## From Primordial Gas to Present-Day Galaxies

ANDREA CIMATTI, FILIPPO FRATERNALI AND CARLO NIPOTI

Shaftesbury Road, Cambridge CB2 8EA, United Kingdom

One Liberty Plaza, 20th Floor, New York, NY 10006, USA

477 Williamstown Road, Port Melbourne, VIC 3207, Australia

314–321, 3rd Floor, Plot 3, Splendor Forum, Jasola District Centre, New Delhi – 110025, India

103 Penang Road, #05–06/07, Visioncrest Commercial, Singapore 238467

Cambridge University Press is part of Cambridge University Press & Assessment, a department of the University of Cambridge.

We share the University's mission to contribute to society through the pursuit of education, learning and research at the highest international levels of excellence.

www.cambridge.org
Information on this title: www.cambridge.org/9781107134768
DOI: 10.1017/9781316471180



First published 2020 (version 2, November 2022)

Printed in the United Kingdom by TJ Books Limited, Padstow Cornwall, November 2022

*A catalogue record for this publication is available from the British Library*

*Library of Congress Cataloging-in-Publication data*
Names: Cimatti, Andrea, 1964– author. | Fraternali, Filippo, 1973– author. | Nipoti, Carlo, 1975– author.
Title: Introduction to galaxy formation and evolution : from primordial gas to present-day galaxies / Andrea Cimatti, Filippo Fraternali, and Carlo Nipoti.
Other titles: Galaxy formation and evolution
Description: Cambridge ; New York, NY : Cambridge University Press, 2020. | Includes bibliographical references and index.
Identifiers: LCCN 2019020556 | ISBN 9781107134768 (hardback : alk. paper)
Subjects: LCSH: Galaxies–Formation. | Galaxies–Evolution.
Classification: LCC QB857 .C56 2020 | DDC 523.1/12–dc23
LC record available at https://lccn.loc.gov/2019020556

ISBN 978-1-107-13476-8 Hardback

Additional resources for this publication at www.cambridge.org/cimatti

# Brief Contents

# Contents

# Preface

## Why This Book?

The study of galaxy formation and evolution is one of the most active and fertile fields of modern astrophysics. It also covers a wide range of topics intimately connected with cosmology and with the evolution of the Universe as a whole. The key to decipher galaxy formation and evolution is to understand the complex physical processes driving the evolution of ordinary matter during its gravitational interplay with dark matter halos across cosmic time. The central theme is therefore how galaxies formed and developed their current properties starting from a diffuse distribution of gas in the primordial Universe. This research field requires major efforts in the observation of galaxies over a wide range of distances, and in the theoretical modelling of their formation and evolution. The synergy between observations and theory is therefore essential to shed light on how galaxies formed and evolved. In the last decades, both observational and theoretical studies have undergone rapid developments. The availability of new telescopes operating from the ground and from space across the entire electromagnetic spectrum opened a new window on distant galaxies. At the same time, major observational campaigns, such as the Sloan Digital Sky Survey, provided huge samples of galaxies in the present-day Universe with unprecedented statistics and allowed one to define the 'zero-point' for evolutionary studies. In parallel, the theoretical models experienced a major advance thanks to the improved performance of numerical simulations of galaxy formation within the cosmological framework.

The idea for this book originated from the difficulties we faced when teaching our courses. We lacked a single and complete Master-level student textbook on how galaxies formed and evolved. This textbook aims to fill a gap between highly specialised and very introductory books on these topics, and enables students to easily find the required information in a single place, without having to consult many sources.

The aim of the book is twofold. The first is to provide an introductory, but complete, description of the key physical processes that are important in galaxy formation and evolution, from the primordial to the present-day Universe. The second is to illustrate what physical and evolutionary information can be derived using multi-wavelength observations. As the research field of galaxy formation and evolution is relatively young and rapidly evolving, we do not attempt to give a complete review of all topics, but rather we try to focus on only the most solid results.

## Readership and Organisation

This textbook assumes a background in general physics at the Bachelor level, as well as in introductory astronomy, fundamentals of radiative processes in astrophysics, stellar evolution and the fundamentals of hydrodynamics. Although this book is primarily intended for students at Master degree level, it can be used as a complement to Bachelor-level courses in extragalactic astrophysics, and we think it can also be a valuable guide to PhD students and researchers.

The content of the chapters is organised as follows. After a general introduction to the field of galaxy formation and evolution (Chapter 1), the book starts with a brief overview on the cosmological framework in which galaxies are placed (Chapter 2). The aim of this chapter is to provide the reader with the key information useful for the rest of the textbook: the Big Bang model, the expansion of the Universe, redshift, the look-back time, the cosmological parameters and the matter–energy cosmic budget. Then, the book continues with a set of four chapters dedicated to the properties of present-day galaxies seen as the endpoint of the evolution that has occurred during the time frame spanned by the age of the Universe ($\approx$13.8 billion years). In particular, Chapter 3 illustrates the statistical properties of galaxies (e.g. morphologies, sizes, luminosities, masses, colours, spectra) and includes a description of active galactic nuclei. The structure, components and physical processes of star-forming and early-type galaxies are presented in Chapter 4 and Chapter 5, respectively. Chapter 4 includes also a description of our own Galaxy seen as a reference benchmark when studying the physics of star-forming galaxies from the 'inside' and with a level of detail not reachable in external galaxies. Chapter 6 deals with the influence of the environment on galaxy properties, and with the spatial distribution of galaxies on large scales. Then, Chapter 7 focuses on the general properties of dark matter halos, and their hierarchical assembly across cosmic time: these halos are crucial because they constitute the skeleton where galaxy formation takes place. Chapter 8 deals with the main 'ingredients' of galaxy formation theory through the description of the key physical processes determining the evolution of baryons within dark matter halos (e.g. gas cooling and heating, star formation, chemical evolution, feedback processes). The subsequent Chapter 9 is dedicated to the evolution of primordial baryonic matter in the early Universe, from the cosmological recombination to the formation of the first luminous objects a few hundred million years after the Big Bang, and the consequent epoch of reionisation. Chapter 10 provides a general description of the theoretical models of the formation and evolution of different types of galaxies, including an introduction to the main methods of numerical modelling of galaxy formation. Finally, Chapter 11 presents a general overview of galaxy evolution based on the direct observation of distant galaxies and their comparison with present-day galaxy types.

**References**

As of writing this book, there are tens of thousands of refereed papers in the literature on galaxy formation and galaxy evolution; not to mention several books on galaxies and cosmology available on the market. This implies that choosing the most significant references for a book like this is really challenging. The difficulty is exacerbated by the very fast evolution of this research field. For these reasons, our choice has been pragmatic and minimalistic. We excluded references before 1900, and we decided to reduce as much as possible the citations to research articles (including our own papers), unless they present a major discovery or a turning point for a given topic, or they are particularly useful for students. Instead, we much preferred to cite recent review articles because they provide an introductory and as much as possible unbiased source of information that is more suitable for students. However, also in this case, it was not feasible to cite all the reviews available in the literature. In the same spirit, the figures selected from the literature were chosen based on their clarity and usefulness to students. Finally, we also suggested a few books where readers can find more details on several topics treated in this textbook. The obvious consequence is that the reference list is unavoidably incomplete. We apologise to any author whose publications may have been overlooked with the selection approach that we adopted.

**Acknowledgements**

This book has benefited from the input of colleagues and students who have helped us in a variety of different and crucial ways. Many of the figures in this book have been produced *ad hoc* for us. We are grateful to the authors of these figures, to whom we give credit in the captions. Here we also wish to explicitly thank our colleagues who have taken the time to read parts of the text, and/or gave us comments and advice that were fundamental to improve the quality of the book. These are: Lucia Armillotta, Ivan Baldry, Matthias Bartelmann, James Binney, Fabrizio Bonoli, Fabio Bresolin, Volker Bromm, Marcella Brusa, Luca Ciotti, Peter Coles, Romeel Davé, Gabriella De Lucia, Mark Dickinson, Enrico Di Teodoro, Elena D'Onghia, Stefano Ettori, Benoit Famaey, Annette Ferguson, Daniele Galli, Roberto Gilli, Amina Helmi, Giuliano Iorio, Peter Johansson, Inga Kamp, Amanda Karakas, Rob Kennicutt, Dusan Kereš, Leon Koopmans, Mark Krumholz, Federico Lelli, Andrea Macciò, Mordecai Mac Low, Pavel Mancera Piña, Antonino Marasco, Claudia Maraston, Federico Marinacci, Davide Massari, Juan Carlos Muñoz-Mateos, Kyle Oman, Tom Oosterloo, Max Pettini, Gabriele Pezzulli, Anastasia Ponomareva, Lorenzo Posti, Mary Putman, Sofia Randich, Alvio Renzini, Donatella Romano, Alessandro Romeo, Renzo Sancisi, Joop Schaye, Ralph Schönrich, Mattia Sormani, Eline Tolstoy, Scott Tremaine, Tommaso Treu, Mark Voit, Marta Volonteri, Jabran Zahid, Gianni Zamorani and Manuela Zoccali.

# 1 Introduction

## 1.1 Galaxies: a Very Brief History

Galaxies are gravitationally bound systems made of stars, interstellar matter (gas and dust), stellar remnants (white dwarfs, neutron stars and black holes) and a large amount of dark matter. They are varied systems with a wide range of morphologies and properties. For instance, the characteristic sizes of their luminous components are from $\sim 0.1$ kpc to tens of kiloparsecs, whereas the optical luminosities and stellar masses are in the range $10^3$–$10^{12}$ in solar units. Roughly spherical halos of dark matter dominate the mass budget of galaxies. As a reference, the size of the stellar disc of our Galaxy[1] is about 20 kpc, whereas the dark matter halo is thought to be extended out to $\approx$300 kpc. The total mass of the Galaxy, including dark matter, is $\sim 10^{12}\ \mathcal{M}_\odot$, whereas the stellar and gas masses amount to only $\approx 5 \times 10^{10}\ \mathcal{M}_\odot$ and $\approx 6 \times 10^9\ \mathcal{M}_\odot$, respectively.

The discovery of galaxies (without knowing their nature) dates back to when the first telescope observations showed the presence of objects, originally called nebulae, whose light appeared diffuse and fuzzy. The first pioneering observations of these nebulae were done with telescopes by C. Huygens in the mid-seventeenth century, and by E. Halley and N.-L. de Lacaille in the first half of the eighteenth century. Interestingly, in 1750, T. Wright published a book in which he interpreted the Milky Way as a flat layer of stars and suggested that nebulae could be similar systems at large distances. The philosopher I. Kant was likely inspired by these ideas to the extent that, in 1755, he explained that these objects (e.g. the Andromeda galaxy) appear nebulous because of their large distances which make it impossible to discern their individual stars. In this context, the Milky Way was interpreted as one of these many stellar systems (island universes).

In 1771, C. Messier started to catalogue the objects which appeared fuzzy based on his telescope observations. These objects were identified by the letter M (for Messier) followed by a number. Now we know that some of these objects are located within our Galaxy (star clusters and emission nebulae; e.g. M 42 is the Orion nebula), but some are nearby galaxies bright enough to be visible with small telescopes (e.g. M 31 is the Andromeda galaxy). However, Messier did not express any opinion about the nature and the distance of these systems. Since late 1700, W. Herschel, C. Herschel and J. Herschel increased the sample of diffuse objects thanks to their larger telescopes, and classified them depending on their

[1] The terms Galaxy (with the capital G) or Milky Way are used to indicate the galaxy where the Sun, the authors and the readers of this book are located.

observed features. In 1850, W. Parsons (Lord Rosse) noticed that some of these nebulae exhibited a clear spiral structure (e.g. M 51).

Since late 1800, the advent of astronomical photography allowed more detailed observations to be performed, and these studies triggered a lively discussion about the nature of the spiral nebulae and their distance. This led to the so-called Great Debate, or the Shapley–Curtis debate referring to the names of the two astronomers who, in 1920, proposed two widely different explanations about spiral nebulae. On the one hand, H. Shapley argued that these objects were interstellar gas clouds located within one large stellar system. On the other hand, according to H. Curtis, spiral nebulae were external systems, and our Galaxy was one of them. Clearly, this debate involved not only the very nature of these objects, but also the size and the extent of the Universe itself. The issue was resolved soon after with deeper observations. In 1925, using the 100-inch telescope at Mount Wilson Observatory, E. Hubble identified individual stars in M 31 and M 33 and discovered variable stars such as Cepheids and novae. In particular, Cepheids are pulsating giant stars that can be exploited as distance indicators. These stars are what astronomers call 'standard candles', i.e. objects whose intrinsic luminosity is known *a priori*, and that therefore can be used to estimate their distance. In 1912, it was found by H. Leavitt that the intrinsic luminosity of Cepheids is proportional to the observed period of their flux variation. Thus, once the period is measured, the intrinsic luminosity is derived and, therefore, the distance can be estimated. Based on these results, Hubble demonstrated that spiral nebulae were at very large distances, well beyond the size of our Galaxy, and that therefore they were indeed external galaxies.

The term 'galaxy' originates from the Greek $\gamma\acute{\alpha}\lambda\alpha$, which means milk, and it refers to the fuzzy and 'milky' appearance of our own Milky Way when observed with the naked eye. Also external galaxies look 'milky' when observed with small telescopes. Discovering that galaxies were external systems also implied that the Universe was much larger than our Galaxy, and this was crucial to open a new window on cosmology in general. In modern astrophysics, the term 'nebula' is still used, but it refers only to objects within the interstellar medium of galaxies. Notable examples are the emission nebulae where the gas is photoionised by hot massive stars, dark nebulae which host cold and dense molecular gas mixed with interstellar dust, and planetary nebulae produced by the gas expelled by stars with low to intermediate mass during their late evolutionary phases. Since the discovery of Hubble, spiral nebulae have therefore been called spiral galaxies. In 1927–1929, based on galaxy samples for which radial velocities and distances were available, G. Lemaître and Hubble found that galaxies are systematically receding from us. In particular, their radial velocity is proportional to their distance: the farther away the galaxies, the higher the redshift of their spectral lines, and therefore the velocity at which they move away from us. This crucial discovery led to the Hubble–Lemaître law[2] which is the experimental proof that the Universe is expanding.

[2] In October 2018, the members of the International Astronomical Union (IAU) voted and recommended to rename the Hubble law as the Hubble–Lemaître law.

## 1.2 Galaxies as Astrophysical Laboratories

Present-day galaxies display a variety of properties and span a very broad range of luminosities, sizes and masses. At first sight, this already suggests that galaxy formation and evolution is not a simple process. However, the existence of tight scaling relations involving galaxy masses, sizes and characteristic velocities (e.g. the Tully–Fisher relation and the fundamental plane) indicates some regularities in the formation and assembly of these systems.

The first distinctive feature of a galaxy is its morphology. The shape of a galaxy as observed on the sky plane is a combination of the intrinsic three-dimensional (3D) structure and its orientation relative to the line of sight. Present-day galaxies show a broad range of shapes. Understanding the physical formation and evolution of the morphological types remains one of the most important, and still open, questions in extragalactic astrophysics. The first systematic study in the optical waveband dates back to 1926, when Hubble started a classification of galaxy morphologies following an approximate progression from simple to complex forms. In particular, Hubble proposed a tuning fork diagram on which the main galaxy types can be placed. Based on this classification, galaxies were divided into three main classes: ellipticals, lenticulars and spirals, plus a small fraction of irregulars. As shown in Fig. 1.1, the Hubble sequence starts from the left with the class of ellipticals (E). This class is further divided into subclasses as a function of their observed flattening.

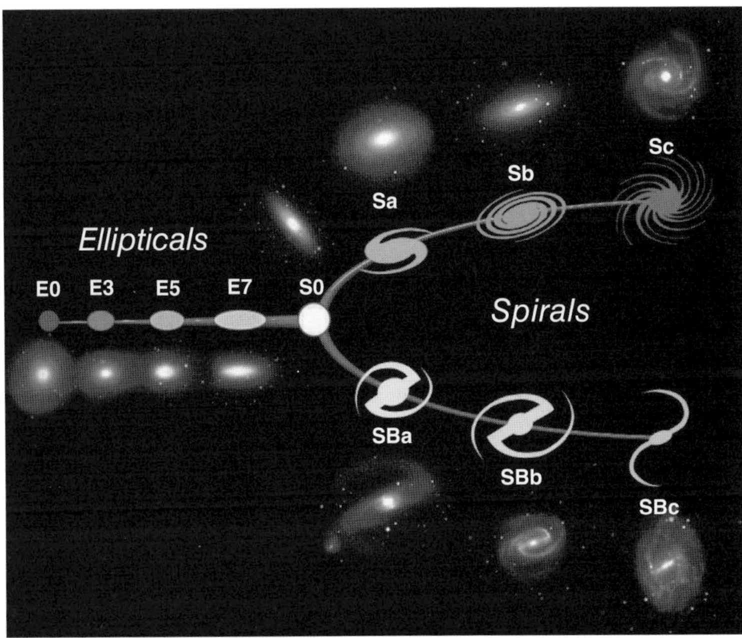


Fig. 1.1 The Hubble classification of galaxy morphology. © NASA and ESA, reproduced with permission.

Perfectly round ellipticals are called E0, whereas the most flattened are the E7. If the observed shape of these galaxies is approximated by ellipses, their flattening is related to the ellipticity $\epsilon = (a - b)/a$, where $a$ and $b$ are the observed semi-major and semi-minor axes, respectively. The number written after the letter E is the integer closest to $10\epsilon$. Proceeding beyond the E7 class, galaxies start to display morphologies with a central dominant spheroidal structure (the so-called bulge) surrounded by a fainter disc without spiral arms. These systems are classified as lenticulars (S0) and represent a morphological transition from ellipticals to spirals. Proceeding further to the right, the tuning fork is bifurcated in two prongs populated by the two main classes of spiral (S) galaxies. In both prongs, spirals have the common characteristic of having a disc-like appearance with well defined spiral arms originating from the centre and extending throughout the outer regions. The top prong includes the so-called normal spirals characterised by a central bulge surrounded by a disc. These spirals are classified Sa, Sb and Sc as a function of decreasing prominence of the bulge (with respect to the disc) and increasing importance of the spiral arms. The bottom prong includes the barred spirals (SB) which show a central bar-like structure which connects the bulge with the regions where the spiral arms begin. Moving further to the right, i.e. beyond Sc types, all galaxies not falling into the previous classes are classified as irregulars (Irr).

Subsequent studies showed that ellipticals and lenticulars are red systems, made of old stars, with weak or absent star formation, high stellar masses, with a wide range of kinematic properties (from fast to absent rotation), and preferentially located in regions of the Universe where the density of galaxies is higher. On the other side of the tuning fork, spirals are bluer, have ongoing star formation, larger fractions of cold gas, stellar populations with a wide range of ages, kinematics dominated by rotation, and are found preferentially in regions with lower density of galaxies. Given the wide range of properties displayed by present-day galaxies, it is crucial to investigate the physical processes which led to their formation and evolution. The study of galaxies involves a wide range of galactic and sub-galactic scales ranging from hundreds of kiloparsecs down to sub-parsec level depending on the processes that are considered. In this respect, galaxies can be seen as 'laboratories' where a plethora of astrophysical processes can be investigated.

## 1.3 Galaxies in the Cosmological Context

Besides their role as astrophysical laboratories, galaxies can be placed in a broader context and exploited as point-like luminous 'particles' which trace the distribution of matter on scales much larger than the size of individual galaxies. This distribution, called large-scale structure, is the 3D spatial distribution of matter in the Universe on scales from tens of megaparsecs to gigaparsecs. Due to its characteristic shape, the large-scale structure is also called the cosmic web. The study of galaxies on these large scales has deep connections with cosmology, the branch of physics and astrophysics that studies the general properties, the matter–energy content and the evolution of the Universe as a whole. Modern cosmology rests on two major observational pillars. The first is the expansion of the Universe (Hubble–Lemaître law). The second is the nearly uniform radiation background observed in the microwaves, the cosmic microwave background (CMB), discovered in 1965 by A. Penzias and R. Wilson. The spectrum of the CMB is an almost perfect black body with a temperature $T \simeq 2.726$ K. The CMB radiation is interpreted as the thermal relic of the Big Bang that occurred about 13.8 Gyr ago when the Universe originated as a hot plasma with virtually infinite temperature and density. Although the detailed properties of the Big Bang itself are unknown, an expanding Universe can be described using the Einstein equations of general relativity together with the Friedmann–Lemaître–Robertson–Walker metric. The current view of the Universe relies on the Big Bang model and on the so-called ΛCDM cosmological framework. In this scenario, also known as standard cosmology, the Universe is homogeneous and isotropic on large scales, and it is made of ordinary matter (i.e. baryonic matter), neutrinos, photons and a mysterious component of cold dark matter (CDM). CDM is dominant with respect to ordinary matter as it amounts to about 84% of the whole matter present in the Universe. CDM is thought to be composed of non-relativistic massive particles that interact with each other and with ordinary matter only through the gravitational force. However, the nature and individual mass of these particles are currently unknown. For this reason, this is one of the main open questions of modern physics. In addition, a further component, called dark energy, is required to explain the current acceleration of the Universe expansion that S. Perlmutter, B. Schmidt and A. Riess discovered in 1998 exploiting distant supernovae as standard candles. In standard cosmology, the space-time geometry is flat (Euclidean), and dark energy is assumed to be a form of energy density (known as vacuum energy) which is constant in space and time. This form of dark energy is indicated by $\Lambda$ and called the cosmological constant. However, other possibilities (e.g. a scalar field) are not excluded, and the nature of dark energy is currently unknown. This represents another big mystery of modern physics.

The ΛCDM model can be fully described by a small number of quantities called cosmological parameters which measure the relative fractions of the matter–energy components and constrain the geometry of the Universe. The ΛCDM model is now supported by a variety of cosmological probes such as the CMB, the Hubble expansion rate estimated from Type Ia supernovae, the properties of the large-scale structure and the mass of galaxy clusters. If the ΛCDM model is assumed, the observational results constrain the cosmological parameters with extremely high accuracy. In particular, in the present-day Universe, dark energy contributes $\approx$70% of the matter–energy budget of the Universe, whereas the contributions of dark matter and baryons amount to $\approx$25% and $\approx$5%, respectively, plus a negligible fraction of photons and neutrinos. The relative uncertainties on these fractions are very small (sub-per cent level). For this reason, modern cosmology is also called precision cosmology. However, it remains paradoxical that the nature of dark matter and dark energy, which together make 95% of the Universe, is still completely unknown despite the accuracy with which we know their relative importance.

Once the cosmological framework has been established, present-day galaxies can be seen as the endpoints that enclose crucial information on how baryonic and dark matter evolved as a function of cosmic time. In this regard, galaxies are also useful to test the ΛCDM cosmology. For instance, the current age of the oldest stars in galaxies should not be older than the age of the Universe itself, estimated to be about 13.8 Gyr based on observational cosmology. This key requirement is met by the age estimates of the Galactic globular clusters based on the Hertzsprung–Russell diagram.

## 1.4 Galaxies: from First Light to Present-Day Galaxies

Galaxies were originated from the primordial gas present in the early Universe. Fig. 1.2 shows a sketch of the main cosmic epochs that are treated in this textbook. Soon after the Big Bang, the baryonic matter was fully ionised and coupled with a 'bath' of black-body photons. In this photon–baryon fluid, the Universe was opaque because photons could not propagate freely due to the incessant Thomson scattering with free electrons. As the Universe expanded, its temperature and density gradually decreased and, about three minutes after the Big Bang, the nuclei of elements heavier than hydrogen (basically only helium and lithium) formed through a process called primordial nucleosynthesis. About 400 000 years after the Big Bang, the temperature and density dropped enough to allow lithium, helium and hydrogen to gradually recombine with electrons and form neutral atoms. This phase is called cosmological recombination. This is the epoch when the Universe became transparent because photons started to propagate freely thanks to the negligible role of Thomson scattering. The CMB radiation observed in the present-day Universe was originated in this phase and therefore represents the earliest possible image of the Universe. After recombination, the Universe was filled of dark matter and diffuse neutral gas composed of hydrogen, helium and lithium only. It is from the evolution of this primordial gas that the first luminous objects and galaxies began to form.

Understanding galaxy formation and evolution is a complex task because it involves several physical processes, their mutual interactions, and their evolution as a function of cosmic time. This is one of the most multi-disciplinary areas of astrophysics as it requires

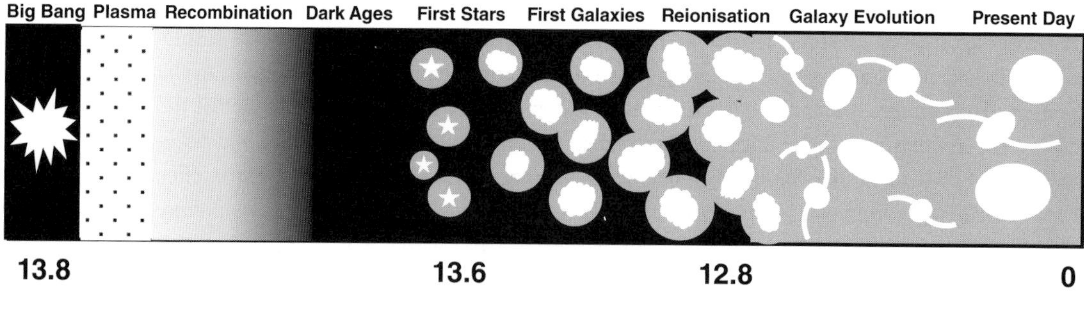


Fig. 1.2 A sketch of the main epochs which characterised the evolution of the Universe, starting from the Big Bang. After the formation of the first stars and galaxies, galaxies followed different evolutionary paths which led to the assembly of the galaxy types that we observe in the present-day Universe.

the cross-talk among a wide range of fields such as cosmology, particle physics (including dark matter) and the physics of baryonic matter. Galaxy formation and evolution is also a relatively young research field because galaxies were recognised as such only about a century ago, and their observation at cosmological distances became possibile only in the mid-1990s thanks to the advent of ground-based 8–10 m diameter telescopes in the optical and near-infrared spectral ranges, in synergy with the *Hubble Space Telescope* (*HST*).

The first step in the study of galaxy formation and evolution requires the definition of a cosmological framework (currently the ΛCDM model) within which galaxies form and evolve. The second step is to include the formation and evolution of dark matter halos which will host the first luminous objects and galaxies. In the ΛCDM model, dark matter halos are the results of the gravitational collapse of CDM in the locations where the matter density is high enough to locally prevail over the expansion of the Universe. As a matter of fact, the competition between the expansion of the Universe and gravity is one of the key processes in galaxy formation. On the one hand, if we take a large volume of the Universe at a given time, the mean matter density decreases with increasing cosmic time due to the expansion of the volume itself. On the other hand, the masses present in the same volume attract each other due to the reciprocal gravitational forces. In the early Universe, the typical masses of these halos were small, but they subsequently grew hierarchically with cosmic time through the merging with other halos and with the accretion of diffuse dark matter. Part of the gas is expected to follow the gravitational collapse of dark matter halos, and then to settle into their potential wells. The possibility to form a galaxy depends on whether this gas can have a rapid gravitational collapse. First of all, gravity must prevail over the internal pressure of the gaseous matter. However, this is not sufficient because the temperature rises as soon as the contraction proceeds. Gas heating is the enemy of galaxy formation because it increases the internal pressure and hampers gravitational collapse. This is why the second key requirement for galaxy formation is that gas cooling prevails over heating. Gas cooling can be produced by the emission of continuum radiation and spectral lines. The emitted photons abandon the gas cloud, carrying energy away, and therefore making the gas cooler and more prone to further gravitational collapse.

The cosmic epoch before the formation of the first collapsed objects (known as first stars or Population III stars) is named the dark ages because the Universe was made only of neutral gas, and luminous sources were completely absent (Fig. 1.2). We think that Population III stars began to form about 100 million years after the Big Bang from the collapse of pristine gas (H, He, Li) within dark matter halos with masses around $10^6\ \mathcal{M}_\odot$. At these early epochs, the main radiative coolants of the gas were primordial molecules such as LiH, HD and $H_2$ previously formed through gas-phase chemical reactions. This collapse led to the formation of protostellar objects and the subsequent ignition of the first thermonuclear reactions in the cores of Population III stars. When these systems started to shine, their strong ultraviolet radiation photoionised the surrounding gas. This was the beginning of the reionisation era. Population III stars ended their life very rapidly and vanished with the expulsion of most of their gas from their dark matter halos by violent supernova explosions. Thus, having lost most of the initial gas, these halos could not host further episodes of star formation. It is therefore thought that the formation of the first galaxies occurred later (a few hundred million years after the Big Bang) in larger dark matter halos with masses around $10^8\ \mathcal{M}_\odot$. These objects are called galaxies because they were massive enough to gravitationally retain a substantial fraction of the gas to prolong star formation without losing and/or heating it excessively due to supernova explosions.

After these early phases, galaxy formation proceeded following a wide range of evolutionary paths depending on the local conditions, the properties of the gas and the interactions with other systems (e.g. merging of their host halos). This is why the full understanding of galaxy formation and evolution is complex and requires a self-consistent treatment of the physical processes of baryonic matter (gas, stars and dust), their kinematics, their evolution within an expanding Universe, and the gravitational interactions with the dark matter component. The physics of baryonic matter is particularly complicated as it involves a variety of ingredients such as radiative processes, multi-phase gas physics and dynamics, gas cooling and heating, radiative transfer, star formation, stellar evolution, metal enrichment and feedback. Moreover, galaxy evolution involves the formation of supermassive black holes, the associated accretion of matter, and also the consequent feedback processes on the surrounding environment. A further complication is that all these processes and their evolution must be investigated on very wide ranges of spatial scales (from sub-parsec to megaparsec) and timescales, say from the lifetime of the most massive stars ($\sim 10^6$ yr) to the age of the Universe ($\sim 10^{10}$ yr). Fig. 1.3 illustrates the main ingredients that need to be included for the physical description of galaxy formation and evolution.

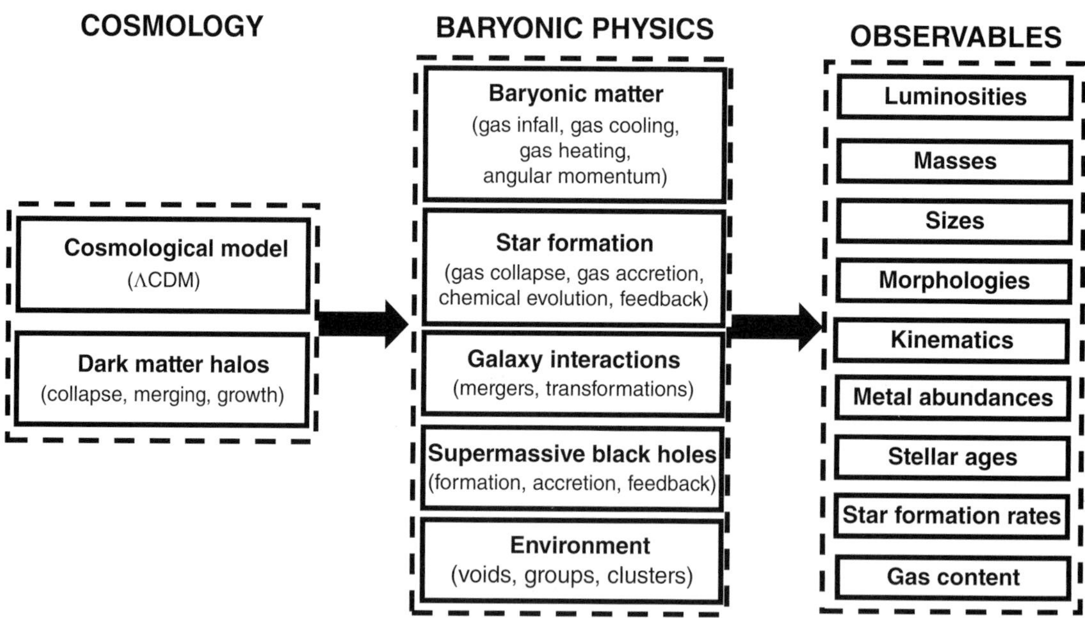


Fig. 1.3 The main ingredients of models of galaxy formation and evolution. *Left*. The cosmological model and the properties of dark matter halos define the 'skeleton' within which galaxies form and evolve. *Centre*. The main processes that drive the evolution of baryonic matter and galaxy formation. *Right*. The predicted properties of galaxies that are compared with the observations to verify the reliability of theoretical models.

## 1.5 Galaxies: Near and Far, Now and Then

Given the above complexity, how can we study galaxy formation and evolution? One approach is through theoretical models which describe coherently the physical processes involved in the formation of galaxies and their subsequent evolution from the smallest to the largest scales. In these models, the $\Lambda$CDM cosmology framework provides the initial conditions (e.g. the dark and baryonic matter fractions, the expansion rate of the Universe as a function of time, the properties of CDM halos and the hierarchical evolution of their masses). Once the cosmological framework is defined, galaxies can be modelled with two main methodologies. The first is based on cosmological hydrodynamic simulations, which follow as much as possible self-consistently the evolution of gas, star formation and feedback processes within dark matter halos. These simulations are very time consuming. This implies that sub-galactic scales can be simulated at the price of not covering large volumes of the Universe due to the limited computational resources. The second approach, called semi-analytic, consists in treating the physics of baryonic matter with a set of analytic prescriptions that, combined with the theoretically predicted evolution of dark matter halos, are tuned to reproduce the observed properties of present-day galaxies. The semi-analytic approach is cheaper from the computational point of view and therefore allows one to simulate large volumes of the Universe up to gigaparsec scales. However, the price to pay is that only the global properties of galaxies can be studied, and limited spatially resolved information is available. For these reasons, the two methods are complementary to each other. A further possibility is to perform analytic/numerical modelling of specific processes which take place within galaxies. An example is given by the chemical evolution models applied to the Milky Way.

The other approach to study galaxy formation and evolution is complementary to the theoretical modelling, and consists in the direct observation of galaxies in order to obtain data (images and spectra) from which the physical and structural properties can be extracted. A first possibility is the so-called archaeological approach where present-day galaxies are exploited as 'fossils'[3] from which it is possible to reconstruct their past history based on what is observed today. For instance, the ages and metal abundances of the stellar populations present in a galaxy allow us to infer how star formation and the enrichment of heavy elements evolved as a function of cosmic time. With this approach, the most reliable results are obtained when the stars within a galaxy can be observed individually and therefore can be placed on the Hertzsprung–Russell diagram. Unfortunately, with the current telescopes, this can be done only within the Milky Way and for galaxies in the Local Group, a $\approx$1 Mpc size region where the Galaxy is located together with its neighbours. The study of our Galaxy is so important as a benchmark of galaxy evolution studies that the *Gaia* space mission has been designed to obtain distances and proper motions of more than a billion stars, with radial velocity measurements for a fraction of them. *Gaia* allowed us to derive a kinematic map of our Galaxy that is essential to investigate its formation and

[3] As present-day galaxies are considered 'fossils', the archaeological approach should be more appropriately called 'palaeontological'.

evolution. Beyond the Local Group, galaxies become rapidly too faint and their angular sizes are too small to observe their stars individually. In these cases, one has to rely on the 'average' information that can be extracted from the so-called integrated light, i.e. the sum of the radiation emitted by the entire galaxy (or by a region of it).

Besides the archaeological studies in the present-day Universe, galaxy formation and evolution can also be investigated with the so-called look-back approach. This consists in the observation of galaxies at cosmological distances. Since light travels at a finite speed, the photons emitted from more distant galaxies reach us after a longer time interval. This means that distant galaxies appear today to us as they were in the past. Thus, it is possible to observe directly the evolution of galaxy properties if we observe galaxies at increasing distances. The fundamental assumption that makes the look-back time approach possible is that the Universe is homogeneous on large scales, so the global properties of the galaxy population on sufficiently large volumes are independent of the position in the Universe. This implies that galaxies in the local volume, in which our Galaxy is located, are representative of the general population of present-day galaxies. Similarly the galaxies observed in a distant volume are assumed to be representative of the past population of galaxies. For instance, if we want to investigate the evolution of spiral galaxies, we need to observe samples of this type of galaxies at increasing distances (i.e. larger look-back times) and to study how their properties (e.g. size, rotation velocity, mass, star formation) change with cosmic time. With this approach, it is truly possible to trace the detailed evolution of galaxies billions of years ago.

The archaeological and look-back approaches are complementary to each other, and their results are essential to build theoretical models and verify their predictions. However, in both cases multi-wavelength data are needed to provide a complete view of galaxy properties and their evolution. The reason is that galaxies are multi-component systems which emit radiation in different regions of the electromagnetic spectrum through diverse processes. For instance, due to the typical temperatures of the stellar photospheres, the starlight is concentrated from the ultraviolet to the near-infrared. Instead, the study of the interstellar molecular gas and dust requires observations from the far-infrared to the millimetre, the atomic hydrogen must be investigated in the radio, and the hot gas and the supermassive black hole activity in the ultraviolet and X-rays. The multi-wavelength approach is limited by the terrestrial atmosphere which is opaque and/or too bright in several spectral ranges. Ground-based telescopes can observe only in the optical, near-infrared and in a few transparent windows of the submillimetre, millimetre and radio. The other spectral ranges are accessible with space-based telescopes. The major advance in multi-wavelength studies of galaxy evolution at cosmological distances became possible thanks to the concurrence of ground- and space-based telescopes which, for the first time, allowed the identification of galaxies at cosmological distances. In the realm of space telescopes, the main contributions to galaxy evolution studies have come from the *Chandra X-ray Observatory*, *XMM-Newton* (X-rays), *Galaxy Evolution Explorer* (*GALEX*; ultraviolet), *HST* (optical/near-infrared), *Spitzer* (mid-infrared) and *Herschel* (far-infrared). In ground-based observations, the look-back approach became a reality with the advent of the 8–10 m diameter Keck telescopes and the Very Large Telescope (VLT) operating since the mid-1990s in the optical and near-infrared, followed by other facilities of comparable size (Gemini, Subaru, Gran Telescopio Canarias and the South African Large Telescope). The James Clerk Maxwell Telescope (JCMT), the telescopes of the Institut de Radioastronomie Millimétrique (IRAM) (such as the NOrthern Extended Millimeter Array; NOEMA), the Atacama Large Millimetre Array (ALMA) and the Karl G. Jansky Very Large Array (VLA) were essential in opening new windows on galaxy evolution at submillimetre, millimetre and radio wavelengths.

The multi-wavelength data provided by these facilities allow us to study galaxy evolution. Imaging observations are crucial to study the morphology and structure of galaxies and their relations with their physical properties. Spectroscopy provides information on the stellar populations, interstellar matter and the presence of supermassive black holes through the analysis of continuum spectra and spectral lines. Furthermore, the Doppler effect allows us to derive the galaxy kinematic properties, to measure dynamical masses and to study scaling relations. Last but not least, the collection of large galaxy samples over wide sky areas allows us to use galaxies as luminous markers to trace the spatial distribution of the underlying dark matter and to exploit them as cosmological probes.

## 1.6 Galaxies: the Emerging Picture and the Road Ahead

What is the picture emerging from the synergy of observations and theory? Most studies seem to converge on a scenario in which the evolution of galaxies is driven across cosmic times by the so-called baryon cycle. Galaxies are thought to accrete gas from the surrounding environment and gradually convert it into stars. The cooling and condensation of neutral hydrogen, and its conversion into molecular hydrogen to fuel star formation, are therefore key processes driving galaxy evolution. In this picture, galaxies grow mainly through gas accretion from the intergalactic medium, while there is a complex equilibrium between gas inflow, the conversion of the available gas reservoir into stars, and gas ejection and heating by feedback processes. In parallel, supermassive black holes form at the centres of galaxies, and trigger the temporary phase of an active galactic nucleus (AGN) whenever the accretion of cold material is efficient enough. Part of the stellar mass is lost by stars during their evolution through winds, planetary nebulae, novae and supernovae. This ejected mass seeds the interstellar medium with metals, molecules and dust grains, while starburst winds and jets from AGNs provide feedback and launch gas outflows. Metal-enriched and pristine halo gas eventually cools and accretes onto the disc to form new stars and feed the central black hole, starting the cycle again. A complex interplay is therefore expected among these processes as a function of galaxy properties, environment and cosmic time. Hence, understanding the evolution of the baryon cycle has become a key question that must be addressed to shed light on the critical steps of galaxy formation and evolution.

Despite the major progress in this research field, the overall picture is still largely incomplete, and several key questions are still open. However, new facilities operating in space such as the *James Webb Space Telescope* (*JWST*), *Euclid*, the *Wide Field Infrared*

*Survey Telescope* (*WFIRST*) in the optical/infrared, and the *Athena X-ray Observatory* in the X-rays have been designed to open new windows through the identification and multi-wavelength studies of galaxies across the entire range of cosmic times since the end of the dark ages. In this landscape, a key role is played by the synergistic studies done with the new generation of gigantic telescopes on the ground such as the Extremely Large Telescope (ELT), the Giant Magellan Telescope (GMT) and the Thirty Meter Telescope (TMT) in the optical/near-infrared, and the Square Kilometre Array (SKA) in the radio. In addition, the development of new numerical models and improved supercomputing facilities allow us to perform new simulations that are essential to investigate how the complex physics of baryonic matter and its interplay with the dark matter halos drove the evolution of different galaxy types as a function of cosmic time.

# 2 The Cosmological Framework

In this chapter we introduce a few fundamental concepts of cosmology that represent the framework for the study of galaxy formation and evolution. For a general introduction to cosmology we refer the reader to specialised textbooks such as Ryden (2017).

## 2.1 The Expanding Universe

The fundamental assumption of the cosmological model is the **cosmological principle**: on sufficiently large scales the Universe is **homogeneous** and **isotropic**. Homogeneity means that the Universe looks the same at any position. Isotropy means that there are not preferred directions: at any position the Universe looks the same in all directions. The cosmological principle is observationally supported by the high degree of isotropy of the cosmic microwave background (§2.4) and by the finding that the observed distribution of galaxies appears homogeneous and isotropic on scales larger than about 100 Mpc. On smaller scales the distribution of matter is neither homogeneous nor isotropic, but in practice we can consider the dynamics of a model Universe that is perfectly homogeneous and isotropic as a framework in which it is possible to study the formation and evolution of structures.

### 2.1.1 Comoving Observers and Scale Factor

In order to study the Universe on large scales it is useful to introduce the concept of **comoving observer** (or **fundamental observer**), that is, a point in the Universe whose motion is negligibly influenced by the local distribution of matter. Comoving observers trace the large-scale dynamics of the isotropic and homogeneous Universe. Thanks to the symmetry implied by the assumption of homogeneity and isotropy, it is possible to label the cosmological events with a single time coordinate $t$. It follows from the cosmological principle that if, at the present time $t_0$, any two comoving observers are separated by a distance $r(t_0)$, at a time $t$ in the past they were separated by $r(t_0)a(t)$, where $a(t)$ is a dimensionless function of time known as the **scale factor**. The usual normalisation of the scale factor, which we adopt in this textbook, is such that $a(t_0) = 1$. Here, with $r$ we have indicated the proper distance (§2.1.4), which is the separation in physical units, computed using the **physical coordinates $\boldsymbol{r}$**. It is also useful to define the **comoving coordinates** $\boldsymbol{x} = \boldsymbol{r}(t)/a(t)$. The comoving distance $x$ (§2.1.4) between two fundamental observers is constant, while the proper distance $r$ between two fundamental observers varies with time, scaling linearly with the scale factor $a(t)$. In other words, the comoving distance between two fundamental observers is their proper distance evaluated at $t_0$.

### 2.1.2 Hubble Flow

The relative speed of two fundamental observers separated at time $t$ by a distance $r(t) = xa(t)$ is

$$v(t)=\dot{r}(t)=x\dot{a}=\frac{\dot{a}}{a}r(t)=H(t)r(t), \tag{2.1}$$

where

$$H(t)\equiv\frac{\dot{a}}{a} \tag{2.2}$$

is the **Hubble parameter**. The value of the Hubble parameter at the present time $H_0 \equiv H(t_0)$ is known as the **Hubble constant**. The reciprocal of the Hubble constant $t_{\rm H} \equiv H_0^{-1}$ is the **Hubble time**, which is an *approximate* estimate of the present-day **age of the Universe**, whose exact value depends on the cosmological model. For the sake of simplicity, throughout the book we use the terms Hubble time and present-day age of the Universe interchangeably.

Eq. (2.1) evaluated at the present time $t_0$ is the **Hubble–Lemaître law** $v = H_0 r$, originally discovered by Lemaître (1927) and Hubble (1929) using measurements of distances and velocities of relatively nearby galaxies. The Hubble constant $H_0$ is empirically found to be positive, which means that, on sufficiently *large scales*, galaxies are receding from each other at speeds proportional to their separations (a phenomenon known as **Hubble flow**). We interpret this recession velocity as a signature of the fact that the Universe is currently expanding ($\dot{a} > 0$). Galaxies receding with the Hubble flow do not move with respect to the space, but are dragged along as the space expands. Given that $r$ is conveniently expressed in Mpc and $v$ in km s$^{-1}$, the Hubble constant is often parameterised as $H_0 = 100h$ km s$^{-1}$ Mpc$^{-1}$ where $h$ is the **dimensionless Hubble constant**. The most recent estimates give $h \simeq 0.7$.

On *small scales*, the relative velocities between neighbouring galaxies are due to local dynamics and *not* to the cosmological expansion. These **peculiar velocities** can be such that, in some cases, neighbouring galaxies are approaching each other. For instance, the Milky Way and Andromeda (§6.3.1) are approaching each other with a relative speed $\approx 110$ km s$^{-1}$.

### 2.1.3 Cosmological Redshift

A fundamental quantity in an expanding homogeneous and isotropic Universe is the cosmological redshift. Let us consider a photon that is emitted by a source (that we assume to be a comoving observer) at time $t$ with rest-frame wavelength $\lambda_{\rm rest}$. If the photon is observed by another comoving observer at time $t_{\rm obs} > t$, the observed wavelength of the photon $\lambda_{\rm obs}$ is given by

$$\frac{\lambda_{\rm obs}}{\lambda_{\rm rest}} = \frac{a(t_{\rm obs})}{a(t)}, \tag{2.3}$$

because between $t$ and $t_{\rm obs}$ the space has expanded by a factor $a(t_{\rm obs})/a(t)$. Specialising to the relevant case in which the photon is observed at the present time $t_{\rm obs} = t_0$, so $a(t_{\rm obs}) = 1$, we define the **cosmological redshift** of the source as

$$z \equiv \frac{\lambda_{\rm obs} - \lambda_{\rm rest}}{\lambda_{\rm rest}} = \frac{1}{a} - 1, \tag{2.4}$$

where $a = a(t)$ is the scale factor at the time of emission. In an expanding Universe, $a < 1$ and therefore $z > 0$: the radiation is shifted by the expansion towards longer wavelengths (hence the name redshift). The photon frequency $\nu$, related to the wavelength by $\nu = c/\lambda$, where $c$ is the speed of light, decreases as $\nu \propto 1/a(t)$ as the Universe expands. From eq. (2.4) it follows that the cosmological redshift $z$ and the scale factor

$$a = \frac{1}{1+z} \tag{2.5}$$

can be used interchangeably to label a given cosmic time $t$ ($a$ increases and $z$ decreases for increasing $t$: at $t = t_0$, $a = 1$ and $z = 0$). Therefore, the cosmological redshift $z$ of a distant galaxy can be determined by measuring, in its observed spectrum, the wavelengths of emission or absorption lines of known rest-frame wavelengths. Galaxies that have higher cosmological redshift are more distant and have emitted a longer time ago the photons that reach us today.

### 2.1.4 Cosmological Distances

The concept of distance between two objects is not trivial in an expanding Universe. Distances are measured between events at different times and the space-time geometry changes in between these times. Moreover, measures of distance in a curved space-time depend on the choice of the lines connecting two events. For these reasons, there are a few different possible definitions of the cosmological distance.

#### Comoving and Proper Distances

Let us consider a photon emitted by a comoving observer at time $t$ and observed by another comoving observer at time $t_{\rm obs}$. In the generic time interval between $t$ and $t + {\rm d}t$ the photon travels $c\,{\rm d}t/a(t)$ in comoving coordinates. Therefore the **comoving distance** between the source and the observer is

$$x \equiv \int_t^{t_{\rm obs}} \frac{c\,{\rm d}t'}{a(t')}. \tag{2.6}$$

By definition, the comoving distance between two fundamental observers is the *same* at *all* times. It is useful to define the comoving distance between us and an object at redshift $z$. Using ${\rm d}t = {\rm d}a/\dot{a}$ in eq. (2.6), for a photon observed at $t_0$ we get

$$x = c\int_{a(t)}^{a(t_0)} \frac{{\rm d}a'}{\dot{a}'a'} = c\int_a^1 \frac{{\rm d}a'}{Ha'^2}, \tag{2.7}$$

where $a = a(t)$ and we have used eq. (2.2). Substituting eq. (2.5) and its differential $\mathrm{d}a = -(1 + z)^{-2}\mathrm{d}z$ in eq. (2.7), we find that the comoving distance of a source at redshift $z$ is

$$x(z) = c\int_0^z \frac{\mathrm{d}z'}{H(z')}, \tag{2.8}$$

where the Hubble parameter is expressed as a function of redshift.

The **proper distance** $r$ is defined as the comoving distance (eq. 2.6) multiplied by the scale factor at the time of observation:

$$r \equiv a(t_{\mathrm{obs}})\int_t^{t_{\mathrm{obs}}} \frac{c\,\mathrm{d}t'}{a(t')}. \tag{2.9}$$

If the photon is observed at the present time ($t_{\mathrm{obs}} = t_0$), the proper distance coincides with the comoving distance, because $a(t_0) = 1$. Because of the expansion of the Universe, the proper distance between two fundamental observers was smaller in the past than it is today. At any time $t$, the **horizon distance**

$$r_{\mathrm{horizon}}(t) = a(t)\int_0^t \frac{c\,\mathrm{d}t'}{a(t')}, \tag{2.10}$$

where $t = 0$ is the time of the Big Bang (§2.3.2), is the maximum distance from which an observer can have received a signal. The horizon distance is an increasing function of cosmic time and represents a measure of the size of the observable Universe at that time ($r_{\mathrm{horizon}} \approx 14\,\mathrm{Gpc}$ at $t_0$).

## Luminosity and Angular Diameter Distances

Let us suppose that we know the intrinsic luminosity $L$ of a cosmological source and that we measure its flux $F$ at $t_0$. In analogy with non-cosmological measures (§C.2), we define the **luminosity distance** of the source $d_{\mathrm{L}}$ such that

$$F = \frac{L}{4\pi d_{\mathrm{L}}^2}. \tag{2.11}$$

The expression that gives the luminosity distance of a source as a function of the source's redshift $z$ depends on the spatial geometry of the Universe (§2.2.1). In the case of a spatially flat Universe this expression is simply

$$d_{\mathrm{L}} = (1 + z)x, \tag{2.12}$$

where $x = x(z)$ is the comoving distance of the source at redshift $z$ (eq. 2.8).

Let us assume that we know the intrinsic size $l$ of an extended cosmological source at redshift $z$ and that we measure for this source an angular size $\theta$. We define the **angular diameter distance** $d_{\mathrm{A}}$ such that

$$\theta = \frac{l}{d_{\mathrm{A}}}. \tag{2.13}$$

It can be shown that $d_{\mathrm{A}}$ and $d_{\mathrm{L}}$ are related by

$$d_{\mathrm{A}} = \frac{d_{\mathrm{L}}}{(1 + z)^2}, \tag{2.14}$$

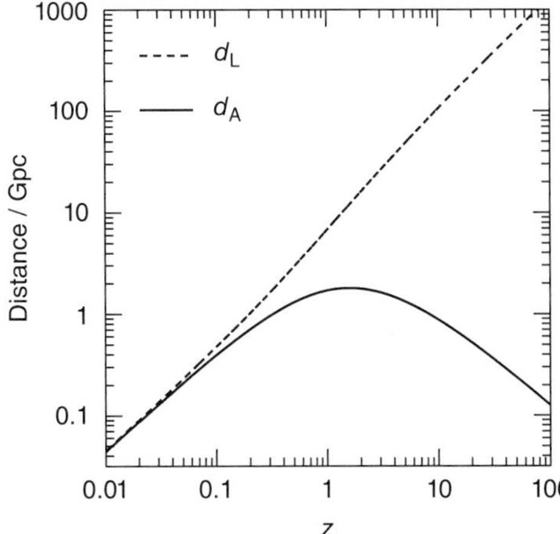


Fig. 2.1 Luminosity distance $d_L$ and angular diameter distance $d_A$ as functions of redshift for the standard cosmological model (§2.3) with parameters as in Tab. 2.1.

so for a spatially flat Universe

$$d_A = \frac{x}{1+z}, \tag{2.15}$$

where $x = x(z)$ (eq. 2.8).

We note that, while $d_L$ is a monotonically increasing function of $z$, this is not necessarily the case for $d_A$: in the currently favoured cosmological model (§2.3) $d_A$ increases with redshift at low $z$, but decreases with redshift at high $z$, peaking at $z \simeq 1.6$ (see Fig. 2.1). The non-monotonic behaviour of the angular diameter distance is a consequence of space-time curvature.

An important consequence of the redshift dependence of $d_L$ and $d_A$ is the cosmological **surface brightness dimming**: the observed surface brightness (§C.2) of a source of given luminosity and size decreases with redshift as $(1 + z)^{-4}$. To understand this, let us take for instance an extended source of given luminosity $L$ and physical area $l^2$, observed at different redshifts. Its average observed surface brightness (eqs. C.7 and C.9) is

$$\langle I \rangle \equiv \frac{4\pi F}{\theta^2}, \tag{2.16}$$

where $F$ is the flux and $\theta^2$ the angular area of the source. Combining eq. (2.16) with eqs. (2.11), (2.13) and (2.14) we get

$$\langle I \rangle = \frac{L}{l^2}\left(\frac{d_A}{d_L}\right)^2 = \frac{L}{l^2}\frac{1}{(1+z)^4}. \tag{2.17}$$

To measure the number density of astronomical objects (for instance galaxies) at cosmological distances, we need to define cosmological volumes. For a spatially flat

Universe, the infinitesimal **comoving volume** element at redshift $z$, delimited by a solid angle $\mathrm{d}\Omega$ and of depth $\mathrm{d}z$, is

$$\mathrm{d}V = x^2\,\mathrm{d}\Omega\,\mathrm{d}x = \frac{c^3}{H(z)}\left[\int_0^z \frac{\mathrm{d}z'}{H(z')}\right]^2 \mathrm{d}z\mathrm{d}\Omega, \tag{2.18}$$

where we have used eq. (2.8). Eq. (2.18) can be integrated to obtain comoving volumes for finite redshift intervals and solid angles.

## 2.2 Dynamics of the Universe

For a given cosmological model, the kinematics of the smooth Universe is fully described by the function $a(t)$ (§2.1.1), or, equivalently, $H(z)$ (§2.1.4). These functions are determined by the dynamics of the isotropic and homogeneous Universe, which is briefly discussed in this section.

### 2.2.1 Friedmann Equations

To describe the expansion of the Universe we need equations for $a(t)$. Such equations must be derived in the context of general relativity. The Universe is modelled as a continuous distribution of material, which we generally refer to as a 'fluid'. In general such a fluid is not a single component, but it is a mixture of different components, characterised by different properties (§2.2.2). The function $a(t)$ is determined by two equations, the **Friedmann equations** (Friedmann, 1922), which are derived from the **Einstein equations** (Einstein, 1916) under the assumption that the distribution of the fluid is homogeneous and isotropic.

The first Friedmann equation, also known as the **Friedmann equation** *tout court*, reads

$$\dot{a}^2 = \frac{8\pi G\rho a^2}{3} - kc^2, \tag{2.19}$$

where $G$ is the gravitational constant, $\rho$ is the mass density of the fluid and $k$ is the **curvature**[1] (with dimensions of the inverse of a length squared), related to the spatial geometry of the Universe, which is spherical for $k > 0$, flat for $k = 0$, and hyperbolic for $k < 0$. The second Friedmann equation, also known as the **acceleration equation**, is

$$\frac{\ddot{a}}{a} = -\frac{4\pi G}{3}\left(\rho + \frac{3P}{c^2}\right), \tag{2.20}$$

where $P$ is the pressure of the fluid. Combining the time derivative of eq. (2.19) with eq. (2.20), we get

$$\dot{\rho} + \frac{3\dot{a}}{a}\left(\rho + \frac{P}{c^2}\right) = 0, \tag{2.21}$$

[1] The curvature $k$ is the 3D analogue of the more familiar **Gaussian curvature** of 2D surfaces, which, for instance, is positive for a sphere, null for a plane and negative for a saddle point.

known as the **fluid equation**. So the dynamics of the Universe is described by any two among eqs. (2.19)–(2.21). When $k = 0$ (spatially flat Universe) the Friedmann equation (eq. 2.19) reads

$$H^2 = \frac{8\pi G\rho}{3}, \tag{2.22}$$

where we have replaced $\dot{a}/a$ with the Hubble parameter (eq. 2.2). It follows that the Universe is spatially flat when $\rho = \rho_{\rm crit}$, where

$$\rho_{\rm crit}(t) \equiv \frac{3H^2(t)}{8\pi G} \tag{2.23}$$

is called the **critical density of the Universe**. At the present time

$$\rho_{\rm crit,0} = \frac{3H_0^2}{8\pi G} \simeq 1.9 \times 10^{-29} h^2 \ {\rm g\ cm^{-3}}, \tag{2.24}$$

roughly corresponding to six protons per cubic metre for $h = 0.7$. It is useful to define the dimensionless **density parameter**

$$\Omega(t) \equiv \frac{\rho(t)}{\rho_{\rm crit}(t)}, \tag{2.25}$$

whose value at the present time $t_0$ is indicated with $\Omega_0 \equiv \Omega(t_0)$. In terms of the density parameter, the Friedmann equation (eq. 2.19) can be written as

$$H^2(1 - \Omega)a^2 = -kc^2, \tag{2.26}$$

so, at the present time, $H_0^2(1 - \Omega_0) = -kc^2$. Therefore, the present-day value of the density parameter $\Omega_0$ determines the sign of $k$ (and then the spatial geometry of the Universe) at *all* times.

### 2.2.2 Components of the Universe

For a given component of the cosmic fluid, the relation between $P$ and $\rho$ is provided by the **cosmological equation of state**

$$P = w\rho c^2, \tag{2.27}$$

where the dimensionless parameter $w$ in general depends on redshift: $w = w(z)$. Combining the equation of state (eq. 2.27) with the fluid equation (eq. 2.21) and eq. (2.5), we obtain the equation for the evolution of the density $\rho$ with redshift:

$$\rho = \rho_0 \exp\left\{3 \int_0^z [1 + w(z')]\, {\rm d}\ln(1 + z')\right\}, \tag{2.28}$$

where $\rho_0 \equiv \rho(t_0)$. When $w$ is a constant (i.e. it does not vary with redshift) the above equation reduces to

$$\rho = \rho_0(1 + z)^{3(1+w)} = \rho_0 a^{-3(1+w)}. \tag{2.29}$$

Different components of the cosmic fluid are characterised by different equations of state. A fundamental distinction is between relativistic and non-relativistic components.[2] For a non-relativistic component, usually referred to as **matter**, which is such that $P \ll \rho c^2$ (in practice, ordinary baryonic matter and non-baryonic cold dark matter; see §7.2), it can be effectively assumed that $w=0$, so $P=0$ and

$$\rho_{\rm m} = \rho_{\rm m,0}(1+z)^3. \tag{2.30}$$

For a relativistic component, usually referred to as **radiation** (for instance photons or, at sufficiently high redshift, neutrinos), $w=1/3$, so

$$\rho_{\rm rad} = \rho_{\rm rad,0}(1+z)^4. \tag{2.31}$$

We note that, for relativistic particles, we define an equivalent mass density $\rho_{\rm rad}$ (in units of $\rm g\,cm^{-3}$) such that the energy density is $\rho_{\rm rad}c^2$, so the quantity $\rho_{\rm rad}$ is defined also for massless particles such as photons.

Another component of the Universe is **dark energy**, which has the property to produce an accelerated expansion ($\ddot{a}>0$). Eqs. (2.20) and (2.27) imply that, for a component to be classified as dark energy, it must have $w<-1/3$. In general, the value of $w$ for a dark energy component is not constant: when $w$ depends on $z$, the evolution of the dark energy density is given by eq. (2.28). A special case of dark energy component is the **vacuum energy** (associated with the **cosmological constant** $\Lambda$), which is characterised by constant $w=-1$, and thus (eq. 2.29) by density

$$\rho_\Lambda = \rho_{\Lambda,0} = {\rm const}, \tag{2.32}$$

such that the energy density is $\rho_\Lambda c^2$. For the above components we define the corresponding density parameters $\Omega_{\rm m} \equiv \rho_{\rm m}/\rho_{\rm crit}$, $\Omega_{\rm rad} \equiv \rho_{\rm rad}/\rho_{\rm crit}$ and $\Omega_\Lambda \equiv \rho_\Lambda/\rho_{\rm crit}$, which are all dependent on time.

Given the definitions of matter and radiation, it is possible that a given species of particles can be classified as radiation at an early time and as matter at a later time. This is the case for very low-mass particles such as neutrinos (with mass $m_\nu$ such that $m_\nu c^2 \lesssim 0.2$ eV). Neutrinos interact very weakly with baryons: these interactions are relevant in the very early Universe, but become negligible at redshifts lower than $z \sim 10^{10}$ (redshift of **neutrino decoupling**). When they decouple, neutrinos are relativistic and behave like radiation. However, the temperature of the neutrino background decreases with time as the Universe expands, so, at the present time, neutrinos are expected to be non-relativistic, and therefore to behave like matter.

## 2.3 Standard Cosmological Model

In the currently favoured cosmological model the cosmic fluid consists of three components: matter, radiation and dark energy. The determination of the cosmological parameters

[2] A particle is **relativistic** if its kinetic energy is of the order of or higher than its rest-mass energy (i.e. its speed is close to the speed of light). Otherwise, the particle is **non-relativistic**.

**Table 2.1** Parameters of the flat ΛCDM cosmological model as obtained by the *Planck* space mission

| $\Omega_{\rm m,0}$ | $\Omega_{\Lambda,0}$ | $\Omega_{\rm b,0}h^2$ | $z_{\rm eq}$ | $h$ |
|---|---|---|---|---|
| $0.315 \pm 0.007$ | $0.685 \pm 0.007$ | $0.0224 \pm 0.0001$ | $3402 \pm 26$ | $0.674 \pm 0.005$ |

Data obtained by Planck Collaboration (2018) fitting a spatially flat ($\Omega_0 = 1$) ΛCDM ($w = -1$) cosmological model to the *Planck* cosmic microwave background data. $\Omega_{\rm m,0}$, $\Omega_{\Lambda,0}$ and $\Omega_{\rm b,0}$ are the present-day matter, vacuum energy and baryon density parameters, respectively. $z_{\rm eq}$ is the redshift of radiation–matter equivalence. The present-day radiation density parameter is $\Omega_{\rm rad,0} = \Omega_{\rm m,0}/(1+z_{\rm eq}) \simeq 9.26 \times 10^{-5}$. $h$ is the dimensionless Hubble constant. We note that when $h$ is measured with a different method, based on the observation of supernovae, the best estimate is $h \simeq 0.732 \pm 0.017$ (Riess et al., 2016). The origin of this discrepancy is unclear.

is derived from different pieces of information, including the analysis and modelling of the cosmic microwave background (§2.4), observations of distant supernovae, measurements of the mass of galaxy clusters, weak gravitational lensing and the galaxy large-scale structure. The specific values of the cosmological parameters and the associated uncertainties depend on the considered set of observations. The present-day total density $\rho_0 = \rho_{\rm m,0} + \rho_{\rm rad,0} + \rho_{\Lambda,0}$ is indistinguishable from the critical density (the Planck Collaboration (2018) find $\Omega_0 = 0.999 \pm 0.002$). The nature (and equation of state) of dark energy is unknown. Though it is not excluded that $w$ depends on redshift, current data appear consistent with constant $w = -1$ (that is, vacuum energy).

A widely adopted cosmological model is such that the Universe is homogeneous, isotropic and spatially flat ($\Omega = 1$ at all times), dark energy is vacuum energy ($w = -1$ at all times), the present-day density parameters are $\Omega_{\rm m,0} = \rho_{\rm m,0}/\rho_{\rm crit,0} \approx 0.3$, $\Omega_{\rm rad,0} = \rho_{\rm rad,0}/\rho_{\rm crit,0} \approx 9 \times 10^{-5}$, $\Omega_{\Lambda,0} = \rho_{\Lambda,0}/\rho_{\rm crit,0} \approx 0.7$, and the dimensionless Hubble constant is $h \approx 0.7$. These values are usually referred to as **standard cosmological parameters**. As we see in §7.2, in order to explain structure formation, most of the matter component of the Universe must be in the form of cold dark matter (CDM). A cosmological model with standard cosmological parameters (and therefore non-zero cosmological constant Λ), in which dark matter is cold, is therefore indicated with the acronym ΛCDM. The **ΛCDM model** has become the reference framework for cosmological studies and is often referred to as the **standard cosmological model** or **concordance cosmological model**. As a reference, in Tab. 2.1 we report estimates of the main cosmological parameters, as obtained with the *Planck* space mission, for a flat ($\Omega = 1$) ΛCDM ($w = -1$) cosmological model.

### 2.3.1 Radiation, Matter and Dark Energy Eras

While the vacuum energy density $\rho_\Lambda$ remains constant with time (eq. 2.32), the density of matter $\rho_{\rm m}$ and the density of radiation $\rho_{\rm rad}$ were higher in the past (§2.2.2). Thus, though vacuum energy is dominant today (Tab. 2.1), before the present **dark energy era**, there has been a **matter era** and, before the matter era, there has been a **radiation era**. The equivalence between matter and dark energy ($\rho_\Lambda = \rho_\mathrm{m}$) occurred at $z \approx 0.30$ (redshift of **Λ–matter equivalence**), while the equivalence between radiation and matter ($\rho_\mathrm{m} = \rho_\mathrm{rad}$) occurred at $z_\mathrm{eq} \approx 3400$, which is known as the redshift of **matter–radiation equivalence** (often referred to as redshift of **equivalence** *tout court*).

In the matter and radiation eras, sufficiently far from equivalence, a good approximation is a one-component model Universe, where the only component is the dominant one at that time. In a matter-only Universe with $\Omega = 1$ (i.e. $k = 0$), known as the **Einstein–de Sitter (EdS) model** (Einstein and de Sitter, 1932), solving the Friedmann equations (eqs. 2.19 and 2.20) gives

$$a = \left(\frac{t}{t_0}\right)^{2/3}, \tag{2.33}$$

or, equivalently,

$$\rho = \frac{1}{6\pi G t^2}. \tag{2.34}$$

In a radiation-only Universe with $\Omega = 1$ (i.e. $k = 0$),

$$a = \left(\frac{t}{t_0}\right)^{1/2}, \tag{2.35}$$

or, equivalently,

$$\rho = \frac{3}{32\pi G t^2}. \tag{2.36}$$

In the most general case, in which the contribution of all components is accounted for, the Friedmann equation (eq. 2.19) for a flat ΛCDM model can be written as

$$H^2(z) = H_0^2\left[\Omega_{\mathrm{m},0}(1+z)^3 + \Omega_{\mathrm{rad},0}(1+z)^4 + \Omega_{\Lambda,0}\right], \tag{2.37}$$

which is a useful expression for the Hubble parameter as a function of redshift.

### 2.3.2 Big Bang, Cosmic Time and Look-Back Time

It follows from eq. (2.35) that, in the radiation era, $a \to 0$ for $t \to 0$, so the energy density of the Universe diverges at early times (eq. 2.36). At $t = 0$ there was a space-time singularity, known as the **Big Bang**. Therefore, we can describe the history of the Universe in terms of **cosmic time**, whose value is $t = 0$ at the time of the Big Bang ($a = 0$ or $z = \infty$) and $t = t_0$ at the present time ($a = 1$ or $z = 0$). The value of $t_0$ is the present-day age of the Universe and can be computed for any given cosmological model. For the standard cosmological model with parameters as in Tab. 2.1, $t_0 \simeq 13.8$ Gyr. The cosmic time increases for decreasing redshift: from eqs. (2.2) and (2.5) it follows that the cosmic time at a given redshift $z$ is given by

$$t(z) = \int_z^\infty \frac{\mathrm{d}z'}{(1+z')H(z')}. \tag{2.38}$$

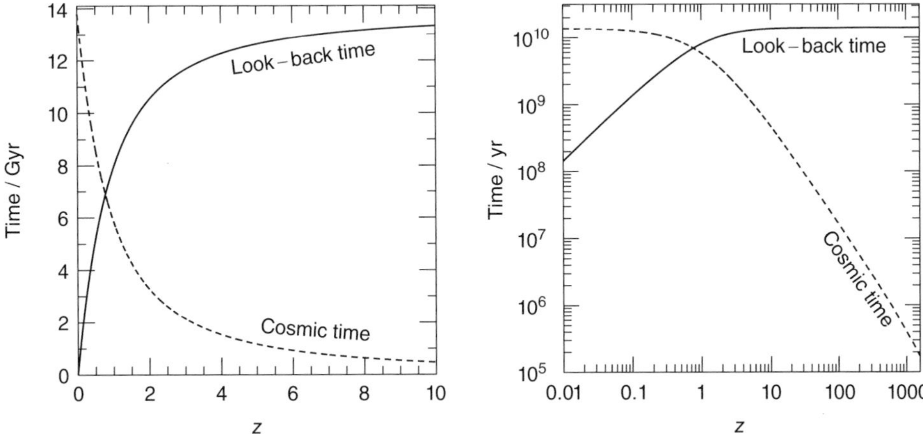


Fig. 2.2 *Left panel.* Cosmic time (i.e. age of the Universe) and look-back time as functions of redshift for a flat ΛCDM model with parameters as in Tab. 2.1. *Right panel.* Same as left panel, but for a different redshift interval and in logarithmic scale.

It is useful to define also the **look-back time** to some redshift $z$ as $t_0 - t(z)$, where $t(z)$ is the cosmic time at redshift $z$: the look-back time is zero at $z = 0$ and increases for increasing redshift. The cosmic and look-back times as functions of redshift for the standard cosmological model are plotted in Fig. 2.2.

The time variation of the expansion rate of the Universe is usually quantified by the **deceleration parameter**

$$q(t) \equiv -\frac{\ddot{a}a}{\dot{a}^2}, \tag{2.39}$$

which is positive if the Universe decelerates ($\ddot{a} < 0$). It follows from eq. (2.20) that, for an EdS model, $q = 1/2$ at all times. In a two-fluid model with matter and vacuum energy,

$$q = \frac{\Omega_{\rm m}}{2} - \Omega_\Lambda, \tag{2.40}$$

so in the standard model the Universe is currently undergoing an accelerated expansion: $q(t_0) \approx -0.5$.

Above we have implicitly assumed that the first cosmological epoch is the radiation era. In fact, strictly speaking, this is not correct when very early times are considered. First, it must be considered that general relativity cannot be applied at cosmic times $t < t_{\rm Planck}$, where $t_{\rm Planck} \sim 10^{-43}$ s is the **Planck time**, such that quantum effects become important on cosmological scales. Second, it has been suggested that between $t \sim 10^{-36}$ s and $t \sim 10^{-34}$ s the Universe experienced a phase of accelerated expansion known as **inflation**. A discussion of inflation is not necessary for the purpose of the present textbook, but the reader should bear in mind that it is not straightforward to extrapolate the standard cosmological model to the very early Universe.

## 2.4 Cosmic Microwave Background, Recombination and Reionisation

Observations in the microwaves have revealed the presence of an almost isotropic radiation with a black-body spectrum (§D.1.3) with temperature $T \simeq 2.726\,\mathrm{K}$, which is in fact the best black-body spectrum known in nature. The photons of this **cosmic microwave background** (CMB) constitute the present-day radiation component of the Universe. We have seen that the radiation density $\rho_{\rm rad}$ was higher in the past, scaling with redshift as $\rho_{\rm rad} \propto (1 + z)^4$ (eq. 2.31). Also the radiation temperature was higher in the past, but scaling with redshift as $T \propto (1 + z)$. Similarly, the matter component was not only denser (eq. 2.30), but also hotter in the past. At very high redshift, when the cosmic plasma is at high temperature, photons and electrons are coupled via **Thomson scattering**: the photon **mean free path** is short, so photons are in thermal equilibrium with the cosmic plasma. At $z \approx 1000$, the mean free path of photons becomes very long (comparable to or longer than the horizon distance; eq. 2.10), so photons and matter decouple. This happens fundamentally for two reasons: (1) the density decreases because of expansion; and (2) the temperature decreases, and, at some point, it becomes low enough that H and He atoms form. The latter process is called **cosmological recombination**, because electrons and nuclei combine to form atoms. Following recombination, the cosmic photons do not interact significantly with atoms, because their energy is lower than the photoionisation energy of H and He. Thus, the redshift of **photon decoupling** (or simply **decoupling**) can be identified with the redshift of recombination $z_{\rm rec}$. Though recombination is not an instantaneous process, we can take as reference value $z_{\rm rec} \approx 1000$ (§9.1). The **surface of last scattering** is the spherical surface in which the CMB photons that reach us now were last scattered. The distance to the surface of last scattering, in redshift space, is $z_{\rm rec}$: this is the highest redshift observable with electromagnetic radiation.

We know that at low redshift the Universe is almost completely ionised, because we observe energetic photons, which would otherwise be absorbed by neutral H and He, emitted by distant sources (§9.9.1). Thus, after recombination, a process called cosmological reionisation (§9.9) must have occurred. The combination of theory and observational data suggests that reionisation started at $z \approx 10$ and ended at $z \approx 6$, and was due to photons emitted by the first stars and galaxies.

We have pointed out above that the CMB is an almost perfect isotropic black-body radiation at temperature $T \simeq 2.726\,\mathrm{K}$. In fact, a dipole (i.e. along a given direction) anisotropy is observed with temperature variations $|\Delta T|/T \sim 10^{-3}$, due to the peculiar motion of the solar system with respect to the Hubble flow (in other words, to the fact that we are not fundamental observers; §2.1.1). When this dipole anisotropy is subtracted, much smaller deviations of the CMB from a perfectly isotropic black-body radiation are detected: in practice, the black-body temperature is not the same all over the sky, but slightly different temperatures are measured at different locations in the sky. Though very small, of the order of $|\Delta T|/T \sim 10^{-5}$, these **CMB temperature anisotropies** contain invaluable information for our understanding of cosmology (§2.3) and galaxy formation (§7.2).

Thanks to microwave space observatories such as the *Wilkinson Microwave Anisotropy Probe* (*WMAP*; Bennett et al., 2013) and *Planck* (Planck Collaboration, 2018), we have now high-resolution observed maps of the CMB temperature fluctuations $\Delta T(\vartheta,\varphi)/T$, where $\vartheta$ and $\varphi$ are spherical angular coordinates that define a position in the plane of the sky. In order to measure how much temperature fluctuation there is on different scales, given that $\Delta T$ is measured on the celestial sphere as a function of angular coordinates, the natural choice is to expand the map of temperature fluctuations in spherical harmonics:

$$\frac{\Delta T(\vartheta,\varphi)}{T} = \sum_{\ell=1}^{\infty} \sum_{m=-\ell}^{\ell} a_{\ell,m} Y_m^\ell(\vartheta,\varphi), \tag{2.41}$$

where $\ell$ and $m$ are the integer indices associated with $\vartheta$ and $\varphi$, respectively ($\ell = 1$ is the dipole term, $\ell = 2$ the quadrupole term, etc.). The corresponding **CMB power spectrum**

$$C_\ell = \frac{1}{2\ell+1} \sum_{m=-\ell}^{\ell} |a_{\ell,m}|^2 \tag{2.42}$$

measures how much fluctuation power we have on the angular scale $\approx \pi/\ell$. The observed CMB power spectrum, shown in Fig. 2.3, is characterised by several peaks, due to acoustic oscillations (analogous to sound waves) of the photon–baryon fluid at the time of decoupling ($z_{\rm rec}$). These are usually referred to as **CMB acoustic peaks** and numbered

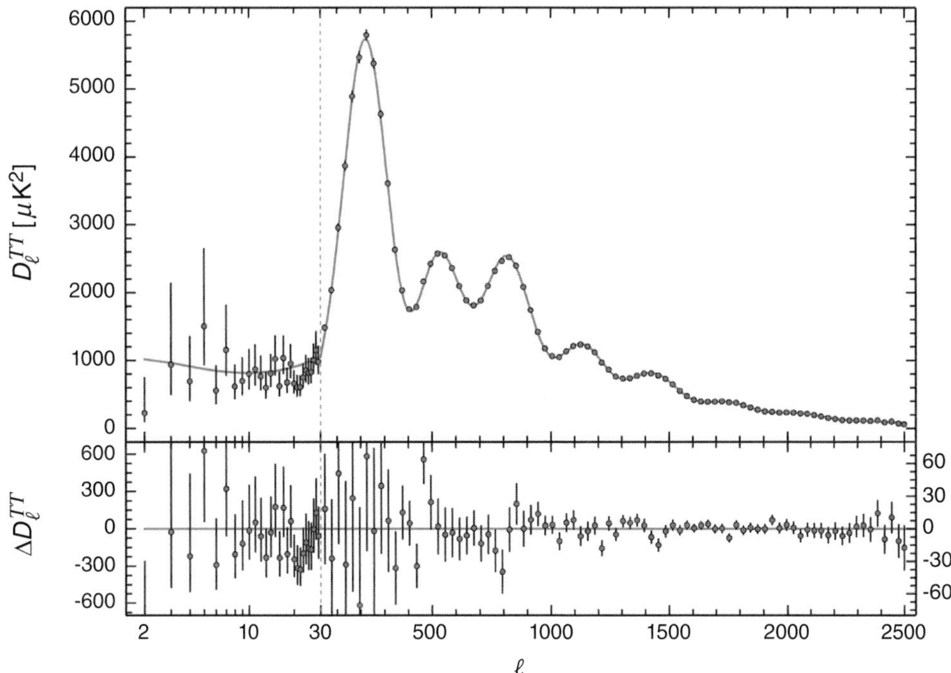


Fig. 2.3 *Top panel*. The observed power spectrum of the CMB (points with error bars) and the model power spectrum (curve) for a flat ΛCDM model with parameters as in Tab. 2.1. *Bottom panel*. Residuals between the model and the data. Here $\mathcal{D}_\ell^{TT} \equiv \ell(\ell+1)C_\ell/(2\pi)$, where $C_\ell$ is given by eq. (2.42). Note that the scale of the horizontal axis is logarithmic for $\ell < 30$ and linear for $\ell > 30$. From Planck Collaboration (2018).

in order of increasing value of $\ell$. In brief, the physical interpretation of the first three peaks is the following.

1. *The first peak* ($\ell \approx 200$), which is the highest and the one on the largest angular scale ($\approx 1°$) is due to acoustic oscillations on the scale of the horizon (§2.1.4) at recombination: since the Big Bang these fluctuations had just enough time to undergo a single compression. The angular scale of the first peak depends on the horizon size at decoupling and on the angular diameter distance to the surface of last scattering, so it is most sensitive to the curvature of the Universe (thus constraining the total density parameter $\Omega$; eq. 2.26).
2. *The second peak* ($\ell \approx 500$, i.e. angular scale $\approx 0.4°$) is due to oscillations that had time to undergo one compression and one rarefaction. The second peak is lower than the first because of the gravitational effect of the baryons on the acoustic oscillations: thus the measure of the height of the second peak relative to the first can be used to constrain the baryon density parameter $\Omega_b \equiv \rho_b/\rho_{crit}$, where $\rho_b$ is the baryon density.
3. *The third peak* ($\ell \approx 800$, i.e. angular scale $\approx 0.2°$) is, as the first and as all odd peaks, a compression peak. It is due to oscillations that have already undergone a compression and a rarefaction: the strength of the third peak depends on the total matter density, so measuring it gives constraints on $\Omega_m$.

Fig. 2.3 also shows the theoretical CMB power spectrum produced by a flat $\Lambda$CDM cosmological model with parameters given in Tab. 2.1, which, as apparent from the residuals, matches very well the observational data.

Before recombination, baryons and photons are coupled, so baryons participate in the acoustic oscillations that give rise to the acoustic peaks of the CMB. After recombination, these **baryon acoustic oscillations** (BAOs) leave an imprint in the large-scale structure of baryonic matter. BAOs are detected in large galaxy surveys as a feature in the matter power spectrum on scales $\approx 150$ Mpc, which is a measure of the comoving **sound horizon** (i.e. the maximum distance travelled by sound waves) at the time of recombination.

## 2.5 Big Bang Nucleosynthesis

Before decoupling, the Universe was composed of a mixture of radiation and fully ionised matter in thermal equilibrium. We have seen that the temperature of this primordial plasma increases with redshift. At sufficiently early times, the temperature was so high ($T \gg 10^{10}$ K) that the reactions transforming protons into neutrons and vice versa were efficient. Later, when the temperature dropped to $T \sim 10^{10}$ K, the timescales of these reactions became longer than the expansion timescale, so the neutron-to-proton number density ratio was frozen to $n_n/n_p \approx 0.2$. Between $T \sim 10^{10}$ K ($t \sim 1$ s) and $T \sim 10^9$ K ($t \sim 10^2$ s) neutrons are converted into protons via $\beta$-decay, so that, at $T \sim 10^9$ K, $n_n/n_p \approx 1/7$. When $T \sim 10^9$ K the temperature is low enough to form light nuclei such as deuterium (D) and $^4$He: this process is known as **Big Bang** (or **primordial**) **nucleosynthesis** and is completed at $t \sim 10^3$ s, when $T \sim 10^8$ K. Assuming that all neutrons end up in $^4$He, one derives a helium mass fraction (§C.6.1) $Y \approx 0.25$ (known as **primordial helium abundance**), consistent with the minimum value of $Y$ measured in observations. The abundance of the other light elements formed together with $^4$He (D, $^3$He and $^7$Li) depends significantly on the baryon density parameter and thus gives very useful constraints on the **cosmic baryon fraction** $f_b = \Omega_b/\Omega_m$. Measurements of the cosmic abundance of light elements (based on observations of low-metallicity interstellar gas and atmospheres of low-mass stars) suggest that the present-day baryon density parameter is $\Omega_{b,0} \simeq 0.05$, in agreement with measurements of the CMB power spectrum (Tab. 2.1). This number is a cornerstone for our understanding of the Universe, not only because it pinpoints the overall mass budget of ordinary matter in the Universe (§6.7), but also because, combined with other pieces of information (such as the properties of the CMB; §2.4), it tells us that most of the matter in the Universe must be non-baryonic, with important implications for theories of structure formation and thus for galaxy formation and evolution (§7.2). We recall that $\Omega_{m,0} \simeq 0.315$ (Tab. 2.1), so the cosmic baryon fraction is $f_b \simeq 0.16$.

## 2.6 Thermal History of the Universe

We have seen that any cosmic time $t$ can be labelled by different quantities. For a given cosmological model, $t$ is monotonically mapped not only into redshift $z$ or the scale factor $a = (1 + z)^{-1}$, but also into the temperature $T$ of the photon background (and, before decoupling, also of the primordial plasma). Especially at early times, it is convenient to use the temperature $T$ to label the cosmic time: for this reason the temporal sequence of the cosmological events and eras is often called the **thermal history of the Universe**. The fundamental stages of this thermal history are reported in Tab. 2.2 for the standard cosmological model (§2.3).

**Table 2.2** Thermal history of the Universe in the standard cosmological model

| $t$ Cosmic time | $T$ Temperature | $z$ Redshift | |
|---|---|---|---|
| $10^{-43}$ s | $10^{32}$ K | $10^{32}$ | Planck time |
| $10^{-36}$–$10^{-34}$ s | $10^{25}$–$10^{28}$ K | $10^{25}$–$10^{28}$ | Inflation |
| 1 s | $10^{10}$ K | $10^{10}$ | Neutrino decoupling |
| 10–1000 s | $10^8$–$10^9$ K | $10^8$–$10^9$ | Big Bang nucleosynthesis |
| $5 \times 10^4$ yr | $9 \times 10^3$ K | 3400 | Matter–radiation equivalence |
| 0.4 Myr | $3 \times 10^3$ K | 1000 | Recombination and photon decoupling |
| 0.5–1 Gyr | 19–30 K | 6–10 | Reionisation |
| 9.5 Gyr | 3.5 K | 0.30 | Λ–matter equivalence |
| 13.8 Gyr | 2.7 K | 0 | Present time |

# 3 Present-Day Galaxies as a Benchmark for Evolutionary Studies

Our understanding of galaxy evolution relies on the possibility to observe galaxies at different look-back times, from the present-day Universe to the highest redshifts accessible with the largest telescopes. The present day refers to when the light travel time from galaxies is small compared to the age of the Universe. In this context, the redshift range $z < 0.1$ is typically considered 'present day' because the light travel time is less than one-tenth of the age of the Universe. The present-day Universe is populated by a variety of galaxy types. These systems display a wide range of luminosities, sizes, structural and kinematic properties, stellar populations and interstellar medium content. Galaxies at $z \approx 0$ are the endpoint of a long process that has lasted $\approx 13.8$ billion years, from the gravitational collapse of the first luminous objects to the formation and differentiation of the galaxy types that we observe today. Present-day galaxies can be observed in the most detail because they can be studied at the highest spatial resolution and their intrinsically faintest features can be detected. The aim of this chapter is to provide an overview of the general properties of present-day galaxies in order to exploit them as a benchmark for understanding their formation and evolutionary processes.

## 3.1 Morphology

Morphology is a key piece of information to understand the structure and properties of galaxies and their evolution across cosmic time. As illustrated in §1.2, Hubble (1926) defined the first classification of morphologies based on four main classes: **ellipticals**, **lenticulars**, **spirals** and **irregulars** (Fig. 1.1). Galaxies on the left and right of the Hubble tuning fork are respectively called early and late types. In particular, the class of **early-type galaxies** (ETGs) includes ellipticals and lenticulars, whereas the ensemble of spirals and irregulars comprises the broad class of **late-type galaxies** (LTGs). Typically, almost all LTGs have ongoing star formation. For this reason, the classification of **star-forming galaxies** (SFGs) is often used rather than basing the classification on morphology (Chapter 4). From now on, the SFG nomenclature will be preferred throughout the book. Instead, ETGs are sometimes called **quiescent** or **passive** galaxies to indicate that they have weak or absent star formation, respectively.

The Hubble classification was revised and expanded by de Vaucouleurs (1959) who proposed additional classes of lenticulars intermediate between S0 and Sa (S0a), spirals (Sd and Sm) and Im irregulars (where the 'm' refers to Magellanic prototypes such as

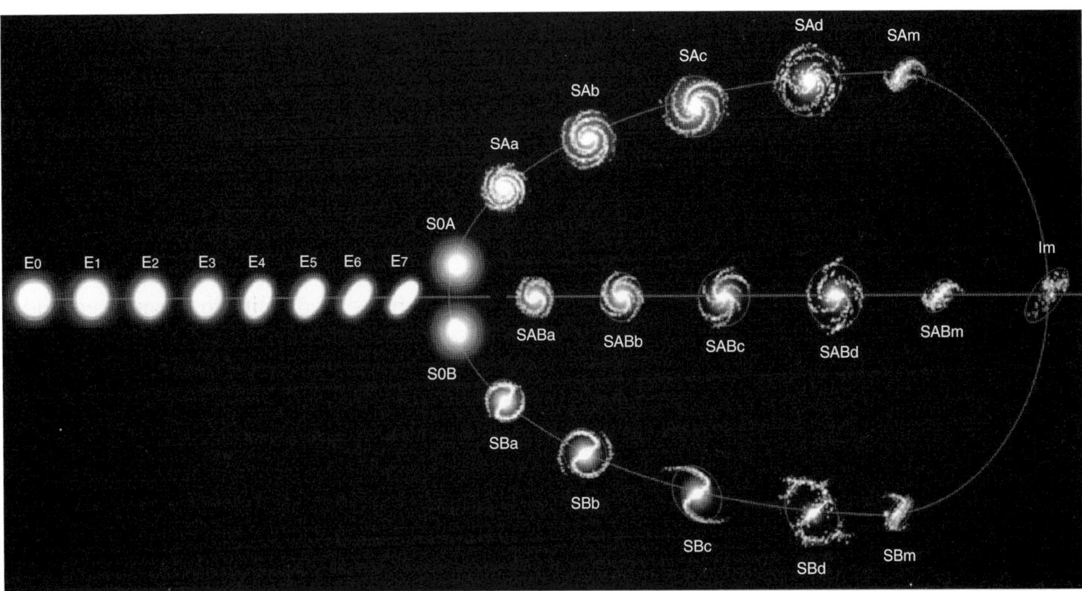


Fig. 3.1 The classification of galaxy morphology according to the revision of de Vaucouleurs. Adapted from Wikipedia. Image by A. Ciccolella and M. De Leo.

the Small Magellanic Cloud; §6.3). Intermediate classes were also introduced (e.g. Sab, SBbc), and the letters 'r' and 's' were adopted to specify if the inner part of a disc galaxy is respectively ring-like or purely spiral. Mixed types such as 'rs' and 'sr' indicate intermediate cases with a prevalence of ring or spiral structure, respectively. Normal, barred and intermediate spirals are called SA, SB and SAB, respectively. The superscripts – and + are used for E and S0 to denote a higher degree of 'earlyness' and 'lateness', respectively. For instance, a class $S0^-$ indicates a lenticular galaxy with a less prominent disc component than in a class $S0^+$. Thus, in the de Vaucouleurs classification (Fig. 3.1), the morphology of a given galaxy is characterised by more detailed information than in the classical Hubble scheme. For example, a galaxy class can be as complex as SAB(rs)bc. Numerical classes (the so-called T-types) are also used to define morphological types. According to the Third Reference Catalogue of Bright Galaxies, the T-types −5, −4, −3, −2, −1, 0, +1, +2, +3, +4, +5, +6, +7, +8, +9, +10 correspond to the de Vaucouleurs classes of E0, $E^+$, $S0^-$, S0, $S0^+$, S0a, Sa, Sab, Sb, Sbc, Sc, Scd, Sd, Sdm, Sm, Im, respectively.

Additional classifications have been proposed based on other galaxy features: luminosity, spectral type, importance of the spheroidal component (bulge; §4.1.2) relative to the disc (bulge-to-disc ratio), structure of spiral arms, peculiarities and others. However, keeping a rather simple classification is important for a comparison with high-redshift galaxies which appear much fainter and have smaller angular sizes (§11.1.6). For this reason, the most popular classification remains the classical Hubble tuning fork extended to the new classes defined by de Vaucouleurs. Galaxies with mass and luminosity more than one order of magnitude lower than our Galaxy are in general called **dwarfs** (Tab. 4.1; Tab. 5.1; §6.3.1).

### 3.1.1 The Morphology of Galaxies Depends on Wavelength

The morphological classifications described above were all historically based on images obtained at optical wavelengths. However, modern observations in the ultraviolet (UV; §C.1) and near-infrared (NIR; §C.1) show that galaxy morphology can strongly vary as a function of wavelength. The change of morphology with wavelength is called **morphological K correction**. The origin of this effect is twofold. The first is that different **stellar populations** with distinct colours are generally located in different regions of a galaxy. For example, the UV morphology is more influenced by hot and massive O and B stars located in spiral arms, where star formation is ongoing. Instead, in the NIR, the bulges are more prominent because of their older (redder) stars, whereas discs appear smoother because long-lived stars are more spread out than young stars. The second reason is **dust extinction** (§4.2.7). Dust is mostly associated with star-forming regions and attenuates preferentially the UV light. This implies that dust extinction can severely influence how a spiral or an irregular galaxy appears as a function of wavelength. ETGs are in general the least affected by these morphological variations thanks to their rather homogeneously old stellar populations and the low amount of dust reddening (§4.2.7). Instead, SFGs can display significantly different morphologies depending on wavelength, and tend to appear more clumpy/irregular in the UV and smoother in the NIR. Fig. 4.2 shows clear examples of these effects.

### 3.1.2 Surface Brightness Profiles

Galaxies appear as extended objects on the sky, and a more quantitative approach to describe galaxy shapes and constrain their structural components is based on how the **surface brightness** $I_\lambda$ (§C.2) varies as a function of the radial distance from the galaxy centre (**surface brightness profile**). The curves connecting points of equal surface brightness are called **isophotes**. Fig. 3.2 shows an example of the surface brightness profile and isophotes of an S0 galaxy. If the observed shape of the isophotes is approximated by

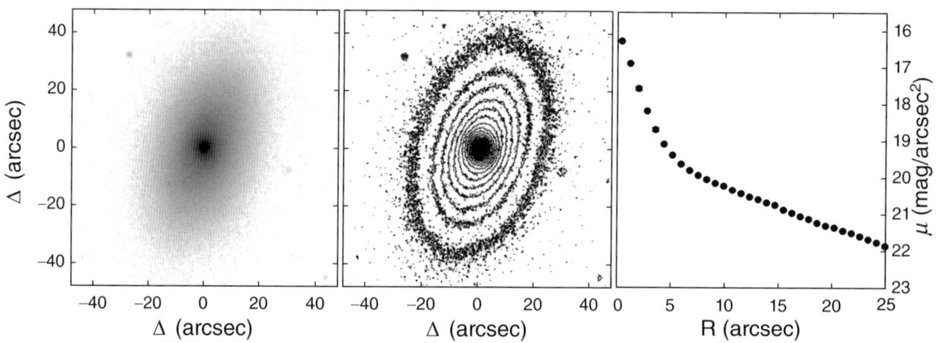


Fig. 3.2 The S0 galaxy NGC 7671 at $z = 0.0138$. *Left panel.* Image in the *r* band (data from Sloan Digital Sky Survey, SDSS). *Central panel.* The isophotes relative to the left image. *Right panel.* The surface brightness profile. At $z = 0.0138$, 5 arcsec correspond to $\approx 1.4$ kpc. Figure courtesy of F. Rizzo.

ellipses, their flattening is related to the **ellipticity** $\epsilon \equiv (a - b)/a$, where $a$ and $b$ are the observed semi-major and semi-minor axes. Either the **elliptical radius** $a$ or the so-called **circularised radius** $R_{\rm circ} \equiv (ab)^{1/2}$ can be used as radius $R$ to label the isophotes and to construct the surface brightness profile $I_\lambda(R)$. The radius of the isophote containing 50% of the total flux is called the **effective radius** or **half-light radius**, and it is indicated with $R_{\rm e}$. From now on, the notation $R_{\rm e}$ will be used irrespectively of being circularised or not. When the surface brightness is expressed in mag arcsec$^{-2}$, it is usually indicated with $\mu_\lambda$ (eq. C.8; right panel of Fig. 3.2). In practice, the $I_\lambda(R)$ or $\mu_\lambda(R)$ radial profiles can be obtained by measuring the flux within concentric annuli at increasing radii. When the surface brightness profiles of the same galaxy are available at two different wavelengths, it is possibile to derive a **colour radial profile**. The study of colour gradients is very important to place constraints on the properties of stellar populations (§4.1.4) and dust extinction within galaxies.

### 3.1.3 The Sérsic Profile

The surface brightness profiles of galaxies of different types can be reproduced with the **Sérsic profile** (Sérsic, 1968)

$$I_\lambda(R) = I_{\lambda,\rm e} \exp\left\{-b(n)\left[\left(\frac{R}{R_{\rm e}}\right)^{1/n} - 1\right]\right\}, \tag{3.1}$$

where $I_{\lambda,\rm e}$ is the surface brightness at a given wavelength $\lambda$ at the effective radius $R_{\rm e}$, $n$ is the **Sérsic index**, and $b$ is a quantity well approximated by $b(n) = 2n - 1/3 + 4/(405n)$. Fig. 3.3 shows the shape of the Sérsic function as a function of $n$. The case

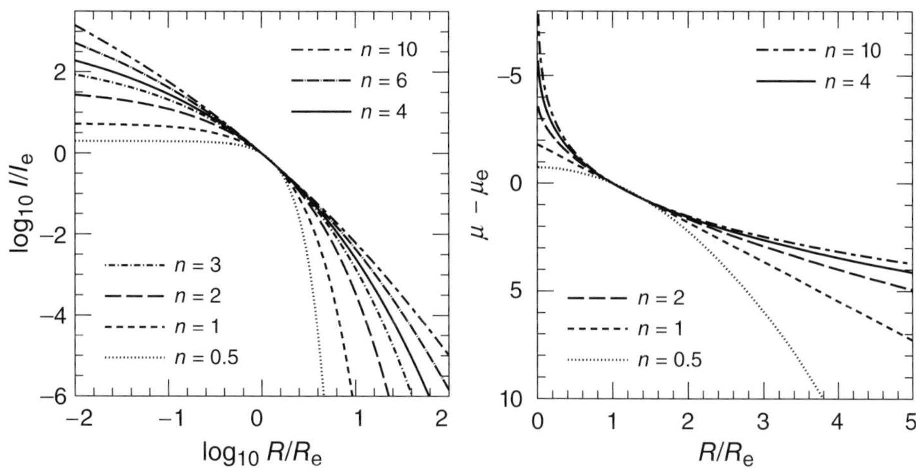


Fig. 3.3 Examples of Sérsic functions for different values of $n$. *Left panel.* Surface brightness normalised to $I_{\rm e}$ as a function of the radius in units of $R_{\rm e}$. *Right panel.* Same as in the left panel, but with the surface brightness expressed in magnitude arcsec$^{-2}$ and adopting a linear scale in abscissa.

of $n = 1$ corresponds to an exponential distribution of surface brightness, whereas for $n = 0.5$ the profile is Gaussian. For increasing $n$, the profile becomes steeper in the centre and shallower in the outskirts (for a detailed study of the Sérsic profile we refer the reader to Ciotti and Bertin 1999). The range of surface brightnesses observed in today's galaxies is very wide, and the different types of galaxies are located in distinct regions of the plane defined by the central surface brightness $I(0)$ and the absolute magnitude, reflecting intrinsic differences in the internal distribution of luminous matter (Fig. 3.4). Galaxies show also a broad distribution of Sérsic index $n$. On average, ellipticals and bulges have the steepest central profiles with $2 < n < 10$, discs have $n \approx 1$, and bars are characterised by flatter central profiles with $n \lesssim 0.5$. When a single Sérsic function is fitted to the observed profile, $n \approx 2.5$ is usually taken as the value separating ETGs from galaxies with later morphological types. However, galaxies are often multi-component systems, and the total observed profile can be the sum of two or more individual Sérsic profiles (e.g. a bulge plus a disc; §4.1). In this regard, two relevant quantities are the **bulge-to-total** ($B/T$) and the **bulge-to-disc** ($B/D$) luminosity ratios, which provide a quantitative measure of the relative importance of the bulge and disc components. Clearly, $B/T$ and $B/D$ correlate with the Hubble classes. For example, the typical values of $B/D$ in the $H$ band for S0/a, Sb and Scd galaxies (corrected for inclination and dust extinction effects) are $B/D \approx 0.3-0.4$, $\approx 0.2-0.3$ and $\lesssim 0.1$, respectively. The $B/T$ ratio can be estimated with the formula $B/T = [1 + (D/B)]^{-1}$, where

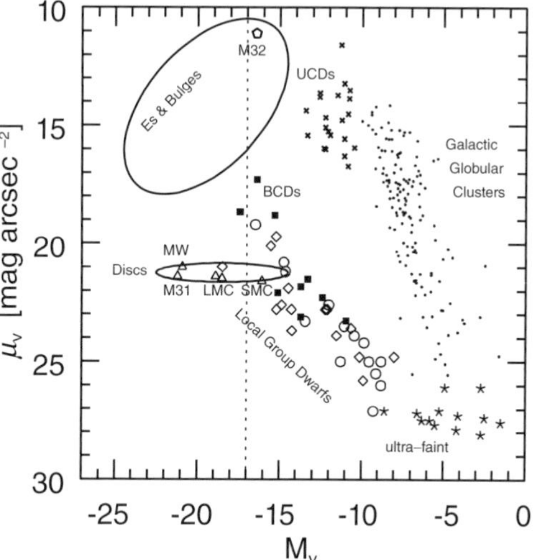


**Fig. 3.4** The distribution of galaxies and globular clusters in the plane defined by the central surface brightness and the absolute magnitude in the $V$ band. The vertical dashed line separates luminous galaxies from dwarfs. Different galaxy types are clearly segregated in well defined regions of this plane. BCDs and UCDs indicate the so-called blue compact dwarfs and ultra-compact dwarfs, respectively. UCDs are dwarf galaxies with very compact sizes and surface brightness comparable with globular clusters. SMC, LMC and MW indicate the Small Magellanic Cloud, the Large Magellanic Cloud and the Milky Way, respectively. M 31 is a spiral galaxy similar to the MW (§6.3). M 32 is an elliptical galaxy satellite of M 31. Data from Tolstoy et al. (2009) and McConnachie (2012). Courtesy of E. Tolstoy.

$(D/B)$ is the disc-to-bulge ratio. In the $H$ band, the typical values for S0/a, Sb and Scd galaxies are $B/T \approx 0.2$–$0.3$, $\approx 0.1$–$0.2$ and $< 0.1$, respectively. Since the bulge is redder than the disc due to the presence of older stars, $B/D$ and $B/T$ depend on wavelength and decrease at bluer wavelengths. Deviations from Sérsic profiles in the innermost regions of galaxies provide information on additional components (e.g. a point-like source or a flat core; §5.1.1). However, the very central regions are difficult to study with ground-based observations because of the blurring effects due to atmospheric turbulence (§11.1.6).

### 3.1.4 Galaxy Sizes

The surface brightness profile can be used also to quantify the size of a galaxy within a given limiting threshold of surface brightness. For example, $R_{25}$, the radius corresponding to the isophote with 25 mag arcsec$^{-2}$, is often used in the case of disc galaxies. However, it is also customary to use the effective radius $R_e$ to indicate the galaxy size. Present-day galaxies have a wide range of sizes. For a fixed luminosity, the size distribution is approximately lognormal. SFGs have sizes ranging from $\sim 1$ kpc in the case of dwarf irregulars, to $\sim 10$ kpc in the case of low surface brightness discs (§4.1). ETGs have effective radii from a few tenths of a kpc to tens of kpc in the case of the most massive systems located at the centre of galaxy clusters (§6.4.1). In general, the galaxy size increases with luminosity and stellar mass ($\mathcal{M}_\star$), but ETGs and SFGs show different behaviours (Fig. 3.5). For SFGs with $\log(\mathcal{M}_\star/\mathcal{M}_\odot) \gtrsim 10.6$ ($M_r \lesssim -20.5$), the dependence of the median half-light radius on mass is $R_e \propto \mathcal{M}_\star{}^\alpha$, where typically $\alpha \approx 0.4$, whereas $\alpha < 0.4$

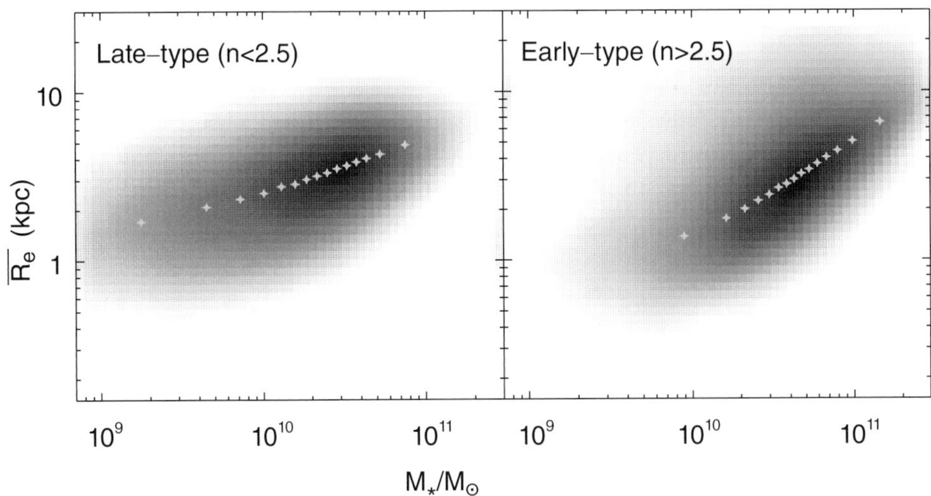


Fig. 3.5 The size–stellar mass relations for late-type (Sérsic index $n < 2.5$; *left panel*) and early-type ($n > 2.5$; *right panel*) galaxies from the SDSS sample. The shaded surfaces show the overall distributions (darker regions indicate higher number density of galaxies). Overplotted on these distributions are the median effective radii in bins of stellar mass. Adapted from Cebrián and Trujillo (2014). Courtesy of M. Cebrián.

for lower masses. In the case of ETGs, the size–mass relation is steeper ($\alpha \approx 0.5$–$0.6$) down to $M_r \approx -20$. Reproducing the distribution of galaxy sizes and its dependence on morphological type, luminosity and mass is a key test for galaxy formation models, as illustrated in §10.10 and §10.11.

## 3.2 Spectral Energy Distributions

The physical properties of galaxies are encoded in the radiation emitted by their stellar and diffuse components. The luminosity or flux of a galaxy across the electromagnetic spectrum is called **spectral energy distribution** (SED). The SEDs of normal (i.e. without AGN; §3.6) galaxies is the sum of three main components.

1. *Radiation from stars*. The dominant stellar contribution is due to the continuum radiation emitted from the photospheres of individual stars. Due to the effective temperatures of stars from O to M spectral types (§C.5), the photospheric black-body emission is concentrated from the UV to the NIR. In the NIR, a spectral bump is present at 1.6 $\mu$m due to the minimum in the opacity of the $H^-$ ion present in the atmospheres of cool stars. This is a nearly ubiquitous feature of all stellar populations with the exception of ages younger than $\approx$1 Myr. In general, the stellar spectra are characterised by a variety of absorption lines (§C.5). In particular, in the case of hot massive stars, stellar winds can produce also significant emission lines. Dust circumstellar shells emit thermal continuum from the mid-infrared (MIR) to the far-infrared (FIR) (§C.1). X-ray and gamma-ray emission can be produced by accretion phenomena in binary stars and in pulsars. The global shape of the stellar continuum from the UV to the NIR is primarily influenced by the ages and metal abundances of the stellar populations, and by dust extinction (Fig. 3.6).
2. *Radiation from diffuse matter*. The emission of photons from the gas can be originated by several processes depending on its temperature, density and ionisation (§4.2). In the spectroscopic notation, I indicates that the atom is neutral, II that it is singly ionised and so on. For instance, H I and H II are the neutral and the (singly) ionised hydrogen, respectively.[1] Note that H II is the same as writing $H^+$. At $T < 1000$ K, vibrational and rotational lines are emitted from molecules (e.g. CO) in the infrared to millimetre spectral range. Neutral atomic hydrogen (H I) with temperatures between tens to 8000 K is responsible for the line emission at 21-cm in the radio. Warm ($T \sim 10^4$ K) gas photoionised by massive OB stars (H II regions) or by white dwarfs (planetary nebulae) emits strong recombination (e.g. the hydrogen Lyman and Balmer series) and **forbidden lines** (e.g. [O II]$\lambda$3726,3729, [O III]$\lambda$4959,5007; §4.2.3) in the UV and optical. Forbidden lines are produced by highly improbable transitions occurring only at low density of the gas. In contrast, emission lines are called **permitted lines** when the probability of the associated transitions is very high (e.g. the hydrogen recombination

[1] The correct way to pronounce H I would be 'H first', given the Roman numeral; however, it has become common practice to use the expression 'H one'.

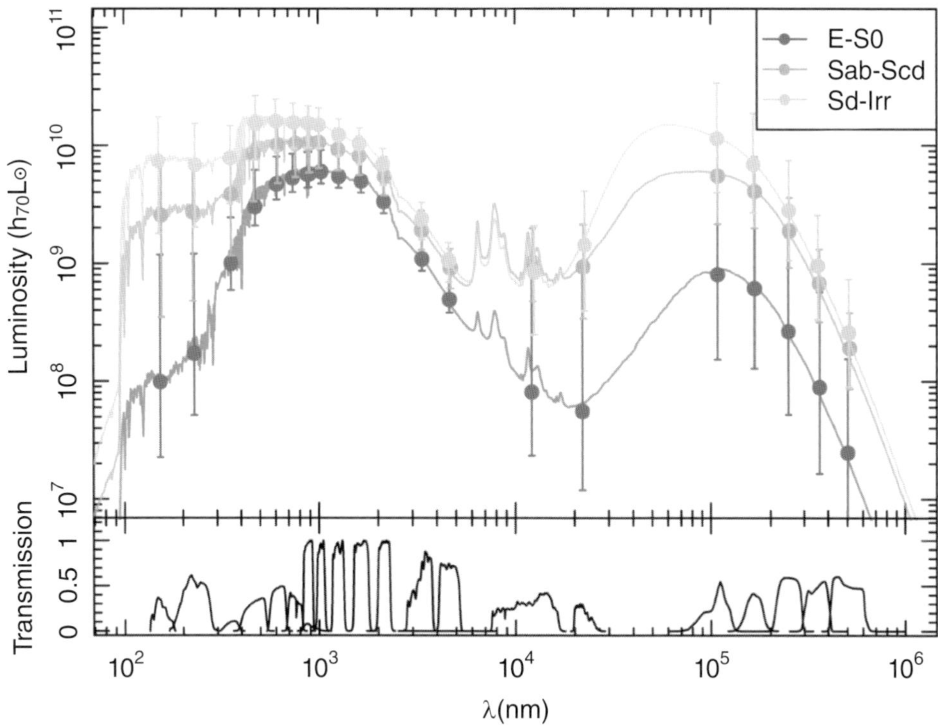


Fig. 3.6

*Top panel*. The average SEDs for three broad morphological classes (E–S0, Sab–Scd and Sd-Irr) based on photometry from the UV to the millimetre. The points are the average fluxes based on 21 filters whose transmission curves (§C.4) are shown in the *bottom panel*. The error bars show the $\pm 1\sigma$ dispersion in the photometry per filter per morphological type. The lines are the best fits to the photometric data obtained with theoretical SEDs including stellar photospheric radiation (dominating from the UV to the NIR range; $\sim 10^2$ nm to $3 \times 10^3$ nm) and the emission from interstellar dust (dominating the MIR to mm region; $\sim 10^4$ nm to $10^6$ nm). The typical double-peaked SED shape is evident for all galaxy types, but the importance of dust emission decreases rapidly for ETGs (E/S0), where star formation is weak or absent. Here $h_{70} \equiv H_0/70$ km s$^{-1}$ Mpc$^{-1}$, where $H_0$ is the Hubble constant. Adapted from Driver et al. (2016). Courtesy of A. Wright.

lines). **Semi-forbidden lines** are produced by transitions with intermediate probabilities and are indicated with a single square bracket (e.g. C II]$\lambda$1907,1909). In addition to emission lines, H II regions emit also bremsstrahlung continuum in the radio at typical frequencies of 1–1000 GHz. Hot ($T \sim 10^{6-7}$ K) shocked gas in supernova remnants (SNRs) emits bremsstrahlung continuum in X-rays and highly ionised lines in the X-ray to optical. Diffuse gamma-ray emission up to energies higher than $\approx$700 GeV has been observed in some SFGs (e.g. M 82) due to the interaction of cosmic rays produced by massive star winds and supernovae with the interstellar medium (ISM). Especially in ETGs, X-ray thermal continuum is emitted by halos of hot ($T \gtrsim 10^6$ K) gas with low density. Finally, a contribution from dust-scattered light can be also present in the UV and optical depending on the properties and geometrical distribution of dust grains with respect to the surrounding stars.

3. *Radiation from dust grains*. Interstellar dust consists of solid particles (**dust grains**) diffused in the ISM, especially within the clouds of molecular and atomic gas. Dust is an important contributor to the MIR–FIR SEDs due to a variety of processes. Three main components play a dominant role. **Polycyclic aromatic hydrocarbons** (PAHs) emit strong spectral bands in the MIR ($\lambda = 3.3$, 6.2, 7.7, 8.6, 11.3, 12.7 $\mu$m). Small dust grains (size $\approx 0.01$ $\mu$m) absorb the UV photons coming from nearby hot stars (OB types) and are heated to $T_d \gtrsim 100$–200 K. This absorbed radiation is re-emitted as grey-body radiation (§D.1.3) mostly in the MIR. Larger grains (size $\approx 0.01$–0.25 $\mu$m) in thermal equilibrium with the stellar radiation field are divided into warm and cold populations depending on the achieved temperature ($T_d \approx 20$–60 K and $T_d \approx 10$–30 K, respectively). In both cases, they emit grey-body continuum in the FIR (e.g. $\lambda \approx$ 100–200 $\mu$m). Although dust emission is stronger in SFGs, a contribution to the MIR continuum can be also present in ETGs due to the grains produced by mass loss in evolving red giant stars.

## 3.3 Integrated Radiation from Galaxies

Beyond distances of a few Mpc, it is not possible to observe the individual stars within a galaxy. This implies that the flux received in a given wavelength interval is the sum of all the emission processes occurring in that spectral range. As galaxies are extended objects on the sky plane, $F_\lambda$ measures the flux at wavelength $\lambda$ **integrated** over a given solid angle or over the entire size of the galaxy (eq. C.2). In the spectral region from the UV to the NIR, the spectrum of a galaxy at a cosmic time $t$ can be modelled as

$$F_\lambda(t) = \sum_{i=1}^{N} D_i(\lambda) S_i(\lambda, t) + \sum_{j=1}^{M} D_j(\lambda) G_j(\lambda, t), \tag{3.2}$$

where $S_i(\lambda, t)$ is the individual spectrum of the $i$th stellar population at time $t$, $G_j(\lambda, t)$ is the spectrum of the $j$th component of the interstellar gas present in the galaxy, and $D_i(\lambda)$ and $D_j(\lambda)$ are the attenuation functions due to dust extinction. For a given colour excess $E(B-V)$ and an extinction curve $k(\lambda)$ (§4.2.7), the attenuation function is

$$D(\lambda) = 10^{-0.4 k(\lambda) E(B-V)}. \tag{3.3}$$

Each component of a galaxy can be subject to its own attenuation function depending on the dust spatial distribution and extinction properties. For instance, in our Galaxy at least three components should be taken into account for $S_i(\lambda, t)$ (disc, bulge and bar stars), plus the photoionised gas emission for $G_j(\lambda, t)$. The spectral components $S_i(\lambda, t)$ and $G_j(\lambda, t)$ are also functions of the abundance of metals (metallicity; §C.6.1). Moreover, $S_i(\lambda, t)$ depends also on the stellar initial mass function (IMF; the distribution of stellar masses at the beginning of star formation; §8.3.9) and the star formation history (i.e. how the star formation evolved as a function of time; §4.1.5; §8.6).

A **photometric SED** is defined as the distribution of the flux obtained with photometric data points as a function of wavelength in a given spectral range. For instance, the ensemble of flux measurements with the optical (*UBVRI*) and NIR (*JHK*) filters (§C.4) can be used to derive the *UBVRIJHK* photometric SED of a given galaxy from $\approx 0.35$ $\mu$m to $\approx 2.5$ $\mu$m. If a galaxy is not spatially resolved, each photometric point of its SED includes the sum of all radiative processes (continuum and lines) contributing to the wavelength interval covered by the filter over the entire galaxy.

The typical SED of a galaxy has a double-peaked shape (Fig. 3.6). The first peak around 0.5–1 $\mu$m is due to the sum of the photospheric black-body emissions of all stars present in the galaxy and to the presence of the 1.6 $\mu$m bump. The second peak around 100 $\mu$m is due to the grey-body continuum emitted by dust grains. The shape and level of the continuum in the transition region between the two peaks depends on the MIR emission from warm dust grains. The level of the UV continuum emission depends strongly on the presence of young, massive, hot stars and therefore on the intensity of star formation activity. However, the UV radiation can be severely attenuated if dust extinction is present. Emission lines from molecular, neutral atomic and ionised gas, as well as emission bands from solid particles, are superimposed onto the continuum spectrum. The 3D structure of the galaxy and its projection along the line of sight also play an important role in the final shape of the emerging spectrum. In principle, photometric SEDs represent a 'gold mine' to infer the physical processes occurring within galaxies. However, in practice it is challenging to disentangle the individual components and avoid the degeneracies between the involved quantities (e.g. age, metallicity, extinction; §8.6.2). Nonetheless, physically motivated models have been developed to decompose the observed photometric SEDs into the stellar and ISM components (§8.6.2).

## 3.4 Galaxy Spectra

Photometric observations allow us to cover broad spectral ranges (e.g. from the UV to the FIR; Fig. 3.6), but the information is limited to the reconstruction of the global shape of an SED and to a general overview of the galaxy emission processes. Instead, spectroscopy is more powerful because it provides more detailed information on galaxy properties, although the covered spectral range is much narrower than in the case of multi-band photometry. In spectroscopy, an important quantity is the **spectral resolution**

$$R \equiv \frac{\lambda}{\delta\lambda}, \tag{3.4}$$

where $\lambda$ is the wavelength of the observation, and $\delta\lambda$ is the smallest difference in wavelength that can be resolved (e.g. $\delta\lambda < 3\text{Å}$ is needed to resolve the [O II]$\lambda$3726,3729 doublet). In general, the higher the spectral resolution, the better the achievable accuracy in the spectral analysis and in the derived physical information. Spectral resolution is also important to measure galaxy kinematics reliably. Kinematic measurements are based on the Doppler effect and consist in measuring the difference ($\Delta\lambda$) between the observed and rest-frame wavelengths of a given spectral line ($\Delta\lambda = \lambda_{obs} - \lambda_{rest}$). Since the radial velocity is $v_r = c\Delta\lambda/\lambda_{rest}$ (§C.7), it is clear that high spectral resolution allows us to measure $\Delta\lambda$ more accurately. A resolution $R$ of a few thousands is considered the typical boundary between low and moderately high resolution in galaxy spectroscopy. The photometric SEDs discussed above can be regarded as 'spectra' with very low spectral resolution ($R \approx 5$ in the optical) because photometric filters are separated from each other by typically $\approx$1000 Å.

As displayed in Fig. 3.7, the spectra of present-day galaxies show a wide range of shapes and features depending on the emission and absorption components, the star formation activity and the ISM properties. In general, there is a broad correlation between the Hubble morphological class and the spectral features in the optical. ETGs show spectra

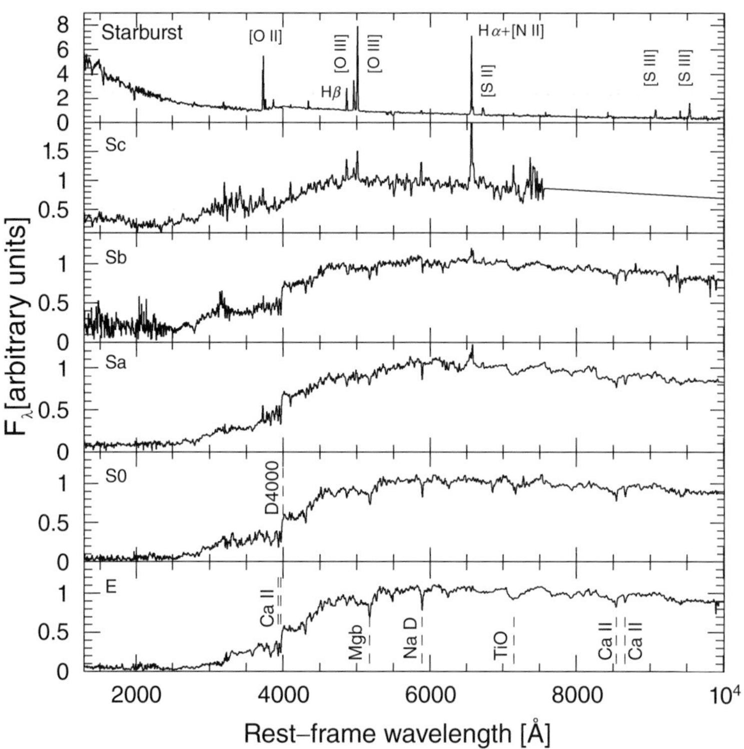


Fig. 3.7 The observed average UV–optical spectra of present-day galaxies of different morphological types (data from Kinney et al., 1996). The main emission and absorption lines are indicated. The starburst spectrum is relative to a case of intense ongoing star formation and low dust extinction with $E(B-V) < 0.1$. The UV continuum and the strength of the emission lines show a rapid decline going from starbursts to ellipticals due to the increasing age of the stellar populations and the gradual decrease of photons capable of photoionising hydrogen and other atomic elements. E/S0 galaxies have red continuum spectra and only absorption lines. The D4000 break is indicated only in the S0 spectrum for clarity, but it is clearly visible in the spectra of galaxies where the bulge component is important (i.e. from E to Sb). The straight line redwards of 7500 Å in the Sc spectrum is the extrapolation of the continuum where data are not available. Some features ($\lambda \approx 3000$–3500 Å and $\lambda \approx 1500$–2500 Å in Sc and Sb spectra, respectively) are due to noise in the spectroscopic data.

characteristic of old systems where star formation ceased a long time ago, and whose stellar populations are evolving passively.[2] In particular, the ETG spectra show (1) red continuum (the older the redder), (2) absorption lines of evolved stars (typically G and K types), (3) no (or very faint) emission lines due to the absent (or very weak) star formation, and (4) a marked continuum discontinuity at 4000 Å. The 4000 Å discontinuity (the so-called **D4000 break**) is due to a sudden onset of absorption features bluewards of 4000 Å that is clearly noticeable in stellar types cooler than G0. A convenient definition is the so-called narrow D4000 break[3]

$$\mathrm{D4000_n} = \frac{\int_{4000}^{4100} F_\nu(\lambda)\mathrm{d}\lambda}{\int_{3850}^{3950} F_\nu(\lambda)\mathrm{d}\lambda}. \tag{3.5}$$

SFGs have typically $\mathrm{D4000_n} \approx 1.1$–1.4, whereas ETGs have $\mathrm{D4000_n} \approx 1.8$–2.1. The $\mathrm{D4000_n}$ amplitude increases with the stellar age (time elapsed since the last star formation activity) and the metal abundance. The 4000 Å break must not be confused with the Balmer jump at 3646 Å corresponding to the sudden decrease in the intensity of the continuum spectrum at the limit of the hydrogen Balmer series. The spectra of ETGs show distinctive breaks also in the UV at 2640 Å, 2900 Å and 3200 Å (Fig. 11.22).

The spectra of early spirals (Sa) are similar to those of E/S0 galaxies, but in addition they show weak emission lines coming from the regions of the disc where star formation is present. The spectra become increasingly bluer and the emission lines stronger for late spirals (Sb–Sd) and irregulars because of the decreasing contribution of the bulge component and the increasing level of star formation.

The strength of a spectral line relative to the underlying continuum is measured by the so-called **equivalent width**, which is defined as

$$W \equiv \int_{\mathrm{line}} \frac{F_{\lambda,\mathrm{cont}}(\lambda) - F_\lambda(\lambda)}{F_{\lambda,\mathrm{cont}}(\lambda)} \mathrm{d}\lambda, \tag{3.6}$$

where $F_\lambda$ is the flux at wavelength $\lambda$, $F_{\lambda,\mathrm{cont}}$ is the average flux of the continuum on either side of the line, and the integral is computed within a wavelength interval which includes the line. The equivalent width is measured in wavelength units (e.g. Å). With this definition, the value of $W$ is positive or negative for an absorption or an emission line, respectively. The equivalent width of absorption lines provides important information because it depends on the transition probability (oscillator strength), the gas column density, the element abundance, the fraction of atoms in a given ionisation state, the velocity distribution of the absorbing gas and the covering factor (i.e. the percentage of the luminous source covered by the clouds of absorbing gas).

The equivalent width should not be confused with the **full width at half-maximum** (FWHM), which is the width (along the wavelength axis) measured at half level between the continuum spectrum and the peak of the line. The units of the FWHM can be

[2] A stellar population evolving in the absence of new events of star formation.

[3] Note that the calculation of $D4000_\mathrm{n}$ has been defined with the spectrum expressed as $F_\nu$, but integrated in $\lambda$ within the ranges indicated in eq. (3.5).

wavelength or velocity. For instance, according to the Doppler formula, a Ly$\alpha$ line ($\lambda = 1216$ Å) with FWHM = 3 Å has a (radial) velocity width of $c(3/1216) \approx 740$ km s$^{-1}$. If the line has a Gaussian profile, the FWHM and the velocity dispersion $\sigma$ are related as FWHM = $2.35\sigma$. Similarly, the **full width at zero intensity** is defined as the base width of a spectral line at the level of the continuum. The measured FWHM or $\sigma$ must be always corrected for the artificial line broadening due to the spectral resolution of the spectrograph. Galaxies with ongoing star formation show the photospheric absorption lines of O and B stars, the emission lines due to the powerful stellar winds produced by these stars, the recombination and forbidden emission lines coming from H II regions, and several ISM absorption lines from elements with a wide range of ionisations (§4.2.3). The extreme cases of galaxies with the highest activity of star formation are called **starbursts** (§4.5). A small fraction of starbursts show strong and broad (with FWHM up to $\approx$ 1000 km s$^{-1}$) emission lines of He II$\lambda$4686, C III$\lambda$4650 and N III$\lambda$4640. These lines are the signatures of Wolf–Rayet stars, i.e. hot ($T \approx$ 25 000–50 000 K), massive ($> 20 M_{\odot}$) stars characterised by very strong stellar winds. **Wolf–Rayet galaxies** represent systems where very massive (i.e. very short-lived) stars are still visible before they explode as Type II supernovae (SNe) within a few Myr. These galaxies are therefore ideal laboratories to study the early phases of starbursts. Fig. 3.8 displays a zoom on the UV spectrum of starburst galaxies and clearly highlights the wealth of spectral lines available for studying the ISM and stellar components in this spectral region.

### 3.4.1 Colours and Bimodality

When photometric SEDs and spectra are not available, colours and luminosities (or absolute magnitudes) are the simplest properties that can be derived for a galaxy because they only require the measurement of the integrated flux and of the distance. Despite their simplicity, colours already provide important indications on galaxy properties because they depend on the ages of stellar populations, the abundance of heavy elements and the dust extinction. However, the consequence of this triple dependence is that colours alone are highly degenerate, and it is impossible to interpret them reliably without additional information (§8.6.2). Optical colours are broadly correlated with the Hubble morphological types. ETGs have the reddest colours, whereas SFGs are bluer. Dust extinction is usually low in ETGs, but it can severely redden the colours of later galaxy types.

The very large samples obtained with massive surveys like the SDSS (Blanton et al., 2017) show that galaxies have a bimodal distribution of colours and are segregated into two main regions of the colour–luminosity (or colour–mass) plane (Fig. 3.9). The first is called the **red sequence** and is mostly populated by ETGs with a wide range of luminosities. However, about 15–30% of red sequence galaxies are not pure ETGs, but SFGs whose colours are red because of a substantial bulge component or dust extinction. The exact percentage depends on mass and redshift range. The second region is called the **blue cloud** and includes the majority of SFGs. The colour bimodality becomes particularly pronounced when the adopted colour index includes a UV filter sensitive to the presence of young stars (e.g. $U - B$ or $u - r$; Fig. 3.9). For a fixed mass or luminosity, the colours in

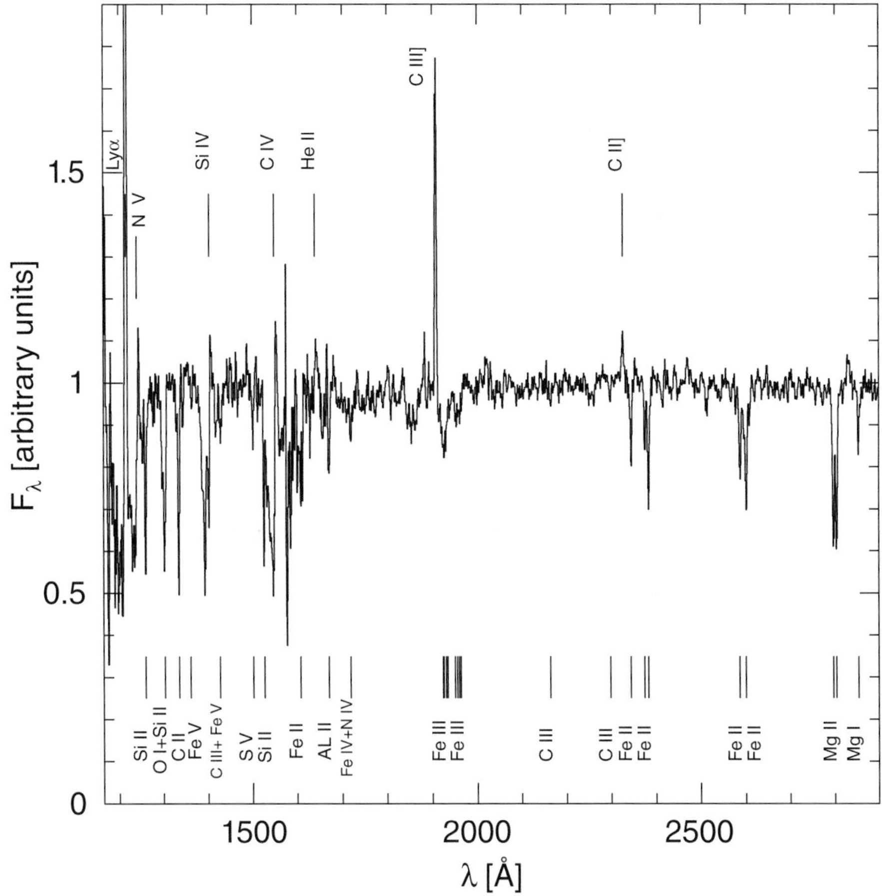


Fig. 3.8 The average UV spectrum of present-day galaxies with intense star formation activity (starbursts). The flux is normalised to the continuum spectrum to facilitate the relative comparison of spectral lines. The emission lines (e.g. He II, C III], C II]) are originated in the photoionised gas. The absorption lines are due to ISM gas (e.g. Si II, C II, O I, Fe II), hot star photospheres (e.g. C III, Fe III, Fe IV, Fe V, S V) or both (e.g. Si IV, C IV). Public data from Leitherer et al. (2011).

the blue cloud are more dispersed than in the red sequence because of the different levels of star formation. Moreover, dust reddening increases the scatter of colours even further. For example, in the case of the $u-r$ colour, the dispersion has root mean square rms$\approx$0.1–0.2 mag and rms$\approx$0.3–0.4 mag for the red sequence and the blue cloud, respectively. The percentage of galaxies within the red sequence and the blue cloud depends on luminosity and mass. The red sequence becomes gradually populated for increasing luminosity or mass. For instance, if the rest-frame $u-r$ colour is considered, the fraction of galaxies located in the red sequence is about 10%, 50% and 100% for $M_r \approx -16$, $-21.5$ and $-23$, respectively. The region between the red sequence and the blue cloud is scarcely populated and is called the **green valley**. Also other quantities show bimodal distributions. Examples are spectral features such as the D4000 break (eq. 3.5), the equivalent width of

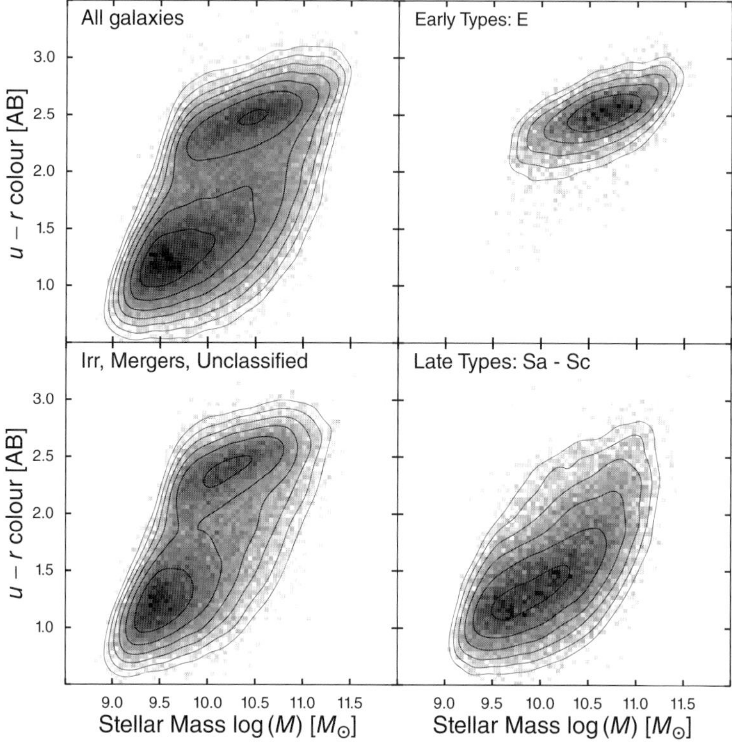


Fig. 3.9 The rest-frame $u - r$ colour as a function of stellar mass and morphology for a sample of 57 160 galaxies with $\log(\mathcal{M}_\star/\mathcal{M}_\odot) > 9$ at $0.02 < z < 0.05$ extracted from the SDSS sample. Darker regions indicate higher number density of galaxies. The lines indicate isodensity contours. Red sequence galaxies have colours in the range of $2 \lesssim u - r \lesssim 3$. Blue cloud galaxies have colours in the range of $0.5 \lesssim u - r \lesssim 1.6$. Figure courtesy of A. Weigel and K. Schawinski.

the H$\delta$ absorption line, the Sérsic index and the effective surface brightness. The galaxy bimodality is an important test for galaxy formation models (§10.10.2).

## 3.5 Statistical Distributions of Galaxy Properties

Galaxies span a wide range of properties. For instance, ultra-faint dwarfs such as Segue 1 ($M_V \approx -1.5$) can be as faint as an individual B star. Instead, giant ellipticals have absolute magnitudes up to $M_V \approx -23$. Similar examples hold for masses, sizes and other galaxy properties. Hence, it is important to understand how the characteristics of galaxies are statistically distributed. These **distribution functions** have a profound meaning because, at a given cosmic epoch, they describe the endpoint of the physical processes that have occurred previously. For these reasons, distribution functions are considered stringent tests for theoretical models of galaxy formation.

### 3.5.1 The Luminosity Function

The **luminosity function** (LF) describes how galaxy luminosities are statistically distributed. The main steps with which the LF of a galaxy sample is derived can be summarised as follows.

1. Measurement of the galaxy redshifts and apparent magnitudes.
2. Calculation of the galaxy luminosity distances and luminosities $L$.
3. Counting the number of galaxies with luminosity between $L - \Delta L$ and $L + \Delta L$.
4. Division of the number of galaxies in each luminosity bin by the volume (in $\mathrm{Mpc}^3$) probed by that bin.

This procedure provides the luminosity function $\Phi(L)$ such that $\Phi(L)\mathrm{d}L$ is the number density of galaxies per cubic Mpc with luminosity between $L$ and $L + \mathrm{d}L$. The same methodology can be applied to galaxy masses to derive the mass function.

Schechter (1976) demonstrated that the distribution of luminosities of galaxies located in clusters can be described by a three-parameter ($\Phi^*$, $L^*$ and $\alpha$) function

$$\Phi(L)\mathrm{d}L = \Phi^* \left(\frac{L}{L^*}\right)^{\alpha} \exp\left(-\frac{L}{L^*}\right) \mathrm{d}\left(\frac{L}{L^*}\right), \tag{3.7}$$

called the **Schechter function**. The function $\Phi(L)$ consists of two components: a power law with slope $\alpha$ for $L \ll L^*$ and an exponential cut-off for $L \gg L^*$, where $L^*$ is the characteristic luminosity and $\Phi^*$ is the normalisation density at $L^*$. $L^*$ marks the transition (the so-called 'knee') where the function $\Phi(L)$ changes from the power law to the exponential regime. The exponential cut-off implies that very luminous galaxies are very rare. The Schechter function can also be expressed as a function of absolute magnitude as

$$\Phi(M)\mathrm{d}M = (0.4 \ln 10)\Phi^* 10^{-0.4(\alpha+1)(M-M^*)} e^{-10^{-0.4(M-M^*)}} \mathrm{d}M, \tag{3.8}$$

where $M^*$ is the absolute magnitude at the knee. The importance of the luminosity function is that it incorporates information on the number density of galaxies as a function of luminosity at a given cosmic epoch. In particular, the LF at $z \approx 0$ is used as the zero-point to investigate how its parameters ($\Phi^*$, $L^*$ and $\alpha$) evolve as a function of redshift and therefore to place constraints on the evolution of the physical processes which shaped present-day galaxies.

The Schechter function allows one also to compute important quantities related to the global properties of galaxies and to investigate which of them play a key role in galaxy evolution (§11.3.4). For instance, the total number density of galaxies above a given luminosity $L_0$ can be calculated as

$$n(L > L_0) = \int_{L_0}^{\infty} \Phi(L)\mathrm{d}L = \Phi^* \Gamma\left(\alpha + 1, \frac{L_0}{L^*}\right), \tag{3.9}$$

where Γ is the incomplete Gamma function.[4] When $L_0 = 0$, eq. (3.9) gives the total number density of galaxies $\Phi^*\Gamma(\alpha + 1)$. The total luminosity density of a given galaxy population (in units of $L_\odot$ Mpc$^{-3}$) above a luminosity $L_0$ is

$$\rho_L(L > L_0) = \int_{L_0}^{\infty} L\Phi(L)\mathrm{d}L = \Phi^* L^* \Gamma\left(\alpha + 2, \frac{L_0}{L^*}\right), \tag{3.10}$$

which, for $L_0 = 0$, gives the total luminosity density of all galaxies of that population, $\rho_{L,\mathrm{tot}} = \Phi^* L^* \Gamma(\alpha + 2)$.

Although the total LF (including all galaxy types, i.e. SFGs + ETGs) can be generally reproduced with a single Schechter function, it has been found that it is better described by a double Schechter function which includes the contributions of the two main galaxy types (ETGs and SFGs):

$$\Phi(L)\mathrm{d}L = \left[\Phi_1^*\left(\frac{L}{L^*}\right)^{\alpha_1} + \Phi_2^*\left(\frac{L}{L^*}\right)^{\alpha_2}\right]\exp\left(-\frac{L}{L^*}\right)\mathrm{d}\left(\frac{L}{L^*}\right), \tag{3.11}$$

where $\Phi_1^*$, $\Phi_2^*$, $\alpha_1$ and $\alpha_2$ are the normalisation densities and power-law exponents of the two Schechter functions, respectively. Instead, the individual luminosity functions of each galaxy morphological type are well described by a single Schechter function. Fig. 3.10 shows an example of a total LF with the relative contributions of ETGs and SFGs. It is evident that the high-luminosity end of the LF is dominated by ETGs, whereas the low-luminosity regime is populated mostly by SFGs. Tab. 3.1 lists the Schechter parameters relative to different galaxy types based on the sample of Fig. 3.10.

**Table 3.1** Optical luminosity functions

| Galaxies | $M_r^*$ (mag) | $\Phi^*$ ($10^{-3}$ mag$^{-1}$ Mpc$^{-3}$) | $\alpha$ | $\rho_{L,\mathrm{tot}}$ ($10^7$ $L_\odot$ Mpc$^{-3}$) |
|---|---|---|---|---|
| All types | −21.71 ± 0.11 | 4.00 ± 0.21 | −1.12 ± 0.03 | 16.02 ± 0.81 |
| Ellipticals | −21.67 ± 0.29 | 1.22 ± 0.19 | −0.65 ± 0.10 | 3.88 ± 0.37 |
| S0–Sa + SB0–SBa | −20.25 ± 0.17 | 2.16 ± 0.29 | 1.08 ± 0.25 | 4.50 ± 0.34 |
| Sab–Scd + SBab–SBcd | −19.83 ± 0.19 | 3.90 ± 0.30 | 0.65 ± 0.20 | 3.81 ± 0.47 |
| Sd–Irr | −18.46 ± 0.16 | 11.56 ± 0.50 | −0.40 ± 0.20 | 1.91 ± 0.06 |

The $r$-band single Schechter function parameters and luminosity density of galaxies at $0.025 < z < 0.06$ and with $M_r < -17.4$ (GAMA survey, Fig. 3.10; Kelvin et al., 2014b). Note that a substantial scatter of the best-fitting Schechter parameters is present in the literature (Kelvin et al., 2014b).

[4] The complete Gamma function $\Gamma(z) = \int_0^\infty t^{z-1}\mathrm{e}^{-t}\,\mathrm{d}t$ can be generalised to the so-called upper incomplete Gamma function, $\Gamma(z, x) = \int_x^\infty t^{z-1}\mathrm{e}^{-t}\,\mathrm{d}t$, such that $\Gamma(z) = \Gamma(z, 0)$.

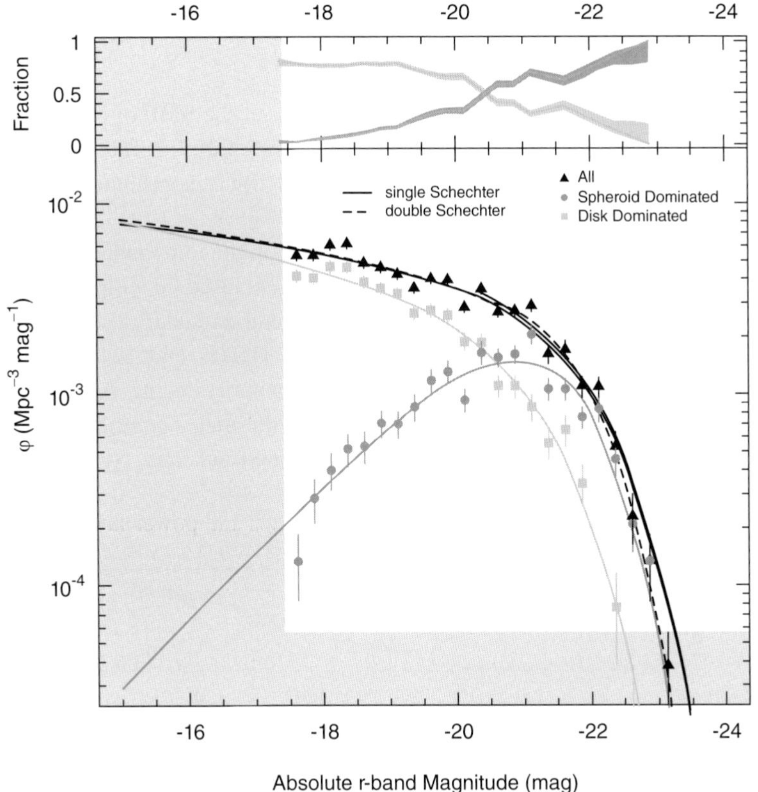


Fig. 3.10

*Bottom panel*. The luminosity function of galaxies at $0.025 < z < 0.06$ in the optical $r$ band. Each data point represents the number density of galaxies per luminosity interval. Error bars show the $1\sigma$ uncertainties. The shaded grey areas indicate those regions where the data are insufficient. The total luminosity function (black) is subdivided into spheroid-dominated (dark grey) and disc-dominated (light grey) galaxies. The total sample has been fitted with a single (solid line) and a double (dashed line) Schechter function. The LFs of the two galaxy subclasses have been fitted with single Schechter functions. *Top panel*. The fractions of spheroid-dominated (dark grey) and disc-dominated (light grey) galaxies with respect to total as a function of absolute magnitude. Tab. 3.1 lists the values of the single Schechter parameters and the luminosity density. Data from the GAMA survey. Adapted from Kelvin et al. (2014b).

## 3.5.2 The Stellar Mass Function

The analysis of stellar light allows one to estimate the galaxy **stellar masses** ($\mathcal{M}_\star$; §11.1.3; eq. 11.4). Thus, it is possible to apply the approach illustrated in §3.5.1 to derive the statistical distribution of galaxy stellar masses, also called the **stellar mass function** (SMF). The Schechter function can successfully reproduce also the SMF. In particular, the SMF displays an exponential cut-off at $\log(\mathcal{M}_\star/\mathcal{M}_\odot) > 11$ and an approximate power-law behaviour at low masses with an exponent $\alpha \approx -1.3$. However, as for the LF, a double Schechter function,

$$\Phi(\mathcal{M})d\mathcal{M} = \left[\Phi_1^*\left(\frac{\mathcal{M}}{\mathcal{M}^*}\right)^{\alpha_1} + \Phi_2^*\left(\frac{\mathcal{M}}{\mathcal{M}^*}\right)^{\alpha_2}\right]\exp\left(-\frac{\mathcal{M}}{\mathcal{M}^*}\right)d\left(\frac{\mathcal{M}}{\mathcal{M}^*}\right), \quad (3.12)$$

provides a better fit to the total SMF of all galaxy types. The SMF of each galaxy type is well represented by a single Schechter function (Fig. 3.11, Tab. 3.2, Tab. 3.3). In particular, the high-mass end is dominated by ETGs, whereas SFGs contribute more at smaller masses.

If galaxies are grouped into two main classes based on the importance of the spheroidal component, about 70% of today's stellar mass is enclosed in ellipticals and S0–Sa galaxies, whereas $\approx 30\%$ resides in disc-dominated and irregular systems (Sb–Sd, Irr). Thus, galaxies with a prominent bulge component (ETGs and Sa galaxies), despite being rarer than late-type systems, contribute substantially to the stellar mass budget of the present-day Universe. As shown in Fig. 3.11, the major contribution to the SMF at $\log(\mathcal{M}_\star/\mathcal{M}_\odot) \gtrsim 10.5$ comes from spheroid-dominated galaxies. At higher masses, in

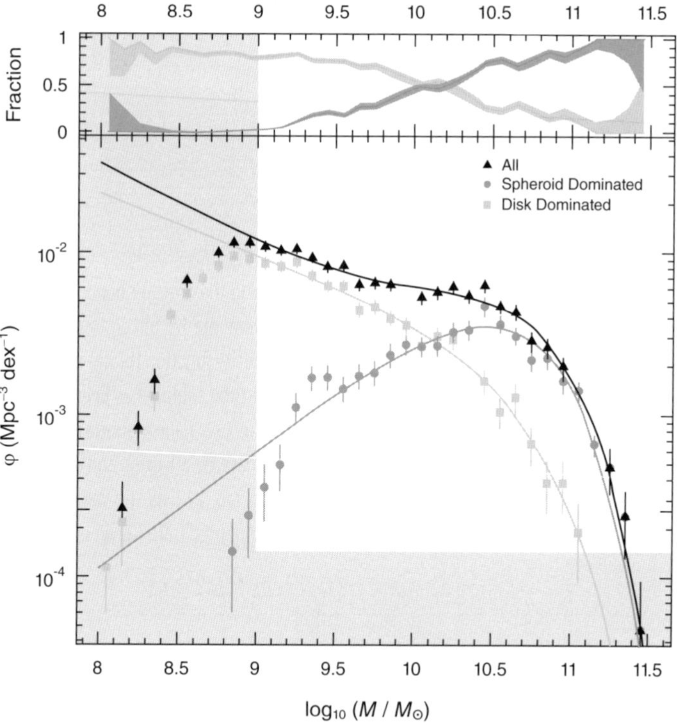


**Fig. 3.11** *Bottom panel.* Similar to Fig. 3.10, but showing the galaxy SMF. The total SMF data (black triangles) are subdivided into spheroid-dominated (dark grey squares) and disc-dominated (light grey circles) galaxies. The SMF of the total sample has been fitted using a double Schechter function (black solid line; Tab. 3.2). The SMFs of spheroids and discs have been fitted each with single Schechter functions (grey solid lines; Tab. 3.3). *Top panel.* The fractions of spheroid-dominated (dark grey) and disc-dominated (light grey) galaxies with respect to total as a function of stellar mass. Tab. 3.3 provides also the values of the stellar mass densities for different galaxy types. Data from the GAMA survey. Adapted from Kelvin et al. (2014a).

**Table 3.2** Total stellar mass function

| $\log(\mathcal{M}^*/\mathcal{M}_\odot)$ | $\alpha_1$ | $\Phi_1^*$ ($10^{-3}$ dex$^{-1}$ Mpc$^{-3}$) | $\alpha_2$ | $\Phi_2^*$ ($10^{-3}$ dex$^{-1}$ Mpc$^{-3}$) |
|---|---|---|---|---|
| $10.64 \pm 0.07$ | $-0.43 \pm 0.35$ | $4.18 \pm 1.52$ | $-1.5 \pm 0.22$ | $0.74 \pm 1.13$ |

Best-fitting parameters of the double-Schechter function which reproduces the total stellar mass function of all galaxy types (Fig. 3.11; Kelvin et al., 2014a).

**Table 3.3** Stellar mass functions per galaxy type

| Galaxies | $\log(\mathcal{M}^*/\mathcal{M}_\odot)$ | $\Phi^*$ ($10^{-3}$ dex$^{-1}$ Mpc$^{-3}$) | $\alpha$ | $\rho_{\mathcal{M},\mathrm{tot}}$ ($10^7 \mathcal{M}_\odot$ Mpc$^{-3}$) |
|---|---|---|---|---|
| Ellipticals | $10.94 \pm 0.10$ | $0.85 \pm 0.27$ | $-0.79 \pm 0.13$ | $6.81 \pm 1.47$ |
| S0–Sa | $10.25 \pm 0.07$ | $2.38 \pm 0.83$ | $0.87 \pm 0.23$ | $7.53 \pm 2.08$ |
| Sab–Scd | $10.09 \pm 0.15$ | $3.57 \pm 0.81$ | $-0.01 \pm 0.31$ | $4.34 \pm 1.62$ |
| Sd–Irr | $9.57 \pm 0.17$ | $3.40 \pm 2.07$ | $-1.36 \pm 0.29$ | $1.77 \pm 1.10$ |
| Spheroid-dominated | $10.60 \pm 0.05$ | $3.96 \pm 1.05$ | $-0.27 \pm 0.16$ | $15.44 \pm 4.31$ |
| Disc-dominated | $10.70 \pm 0.23$ | $0.98 \pm 0.42$ | $-1.37 \pm 0.11$ | $7.08 \pm 3.91$ |

Single Schechter function parameters and stellar mass density for different galaxy types (Fig. 3.11; Kelvin et al., 2014a).

the exponential cut-off part of the SMF at $\log(\mathcal{M}_\star/\mathcal{M}_\odot) > 11.5$, the number density of ETGs is about one order of magnitude larger than that of late-type systems. The general features of the SMF, the relative contributions of ETGs and SFGs, and the absence of galaxies with $\log(\mathcal{M}_\star/\mathcal{M}_\odot) > 12$ are critical tests for galaxy formation models (§10.10.1; §10.11).

### 3.5.3 The Baryonic Mass Function

More complete information on the matter content of galaxies comes from the so-called **baryonic mass function**. Different gas phases can be present within galaxies, from the hot ionised gas located in the halos to the cold molecular gas distributed in the discs (§4.2). Stellar mass is not always the dominant baryonic component because the gas-to-stellar-mass ratio increases with decreasing stellar mass. For instance at $z \approx 0$, neutral atomic hydrogen (H I) dominates the gas mass in galaxies, and it can provide the major contribution to the total baryonic content of low-mass galaxies. The transition from star-dominated to H I-dominated systems takes place at $\log(\mathcal{M}_\star/\mathcal{M}_\odot) \lesssim 9.5$ (Fig. 4.8, top left). The baryonic mass function can be relative to all baryons (stars + all gas phases + dust), or to specific components (e.g. stars + atomic or molecular gas). The mass of a specific gaseous component can be derived from the direct observation of the gas (e.g. the 21 cm line for H I; §4.2.1), or estimated indirectly through statistical scaling relations between gas and dust masses (§11.1.5). The stellar + neutral atomic gas mass function is reproduced very well with a Schechter function. Galaxies with $\log(\mathcal{M}_{\rm HI}/\mathcal{M}_\odot) > 10.0$ are rare in the present-day Universe as their number densities are about 2–3 orders of magnitude lower than galaxies with the same cut in stellar mass, whereas the number densities become comparable for $\log(\mathcal{M}_{\rm HI}/\mathcal{M}_\odot) < 9.5$.

## 3.6 Active Galactic Nuclei

A fraction of present-day galaxies show phenomena in their nuclear regions that cannot be explained with normal stellar activity. These systems are called **active galactic nuclei** (AGNs). The main observed features of AGNs can be summarised as follows.

- Very high bolometric luminosities (§C.2), up to $L \sim 10^{48}$ erg s$^{-1}$.
- Non-stellar UV and optical continuum.
- Strong X-ray and gamma-ray emission. An X-ray luminosity higher than $3 \times 10^{42}$ erg s$^{-1}$ in the rest-frame band 2–10 keV is usually taken as the threshold above which the X-rays are dominated by AGN emission. Below this threshold, it has been found that ongoing star formation activity contributes significantly to the X-ray luminosity due to the presence of high-mass binary stars (§11.1.2).
- Non-thermal radio emission with a wide range of luminosities.
- Broad permitted emission lines with FWHM up to >10 000 km s$^{-1}$.
- Narrow (FWHM $\approx$ 1000 km s$^{-1}$) emission lines (forbidden and permitted) characterised by ionisation higher than in SFGs.
- Emission of relativistic plasma jets departing from the nucleus with velocities near the speed of light.
- Time variability of the continuum and broad emission-line luminosities on a wide range of timescales (from minutes to years).

Not every AGN shares all of these characteristics, and a complex taxonomy has been developed to define classes and subclasses of these objects depending on their specific observed properties (see Padovani et al., 2017, for a review). However, all AGN types can be broadly grouped into four main classes.

1. *Quasars or QSOs*. These objects were initially discovered as optical counterparts of radio sources at cosmological distances (Schmidt, 1963). Their large distances were derived from the redshifted emission lines observed in their optical spectra. These objects were called **quasi-stellar radio sources (quasars)**. The quasi-stellar adjective refers to the compact/star-like morphology in the optical. Later, a larger number of objects were found to have similar optical properties, but weaker or absent radio emission. These objects were named **quasi-stellar objects (QSOs)**. Today, the two names are considered interchangeable. Strong radio emission is present only in a small

fraction of these systems. An AGN is called **radio-loud** when the radio-to-optical flux ratio is larger than a given value. A typical threshold is $F(5\ \mathrm{GHz})/F_R \geq 10$, where $F(5\ \mathrm{GHz})$ and $F_R$ are the fluxes at 5 GHz (radio) and in the $R$ band (optical; $\approx 680$ THz), respectively. If an AGN does not satisfy the radio-loudness criterion, it is called **radio-quiet**. The radio emission is originated by relativistic electrons moving in a magnetic field and thus emitting synchrotron radiation (§D.1.7). Synchrotron emission produces power-law continuum spectra that can be described by $F_\nu \propto \nu^\alpha$, where $F_\nu$ is the observed flux at a given frequency $\nu$, and $-0.5 \lesssim \alpha \lesssim -1$. Radio-quiet QSOs outnumber the radio-loud ones by a factor of $\sim 10$. The bolometric luminosities of QSOs are the highest of all AGNs ($L \sim 10^{45}$–$10^{48}$ erg s$^{-1}$). The optical spectra show broad permitted and narrow forbidden emission lines. The continuum and broad lines are often variable on timescales of months. Thanks to their high luminosity, QSOs have been identified up to very high redshifts ($z > 7$).

2. *Seyfert galaxies*. Named after their discoverer (Seyfert, 1943), these systems are disc galaxies hosting a high surface brightness nuclear region emitting strong emission lines. Their bolometric luminosities ($L \sim 10^{43-45}$ erg s$^{-1}$) are lower than those of QSOs. Based on their optical spectra, Seyferts have been classified Seyfert 1 (where both broad and narrow emission lines are present) and Seyfert 2 (where only narrow lines are visible).
3. *Radio galaxies*. These galaxies emit synchrotron radiation with radio luminosities similar to those of radio-loud QSOs. However, while in QSOs the host galaxy is almost invisible due to the outshining luminosity of the active nucleus, in radio galaxies it is possible to observe directly the galaxy hosting the AGN activity. Based on the Fanaroff–Riley (FR) classification (Fanaroff and Riley, 1974), two main classes, **FR I** and **FR II**, have been defined depending on whether the radio source luminosity at 178 MHz is lower or higher than $2 \times 10^{32}$ erg s$^{-1}$ Hz$^{-1}$. This critical luminosity corresponds to a marked change in the properties of radio galaxies. For a fixed radio luminosity, FR IIs are on average more luminous in the optical continuum and emission lines than FR Is. While the structure of FR I radio sources shows a variety of radio morphologies, FR IIs are easily recognised by their well defined radio lobes with evident regions of high surface brightness (called hot spots) often connected to the galaxy nucleus through **jets** (collimated outflows of relativistic electrons departing from the galaxy nucleus) and/or bright edges. The prototype of an FR II is Cygnus A. The sizes of radio sources range from tens of kpc to Mpc scales. As in Seyferts, the optical spectra can show broad and narrow emission lines (**broad-line radio galaxies**), or only narrow lines (**narrow-line radio galaxies**). The host galaxies are preferentially early types, although sometimes the morphologies appear rather complex and disturbed. Typically, FR Is are located in moderately rich cluster environments, whereas FR IIs tend to be more isolated.
4. *Blazars*. This class contains two types of objects both very luminous in the radio. The first includes **flat-spectrum radio-loud quasars** (FSRQs) characterised by radio synchrotron emission spectra $F_\nu \propto \nu^\alpha$, with $\alpha > -0.5$. The optical spectra of FSRQs are similar to those of radio-loud QSOs. The second class includes the **BL Lacertae (BL Lac) objects**. The name comes from the designation used for variable stars, as

this object has a star-like morphology and was thought to be a variable star within our Galaxy at the time of its discovery in the constellation of the Lacerta. Differently from FSRQs, the optical spectra of BL Lac objects show nearly featureless continua and, rarely, weak emission and/or absorption lines. FSRQs and BL Lac objects belong to the same class of **blazars** because they are AGNs whose jets are oriented at small angles with respect to the line of sight of the observer (say $< 20°$). This geometry originates relativistic effects such as Doppler boosting of the flux, strong and random flux variability with timescales ranging from hours at high energies (e.g. X-rays) to weeks/months at longer wavelengths (e.g. optical), and strong linear polarisation. Blazars are also very luminous gamma-ray sources.

In general, objects with broad emission lines are called **Type 1 AGNs**, whereas **Type 2 AGNs** means that only the narrow lines are visible. For the classes of AGNs listed above, Fig. 3.12 shows representative spectra in the UV–optical region. The fraction of galaxies hosting a radio-loud AGN increases strongly with the stellar mass of the host galaxy and scales approximately as $\propto \mathcal{M}_\star^{2.5}$. This fraction increases to $\approx 30\%$ for $\log(\mathcal{M}_\star/\mathcal{M}_\odot) \gtrsim 11.7$.

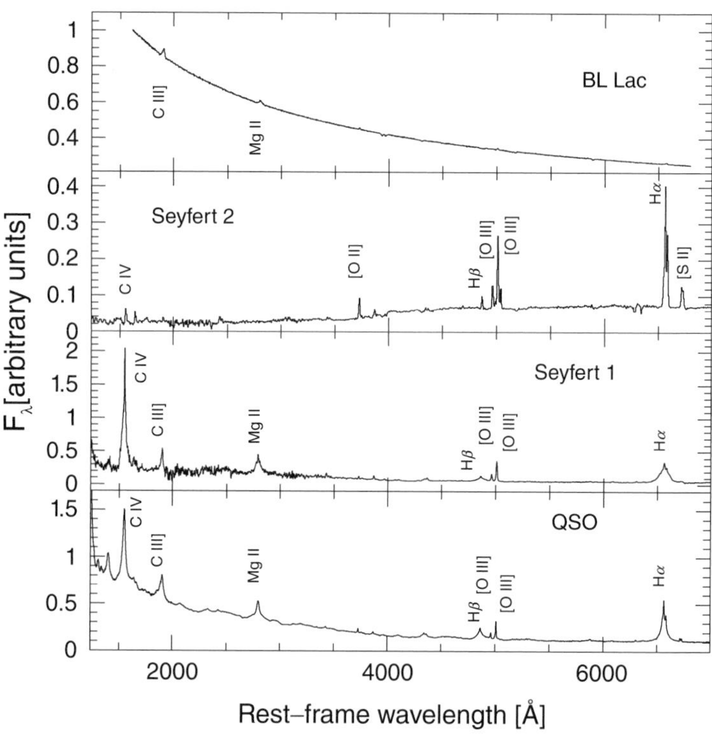


Fig. 3.12 The average rest-frame UV–optical spectra of QSOs (data from Telfer et al., 2002), Seyfert 1 and Seyfert 2 galaxies (data from Francis et al., 1991) and BL Lacertae objects (data from Landoni et al., 2013). Broad emission lines are present only in QSOs and Seyfert 1s. BL Lac objects are nearly featureless due to the high level of the continuum which outshines the spectral lines.

### 3.6.1 The Supermassive Black Hole Paradigm

Following the seminal works of Salpeter (1964) and Zeldovich and Novikov (1964), the general consensus is that the AGN energy is powered by accretion onto a **supermassive black hole** (SMBH) with $\log(\mathcal{M}_\bullet/\mathcal{M}_\odot) \approx 6$–$10$ located at the centre of the host galaxy. The prerequisite is the presence of matter (gas and stars) close enough to the SMBH to be accreted. The radiative energy ($E$) is generated by the conversion of gravitational energy of the matter infalling with an accretion rate $\dot{\mathcal{M}}_{\rm acc}$, and heated to high temperatures in an **accretion disc** surrounding the SMBH (Pringle and Rees, 1972; Shakura and Sunyaev, 1976). The accretion process can take place only if the inward gravitational force acting on the gas is stronger than the outward radiation pressure force (§8.8). The main AGN fuelling equation can be written as

$$L = \frac{\mathrm{d}E}{\mathrm{d}t} \approx \frac{\mathrm{d}\mathcal{M}_{\rm acc}}{\mathrm{d}t}\epsilon_{\rm rad}c^2, \tag{3.13}$$

where $L$ is the radiative bolometric luminosity and $\epsilon_{\rm rad}$ is the **radiative efficiency** of the process. A fiducial value adopted for the efficiency is $\epsilon_{\rm rad} \approx 0.1$. In comparison, the efficiency of thermonuclear reactions in stellar nuclei is much lower (0.007 or less). Based on the previous equations, an accretion rate $\mathrm{d}\mathcal{M}_{\rm acc}/\mathrm{d}t \approx 2\mathcal{M}_\odot$ yr$^{-1}$ is sufficient to explain a typical QSO luminosity of $L \sim 10^{46}$ erg s$^{-1}$. The growth of the SMBH is

$$\frac{\mathrm{d}\mathcal{M}_\bullet}{\mathrm{d}t} \approx \frac{\mathrm{d}\mathcal{M}_{\rm acc}}{\mathrm{d}t}. \tag{3.14}$$

The total energy radiated by the AGN during its net lifetime $\tau$, assuming that $\epsilon_{\rm rad}$ is constant with time, is

$$E = L\tau = \epsilon_{\rm rad}\mathcal{M}_{\rm acc}c^2 \approx \epsilon_{\rm rad}\mathcal{M}_\bullet c^2, \tag{3.15}$$

where $\mathcal{M}_{\rm acc}$ is the total mass accreted. Thus, the final mass of the SMBH can be estimated as

$$\mathcal{M}_\bullet \approx \frac{L\tau}{\epsilon_{\rm rad}c^2}. \tag{3.16}$$

The accretion lifetime $\tau$ is uncertain, but estimates based on statistical analysis, such as the so-called Soltan argument (Soltan, 1982), suggest that $\tau \sim 10^{7-9}$ yr. This approach is based on comparing the relic SMBH mass density in the present-day Universe (due to the past AGN activity) with the integrated energy output of all AGNs. Nonetheless, this method is affected by significant systematic uncertainties because it requires knowledge of the bolometric luminosity of the AGN and the efficiency $\epsilon_{\rm rad}$. Despite these uncertainties, for $L = 10^{46}$ erg s$^{-1}$ and $\epsilon_{\rm rad} = 0.1$, the expected SMBH masses at the end of the accretion process are of the order of $\mathcal{M}_\bullet \sim 10^{7-9}\ \mathcal{M}_\odot$, in agreement with those of SMBHs found in nearby bulges and ETGs. These fossil SMBHs in present-day galaxies, such as the SMBH of our Galaxy ($\mathcal{M}_\bullet \simeq 4.1 \times 10^6\ \mathcal{M}_\odot$), could be the remnants of the accretion that occurred in the past when the AGN process was active. It is still unclear how the AGN phenomenon ceases. Some possibilities include the exhaustion of gas and stars around the SMBH and/or a decline of the efficiency $\epsilon_{\rm rad}$.

### 3.6.2 The Structure of Active Galactic Nuclei

The widely accepted scenario of the AGN structure is that the central source (the accretion disc around the accreting SMBH) is very small ($\sim 10^{14}$–$10^{15}$ cm, about the size of the solar system!). The radius of the accretion disc ($R_{\rm acc}$) depends on the mass of the SMBH, and scales approximately as $R_{\rm acc} \propto \mathcal{M}_\bullet^{2/3}$ according to theoretical predictions. This central engine is thought to be surrounded by the so-called **broad-line region** (BLR), where dense ($n \sim 10^8$–$10^{11}$ cm$^{-3}$) photoionised ($T \sim 10^4$ K) gas clouds move fast (FWHM up to >10 000 km s$^{-1}$) and are responsible for the emission of the broad permitted lines. The central source and the BLR are thought to be surrounded by a toroidal structure made of colder gas and dust (called **dusty torus**) that induces the radiation emitted by the central regions to escape anisotropically with a biconical pattern. The narrow (FWHM $\approx$ 500–1000 km s$^{-1}$) emission lines originate from the so-called **narrow-line region** (NLR) where the gas has a lower density ($n \sim 10^{2-4}$ cm$^{-3}$), as demonstrated by the presence of several forbidden lines. The typical size of the NLR is of the order of 100 pc or larger, and often shows a biconical morphology that is consistent with the presence of an anisotropic radiation field emerging from the inner regions inside the torus. In this scenario, many of the observed differences amongst the AGN types (e.g. broad versus narrow emission lines) are ascribed to geometrical effects depending on the orientation of the system with respect to the line of sight of the observer. Based on this **unification scheme**, AGNs with broad emission lines (e.g. QSOs, Seyfert 1 and broad-line radio galaxies) represent cases where the region inside the torus is directly visible, whereas the properties of Seyfert 2 and narrow-line radio galaxies can be explained if the central engine and the BLR are obscured by the dusty torus (oriented edge-on), and only the emission from the NLR (which is more extended than the torus) is visible. In this scenario, blazars represent extreme cases with a pole-on line of sight nearly aligned with the relativistic jets ejected from the nucleus.

### 3.6.3 The Spectral Energy Distribution of Active Galactic Nuclei

The radiation emitted by AGNs is due to several processes. The inner structure of the accretion disc has temperatures $T \sim 10^5$ K and produces a thermal continuum bump peaking in the UV and soft X-rays. Another source of thermal emission is the dusty torus, where the dust grains are heated up to $T_{\rm d} \approx 1000$ K producing a continuum peak in the MIR spectrum. In addition, synchrotron and inverse Compton emission play a key role in producing a non-thermal continuum extended across the entire electromagnetic spectrum. This multi-component SED can be approximated with a power law ($L_\nu \propto \nu^\alpha$, with $\alpha \approx -1$) over several decades of frequency, but with significant deviations (bumps, depressions, breaks) due to the thermal components and the relative weights of the different radiation processes. Because of the much higher luminosity, the SEDs of AGNs outshine those of the host galaxies by orders of magnitude, especially in the case of Type 1 AGNs. Fig. 3.13 shows the SEDs of representative AGNs compared to those of normal galaxies. Some AGNs are so heavily obscured by dust and dense gas that it is difficult or even impossible to unveil them, with the exception of the spectral regions not affected by absorption and

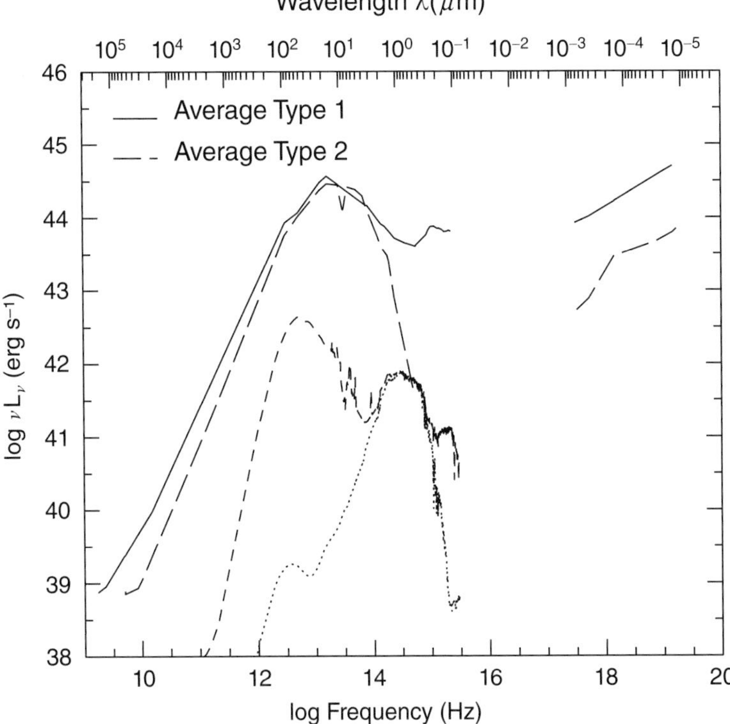


Fig. 3.13 The average SEDs of Type 1 (solid line) and Type 2 (long-dashed line) AGNs (data from Prieto et al., 2010) compared with those of normal elliptical (dotted) and spiral galaxies (short-dashed) (data from Polletta et al., 2007). The differences in the SED shape and luminosity between AGNs and normal galaxies are evident. Courtesy of M. Brusa.

attenuation processes (i.e. hard X-rays, FIR, radio). However, also the X-ray photons can be absorbed if the column density of neutral hydrogen is high enough. Two processes can cause the attenuation of X-ray photons: photoelectric absorption by electrons bound to atoms, and Compton scattering by free electrons. The former is significant at column densities of neutral hydrogen $N_{\mathrm{H}} > 10^{21}$ cm$^{-2}$, but irrelevant if X-ray photons have energy higher than $\approx$10 keV. The latter, consisting in an inelastic scattering where part of the photon energy is absorbed by the recoiling electron, is important at $N_{\mathrm{H}} > 10^{24}$ cm$^{-2}$: in this high-density regime an AGN is called **Compton-thick**. In the extreme cases with $N_{\mathrm{H}} > 10^{25}$ cm$^{-2}$, the high-energy spectrum becomes completely depressed over the whole X-ray spectral range.

# 4 Present-Day Star-Forming Galaxies

The term 'star-forming galaxies' (SFGs) is used to indicate galaxies that are actively forming their stellar component at the time of the observation. In the local Universe, large SFGs are called **disc galaxies**, spiral galaxies or late-type galaxies (see the morphological classification in §3.1). SFGs of lower masses (typically below a stellar mass $\mathcal{M}_\star \sim 10^9\ \mathcal{M}_\odot$) are considered dwarf galaxies: we refer to a galaxy in this category as **dwarf irregular** (dIrr). Note that dIrrs include both Sm and Im types described in §3.1. When the star formation activity is particularly vigorous, a galaxy is called a **starburst galaxy** (§3.3). Elliptical and lenticular (S0) galaxies (Chapter 5) can also exhibit low levels of star formation, but typically, for a given $\mathcal{M}_\star$, at least one order of magnitude lower than the cases discussed in this chapter.

Within the class of SFGs there are a number of subclasses that have been already discussed in §3.1. Mostly for historical reasons, this nomenclature reflects the optical morphology of these systems and thus specifically relates to their stellar content. In addition to this, the most fundamental feature of SFGs is that they have *cold* gas, at typical temperatures from tens to hundreds kelvin. This gas, detected via various emission and absorption processes across the electromagnetic spectrum, is closely related to the presence of star formation. In this chapter, we discuss the main properties of present-day SFGs and their scaling relations. At the end (§4.6), we describe the stellar and gaseous contents of our own Milky Way. Tab. 4.1 summarises the physical properties of SFGs as they can be inferred from observations in the nearby Universe.

## 4.1 Stars

The stellar constituents of SFGs are primarily two: the disc and the bulge. The relative prominence of discs and bulges is one of the main drivers of the Hubble/de Vaucouleurs classifications (Chapter 1 and §3.1). Spiral galaxies of earlier types (Sa) have large bulges and the prominence of the bulge component progressively decreases moving to Sb, Sc and Sd spirals. Roughly 60% of all spiral galaxies host stellar bars (§4.1.2). Moreover, disc galaxies are embedded in extended stellar components, called stellar halos (§4.1.3). Low-mass SFGs have a more irregular morphology and the irregularity often correlates with the level of star formation. Stars belonging to the different galaxy components tend to have different physical properties. In the following we give a description of the various stellar components of present-day SFGs. Other reference textbooks on this topic are Sparke and Gallagher (2006) and Binney and Merrifield (1998).

**Table 4.1** Typical properties of present-day star-forming galaxies

| | Spirals[a] | Dwarf Irrs[b] | Milky Way[c] |
|---|---|---|---|
| Luminosity ($B$ band) ($L_\odot$) | $10^9$–$10^{11}$ | $\lesssim 10^9$ | $3 \times 10^{10}$ |
| Magnitude ($B$ band) | −17 to −23 | $\gtrsim -17$ | −20.7 |
| Disc stellar mass ($\mathcal{M}_\odot$) | $10^9$–few $\times 10^{11}$ | $10^5$–$10^9$ | $\approx 4 \times 10^{10}$ |
| Bulge mass ($\mathcal{M}_\odot$) | $10^8$–$10^{11}$ | absent | $1$–$1.5 \times 10^{10}$ |
| Stellar disc scalelength (kpc) | 1–10 | $< 5$ | 2–3 |
| Presence of bars | 50–70% | rare | yes |
| H I mass ($\mathcal{M}_\odot$) | $10^8$–$5 \times 10^{10}$ | $10^6$–$10^9$ | $4 \times 10^9$ |
| Extent of the H I disc (kpc) | 10–100 | 1–10 | $\approx 25$ |
| Molecular gas mass ($\mathcal{M}_\odot$) | few $\times 10^7$–$10^{10}$ | uncertain | $\approx 1 \times 10^9$ |
| Rotation velocity[d] (km s$^{-1}$) | 100–400 | $< 100$ | 220–250 |
| Virial (total) mass ($\mathcal{M}_\odot$) | $10^{11}$–$10^{13}$ | $10^8$–$10^{11}$ | $1$–$1.6 \times 10^{12}$ |
| Star formation rate ($\mathcal{M}_\odot$ yr$^{-1}$) | 0.1–few tens | $10^{-4}$–1 | 1–3 |
| $\langle \Sigma_{\rm SFR} \rangle$[e] ($\mathcal{M}_\odot$ yr$^{-1}$ kpc$^{-2}$) | $10^{-4}$–1 | $10^{-4}$–1 | $\approx 0.003$ |

[a] Indicative ranges from Courteau et al. (2007), Lelli et al. (2016a) and Catinella et al. (2018).
[b] Values mostly from Tolstoy et al. (2009) and Lelli et al. (2016a).
[c] Values mostly from Bland-Hawthorn and Gerhard (2016).
[d] Peak velocity of the rotation curve (§4.3.3).
[e] Star formation rate (SFR) surface density from Kennicutt and Evans (2012).
Starburst galaxies (§4.5) have SFRs up to a few hundreds $\mathcal{M}_\odot$ yr$^{-1}$ and SFR surface densities up to $10^3$ $\mathcal{M}_\odot$ yr$^{-1}$ kpc$^{-2}$.

### 4.1.1 Stellar Discs

The presence of a rotating disc of stars is the main feature of today's SFGs. Typically, the disc hosts stars of different ages ranging from old ($\gtrsim$ 10 Gyr) stars that formed in the early Universe down to very young (currently forming) stars. Galaxy discs, and the locations of the spiral arms in particular, are the regions where the vast majority of the star formation takes place today.

Stellar discs are supported by coherent rotation around the galaxy centre. Typical values for the rotation velocity can vary from a few tens to a few hundreds of km s$^{-1}$ from dwarf to Sa spirals (Tab. 4.1). The velocity dispersion of stars in discs is of the order of ten to several tens of km s$^{-1}$. A common way to describe the type of dynamical support in galaxies is to measure the ratio between rotation velocity and velocity dispersion: the **$V/\sigma$ ratio**. This ratio is often of the order of 10 for the stellar discs of large spiral galaxies (fully rotation-dominated) but it can be as low as $\sim 1$ in small dIrrs. In the former case (high $V/\sigma$) the kinematics is dominated by circular motions and discs are relatively thin (§4.6.1), while they can be thicker in the latter case.

The radial distribution of the stellar light in a galaxy disc, seen face-on, is well described by an **exponential profile** of surface brightness (§C.2),

$$I(R) = I_0 \exp\left(-\frac{R}{R_{\rm d}}\right), \tag{4.1}$$

where $I_0$ is the central ($R = 0$) surface brightness, usually expressed in $L_\odot\,\mathrm{pc}^{-2}$, and $R_\mathrm{d}$ is the exponential scale radius called **disc scalelength** (the half-light or effective radius, §3.1.4, of a pure exponential disc is $R_\mathrm{e} \simeq 1.678\,R_\mathrm{d}$). Typical values of the disc scalelengths of spiral galaxies are in the range 1–10 kpc. Eq. (4.1) can be seen as a particular Sérsic function (§3.1.3) with $n = 1$ and can be conveniently rewritten as a linear relation using magnitudes,

$$\mu(R) = \mu_0 + 1.0857\left(\frac{R}{R_\mathrm{d}}\right), \tag{4.2}$$

where $\mu(R)$ and $\mu_0$ (central surface brightness) are expressed in mag arcsec$^{-2}$ (eq. C.8).

The central surface brightness *of the discs* of spiral galaxies was originally proposed to be always around $\mu_0 \approx 21.7$ mag arcsec$^{-2}$ in the $B$ photometric band (§C.4), a property called the **Freeman law** (Freeman, 1970). In the subsequent years, it became clear that this observation was hampered by selection effects that caused discs at lower central surface brightness to be missed in those first surveys. Today, the Freeman value remains a reference for the discs of large spirals. Galaxies with central surface brightness in the $B$ band $\mu_0 \gtrsim$ 23 mag arcsec$^{-2}$ are called **low surface brightness** (LSB) galaxies as opposed to **high surface brightness** (HSB) galaxies. LSB galaxies have typically also lower luminosities.

Galaxy discs are seen, on the sky, at some **inclination angle** $i$ with respect to the line of sight: the convention is to have $i = 0$ for face-on and $i = 90°$ for edge-on systems. The apparent flattening of the galaxy is thus due to the inclination of the disc if the disc thickness (see below) can be neglected: a circle inclined with respect to the line of sight appears as an ellipse. Thus, the surface brightness profile is extracted by averaging the observed surface brightness ($I_\mathrm{obs}$) along ellipses with **axis ratio**[1] given by the disc inclination and then by converting it to a *face-on* ($I$) surface brightness $I = I_\mathrm{obs}\cos i$. To understand this correction, recall that the projection of an 'intrinsic' area in the plane of the disc increases by a factor $1/\cos i$ going from an inclined to a face-on view.

Real galaxy discs do have non-zero **thickness** and thus the distribution of their stellar light also changes in a direction perpendicular to the plane of the disc. This so-called **vertical profile** can be studied in spiral galaxies seen edge-on. In the study of disc galaxies, one typically uses cylindrical coordinates $(R, \phi, z)$ centred at the galaxy nucleus and with the galaxy disc lying on the $z = 0$ plane. In this geometry, the stellar luminosity density (or emissivity; §D.1.2) can be written as

$$j_\star(\boldsymbol{x}) = j_{\star,0}\,\exp\left(-\frac{R}{R_\mathrm{d}}\right)\exp\left(-\frac{|z|}{h_\star}\right), \tag{4.3}$$

where $j_{\star,0}$ is the central stellar emissivity (typically in $L_\odot\,\mathrm{pc}^{-3}$) and $h_\star$ is the **scaleheight** of the stellar disc. The study of edge-on galaxies observed in IR bands and thus mildly affected by dust extinction (§4.2.7) reveals that $h_\star$ is much smaller than the disc scalelength (typically $h_\star \approx 0.1R_\mathrm{d}$). This justifies neglecting the thickness for the deprojection of the surface brightness of inclined galaxies, as explained above. Differently from the gaseous

[1] The axis ratio of a disc projected on the plane of the sky is defined as the ratio between its projected semi-minor axis $b$ and its semi-major axis $a$. This is related to the inclination $i$ of the disc as $\cos i = b/a$.

disc (§4.6.2), the thickness of a stellar disc does not appear to strongly vary with radius within the same galaxy. Finally, most galaxy vertical profiles are not fitted by a single exponential, but a second disc component with larger scaleheight (thick disc) materialises at large distances from the midplane. We discuss this component in the context of the Milky Way (§4.6.1).

The central surface brightness of galaxy discs does not usually significantly exceed the Freeman value. However, when a prominent bulge is present, typically for Hubble types Sa and Sb, the *observed* central surface brightness can be much brighter (Fig. 3.4). In order to derive the disc surface brightness in the case of prominent bulges one needs to perform the so-called **bulge–disc decomposition**, an example of which is shown in Fig. 4.1 (left). The modern way to perform this decomposition is by fitting the two-dimensional (2D) surface brightness in one or more optical/IR band(s) with codes that generate mock galaxies from multi-component models. Typically, one includes an exponential disc (eq. 4.1; two free parameters: $I_0$ and $R_\mathrm{d}$) and a bulge with a Sérsic profile (eq. 3.1; three free parameters: $I_\mathrm{e}$, $R_\mathrm{e}$ and $n$). Bars, other disc components, spiral arms and other features can also be added to the fit.

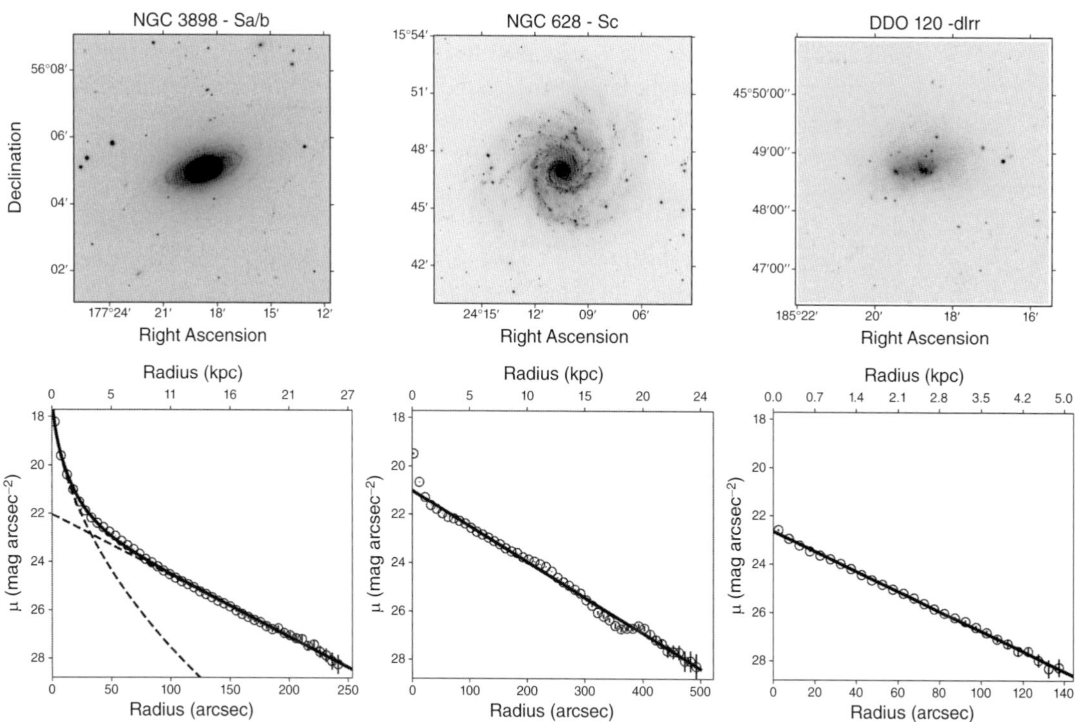


Fig. 4.1 *Top panels*. Optical images (SDSS *g* band) of three representative SFGs of Hubble types Sa/b, Sc and dIrr, respectively, from left to right. *Bottom panels*. Surface brightness (*g* band) profiles of the galaxies in the corresponding above images. The solid curves are fits to the profiles with exponential discs and with the sum of an exponential disc and a Sérsic profile for the bulge (both dashed curves) for the Sa/b galaxy (*left*). Note that also the Sc galaxy (*middle*) shows the presence of a small bulge (inner tens of arcsec). Courtesy of A. Marasco.

We can integrate eq. (4.1) out to a generic radius $R$ to obtain the luminosity within that radius,

$$L(R) = 2\pi \int_0^R I_0 \exp\left(-\frac{R'}{R_\mathrm{d}}\right) R'\,\mathrm{d}R'$$
$$= 2\pi I_0 R_\mathrm{d}^2 \left[1 - \exp\left(-\frac{R}{R_\mathrm{d}}\right)\left(1 + \frac{R}{R_\mathrm{d}}\right)\right] \xrightarrow[R\to\infty]{} 2\pi I_0 R_\mathrm{d}^2, \tag{4.4}$$

where the last expression gives the total luminosity of the exponential disc. Deep photometry of nearby galaxies shows that a significant fraction of discs remain exponential out to the outermost radius reached. Some stellar discs (the fraction of which is much debated) show **breaks** that can take the form of either a **truncation**, i.e. the surface density falls below the exponential extrapolation, or an **up-bending profile**. These breaks are potentially caused by tidal interactions with companion galaxies or by the migration of stars from the inner disc (§10.7.5).

An almost ubiquitous property of galactic discs is the presence of **spiral arms** (left and middle top panels in Fig. 4.1). The spiral often develops over the whole disc with two or four distinct arms. We call this large-scale structure a **grand design** spiral. Spiral patterns that do not cover the whole disc or lose coherence are called **intermediate spirals**. Finally, in a few disc galaxies, the spiral pattern is nearly absent and the disc appears as a 'sea' of small arm-like structures; these are called **flocculent spirals**. For regular spirals, we can define the **pitch angle** as the angle between the tangent to the arm at a certain location and a circle centred at the galaxy's nucleus and intersecting that location. Spiral arms with very low pitch angles are called **wound-up spirals** (this can be seen in Fig. 8.7): a ring of stars would have null pitch angle. The pitch angle tends to increase from Sa galaxies to Sc/Sd galaxies (Fig. 3.1).

The typical rotation velocity ($v_\mathrm{rot}$)[2] of stars and gas in disc galaxies, excluding the central regions, is nearly constant with radius (flat rotation curve, see §4.3.3). As a consequence, the **angular speed** $\Omega = v_\mathrm{rot}/R$ (also called **angular frequency**) decreases as $\approx 1/R$ at large radii. In these conditions, the existence of a grand design (or even an intermediate) spiral structure is surprising. Indeed, the inner disc completes a full rotation much faster than the outer disc and this should distort the spiral pattern very quickly and make it disappear. We say that the arms should 'wind up' rapidly. Solutions to this **winding problem** are discussed in §10.1.6.

### 4.1.2 Bulges and Bars

The central regions of spiral galaxies (in particular Hubble types Sa and Sb) host a spheroidal component called the **stellar bulge**. The properties of bulges are not homogeneous, as it appears that there are two distinct types of structures that are likely to have formed in different ways. The first are called **classical bulges** and have properties in common with the stellar component of elliptical galaxies (§5.1). These bulges tend to have

[2] Hereafter, we indicate the rotation velocity with $v_\mathrm{rot}$ but keep using $V$ for the $V/\sigma$ ratio, as it is a much used convention.

radial profiles with high ($2 < n \lesssim 4$) Sérsic index (§3.1.3) and thus very high central surface brightnesses and stellar densities. The kinematics is characterised by random motions (measured by the stellar velocity dispersion) and a modest or absent rotation. When bulges are flattened they appear consistent with being flattened by rotation (see §5.1.2). We discuss the formation mechanisms of these structures in §10.3. The second type of bulges are called **pseudobulges**. They are characterised by a larger amount of rotation and by stellar profiles with low ($n \lesssim 2$) Sérsic indices. They are more flattened than classical bulges and less concentrated. Cold gas is observed in the region of a pseudobulge with significant star formation in the central parts. These structures are thought to arise from the internal evolution of the disc (§10.2).

The prominence of the bulge is quantified by the $B/T$ ratio (§3.1.3), also called **bulge fraction**. This ratio usually refers to the luminosity in an optical or IR band. Later Hubble types (Sb and Sc) have lower $B/T$ ratios (typically below 0.2) than Sa spirals and they tend to have a larger fraction of pseudobulges. A fraction of Sc (or Sc and Sd in the de Vaucouleurs classification, §3.1) galaxies appear **bulgeless** in the sense that no spheroidal component is observed in their inner parts. Conversely, there are galaxies where bulges of different types coexist.

In general, discs and bulges do not have the same average stellar populations (§3.3) as the stars in the bulges tend to be older. This implies that the *luminosity* $B/T$ ratios do not straightforwardly correspond to fractions of the bulge mass to the total stellar mass. The ratio between stellar mass and luminosity $L_X$ in a given photometric band $X$ (§C.4), normalised to the solar values, is called the **stellar mass-to-light ratio** and is indicated with $\Upsilon_\star^X \equiv (\mathcal{M}_\star/L_X)/(\mathcal{M}_\odot/L_{X,\odot})$, where $L_{X,\odot}$ is the Sun's luminosity in that band. A way to estimate the mass of bulge and disc components is then to multiply their luminosities by two different stellar mass-to-light ratios. Mass-to-light ratios can be motivated by fits of the stellar spectra with population synthesis models (§8.6.2). We discuss this further in §4.3.4. An indicative ratio between the mass-to-light ratios of the bulge ($\Upsilon_{\star,\mathrm{bu}}$) and the disc ($\Upsilon_{\star,\mathrm{d}}$) components in NIR bands is $\Upsilon_{\star,\mathrm{bu}} \approx 1.5\Upsilon_{\star,\mathrm{d}}$. In Tab. 4.2 we list the main properties of the stellar bulges of disc galaxies.

About 60% of disc galaxies host central **stellar bars**. These are prolate structures with their major axis aligned along the plane of the galactic disc and characterised by axis ratios that can reach 1 : 5. They tend to have a disc-like, although relatively old, stellar population. For this reason, they are more easily observed at NIR wavelengths (see also §3.1.1). Fig. 4.2 shows this clearly for the spiral galaxy NGC 253, a member of the nearby group of galaxies called the Sculptor Group. The disc of this galaxy has an inclination of about 78° (close to edge-on) and the near side is the one towards north-west. If we had only the UV emission (tracing young stars) and the H$\alpha$ emission (tracing photoionised gas, §4.2.3), we would not be able to say that this galaxy is barred. However, at NIR wavelengths the central bar appears very bright and dominates the stellar light emission.

The distinction between bars and bulges is not always straightforward, especially for bars with low axis ratios, so some disc galaxies may be misclassified. Possibly also as a consequence of these uncertainties in the classification, the general properties of bars show similarities with those of pseudobulges (Tab. 4.2). A clear case in which a bar and a bulge

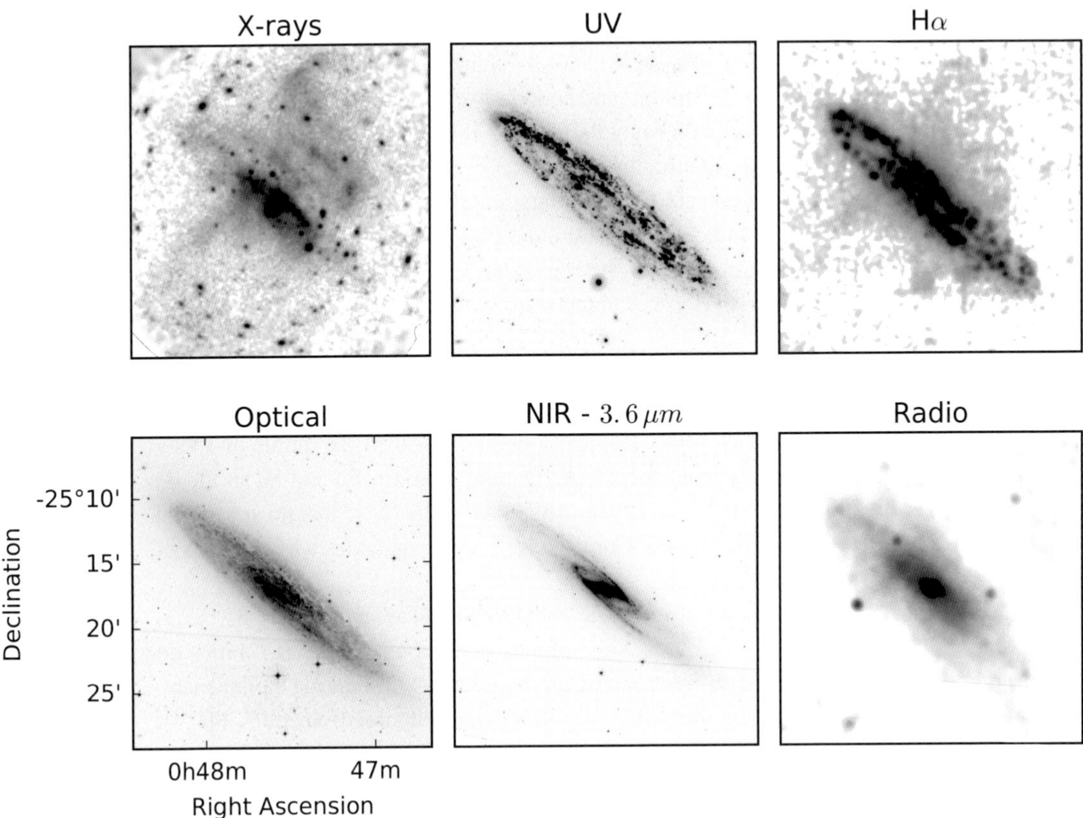


Fig. 4.2 The nearby barred and starburst spiral galaxy NGC 253, at a distance of about 3.5 Mpc ($1' \simeq 1$ kpc), seen at different wavelengths. *Top row from left to right*. X-rays (soft band: 0.2–1 keV) from *XMM-Newton*; near-UV (1800–2800 Å) from *GALEX*; H$\alpha$ emission line (narrow filter centred at 6563 Å) from the Cerro Tololo Inter-American Observatory. *Bottom row from left to right*. Optical (red band: 6300–6900 Å) from the Digital Sky Survey (DSS); NIR (3.2–3.9 $\mu$m) from *Spitzer*; radio continuum (1.4 GHz) from the VLA. Broadly speaking, UV light shows the location of very young stars, whereas the optical is contributed by stars of different ages and the NIR band mostly shows the old stellar population. H$\alpha$ emission is produced by photoionised gas (§4.2.3), X-rays come from both X-ray binaries and collisionally ionised gas (§4.2.4), while the radio emission is mostly due to relativistic electrons spiralling in NGC 253's magnetic field (§4.2.8). Note that H$\alpha$ and X-rays show a prominent galactic wind (§4.2.10) above and below the disc of the galaxy. Data credits: X-ray from Bauer et al. (2008), data courtesy of W. Pietsch; H$\alpha$ from Hoopes et al. (1996), data courtesy of J.M. van der Hulst; the 'Second Epoch Survey' (DSS2) was produced by the AAO using the UK Schmidt Telescope, plates have been digitised and compressed by the Space Telescope Science Institute (STScI); the *Spitzer Space Telescope* is operated by the Jet Propulsion Laboratory, Caltech, under a contract with NASA; radio continuum from Carilli et al. (1992), data courtesy of C. Carilli.

coexist is shown in Fig. 4.3 (left). This galaxy is nearly face-on and the central part of its disc is fully dominated by the highly elongated, non-axisymmetric structure of the stellar bar. The central regions of the bar, however, appear very round and this is the bulge.

The vertical structure of bulges is better observed in edge-on galaxies. Fig. 4.3 (right panels) shows three examples of edge-on galaxies with markedly different bulges. In the top panel we have a galaxy with a classical bulge, while the other two galaxies

**Table 4.2** Properties of bulges and bars

| | Classical bulges | Pseudobulges | Bars |
|---|---|---|---|
| Sérsic index | $>2$ | $\lesssim 2$ | $\lesssim 1$ |
| Colours[a] | redder | bluer | bluer + dust lanes |
| Stellar population | old | disc-like/old | disc-like/old |
| Fractional stellar mass[b] | $\sim 10\%$ | $\sim 10\%$ | $\sim 10\%$ |
| Kinematics | $V/\sigma < 1$ | $V/\sigma \gtrsim 1$ | complex |
| ISM | scarce | present | present |

[a] Typical colours of bulges are $1.2 \lesssim B - R \lesssim 2$. Redder bulges are at the top of this range and vice versa for bluer bulges.

[b] This is an average fractional mass estimated for massive ($\mathcal{M}_\star \gtrsim 10^{10}\,\mathcal{M}_\odot$) spirals only; in these, the discs contain about 70% of the stellar mass. In smaller galaxies the bulge and bar components become less important. Values from Weinzirl et al. (2009).

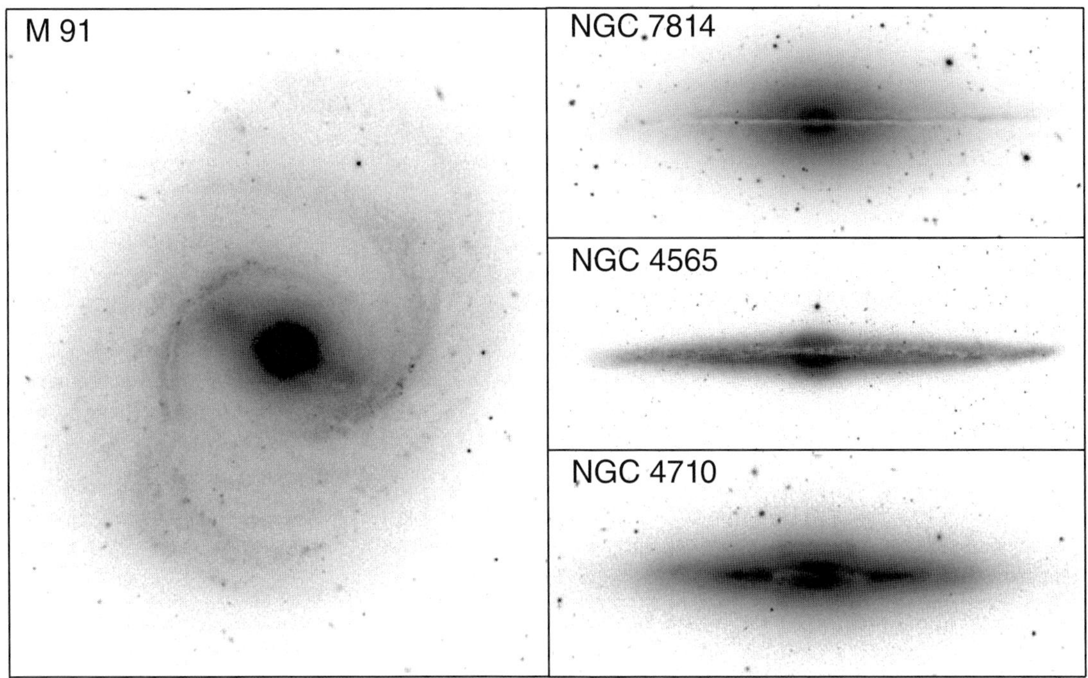


**Fig. 4.3** *Left panel*. Optical image of the nearly face-on SBb spiral galaxy M 91 with a stellar bar and a bulge. *Right panels*. Optical images of three edge-on galaxies showing the different types of stellar bulges. The *top panel* shows an Sa/b spiral with its classical bulge. The *middle panel* shows an Sb type with a boxy (mildly peanut-shaped) bulge. The galaxy in the *bottom panel* is an S0/a with a strongly peanut-shaped bulge. The sizes of the fields of view are approximately as follows. M 91: $5' \times 6' \simeq 18$ kpc $\times$ 23 kpc; NGC 7814: $8' \times 3' \simeq 34$ kpc $\times$ 13 kpc; NGC 4565: $17' \times 6' \simeq 59$ kpc $\times$ 22 kpc; NGC 4710: $5' \times 2' \simeq 25$ kpc $\times$ 10 kpc. Data credit: SDSS and Aladin Sky Atlas, developed at CDS, Strasbourg Observatory, France (see Bonnarel et al., 2000, for details).

have **boxy** or **peanut-shaped bulges**. These types of bulge shapes are typical of the category that we have called pseudobulges, although note that this classification may vary in the literature. In §10.2 we see how pseudobulges can form via instabilities occurring in the bar, thus underlining the tight connection between the two structures. Boxy or peanut-shaped bulges in edge-on galaxies are considered evidence of the presence of bars, which are otherwise difficult to detect due to projection effects. Lurking bars in edge-on galaxies can also be unveiled by the peculiar kinematic pattern of gas and stars (§10.2.1). However, kinematic data of the required quality are not always available.

### 4.1.3 Stellar Halos

All SFGs, with the possible exception of low-mass ones, are embedded in extended **stellar halos** made of diffuse stars, star clusters (called globular clusters; §4.6.1) and stellar streams. Globular clusters are aggregations of nearly coeval stars that are found around both star-forming and early-type galaxies. Their number correlates with the mass of the host galaxy or with the galaxy type. Spiral galaxies can have a few hundred globular clusters while massive ETGs can have up to thousands. They are only seldom found around dwarf galaxies, a notable exception being the Fornax dwarf spheroidal galaxy (§6.3) that has five globular clusters.

Other than star clusters, diffuse stars and stellar streams are also observed in galactic halos. We refer to halo stars as those that are clearly not part of any of the components mentioned above, namely disc, bulge and bar. However, this distinction is not always straightforward (see in particular the case of the Milky Way in §4.6.1). In external galaxies, there are two techniques to detect stellar halos: one can either detect resolved individual stars or image the diffuse stellar light at low surface brightness. The first technique consists in extracting the position of all stars above the detection limit of the observations in a field around a galaxy. This is a very powerful way to unveil faint structures, but can only be applied to nearby objects, often with the use of either large ground-based telescopes like the VLT or with the *HST*. The second technique can be carried out with relatively small telescopes and very long exposure times, and it can reach larger distances. Both techniques show that stellar halos are very extended although much less massive than the disc components of the host galaxies. **Stellar streams** appear frequent although a fraction of stars can belong to a more diffuse component.

Fig. 4.4 shows two examples of stellar halos mapped with the above two techniques. The left panel shows the Andromeda galaxy (M 31), the other large spiral of the Local Group other than the Milky Way (§6.3) at a distance $d \approx 780$ kpc from us. A number of campaigns with different telescopes have revealed a series of stellar streams that are thought to be the result of minor merger events and fly-bys of companion galaxies (§6.1 and §10.4). The right panel of Fig. 4.4 shows the more distant galaxy NGC 5907 ($d \approx 14$ Mpc). Deep photometry reveals extended loops of stars surrounding the edge-on disc. These streams are likely produced by one satellite wrapping around the main galaxy more than once (the orbital timescale of the progenitor exceeds 1 Gyr).

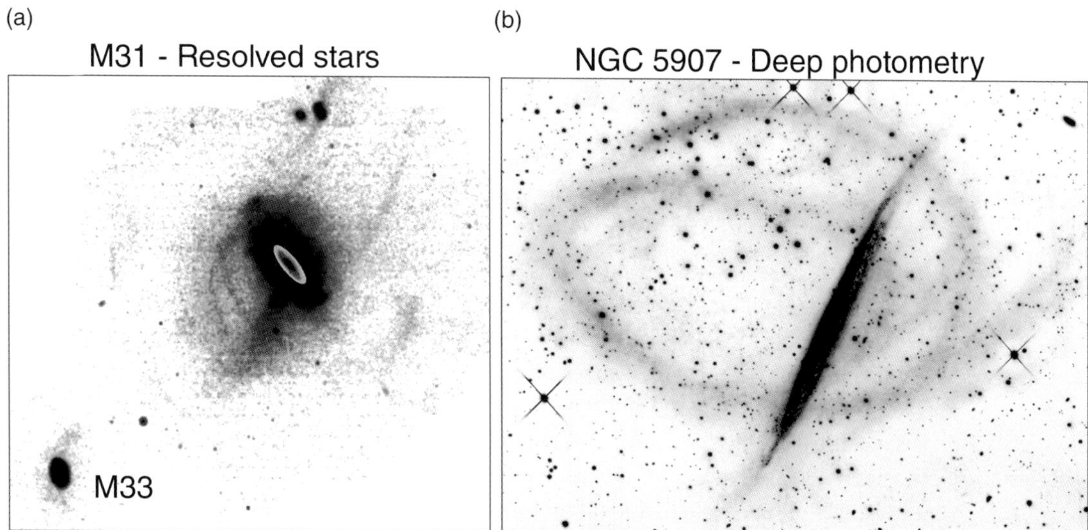


Fig. 4.4 *Left panel.* Imaging of resolved individual stars in the large field of view surrounding the Local Group galaxies M 31 (centre, the ellipse outlines its bright stellar disc) and M 33 (bottom left). The projected distance between the two galaxies is about 200 kpc. From Martin et al. (2013). © AAS, reproduced with permission. *Right panel.* Deep diffuse-light photometry of the nearby edge-on galaxy NGC 5907 (see Martínez-Delgado et al., 2008, for details). The size is about $24' \times 18'$ corresponding to 98 kpc $\times$ 74 kpc at the distance of NGC 5907. Image credit: R. J. Gabany.

### 4.1.4 Colours and Metallicity

The photometric properties of SFGs change with the wavelength of the observations (§3.1.1). Properties like the scalelength of the stellar disc, the $B/T$ ratio (§3.1.3) and the prominence of the bar are particularly affected. As a general trend, exponential discs tend to have smaller scalelengths at long wavelengths towards the NIR bands with respect to blue optical bands. This property can be seen as the fact that discs are *bluer* in the outer parts than in the inner parts. A visualisation of this is provided, for instance, by Fig. 4.2, where it is clear that the outer disc is brighter in UV than in the NIR. As a reference, the optical $B - R$ colour goes from inner values $1.2 \lesssim B - R \lesssim 2$ (in particular if a bulge is present) to values $B - R \lesssim 1$ in the outer parts of the discs. The interpretation of these **colour gradients** in galaxy discs is that the metal content and/or the average age of the stellar populations decrease with increasing radius.

Estimates of the metal abundances of stars in the outer parts of nearby galaxy discs consistently show that outer stars are more metal-poor than central stars. These *negative* **metallicity gradients** are also observed in the gas, in particular using H II regions (see Fig. 4.21 for the Milky Way). Note that, in the case of stellar and gas abundances, one does not typically use the same elements, and conversions must be applied (§C.6.1). The presence of metallicity gradients is of fundamental importance to understand the evolution of discs. The general idea is that discs form *inside-out*. In this scenario, the inner stars have more time to pollute the surrounding gas with metals and thus to increase its metallicity (§10.7.4). Finally, we mention that the average metallicity of the entire galaxy correlates with its stellar mass as we see in §4.4.3.

### 4.1.5 Star Formation Rates and Histories

Star formation is a fundamental process of galaxies that builds up their stellar component. The distinctive feature of SFGs with respect to ETGs is that stars are still forming today at a non-negligible rate. This latter is called **star formation rate** (SFR) and it is usually given in units of $\mathcal{M}_\odot\,\mathrm{yr}^{-1}$. Typical values of the SFRs for different classes of SFGs are given in Tab. 4.1 (to know how the SFRs are estimated, see §11.1.2). When the SFR of a galaxy is several times larger than most galaxies of the same stellar mass (§4.4.4), the galaxy is considered a starburst. The SFR of a galaxy changes with time and the reconstruction of its past evolution is called **star formation history** (SFH).

In very nearby galaxies (typically within a few Mpc), the past SFH can be studied using observations of individual stars. Having the photometry of a large number of stars allows us to build colour–magnitude diagrams (§C.5) and reconstruct the SFHs in detail (§6.3.3). These studies show the ubiquitous existence in SFGs of old stellar populations, dating back to 10–13 Gyr ago, together with a variety of subsequent generations of stars. Even local starburst galaxies, which are currently experiencing a peak of star formation activity, have old stars. As a consequence, an SFG hosts a very complex mixture of stellar populations with different ages.

Leaving aside the very nearby galaxies, for most present-day SFGs stars cannot be observed individually as they are too faint and/or their emissions cannot be separated by the resolution of the telescope (§11.1.6). This effect is called crowding. One then resorts to integrated light observations, in the form of SEDs (§3.3) or spectra (§3.4), where the light of the entire galaxy (or a fraction of it) is captured together. In practice, the UV, optical and IR emissions of the galaxy gather contributions from stars of a variety of ages and metallicities that one has to disentangle in order to derive its SFH. This can be done by fitting composite stellar population models that we describe in §8.6.2 and in §11.1.3. Both the reconstruction done with single stars and with integrated light are types of the so-called 'archaeological' approach to gather past information about galaxy evolution (§11.3.10). The main result of this approach for SFGs is that they, unlike ETGs, tend to have relatively *flat* (slowly evolving) SFHs. This is particularly true for the outer parts of the stellar discs, while the inner discs may have steeper SFHs. Pseudobulges and bars have SFHs similar to the inner discs, whereas classical bulges have SFHs more similar to those of ETGs.

A simple calculation can help elucidating why the SFHs are nearly flat. Take a current SFR of a few $\mathcal{M}_\odot\,\mathrm{yr}^{-1}$ and assume that it has remained constant from the time of formation (of order $\sim 10$ Gyr ago) until today. This would build up a stellar mass of a few $\times 10^{10}\,\mathcal{M}_\odot$, comparable to the masses of today's stellar discs (Tab. 4.1). For a more precise estimate, one should consider that part of the stellar mass is eventually returned to the ISM by stellar evolution processes (§8.4.1), but the basic result does not change, i.e. a relatively flat SFH produces roughly the amount of stars that are observed in discs today (§10.7). However, the exact shape of the SFH does depend on the stellar (or the total) mass of the galaxy. Massive spirals tend to have a more steeply declining SFH while smaller SFGs have SFHs that are flat or even rising with time. The SFHs of dIrrs also appear characterised by distinct episodes of activity (§6.3.3).

## 4.2 The Interstellar Medium of Star-Forming Galaxies

The main property that characterises SFGs is that they host large amounts of interstellar gas and dust. We call this component **diffuse matter** or **interstellar medium** (ISM). For reasons that are discussed in the next sections and in §8.1.4, it is customary to classify the gas as *cold* at temperatures[3] $T \lesssim 10^2$ K, *warm* ($10^3$ K $\lesssim T \lesssim 10^4$ K) and *hot* ($T \gtrsim 10^6$ K). The main mass components of the ISM of an SFG are the neutral atomic gas, the photoionised gas and the molecular gas. Of these, the neutral atomic gas is typically dominant in mass in present-day galaxies and more extended radially, whereas the molecular gas is more concentrated in inner parts, and in particular in the spiral arms, where stars are forming more actively. Although not dominant in mass, other diffuse components are crucial for the complex physical and chemical processes that take place in the ISM, in particular the dust (§3.2). In the next sections, we describe the general properties of the ISM of a present-day SFG. The interested reader can find further details in Spitzer (1978) and Binney and Merrifield (1998).

### 4.2.1 Neutral Atomic Gas

Most of the mass in the ISM of a present-day SFG is in the form of **neutral atomic gas**. This is mostly composed of hydrogen, with a percentage in atoms of helium of about 9%. Metals are rare and do not significantly contribute to the mass (§C.6), but are very important for the physics and the chemistry of the gas. The average mass per particle in the neutral atomic ISM in units of the proton mass (mean atomic weight; §D.2.2) is $\mu \approx 1.3$. The neutral gas is mainly observed via the **21 cm hyperfine line** of the hydrogen atom: H I. This transition occurs when the electron's spin flips from a state of parallel to anti-parallel orientation with respect to the spin of the proton in a hydrogen atom situated in the lowest energy state.

A general property of the interstellar gas is that its particles are continuously moving and interact with each other through (elastic) 'collisions': encounters between atoms, electrons or molecules that have the effect of changing the particle velocities. In most of the ISM, the velocities of the particles, and thus the kinetic energies of the collisions, are linked to the gas temperature: the higher the temperature, the higher the speeds (§4.2.3 and eq. 4.9). However, collisions can, in some situations, also have the effect of exciting upper energy levels of atoms and molecules, and, in extreme cases, of ionising them (§4.2.4). These can be seen as *inelastic* collisions. Given a generic transition (electronic, rotational, etc.)

[3] An intermediate temperature range at $T \sim 10^5$ K is sometimes called *warm–hot*. In general, note that definitions of temperature ranges may vary between texts.

between a lower level $i$ and upper level $j$, we can define the equivalent temperature of the transition energy as $T_* \equiv \Delta E_{ji}/k_B$, where $\Delta E_{ji}$ is the energy jump between the two levels. If the gas temperature is of the same order of or higher than this temperature and the gas density is high enough (see §4.2.3), we can expect the upper level of the transition to be well populated.

In the case of H I, the energy difference ($\Delta E_{10}$) between the two hyperfine levels is extremely small ($5.9 \times 10^{-6}$ eV) corresponding to a temperature $T_* \simeq 0.068$ K. This temperature is so low (much lower than the CMB temperature!) that the upper level is always populated by collisions (§D.1.4 and §C.2.1). The spontaneous emission from upper to lower level (necessary to emit a photon) is extremely improbable (Einstein coefficient $A_{10} \simeq 2.88 \times 10^{-15}\,\mathrm{s}^{-1}$ or mean lifetime of 11 Myr) and one may expect the transition not to be observable. Instead, given the very high number of neutral hydrogen atoms in the ISM of galaxies, the transition can be easily detected. What happens in practice is that most excitations and de-excitations of H I are collisionally induced. However, about 1/1000 of the de-excitations are spontaneous and this produces the emission of a photon at a frequency $\nu \simeq 1420.406$ MHz corresponding to $\lambda \simeq 21.106$ cm.

The atomic neutral hydrogen gas is the most abundant component of the ISM in any present-day SFG (Fig. 4.8). H I **column densities** in galaxies have typically values of $N_{\mathrm{HI}} \approx 10^{20}$–$10^{21}\,\mathrm{cm}^{-2}$. Below $N_{\mathrm{HI}} \approx$ a few $\times 10^{19}\,\mathrm{cm}^{-2}$, H I becomes difficult to observe with current radio telescopes and its emission is expected to drop significantly as a larger and larger fraction of the gas is photoionised (§4.2.3). At column densities $N_{\mathrm{HI}} >$ a few $\times 10^{21}\,\mathrm{cm}^{-2}$, H I starts to become optically thick (§D.1.1). Except for these very high densities, the H I emission can be, to a first approximation, considered optically thin.

Under the assumption of an optically thin medium, and following a derivation that is described in §C.2.1, one can find a simple expression that links the H I mass of a source to its observed flux (§C.2) at 21 cm as

$$\mathcal{M}_{\mathrm{HI}} \simeq 2.343 \times 10^5 (1+z)^{-1} \left(\frac{d_{\mathrm{L}}}{1\,\mathrm{Mpc}}\right)^2 \left(\frac{F_{\mathrm{HI}}}{\mathrm{Jy\,km\,s^{-1}}}\right) \mathcal{M}_{\odot}, \tag{4.5}$$

where $d_{\mathrm{L}}$ is the luminosity distance (§2.1.4) or simply the distance for nearby sources and $F_{\mathrm{HI}}$ is the total H I flux. Due to the fact that different parts of a galaxy move with respect to us at different velocities (§4.3.2), which correspond to different observed frequencies (§C.7), to obtain the total flux one needs to integrate the monochromatic flux $F_\nu$ over these line-of-sight velocities covered by the source so $F_{\mathrm{HI}} = \int F_\nu(v)\mathrm{d}v$. Thus, the total flux is in units of jansky (Jy)[4] multiplied by $\mathrm{km\,s^{-1}}$. In the end, under the above assumptions, the H I mass of a galaxy is a rather precise measurement. The main uncertainty is typically the distance of the galaxy, which must be determined, for nearby galaxies, using distance estimators (§4.4.1), while one can use the Hubble flow at cosmological distances (§2.1.2). Although H I is often considered optically thin, it has been estimated that up to $\approx 30\%$ of the emission could be self-absorbed, so $\mathcal{M}_{\mathrm{HI}}$ as given in eq. (4.5) could be larger by a factor up to 1.3, depending on the inclination of the galaxy (more inclined galaxies are

[4] $1\,\mathrm{Jy} = 10^{-23}\,\mathrm{erg\,s^{-1}\,cm^{-2}\,Hz^{-1}}$.

more self-absorbed). Finally, note that eq. (4.5) gives only the *hydrogen* mass and must be multiplied by $\approx 1.4$ to obtain the *total mass* (including helium) of atomic neutral gas.

The 'size' of an H I disc ($R_{\rm HI}$) is usually defined as the radius at which the column density is $1\,\mathcal{M}_\odot\,{\rm pc}^{-2}$ ($N_{\rm HI} \simeq 1.25 \times 10^{20}\,{\rm cm}^{-2}$). The H I column density is a measure of the surface brightness and thus $R_{\rm HI}$ is analogous to the definition of the characteristic radius $R_{25}$ of the optical disc (§3.1.4). Note that for galaxies not at cosmological distances ($z < 0.1$) the surface brightness is independent of the distance (§C.2). The typical value of the ratio $R_{\rm HI}/R_{25}$ is in the range 1.5–2 in present-day disc galaxies. Fig. 4.5 shows the

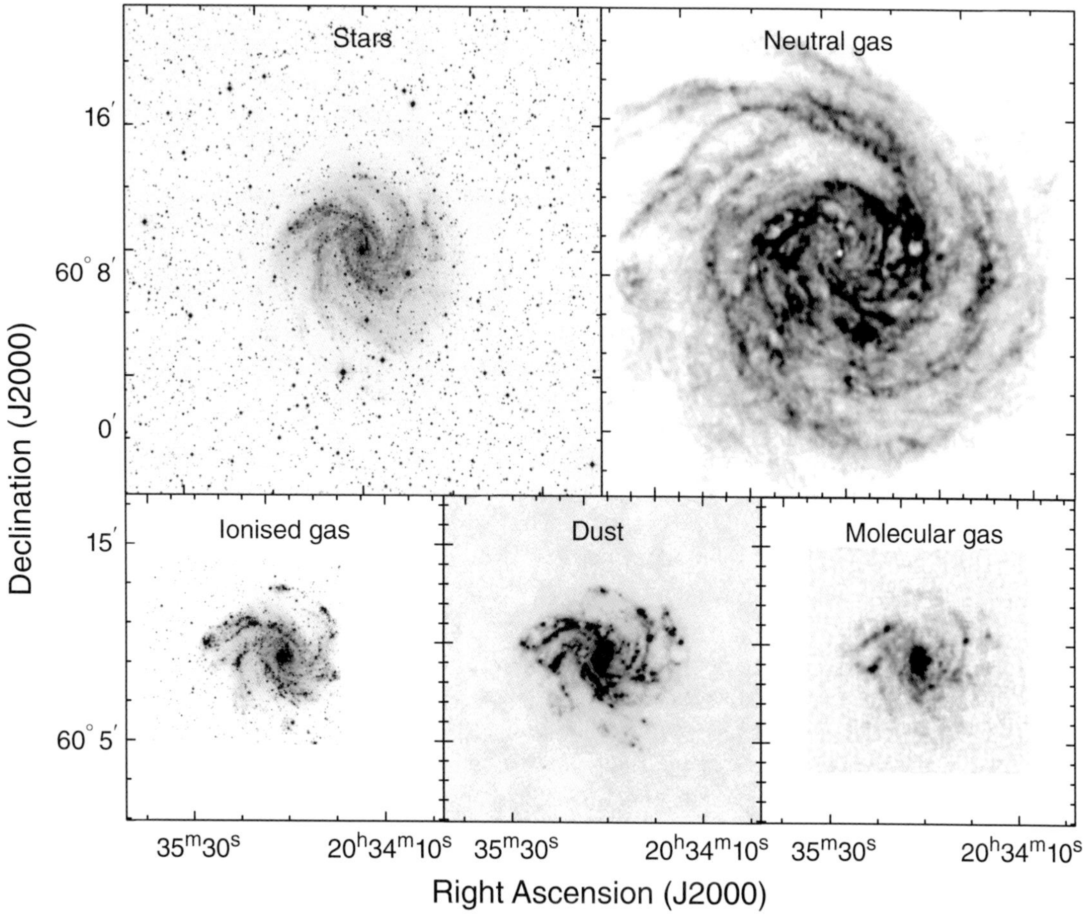


**Fig. 4.5** Gas components in the nearby spiral galaxy NGC 6946 compared to an optical image (DSS) showing the bright stellar component ($R_{25} = 5.6$ arcmin $\simeq 9.8$ kpc) (*top left panel*). *Top right*. 21 cm emission (from the Westerbork Synthesis Radio Telescope) showing the atomic neutral gas (data from Boomsma et al., 2008), $R_{\rm HI} = 13$ arcmin $\simeq 23$ kpc. *Bottom left*. H$\alpha$ emission (from the Kitt Peak National Observatory) showing the ionised gas. *Bottom middle*. 100 $\mu$m continuum (from *Herschel*) tracing the dust. *Bottom right*. CO ($J = 2 \rightarrow 1$) emission (from IRAM) tracing the molecular gas. All images are on the same scale. Data credit for the bottom panels: NASA/IPAC Extragalactic Database.

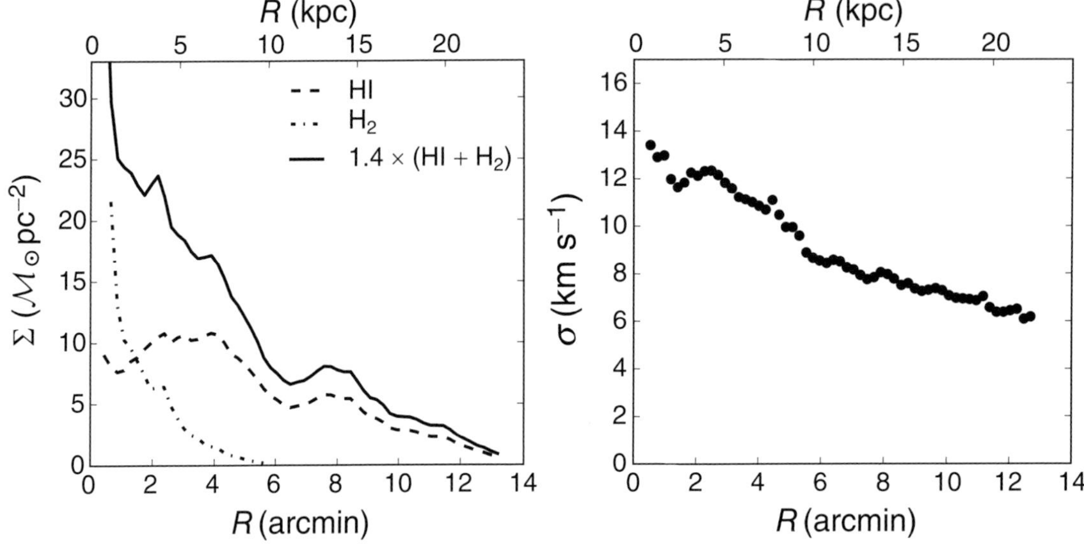


Fig. 4.6 Radial trends of surface density and velocity dispersion of the cold/warm ISM in the nearby spiral galaxy NGC 6946 (see Fig. 4.5). *Left panel.* Surface density profiles of H I, $H_2$ and the total gas density. Note how the atomic gas extends much further out than the molecular gas: this result is general for present-day disc galaxies. The multiplication by 1.4 is often employed to correct for the presence of helium and metals (although, rigorously, one should use slightly different factors for the two components). *Right panel.* Velocity dispersion of the neutral atomic gas averaged in annuli at different radii (data from Boomsma et al., 2008). The dispersion has an average value of $\sigma_{\rm HI} \approx 10$ km s$^{-1}$ and decreases with radius, reaching about $\sigma_{\rm HI} \approx 6$–7 km s$^{-1}$ in the outer parts: this is also a general trend.

comparison between optical, H I and other gas components in the spiral galaxy NGC 6946. The radial profile of the H I surface density is shown in Fig. 4.6 (left).

The amount of H I in SFGs changes with the Hubble type: the later types are more gas-rich. In particular the ratio between stellar and H I masses goes from about 10 in massive spiral galaxies to values of order unity or less for dwarf galaxies (§3.5.3 and §4.2.6). The smallest dwarfs can be totally dominated by H I in terms of their baryonic component.

The neutral atomic gas can also be studied with **fine-structure transitions**, in particular the [C II] line at 158 $\mu$m. This line results from transitions between the $^2P_{3/2}$ and $^2P_{1/2}$ fine-structure states of the singly ionised carbon atom $C^+$. It is important to realise that ISM regions of neutral atomic gas (sometimes referred to as H I regions) are never *fully* neutral. In particular, metals with ionisation potentials below that of hydrogen ($\chi \simeq 13.6$ eV; §4.2.3) can be easily ionised by the stellar radiation field. Among them, carbon (ionisation potential $\chi \simeq 11.3$ eV) is the most important, mainly due to its high abundance (§8.1.1). The separation between the two above fine-structure energy levels corresponds to a temperature of only $T_* \simeq 92$ K, which makes $C^+$ easily collisionally excited. $C^+$ is typically optically thin and it is also observed in regions of molecular gas (that we describe in §4.2.5) illuminated by radiation from massive stars. In these circumstances the highly abundant CO molecules can be dissociated by UV photons (see §8.3.1). These are called **photodissociation regions** and are also characterised by emission of other fine-structure transitions, in particular the [O I] line at 63 $\mu$m.

### 4.2.2 Turbulence of the ISM

A fundamental property of the ISM of SFGs is turbulence. If we observe the H I emission line with a radio telescope in a small region of a galaxy (not affected by large-scale motions) we can derive a precise measurement of the line broadening and thus of the velocity dispersion of the gas. The lines are, to a first approximation, Gaussian with observed dispersion $\sigma_{\rm HI}$ between about 6 and 20 km s$^{-1}$, with an average value $\sigma_{\rm HI} \approx 10$ km s$^{-1}$. The velocity dispersion in galaxies is usually declining with radius (Fig. 4.6, right). In §4.6.2, we see that the neutral atomic gas is divided into two gas phases: the cold and the warm neutral medium with characteristic temperatures of $T \sim 10^2$ K and $T \sim 10^4$ K. The observed line broadening is almost always larger than what one would expect from thermal broadening (an order of magnitude larger for the cold neutral medium; §4.6.2); for a description of the different types of line broadening see §9.8.2. Thus, the neutral atomic gas in a galactic disc has random motions that are *suprathermal*. This is the main observational evidence of **ISM turbulence**, which, in essence, constitutes a disordered kinetic energy not ascribable to the gas temperature and thus to the random thermal motions of elementary particles. In fact, a way to visualise turbulence is to imagine the motion of portions of the gas (rather than elementary particles) with respect to each other. We return to this in §8.3.7.

### 4.2.3 Photoionised Gas

In regions of the ISM exposed to UV photons with energies $h\nu \geq \chi$, where $\chi \simeq 13.6$ eV is the ionisation potential of the hydrogen atom, the gas becomes largely ionised. We call this process **photoionisation**. The ionising photons have wavelengths shorter than 912 Å, thus they lie in the far/extreme-UV (§C.1). They are produced by young massive stars of spectral types O and B (§C.5) or by AGNs (§3.6). The photoionised regions around massive stars are called **H II regions**.

#### Emission Processes

Photoionised gas is observed in emission through **recombination lines** and forbidden lines. Recombination lines are produced by electrons cascading down to lower electronic states after they have been (re)captured by an ion (§D.1.5). The most commonly observed in present-day galaxies, as they occur at optical wavelengths, are the **Balmer series** lines of the hydrogen atom (transition from the $n$th state to the second state, $n > 2$), called H$\alpha$ ($3 \rightarrow 2$, $\lambda \simeq 6563$ Å), H$\beta$ ($4 \rightarrow 2$, $\lambda \simeq 4861$ Å) and so forth. The **Lyman series** lines ($n$th to first state, with $n > 1$) are often used for galaxies at higher redshifts as their rest-frame UV emission falls in the optical band at $z > 2$ (§11.2.7). In disc galaxies, the H$\alpha$ emission traces the spiral arms (Fig. 4.5) and, occasionally, the gas outflows (Fig. 4.2). Unlike the H I gas, it is concentrated in the inner star-forming regions (Fig. 4.5). Because of their relation to massive (young) stars, recombination lines, and in particular Balmer lines, are used as SFR estimators (§11.1.2). Recombination lines are a type of electric dipole transition, i.e. they satisfy a set of quantum-mechanical selection rules (for details see e.g. Draine, 2011).

These transitions are called permitted (§3.2) as they have very high spontaneous transition probabilities (Einstein coefficients $5\times10^8\ \mathrm{s}^{-1}$ and $4\times10^7\ \mathrm{s}^{-1}$ for Ly$\alpha$ and H$\alpha$, respectively), corresponding to very short lifetimes (inverse of the coefficients).

The second class of emission lines, used to detect and study the photoionised gas in the ISM, are the forbidden lines (§3.2), so-called because they violate some of the rules for electric dipole transitions and are thus very improbable (Einstein coefficients $\lesssim 1\ \mathrm{s}^{-1}$). Forbidden lines are excited by collisions, typically electrons colliding on ions. At gas densities of Earth laboratory experiments the de-excitation of the ions also occurs by collisions, whereas the spontaneous decay (and consequent emission of these lines) is essentially impossible. However, the ISM is much less dense and collisional de-excitation is less probable, making the spontaneous decay possible and detectable. This occurs below the so-called critical density (see below) of the transition, a condition that is typically fulfilled for a large number of forbidden lines in an ISM nebula (i.e. an H II region).

Prime examples of these transitions are represented by the lower ions of the oxygen atom. The first ionisation potential of oxygen is similar to that of hydrogen, so oxygen in photoionised regions is largely in the form of $O^+$. The ground levels of $O^+$ can then be collisionally excited and the spontaneous decay produces two lines at similar wavelengths ($\lambda \simeq 3726$ Å and $\lambda \simeq 3729$ Å) due to the fact that the electrons are decaying from two very close energy levels ($^2D_{3/2}$ and $^2D_{5/2}$) to the same $^4S_{3/2}$ level. Such an emission is called a **doublet** and is indicated as [O II]$\lambda\lambda$3726, 3729. The [O II] doublet is often the second brightest emission feature after H$\alpha$ in the optical spectrum of an SFG (Fig. 3.7).

The critical densities of [O II]$\lambda$3729 and $\lambda$3726 are of the order of $10^3\ \mathrm{cm}^{-3}$ and $10^4\ \mathrm{cm}^{-3}$, respectively, while the gas in H II regions is often (but not always) at lower densities (see below). In general, the **critical density** ($n_{\mathrm{crit}}$) of a transition constitutes a threshold that separates two regimes. At densities[5] $n \ll n_{\mathrm{crit}}$, collisional de-excitation is rare and spontaneous decay (and emission) is favoured. However, at these densities, collisional excitation is also less frequent than spontaneous decay and so ions spend very little time in their upper levels, which are quickly depopulated. Thus, an approximation that is often made is that all ions are in the lowest energy level (we return to this in §8.1.1). When the density approaches $n \approx n_{\mathrm{crit}}$, the upper level becomes amply populated (by collisional excitation) but also depopulated by collisional de-excitation. Spontaneous emission is still possible, but it becomes highly improbable for $n \gg n_{\mathrm{crit}}$ as the upper levels mostly decay by collisional de-excitations. These are conditions that approach thermodynamic equilibrium, where the energy levels are populated according to the Boltzmann law (§D.1.4).

Other than emission lines, photoionised gas emits continuum radiation due to the process of **bremsstrahlung**, also called **braking** or **free–free radiation** (§D.1.6). This continuum emission is due to close encounters between electrons and ions that deviate the electrons

[5] Note that $n$ is the density of the *colliding* particles that are exciting the particular transition, e.g. the electrons in the case of the $O^+$ transitions.

from their course. At $T \sim 10^4$ K the bremsstrahlung shines at radio wavelengths and a significant fraction of the radio emission of SFGs (especially at high frequencies) is due to this thermal emission.

## The Strömgren Sphere

Given a medium consisting of neutral particles, ions and electrons, we define the **ionisation fraction** (or degree of ionisation) as

$$x \equiv \frac{n_\mathrm{i}}{n_\mathrm{t}}, \tag{4.6}$$

where $n_\mathrm{i}$ is the number density of ions and $n_\mathrm{t} = n_\mathrm{i} + n_\mathrm{n}$ is the density of atoms (whether neutral or ionised); $1-x$ is the **neutral fraction**. Note that the above definition is sometimes given for a gas of pure hydrogen, in which case it can also be given as $n_\mathrm{e}/n_\mathrm{t}$, with $n_\mathrm{e}$ the electron density. A gas is fully ionised when $x = 1$. As mentioned, hydrogen atoms in their ground state are ionised by photons at $\lambda < 912$ Å and, given that this is the endpoint of the Lyman series ($n = 1 \rightarrow \infty$), radiation at these wavelengths is called the **Lyman continuum**.

The classical theory of H II regions predicts that the region around a massive star should be almost fully ionised ($1 - x \approx$ few $\times\, 10^{-5}$) out to a characteristic distance called the **Strömgren radius** (Strömgren, 1939),

$$R_\mathrm{S} = \left[\frac{3}{4\pi}\frac{Q(H)}{n_\mathrm{e}^2 \alpha_\mathrm{H}}\right]^{1/3} \simeq 0.7 \left[\frac{Q(H)}{10^{49}\ \mathrm{s}^{-1}}\right]^{1/3} \left(\frac{n_\mathrm{e}}{10^3\ \mathrm{cm}^{-3}}\right)^{-2/3} \mathrm{pc}, \tag{4.7}$$

where $Q(H)$ is the emission rate of hydrogen ionising photons from the star and $\alpha_\mathrm{H}$ is the recombination coefficient (rate at which an electron is recaptured by a hydrogen ion in units of $\mathrm{cm}^3\ \mathrm{s}^{-1}$),[6] which is a weak function of the electron temperature. Eq. (4.7) is obtained by assuming a gas of pure hydrogen with a uniform gas density and by imposing equilibrium between photoionisation and recombination rates (a condition called **ionisation balance**) in a spherical volume around the star: $Q(H) = (4\pi/3) R_\mathrm{S}^3 n_\mathrm{e} n_\mathrm{p} \alpha_\mathrm{H}$ ($n_\mathrm{e} = n_\mathrm{p}$ for pure hydrogen). This is the volume that the star *can keep* ionised (against the attempt of the gas to recombine) thanks to its rate of ionising photons. Given the high rate of ionisation within this sphere ($x \approx 1$), the electron density in eq. (4.7) is essentially equal to the number density of the medium (if we assume it atomic) before the ionisation began ($n_\mathrm{e} \approx n$). The Strömgren sphere has typical sizes that can reach a few parsecs for the brightest O stars ($Q(H) \simeq 7 \times 10^{49}\ \mathrm{s}^{-1}$ for an O4 star; §C.5). It is expected to have a sharp edge and no Lyman continuum photons should escape from it. Very massive stars have a fraction of their photons powerful enough to also ionise helium once ($h\nu \geq 24.6$ eV), but not twice ($h\nu \geq 54.4$ eV). The ionisation of He occurs in a region close to the star out to a characteristic radius that lies within the Strömgren radius.

In realistic conditions, H II regions are larger than a theoretical Strömgren sphere and also the fraction of photons that escape is not null, mostly due to the fact that the ISM is clumpy and inhomogeneous (**porosity** of the ISM). This is largely caused by stellar winds

[6] Note that, for this specific problem, the rate $\alpha_\mathrm{H}$ is calculated so as to exclude recombinations from infinity to $n = 1$, as they will produce another ionising photon that is promptly reabsorbed by the medium.

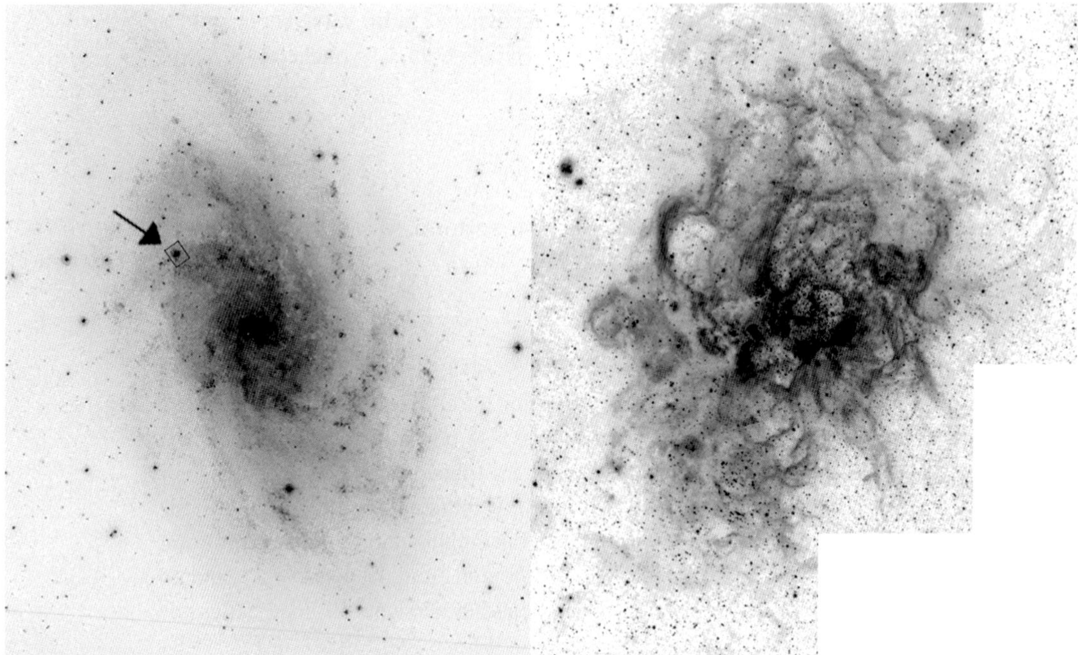

Fig. 4.7 Giant H II region in the Local Group galaxy M 33 (§6.3). *Left panel.* DSS image of this galaxy with an arrow indicating the location of the giant H II region NGC 604. Image size: $55' \times 65' \simeq 13$ kpc $\times$ 15 kpc. Credit: DSS and Aladin Sky Atlas. *Right panel.* Blow-up of the squared area obtained with *HST* (combination of optical wide and narrow filters), showing the H II region. Image size: $1.9' \times 2.2' \simeq 450$ pc $\times$ 520 pc. Credit: NASA/ESA and the Hubble Heritage Team (AURA/STScI).

(§8.7.2) and SN explosions (§8.7.1) that make these regions rather chaotic and carved with shells and bubbles. We can appreciate this in the *HST* image in the right panel of Fig. 4.7, showing a giant H II region in M 33. Note that, in this case, the UV emission is produced by the combined action of several O/B stars.

## Estimates of Temperature and Density

As explained in §4.2.1 and later in §4.6.2, the physics of the thermal part of the ISM of a galaxy is fully dominated by collisions between particles. This is an important property and has the fundamental consequence that all ISM particles at *some location* tend to share a Maxwellian distribution (§D.1.4) with the *same* temperature. It is because of this that we are allowed to give only one value for the temperature of a certain gas structure. Thus, for instance, in an H II region we do not speak of electron temperature ($T_e$) as opposed to ion temperature ($T_i$), but instead we assume a single gas temperature $T \approx T_e \approx T_i$ because both electrons and ions obey the same Maxwellian. This approximation is usually correct within 1% or better. Note, however, that the thermal speeds (eq. 4.9) of electrons ($v_e$) and ions ($v_i$) differ greatly ($v_e \gg v_i$). Note also that, in some cases, the gas temperature is in fact given as electron temperature $T_e$, but this mostly reflects the method used to determine it and does not imply that ions or neutral particles have different temperatures.

The temperature of an H II region can be determined using the ratios between different emission lines of the same element. In general, these ratios depend on both temperature and density but, in certain cases, the density dependence can be neglected. One such case is the doubly ionised oxygen $O^{2+}$, which is the second (and last) oxygen ion that can be produced by photoionisation from massive stars (ionisation potential 35.1 eV). This ion has three relevant energy levels: $^3P$, $^1D$ and $^1S$ in order of increasing energy. As mentioned, collisional excitations populate the upper levels making forbidden transitions possible: the critical densities of these transitions are $n_{\rm crit} > 5 \times 10^5\ {\rm cm}^{-3}$, always higher than typical gas densities in H II regions. The main [O III] forbidden lines are produced by the decay between $^1D_2$ level and levels $^3P_2$ ($\lambda = 5007$ Å) and $^3P_1$ ($\lambda = 4959$ Å), and the decay between $^1S_0$ and $^1D_2$ levels at $\lambda = 4363$ Å.

The excitation energy of the [O III]$\lambda$5007 and the [O III]$\lambda$4959 lines is about 2.5 eV, while that of [O III]$\lambda$4363 is 5.3 eV. The relative population of the levels, determined by the collisions between ions and electrons, depends mostly on the electron temperature (at higher temperature the upper level is more populated). This makes the ratio between the line emissivities (or the observed line fluxes) a 'thermometer' of the H II region. The treatment of the problem requires other approximations and the knowledge of coefficients for spontaneous emission and collision strengths. In the end, it can be shown that the line flux ($F$) ratio between the sum of the $^1D_2 \rightarrow {}^3P$ transitions and the $^1S_0 \rightarrow {}^1D_2$ transition is

$$\frac{F(4959\ \text{Å}) + F(5007\ \text{Å})}{F(4363\ \text{Å})} \simeq 7.9 \exp\left(\frac{32\,900\ \rm K}{T_e}\right), \tag{4.8}$$

where 32 900 K is the temperature corresponding to the energy difference between the levels: 2.8 eV/$k_B$. If all these lines (in particular the one at $\lambda = 4363$ Å, which is the weakest) are detected, then we can estimate the (electron) temperature of the region.

The typical temperatures of H II regions determined using the forbidden line ratios or other methods turn out to be $T \sim 10^4$ K. Surprisingly, this is much lower than the temperature of the photosphere of the O star (Tab. C.4), which is responsible for the ionisation of the medium. This suggests that the gas surrounding these massive stars is not simply heated by the stars, but it also cools efficiently. It is the emission of the forbidden lines, in particular [O II], [O III], [N II] and [S III], that causes the cooling of H II regions. The forbidden line cooling is one contributor to the so-called radiative cooling that we describe in detail in §8.1.1. The basic process is simple: an atom (ion in these cases) is excited by a collision with another particle (typically, but not always, an electron). The excited state decays emitting a photon that, given its low transition probability, has also a very low probability of being reabsorbed by the gas and thus leaves the region carrying away energy. In the end, this process drains kinetic energy from the incident electrons and produces a decrease of the gas temperature.

The electron density ($n_e$) of H II regions can be estimated in different ways, for instance from radio observations using the bremsstrahlung radiation given the dependence of its emissivity on the density squared (eq. D.18) and from optical recombination lines, given the same dependence of the recombination rate (§D.1.5). Note that in both cases one has to make assumptions about the temperature or measure it independently. Finally, an accurate estimate of $n_e$ can also be obtained from line flux ratios between forbidden lines, in particular lines that have different transition probabilities. In contrast to what was done for the temperature above, one now concentrates on transitions that are sensitive to the density and where the temperature dependence can be neglected. A prime example is the [O II] doublet with the two lines at $\lambda = 3726\,\text{Å}$ and $\lambda = 3729\,\text{Å}$. With such a small difference in wavelength (and excitation energy), the line flux ratio $F(3726\,\text{Å})/F(3729\,\text{Å})$ is rather insensitive to the gas temperature, while it strongly depends on $n_e$. This is due to the fact that the critical density of [O II]$\lambda$3729 is $n_{crit} \sim 10^3\,\text{cm}^{-3}$ and the line emission changes significantly between the regime $n < n_{crit}$ (little collisional de-excitation) and $n > n_{crit}$ (collisional de-excitation dominates; §D.1.4). As a consequence, having predictions for the line ratio from theory, the *observed* flux ratio can be used to estimate the electron density. Another doublet flux ratio frequently used to determine the electron densities of H II regions is [S II] $F(6716\,\text{Å})/F(6731\,\text{Å})$ with again $n_{crit} \sim 10^3\,\text{cm}^{-3}$. Typical electron densities of H II regions are found to be in the range $10$–$10^4\,\text{cm}^{-3}$, reaching up to $10^5$–$10^6\,\text{cm}^{-3}$ in very compact H II regions.

### 4.2.4 Collisionally Ionised Gas

When the velocities of the gas particles are high enough, the ionisation of an atom can occur by collisions. If these velocities are due to thermal motions, we can estimate the temperature above which atoms should be fully ionised. This is the equivalent ionisation temperature $T_{ion} \equiv \chi/k_B$, where $\chi$ is the ionisation potential. For hydrogen, $\chi \simeq 13.6\,\text{eV}$ and $T_{ion} \simeq 1.58 \times 10^5\,\text{K}$. In practical situations, however, the minimum temperature for collisional ionisation is significantly lower ($T \approx 2 \times 10^4$ K, see details in §8.1.1). This is due to the fact that the competing process (radiative recombination) has a relatively small cross section, so that rare electrons in the high-energy tail of the distribution are sufficient to compensate the relatively slow pace of radiative recombinations and keep the medium ionised. We recall that the Maxwellian distribution of particle speeds (§D.1.4) peaks around the **thermal speed**[7]

$$v_T \equiv \sqrt{\frac{k_B T}{m}}, \tag{4.9}$$

where $m$ is the mass of the gas particle, but has a tail at higher speeds.

Collisionally ionised gas is typically produced by **shock waves** or **shocks** that occur when a portion of the ISM is accelerated to speeds that greatly exceed the sound speed (Mach number much larger than unity; eq. D.32). A shock wave propagates into the medium at supersonic speeds and perturbs it (we say that it *shocks* the previously unperturbed gas). A fundamental consequence of the passage of a shock wave is that the particles in the perturbed medium acquire fast random motions and thus the temperature of the gas is raised to values much higher than the unperturbed temperature. For fast shocks,

[7] The peak is at $\sqrt{2}v_T$.

the Rankine–Hugoniot jump conditions (§D.2.4) predict that the temperature increases with the square of the shock speed ($v_{\rm sh}$),

$$T = \frac{3}{16}\frac{\mu m_{\rm p}}{k_{\rm B}} v_{\rm sh}^2 \simeq 1.4 \times 10^5 \left(\frac{\mu}{0.62}\right)\left(\frac{v_{\rm sh}}{100\,{\rm km\,s^{-1}}}\right)^2 \,{\rm K}, \tag{4.10}$$

where $m_{\rm p}$ is the proton mass and $\mu \simeq 0.62$ is the mean atomic weight of a fully ionised plasma at solar metallicity.

In the ISM of an SFG, where the typical sound speeds are of the order of a few $\rm km\,s^{-1}$ (§4.6.2), these high speeds are easily reached as a consequence of stellar processes. Stellar winds from massive stars release material with speeds $v_{\rm w} \sim 10^3\,\rm km\,s^{-1}$ and produce strong shocks (§8.7.2). SN explosions create large shells of shocked ISM that expand at speeds of hundreds of $\rm km\,s^{-1}$. These are called supernova remnants (SNRs) and a large fraction of their kinetic energy is expected to heat the ISM and produce collisionally ionised gas (§8.7.1). In galaxies with high SFR surface densities (starburst galaxies) some hot ISM escapes the disc via powerful galactic winds (Fig. 4.2, top left, and §4.2.10).

The collisionally ionised gas in SFGs has a typical temperature of $T \sim 10^6$ K and emits in the soft X-ray band (§C.1) as $k_{\rm B}T = 0.1$ keV corresponds to $T \simeq 1.16 \times 10^6$ K. The base emission is thermal bremsstrahlung (§D.1.6), given that the medium is largely ionised, in particular for hydrogen and helium. Emission lines from metallic ions, however, provide a very important contribution (§8.1.1).

### 4.2.5 Molecular Gas

At temperatures $T < 40$ K and column densities $N_{\rm H} > 10^{21}\,\rm cm^{-2}$, hydrogen atoms combine to form molecules of $H_2$. There are different formation processes that we discuss in §8.3.1 and §9.4.1. Here we describe the distribution of molecular gas in present-day SFGs and its relation to star formation. Unfortunately, **molecular hydrogen** does not have appreciable emissions in any portion of the electromagnetic spectrum at low temperatures. In particular, being made of two identical atoms, it does not have an electric dipole and rotational transitions are limited to quadrupoles. The brightest line would be the $J = 2 \to 0$ emission at 28.2 $\mu$m, but the long lifetime of the $J = 2$ level of 1000 yr and the high excitation energy (equivalent temperature $T_* \simeq 510$ K) make it quite hard to detect. Rovibrational transitions, for instance $(v = 1, J = 3) \to (v = 0, J = 1)$ at 2.1 $\mu$m, and electronic transitions require very high temperatures, of 6600 K and $\sim 10^5$ K, respectively. These temperatures are achieved in shocks where the density is also high and $H_2$ can be collisionally dissociated. The binding energy of an $H_2$ molecule is $E_{\rm b} = 4.5$ eV. However, the main channel for $H_2$ **dissociation** is photodissociation, which takes place through two steps. First the molecule is excited by a photon to levels above the ground level (excitation to the first level requires an energy of 11.2 eV). Then, the subsequent de-excitation has a non-negligible probability to result in a state where the two atoms are unbound, hence the molecule is dissociated (§9.5.3). Note that UV photons with $11.2\,{\rm eV} < h\nu < 13.6\,{\rm eV}$, which cannot ionise a hydrogen atom, can photodissociate an $H_2$ molecule.

Most of the molecular gas in SFGs appears located in large **molecular clouds**. These are structures with masses up to $\sim 10^7\,\mathcal{M}_\odot$ and typical temperatures of $\approx 10$ K, where most of the star formation in a galaxy takes place (see Tab. 4.4 for the Milky Way). For the gas to be maintained in a molecular form, molecules must be protected from UV photons that could dissociate them. This happens naturally because molecular hydrogen becomes optically thick already at relatively low densities, lower than the typical densities of molecular clouds, which are $n_{H_2} \gtrsim 10^2$ cm$^{-3}$ (or column densities of $N_{H_2} > 10^{21}$ cm$^{-2}$). As a consequence, the outer molecules of a cloud absorb the incoming radiation, effectively shielding the inner gas, which is then protected from photoionisation, photoelectric heating and photodissociation. This process is called **self-shielding** and is very important for the survival not only of $H_2$ but also of other molecular species inside molecular clouds.

Given that $H_2$ is nearly undetectable in emission, most of the information about molecular gas in the ISM of galaxies comes from **carbon monoxide** (CO). CO is the second most abundant molecule in the ISM, but much less abundant (a factor $\sim 10^{-4}$) than $H_2$. It has useful rotational transitions, such as $J = 1 \rightarrow 0$ at 2.6 mm and $J = 2 \rightarrow 1$ at 1.3 mm. For CO ($J = 1 \rightarrow 0$), the transition energy is low (equivalent temperature $T_* \simeq 5.5$ K), thus the upper levels are very easily excited. Its spontaneous emission is fairly probable (lifetime $\approx 1 \times 10^7$ s) and the critical density is relatively high, $n_{\rm crit} \approx 3 \times 10^3$ cm$^{-3}$. These transitions are excited by collisions with molecules of $H_2$ (given that these are the vast majority): for this reason the CO emission is considered a good *tracer* of all molecular gas. The study of the CO emission is the most common way to determine the molecular gas content in present-day and distant galaxies.

## From CO Emission to Molecular Gas Mass

In §4.2.1 we have seen that the H I mass of a source can be easily derived from its 21 cm flux given that it is usually optically thin. The situation is rather different for CO as the lower rotational transitions ($J = 1 \rightarrow 0$ and $J = 2 \rightarrow 1$) of the most common isotope $^{12}C^{16}O$ are essentially always optically thick (§D.1.1), because the column densities that produce unit optical depth are always reached in molecular clouds. Other species such as $^{12}C^{18}O$ ($\sim 10^{-7}$ times the abundance of $H_2$) could be considered optically thin but, being detected only in very high-density regions, they are not necessarily good tracers of the total gas mass. Emission from an optically thick homogeneous cloud only comes from the *surface* of the cloud and there is no way for the observed flux to be proportional to the number of emitting particles in its interior and so its total mass. Luckily, molecular clouds are not homogeneous structures and this is the reason why we can still use the CO flux to estimate their mass, as we now see.

When molecular clouds are observed at very high spatial resolution they consist of a large number of substructures in the form of smaller clouds (cloudlets) and filaments (§4.6.2). If we assume that these cloudlets have the same mass or a similar mass spectrum in all molecular clouds, then the measured CO flux becomes proportional to the total number of emitting cloudlets. In practice, in this reasoning, we have substituted the emission from CO molecules with emission from cloudlets (each containing large numbers of CO molecules). The fact that a single cloudlet is optically thick is not important, because, in this picture, the CO flux only *counts* the number of cloudlets. What is important is that they do not significantly hide one another, i.e. that 'shadowing' is negligible. Simple arguments based on the sizes of cloudlets in spatial and velocity space show that shadowing is likely never an important effect.

The above considerations lead to an important consequence. We started by saying that, unlike H I, the CO flux cannot be proportional to the gas mass, but we now find that it is potentially proportional to the number of cloudlets in a molecular cloud. This effectively makes the CO flux proportional to the total gas mass of the molecular cloud. The proportionality however is not known *a priori* like for H I and it has to be calibrated. This has been done in the Milky Way by estimating the column density of $H_2$ in a large number of clouds or lines of sight using different methods. These include (1) the observations of gamma-rays produced by cosmic rays (§4.6.2) interacting with the gas in the cloud (their emissivity is proportional to the hydrogen density), (2) dust extinction/emission assuming a certain gas-to-dust ratio (§4.2.7) and (3) the virial theorem assuming that the clouds are gravitationally bound and in equilibrium (§8.3.5). The final result is the $\boldsymbol{X_{\mathrm{CO}}}$ **factor**

$$X_{\mathrm{CO}} \equiv \left(\frac{N_{\mathrm{H}_2}}{\mathrm{cm}^{-2}}\right)\left(\frac{\int T_{\mathrm{b}}(v)\mathrm{d}v}{\mathrm{K\,km\,s}^{-1}}\right)^{-1} \approx 2\times 10^{20}, \tag{4.11}$$

which is used to obtain the column density of molecular hydrogen ($N_{\mathrm{H}_2}$) from the velocity-integrated CO ($J=1\rightarrow 0$) brightness temperature ($T_{\mathrm{b}}$; eq. C.10), which is a measure of the CO surface brightness (§C.2). Such a conversion factor is also called the $\boldsymbol{\alpha_{\mathrm{CO}}}$ **factor**. Note, however, that $X_{\mathrm{CO}}$ converts from CO flux to hydrogen mass, while in the $\alpha_{\mathrm{CO}}$ factor a correction for helium is also included (usually multiplying by $\approx 1.3$–$1.4$) and thus it should convert from CO flux to *total* molecular mass.

In present-day galaxies, CO emission tends to be observed in the central parts and is less radially extended than H I (Fig. 4.5). As a consequence, the surface density of the molecular gas derived using the CO brightness temperature usually peaks close to the centre and falls with radius much faster than that of the neutral atomic gas. Fig. 4.6 (left) shows the distribution of molecular and neutral atomic gas as a function of radius for the galaxy shown in Fig. 4.5. Interestingly, the H I density shows a depression in the inner parts that is filled by the molecular gas. This is a feature quite commonly observed in nearby galaxies.

Although the calibration in eq. (4.11) is widely used to determine the molecular gas mass, it has been tested properly only in the Milky Way, where it holds within a factor of 2. Complications may arise if used in galaxies whose ISM has very different properties; for instance, in metal-poor galaxies such as low-mass galaxies (§4.4.3), where there are less metals and thus less CO. As a consequence, the ratio between $H_2$ and CO molecules may be higher and one would need to take this into account. Conversely, there are indications using estimates from the dust (§4.2.7) that $X_{\mathrm{CO}}$ can be significantly lower than the value in eq. (4.11) in the very central (high-density) regions of galaxies. Sometimes, CO ($J=1\rightarrow 0$) is not available and one has to resort to the use of other transitions and yet different calibrations (§11.1.4).

### 4.2.6 Global Properties of the Neutral Gas in Star-Forming Galaxies

In this section, we summarise the global properties of the neutral atomic and molecular gas in present-day SFGs. We do this with the help of Fig. 4.8, where the total masses of H I ($\mathcal{M}_{\rm HI}$) and $H_2$ ($\mathcal{M}_{\rm H_2}$) of a sample of local galaxies are compared with other global properties, in particular the stellar mass ($\mathcal{M}_\star$) and the SFR. The top left panel of this figure shows the total gas content ($\mathcal{M}_{\rm HI} + \mathcal{M}_{\rm H_2}$, corrected for helium) as a function of the stellar mass of the galaxies. There is a clear tendency for more massive SFGs to have more gas, although the relation is shallow and with large scatter. We can define the **gas-to-stellar-mass ratio** and **gas fraction** respectively as

$$f_{\rm gas,\star} \equiv \frac{\mathcal{M}_{\rm gas}}{\mathcal{M}_\star} \quad \text{and} \quad f_{\rm gas} \equiv \frac{\mathcal{M}_{\rm gas}}{\mathcal{M}_{\rm b}}, \tag{4.12}$$

where $\mathcal{M}_{\rm b}$ is the total baryonic mass ($\mathcal{M}_{\rm b} = \mathcal{M}_\star + \mathcal{M}_{\rm gas}$). The dashed lines in the top left panel of Fig. 4.8 show that $f_{\rm gas,\star}$ goes from values larger than unity for low-mass galaxies to $\lesssim 10\%$ for very massive spirals. The baryonic component of low-mass (dwarf) galaxies can be totally dominated by the gas ($f_{\rm gas} \approx 1$). Finally, the top right panel shows that the ratio between H I mass and $H_2$ mass is, in general, larger than unity: present-day galaxies have more atomic than molecular gas.

The bottom panels of Fig. 4.8 show relations between the gas content and the SFR in present-day galaxies. We discuss the importance of these relations in §4.2.9 and §4.4.4. Here, we just note that if a galaxy has more gas, it tends to form stars at a higher rate. This is true for both $H_2$ and H I, as we can appreciate from the relations in the two bottom right panels.

### 4.2.7 Interstellar Dust

A key component of the ISM in galaxies is **interstellar dust** (see Draine, 2003, for a review). Despite constituting a small fraction of the total ISM mass (of order 1%), dust is important in a number of key processes, including $H_2$ formation (§8.3.1) and the reprocessing of stellar light (§3.3). Dust grains typically absorb UV and optical radiation from stars and re-emit it at longer (IR) wavelengths. Dust can therefore be *detected* both from the absorption of background UV/optical light and from the emission in mid/far-IR bands. On average, in present-day SFGs, one-quarter to one-third of the stellar radiation is turned into dust emission. Radiation scattering by dust is also an important effect and produces structures around massive stars called **reflection nebulae**. The combined contribution of absorption and scattering is called dust extinction.

Consider a source along a line of sight that goes through an intervening dusty medium that attenuates part of its radiation. The extinction in magnitudes at a certain wavelength can be defined as

$$A_\lambda \equiv -2.5 \log\left(\frac{F_\lambda}{F_{\lambda,0}}\right) = 2.5 \log(e)\tau_\lambda \simeq 1.086\tau_\lambda, \tag{4.13}$$

where $F_\lambda$ and $F_{\lambda,0}$ are, respectively, the measured and intrinsic fluxes of the source, and $\tau_\lambda$ is the optical depth due to dust extinction. Eq. (4.13) comes from the solution of the

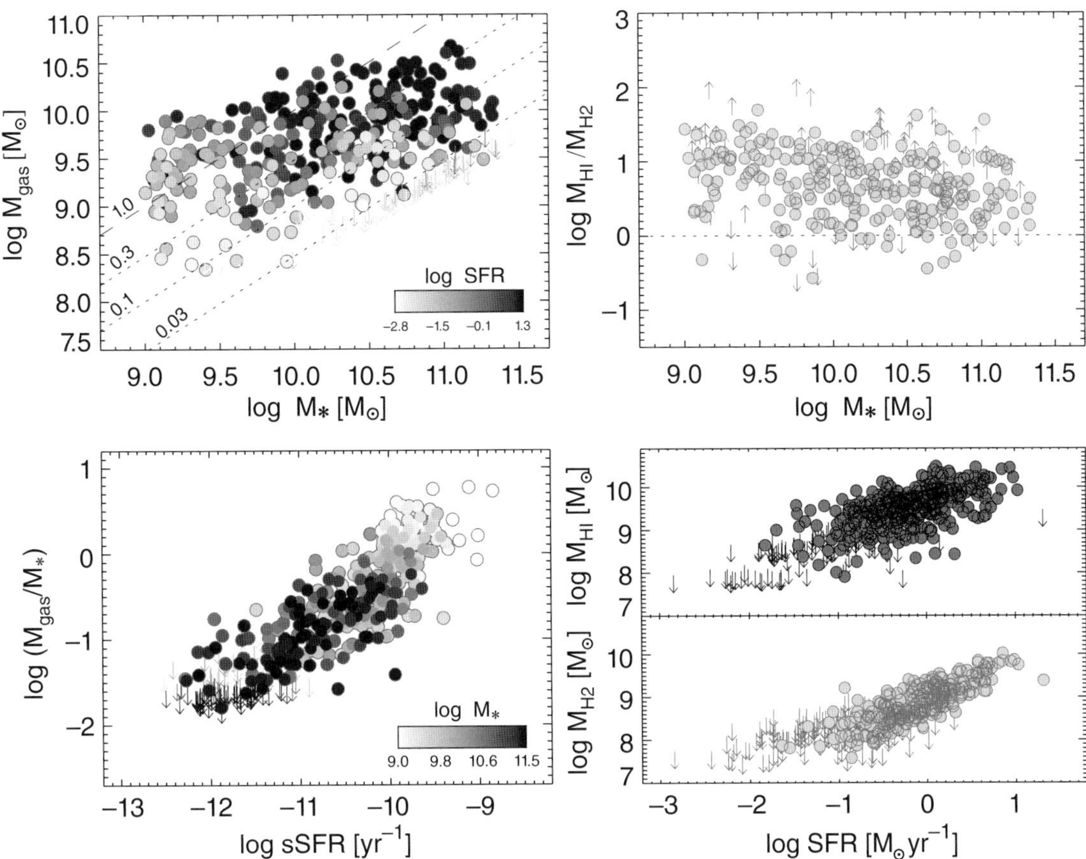


Fig. 4.8

Relations between global gas and stellar properties in present-day ($0.01 < z < 0.05$) galaxies. *Top left*. Total (atomic neutral + molecular) gas mass as a function of the stellar mass of the galaxy. The points are grey-coded with the SFR of the galaxy and the dashed lines show different values of the gas-to-stellar-mass ratio ($f_{gas,\star}$; eq. 4.12). *Top right*. Ratio between neutral atomic and molecular hydrogen mass as a function of the stellar mass of the galaxy. *Bottom left*. Gas-to-stellar-mass ratio $f_{gas,\star}$ versus specific SFR (eq. 4.44). The points are grey-coded with the stellar mass of the galaxy. *Bottom right*. Hydrogen mass (atomic above and molecular below) versus SFR. In all panels lower and upper limits are shown as arrows. Note that the galaxies in this sample have been selected only in stellar mass ($10^9 < \mathcal{M}_\star < 10^{11.5}$) and thus include also ETGs (most of the upper limits in the gas masses). Data from Catinella et al. (2018). Figure courtesy of B. Catinella.

equation of radiative transfer (eq. D.4) assuming that the intervening medium does not emit at the wavelength $\lambda$ and thus $F_\lambda = F_{\lambda,0}\exp(-\tau_\lambda)$. $A_\lambda$ gives the amount of extinction as a function of wavelength and it has been precisely determined in the Milky Way and in the Magellanic Clouds (§6.3) along lines of sight towards stars of known luminosity. The outcome is called an **extinction curve**, usually given as a ratio $A_\lambda/A_X$ versus wavelength, with $A_X$ the extinction in some spectral band $X$. Its exact shape varies across the Galaxy and in different galaxies, although, in general, it falls roughly as $A_\lambda \propto \lambda^{-1}$ from $\lambda \approx 1000$ Å (UV) down to FIR wavelengths, where it approaches zero extinction. Fig. 4.9 shows some average observed extinction curves normalised to the $V$ band. For $\lambda < 1000$ Å (not shown in Fig. 4.9) the extinction eventually falls but not too steeply as soft X-rays are also absorbed and scattered by dust.

A useful quantity to characterise extinction is the **colour excess**, defined as the **reddening** of a source due to extinction. For instance, considering the optical $B$ and $V$ bands, the colour excess is

$$E(B-V) \equiv (B-V) - (B-V)_0, \tag{4.14}$$

where $(B-V)_0$ is the colour that the source would have in the absence of extinction. The colour excess is linked to the extinction as $E(B-V) \equiv A_B - A_V$ and this can be used to define the parameter

$$R_V \equiv \frac{A_V}{A_B - A_V} = \frac{A_V}{E(B-V)}, \tag{4.15}$$

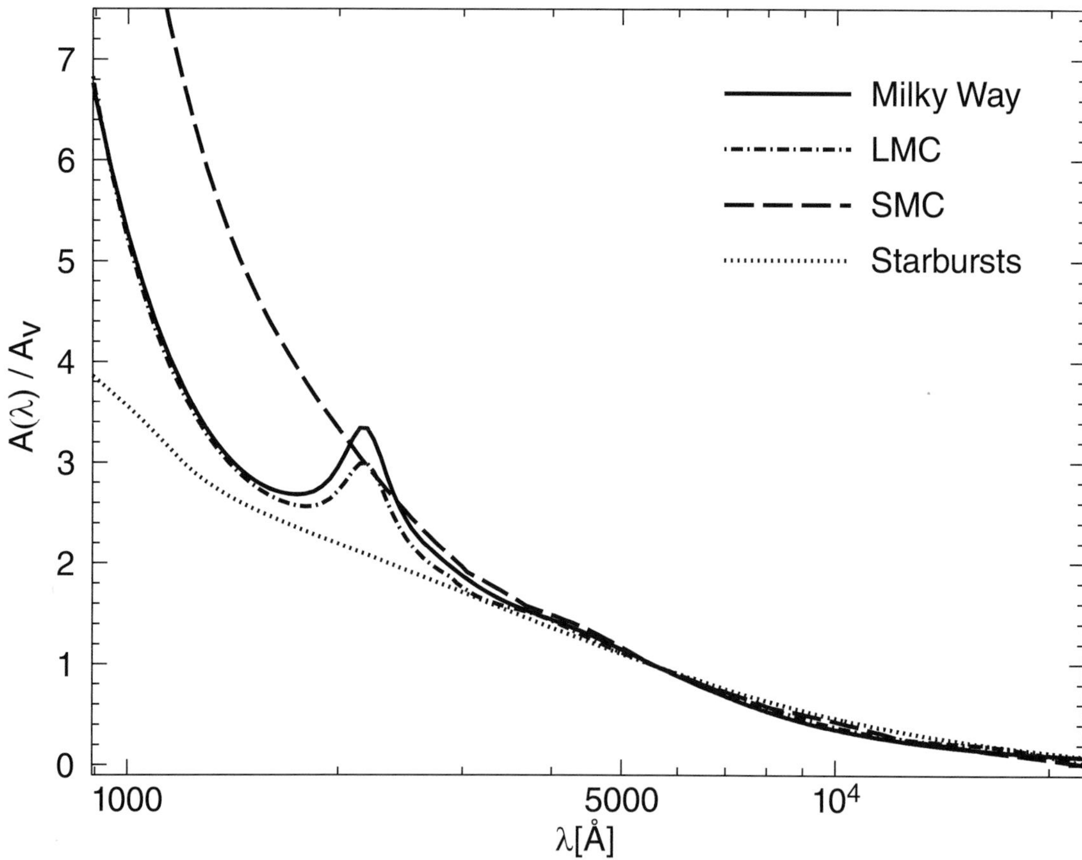


Fig. 4.9 Extinction curves normalised to the $V$ band in the Milky Way, in the Large Magellanic Cloud (LMC), in the Small Magellanic Cloud (SMC) and in starburst galaxies. The curves have been determined by Fitzpatrick and Massa (2007) (Milky Way), Gordon et al. (2003) (SMC and LMC) and Calzetti et al. (2000) (starbursts). Figure courtesy of M. Bolzonella.

which is a measure of the steepness of the extinction curve in the region of the $V$ band ($\approx 5500$ Å). Values of $R_V$ for the Milky Way vary from $\simeq 2.4$ (diffuse ISM) to $\simeq 5.5$ (dense molecular clouds); the typically used value is $R_V = 3.1$. In Fig. 4.9 larger values of $R_V$ normally correspond to flatter curves. The inverse $1/R_V$ is called the **normalised extinction** in $V$ band. All the above definitions can be given for bands different than $B$ and $V$. If we rewrite eq. (4.15) for a generic wavelength $\lambda$, it reads $A_\lambda/E(B-V) \equiv k(\lambda)$, where $k(\lambda)$ is another definition of extinction curve (see eq. 3.3).

The amount of interstellar dust extinction appears to be tightly related to the amount of gas. The extinction at a certain wavelength ($A_V$ for instance) is proportional to the observed gas column density. This proportionality is calibrated in the Milky Way as

$$N_{\rm H} \approx 2 \times 10^{21} \left(\frac{A_{\rm V}}{\rm mag}\right) \, {\rm cm}^{-2}, \qquad (4.16)$$

where $N_{\rm H}$ represents the column density of hydrogen atoms in both atomic and molecular form. The above relation can be converted into a relation between the gas and dust mass (gas-to-dust ratio, §3.5.3), which turns out to be about 100. This ratio is considered fairly constant at least in galaxies with metallicities similar to the Milky Way.

The extinction curve $A_\lambda$ gives us information about the dust particles as well. The classical Mie theory (Mie, 1908; Tielens, 2005) approximates dust grains as spheres with different properties subject to plane-parallel incident light. Its application to the ISM tells us that interstellar dust grains have a wide range of sizes with an upper limit at $\sim 1\,\mu$m. The main constituents of dust are silicates and carbon; the silicates also contain magnesium and iron, in addition to silicon and oxygen. Carbon aggregates into graphites, PAHs (§3.2) or amorphous structures. In the Milky Way and the Large Magellanic Cloud, $A_\lambda$ shows a clear peak at $\lambda \simeq 2175$ Å (Fig. 4.9), potentially due to graphite grains or other forms of ordered carbon material. This feature does not seem to be present in the Small Magellanic Cloud, probably due to the low metallicity of its ISM. Dust grains can lock up a significant fraction of heavy elements and deplete the gaseous phase of the ISM from these elements. This phenomenon is called **depletion into dust** and one must take it into account when trying to determine the ISM metallicity (§4.1.4).

Stellar light typically heats up the dust to temperatures of tens of kelvin. A grey body (§D.1.3) emitting at these temperatures shines in the FIR at about $100\,\mu$m (eq. D.11): indeed this is where a large fraction of the emission of SFGs takes place (Fig. 3.6). In Fig. 4.5 we show the $100\,\mu$m emission for the nearby spiral galaxy NGC 6946 tracing the location of the warm dust in this galaxy. It follows very well the H$\alpha$ emission, testifying to the close link with star-forming regions (young stars photoionise the gas in H II regions and also heat the dust grains). Most of the emission in the FIR is due to grains with typical sizes $\sim 0.1\,\mu$m, which are the most common. However, smaller grains ($\lesssim 0.01\,\mu$m) also exist in the ISM. They can be heated to higher temperatures (hundreds of K) and produce substantial thermal emission at NIR wavelengths (§3.2).

### 4.2.8 Magnetic Field

The radio continuum emission of SFGs is produced by two main mechanisms: the (thermal) bremsstrahlung emission (§D.1.6) and (non-thermal) synchrotron emission (§D.1.7), see

Fig. 11.3. The bremsstrahlung emission is generally associated with H II regions and the presence of photoionised gas (§4.2.3). The synchrotron emission is produced by relativistic electrons that are spiralling around the lines of the magnetic field in the diffuse ISM or in individual SNRs (§8.7.1). The relative importance of thermal and non-thermal emission depends on frequency. Typically, at high frequencies (say $\nu > 5\,\mathrm{GHz}$) thermal emission prevails while the contribution of synchrotron emission is dominant at lower frequencies. The shape of the radio spectrum is often used to discriminate between thermal and non-thermal emission (§D.1.7). Another property of the synchrotron emission is to be polarised in a direction perpendicular to the local magnetic field $\boldsymbol{B}$. Thus, the presence and the orientation of the magnetic field can be directly probed through the linearly polarised radio continuum. When galaxies are observed face-on, magnetic field lines tend to be stretched radially along the plane of the disc. When prominent spiral arms are present, the magnetic field is aligned along the arms.

The synchrotron emission also allows us to estimate the magnetic field strength. This is often done using the assumption of **equipartition** between the energy density of the magnetic field and that of the relativistic electrons (see §4.6.2) that produce the synchrotron radiation. Under this assumption, one estimates the equipartition field strength $B_{\mathrm{eq}} \propto I_\nu^{1/(3+\alpha)}$, where $I_\nu$ and $\alpha$ are the intensity (§C.2) and the spectral index of the synchrotron emission, respectively. The intensity of the synchrotron emission and of the magnetic field is highest in the discs of SFGs, while a more tenuous magnetic field permeates the halo region, as shown by observations of edge-on galaxies. In some galaxies the radio halos are quite prominent and extended (Fig. 4.2, bottom right).

The strength of the magnetic field can also be measured using the **Faraday rotation** towards pulsars and AGNs. Faraday rotation occurs when linearly polarised radiation goes through a magnetised medium. The plane of polarisation tends to rotate by an angle $\beta = R_{\mathrm{M}}\,\lambda^2$, where $\lambda$ is the wavelength of the observations and $R_{\mathrm{M}} \propto \int n_{\mathrm{e}} B_{||}\,\mathrm{d}x_{\mathrm{los}}$ is the **rotation measure** and the integration is along the path to the source. With this technique one obtains a measure of $\langle n_{\mathrm{e}} B_{||}\rangle$, where $n_{\mathrm{e}}$ is the electron density and $B_{||}$ is the component of magnetic field parallel to the line of sight. Thus $B_{||}$ can be estimated by making appropriate guesses for the average $n_{\mathrm{e}}$ of the medium and the size of the emitting region. Ideally, one can use the **dispersion measure**, which gives the electron density integrated along the line of sight, $D_{\mathrm{M}} \equiv \int n_{\mathrm{e}}\,\mathrm{d}x_{\mathrm{los}}$. This requires an object that emits a pulse of radiation, e.g. a pulsar, and thus it has been used, in particular, in the Milky Way and in the Magellanic Clouds. The delays of the pulsar signals are functions of the observed frequency and of $D_{\mathrm{M}}$ and this allows us to obtain the mean electron density $\langle n_{\mathrm{e}}\rangle$. By combining $R_{\mathrm{M}}$ and $D_{\mathrm{M}}$ for the same object, one finally estimates the electron-density-weighted average value of the magnetic field along the line of sight $\langle B_{||}\rangle$.

The Faraday rotation and the equipartition methods applied to nearby spiral galaxies including the Milky Way return consistent values of the ISM magnetic field strength in the range

$$\langle B\rangle \approx 1\text{–}10\,\mu\mathrm{G}. \tag{4.17}$$

The strength of the field is higher along spiral arms and in dense regions in general. This amplification of the magnetic field strength is explained in §8.3.6. In these dense regions and in the inner parts of some galaxies the magnetic field strength can reach values of a few tens of $\mu$G.

### 4.2.9 Star Formation Laws

We have seen that, on average, present-day spiral galaxies form stars at typical rates of 0.1 to a few tens $\mathcal{M}_\odot\,\mathrm{yr}^{-1}$ (Tab. 4.1). Dwarf irregulars form stars at lower rates mostly because of their smaller sizes, but they can also be starbursting (§4.5). Given the very different sizes of SFGs, a useful quantity to measure the level of star formation activity is the **star formation rate surface density** ($\Sigma_{\rm SFR}$), which is the SFR per unit area of the galaxy disc.

A fundamental relation for star formation in galaxies is the **Schmidt–Kennicutt law** that links the SFR density to the gas density. It was originally discovered in the Milky Way using volume densities (Schmidt, 1959), but then given, for external galaxies, in terms of surface densities, which are much easier to measure (Kennicutt, 1998). The surface density relation reads

$$\Sigma_{\rm SFR} = B\left(\frac{\Sigma_{\rm gas}}{1\,\mathcal{M}_\odot\,\mathrm{pc}^{-2}}\right)^{\alpha}\,\mathcal{M}_\odot\,\mathrm{yr}^{-1}\,\mathrm{kpc}^{-2}, \tag{4.18}$$

where $\Sigma_{\rm gas}$ is the total gas (molecular and atomic) surface density corrected for the helium fraction, $B \approx 1\times10^{-4}$ and $\alpha \approx 1.4$. This relation is shown in Fig. 4.10, where each point represents an SFG and both $\Sigma_{\rm gas}$ and $\Sigma_{\rm SFR}$ are averaged across the galaxy discs. The relation holds for very different galaxy types, including normal spirals and starbursts. The scatter appears to increase if one includes metal-poor galaxies and towards low column densities.

A source of uncertainty for the star formation law comes from the need to convert from CO observations to molecular gas densities for galaxies of very different types (§4.2.5). Indeed a large fraction of the galaxies that appear in Fig. 4.10 have very high gas column densities ($\Sigma_{\rm gas} > 10\,\mathcal{M}_\odot\,\mathrm{pc}^{-2}$) and they are dominated by molecular gas, i.e. $\Sigma_{\rm H\,I}$ is negligible. Therefore, the chosen value of $X_{\rm CO}$ (eq. 4.11) is quite crucial and using different calibrations can give different slopes or even different normalisations, in particular in the starburst galaxy regime.

Eq. (4.18) describes an empirical relation that may have a rather straightforward explanation. Suppose that the SFR volume density ($\rho_{\rm SFR}$) depends on the amount of gas locally available (the gas volume density $\rho_{\rm gas}$) divided by a typical timescale for star formation that we can take to be proportional to the free-fall time $t_{\rm ff} \propto \rho_{\rm gas}^{-1/2}$ (eq. 8.45). If we further assume that the thickness of the gas disc and of the layer where star formation takes place are *constant* in all galaxies, we have that

$$\rho_{\rm SFR} \propto \frac{\rho_{\rm gas}}{t_{\rm ff}} \propto \rho_{\rm gas}^{3/2} \quad\Longrightarrow\quad \Sigma_{\rm SFR} \propto \Sigma_{\rm gas}^{3/2}, \tag{4.19}$$

and thus a proportionality very similar to the Schmidt–Kennicutt law.

Other than using average quantities over the whole star-forming disc like in Fig. 4.10, the star formation law has also been determined in spatially resolved regions of galaxies down to sub-kpc sizes. Despite a general agreement with the global law, indications have been found that, of the two gas components (atomic and molecular), the one that best

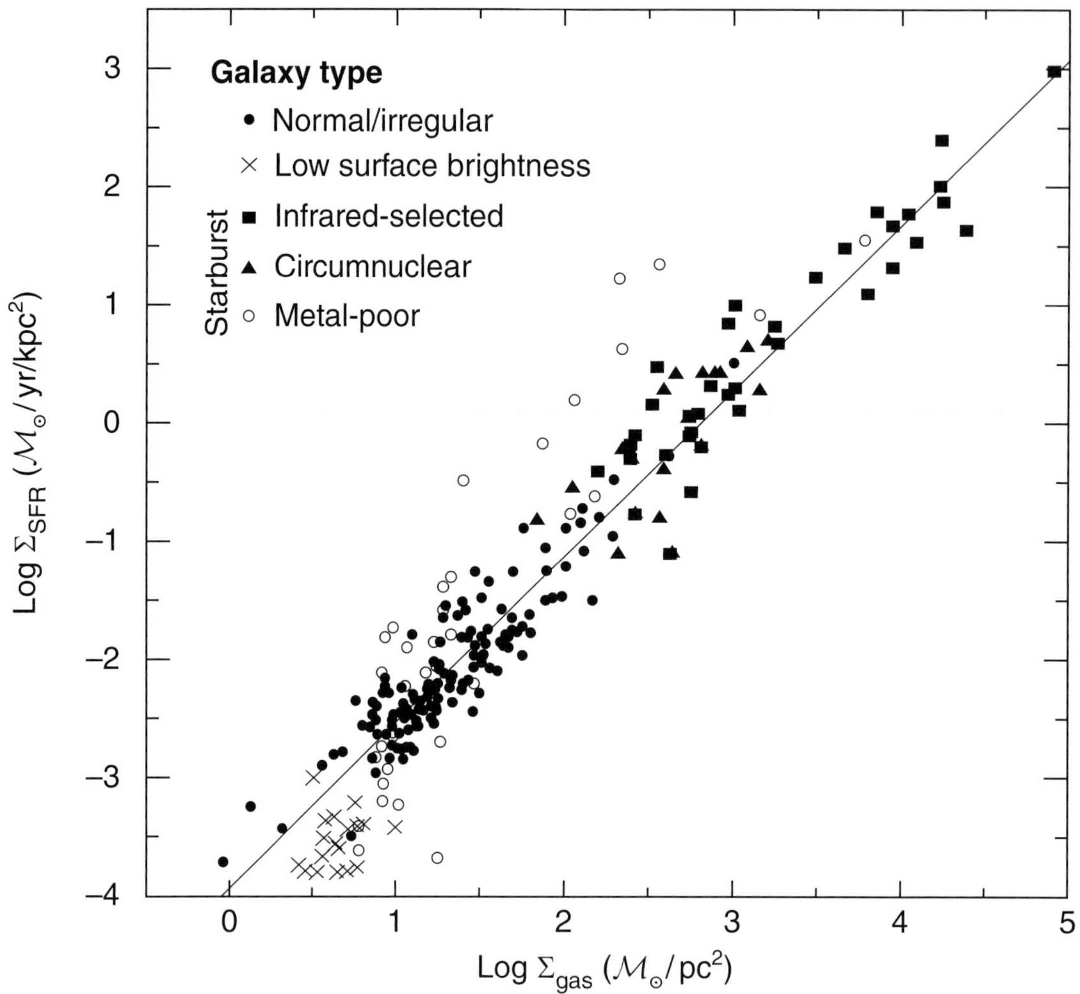


Fig. 4.10 Total (atomic and molecular) gas surface density versus SFR surface density (Schmidt–Kennicutt) relation for a sample of nearby spirals and starburst galaxies. Updated version of the original from Kennicutt (1998). © AAS, reproduced with permission. Courtesy of R. Kennicutt.

correlates with the SFR is the molecular gas density. This led to the proposal of a relation between SFR and molecular (only) gas surface densities: this relation has a slope shallower than 1.4. However, the SFR does correlate with the mass of atomic gas as we can see by simply looking at the global quantities (Fig. 4.8, bottom right). Therefore, it is likely that molecular and atomic gas trace two different regimes of high (molecular gas) and low (H I) SFRs. The low-density regime is typical of low-mass galaxies and external parts of galaxy discs, where there is very little molecular gas detected (Fig. 4.6). The slope of the Schmidt–Kennicutt law at these low densities is still uncertain and a number of parameterisations have been proposed. Finally, to add to the complex picture, other relations have been found between the *mass* of the very dense molecular gas (only a small fraction of the total molecular gas; Tab. 4.4) and the SFR. The interested reader is referred to Kennicutt and Evans (2012) for more details.

A possible way out of some of the complexities outlined above is that the fundamental star formation law should be expressed in terms of volume densities ($\rho_{gas}$ versus $\rho_{SFR}$) instead of column densities. These latter are indeed quantities projected along the line of sight and are strongly influenced by the thickness of the gas and of the star-forming layers. In §4.6.2 we see that the thickness of gas discs in galaxies is not constant with radius and it varies from galaxy to galaxy. Thus, the implication in eq. (4.19) is probably valid only in a restricted regime of high densities (inner region of discs). Low-mass galaxies and H I-dominated regions of galaxy discs (outer parts) tend to have larger thicknesses and thus the measured surface densities cannot be straightforwardly linked to the volume densities. The large scatter of the star formation law at low column densities and the lack of apparent correlation with $\Sigma_{\rm H\,I}$ could be explained by this effect.

### 4.2.10 Gas Inflow and Outflow

SFGs continuously exchange gas with the surrounding environment. This exchange occurs through **gas inflows** (often called **gas accretion**) and **gas outflows**. Gas accretion is quite difficult to observe directly. Deep observations of neutral atomic gas in emission (§4.2.1) have not revealed a significant population of 'floating' gas clouds that could be evidence of ongoing gas accretion. Thus most of the accretion is thought to take place at lower column densities ($< 10^{20}$ cm$^{-2}$) than those probed by H I emission. This gas must be highly ionised and can be seen only in absorption towards distant QSOs; we describe this medium in §9.8. The study of gas in emission, however, can efficiently probe gas accretion indirectly. Pieces of evidence of this kind are provided by disturbances of the kinematics and the morphology of the outer H I discs, by the H I warps (§4.3.3) and by the peculiar kinematics of the gas located at kpc distances above and below the disc galaxy planes, called extraplanar gas (see below). For an in-depth description of these issues see Sancisi et al. (2008).

Gas accretion must have occurred abundantly at the epoch of galaxy formation (§8.2), but also later accretion is an important phenomenon for the evolution of SFGs, as we describe in §10.7.2. Gas has the fundamental role of *feeding* the star formation in galaxies. Given the SFR of a galaxy and the (cold) gas available in galactic discs, one can estimate the gas **depletion time** (or **consumption time**)

$$t_{\rm depl} \equiv \frac{\mathcal{M}_{\rm gas}}{\rm SFR}, \tag{4.20}$$

which is a measure of the time a galaxy can continue forming stars at the current rate given its present gas supply. If one considers the molecular gas available in present-day spiral galaxies, this depletion time is $\sim 1$ Gyr (Tab. 4.1). Including H I in the estimate increases it to a few Gyr. Note, however, that most of the H I typically lies at large radii with respect to the region where star formation takes place (Fig. 4.5) and thus an inflow through the disc would be needed to bring it to the central regions (§10.7.2).

Stellar evolution processes return to the ISM part of the gas initially locked up into stars. This happens in two stages. Massive stars return gas fairly quickly with stellar winds and Type II (core-collapse) SN explosions. The typical timescale for this return is less than $\tau_{\rm MS}(8\,\mathcal{M}_\odot) \approx 30\,{\rm Myr}$, which is the main-sequence time of a $8\,\mathcal{M}_\odot$ star: the last to explode as a core-collapse SN. This time can be slightly longer if the stars are in binaries. Intermediate-mass stars evolve more slowly and return part of their material to the ISM via mass loss from giants, supergiants, novae and planetary nebulae (§8.5.1 and §C.5). The returned gas has two important effects: it prolongs the depletion times calculated above and, most importantly, it pollutes the ISM with metals.

Feedback from SN explosions is thought to have a major impact on the evolution of SFGs. SN ejecta travel at supersonic speeds producing shock waves (§D.2.4) and collisionally ionised gas (§4.2.4). The expansion of SN shells also compresses the ISM promoting new star formation in the disc, a phenomenon referred to as **self-regulated** (or **self-propagating**) **star formation**. Around star clusters containing O/B stars or around O/B associations, large interstellar bubbles (superbubbles) form and expand (§8.7.3). These bubbles can reach sizes larger than the typical disc thickness of a few hundred parsecs (§4.6.2). When this happens they *blow out* of the galactic disc ejecting large amounts of gas into the galactic halo. This sets in motion a circulation called **galactic fountain** where disc material is temporarily removed from star-forming regions, travels through the halo region and falls back to another area of the disc (§10.7.3). This process produces thick (several kpc) gas layers of neutral and ionised **extraplanar gas** that contain $\sim 10\%$ of the total mass of cold/warm gas in a disc galaxy. The galactic fountain blow-outs also leave large (up to a few kpc in diameter) *holes* in the distribution of the cold ISM in the disc: we observe these in face-on galaxies like the one shown in Fig. 4.5 (top right).

Intense star formation, in particular in starburst galaxies, can also cause the onset of the so-called **galactic winds** (§8.7.3). These are powerful ejection processes powered by multiple stellar winds and SN explosions that take place in a short time and in a relatively small region of the galaxy disc characterised by very high SFR surface density. We see an example in Fig. 4.2 both in the X-ray and in the H$\alpha$ emission for the nearby starburst galaxy NGC 253. This galaxy has a global SFR exceeding $10\,\mathcal{M}_\odot\,{\rm yr}^{-1}$. The star formation appears widespread in the disc (see UV image) but in the central few kpc the SFR density reaches values so high ($\sim 1\,\mathcal{M}_\odot\,{\rm yr}^{-1}\,{\rm kpc}^{-2}$) that a large-scale roughly biconical wind is driven out of the disc in the perpendicular direction. Galactic winds are typically observed in emission using recombination lines (H$\alpha$) and forbidden lines like [O III]. They can also be seen in absorption as blueshifted spectral lines towards the optical/UV continuum of the galaxy. The typical speeds of the ejected material are up to hundreds of $\rm km\,s^{-1}$.

A recurrent question is whether the material expelled through galactic winds can escape the gravitational potential well of a galaxy. This would deplete the galaxy of gas and inject metal-rich (as it comes from a region of intense star formation) material into the surrounding medium. The **escape speed** at a position $\boldsymbol{r}$ is defined as

$$v_{\rm esc}(\boldsymbol{r}) \equiv \sqrt{2|\Phi(\boldsymbol{r})|}, \tag{4.21}$$

where $\Phi(\boldsymbol{r})$ is the gravitational potential at that location. For a spherical potential with a power-law density distribution (eq. 4.25) of slope $2 < \delta < 3$, the escape speed at the radius $r$ is

$$v_{\rm esc}(r) = \sqrt{\frac{2}{\delta - 2}}\, v_{\rm c}(r), \tag{4.22}$$

where $v_{\rm c}(r)$ is the circular speed at $r$ (eq. 4.23). In §4.3.1, we see that the slope of the potential in disc galaxies is $\delta \approx 2$ and approaches $\delta \approx 3$ in the outer halo, thus the escape speed is typically larger than $v_{\rm c}$ and its value can be rather uncertain. Including the disc potential, one finds that the escape speed from the centre of the Milky Way is $v_{\rm esc} \approx 800\,{\rm km\,s^{-1}}$, higher than typical velocities measured in galactic winds powered by stellar feedback. We discuss the consequences of this in §10.7.3.

## 4.3 Mass Distribution

Disc and dwarf irregular galaxies are ideal systems in which to trace the total distribution of matter. This is due to the fact that the kinematics of most of the visible matter in their discs is dominated by rotation. Some matter components (e.g. the cold ISM) can be considered *kinematically cold*, which means that they have low velocity dispersions compared to rotation ($V/\sigma \gg 1$). In these cases, if the gravitational potential is nearly axisymmetric, we can expect them to follow almost perfect circular orbits. This allows us to derive the so-called **rotation curve**, i.e. the rotation velocity as a function of radius, $v_{\rm rot}(R)$, which, as we see below, is readily related to the galactic potential. Neutral hydrogen is one of these cold components. Moreover, it is insensitive to dust extinction and it also has the key advantage to extend out to large distances (beyond the optically bright regions; Fig. 4.5). Thus we can probe the potential at large radii, where the stellar and other baryonic components fade away. The derivation and analysis of the first high-quality H I rotation curves in the 1970s and 1980s have revealed that the rotation velocity in disc galaxies tends to remain constant out to the outermost measured radii. The fact that H I *rotation curves are flat* became the key observational evidence that established the presence of a large amount of dark matter in spiral galaxies. In the next sections, we describe the theoretical link between gas rotation and the gravitational potential and what we can infer about the dark matter distribution by studying galaxy rotation curves.

### 4.3.1 Gravitational Potential

Let us first consider a system with a spherical mass distribution. The **circular speed** of such a system is defined as

$$v_{\rm c}(r) \equiv \sqrt{r\frac{{\rm d}\Phi(r)}{{\rm d}r}} = \sqrt{\frac{GM(r)}{r}}, \tag{4.23}$$

where $\Phi(r)$ is the gravitational potential and $M(r)$ is the mass contained within the radius $r$. Thus for a spherical system, the circular speed at a certain radius is readily related (through eq. 4.23) to the mass enclosed within that radius. To a first approximation, this is valid also for flattened systems like discs. Note that the circular speed is a property of any system with a generic mass distribution $\mathcal{M}(r)$ *independently* of whether the system is rotating or not. It should be simply thought of as the speed of a test particle in pure circular orbit at radius $r$.

Let us now suppose that the rotation velocity ($v_{\rm rot}$) that we measure in disc galaxies with, for instance, H I observations is such that $v_{\rm rot} = v_{\rm c}$ (we discuss this in §4.3.4). In §4.2.6 we have seen that in large spiral galaxies the stellar component is typically dominant in mass over the gas component ($\mathcal{M}_\star \gg \mathcal{M}_{\rm gas}$). Thus in the absence of any other mass component (suppose that there is no dark matter), in the outer regions of a galaxy disc where the stellar component fades exponentially, eq. (4.23) would imply that the rotation should decrease. In particular, if the mass does not grow significantly beyond a certain radius and stays constant to a value $\mathcal{M}$, at these radii we would expect

$$v_{\rm rot}(r) \simeq \sqrt{\frac{G\mathcal{M}}{r}} \propto r^{-1/2}, \tag{4.24}$$

which is called **Keplerian fall**, given the analogy with the solar system. This behaviour is not observed in galactic rotation curves (§4.3.3), implying that the outer parts of galaxies must contain much more matter than observed. In fact, the inferred profile of $v_{\rm c}$ is approximately flat at large radii, which implies (from eq. 4.23) that the mass increases linearly with radius.

It is instructive to derive the circular speed of a spherical system with a power-law density profile

$$\rho(r) = \rho_0 \left(\frac{r}{r_0}\right)^{-\delta}, \tag{4.25}$$

where $r_0$ is a reference radius and $\rho_0$ is the density at that radius. For $\delta \neq 3$,

$$\mathcal{M}(r) = 4\pi \int_0^r \rho(r') r'^2 \, {\rm d}r' = \frac{4\pi\rho_0 r_0^3}{3-\delta}\left(\frac{r}{r_0}\right)^{3-\delta}, \tag{4.26}$$

so the circular speed (eq. 4.23) is such that

$$v_{\rm c}^2(r) = \frac{4\pi G\rho_0 r_0^2}{3-\delta}\left(\frac{r}{r_0}\right)^{2-\delta}. \tag{4.27}$$

Thus, to obtain a flat rotation curve, one would need $\delta = 2$, which is the density profile of a singular isothermal sphere (eq. 5.37).

Moving from spherical systems to flattened systems, like discs, requires a certain amount of involved calculations, which are fully covered in other texts (e.g. Binney and Tremaine, 2008). For axisymmetric systems, a standard approach is to consider an oblate spheroidal system with axis ratio $q$ and take the limit $q \to 0$, while keeping constant the face-on surface density. The potential of this flattened system is then calculated as the sum of (infinite) concentric **homoeoids**.[8] The most convenient sets of coordinates to describe flattened

[8] A homoeoid is a shell bounded by two similar surfaces, in this case the surfaces of two flattened spheroids with infinitesimal difference in size.

systems are the cylindrical coordinates, where the plane $z = 0$ is taken to coincide with the plane of the disc. The circular speed can then be defined such that

$$v_c^2(R) \equiv R \left[ \frac{\partial \Phi(R, z)}{\partial R} \right]_{z=0}, \tag{4.28}$$

where the partial derivative is evaluated in the $z = 0$ plane.

If we consider an exponential razor-thin (zero-thickness) disc with surface density profile $\Sigma(R)$ given by eq. (4.1) with $I$ replaced by $\Sigma$, eq. (4.28) gives

$$v_c^2(R) = 4\pi G \Sigma_0 R_d\, y^2 \left[ I_0(y) K_0(y) - I_1(y) K_1(y) \right], \tag{4.29}$$

where $\Sigma_0$ is the central surface density, $K_n$ and $I_n$ are modified Bessel functions of $n$th order and $y \equiv R/(2R_d)$. The function in eq. (4.29) rises from the centre to a maximum at $R \simeq 2.2\, R_d$, beyond which it declines quite fast approaching the Keplerian fall of a point mass with the same mass of the exponential disc for $R \gg R_d$. A plot of this function can be seen in the bottom left panel of Fig. 4.15.

### 4.3.2 Spectroscopic Data

Rotation curves of disc galaxies are typically derived from emission-line observations (H I, CO or optical recombination lines). Stellar absorption lines are also used to derive the rotation velocities of the stellar disc. The basic physical effect that one aims to measure is the Doppler shift of a line (velocity along the line of sight) due to the rotation of the galaxy. The type of observations used for this kind of work are spectroscopic data, in the form of either **datacubes** or **long-slit observations** (Fig. 4.11).

A datacube can be seen as a collection of images taken at different wavelengths (frequencies or line-of-sight velocities). Data are stored in a 3D array with two spatial dimensions and one spectral dimension (Fig. 4.11, bottom). The first datacubes have been obtained for radio line observations, in particular H I and CO data. A radio telescope can divide the observational band into **velocity channels** (or frequency channels), typically with widths $\Delta V$ in the range 1–10 km s$^{-1}$ and images of the source are collected in each of these channels. The image of a channel, called a **channel map**, shows the gas that is moving at line-of-sight velocities $V_{chan} - \Delta V/2 < v < V_{chan} + \Delta V/2$, with $V_{chan}$ the velocity at the centre of the channel (see also §C.7). Today, datacubes are used across the whole electromagnetic spectrum, in particular at optical/NIR wavelengths; they are the outcome of **integral field unit** (IFU) spectrographs (a technique called **integral field spectroscopy**). IFUs (e.g. MUSE and KMOS at the VLT) are instruments that spread the incoming light and record a flux at different wavelengths, but they do not collect a spectrum from a single aperture on the sky. Instead, they can simultaneously obtain spectra at different locations in the sky plane (called **spaxels**) with the aim of *mapping* at best the astronomical object under scrutiny. This technique is also called 2D spectroscopy, where 2D refers to the spatial covering on the sky.[9] In contrast, more traditional optical observations used

[9] Note that 2D spectroscopy produces a 3D datacube, as one dimension is along the spectral axis.

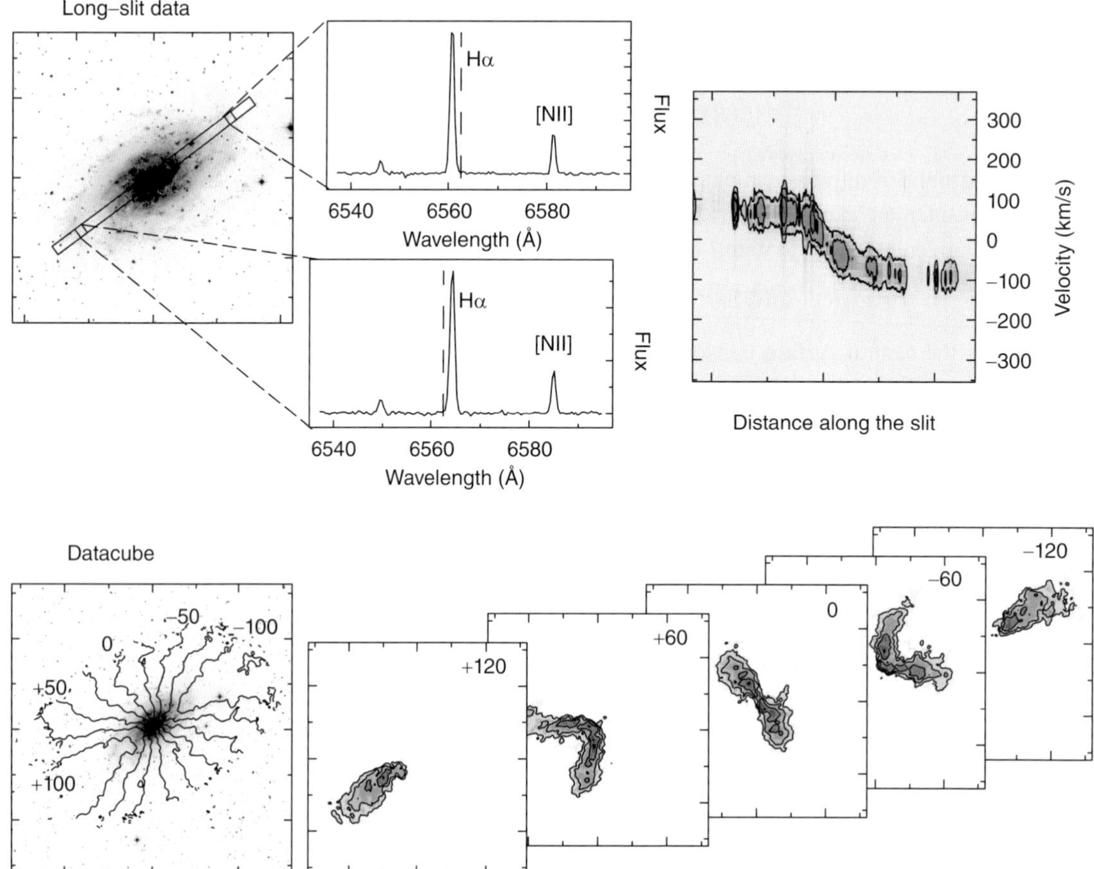


**Fig. 4.11** *Top panels.* Long-slit observations in the optical band of the nearby galaxy NGC 2403. The slit is positioned along the major axis of the galaxy (optical DSS image, $21' \times 21' \simeq 20$ kpc $\times$ 20 kpc) to maximise the detection of the rotation signal (see eq. 4.30). Two spectra, taken at opposite locations with respect to the centre, reveal blueshifted and redshifted emission lines (the dashed vertical lines show the location of the H$\alpha$ line at rest in this galaxy). If we focus on the H$\alpha$ line and visualise the intensity in all the spectra along the slit, we obtain a diagram of the line-of-sight velocity along the major axis showing a pattern of rotation (rightmost panel). Data from Fraternali et al. (2004). *Bottom panels.* A visualisation of the H I datacube of the same galaxy as in the top panel. The left panel shows the velocity field (spider diagram) overlaid on the optical DSS image (zoomed out by a factor of 2 with respect to the top panel). The numbers indicate line-of-sight velocities (in km s$^{-1}$) with respect to the systemic velocity of the galaxy. The other panels show five channel maps at representative line-of-sight velocities (top right corners in km s$^{-1}$). Data from Fraternali et al. (2002).

long slits, which can be seen as a collection of spectra along a segment on the sky (Fig. 4.11, top).

Optical and NIR spectrographs tend to have lower spectral resolutions than radio telescopes, of order of tens of km s$^{-1}$. The angular resolution in optical/IR is limited by the Earth's atmosphere for ground-based telescopes, but it can improve dramatically for space-based instruments or when the telescope is equipped with adaptive optics (§11.1.6). Most radio or mm/sub-mm telescopes are instead **interferometers**, i.e. several antennas that work simultaneously and use **fringe interference** to reconstruct the image of the observed sources; for a reference on this topic see Thompson et al. (2017). This allows us to reach much higher angular resolutions than with **single-dish** observations. Typical angular resolutions achievable for kinematic studies in the radio at 21 cm (1.4 GHz) with the VLA are 4–14 arcseconds, while we can reach resolutions of $\sim 0.1$ arcseconds at millimetre wavelengths with ALMA.

### 4.3.3 Rotation Curves

In this section, we describe how to derive rotation curves from emission line datacubes. The use of these rotation curves to infer the dark matter content of a galaxy is discussed in §4.3.4. From a datacube, it is customary to obtain the so-called **velocity field**, which is a map showing the characteristic velocity at every position on the sky over the galaxy disc (Fig. 4.11, bottom left). These velocities can be extracted as the intensity-weighted mean velocities of emission/absorption lines or by performing fits to the line profiles with Gaussians or other functions (e.g. eq. 5.4). If a galaxy disc is in regular rotation, velocity fields show the pattern of blueshifted and redshifted velocities on the two opposite sides of the galaxy, also called **approaching** and **receding sides**. Note that the blueshift and the redshift are meant with respect to the so-called **systemic velocity** ($v_{\rm sys}$) of the galaxy, which is the measured line-of-sight velocity shift of the galaxy as a whole. This is due to the Hubble flow (§2.1.2) plus the peculiar motion of the galaxy (§C.7). Examples of velocity fields are shown in Fig. 4.12 (second row) for three representative galaxies of different Hubble types: Sa, Sc and dIrr.

In §4.1.1 we have seen that the three types of galaxies shown in Fig. 4.12 have different surface brightness profiles. We see now that this difference in surface brightness translates into markedly different rotation curves. These rotation curves are representative of three classes of galactic rotation curves that we now describe.

1. The most typical shape for a spiral galaxy is that displayed in the central column of Fig. 4.12 with an inner rise and then a flattening at large radii out to the outermost measured value. The inner rise is an almost rigidly rotating **solid-body** regime (out to $R \simeq 5$ kpc in our example), in which the *angular speed* (§4.1.1) is roughly constant and the rotation velocity grows linearly with radius: $\Omega(R) = v_{\rm rot}(R)/R \simeq$ const. The flat part is instead a region of **differential rotation** (varying angular speed), in which the *rotation velocity* is nearly constant $v_{\rm rot}(R) \simeq v_{\rm flat}$, which is very typical of present-day disc galaxies.
2. Massive HSB galaxies (§4.1.1) with large bulges have rotation curves that rise steeply and then *decline* out to a radius where they also flatten. In these galaxies, the solid-body regime is often not traced (Fig. 4.12, top left) as it occurs within a radius that is smaller than the angular resolution of the data. Note that the flat part of the rotation curve can be seen only if the kinematic observations reach far out in radius ($R > 40$ kpc in our example), which is typically the case with H I data.

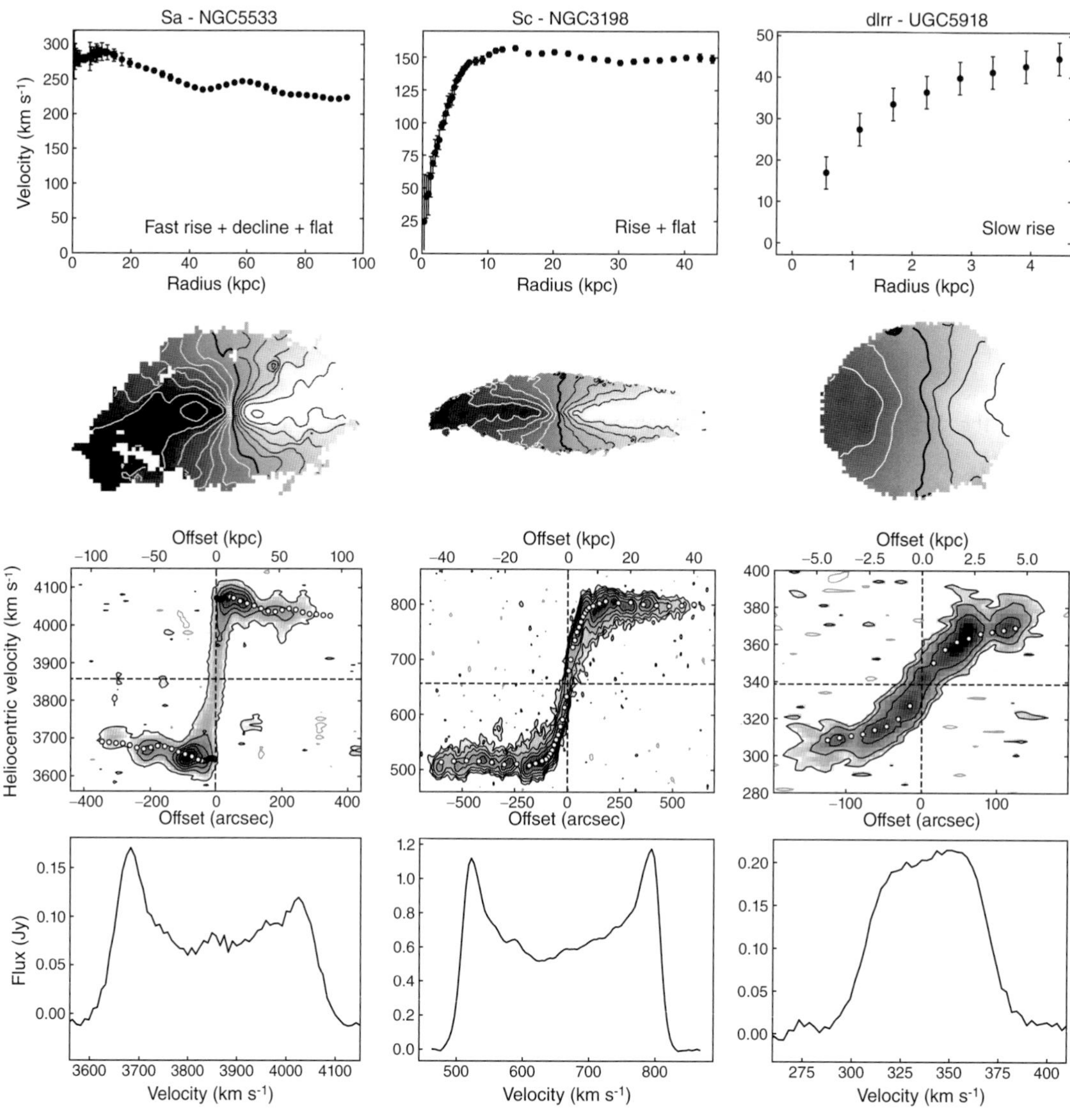


Fig. 4.12 *Top row*. Three classes of rotation curves of present-day galaxies of Hubble types Sa, Sc and dIrr. *Second row*. H I velocity fields of the same galaxies rotated so as to have the major axes aligned horizontally. The isovelocity contours are separated by 30 km $s^{-1}$ in NGC 5533 and NGC 3198, and by 10 km $s^{-1}$ in UGC 5918. The approaching sides are darker. *Third row*. Position–velocity diagrams extracted from the H I datacubes along the major axis of the galaxies. The points show the projected rotation curves from the top panels. *Bottom row*. Global H I line profiles extracted from the datacubes of the three galaxies. H I data from the WHISP survey (van der Hulst et al., 2001) and from Gentile et al. (2013) (NGC 3198); rotation velocities from Lelli et al. (2016a). Figure courtesy of C. Bacchini.

3. Dwarf irregular galaxies tend to have slowly rising (nearly solid-body) rotation curves that may not reach a flat part within the observed radii. This is, however, a rule with exceptions. Some dwarf galaxies have fast-rising rotation curves of the shape shown in the second column of Fig. 4.12. Others appear as nearly perfect solid bodies. In

general, we say that low-mass galaxies show a considerable *diversity* in their rotation curve shapes. Interestingly, also in the dwarf galaxy regime, there appears to be a close link between the shape of the rotation curve and the surface brightness of the galaxy, with LSB galaxies having slowly rising curves and HSB galaxies having steeply rising curves.

Let us now look at the velocity fields from which the above rotation curves have been derived, shown in the second row of Fig. 4.12. The velocity fields of the Sa galaxy and of the Sc galaxy have a characteristic shape that has been termed a **spider diagram**, with isovelocity contours that tend to crowd towards the centre. This is reflected in the shape of the respective rotation curves, which can be seen as a plot of the (deprojected) velocities along the major axis of the velocity field (here aligned horizontally). The velocity field of the dIrr galaxy shows instead contours that tend to be parallel to each other and perpendicular to the major axis. Contours *exactly* parallel to each other are the pattern of a perfect solid body.

We briefly describe how rotation curves can be derived from the fitting of velocity fields. The main assumption is that the kinematics is totally dominated by rotation and that the different velocities across the velocity fields are simply due to the orientation of the rotation velocity vector with respect to the line of sight. The galaxy disc is decomposed into concentric rings whose shape on the sky can be uniquely described by the radius ($R$) of the ring and two angles: the inclination angle ($i$) and the **kinematic position angle** ($\phi$). This latter is defined as the angle in the anticlockwise direction between the north and the major axis of the receding side of the galaxy. This decomposition in concentric rings is called the **tilted-ring model**. The rotation curve $v_{\rm rot}(R)$ is found by fitting the velocity field with the relation

$$v_{\rm los}(x, y) = v_{\rm sys} + v_{\rm rot}(R) \sin i \cos \theta, \tag{4.30}$$

where $v_{\rm los}(x, y)$ is the measured velocity at a certain projected location $(x, y)$ on the sky and $\theta$ is the azimuthal angle in the plane of the galaxy. This angle is taken with respect to the major axis and is such that $\cos \theta = [-(x - x_0) \sin \phi + (y - y_0) \cos \phi]/R$ and $\sin \theta = [-(x - x_0) \cos \phi - (y - y_0) \sin \phi]/(R \cos i)$, with $x_0$ and $y_0$ the sky coordinates of the (kinematic) centre of the galaxy. The tilted-ring fit can be performed either by keeping the inclination and position angles fixed with radius or by letting them vary. In the latter case, one can trace the **warps**, which are common features of the outer galaxy discs, frequently observed in H I, but sometimes also in the stellar component. A warp can be envisioned as the tilt of the rotation axis of the outer galaxy rings. In edge-on galaxies, this leads to an 'integral sign' shape as the position angle of the outer disc twists with respect to the inner disc. In galaxies at intermediate inclinations, one observes a distortion of the outer velocity field: these are called kinematic warps. For instance, the indication of a warp is visible in NGC 5533 (Fig. 4.12) as an anticlockwise rotation of the outer isovelocity contours, especially on the approaching side.

The third row of Fig. 4.12 shows the **position–velocity diagrams** extracted along the major axes of these galaxies. These diagrams can be seen as a *collection of spectra* along a spatial direction on the sky and they are akin to the result of a long-slit observation (Fig. 4.11, top). In galaxies where the kinematics is regular and symmetric, position–velocity diagrams along the major axis ($\cos\theta = 1$ in eq. 4.30) well describe the rotation of the whole disc. The white points overlaid on these diagrams in Fig. 4.12 show the projected rotation velocities obtained by the fitting of the velocity field, as explained above. Note how they lie, at nearly every radius, close to the peaks of the H I emission.

We conclude by mentioning that all the data described in this section are taken at high angular resolution, which translates into having several independent points in the final rotation curves. However, observations of distant galaxies (Fig. 11.16) have typically only a few resolution elements across the entire disc and the derivation of the rotation curve requires more caution. The low angular resolution causes parts of the disc with very different velocities to end up in the *same* resolution element, especially in the inner disc where the isovelocity contours crowd (second row of Fig. 4.12). This produces two effects: (1) the resulting line-of-sight velocity is the intensity-weighted average of the velocities contained in the resolution element; (2) the line profiles become broader due to the contribution of different velocities coming from different parts of the disc. The problem is aggravated by the non-homogeneous and clumpy distribution of the gas density. This effect, referred to as **beam smearing**,[10] needs to be corrected for to derive reliable rotation curves and velocity dispersions from low-resolution data.[11] If not properly taken into account, the typical consequences are an underestimate of the rotation velocity and an overestimate of the velocity dispersion. In optical/UV observations, one should also consider the effect of dust extinction (§4.2.7) that may conceal kinematic information from some parts of the galaxy.

The most extreme case of beam smearing occurs when the beam size (angular resolution) of the observations is larger than the galaxy itself: the object is spatially *unresolved*. The line profiles resulting from these observations are called **global profiles** (bottom row of Fig. 4.12). If a galaxy has no rotation, the peak of the global profile would be at the systemic velocity ($v_{\rm sys}$) and the broadening would be given only by the turbulent velocity dispersion of the gas (§4.2.2). Instead, in the presence of rotation, the gas emits at velocities $v_{\rm sys} \pm v_{\rm rot}(R)\sin i\cos\theta$, where $v_{\rm rot}(R)\sin i\cos\theta$ is the line-of-sight component of the rotation velocity (eq. 4.30), and + and − refer to the receding and the approaching sides, respectively. If, in a large fraction of the disc, we have $v_{\rm rot} \simeq v_{\rm flat}$ (and $i$ does not change appreciably) then the emission accumulates around $v_{\rm sys} \pm v_{\rm flat}$ and one obtains the 'double-horn' shape visible in the first two columns of the bottom row of Fig. 4.12. These shapes are typical of H I observations of present-day spiral galaxies. On the other hand, if the rotation velocity changes with radius, like in the dIrr galaxy, the horns tend to vanish.

### 4.3.4 Mass Decomposition

If the tracer component that we use to derive the rotation curve is kinematically cold ($V/\sigma \gg 1$), the measured rotation velocity at a certain galactocentric radius can be

[10] This term has been initially used by the radio community, in particular for H I observations, hence the term 'beam', which is the analogue of the point spread function (§11.1.6) for a radio telescope.

[11] An efficient way to correct for beam smearing is to produce artificial observations from tilted-ring models (including observational effects) and fit them directly to the emission-line datacube.

considered equal to the circular speed at that radius: $v_{\rm rot}(R) = v_{\rm c}(R)$. For H I rotation curves of disc galaxies this is always the case, since $\sigma_{\rm HI} \approx 10\,{\rm km\,s^{-1}}$ (Fig. 4.6). Instead, for dIrrs a correction may be required as we see at the end of this section. Using mathematical tools like the one described in §4.3.1 one can determine from observations the circular speed contributed by any of the main baryonic components of a galaxy. These are: the stellar bulge, the stellar disc and the gaseous disc. The stellar halo (§4.1.3) is a very minor component in terms of mass and it is always neglected in these calculations.

Let us first consider the bulge and the stellar disc. From optical/NIR observations, we measure the surface brightness and we must, first, convert it to a mass surface density and, second, to a circular speed, using eq. (4.28), where $\Phi$ is, in this case, the potential of the bulge or the stellar disc. The second step involves the assumption of a geometry for the mass component: usually a spheroid for the bulge and a thin structure with a scaleheight constant with radius for the disc. The first step is more critical as it involves the choice of stellar mass-to-light ratios ($\Upsilon_\star$; §4.1.2) for the two components. Here, we have to deal with uncertainties due to the derivation of stellar masses (or surface densities in this case) from optical/NIR observations. We can consider two approaches.

1. A useful approach that gives an upper limit to the stellar mass is to take the largest $\Upsilon_\star$ allowed by the rotation curve in the inner parts. This is called the **maximum-disc hypothesis**.[12]
2. A more physically motivated way to proceed is to use $\Upsilon_\star$ compatible with stellar population synthesis models (§8.6). This gives typical values of order one (with significant spread) in the optical bands and $\Upsilon_\star \approx 0.2$–$0.7$ in the NIR bands.

These two approaches return comparable $\Upsilon_\star$ for large disc galaxies, but very different values for small spirals or dwarfs. Fig. 4.13 illustrates this point for the same representative rotation curves used in Fig. 4.12. The circular speed contributions of the stellar discs in the top row have been obtained forcing a maximum disc, while, in the second row, $\Upsilon_\star$ has been fixed to 0.45, a realistic value for present-day SFGs in the 3.6 $\mu$m band.

The gas contribution to the rotation curve is calculated assuming that the gas is distributed in a thin disc. Typically one only considers neutral atomic gas (H I) for two reasons. First, H I dominates the gas mass (Fig. 4.8). Second, both molecular and ionised gas tend to have similar distributions to that of the stellar component (see Fig. 4.5) and their contributions are effectively incorporated into the uncertain $\Upsilon_\star$. On the contrary, H I is more radially extended and it is the dominant baryonic component in the outer parts. In dwarf galaxies, it is possible that H I dynamically dominates everywhere. Another important property of H I is that its mass and surface density are known with precision (eqs. 4.5 and C.16), so uncertainties analogous to that of the stellar mass-to-light ratio are not present.

[12] The term 'maximum disc' or 'maximal disc' is technically correct for galaxies where the disc fully dominates the stellar light, e.g. Hubble types Sc (Fig. 4.1), in which case it is customary to use a single value of $\Upsilon_\star$ for the entire stellar surface density profile. In galaxies with prominent bulges, a more correct terminology, however scarcely used, would be 'maximum light' or 'maximum stellar mass', as both the bulge and the disc are maximised and, potentially, with different $\Upsilon_\star$.

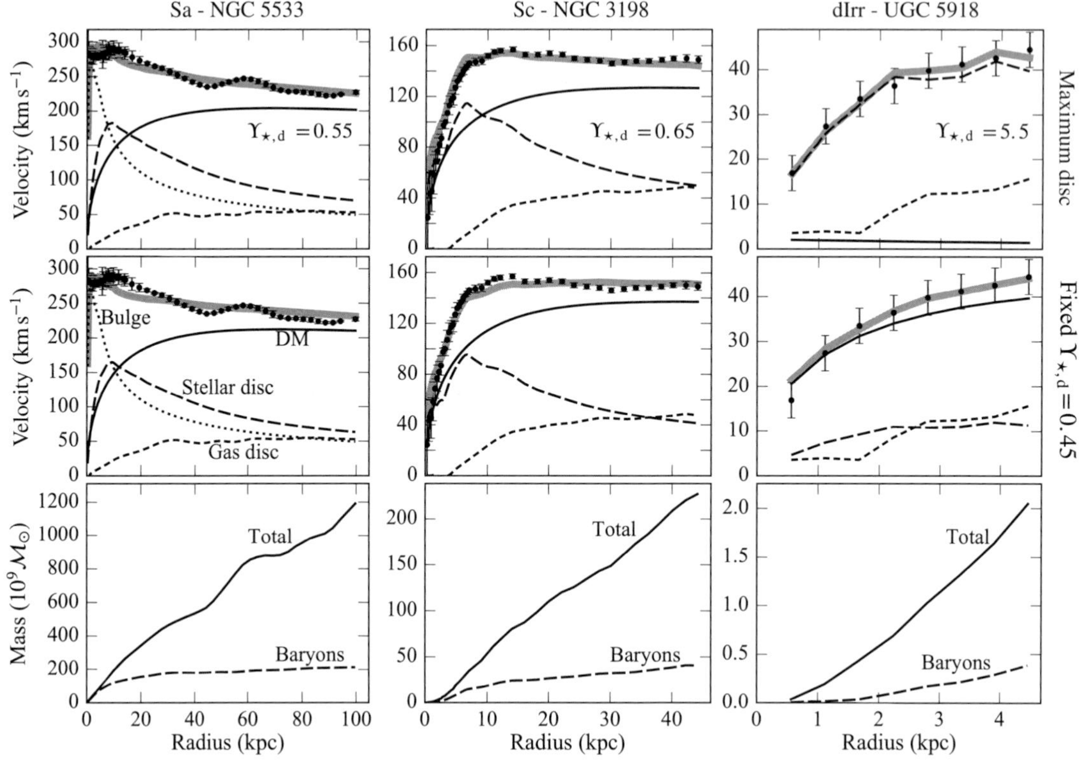


**Fig. 4.13** *Top and middle rows*. Rotation curves and corresponding decompositions for the three representative galaxies shown in Fig. 4.12. The contributions of the bulge, stellar disc, gaseous disc and of the dark matter (DM) are shown respectively as dotted, long-dashed, short-dashed and solid lines. The *top panels* show decompositions that give the maximum importance to visible components (maximum disc). The values of the mass-to-light ratios ($\Upsilon_\star$) of the stellar discs are indicated. In the *middle panels*, $\Upsilon_\star$ are fixed to values in agreement with stellar population models ($\Upsilon_\star = 0.7$ for the bulge and $\Upsilon_\star = 0.45$ for the discs in the three galaxies). All $\Upsilon_\star$ are given in the 3.6 $\mu$m band. The contributions of the baryonic components have been calculated numerically from the surface brightness profiles. The concentration of the dark matter halos (§7.5.2) are fixed to $c_{200} = 8$, 9 and 10.5, respectively for the Sa, the Sc and the dIrr (Fig. 7.9). The final fit is shown in grey. *Bottom panels*. Mass profiles: the solid curve shows the total observed mass, the long-dashed curve shows the contribution of all the baryonic components using the decompositions in the middle row. Rotation curves and photometric data from Lelli et al. (2016a).

Once we have the contributions of stellar and gas components, they are summed quadratically to give the full baryonic contribution to the circular speed,

$$v_{\rm c,b} = \left(\Upsilon_{\star,\rm bu} v_{\rm c,bu}^2 + \Upsilon_{\star,\rm d} v_{\rm c,d}^2 + v_{\rm c,gas}^2\right)^{1/2}, \tag{4.31}$$

where $v_{\rm c,bu}$ and $v_{\rm c,d}$ are, respectively, the contributions of the stellar bulge and disc calculated for $\Upsilon_\star = 1$, while $\Upsilon_{\star,\rm bu}$ and $\Upsilon_{\star,\rm d}$ are the actual (assumed or fitted) stellar mass-to-light ratios of bulge and disc. The gas contribution ($v_{\rm c,gas}$) is for the total gas mass, including helium. In the next section, we see how the baryonic circular speed fails to explain the observed rotation curve of *any* galaxy.

We conclude this section by mentioning that in low-mass galaxies the initial assumption of $v_{\rm rot}(R) = v_{\rm c}(R)$ may occasionally break down, especially if we are using tracers with higher velocity dispersions than H I like ionised gas or stars. We say that these discs are kinematically hot, i.e. they have $V/\sigma \sim 1$ to a few. The material rotating with these $V/\sigma$ ratios experiences a phenomenon called **asymmetric drift** that implies a lower average rotation velocity, because also the velocity dispersion contributes to the support against gravity. The asymmetric drift can be described in terms of the Jeans equations (§5.3.2) and corrected for. In particular, if we are considering the gas component, we can make the simplifying assumption that the velocity dispersion is isotropic[13] and only a function of the radius, $\sigma_{\rm gas} = \sigma_{\rm gas}(R)$ (see also §4.6.2). Then, given the face-on surface density of the gas, $\Sigma_{\rm gas}(R)$, it can be shown that its circular speed is obtained from the equation

$$v_{\rm c}^2(R) \simeq v_{\rm rot}^2(R) - R\sigma_{\rm gas}^2 \frac{\partial \ln\left(\Sigma_{\rm gas} h_{\rm gas}^{-1} \sigma_{\rm gas}^2\right)}{\partial R}, \tag{4.32}$$

where $h_{\rm gas}(R)$ is the scaleheight of the gas disc and the second term on the right-hand side (r.h.s.) is the asymmetric drift correction. Note that the argument of the logarithmic derivative typically decreases with radius leading to a *positive* correction on $v_{\rm rot}$. In the end, for the gas component, the asymmetric drift correction is essentially a correction for pressure support, if the pressure is written as in eq. (8.50). Instead, the analogous correction for the stellar component, whose velocity dispersion is not isotropic in general, is more complex (see Binney and Tremaine, 2008, for a full derivation).

### 4.3.5 Dark Matter

The combined contribution to the circular speed of the baryonic matter components calculated in §4.3.4 is insufficient to explain the measured H I rotation curves of present-day SFGs. There is, in fact, a dramatic *discrepancy* between the predicted and the observed velocities. We can see this clearly in the bottom row of Fig. 4.13, where we compare the observed mass contained within $R$, inferred from the rotation curve, with the sum of the masses contributed by all the baryonic components (stellar bulge, stellar disc and gas disc), obtained using $\Upsilon_{\star,\rm d} = 0.45$ for the stellar disc (second row of Fig. 4.13). The difference is striking: there is a clear need[14] for a large amount of 'undetected' matter, called **dark matter**, whose contribution is particularly necessary in the outer parts of the galaxies. As shown in §7.5, dark matter is expected to be distributed in ellipsoidal structures called **dark matter halos**. In the following, we show how H I rotation curves allow us to constrain the density profiles of these halos, assuming for simplicity that they are spherical.

[13] In the ISM, two-body interactions between gas particles (§4.2.4) and/or between gas clouds are expected to make the system isotropic. For instance, if a phenomenon, such as a disc instability (§10.2.1), produces an increase of one component (e.g. the radial one) of the velocity dispersion, ISM collisions can quickly transfer this kinetic energy to the other components.

[14] Alternative explanations to avoid this discrepancy have been suggested, most notably the possibility that Newtonian dynamics is modified in the low-acceleration regime (e.g. modified Newtonian dynamics, MOND; Milgrom, 1983).

The first dark matter profile proposed has been the **cored pseudo-isothermal profile**[15] (van Albada et al., 1985), whose density can be written as

$$\rho(r) = \frac{\rho_0}{1 + (r/r_c)^2}, \tag{4.33}$$

where $\rho_0$ is the central density and $r_c$ is the core radius. This profile leads to a flat rotation curve at large radii (where $\rho \propto r^{-2}$; see eq. 4.27) and has a core in the centre ($\rho \approx \rho_0$ for $r \ll r_c$). The circular speed of the pseudo-isothermal halo is given by

$$v_c^2(r) = v_\infty^2 \left[1 - \frac{r_c}{r} \arctan\left(\frac{r}{r_c}\right)\right], \tag{4.34}$$

with asymptotic value $v_\infty = \sqrt{4\pi G \rho_0 r_c^2}$.

In §7.5.1 we see that $N$-body cosmological simulations predict a characteristic dark matter density profile called, after its discoverers, a Navarro–Frenk–White (NFW) profile (Navarro et al., 1996). A spherical NFW model has density distribution given by eq. (7.50) and circular speed given by

$$v_c^2(r) = v_\Delta^2 \frac{1}{x} \frac{\ln(1 + c_\Delta x) - \frac{c_\Delta x}{1+c_\Delta x}}{\ln(1 + c_\Delta) - \frac{c_\Delta}{(1+c_\Delta)}}, \tag{4.35}$$

where $v_\Delta$ is the circular speed at $r_\Delta$, the radius inside which the average density equals $\Delta$ times the critical density of the Universe (§7.3.2), $c_\Delta \equiv r_\Delta / r_s$ is the concentration of the halo ($r_s$ is the scale radius; §7.5.2) and $x = r/r_\Delta$. In principle, as for the cored pseudo-isothermal profile, an NFW halo has two free parameters: $v_\Delta$ (or $\mathcal{M}_\Delta$ using eqs. 7.23 and 7.24) and $c_\Delta$. However, the concentration $c_\Delta$ of a dark matter halo is predicted to be a function of the halo mass (Fig. 7.9) and it should not be considered as a completely free parameter. In §7.3.2 we see that a typical choice for the overdensity $\Delta$ is $\Delta = 200$, which we use in the following.

Fig. 4.13 shows fits with NFW dark matter profiles to the rotation curves of our representative galaxies. The first row shows fits with $\Upsilon_\star$ chosen to maximise the stellar contribution (maximum disc). In the second row, $\Upsilon_\star$ are fixed to realistic values compatible with stellar population synthesis models. In both fits, the dark matter halos have only one free parameter ($v_{200}$) as the concentrations ($c_{200}$) have been fixed to realistic values for these galaxies. As anticipated, the maximum-disc $\Upsilon_\star$ of large spiral galaxies (e.g. the Sa galaxy) are compatible with those from stellar evolution models. This shows that the inner gravitational potential of a large spiral is dominated by the stellar component. The stellar bulge of the Sa galaxy in Fig. 4.13 is driving the shape (decline) of the inner rotation curve and the dark matter dominates beyond about 12–15 kpc from the centre. The situation is very different when we look at the dwarf irregular galaxy. If we choose a realistic $\Upsilon_\star$, the contribution of the stellar disc is completely negligible and the dark matter dominates the potential everywhere (middle row, right). If we force a maximum-disc $\Upsilon_\star$, its value turns out to be 10 times larger than what would be justifiable by stellar population synthesis models (top right). The Sc galaxy is somewhat in the middle of these extremes, with a

[15] This profile, sometimes simply called isothermal, is similar to the singular isothermal sphere (eq. 5.37), but with a central core.

marginally 'submaximal' stellar disc and a similar contribution of the dark matter and baryons also in the central regions (middle column). Note that the photometry used in these decompositions is in the NIR (3.6 $\mu$m) and it is considered optimal to trace the stellar mass in galaxies (§8.6).

In the end, the fitting of rotation curves like those presented here (middle row in Fig. 4.13) allows us to estimate the total (virial) mass $M_{200}$ (see §7.3.2) of the dark matter halos. The values one obtains for the galaxies shown in Fig. 4.13 are about $2 \times 10^{12}\ \mathcal{M}_{\odot}$, $5 \times 10^{11}\ \mathcal{M}_{\odot}$ and $1.4 \times 10^{10}\ \mathcal{M}_{\odot}$, respectively, for the Sa, the Sc and the dIrr. These values are typical for galaxies in these three classes. Note that these masses are significantly higher than those derived within the outermost radii of the rotation curves (bottom panels of Fig. 4.13), as a significant fraction of dark matter is expected to lie beyond the galaxy discs. In general, the part of the potential probed by observed rotation curves is at radii $r \ll r_{\rm vir}$, with $r_{\rm vir}$ the virial radius (§7.3.2); thus calculating the total mass requires an extrapolation. In Fig. 4.15 (bottom right) we show the full circular speed of an NFW halo out to the virial radius; this curve peaks at $r \simeq 2r_{200}/c_{200}$.

We conclude with a brief comparison between NFW and pseudo-isothermal profiles. In general, they tend to return comparably acceptable fits for most rotation curves of galaxies. This is due to the presence of two free parameters for the halo profiles (but see the above discussion about the concentration of the NFW) and some freedom in the choice of $\Upsilon_{\star}$. Some discrepancies between the data and the prediction of the NFW profiles have been found in the inner parts of LSB/dwarf galaxies. Indeed, the NFW density distribution leads to $\rho(r) \propto r^{-1}$ for $r \to 0$, while the rotation curves of several dwarf galaxies show a preference for a core, $\rho(r) \approx$ const for $r \to 0$ (see Fig. 7.10). This finding is a manifestation of the so-called cusp/core problem (§7.5.2), which suggests that the profiles of the dark matter halos can be significantly modified by the baryons (§7.5.5).

## 4.4 Scaling Relations

A number of physical properties of present-day SFGs correlate with each other across a broad range of scales, from dwarf to massive galaxies. These correlations are called **scaling relations** or **scaling laws**. They are relatively tight relations between fundamental quantities whose existence helps us understand the physics of galaxy formation. In general, they portray a remarkable regularity in galaxies despite the complex physical processes that are at play (Chapter 8). In the next sections, we describe the main scaling relations of SFGs (those of ETGs are presented in §5.1.3 and §5.4). The relevance of these relations and their connection to theory are discussed in §10.10.3.

### 4.4.1 Tully–Fisher Relation

This relation was first reported by Tully and Fisher (1977) as a correlation between the velocity widths $W$ of the global H I profiles (Fig. 4.12, bottom row) and the absolute magnitudes $M$ of spiral galaxies. As we have seen in §4.3.3, the velocity width[16] of a global profile of a rotating disc is roughly $W \approx 2v_{\rm rot} \sin i$. If we know the inclination of the galaxy independently (for instance from the fitting of the optical isophotes) we can derive an inclination-corrected width $W^i \approx 2v_{\rm rot}$. The **Tully–Fisher relation** (TFR) then reads

$$M = A - \alpha \left(\log W^i - 2.5\right), \tag{4.36}$$

with zero-point $A$ and slope $\alpha$ (in the range $\approx 7$–$10$) that depend on the observational band. We show a version of this relation at 3.6 $\mu$m in Fig. 4.14 (left).

More intelligibly, the TFR can be expressed as a relation between the luminosity of the galaxy and its rotation velocity,

$$\log\left(\frac{L}{10^{10}L_\odot}\right) = B + \beta \log\left(\frac{v_{\rm rot}}{200\,{\rm km\,s^{-1}}}\right), \tag{4.37}$$

with zero-point $B$ depending on the observational band (indicative values are $B \approx 0.5$ and $\approx 1$ in the $V$ and 3.6 $\mu$m bands, respectively) and slope $\beta \approx 3$–$5$. The stellar luminosity can also be seen as a proxy for the stellar mass, choosing an appropriate $\Upsilon_\star$ (§4.3.4). The rotation velocity is, in general, a function of radius (see rotation curves in Fig. 4.12), so one must choose where to measure the velocity. A preferred choice, when available, is to choose the velocity of the flat part of the rotation curve. Indeed the TFR becomes tighter (it has less intrinsic scatter) if one uses $v_{\rm flat}$ in eq. (4.37).

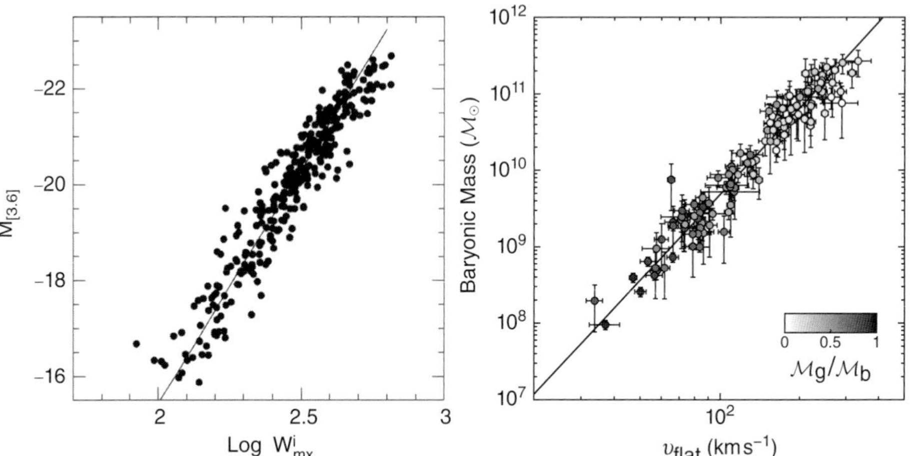


Fig. 4.14 *Left panel.* The Tully–Fisher relation between the absolute magnitudes of a sample of local disc galaxies and the widths of their global H I profiles. Here the magnitude is taken at 3.6 $\mu$m and the widths have been corrected for the inclination of the galaxies. The black line is a fit with eq. (4.36) with $A \simeq -20.3$ and $\alpha \simeq -9.8$. Adapted from Sorce et al. (2013). Courtesy of J. Sorce. *Right panel.* Baryonic Tully–Fisher relation between the total baryonic mass (stars and H I) of SFGs and their rotation velocity taken in the flat part of the rotation curves. The greyscale gives the gas fraction (eq. 4.12). Adapted from Lelli et al. (2016b). Courtesy of F. Lelli.

[16] These widths are often taken at some percentage of the peak flux of the global line profile, usually 20% or 50% and called $W_{20}$ and $W_{50}$, respectively.

Given the tightness of the relation (intrinsic scatters $\sigma \approx 0.15$–$0.2$ dex and $\sigma_\perp \lesssim 0.1$ dex[17]), the TFR is often used, for nearby SFGs, as a **distance indicator**. In practice, one simply has to determine the rotation velocity of a disc galaxy (or, even more straightforwardly, the width of the global H I profile and the disc inclination) to have an estimate of the absolute magnitude or luminosity from eq. (4.36) or (4.37), which then gives the distance by measuring the apparent magnitude or the flux (eqs. C.5 and C.24). This method allows us to reach larger distances than other (usually more precise) indicators like the period–luminosity relation of Cepheid variable stars (§4.6.1). However, distances determined using the TFR are uncertain due to the intrinsic scatter of the relation and the fact that there is not a unique calibration. Note that the TFR cannot be used as a distance indicator at cosmological distances because scaling relations like the TFR may evolve with time (§11.3.13).

A further version of the Tully–Fisher relation takes into account the total baryonic mass $\mathcal{M}_\mathrm{b}$. In order to construct this relation one needs to derive a stellar mass, obtained by multiplying the galaxy luminosity by $\Upsilon_\star$ or with other methods, and add the mass of the gas, usually only the H I mass being the dominant gas component (see Fig. 4.8, top right). The resulting relation, called the **baryonic Tully–Fisher relation** (McGaugh et al., 2000), can be written as

$$\log\left(\frac{\mathcal{M}_\mathrm{b}}{10^{10}\,\mathcal{M}_\odot}\right) = C + \gamma \log\left(\frac{v_\mathrm{rot}}{200\,\mathrm{km\,s^{-1}}}\right), \tag{4.38}$$

where $C$ and $\gamma$ are the normalisation and the slope. Considering the gas in the mass budget makes little difference for massive galaxies, whereas it changes the location of the smallest dwarf galaxies, which are largely gas-dominated, by up to an order of magnitude. A version of the baryonic Tully–Fisher relation with $v_\mathrm{rot} = v_\mathrm{flat}$ for a sample of present-day SFGs is shown in Fig. 4.14 (right panel). To convert the luminosities to stellar masses, $\Upsilon_\star = 0.5$ at 3.6 $\mu$m has been used for all galaxies. Fitting eq. (4.38) (solid curve) returns a slope $\gamma \simeq 3.7$ and a normalisation $C \approx 0.8$. The relation is very tight: the intrinsic scatter at fixed $v_\mathrm{flat}$ is $\sigma \sim 0.1$ dex. Note that, unlike for the TFR, where a change in $\Upsilon_\star$ only affects the normalisation of the relation, in the baryonic Tully–Fisher relation a different $\Upsilon_\star$ produces a different slope as well. For instance, using the galaxy sample in Fig. 4.14, the relation would have slopes $\gamma = 3.5$ and $\gamma = 3.9$ for $\Upsilon_\star = 0.3$ and $\Upsilon_\star = 0.7$, respectively.

### 4.4.2 Specific Angular Momentum–Stellar Mass Relation

A fundamental property of galaxies is their angular momentum content (§10.1). If we consider a disc of stars or gas with a surface density $\Sigma(R)$ and rotating at $v_\mathrm{rot}(R)$, we can define the **specific angular momentum** (angular momentum per unit mass) within a certain radius $R$ as

$$j(<R) \equiv \frac{2\pi \int_0^R \Sigma(R') v_\mathrm{rot}(R') R'^2\,\mathrm{d}R'}{2\pi \int_0^R \Sigma(R') R'\,\mathrm{d}R'}. \tag{4.39}$$

[17] $\sigma$ is the scatter at fixed $v_\mathrm{rot}$, while $\sigma_\perp$ is the perpendicular scatter between the data and the model fit.

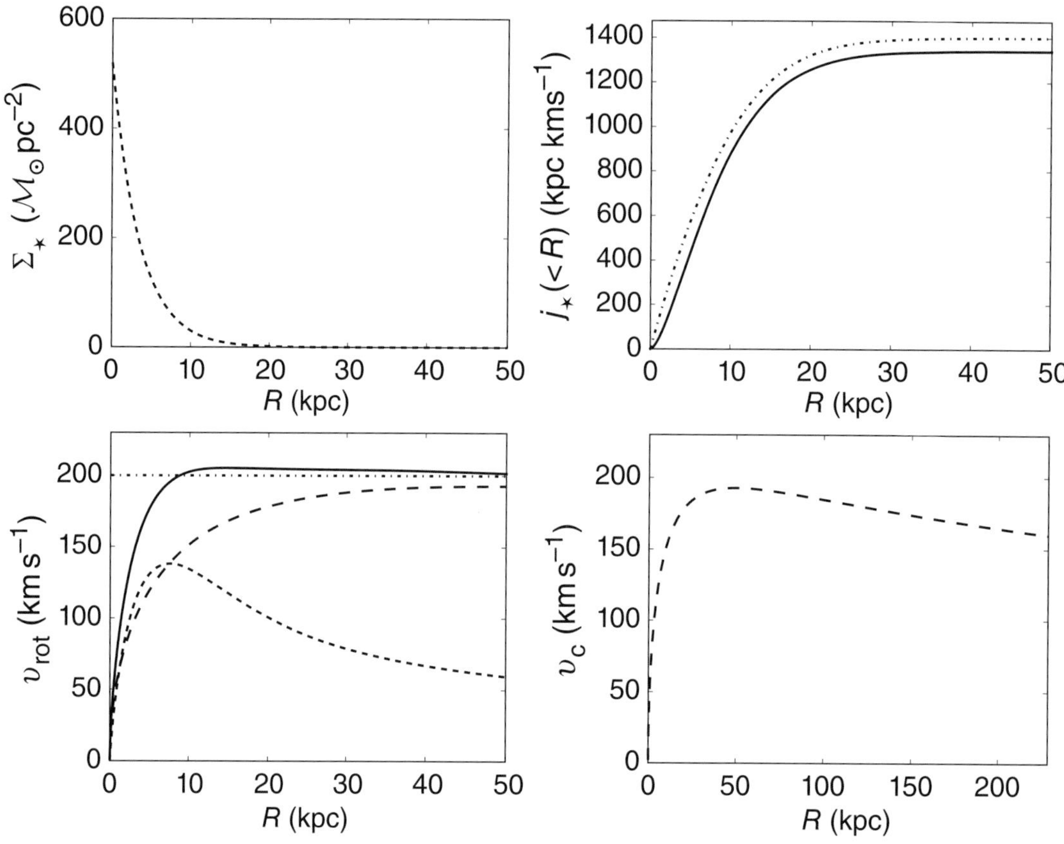


Fig. 4.15 *Left panels*. Simple model of a galaxy's disc with an exponential density profile with $R_d = 3.5$ kpc and $\mathcal{M}_d = 4 \times 10^{10}\ \mathcal{M}_\odot$ (*top*) and two types of rotation curves, flat at $v_{\rm flat} = 200$ km s$^{-1}$ and realistic (*bottom*). The realistic curve is given by the combined contributions of the circular speeds of an exponential disc (short dashed, eq. 4.29) and a spherical NFW dark matter halo (long dashed, eq. 4.35) with $c_{200} = 10$ and $v_{200} = 160$ km s$^{-1}$. Note that we are assuming that the stars are rotating at the circular speed $v_{\star,\rm rot}(R) = v_c(R)$. *Top right panel*. Specific angular momentum of the stellar disc contained within the radius $R$, $j_\star(<R)$, for the flat and the realistic rotation curves. The asymptotic value reached by the flat curve is given by eq. (4.40) with $\alpha = 1$. *Bottom right panel*. Circular speed of the NFW dark matter halo out to $r_{200} \simeq 229$ kpc.

Note that eq. (4.39) is simply a measure of $R\, v_{\rm rot}(R)$, weighted by the surface density. In the following, we will mainly deal with the stellar disc and take $\Sigma(R) = \Sigma_\star(R)$, i.e. the stellar surface density, which we can assume to be exponential (obtained from the surface brightness in eq. 4.1 assuming an $\Upsilon_\star$), and $v_{\rm rot}(R) = v_{\star,\rm rot}(R)$, i.e. the rotation curve of the stellar component. Integrating eq. (4.39) to infinity we obtain the total specific angular momentum of an exponential stellar disc

$$j_\star = 2\alpha R_d v_{\rm flat} = \alpha\, j_{\rm flat}, \tag{4.40}$$

where $j_{\rm flat} \equiv 2R_d v_{\rm flat}$ and $\alpha$ is a dimensionless factor of order one that accounts for the exact shape of the rotation curve (§4.3.3); $\alpha = 1$ if $v_{\star,\rm rot}(R) = v_{\rm flat}$ at every $R$.

Fig. 4.15 shows the specific angular momentum within $R$ for a disc with exponential surface density profile and either a flat or a more realistic rotation curve. This disc has a scalelength $R_d = 3.5$ kpc and we can see that most of the specific angular momentum is contained within a few scalelengths reaching an asymptote at $R \approx 5R_d$. The asymptotic value given by the disc with a flat rotation curve is exactly $j_{flat}$ of eq. (4.40), while the disc with a realistic curve is slightly below. Thus the approximation of a flat rotation curve is reasonably good and eq. (4.40) with $\alpha = 1$ allows us to estimate the total specific angular momentum $j_\star$ of an exponential stellar disc using only two quantities that are relatively easy to measure.

Accurate measurements of the specific angular momentum of SFGs reveal a tight relation between $j_\star$ and the stellar mass. This relation, originally proposed by Fall (1983), can be written as

$$\log\left(\frac{j_\star}{10^3\,\text{kpc km s}^{-1}}\right) = D + \zeta \log\left(\frac{\mathcal{M}_\star}{10^{11}\,\mathcal{M}_\odot}\right), \tag{4.41}$$

where $D$ is a normalisation and $\zeta \approx 0.6$ is the slope of the relation. A version of this relation for present-day disc galaxies and dIrrs is shown in Fig. 4.16. Here $j_\star$ has been estimated using a discretised version of eq. (4.39) with $\Sigma_\star$ from 3.6 $\mu$m photometry. In §10.1 and §10.10.3, we discuss the importance of angular momentum assembly for galaxy formation and the significance of the $j_\star$–$\mathcal{M}_\star$ relation.

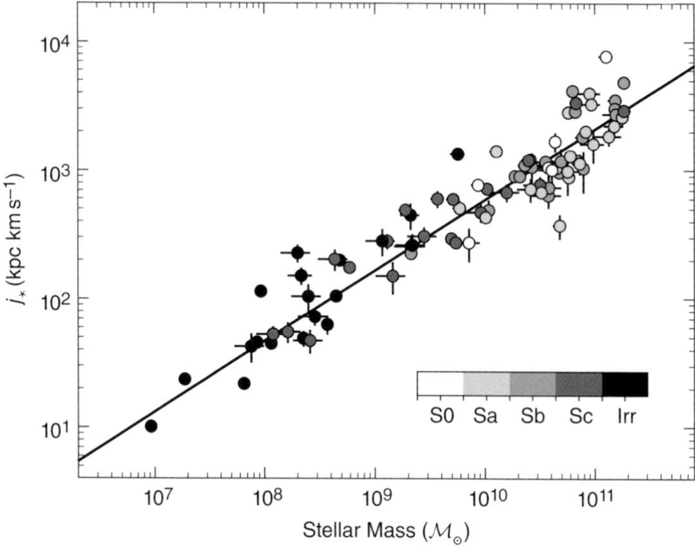


Fig. 4.16 Specific angular momentum of the stars ($j_\star$) versus the stellar mass in a sample of present-day spirals and dIrrs. The greyscale shows the Hubble types of the galaxies. The curve is a fit with eq. (4.41) with $\zeta \simeq 0.55$, $D \simeq 0.34$ and perpendicular intrinsic scatter $\sigma_\perp \simeq 0.17$ dex. Adapted from Posti et al. (2018).

### 4.4.3 Mass–Metallicity Relation

If we measure the global metallicity of a galaxy we find that galaxies with higher stellar masses tend to have higher metallicities. This is the essence of the **mass–metallicity relation**, which spans several orders of magnitude in mass and more than one in metallicity. In Fig. 4.17, we display two versions of the mass–metallicity relation of SFGs. The left panel shows *stellar* metallicities obtained either using very bright stars hosted by nearby galaxies or by fitting the integrated optical spectrum of the whole galaxy. Note that the latter yields a *luminosity-weighted* average metallicity of the stars. The relation appears tight and linear (on logarithmic scales) up to stellar masses of $\mathcal{M}_\star \sim 10^{10}\,\mathcal{M}_\odot$, beyond which it flattens. The right panel of the same figure shows the relation one obtains using *gas abundances*, i.e. the metallicities of the ionised ISM (§4.2.3 and §11.1.4). Here, one typically determines the oxygen abundance, expressed in units of $12 + \log(\mathrm{O/H})$, and converts it to $Z/Z_\odot$ using a value for the solar oxygen abundance (see §C.6.1). On the whole, the gas mass–metallicity relation of present-day SFGs shows a shape very similar to the stellar relation, although with a large scatter.

The linear part of the mass–metallicity relation can be expressed as

$$\log\left(\frac{Z}{Z_\odot}\right) = F + \eta \log\left(\frac{\mathcal{M}_\star}{10^{10}\,\mathcal{M}_\odot}\right), \tag{4.42}$$

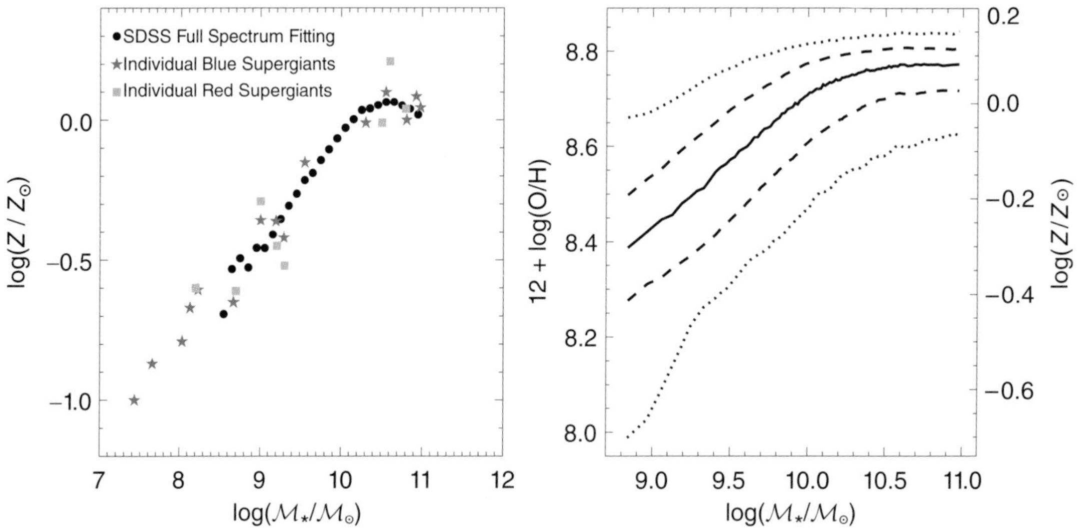


Fig. 4.17 *Left panel.* Mass–metallicity relation of present-day SFGs obtained from stellar abundances. The three determinations used either single stars (blue and red supergiants) or the fitting of the integrated optical spectra of the galaxies. *Right panel.* Mass–metallicity relation of present-day SFGs in the gas phase (solid curve). Here the metallicities are obtained from the emission lines (mostly oxygen) of the ionised gas. The dashed and dotted curves contain 68% and 95% (respectively) of the SFG population. Both figures are adapted from Zahid et al. (2017). © AAS, reproduced with permission. Courtesy of J. Zahid.

with $\eta \approx 0.35$ and $F \approx -0.1$, depending on the determination method. To take into account the flattening at high masses, different functional forms have been proposed such as, for instance, second-order polynomials. Dwarf irregular galaxies smaller than those shown in Fig. 4.17 can be extremely metal-poor with cases where the abundance of metals approaches 1% of the solar value. For instance, the starburst dwarf I Zw 18 has a metallicity $Z \simeq 0.02 Z_\odot$, as determined from its H II regions.

### 4.4.4 Star Formation Main Sequence

The global SFR of SFGs correlates with their total stellar mass. This relation is called **star formation main sequence**[18] (SFMS) and reads

$$\log\left(\frac{\mathrm{SFR}}{\mathcal{M}_\odot\,\mathrm{yr}^{-1}}\right) = Y + \kappa \log\left(\frac{\mathcal{M}_\star}{10^{11}\,\mathcal{M}_\odot}\right), \tag{4.43}$$

with a slope in the range $0.6 \lesssim \kappa \lesssim 1.0$, depending on the sample selection, and a normalisation $Y \approx 1$. The SFMS of present-day SFGs is shown in Fig. 4.18. The scatter of the relation is quite large ($\approx$ 0.2–0.3 dex), but both galaxy stellar masses and SFRs may have large uncertainties (§8.6 and §11.1.2), so part of this scatter is observational and the

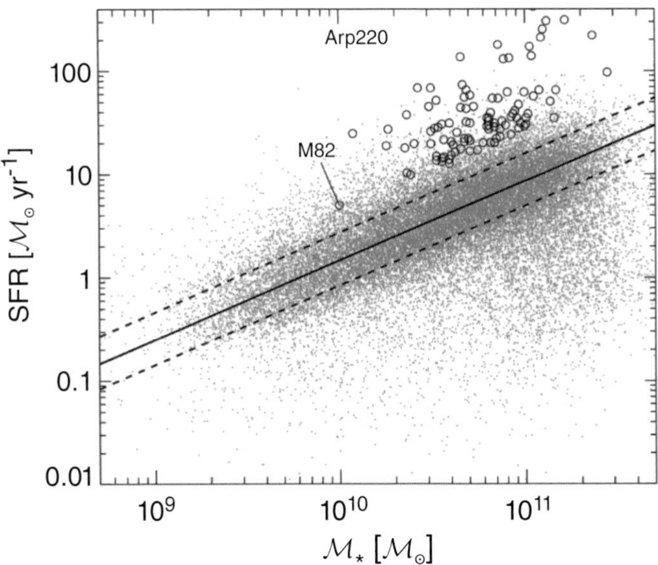


Fig. 4.18 The correlation between stellar mass and SFR in present-day ($0.04 < z < 0.1$) SFGs, referred to as the SFMS. Every dot in this plot is an SFG (passive galaxies have been excluded). The solid line shows a fit to the points with eq. (4.43) with $\kappa \simeq 0.8$, while the dashed lines show the 16th and 84th percentiles of the distribution. The large circles mark the location of known starburst galaxies. Adapted from Elbaz et al. (2007). Courtesy of D. Elbaz.

[18] Note that this term does not have any relation with the *stellar* main sequence that one obtains by plotting absolute magnitudes versus colours of individual stars (Hertzsprung–Russell diagram; §C.5).

intrinsic relation may be tighter. Starburst galaxies (§4.5) lie above this relation by at least a factor of a few in SFR. Galaxies that are in the process of quenching their star formation (§10.6) lie below this relation.

A useful quantity is the **specific star formation rate**

$$\mathrm{sSFR} \equiv \frac{\mathrm{SFR}}{\mathcal{M}_\star}, \tag{4.44}$$

whose inverse is a characteristic time, related to the build-up of the galaxy stellar mass (§10.10.3), of the order of 10 Gyr in main-sequence galaxies and much shorter in starbursts. The sSFR shares similarities with the **Scalo birthrate parameter** (Scalo, 1986)

$$b \equiv \frac{\mathrm{SFR}}{\langle \mathrm{SFR} \rangle}, \tag{4.45}$$

i.e. the ratio between the current and the past (averaged) SFR in a galaxy. Indeed, the mass locked up in stars in a galaxy is simply the integral of its SFR over time corrected for the fraction of mass returned to the ISM due to SN explosions and mass losses (§8.4.1). Given the slope of the main sequence, both the sSFR and $b$ are weak functions of the galaxy mass.

### 4.4.5 Size–Mass Relations

The stellar discs of SFGs follow a size–mass relation that links their scalelength $R_\mathrm{d}$ to their mass $\mathcal{M}_\mathrm{d}$ as

$$\log\left(\frac{R_\mathrm{d}}{1\,\mathrm{kpc}}\right) = H + \xi \log\left(\frac{\mathcal{M}_\mathrm{d}}{10^{10}\,\mathcal{M}_\odot}\right), \tag{4.46}$$

with $H \approx 0.4$ and $\xi \approx 0.2$–$0.3$. In §3.1.4 and Fig. 3.5 we have seen this relation expressed in terms of effective radii $R_\mathrm{e}$. Note, however, that there the whole stellar component of the galaxies has been considered and not only the stellar disc. Thus the two slopes cannot be straightforwardly compared especially at high masses where prominent bulges are present. The scatter in the stellar size–mass relation is quite large: galaxies of the same stellar mass can have disc scalelengths that differ from each other by a factor of 5.

A much tighter size–mass relation is obtained if one uses the neutral atomic gas. This relation reads

$$\log\left(\frac{R_\mathrm{H\,I}}{10\,\mathrm{kpc}}\right) = K + \chi \log\left(\frac{\mathcal{M}_\mathrm{H\,I}}{10^{10}\,\mathcal{M}_\odot}\right), \tag{4.47}$$

where $\mathcal{M}_\mathrm{H\,I}$ is the total H I mass of the galaxy, $R_\mathrm{H\,I}$ is defined in §4.2.1 and the parameters are $K \approx 0.5$ and $\chi \approx 0.5$–$0.55$. This relation has an intrinsic scatter smaller than 0.1 dex at fixed $R_\mathrm{H\,I}$. Its slope simply implies that the typical H I surface density is nearly the same for all disc galaxies: if the average surface densities were *exactly* the same one would have $\mathcal{M}_\mathrm{H\,I} \propto R^2_\mathrm{H\,I}$, which is very close to the observed value.

## 4.5 Starburst Galaxies

A small fraction (a few per cent) of SFGs have SFRs that are significantly higher (up to 1–2 orders of magnitude) than most galaxies of the same stellar mass (Fig. 4.18). These are **starburst galaxies**, an example of which is shown in Fig. 4.2. Starbursts are episodes of intense star formation that may last a relatively short time ($t_{\rm burst} \lesssim 10^8$ yr). The high star formation activity is typically concentrated in the central parts (few kpc or less) of the galaxies. There, the SFR densities can exceed those of a normal SFG by various orders of magnitude (Fig. 4.10). The intense star formation causes important dust thermal emission. As a consequence, starbursts have an extremely high IR luminosity $L_{\rm IR}$. When selected at IR wavelengths, they are called **luminous IR galaxies** if $\log(L_{\rm IR}/L_\odot) > 11$ or **ultra-luminous IR galaxies** (ULIRGs) if $\log(L_{\rm IR}/L_\odot) > 12$.

Fig. 4.19 shows the luminosity functions (§3.5.1) of a sample of IR-selected sources at low redshift divided into normal SFGs and starbursts. The starburst galaxies have been selected as those at 0.6 dex or more above the SFMS (§4.4.4). They are only $\sim 1\%$ at *normal* IR luminosities, but they dominate at luminosities above a few $\times 10^{11}\,L_\odot$. In general, ULIRGs are orders of magnitude less numerous than normal SFGs.

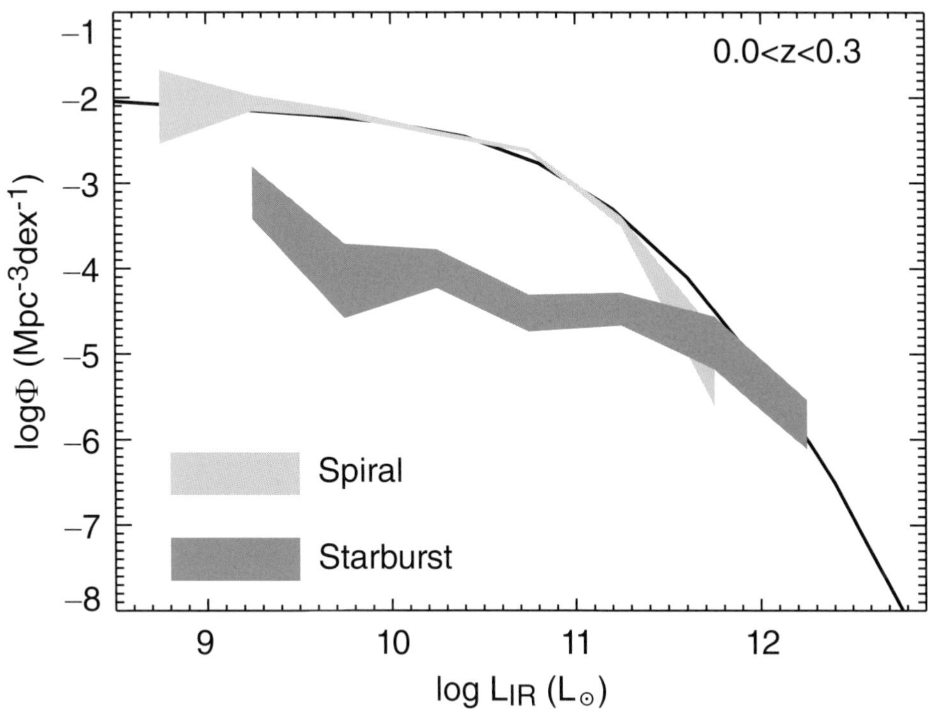


**Fig. 4.19** Total IR ($\lambda_{\rm rest} = 8$–$1000\,\mu$m) luminosity function of low-redshift ($z < 0.3$) sources from observations taken with *Herschel*, divided into normal spiral (light grey area) and starburst (dark grey area) galaxies. The black curve shows the total luminosity function. Data from Gruppioni et al. (2013). Figure courtesy of C. Gruppioni.

The mechanisms that can trigger the starburst are not fully established, but it is likely that strong interactions between galaxies (e.g. mergers; §6.1) play an important role. These phenomena can gather a large amount of gas in the centre of the merging system (§10.3.2). The central parts can reach very high gas densities causing an extremely intense star formation activity. This enhancement of the SFR is observed in nearby merging galaxies, for example in the pair of interacting galaxies (NGC 4038 and NGC 4039) known as the Antennae. One of the closest and best known starburst galaxies is M 82, which is located at about 3 Mpc in the M 81 group. Its starburst activity appears to have been triggered by an interaction (fly-by) between the three main galaxies of the group that produced a copious exchange of gas (see the front cover of this book).

Starbursts can have a variety of masses. In fact there are galaxies with masses typical of dIrrs that are currently starbursting (they are not represented in Fig. 4.18): these are called **blue compact dwarfs** (or BCDs). They have enhanced SFR with respect to dIrrs from a few to 10 times and enhanced SFR densities by one or more orders of magnitude (see also §10.9.1).

## 4.6 The Milky Way

The Milky Way gives us the privileged view of a disc galaxy seen from within. On the one hand, this is a great advantage because it allows us to study in detail the distribution and the kinematics of the various stellar components, the complexity of the ISM and the chemical composition. On the other hand, our large-scale view of the Galaxy suffers from us being located inside the thin disc. The *Gaia* mission has greatly improved our knowledge of the stellar component although optical observations towards the inner disc are problematic due to the high dust extinction (§4.2.7). As a consequence, some uncertainties remain such as the exact scalelength of the stellar disc, the number and the location of the spiral arms and the properties of the stellar bar. The large-scale view of the ISM of the Milky Way is also problematic due to the difficulties in determining distances of gas clouds. In this section, we focus on physical properties that we can study better in the Milky Way than in external disc galaxies.

### 4.6.1 Stellar Components

The Milky Way is a barred spiral galaxy of Hubble type presumably SBbc (for a review see Bland-Hawthorn and Gerhard, 2016). The very existence of the stellar bar is a relatively recent discovery, again stressing the difficulties in studying the large-scale stellar distribution from inside. This is mostly due to the fact that the vast majority of the optical light coming from disc stars is heavily extincted by dust: $A_V$ (eq. 4.13) reaches 40 magnitudes towards the Galactic centre! The only useful wavelength to study stars in the inner Milky Way is the NIR, where dust extinction is minimised (§4.2.7), stellar emission is still significant and dust emission is not important. At longer wavelengths the latter

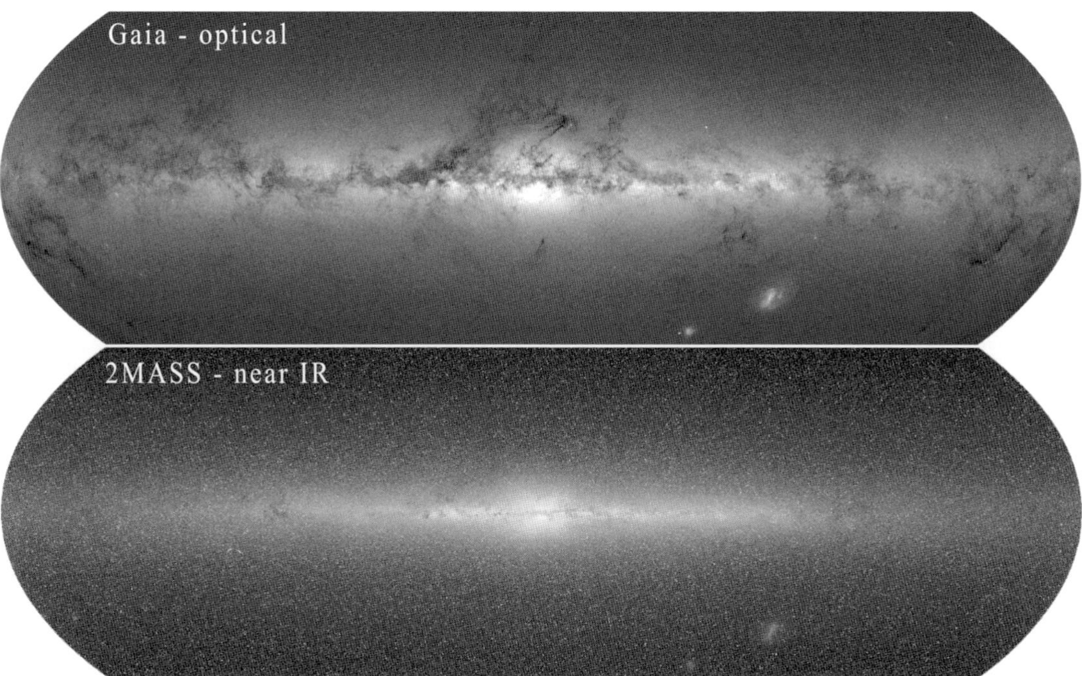


Fig. 4.20 The Milky Way seen at optical and IR wavelengths. *Top panel*. Map of nearly all the sky reconstructed from the fluxes in the $G$, $G_{BP}$ and $G_{RP}$ bands (§C.5) of about $1.7 \times 10^9$ stars observed by *Gaia*. *Bottom panel*. Map of nearly all the sky at IR wavelength obtained combining The Two Micron All Sky Survey (2MASS) fluxes in the $J$, $H$ and $K$ bands (§C.4) of half a billion stars. Both images are centred at the Milky Way's nucleus (Galactic longitude $\ell = 0$ and latitude $b = 0$) and use similar projections in Galactic coordinates; the uppermost and lowermost latitudes (roughly at $|b| > 45°$) have been cut out. The two bright sources on the right below the Galactic plane are the Magellanic Clouds (§6.3.1). Credit: *Gaia* map from Gaia Collaboration (2018b); Atlas Image mosaic obtained as part of 2MASS/J. Carpenter, T. H. Jarrett and R. Hurt. 2MASS is a joint project of the University of Massachusetts and the Infrared Processing and Analysis Center/California Institute of Technology, funded by the National Aeronautics and Space Administration and the National Science Foundation.

two conditions are not satisfied. In the following we describe the properties of the stellar components of the Milky Way mostly reconstructed observing *individual stars*, which is a technique that we can fully exploit at best in our Galaxy. Fig. 4.20 shows two panoramic views obtained from the detection of about a billion individual stars in the optical and in the NIR. Note the prominent (dark) dust lanes that obscure the view of the disc at optical wavelengths. For general properties of the stellar components of disc galaxies see §4.1.

## Stellar Disc

Our solar system is located within the disc of the Milky Way at $R_\odot \simeq 8.2\,\mathrm{kpc}$ from the centre and about 20 pc above the midplane. The region surrounding the Sun (within about

500 pc) is referred to as the **solar neighbourhood**. The surface brightness at the solar location is $\mu_B \simeq 23.75$ mag arcsec$^{-2}$ in the $B$ band corresponding to a total stellar surface density $\Sigma_\star \simeq 30\, \mathcal{M}_\odot\, \mathrm{pc}^{-2}$, to which we should add about $7\, \mathcal{M}_\odot\, \mathrm{pc}^{-2}$ of stellar remnants (white dwarfs, neutron stars and black holes). The solar neighbourhood is where we have historically collected most of the information about the chemical composition and the kinematics of the Milky Way's stars. *Gaia* has greatly enlarged this area by determining accurate distances and proper motions of more than one billion stars across the Galaxy's disc and halo (Fig. 4.20). This information can be combined with spectroscopic surveys to determine the accurate chemical composition of the stars as we see in the following sections.

One aspect that has been studied in great detail is the local vertical distribution of stars. In general, the vertical profile of a self-gravitating system with constant velocity dispersion is expected to take the form of a hyperbolic secant square.[19] This function falls like an exponential at large heights and, in practice, exponential profiles are often used (eq. 4.3, but with the volume density of stars $\rho_\star$ in place of the emissivity $j_\star$). As for external galaxies, the vertical distribution of the solar neighbourhood's stars requires at least two exponentials to be fitted: the **thin disc** and the **thick disc**. The local thin disc has a scaleheight $h_\star$ of about 300 pc whilst the thick disc has $h_\star \approx 1$ kpc. At our location, roughly 15% of the stars appear to belong to the thick disc. The scalelength of the thin disc is in the range $2\,\mathrm{kpc} < R_{\mathrm{d,MW}} \lesssim 3\,\mathrm{kpc}$, whilst that of the thick disc is $\approx 2$ kpc.

Stars belonging to one or the other disc tend to have different properties, in particular different kinematics, chemical compositions and ages. However, the separation is not always straightforward as we see below. Thick disc stars tend to be more metal-poor ($[\mathrm{Fe/H}] \lesssim -0.3$) and older (age $\gtrsim 8$ Gyr) than those that are part of the thin disc. Moreover, thick disc stars are richer in $\alpha$-elements (see below), at least in the region inside the **solar circle**.[20] Finally, the thick disc is, not surprisingly, kinematically *hotter* and thus its stars have higher velocity dispersions. In particular the average vertical dispersion of thick disc stars ($\sigma_z \approx 40$–$50\,\mathrm{km\,s^{-1}}$) is larger than that of the thin disc ($\sigma_z \approx 20$–$25\,\mathrm{km\,s^{-1}}$)[21] and, due to this, thick disc stars can reach larger distances above and below the midplane. The other components of the velocity dispersion are also higher than in the thin disc. The above values are given as averages over the 'old' stellar population at the solar circle, i.e. excluding young stars. The stellar velocity dispersion is, in fact, an increasing function of stellar age. The total (3D) velocity dispersion of disc stars in the solar neighbourhood goes from $\sigma_{\mathrm{tot}} \approx 25\,\mathrm{km\,s^{-1}}$ for the youngest stars to $\sigma_{\mathrm{tot}} \approx 70\,\mathrm{km\,s^{-1}}$ for the oldest (10–12 Gyr) stars.

[19] It can be shown that this is the vertical profile of a self-gravitating slab in 'hydrostatic equilibrium' (§4.6.2), where for a collisionless system the 'pressure' is $\rho\sigma_z^2$. The solution reads $\rho(z) = \rho_{\mathrm{s},0}\, \mathrm{sech}^2(z/h_\mathrm{s})$, where $\rho_{\mathrm{s},0}$ and $h_\mathrm{s}$ are the midplane density and scaleheight. At large $z$, this function can be approximated with an exponential $\rho(z) = \rho_{\mathrm{e},0} \exp(-z/h_\mathrm{e})$ provided that $h_\mathrm{e} = h_\mathrm{s}/2$ and $\rho_{\mathrm{e},0} = 4\rho_{\mathrm{s},0}$.

[20] We define the solar circle as the circle centred on the Galactic centre, lying on the Galactic plane and with a radius equal to the distance between the Sun and the Galactic centre $R_\odot \simeq 8.2$ kpc.

[21] We remind the reader that, in general, in collisionless systems like the stellar discs the three components of the velocity dispersion are not equal (§C.8). For instance, the average radial velocity dispersion of the thin disc is $\sigma_R \approx 35\,\mathrm{km\,s^{-1}}$.

The motion of stars in the thin disc of the Milky Way is largely dominated by coherent rotation at nearly the circular speed. However, due to the non-negligible velocity dispersion, there is also a small asymmetric drift effect that lowers the average rotation velocity (§4.3.4). The effect becomes more pronounced for thick disc stars that, as a consequence, rotate, on average, more slowly than thin disc stars. The circular speed at different radii (rotation curve; §4.3.3) has been determined using, in particular, H I observations for the inner Galaxy (inside the solar circle) and other (mostly stellar) tracers for the outer disc. This rotation curve is important as it allows us to estimate the total (baryonic and dark) mass as a function of radius (§4.3.5). The local circular speed at the distance of the Sun from the Galactic centre is in the range $v_c(R_\odot) \approx 220\text{–}250\,\mathrm{km\,s^{-1}}$. The **local standard of rest** (LSR) is defined as the inertial rest frame centred on the Sun and moving in the direction of the disc rotation at $v_c(R_\odot)$.

The study of nearby stars has also revealed the presence of different types of peculiar motions. The Sun itself has a peculiar motion with respect to the LSR of about $15\,\mathrm{km\,s^{-1}}$. Moreover, there are groups of stars called **stellar groups** (or streams) observed in and around the solar neighbourhood that are moving at velocities that depart from pure rotation by several tens of $\mathrm{km\,s^{-1}}$. Some of these are remnants of old stellar clusters or associations. Others are considered perturbations of the local kinematics produced by the inner stellar bar (see next section) and by the spiral arms. Finally, the interaction between the Milky Way disc and its satellites (§6.3) can also be responsible for the peculiar motions of some of these stellar groups.

As we see in §4.6.2, most of the star formation in the Milky Way occurs in giant molecular clouds where stars do not form in isolation but in star clusters and associations. Once formed, these clusters can remain visible as aggregations of nearly *coeval* stars for a long time. We call them **open clusters**. Several of the open clusters that we observe today are young (a few hundred Myr), but a small percentage of them are old, up to about 8 Gyr. They display a variety of luminosities ranging from a few hundreds to tens of millions solar luminosities. The most luminous ones are the youngest clusters where massive stars have not yet exploded as SNe. Young clusters tend to lie in the thin disc overlapping with the thin layer of molecular gas. As time passes, they are thought to disperse and thus become more difficult to observe.

Open clusters can be used to determine the metallicity of stars in different regions of the Milky Way as their distances are often very well constrained by their colour–magnitude diagrams. In §4.1.4 we have seen that the discs of external galaxies have negative metallicity gradients from the centre to the outer regions. This property is also observed in the disc of the Milky Way. Fig. 4.21 shows the metallicity obtained from stellar and gas (H II regions) tracers in the disc of the Milky Way as a function of Galactocentric radius. The stellar tracers include open clusters, Cepheid and O/B stars. The metallicities are measured from the stellar spectra, while the distances of these objects can be obtained using independent estimates. For instance, Cepheid stars are well known **standard candles** as their period of pulsation correlates with their luminosity. Altogether, these tracers show a coherent decrease of the metallicity from the inner to the outer disc of the Milky Way.

Finally, star clusters, and open clusters in particular, are used to observationally determine the shape of the stellar initial mass function (IMF; §8.3.9). In general, when stars

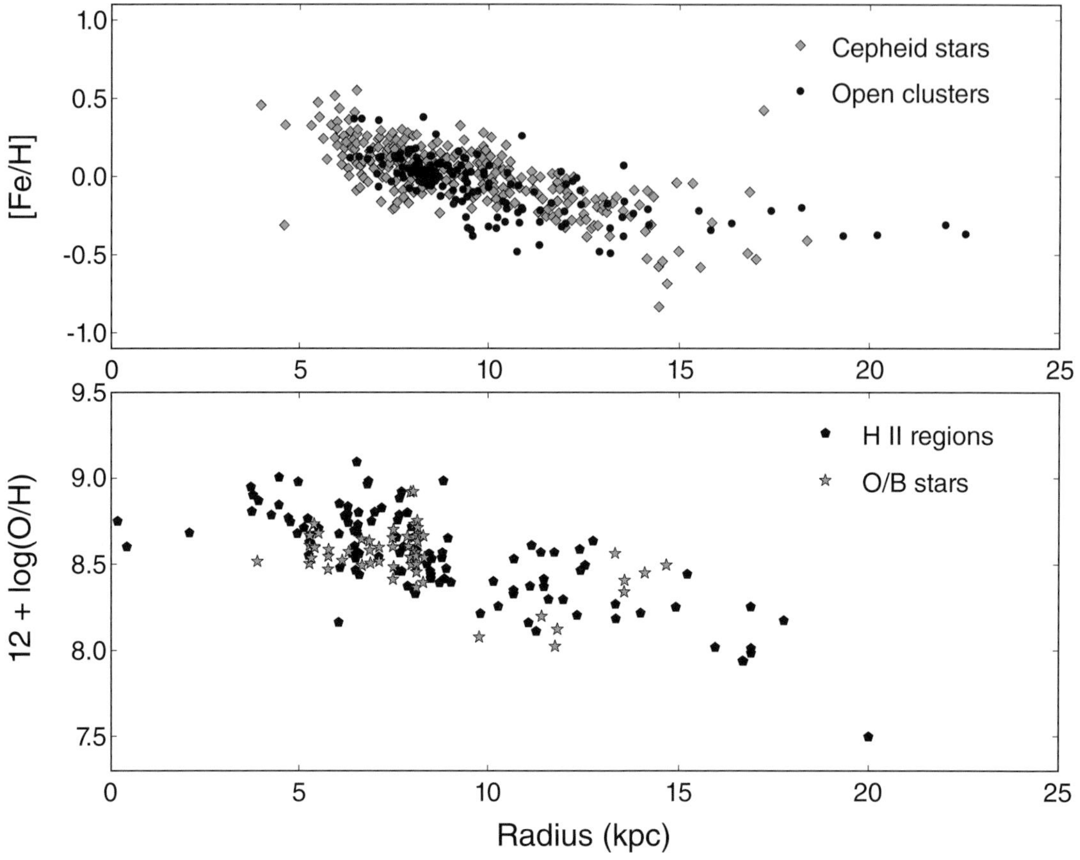


Fig. 4.21 Metallicity of stars and gas as a function of Galactic radius. *Top panel.* [Fe/H] obtained from stellar spectra of open clusters of different ages (data from Magrini et al., 2010; Frinchaboy et al., 2013; Heiter et al., 2014) and Cepheid stars (data from Genovali et al., 2014). *Bottom panel.* Abundance of oxygen over hydrogen from Galactic H II regions (data from Rood et al., 2007; Deharveng et al., 2010) and young O/B stars (data from Daflon and Cunha, 2004). For the conversion between the two metallicity scales see §C.6.1.

form in large number, in a cluster or in a galaxy, their mass distribution is such that many more stars form at low masses ($\lesssim 1\,\mathcal{M}_\odot$) than at high masses. By observing *individual stars* in a young cluster (being young is important, as it also contains massive stars) we can, in principle, determine these distributions precisely and also test whether different clusters have the same distribution. These determinations generally follow two steps: first, one determines a luminosity function of the stars; second, one converts from luminosity to stellar masses. The second passage is quite critical and it requires the use of a mass–magnitude relation that must be determined theoretically and/or calibrated observationally. Observations of several star clusters (see the Trapezium cluster in Fig. 8.10) and also in the field of the Galactic disc show distributions that are reasonably similar to each other. A number of fitting functions for these distributions have been proposed, as we discuss in §8.3.9.

## The Inner Milky Way: Bar and Bulge

In the inner regions of our Galaxy, both photometry and the study of individual stars show the presence of a prominent bulge and a bar. The bulge is easily observed at different wavelengths as it extends to high latitudes above and below the disc, and thus out of the dust lane that absorbs most of the midplane light (Fig. 4.20). The visualisation of the bar is instead more subtle. If there were only a bulge, i.e. an axisymmetric structure, and we can assume to have properly accounted for the dust extinction, we should not expect to see any asymmetries between positive and negative longitudes around the Galactic centre. Instead, an asymmetry in the stellar surface density is clearly visible as we can appreciate by looking at Fig. 4.22, where we see that the stellar distribution is brighter and thicker at positive longitudes (left) with respect to negative longitudes. This is expected if there is an accumulation of stars close to us on one side of the Galactic centre. Such a configuration is typical of a bar structure if the near part of the bar is at positive latitudes and we can use this asymmetry to estimate its orientation. The angle between the major axis of the bar and a line connecting us to the Galactic nucleus is between 20 and 40 degrees and the bar extends spatially to $R \approx 4$–5 kpc from the centre. This structure is referred to as the **long bar**. The innermost structure, which overlaps with the location of the bulge, is sometimes called the **bulge/bar**, given the difficulties to precisely disentangle the two components. The bar in this region is also called the **main bar**. Note that the thick disc also broadly overlaps with the bulge to add further complications. The inner thin disc is instead mostly seen in the gas component and in very young stars. Overall, the stellar distribution of the inner Milky Way resembles that of edge-on external galaxies with boxy/peanut-shaped bulges (Fig. 4.3, right).

The stellar population of the bulge has been studied extensively and it has been shown that a significant fraction of its stars are old ($\sim 10$ Gyr) and metal-rich ([Fe/H] $\gtrsim 0$). A population at lower metallicity ([Fe/H] $\lesssim 0$) is also present, but the quantification of its fraction is not certain. In general, the stars in the bulge have higher velocity dispersion

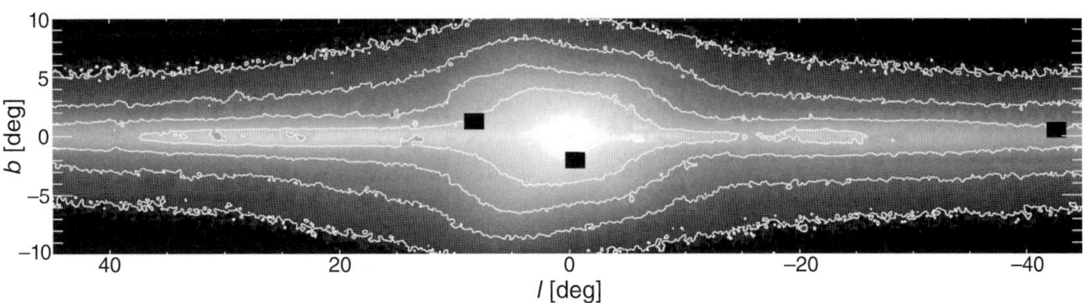


Fig. 4.22 Stellar surface density (greyscale and contours) in the $K_S$ band (§C.4) towards the inner Milky Way after correcting for dust extinction: white indicates higher densities. The axes show Galactic longitude ($l$) and latitude ($b$) in degrees. The black rectangles are regions without data of sufficient depth. The asymmetry with respect to the Galactic centre is due to the orientation of the bar with respect to us. Adapted from Wegg et al. (2015). Courtesy of C. Wegg.

than those in the disc, with values $\sigma \approx 100\,\mathrm{km\,s^{-1}}$. They however also show coherent (disc-like) rotation with a velocity that rises to values $v_{\rm rot} \gtrsim 100\,\mathrm{km\,s^{-1}}$. Only the most metal-poor stars ([Fe/H] $< -1$) appear to have a different kinematics, more dominated by velocity dispersion than rotation $V/\sigma < 1$. These stars are however a small fraction ($\approx$ 5–8%) and their association with the bulge is not certain: they could belong to the inner stellar halo. All these measurements give important clues to understand the origin of the Galactic bulge (§10.2).

The presence of the bar in the central region is also clearly visible in the kinematics of the gas. In §4.3.4 we have seen that, in galaxies like the Milky Way, the stellar component tends to dominate the gravitational potential in the inner regions. A stellar bar causes the potential to lose axisymmetry and become triaxial. Thus one may expect non-circular motions in the regions where the bar dominates. Indeed both CO and H I observations of the inner Milky Way show patterns that are not compatible with pure circular orbits.

## Stellar Halo

The thin and the thick discs of the Milky Way appear surrounded by an extended stellar component: the stellar halo. Historically, stars in the disc have been called **Population I stars**, while halo stars **Population II stars**. These latter were discovered because they were not participating in the rotation of the disc as they were moving at very high velocities in the heliocentric rest frame. As for external galaxies, the stellar halo of the Milky Way is composed of globular clusters, stellar streams and diffuse stars.

The Milky Way hosts about 160 **globular clusters** (see also §4.1.3), with stellar masses from $\sim 10^5\,\mathcal{M}_\odot$ up to a few $\times 10^6\,\mathcal{M}_\odot$. Despite their masses and luminosities being very similar to those of faint dwarf galaxies, they are much more concentrated (Fig. 3.4) and they do not appear to have dark matter. The majority of the globular clusters are associated with the Milky Way's halo and they have metal-poor ($-2.5 \lesssim$ [Fe/H] $\lesssim -0.8$) and old stellar populations, with ages that approach the Hubble time. A second population of globular clusters is, instead, associated with the Galactic thick disc and they are characterised by metallicities higher ([Fe/H] $\gtrsim -0.8$) than the halo population. A few globular clusters are associated with the Galactic bulge. For decades, every globular cluster has been thought to host a single (coeval) stellar population, but we now know that they have multiple populations. These populations can differ in age by at most a few hundred Myr and show distinctive chemical patterns.

Observations of resolved stars in wide-field imaging surveys have revealed the presence of several streams with coherent structure (and kinematics) in the Galactic halo. These are likely the result of the disruption of dwarf galaxies and/or globular clusters due to tidal interactions with the Milky Way (§8.9.4). The most prominent of these structures is the **Sagittarius stream**, a huge, highly extended, stellar structure produced by the infall of a relatively massive dwarf galaxy (the Sagittarius dwarf) into the Milky Way's potential. The central region of the remnant of the dwarf galaxy is currently located behind the Galactic bulge with respect to us. The Sagittarius stream can be traced wrapping over the whole sky and, to an external viewer, it would look similar to the stream shown in Fig. 4.4 (right).

The diffuse stars in the halo are refered to as 'field' halo stars, the origin of which is undergoing detailed scrutiny thanks to the new data obtained with *Gaia*. These stars may stem from disrupted satellites whose streams are no longer spatially distinguishable (§10.4). Historically halo stars have been identified according to: (1) their kinematics, which is characterised by very little or no rotation (high apparent velocities in the LSR), (2) low metallicities (typically $[Z/Z_{\odot}] < -0.6$) and (3) high $[\alpha/\mathrm{Fe}]$ ratio (see below). These criteria are somewhat arbitrary and open to misclassification. Today, thanks to the precise determination of distances and proper motions obtained with *Gaia*, we are able to reconstruct the 3D motions of very large samples of stars. Fig. 4.23 reports the location of disc and halo stars within 2 kpc from the Sun in a plane showing ordered rotation versus random motions. Not surprisingly, the vast majority of nearby stars belong to the thin disc and rotate at velocities close to the local circular speed. Thick disc stars appear to significantly depart from the thin disc kinematics, while the halo population is visible at negligible rotation velocity and large radial and vertical motions. It has been suggested

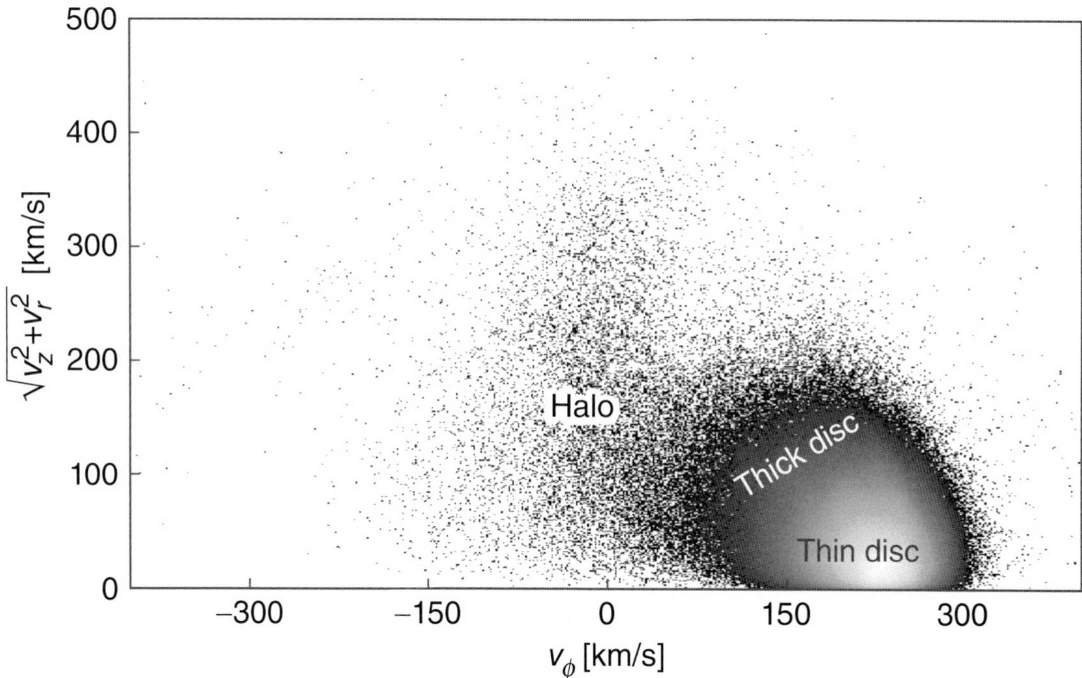


Fig. 4.23 Kinematics of an unbiased sample of stars with precise distance determinations observed by *Gaia* within 2 kpc from the Sun (see Koppelman et al., 2018, for details). The *x*-axis shows the azimuthal velocities $v_\phi$ (rotation), while the *y*-axis is a proxy for random motions ($v_r$ and $v_z$ are the radial and vertical velocities, respectively). Light grey (thin disc) corresponds to the highest density of stars. The vast majority of stars in our vicinity are thin disc stars, located at high azimuthal velocities ($v_\phi \approx 200$–$250$ km s$^{-1}$) and having relatively small random motions. The tail towards lower rotation and larger random motions is characteristic of thick disc stars; however note that there is not a clear separation between the two components. The 'vertical' structure in the centre contains most of the local halo stars, it does not show rotation (possibly a slight retrograde motion) and has very large random motions. Courtesy of H. Koppelman.

that the halo stars in Fig. 4.23 are mostly contributed by a relatively massive satellite that merged with the Milky Way about 10 Gyr ago. From Fig. 4.23 it is also clear that the distinction between the various components (halo stars versus thick disc and thick versus thin disc) is not totally straightforward.

Globally, the overall stellar profile of the halo appears to decline as a power law with radius ($\rho \propto r^{-\alpha}$ with $\alpha \approx 3$ in the inner 30 kpc), starting from a relatively flat distribution close to the disc and becoming progressively rounder at larger radii. The total mass of the stellar Galactic halo is $\mathcal{M}_\star \lesssim 1 \times 10^9\, \mathcal{M}_\odot$, including globular clusters and stellar streams.

## The Chemical Composition of the Stars

A useful approach to investigate the origin and the formation timescales of the stellar components of the Milky Way is shown in Fig. 4.24. Here, we see the ratio between the abundance of magnesium (one of the **$\alpha$-elements**[22]) and of iron ([Mg/Fe] ratio or, generically, **[$\alpha$/Fe] ratio**) versus metallicity [Fe/H] (abundance of iron versus hydrogen; see also §C.6) in the photospheres of stars that belong to different components of the Milky Way: the stellar halo, the thick disc and the thin disc.

Stars form from the collapse of interstellar gas that has been enriched by previous stellar generations, and when a star is born, a sample of this medium is locked up in its photosphere. Low-mass stars can live for very long times, exceeding the Hubble time for $\mathcal{M} < 0.8\, \mathcal{M}_\odot$, and observing these stars in the Milky Way allows us to reconstruct the enrichment of the ISM throughout time. A large fraction of the metal enrichment is due to the two types of SNe. **Type II SNe**, generated by the explosion of young massive stars, release predominantly $\alpha$-elements but also some iron (Tab. 8.1). Thus, if we start from a certain low metallicity, the explosion of Type II SNe will increase the metallicity without significantly changing the [$\alpha$/Fe] ratio. This is what we see happening for stars associated with the Galactic halo and some of the thick disc stars that populate the plateau[23] at [Mg/Fe] $\approx$ 0.4 in Fig. 4.24.

This pattern changes when the first **Type Ia SNe** start to explode. These SNe are produced by the explosion of a white dwarf exceeding the Chandrasekhar limit (§8.4.1) due to mass transfer from a red giant branch companion (§C.5). They release large amounts of iron and only a few $\alpha$-elements (Tab. 8.1). This produces a characteristic 'knee' and a shift of the stars towards lower [Mg/Fe] and higher [Fe/H] (see location of the thick and thin disc stars in Fig. 4.24). In general, the [$\alpha$/Fe] versus [Fe/H] diagram is considered a 'clock' for the enrichment of the ISM: the horizontal axis of Fig. 4.24 can be seen as a temporal axis. The typical timescale for the formation of a knee in this plot is the time that it takes for the explosion of Type Ia SNe, estimated to be, on average, $\sim$ 1 Gyr.

From Fig. 4.24, we also see that halo stars are observed with metallicities of less than 0.01% solar. In fact, a few stars have been found with metallicity as low as [Fe/H] $\lesssim -6$, where Fe lines are not seen, but other elements like C and Mg are. These extremely low

[22] The $\alpha$-elements are those synthesised in stars by $\alpha$-processes (an $\alpha$-particle being the nucleus of helium) and include Ne, Mg, Si, S, Ar, Ca and Ti. Oxygen is often considered an $\alpha$-element as well. They are mostly produced by core-collapse (Type II) SNe (§8.5.1).

[23] The exact location of this plateau in [$\alpha$/Fe] depends on the shape of the IMF (§8.3.9)

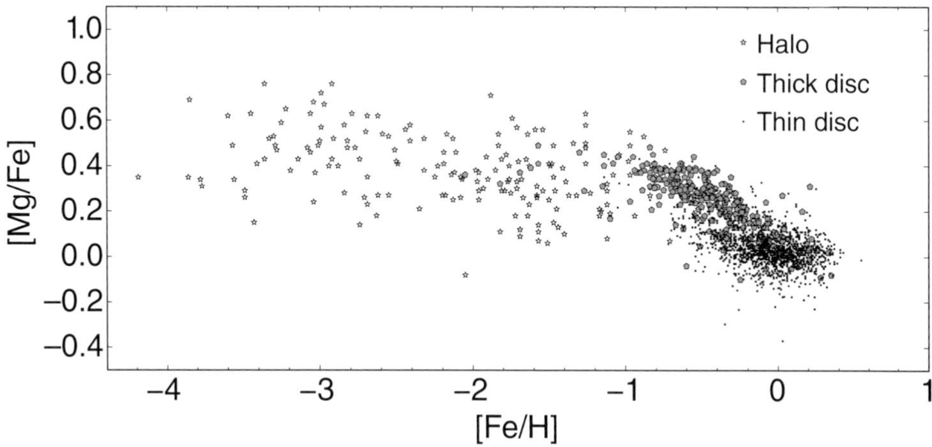


Fig. 4.24 The ratio between the abundances of the $\alpha$-element Mg and Fe [Mg/Fe] versus metallicity [Fe/H] in stellar photospheres. The different symbols show Milky Way stars that are likely to belong to the stellar halo (open stars), the thick disc (grey pentagons) and the thin disc (black dots). Note that the location of the thick disc stars at high [$\alpha$/Fe] values shown here is typical of the inner thick disc (within $R_{\odot}$). Data for the halo stars from Venn et al. (2004) and Cayrel et al. (2004). Data for the thin and thick discs from Adibekyan et al. (2012) and Bensby et al. (2014). Typical errors in the metallicity are $\pm 0.1$ dex, increasing at lower metallicities.

Fe stars often show strongly enhanced carbon. Stars with *primordial* abundance, i.e. completely without any metals, have not been found despite extensive searches. However, the first stars that formed in the very early Universe are expected to have had this composition and the fact that we do not see them around today provides clues as to their properties and initial masses. As extension of the historical classification scheme, primordial zero-metallicity stars are called Population III stars and we describe them in §9.5.

Finally, we mention that the samples of stars shown in Fig. 4.24 are heterogeneous, as they have been selected with different criteria: e.g. using either their kinematics or their chemistry. As a consequence, some stars may be misclassified as thin/thick disc or halo. Most importantly, the number of stars in each part of the diagram are *not* representative of the relative percentages of the three populations in the Milky Way. In particular, the lowest-metallicity stars are extremely rare and constitute only a small fraction of the halo population, which in turn constitutes a small fraction of all the stars present in the Milky Way.

### 4.6.2 The Galactic Interstellar Medium

Globally, the average density of the Galactic ISM in the plane of the disc of the Milky Way is $\sim 1$ particle cm$^{-3}$. The properties of the ISM of our Galaxy are similar to those that we have outlined for SFGs in general (§4.2). However, as for the stellar component, there are some physical properties that can be better studied in the Milky Way than in external galaxies. More details on this topic can be found in Tielens (2005) and Stahler and Palla (2005).

## Neutral Atomic Gas

As for external galaxies, the main gaseous component of the Galactic disc is neutral atomic gas with a total mass of about $4 \times 10^9\ M_\odot$. The H I in the Milky Way is thought to separate into two **gas phases**, the **cold neutral medium** (CNM) and the **warm neutral medium** (WNM). The CNM is made of 'clouds' with typical volume density $n_{\rm CNM} \approx 30\,{\rm cm}^{-3}$ (Tab. 4.3) and temperatures in the range $T_{\rm CNM} \approx 30$–$120$ K, whilst the WNM is much more rarefied with a temperature $T_{\rm WNM} \approx 6000$–$10\,000$ K and a density $n_{\rm WNM} \approx 0.3\,{\rm cm}^{-3}$. Note that the temperatures of the WNM are comparable with those of the photoionised gas in H II regions (§4.2.3).

The physical process that shapes the ISM into distinct phases is called thermal instability and we describe it in §8.1.4. The cold and warm phases mentioned above are roughly in **pressure equilibrium** ($n_{\rm CNM}T_{\rm CNM} \approx n_{\rm WNM}T_{\rm WNM}$). This equilibrium is an important property of the gaseous ISM. The average ISM pressure in the disc of the Milky Way is $\langle P_{\rm ISM}\rangle \approx 3 \times 10^{-13}$ dyne cm$^{-2}$.

Roughly 50–60% of the neutral atomic gas is thought to be at *warm* temperatures characteristic of the WNM or intermediate between the two phases. The WNM has a larger volume filling factor[24] ($f \approx 25\%$) than the CNM ($f \approx 1\%$). In general, the gas thermal speed is given by eq. (4.9). Using typical values for the temperatures of the two phases ($T_{\rm CNM} = 70$ K and $T_{\rm WNM} = 8000$ K), we can calculate the thermal broadening speeds that we should expect to observe as $v_{\rm T,\,CNM} \simeq 0.75$ km s$^{-1}$ and $v_{\rm T,\,WNM} \simeq 8$ km s$^{-1}$. These speeds are generally lower than the observed H I broadening of $\sigma_{\rm HI} \approx 10$ km s$^{-1}$, which implies, as for external SFGs, a significant contribution from gas turbulence (§4.2.2).

**Table 4.3** Properties of the gaseous ISM of the Milky Way

| Property | Molecular | CNM | WNM | H II regions | WIM | HIM |
|---|---|---|---|---|---|---|
| Temperature (K) | $\sim 10$ | 30–120 | 8000 | $10^4$ | $10^4$ | $\gtrsim 10^6$ |
| Density (cm$^{-3}$) | $\gtrsim 10^2$ | 30 | 0.3 | 10–$10^4$ | 0.1 | $\sim 10^{-3}$ |
| Thermal speed (km s$^{-1}$) | $\sim 0.1$ | 0.5–1 | 8 | $\sim 10$ | $\sim 10$ | $\gtrsim 100$ |
| $\langle\Sigma_{\rm gas}\rangle^a$ ($M_\odot$ pc$^{-2}$) | 3 | $4.5^b$ | | — | 2 | — |
| Mass ($10^9\ M_\odot$) | 1 | 2 | 2 | 0.1 | 1 | $\lesssim 0.1$ |
| Scaleheight$^c$ (kpc) | 0.06 | 0.1 | 0.2–0.3 | — | 1 | 3 |
| Filling factor (%) | $\sim 0.1$ | $\sim 1$ | 25 | — | 25 | 50 |

Most values from Tielens (2005), Marasco et al. (2017) and Heiles and Troland (2004). Note that most values in this table are approximate.

$^a$ For neutral molecular and atomic gas, these are average hydrogen surface densities in the range 3 kpc $\lesssim R \lesssim R_\odot$.

$^b$ Sum of both CNM and WNM.

$^c$ In the inner Galaxy, $R < R_\odot$, the H I values do not include extraplanar gas (§4.2.10).

[24] The filling factor is the fraction of a volume occupied by a certain component. In the case of the Galactic disc one assumes a certain outer radius and thickness of the disc to calculate the volume. A filling factor of 100% corresponds to the complete coverage of the entire volume.

The random motion (velocity dispersion) of the gas also determines the thickness of the gas layer. We can estimate this thickness by considering a gaseous disc in vertical **hydrostatic equilibrium**, meaning that the gas pressure gradient balances the gravitational pull given by all the matter components of the galaxy (§4.3.1). For the pressure, we consider a generic definition $P = \sigma_{\rm gas}^2 \rho_{\rm gas}$, where $\rho_{\rm gas}$ and $\sigma_{\rm gas}$ respectively are the gas density and velocity dispersion. This pressure can have contributions from the thermal pressure and, given the importance of turbulence in the ISM (§4.2.2), from the 'turbulent pressure' (eq. 8.50). For simplicity, we assume that this velocity dispersion does not vary with $z$: $\partial \sigma_{\rm gas}/\partial z = 0$. Using the Euler equation (eq. D.24), written in cylindrical coordinates, and imposing equilibrium ($\mathrm{D}\boldsymbol{u}/\mathrm{D}t = 0$), we get

$$\frac{1}{\rho_{\rm gas}}\frac{\partial P}{\partial z} = \sigma_{\rm gas}^2 \frac{\partial \ln \rho_{\rm gas}}{\partial z} = -\frac{\partial \Phi}{\partial z}, \tag{4.48}$$

where $\Phi(R, z)$ is the total gravitational potential. Eq. (4.48) can be integrated in $z$, for every $R$, to obtain

$$\rho_{\rm gas}(R, z) = \rho_{\rm gas}(R, 0) \exp\left[-\frac{\Phi(R, z) - \Phi(R, 0)}{\sigma_{\rm gas}^2(R)}\right], \tag{4.49}$$

where $\rho_{\rm gas}(R, 0)$ and $\Phi(R, 0)$ are the gas density and the potential, respectively, in the midplane ($z = 0$) of the disc. If we remain close to the midplane, we can expand the gravitational potential as a Taylor series

$$\Phi(R, z) = \Phi(R, 0) + z\left(\frac{\partial \Phi}{\partial z}\right)_{(R,0)} + \frac{z^2}{2}\left(\frac{\partial^2 \Phi}{\partial z^2}\right)_{(R,0)} + \cdots. \tag{4.50}$$

The first derivative in eq. (4.50) is null as the vertical component of the gravitational acceleration changes sign in the midplane; thus the combination of this equation with eq. (4.49) leads to a Gaussian shape of the density

$$\rho_{\rm gas}(R, z) \simeq \rho_{\rm gas}(R, 0) \exp\left[-\frac{z^2}{2h_{\rm gas}^2(R)}\right], \tag{4.51}$$

where we have defined the scaleheight as

$$h_{\rm gas}(R) \equiv \frac{\sigma_{\rm gas}(R)}{\sqrt{\left(\frac{\partial^2 \Phi}{\partial z^2}\right)_{(R,0)}}} = \frac{\sigma_{\rm gas}(R)}{\sqrt{4\pi G \rho_{\rm m}(R, 0) - \frac{1}{R}\frac{\partial}{\partial R}\left(R\frac{\partial \Phi}{\partial R}\right)_{(R,0)}}}, \tag{4.52}$$

where $\rho_{\rm m}(R, 0)$ is the total matter density in the midplane and the last equality comes from the vertical Poisson equation (eq. 7.3) written in cylindrical coordinates for an axisymmetric system.

Eq. (4.52) allows us to estimate the scaleheight of any gaseous component of the disc of the Milky Way or an external galaxy. Note that the second term in the square root is a radial derivative of the square of the circular speed (eq. 4.23), which is null if the rotation curve is flat and thus in a significant fraction of a typical galaxy disc (§4.3.3). At the solar circle, this is the case, $(\partial v_{\rm c}/\partial R)_{R=R_\odot} \approx 0$, and we can also assume that the dominant matter density in the midplane is the baryonic component (stars and gas). Using the measured local baryon surface density $\Sigma_{\rm b} \approx 47\, \mathcal{M}_\odot\, {\rm pc}^{-2}$ and a thickness for the stellar disc of $\approx 400$ pc (intermediate between thin and thick discs), we obtain a thickness for the H I layer of about[25] $h_{\rm H\,I} \approx 135$ pc, which is compatible with the observed value (Tab. 4.3). In the outer parts of the disc, the gravitational pull of the stars decreases exponentially and the potential becomes dominated by the dark matter. Given that the H I velocity dispersion in a galaxy disc decreases slowly (Fig. 4.6), the scaleheight increases steadily with the radius. As a consequence, the H I layer can become much thicker at large radii, a phenomenon called **disc flaring**. In the Milky Way, the H I disc flares to $h_{\rm H\,I} \sim 1$ kpc at $R \approx 25$ kpc, the outermost radius at which H I is detected. The outer parts of the Galactic H I disc are also warped with respect to the inner disc (§4.3.3).

About 10% of the H I in the Milky Way appears to be out of hydrostatic equilibrium, as it is observed at much larger heights from the plane (a few kpc) than those estimated above. As we mentioned in external galaxies, this is called extraplanar H I and it is thought to be mostly produced by the galactic fountain (§4.2.10). The regions outside the disc also offer the possibility to detect gas that is accreting onto the Milky Way from the external environment (§4.2.10). For instance, observations carried out in H I emission, but also in absorption towards QSOs (§9.8.3) and halo stars, have revealed the presence of the so-called **high-velocity clouds**. These are gas clouds that are easily singled out in the data given their anomalous velocities with respect to the normal rotation of the disc gas. They seem to have a global motion of infall towards the Galaxy disc and tend to have low metallicities. Their rate of accretion is however difficult to estimate and so is their precise origin (see Putman et al., 2012, for a review). We discuss these issues further in §10.7.2.

## Photoionised and Collisionally Ionised Gas

The photoionised gas constitutes the second component of the warm phase of the Milky Way's ISM. Like in external SFGs, the recombination line luminosity is dominated by H II regions while most of the mass of photoionised gas is instead in a diffuse component permeating the whole disc. In the Milky Way, the total mass of photoionised gas in H II regions is of the order of $10^8\,\mathcal{M}_\odot$. The diffuse **warm ionised medium** (WIM) has instead a total mass of $\sim 10^9\,\mathcal{M}_\odot$ and a large volume filling factor of about 25% (Tab. 4.3). The ionising radiation for the WIM can come from different sources. Some UV photons produced by massive stars propagate much further than the typical Strömgren radius (§4.2.3). This is due to the fact that the ISM is not homogeneous and UV photons can travel more efficiently in directions with low density of neutral hydrogen. Additional UV flux can come from those B stars that are not powerful enough to produce H II regions and highly evolved stars. Finally, in the outer parts of the disc, or at relatively large heights from the plane, H I can be photoionised by the extragalactic ultraviolet background (UVB; §8.1.2). The typical densities and scaleheight of the WIM disc in the Milky Way are $n_{\rm WIM} \approx 0.1\,{\rm cm}^{-3}$ and $h_{\rm WIM} \approx 1$ kpc, respectively.

[25] The exact value depends on the velocity dispersion, assumed to be $\sigma_{\rm gas} = \sigma_{\rm H\,I} = 8\,{\rm km\,s}^{-1}$ here, and on different choices for the thickness of the stellar thin and thick discs (§4.6.1). The dark matter contribution at the solar circle is negligible; indeed a spherical dark matter halo has a much lower density ($\rho_{\rm DM} \approx 0.005$–$0.01\,\mathcal{M}_\odot\,{\rm pc}^{-3}$) than the baryons in the midplane of the disc.

The collisionally ionised gas constitutes the third phase of the ISM. In the Milky Way it is called the **hot intercloud medium** (HIM) and its total mass is very uncertain, probably between $10^7$ and $10^8\ \mathcal{M}_\odot$. The filling factor approaches unity and the scaleheight is a few kpc, depending on the temperature of the gas. This phase is produced mostly by shock waves generated by SN explosions and stellar winds (§8.7). The three-phase model of the Galactic ISM (McKee and Ostriker, 1977) states that these phases (CNM, WNM/WIM and HIM) live together in nearly pressure equilibrium as gas is continuously exchanged from one phase to another by the interplay between cooling and heating (§8.1). The region outside the disc is permeated by the Galactic hot corona ($T \simeq 2 \times 10^6$ K and $n \lesssim 10^{-3}$ cm$^{-3}$) that is thought to extend to very large distances in the Galactic halo (§6.7).

## Molecular Clouds

Most of the molecular gas in the Milky Way ($\mathcal{M}_{H_2} \approx 1 \times 10^9\ \mathcal{M}_\odot$) appears located in **giant molecular clouds** (GMCs) that are spread all around the disc mostly along the spiral arms where compression of the ISM is facilitated (§8.3.3). The typical masses of Galactic GMCs are of order $10^5\ \mathcal{M}_\odot$ up to a few $\times 10^6\ \mathcal{M}_\odot$, while their sizes can reach $\approx 100$ pc. Giant molecular clouds are not homogeneous structures; instead they host a variety of complex and often filamentary substructures down to very small scales (Fig. 4.25). These substructures have progressively decreasing masses and sizes in an almost 'fractal' hierarchy. The smallest features are called **dense cores** and have sizes $\lesssim 1$ pc and masses of a single star. Tab. 4.4 summarises the main physical properties of GMCs and dense cores. Newborn stars are often observed inside dense cores. To study these very dense and compact features, molecules different from CO are often employed. Indeed, molecular clouds contain a variety of different molecules such as OH, $NH_3$, CS, $H_2CO$ and HCN. The last three in particular have transitions with high critical densities ($n_{crit} \gtrsim 10^6$ cm$^{-3}$; §4.2.3), which make them excellent probes of high-density environments. A fraction, whose value is uncertain (possibly around 20% or more), of the molecular gas is in a more diffuse component.

The temperatures of molecular clouds are very low, typically $T \approx 10$–$15$ K. We can see this in Fig. 4.25, which shows a portion of the Orion molecular cloud as $H_2$ column density (left) and dust temperature (right). The Orion cloud is the nearest GMC, located at a distance of a mere 400 pc from us with a total gas mass of about $10^5\ \mathcal{M}_\odot$. Fig. 4.25 shows a region called Orion B, characterised by complex filamentary structures and dense

**Table 4.4** Physical properties of Galactic molecular clouds

| Type | $n_{H_2}$ (cm$^{-3}$) | Size (pc) | Mass ($\mathcal{M}_\odot$) | $T$ (K) | $c_s$ (km s$^{-1}$) | $\sigma_{gas}$[a] (km s$^{-1}$) | $\langle B \rangle$ ($\mu$G) |
|---|---|---|---|---|---|---|---|
| GMCs | $10^2$–$10^3$ | 30–100 | $10^5$–$10^6$ | 15 | $\sim 0.1$ | 1–2 | 10 |
| Dense cores | $10^4$–$10^6$ | 0.1–1 | 0.1–$10^2$ | 10 | $\sim 0.1$ | 0.1–0.3 | 30–$10^2$ |

[a] Typical velocity dispersion observed from the broadening of emission lines.

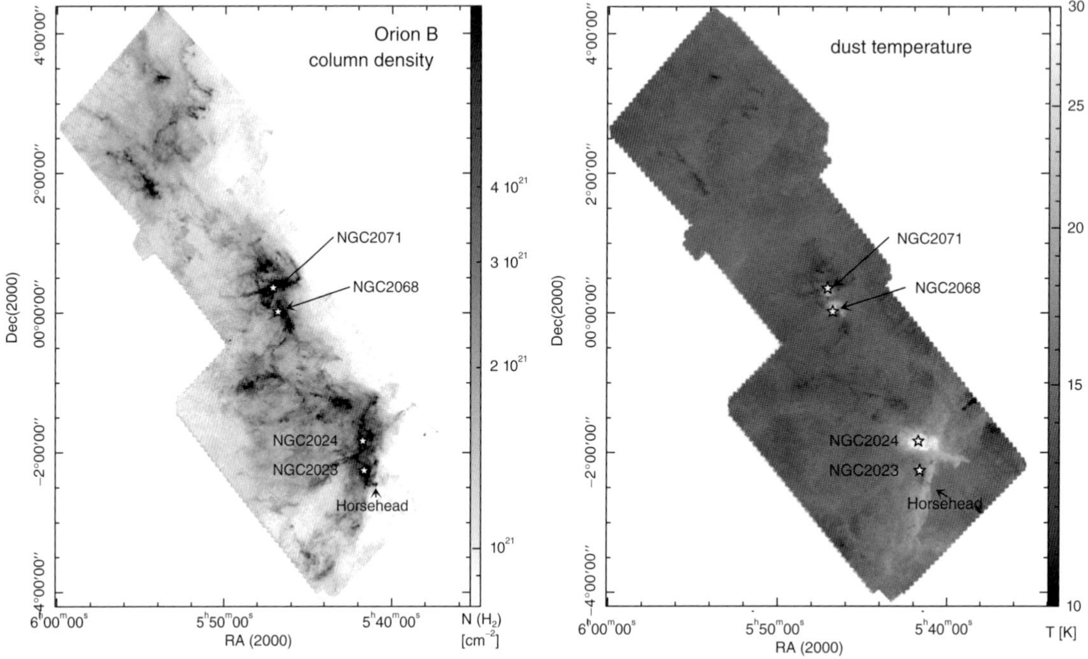


Fig. 4.25 Maps of $H_2$ column density (estimated from a fit of the FIR SED; *left panel*) and dust temperature (*right panel*) of the Orion B molecular cloud. The physical size of this image in the vertical direction is about 60 pc. The darkest (densest) regions (*left panel*) are sites of ongoing star formation and the feedback from these protostars is locally raising the dust temperature (*right panel*). Adapted from Schneider et al. (2013). © AAS, reproduced with permission. Courtesy of N. Schneider.

agglomerations, which are the sites of ongoing star formation. Most of the gas in this cloud is at a temperature[26] $T \simeq 15\,\mathrm{K}$, with the densest regions (black) going down to $T \simeq 10\,\mathrm{K}$. The notable exceptions are the dense cores where new stars have just formed (indicated with names from the NGC catalogue). There, stellar feedback from the young stars raises the temperature up to $T \approx 30\,\mathrm{K}$.

The stars formed inside a GMC allow us to estimate the *age* of the GMC itself, for instance by dating an entire star cluster. Typical ages are found up to a few $\times 10^7$ yr. Fig. 4.25 also shows that on the large scale, where most of the gas is contained, the GMC appears rather 'inactive'. Given the low temperatures of the gas, we can expect molecular clouds to have relatively low internal energy (and thus thermal pressure). However, due to their large masses, they have strong self-gravity. Thus, to avoid the precipitous gravitational collapse of the whole cloud, the gravitational pull must be counteracted by some energy other than internal energy. As we see in §8.3, the best candidates for this role are turbulence and magnetic fields. The fact that molecular clouds are highly turbulent structures is indeed observed. Typical line broadening of molecules (CO for instance)

[26] We remind the reader that, due to the frequent collisions (§4.2.1), we do not expect large differences between the dust and gas temperatures.

observed in GMCs are 1–2 km s$^{-1}$. This strongly contrasts with the thermal broadening of order $v_T \lesssim 0.1\,\mathrm{km\,s^{-1}}$ expected for CO molecules at $T \lesssim 15\,\mathrm{K}$ thus GMCs can be considered highly turbulent.

### Other Components of the ISM

The ISM of the Milky Way contains dust with properties that have been described in §4.2.7. The magnetic field has been discussed in §4.2.8 for both external galaxies and the Milky Way. Here we add that the Galactic magnetic field also affects the orientation of elongated dust grains by making them align perpendicularly to the field lines. This occurs for two reasons: (1) dust grains tend to spin about their shortest axis and (2) the magnetic field $\boldsymbol{B}$, in time, torques the dust grains to align their rotation axes in the direction of $\boldsymbol{B}$. This latter effect takes place because most dust grains are charged. A region of space with dust grains preferentially aligned in one direction tends to absorb and scatter background radiation more in that direction and let through orthogonally polarised photons. Thus, if the region is threaded by a coherent magnetic field we observe a regular pattern in the polarised radiation. This polarisation of the stellar light is classical evidence for the presence of a, at least partially, coherent magnetic field in our Galaxy.

Complementary information is given by the polarisation of dust *emission* measured by the *Planck* mission. This gives an all-sky image of the orientation of the Galactic magnetic field and provides insights on both its large-scale distribution and small-scale structure. Moreover, it testifies to the role of the magnetic field in shaping the filamentary structures in the ISM and to the importance of turbulence. We discuss these issues further in §8.3.6 and §8.3.7.

Measurements of magnetic field strengths in the Milky Way have been obtained using the **Zeeman effect** of ISM emission lines. This effect occurs when the electron magnetic moment interacts with an external magnetic field. The result is a splitting of the emission/absorption line into sublevels (typically three), whose difference in frequency with respect to the unperturbed level is proportional to the magnetic field strength $\Delta\nu = (b/2)B$, with a $b$ parameter different for different lines. Measurements of the Zeeman effect from 21 cm H I absorption return values of the magnetic field $B \approx$ few $\times\, 10^{-5}$ G, higher than the average field in the Milky Way: $\langle B_{\mathrm{MW}} \rangle \approx 3\,\mu\mathrm{G}$. This is due to the fact that this technique is biased towards dense clouds, where the magnetic field tends to be more intense (see §8.3.6). The Zeeman effect is also used to determine the magnetic field strength in molecular clouds, using typically molecules of OH, CN and CH. In general, the frequency shift that one needs to measure is quite small; for instance for OH, $b \simeq 2.0\,\mathrm{Hz}\,\mu\mathrm{G}^{-1}$ and $\simeq 3.3\,\mathrm{Hz}\,\mu\mathrm{G}^{-1}$ for the two main transitions. Typical values of the magnetic field strength obtained for molecular clouds are shown in Tab. 4.4.

Another component that we can study in great detail in the Milky Way are the so-called **cosmic rays**. These are high-energy (relativistic) particles, in number mostly protons with $\approx 10\%$ of He nuclei, and about 1% of heavy nuclei and electrons. They impact on the Earth's atmosphere producing showers of secondary particles that can be detected from the ground using different techniques. Depending on their energies they are likely to have different origins. Low-energy ($E \lesssim 10^{15}$ eV) cosmic rays are thought to be generated and accelerated in SNRs (§8.7.1) and then 'swirl' around the Galactic magnetic field (§4.2.8). At higher energies ($E \gtrsim 10^{15}$ eV) also nearby AGNs may contribute to their production. For $E \gtrsim 10^{19}$ eV (ultra-high-energy cosmic rays) the origin is probably extragalactic. In the 'ecosystem' of the Galactic ISM, the relativistic particles of cosmic rays provide a key contribution to the thermal balance (§8.1.2). In particular in dense (cold) atomic clouds and molecular clouds, where shielding of radiation is very strong, they are the most important heating mechanism (§8.1.4). Cosmic rays can also be important contributors to stellar feedback from galaxies (§8.7.4).

### 4.6.3 The Galactic Centre

The centre of the Milky Way hosts an SMBH of $\mathcal{M}_\bullet \simeq 4.1 \times 10^6\,\mathcal{M}_\odot$. This black hole mass is allegedly the best determined in galaxies because it comes from the study of the proper motions of stars around the black hole and the fitting of their orbits. The Galactic SMBH is identified with a source called **Sagittarius A$^*$** (Sag A$^*$) emitting radiation throughout the whole electromagnetic spectrum and experiencing strong variability, from the X-rays (X-ray flares) to the radio. Sag A$^*$ is embedded in an extremely active region (on a scale of hundred pc) also characterised by intense star formation (5–10% of the SFR of the Milky Way is there). At optical wavelengths, this region is heavily obscured, but IR observations reveal the presence of several star clusters containing massive recently born stars. A large number of SNRs, seen in radio, and a very hot plasma with temperatures of $\approx 10^7$ K are also observed. The molecular gas and the atomic gas are distributed in a massive ($10^7$–$10^8\,\mathcal{M}_\odot$) rotating ring of a few hundred parsec diameter, tilted with respect to the orientation of the outer Galactic disc. This is referred to as the **central molecular zone**.

The *Fermi* telescope observing gamma-rays has discovered a couple of large bubbles, called **Fermi bubbles**, clearly originating from the Galactic centre and extending up to almost 10 kpc above and below the Galactic plane. These huge structures are possibly produced by inverse Compton scattering (§8.1.3) of the interstellar radiation field by the relativistic electrons ejected from the Milky Way's nuclear region. The origin of these particles remains unclear: they can be produced either by intense nuclear star formation or by relativistic jets from the past activity (roughly $\lesssim$ 10 Myr ago) of the SMBH.

# 5 Present-Day Early-Type Galaxies

Based on the Hubble classification (§1.2 and §3.1), galaxies with smooth light distributions and approximately elliptical isophotes are globally classified as early-type galaxies (ETGs; §3.1). Because of their observed shape, ETGs are sometimes called **spheroids**, though strictly speaking this is a misnomer because ETGs do not necessarily have intrinsic spheroidal shape (§5.1.1). The family of ETGs comprises several types of galaxies, ranging from dwarf galaxies such as **dwarf ellipticals** (dEs) to luminous galaxies such as lenticulars (S0s), ellipticals, up to giant ellipticals, which are often the central and most luminous galaxies in clusters of galaxies (usually referred to as brightest cluster galaxies; §6.4.1). The optical luminosity of ETGs spans a wide range: for instance in the $B$ band, $10^7 \lesssim L_B/L_{\odot,B} \lesssim 10^{12}$ ($-25 \lesssim M_B \lesssim -13$). The characteristic properties of different classes of ETGs are summarised in Tab. 5.1.

Present-day ETGs have weak or absent ongoing star formation, little cold gas and do not have prominent stellar discs and spiral arms. For these characteristics, ETGs are also called quiescent or passive galaxies, in contrast to LTGs, which are star-forming (Chapter 4). Dwarf spheroidal galaxies (dSphs[1]), though sharing a few properties with ETGs (they are quiescent and have spheroidal morphology), represent a distinct class of objects and are not included in the family of ETGs: they are described in §6.3. Observations of ETGs across the electromagnetic spectrum have revealed the existence of important emission that is not due to stars. Ellipticals have hot ($10^6\,\mathrm{K} \lesssim T \lesssim 10^7\,\mathrm{K}$) gaseous halos that, especially in the brightest systems, are detected as smooth extended sources in X-rays. Ellipticals can also host, in their centre, an AGN (§3.6), which is believed to be powered by accretion onto an SMBH and gives an important contribution to the galaxy emission, especially in the radio band.

Assessing the properties of the present-day quiescent galaxies, and in particular of massive ellipticals, is important for understanding galaxy evolution, because these galaxies are believed to be the end product of a complex formation process. Given their regular structure and their brightness, luminous ellipticals have been studied in great detail and represent in a sense the prototypes of ETGs. Therefore in this chapter we will often refer specifically to ellipticals, though many of the observational and modelling techniques described for ellipticals also apply to other ETGs and, to some extent (for instance in the analysis of the stellar component), also to SFGs.

[1] For the classification of dSphs and dEs we follow Kormendy and Bender (2012). In practice, centrally concentrated dwarfs following the sequence of luminous ellipticals in the surface brightness–magnitude diagram of Fig. 3.4 are classified as dEs.

**Table 5.1** Properties of dwarf ellipticals (dEs), ellipticals (Es) and brightest cluster galaxies (BCGs)

| | dEs | Es | BCGs |
|---|---|---|---|
| Luminosity ($B$ band) ($L_\odot$) | $10^7$–$10^9$ | $10^9$–$10^{11}$ | $10^{11}$–$10^{12}$ |
| Magnitude ($B$ band) | −13 to −17 | −17 to −22 | −22 to −25 |
| Stellar mass ($\mathcal{M}_\odot$) | $10^7$–$10^9$ | $10^9$–$10^{11}$ | $10^{11}$–$10^{12}$ |
| Effective radius (kpc) | 0.1–3 | 1–10 | 3–30 |
| Stellar velocity dispersion (km s$^{-1}$) | 20–100 | 100–300 | 250–400 |
| Hot gas mass ($\mathcal{M}_\odot$) | — | $10^8$–$10^{10}$ | $> 10^{10}$ |
| Black hole mass ($\mathcal{M}_\odot$) | — | $10^7$–$10^9$ | $10^9$–$10^{10}$ |
| Virial (total) mass ($\mathcal{M}_\odot$) | $10^8$–$10^{10}$ | $10^{10}$–$10^{13}$ | $> 10^{13}$ |

This table reports indicative ranges of a selection of physical quantities for three ETG subclasses. We refer the reader to Kormendy et al. (2009) and Lauer et al. (2014) for overviews of the ETG properties.

## 5.1 Stars

We start our description of present-day ETGs by reviewing the main structural and kinematic properties of their stellar components, and the characteristics of their stellar populations.

### 5.1.1 Spatial Distribution

At visible wavelengths the light emitted by stars is the dominant component of ETGs. In general the light emitted by individual stars of ETGs cannot be resolved, due to the large distances (an exception is the dE M 32 in the Local Group; §6.3; Tab. 6.2): the stellar emission appears as a diffuse unresolved smooth light distribution.

#### Shape of Isophotes

To a first approximation each isophote of the stellar emission of an ETG can be described as an ellipse with semi-major axis $a$ and semi-minor axis $b$. Each isophote is labelled by its radius $R$, which is defined as either the elliptical radius $a$ or the circularised radius[2] $R_{\rm circ} = (ab)^{1/2}$ (§3.1.2), and is characterised also by its **position angle**, defined as the angle in the anticlockwise direction between the north and the major axis of the galaxy (see also §4.3.3). The position angle can vary systematically with $R$, a phenomenon known as **isophote twist**.

In general, the ellipticity $\epsilon = 1 - b/a$ (§3.1.2) is not the same for all isophotes: an interesting property of an ETG is therefore its **ellipticity profile** $\epsilon(R)$. In practice, in many

[2] When comparing galaxies with different ellipticities, the circularised radius may be more meaningful than the elliptical radius, because the area enclosed within an elliptic isophote $\pi ab = \pi R^2_{\rm circ}$ is independent of the axis ratio $b/a$ at fixed $R_{\rm circ}$.

cases the ellipticity profile is constructed defining $\epsilon(R)$ as the luminosity-weighted average ellipticity within the isophote of radius $R$. Often $\epsilon$ increases with radius, but the shape of the ellipticity profile varies a lot from galaxy to galaxy. It is sometimes useful to characterise a galaxy with a single value of the ellipticity. In some cases this value is obtained by simply fitting a 2D elliptical model to the galaxy surface brightness distribution. Otherwise, when the ellipticity profile is available, the galaxy ellipticity is defined as the value of $\epsilon$ measured within a reference radius: for instance the ellipticity $\epsilon_e \equiv \epsilon(R_e)$ measured within the effective radius $R_e$ (§3.1.2). Ellipticals have observed ellipticities in the range $0 \lesssim \epsilon_e \lesssim 0.7$.

The observed shape of an ETG is given by a combination of intrinsic shape and projection effects. ETGs are in general ellipsoids, often significantly triaxial. The intrinsic 3D distribution of stars of ETGs can be represented by concentric ellipsoids with principal semi-axes $c_{int} \leq b_{int} \leq a_{int}$. An isodensity surface is a **triaxial ellipsoid** when $c_{int} < b_{int} < a_{int}$, an **oblate spheroid** when $c_{int} < b_{int} = a_{int}$, and a **prolate spheroid** when $c_{int} = b_{int} < a_{int}$. The simplest models are those in which the concentric ellipsoids are similar and their principal axes are aligned. However, in general the direction of the principal axes and the axis ratios depend on the distance from the system's centre. It can be shown that the presence of isophote twist implies that the intrinsic distribution is triaxial with varying axis ratios (see Binney and Merrifield, 1998, for a detailed treatment). Observations show that the brightest ellipticals are typically triaxial systems, while lower-luminosity ellipticals tend to be oblate spheroidal systems.

In general, the isophotes of ETGs deviate from ellipses: if, in polar coordinates $(R, \theta)$, the isophote is described by $R(\theta)$ and the best-fitting ellipse by $R_{ell}(\theta)$, the **isophote deviation** is measured by the residual $\Delta R(\theta) \equiv R(\theta) - R_{ell}(\theta)$. To quantitatively describe the deviations from the elliptic shape, it is useful to write the Fourier expansion of the residual as a function of $\theta$:

$$\Delta R(\theta) = \sum_{j=1}^{\infty} \left[ a_j \cos(j\theta) + b_j \sin(j\theta) \right], \qquad (5.1)$$

which is a way to determine the characteristic angular scales on which the deviation is apparent. The dominant term in the expansion is usually $a_4$, which is the first coefficient in the series corresponding to a deviation from an ellipse symmetric with respect to both the major and the minor axes. When $a_4 > 0$ the isophote is **discy**, and when $a_4 < 0$ the isophote is **boxy**, as illustrated by Fig. 5.1. It follows that, based on the analysis of the isophote deviation, ETGs are classified as either **discy** or **boxy galaxies**. Among massive ellipticals, more luminous galaxies tend to be boxy and less luminous galaxies are in general discy.

## Surface Brightness Profiles

ETGs are extended systems without a well defined edge. A convenient measure of the size of an ETG is a radius containing a given fraction of the total luminosity, and in particular the effective radius $R_e$, which contains half of the total luminosity (§3.1.2). From the observed surface brightness (§C.2) distribution $I(\boldsymbol{R})$, where $\boldsymbol{R}$ is a 2D vector in the plane of the sky, it is possible to derive the surface brightness profile $I(R)$, where $R$ is either

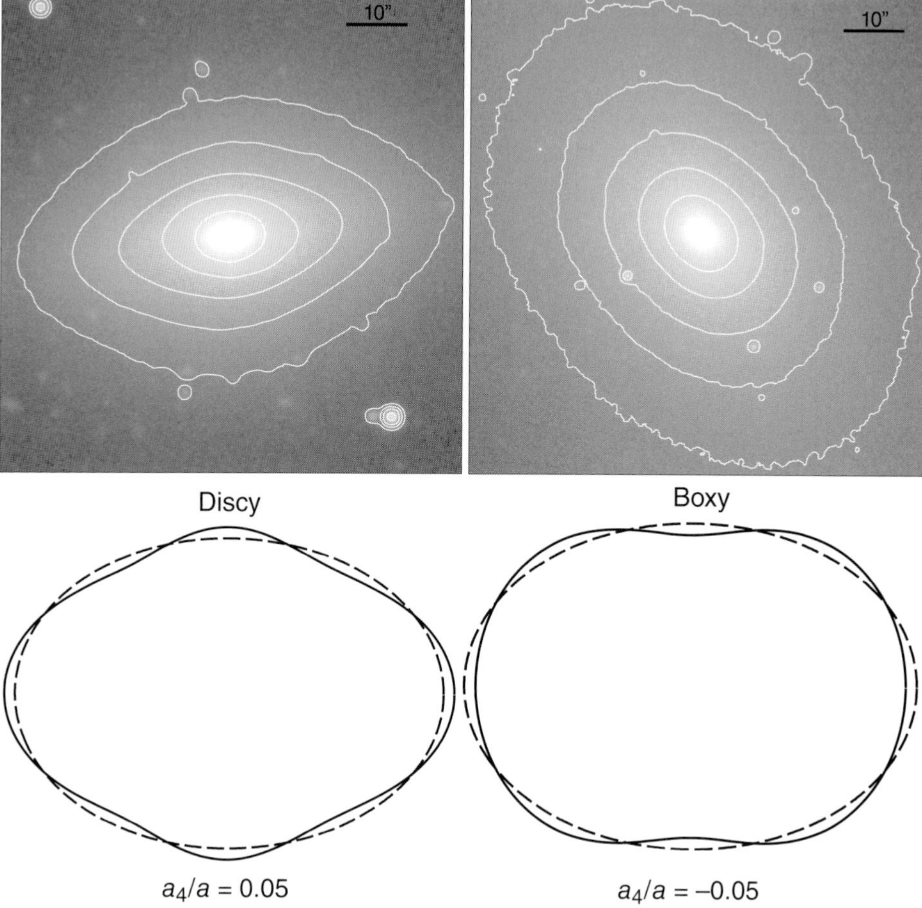


Fig. 5.1 *Top panels.* SDSS *r*-band images of the discy galaxy NGC 4660 (*left panel*, 10 arcsec $\simeq$ 1.2 kpc) and of the boxy galaxy NGC 4365 (*right panel*, 10 arcsec $\simeq$ 1.8 kpc). The contours indicate isophotes. Adapted from Krajnović (2011). Courtesy of D. Krajnović. *Bottom panels.* The solid curves show examples of discy (*left panel*) and boxy (*right panel*) isophotes. These curves are constructed using eq. (5.1) with $a_4 \neq 0$ and all the other coefficients set to zero. Here $a$ is the semi-major axis of the corresponding ellipse (dashed curves).

the elliptical or the circularised radius. The surface brightness profiles of ETGs are well described by the Sérsic law (eq. 3.1), which is characterised by the Sérsic index $n$. The case of $n=4$ corresponds to the **de Vaucouleurs $R^{1/4}$ law**,

$$I(R) = I_\mathrm{e} \exp\left\{-b(4)\left[\left(\frac{R}{R_\mathrm{e}}\right)^{1/4} - 1\right]\right\}, \tag{5.2}$$

with $b(4) \simeq 7.67$, which was first found by de Vaucouleurs (1948) as able to reproduce the profiles of present-day ellipticals. However, when $n$ is left as a free parameter, better fits are found for the surface brightness profiles of ellipticals: the best-fitting Sérsic index tends to increase with galaxy luminosity, from $n \approx 2$ for low-luminosity ellipticals to $n \approx 10$ for the brightest ellipticals. For higher values of $n$ the central profile is steeper and the outer profile is shallower (Fig. 3.3).

In some cases the very central surface brightness profiles of ellipticals deviate significantly and systematically from a Sérsic profile. The brightest ellipticals are characterised by a central light deficit: inside a **break radius** $R_b$ the light profile is shallower than the inward extrapolation of the Sérsic law that best fits the outer profile. These galaxies are therefore called **cored** (or **core-Sérsic**) **galaxies** and usually have $R_b/R_e \sim 0.01$. In general, less luminous ellipticals have no central light deficit: these galaxies are called **coreless** (or **power-law** or **Sérsic**) **galaxies**. Examples of light profiles of cored and coreless ellipticals are shown in Fig. 5.2. In fact, coreless ellipticals are often characterised by the presence of central extra light, in the sense that the central light profile is systematically above the inward extrapolation of the outer Sérsic profile, as apparent from Fig. 5.3. We see below that the presence or absence of a core in the light distribution correlates with several other observed properties of ETGs (Tab. 5.2).

## Lenticular Galaxies

The structure of lenticulars (S0s) differs from that of ellipticals in many respects. S0s are more similar to spiral galaxies, because they have two distinct stellar components: a central bulge and an extended disc. However, unlike the discs of spirals, the discs of S0s have extremely weak or no spiral arms. The surface brightness profile of an S0 is not described well by a single Sérsic law, but can be well represented by a two-component

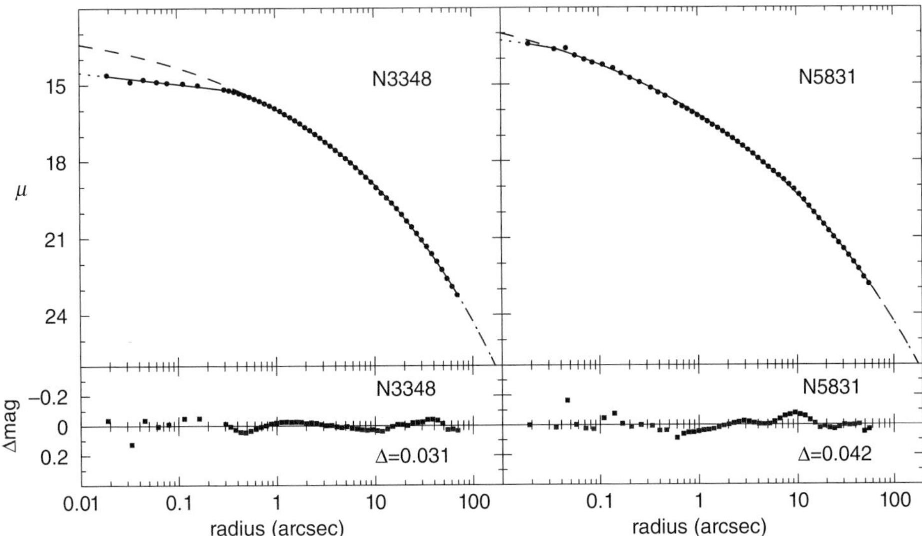


Fig. 5.2 *Top panels.* The points indicate the surface brightness profiles (in units of mag arcsec$^{-2}$) of the cored elliptical galaxy NGC 3348 (*left panel*, effective radius $R_e \simeq 21.8$ arcsec $\simeq 4.1$ kpc) and of the coreless elliptical galaxy NGC 5831 (*right panel*, effective radius $R_e \simeq 28.3$ arcsec $\simeq 2.9$ kpc). In each panel the dashed curve is the Sérsic fit to the profile, excluding the core. The solid curve is the best fit to the entire profile with a function obtained by combining a Sérsic profile with a central power law. *Bottom panels.* Residuals between the observed profiles and the best fits (solid curves in the upper panels). From Graham et al. (2003). © AAS, reproduced with permission.

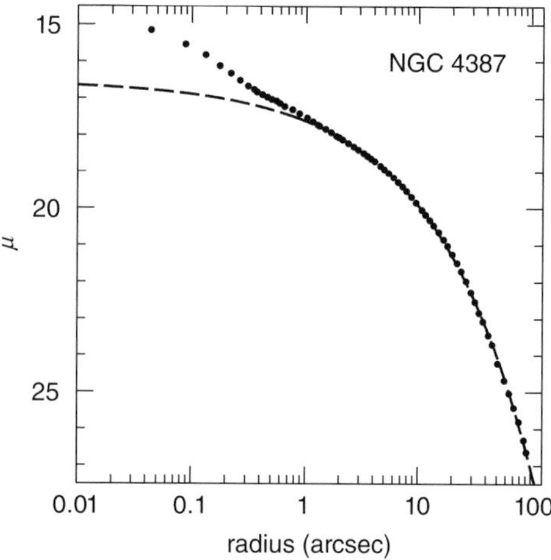


Fig. 5.3 Surface brightness profile (in units of mag arcsec$^{-2}$) of NGC 4387, a coreless elliptical galaxy with central extra light (the effective radius is $R_e \simeq 14.4$ arcsec $\simeq 1.25$ kpc). The dashed curve is the Sérsic fit to the outer parts of the profile. Data from Kormendy et al. (2009).

system: the bulge, described by a Sérsic law with $n \gtrsim 2$, dominates in the centre, while the disc, described by an exponential profile ($n \approx 1$), dominates in the outer parts. These two components of the S0s are apparent also in their ellipticity profiles (relatively low values of $\epsilon$ in the centre and the highest values of $\epsilon$ in the outskirts) and in their isophote deviation profiles (large positive values of $a_4$, signatures of discy isophotes, are observed in the outer disc-dominated parts). S0 galaxies with no evidence of a bulge are believed to be the quiescent counterparts of bulgeless spiral galaxies or Irrs (§3.1 and §4.1.2). The origin of S0s is discussed in §10.8.2.

### Early-Type Galaxies with Substructures

Though most ETGs appear regular and featureless, it is important to mention that a fraction of the present-day quiescent galaxies have **shells** (or **ripples**) probably due to accretion of satellite galaxies. The fraction of galaxies with such substructures is hard to determine, because the estimate depends critically on the availability of deep photometry to detect low surface brightness features (see also §4.1.3). A magnificent example of an ETG with complex substructures in the outskirts is shown in Fig. 5.4.

## 5.1.2 Kinematics

The study of the motion of stars within ETGs gives complementary information with respect to the study of the spatial distribution of stars. This information can be used to estimate the dark matter content and distribution of ETGs, and to constrain their formation and evolution history.

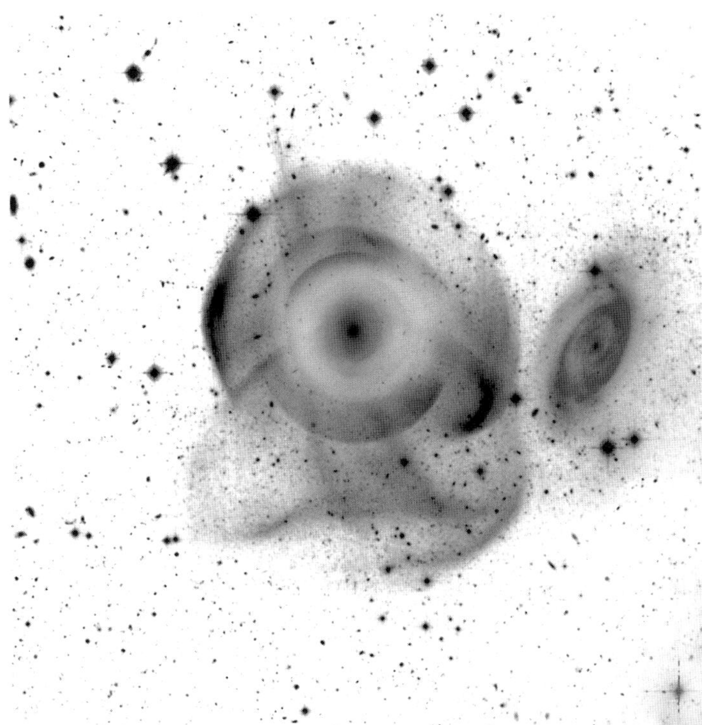

Fig. 5.4 Deep optical image of the S0 galaxy NGC 474, characterised by a complex structure of shells and stellar streams. The smaller galaxy in the right part of the image is the companion spiral NGC 470. The image, taken with the Canada–France–Hawaii Telescope, was processed to enhance the inner shells of NGC 474. Courtesy of P.-A. Duc and J.-C. Cuillandre.

## Line-of-Sight Velocity Distribution

The kinematics of stars in ETGs is studied by means of spectroscopy, mainly in the optical band. In general, velocities of individual stars cannot be measured, but it is possible to obtain spectra of integrated light from all stars within the aperture of the spectrograph. Taking a spectrum at a given position, we can measure the properties of the stellar absorption lines (for instance, Ca, Mg and Na lines), which are broadened by the Doppler effect due to the motion of the stars within the galaxy. Therefore, the **line profile** provides information about the local line-of-sight velocity distribution (LOSVD) $F(v_{\rm los}, \boldsymbol{R})$, that is, the distribution of the stellar line-of-sight velocity $v_{\rm los}$ (relative to the systemic velocity; §4.3.3) at position $\boldsymbol{R}$ in the plane of the sky (eq. C.43). The LOSVD at a given position can be characterised by measuring its moments: the mean line-of-sight velocity $\overline{v}_{\rm los}$ (eq. C.44) and the line-of-sight velocity dispersion $\sigma_{\rm los}$ (eq. C.45).

With the so-called integral field spectroscopy (§4.3.2) it is possible to take several spectra at different positions of the galaxy and to construct 2D maps of $\overline{v}_{\rm los}(\boldsymbol{R})$ and $\sigma_{\rm los}(\boldsymbol{R})$ (see Cappellari, 2016, for a review). The kinematic maps of two representative ETGs are shown in Fig. 5.5. When $\overline{v}_{\rm los}(\boldsymbol{R})$ is positive on the one side of the galaxy and negative

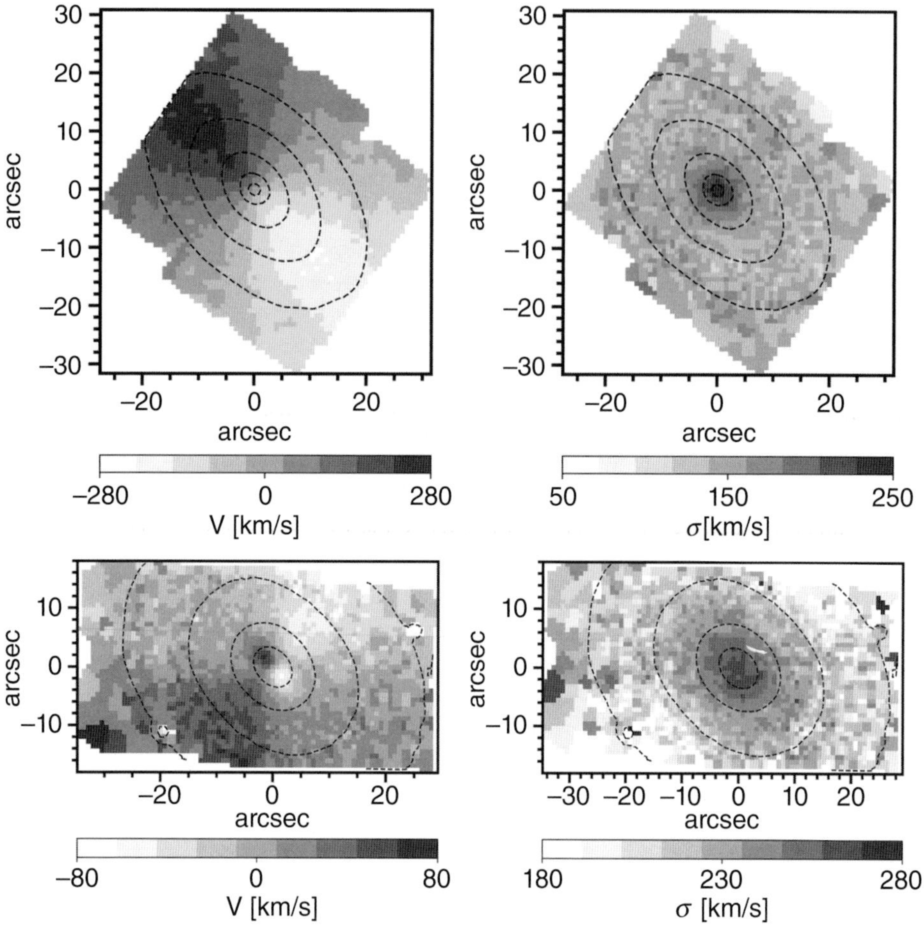


Fig. 5.5 *Top panels*. Line-of-sight velocity (*left*) and velocity dispersion (*right*) maps for the fast rotator ETG NGC 2974 (1 arcsec $\simeq$ 0.1 kpc). *Bottom panels*. Same as top panels, but for the slow rotator ETG NGC 4365 (1 arcsec $\simeq$ 0.1 kpc). Note that the greyscales in the bottom panels are very different from the corresponding greyscales in the top panels. Data from Cappellari et al. (2011). Courtesy of D. Krajnović.

on the other side (top left panel of Fig. 5.5), we are detecting a clear signature of rotation (§4.3.3). The absence or weakness of this feature (bottom left panel of Fig. 5.5) might indicate that rotation is negligible, but it can also be the case that the galaxy rotates significantly, with rotation axis almost aligned with the line of sight (i.e. the galaxy is face-on). For sufficiently symmetric systems it is possible to define the line-of-sight velocity profile $\overline{v}_{\rm los}(R)$ and velocity dispersion profile $\sigma_{\rm los}(R)$, where $R$ is a coordinate labelling, in the plane of the sky, the kinematic major axis (§4.3.3), along which the velocity gradient is maximum ($R$ is the elliptical radius only if the kinematic major axis coincides with the photometric major axis).

In order to characterise the velocity dispersion of an ETG with a single number, it is useful to introduce the **central velocity dispersion** $\sigma_0$, which is the line-of-sight projected velocity dispersion measured within a given aperture (i.e. angular area) $S_{\rm ap}$, centred in the galaxy centre, defined by

$$\sigma_0^2 = \frac{\displaystyle\int_{S_{\rm ap}} \sigma_{\rm los}^2(\boldsymbol{R}) I(\boldsymbol{R}) {\rm d}^2\boldsymbol{R}}{\displaystyle\int_{S_{\rm ap}} I(\boldsymbol{R}) {\rm d}^2\boldsymbol{R}}, \tag{5.3}$$

where $I$ is the surface brightness. The details of the definition of $\sigma_0$ depend on the instrument used for the spectroscopic observation. It is convenient to associate $S_{\rm ap}$ with an aperture radius $R_{\rm ap}$, which gives a measure of the size of the region probed by the spectrograph, so that, for instance, in the idealised case of a circular aperture the area of $S_{\rm ap}$ is $\pi R_{\rm ap}^2$. In most cases, at least in low-redshift observations, $R_{\rm ap}$ is in the range $0.1 \lesssim R_{\rm ap}/R_{\rm e} \lesssim 1$, hence the name central velocity dispersion. When the aperture is $R_{\rm e}$, the central velocity dispersion is often indicated with $\sigma_{\rm e}$.

The LOSVDs of ellipticals can deviate significantly from a Gaussian. A convenient way to quantify such deviations is to parameterise the LOSVD $F(v_{\rm los}, \boldsymbol{R})$ as a truncated **Gauss–Hermite series** (see Gerhard, 1993, and van der Marel and Franx, 1993), which has been demonstrated to be able to reproduce sufficiently well the observed line profiles of spheroids. Taking only the first terms of the series, the LOSVD at a given position $\boldsymbol{R}$ can be written as

$$F(v_{\rm los}) = \frac{1}{\sqrt{2\pi}\sigma_{\rm los}} \exp\left(-\frac{w^2}{2}\right) [1 + h_3 H_3(w) + h_4 H_4(w)], \tag{5.4}$$

where $w \equiv (v_{\rm los} - \overline{v}_{\rm los})/\sigma_{\rm los}$, $H_3(w) = (2w^3 - 3w)/\sqrt{3}$ and $H_4(w) = (2w^4 - 6w^2 + 3/2)/\sqrt{6}$ are **Hermite polynomials**, and $h_3$ and $h_4$ are dimensionless coefficients. The effect of $h_3$ and $h_4$ on the LOSVD is illustrated in Fig. 5.6. The parameter $h_3$ (related to the skewness) measures asymmetric deviations from a Gaussian, while $h_4$ (related to the kurtosis) measures symmetric deviations. The coefficient $h_3$ is sensitive to the presence of ordered motions: if there is rotation with a velocity component along the line of sight, $h_3 \neq 0$, the profile is asymmetric and the peak of $F(v_{\rm los})$ is shifted from $\overline{v}_{\rm los}$. The coefficient $h_4$ is sensitive to the anisotropy of the velocity distribution: $F(v_{\rm los})$ is more centrally peaked than a Gaussian when $h_4 > 0$ and more flat-topped than a Gaussian when $h_4 < 0$. With accurate measurements of $h_3$ and $h_4$ for observed stellar absorption line profiles it is possible to constrain the intrinsic stellar velocity distribution and therefore to break the so-called mass–anisotropy degeneracy (§5.3.2).

Observationally, a full characterisation of the kinematics of stars in ETGs is therefore given by the maps $\overline{v}_{\rm los}(\boldsymbol{R})$, $\sigma_{\rm los}(\boldsymbol{R})$, $h_3(\boldsymbol{R})$ and $h_4(\boldsymbol{R})$. When assumptions on the intrinsic shape and on the inclination of the galaxy are made, these kinematic maps can be compared with 3D galaxy models: such a comparison can provide information on the phase-space distribution of stars (§C.8) and on the intrinsic mass distribution (§5.3).

## Fast and Slow Rotators

While the kinematics of present-day spiral galaxies is dominated by the ordered motions of the disc stars (§4.1.1), random motions are a key ingredient of the kinematics of ETGs.

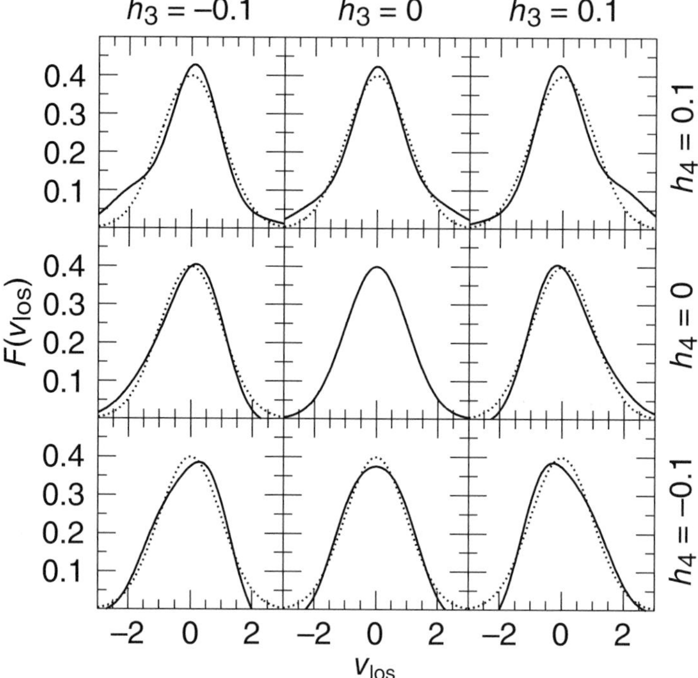


Fig. 5.6 Effect of $h_3$ and $h_4$ on the LOSVD. Here the LOSVD $F(v_{\rm los})$ is defined by eq. (5.4), taking $\overline{v}_{\rm los} = 0$ and $\sigma_{\rm los} = 1$. Figure inspired by van der Marel and Franx (1993).

A first estimate of the relative importance of ordered and random motions is given by the $V/\sigma$ ratio (§4.1.1), where $V$ and $\sigma$ are characteristic values of the line-of-sight velocity and velocity dispersion, respectively. In early studies of the kinematics of ETGs, it was often assumed that $V = v_{\rm max}$ and $\sigma = \sigma_0$, where $v_{\rm max}$ is the maximum value of $|\overline{v}_{\rm los}|$, and $\sigma_0$ is the central velocity dispersion (eq. 5.3). Measurements of $v_{\rm max}/\sigma_0$ have shown that ellipticals are **pressure-supported stellar systems**:[3] $\sigma_0$ is at least of the order of and often much higher than $v_{\rm max}$, so support against gravity is given more by velocity dispersion than by ordered motions. In this respect ellipticals are very different from massive SFGs, which have $V/\sigma \gg 1$ (§4.1.1). The difference is apparent also in terms of specific angular momentum $j_\star$ (§4.4.2): at given stellar mass, ETGs are found to have a factor of $\approx 5$ lower $j_\star$ than SFGs.

With the advent of integral field spectroscopy, $v_{\rm max}/\sigma_0$ has been replaced by a new definition of the ratio $V/\sigma$, with $V \equiv \left(\left\langle \overline{v}_{\rm los}^{2} \right\rangle_{\rm e}\right)^{1/2}$ and $\sigma = \sigma_{\rm e}$, where $\langle \cdots \rangle_{\rm e}$ indicates the luminosity-weighted average within the effective ellipse with semi-major axis $R_{\rm e}$, and $\sigma_{\rm e}$ is the central velocity dispersion measured within $R_{\rm e}$. Alternatively, the importance of rotation is quantified by the parameter

[3] The term 'pressure-supported' is used in analogy with gaseous systems: in collisionless stellar systems (§C.8) the role of 'pressure' is played by the kinetic energy density associated with the random motions of the stars.

$$\lambda_{R_e} \equiv \frac{\langle R|\overline{v}_{\rm los}|\rangle_{\rm e}}{\left\langle R\sqrt{\overline{v}^2_{\rm los} + \sigma^2_{\rm los}}\right\rangle_{\rm e}}. \tag{5.5}$$

In practice, when a binned 2D map of integral field spectroscopy data is available, $\lambda_{R_e}$ is computed as

$$\lambda_{R_e} = \frac{\sum_i F_i R_i |\overline{v}_{{\rm los},i}|}{\sum_i F_i R_i \sqrt{\overline{v}^2_{{\rm los},i} + \sigma^2_{{\rm los},i}}}, \tag{5.6}$$

where $F_i$, $R_i$, $\overline{v}_{{\rm los},i}$ and $\sigma_{{\rm los},i}$ are, respectively, the flux, distance to the centre, line-of-sight velocity and line-of-sight velocity dispersion of the $i$th bin, and the sum is extended over all bins within the effective isophote.

The observed shape of the isophotes of ETGs is related to the kinematics of stars. For instance, oblate ETGs can be flattened by rotation, but this is not necessarily the case, because also the anisotropy of the stellar velocity distribution can produce a flattening, even in the absence of rotation. We now see how it is possible to determine whether an ETG is flattened by rotation or by anisotropy, based on observable quantities.

Observationally determined values of $V/\sigma$ and $\lambda_{R_e}$ for a sample of ETGs are shown as functions of the ellipticity $\epsilon_{\rm e}$ in Fig. 5.7. In order to interpret the distribution of the observed ETGs in the $\epsilon_{\rm e}$–$(V/\sigma)$ and $\epsilon_{\rm e}$–$\lambda_{R_e}$ planes, we ask ourselves under what conditions a spheroid is consistent with being flattened by rotation. Let us consider for simplicity an axisymmetric system, which is observed edge-on, and let us take a Cartesian coordinate system $(x, y, z)$, in which $z$ is the symmetry axis and $z = 0$ defines the plane of symmetry.

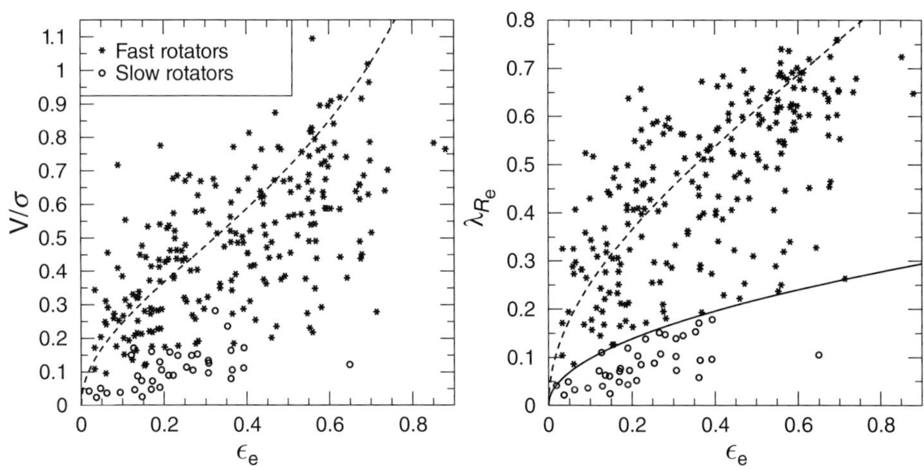


Fig. 5.7 Ordered-to-random motion ratio $V/\sigma$ (*left panel*) and $\lambda_{R_e}$ parameter (eq. 5.5; *right panel*) as functions of the ellipticity $\epsilon_{\rm e}$ for fast (asterisks) and slow (circles) rotators of the ATLAS$^{\rm 3D}$ sample of ETGs. Here $V = \left(\langle\overline{v}^2_{\rm los}\rangle_{\rm e}\right)^{1/2}$ and $\sigma = \sigma_{\rm e}$. The dashed curves indicate the loci of edge-on isotropic rotators. The solid curve corresponds to $\lambda_{R_e} = 0.31\sqrt{\epsilon_{\rm e}}$, commonly adopted to separate fast and slow rotators. Data from Emsellem et al. (2011).

For such a system the parameter $\lambda_{R_e}$ and the ratio $V/\sigma$ are related to the ellipticity $\epsilon_e$ essentially only through the **global anisotropy parameter**

$$\delta \equiv 1 - \frac{\Pi_{zz}}{\Pi_{xx}}, \tag{5.7}$$

where $\Pi_{xx}$ and $\Pi_{zz}$ are the diagonal components of the random motion kinetic energy tensor along $x$ and $z$, respectively (eq. C.41). A stellar system with isotropic velocity distribution has $\delta = 0$, because the components of its velocity dispersion tensor (eq. C.40) are $\sigma_{xx}^2 = \sigma_{yy}^2 = \sigma_{zz}^2$. If such an isotropic system does not rotate, it is spherical; if it rotates, it is an oblate spheroid, with symmetry axis given by the rotation axis, called the **isotropic rotator**. The dashed curves in Fig. 5.7 indicate, in the planes $\epsilon_e$–$(V/\sigma)$ and $\epsilon_e$–$\lambda_{R_e}$, the loci of model isotropic ($\delta = 0$) rotators observed edge-on. Isotropic rotators that are not observed edge-on lie in the regions to the left of these curves. Edge-on anisotropic ($\delta > 0$) rotators lie to the right of these curves.

It follows that systems close to the dashed curves in Fig. 5.7 (**fast rotators**) are flattened by rotation, while systems that lie substantially below these curves (**slow rotators**) are flattened by orbital anisotropy. A curve that is conventionally used to separate fast and slow rotators in the $\epsilon_e$–$\lambda_{R_e}$ plane is $\lambda_{R_e} = 0.31\sqrt{\epsilon_e}$ (Fig. 5.7, right). These two families of objects have markedly different kinematic maps, as illustrated by Fig. 5.5, which shows the line-of-sight velocity and velocity dispersion maps of a fast rotator (top panels) and a slow rotator (bottom panels).

The distributions of bright ETGs in the $\epsilon_e$–$(V/\sigma_e)$ and $\epsilon_e$–$\lambda_{R_e}$ planes are correlated with galaxy luminosity: less luminous systems tend to be fast rotators, while more luminous systems are typically slow rotators. Though there is not a clear-cut distinction between the two populations, $L_V \approx 3 \times 10^{10} L_{\odot,V}$ can be taken as a reference luminosity separating fast and slow rotators. Such a trend does not appear to extend to the lowest luminosities: dEs present a variety of kinematic behaviours. The aforementioned kinematic classification is often used more generally to define two families of ETGs: fast rotators ($\approx 80\%$ of nearby ETGs) and slow rotators (the remaining $\approx 20\%$). Galaxies belonging to each of these two families share similar properties not only from the point of view of kinematics and structure, but also in terms of stellar populations, gas content, X-ray and radio emission. The main properties of fast and slow rotators are summarised in Tab. 5.2. Explaining the origin of the segregation of present-day ETGs in these two groups is an important goal of any galaxy formation model, as we see in §10.8.1 and §10.8.2.

The kinematic maps of ETGs are not always regular. In some cases the kinematic behaviour of the stellar galaxy core is different from that of the main body of the galaxy, in the sense that the rotation axes of the main body and of the core are misaligned. When this feature is observed the galaxy is said to have a **kinematically distinct** (or **kinematically decoupled**) **core**. A clear example of a galaxy with kinematically decoupled core is NGC 4365 (bottom left panel of Fig. 5.5). In some special cases the core rotates with roughly the same rotation axis as the main body of the galaxy, but with opposite direction of the angular momentum: this feature is called a **counter-rotating core**. Kinematically distinct cores are frequent among slow rotators and seldom found among fast rotators. Counter-rotating cores are rare, but almost all known cases are in slow rotators.

**Table 5.2** Properties of fast and slow rotators

| Fast rotators | Slow rotators |
|---|---|
| Flattened by rotation | Flattened by anisotropy |
| Oblate spheroidal | Triaxial |
| Isotropic velocity distribution | Anisotropic velocity distribution |
| More flattened ($\langle\epsilon_e\rangle \approx 0.4$) | Less flattened ($\langle\epsilon_e\rangle \approx 0.2$) |
| Discy isophotes | Boxy isophotes |
| Sérsic index $n \lesssim 4$ | Sérsic index $n \gtrsim 4$ |
| Coreless surface brightness profile | Cored surface brightness profile |
| Lower optical luminosity | Higher optical luminosity |
| Younger stars | Older stars |
| No $\alpha$-element enhancement | $\alpha$-element enhancement |
| Weak radio emission | Strong radio emission |
| Weak diffuse X-ray emission | Strong diffuse X-ray emission |
| No kinematically distinct core | Kinematically distinct core |

Table inspired by Kormendy and Bender (2012).

## 5.1.3 Stellar Populations

Early-type galaxies are key probes of galaxy evolution because their SFHs allow us to constrain how massive galaxies formed and evolved. However, the individual stars of ETGs at distances beyond a few Mpc cannot be resolved (§5.1.1). This implies that the information on the stellar populations can be inferred only through the analysis of their integrated colours and spectra (§3.3).

Regarding colours, as described in §3.4.1 and shown in Fig. 3.9, ETGs are in general redder than SFGs. As ETGs have a low content of cold gas and interstellar dust (§5.2.2) and dust extinction is usually low (typically $A_V < 0.3$ mag), this implies that these galaxies are intrinsically red. Since old or metal-rich stars are red (Fig. 8.16), the ETG colours can be due to evolved stellar populations and/or a high abundance of metals (§C.6.1).

In the 1980s, it was found that the colours of cluster ETGs (§6.4) show tight correlations with luminosity and the central velocity dispersion $\sigma_0$. In particular, redder ETGs have higher luminosities (**colour–magnitude relation**) and $\sigma_0$ (**colour–velocity dispersion relation**). Since galaxies with higher stellar mass are found to have higher $\sigma_0$ (§5.4.3; eq. 5.16 and eq. 5.17), the colour–$\sigma_0$ relation indicates also that redder ETGs are more massive. More recently, it has been found that all ETGs (i.e. not only those located in clusters) follow a correlation between colour and mass. The red sequence illustrated in §3.4.1 (Fig. 3.9) is a manifestation of the correlation between the colour and the stellar mass, although the scatter of this correlation is larger than that of cluster ETGs. To summarise, the general implication of these results is that massive ellipticals are older and/or more metal-rich. However, the colours alone are not sufficient to derive the physical and evolutionary interpretation of these correlations because of the so-called

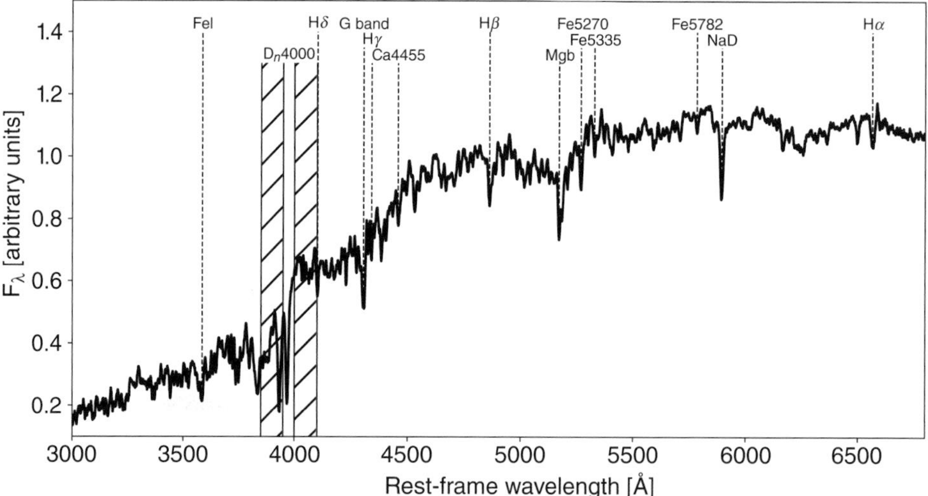


Fig. 5.8 The average spectrum of 3425 ETGs with $\log(\mathcal{M}_\star/\mathcal{M}_\odot) > 11$ ($\bar{z} \simeq 0.4$) extracted from the SDSS-III BOSS public survey data (Dawson et al., 2013). The main absorption lines are indicated. The hatched regions indicate the wavelength intervals used for the calculation of the $D4000_n$ break (eq. 3.5). Courtesy of M. Moresco.

**age–metallicity degeneracy** which does not allow us to understand whether a red colour alone is due to old stars or to high metallicity (§11.1.3).

To overcome this limitation, more stringent constraints on the stellar content and SFHs of ETGs can be extracted from their spectra. Fig. 5.8 shows the average spectrum of ETGs from the near-UV to the red optical. The shape of the continuum, the D4000 break and the absorption lines are typical of evolved stars (§3.4). This kind of data is rich in information on the stellar populations of ETGs. The extraction of such information can be done following two main approaches. The first is to fit the whole continuum and absorption lines with theoretical models of stellar populations (**full spectral fitting**; §8.6 and §11.1.3). The second is to measure the equivalent widths of individual absorption lines (e.g. the **Lick indices**; Fig. 11.21) that are sensitive to age and metallicity separately, and allow us to break the degeneracy between these two quantities. Within the uncertainties, these two methods give consistent results and provide a rather coherent picture on the stellar content of ETGs. In general, spectroscopic studies have revealed that the stellar populations of present-day ETGs are *both* old (from a few Gyr to almost the age of the Universe) *and* metal-rich (with metallicities from solar to supersolar). Moreover, it has been found that the stellar population properties depend strongly on galaxy mass: massive ETGs are older and more metal-rich than less massive ones. In particular, three main correlations have been found with the central velocity dispersion $\sigma_0$ (Fig. 5.9). These relations are extremely important to reconstruct the SFHs of ETGs (§11.3.10), and the main findings can be summarised as follows.

- *Age–velocity dispersion relation*. The correlation indicates that the age of stellar populations increases with $\sigma_0$ (and thus with mass). The ages are in general old

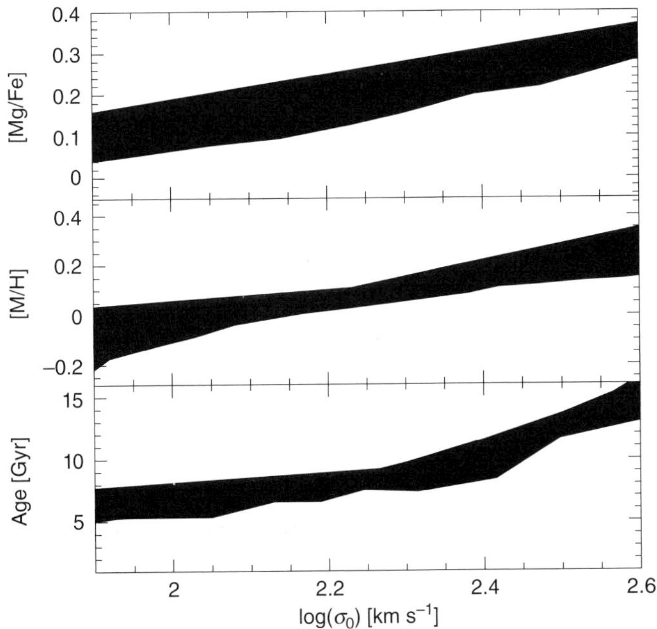


Fig. 5.9 The scaling relations between central velocity dispersion $\sigma_0$ and the stellar population properties of ETGs at $z \approx 0$ based on full spectral fitting or Lick indices. From bottom to top: stellar age, total metallicity (in units of [M/H]) and the $\alpha$-element magnesium abundance relative to iron ([Mg/Fe]). The shaded regions enclose the range of the correlations found in different works (data from Gallazzi et al., 2006; Thomas et al., 2010; Conroy et al., 2014; Johansson et al., 2012).

(above $\approx$ 5 Gyr), and the most massive ETGs at $z \approx 0$ can be nearly as old as the Universe. It has also been found that, for a fixed mass, ETGs with smaller sizes are older, more metal-rich and more $\alpha$-enhanced (see below) than their larger counterparts.

- *Metallicity–velocity dispersion relation.* This correlation shows that the stellar metallicity ($Z$) increases with galaxy $\sigma_0$. The metallicity of ETGs is generally high and reaches supersolar values for the most massive systems.
- *$\alpha$-element abundance–velocity dispersion relation.* This correlation indicates that the abundance of $\alpha$-elements relative to iron ([$\alpha$/Fe]) also increases with mass. In general, ETGs have supersolar values of [$\alpha$/Fe], with the most massive reaching abundances up to [$\alpha$/Fe] $\approx 0.3$.

These scaling relations do not strongly depend on the environmental density (§6.5.1), and are mostly (if not completely) driven by galaxy mass.

The physical interpretation of these results will be given in the general context of ETG evolution (§11.3.10). Here it is relevant to anticipate that the age–$\sigma_0$ relation (bottom panel of Fig. 5.9) implies that massive ETGs formed their stars earlier (i.e. at higher redshifts) as they are older than the lower-mass ETGs in the present-day Universe. Another key result is the high abundance of $\alpha$-elements relative to iron. As described in

§4.1.4 (Fig. 4.24), the [$\alpha$/Fe] ratio allows us to estimate the duration of star formation ($\Delta t$) because the $\alpha$-elements are produced mostly by the (short-lived) Type II SNe. If the duration is relatively short (e.g. $\Delta t \lesssim 0.5$–1 Gyr), [$\alpha$/Fe] is high, and its absolute value depends mostly on the IMF, i.e. on the number of massive stars which exploded as Type II SNe ($\alpha$-enhancement). Instead, [$\alpha$/Fe] gradually decreases with increasing $\Delta t$ due to the later onset of Type Ia SNe which increase the Fe abundance relative to the $\alpha$-elements. The values of [$\alpha$/Fe] (top panel of Fig. 5.9) imply that the star formation timescales are in general short (say $< 0.5$–1 Gyr), and that they are shortest (down to 0.1–0.3 Gyr) for the most massive ETGs. Based on these results, it has been estimated that ETGs with $\log(\mathcal{M}_\star/\mathcal{M}_\odot) > 10.5$ formed about 80–90% of their stellar mass at $z > 2$ (i.e. more than 10 Gyr ago), whereas ETGs with $\log(\mathcal{M}_\star/\mathcal{M}_\odot) < 10$ formed the same fraction of mass much later. This mass-dependent evolution (massive galaxies formed their stars earlier and faster) is called downsizing and it will be illustrated in §11.3.10.

The results of Fig. 5.9 are relative to the properties integrated over the entire galaxy size (§3.3). However, it is also possible to derive spatially resolved information, such as radial gradients of stellar properties, through integral field spectroscopy (§4.3.2). Typically, ETGs display gradients where the colours are redder in the inner regions. These gradients are usually ascribed to an increasing metallicity and $\alpha$-element abundance in the galaxy inner regions. This information is important to investigate how and when ETGs assembled their internal stellar component.

## Young Stars in Old ETGs

Even if old stars are the dominant population in most ETGs at $z \approx 0$, young stars are sometimes present. Space observations in the UV with *GALEX* showed that $\approx$ 15–30% of galaxies with elliptical morphology have colours which require rather recent episodes of star formation occurring $< 1$ Gyr ago. However, these cases, still consistently with the downsizing scenario, are mostly limited to lower-mass ellipticals, and the contribution of young stars to the total stellar mass is usually small ($\approx$ 1–3%). The presence of younger stars in a fraction of ETGs has been inferred also with optical spectroscopy in the so-called **E+A** (or **K+A**) **galaxies**. These are characterised by the simultaneous presence of K-star-like spectra (typical of old stars) and the strong Balmer line absorptions typical of A-type stars. The Balmer absorptions suggest that a secondary episode of star formation ceased less than about 0.5–1 Gyr ago, and therefore these galaxies represent cases with a younger population superimposed onto the underlying older one. E+A galaxies represent only $\approx$ 3% of the whole galaxy population with $\log(\mathcal{M}_\star/\mathcal{M}_\odot) > 10.2$ at $z \approx 0$. Also the radial gradients allow us to investigate the presence of multiple stellar populations within ETGs. The results show that a small fraction of ETGs have inverted gradients with blue cores due to the presence of younger stars in the central regions. In general, the fraction of ETGs with signatures of younger stars (sometimes called rejuvenated ETGs) seems to be larger in dense environments where galaxy mergers and interactions are thought to play an important role (§6.1).

### The UV Upturn

Observations of present-day ETGs showed also an unexpected phenomenon, called **UV upturn** (not to be confused with the above case of young stars in ETGs). This consists in an increase of the continuum flux from $\lambda \approx 2500$ Å to the Lyman limit ($\lambda = 912$ Å), and it is present in the majority of luminous ETGs. This has been a mystery ever since its first detection by the *Orbiting Astronomical Observatory* (*OAO-2*) because hot ($T \approx 25\,000$ K) stars are not expected in these galaxies, where the stellar populations are old, metal-rich and do not show signatures of ongoing star formation. The extended and smooth spatial distribution of the UV excess rules out the possibility of non-thermal radiation from an AGN (§3.6), whereas the lack of spectral features of hot stars excludes the presence of young massive stars as the origin of this phenomenon. Several explanations have been proposed in the context of old stars: low-metallicity old stars, extreme horizontal-branch stars, extreme helium enrichment and binary stars. However, the explanation of the phenomenon is still a matter of debate.

### The Initial Mass Function of ETGs

To conclude this overview on the stellar populations in ETGs, it is important to mention the stellar IMF (§8.3.9). This is an essential ingredient of galaxy evolution because of its deep links with star formation, chemical evolution and SN feedback processes. However, as explained in §4.6.1, the IMF can be reliably derived only within the Milky Way, where individual stars can be counted and their mass distribution derived. Due to this major limitation, it is customary to extrapolate IMFs that have been determined in the Milky Way (§8.3.9) to other galaxies. This might be inappropriate, especially in the case of ETGs that are known to have SFHs markedly different from that of the Milky Way. Based on indirect observational contraints such as stellar absorption lines sensitive to low-mass stars or other methods, some studies suggest that the IMF of some ETGs might indeed be different from the Galactic one.

## 5.2 Diffuse Matter

Compared to SFGs, ETGs are generally gas-poor. Nevertheless, thanks to the growing observational capabilities at different wavelengths, it is now well established that gas in all phases (molecular, neutral atomic and ionised) is typically present in ETGs. Here we review the main properties of this gas.

### 5.2.1 Hot Gas

By far the dominant gaseous component in the most massive ellipticals is the almost completely ionised plasma nearly at the system's virial temperature (§8.2.1), $10^6$–$10^7$ K. Such **hot gas halos** are detected in X-rays: their emission is thermal, mainly due to free–free (bremsstrahlung; §D.1.6) and free-bound (recombination; §D.1.5) continuum, and emission lines produced by radiative de-excitation of collisionally excited inner-shell electrons of highly ionised metals (§4.2.4 and §8.1.1). At a given position $\boldsymbol{x}$ in the galaxy, the combined emissivity (luminosity per unit frequency per unit volume, in units of erg s$^{-1}$ Hz$^{-1}$ cm$^{-3}$) from these mechanisms can be expressed as

$$j_\nu(\boldsymbol{x}) \equiv \frac{\mathrm{d}L}{\mathrm{d}\nu \mathrm{d}V} = n_\mathrm{e}(\boldsymbol{x}) \sum_{\mathrm{E}^k} n_{\mathrm{E}^k}(\boldsymbol{x}) \Lambda_\nu(\mathrm{E}^k, T), \tag{5.8}$$

where $\mathrm{E}^k$ is a $k$-times ionised ion of the generic element E, $n_\mathrm{e}$ is the electron number density, $n_{\mathrm{E}^k}$ is the number density of ion $\mathrm{E}^k$, $\Lambda_\nu$ is the emissivity per ion per electron at the frequency $\nu$, and the sum is performed over all ions of all elements. The X-ray emissivity (luminosity per unit volume, in units of erg s$^{-1}$ cm$^{-3}$) is

$$j_\mathrm{X}(\boldsymbol{x}) = \int_\mathrm{X} j_\nu(\boldsymbol{x}) \mathrm{d}\nu = n_\mathrm{e}(\boldsymbol{x}) n_\mathrm{t}(\boldsymbol{x}) \Lambda, \tag{5.9}$$

where $n_\mathrm{t}(\boldsymbol{x}) \equiv \sum_{\mathrm{E}^k} n_{\mathrm{E}^k}$ is the total number density of ions, $\Lambda$ is the radiative cooling function, depending on temperature $T$ and metallicity $Z$ (§8.1.1), and the integral is over an X-ray spectral band (§C.1). Assuming that the medium is everywhere optically thin, the X-ray surface brightness $I_\mathrm{X}$ is obtained by integrating $j_\mathrm{X}$ along the line of sight $x_\mathrm{los}$: for a spherical system with emissivity $j_\mathrm{X}(r)$,

$$I_\mathrm{X}(R) = 2 \int_R^\infty \frac{j_\mathrm{X}(r) r \, \mathrm{d}r}{\sqrt{r^2 - R^2}}, \tag{5.10}$$

where $R$ is the projected distance from the centre of the system and we have used $x_\mathrm{los}^2 = r^2 - R^2$ to change the integration variable from $x_\mathrm{los}$ to $r$. The total X-ray luminosity of the gaseous halo is

$$L_\mathrm{X} = \int j_\mathrm{X}(\boldsymbol{x}) \mathrm{d}^3\boldsymbol{x} = 2\pi \int_0^\infty I_\mathrm{X}(R) R \, \mathrm{d}R, \tag{5.11}$$

where the first integral is performed over the entire volume occupied by the hot gas and the second equality holds in circular symmetry. For ETGs, $L_\mathrm{X}$ is found in the range $10^{39}$–$10^{43}$ erg s$^{-1}$.

The most prominent feature of the X-ray spectrum continuum is the exponential cut-off at energy $h\nu \approx k_\mathrm{B}T$ (eq. D.18): this feature is very useful to determine the plasma temperature $T$. At the temperatures relevant to elliptical galaxies ($10^6$–$10^7$ K), the contribution from emission lines from highly ionised metals (in particular O, Ne, Mg, Si, S and Fe) is very important and often dominant (as happens for relatively low-temperature clusters of galaxies; see Fig. 6.9).

When spatially resolved spectroscopic X-ray data are available, it is possible to obtain profiles of gas temperature $T$ and density $n = n_\mathrm{t} + n_\mathrm{e}$ by deprojection, typically assuming spherical symmetry. Given that the X-ray surface brightness depends not only on density, but also on temperature and metallicity (through $\Lambda$; see eq. 5.9), to obtain $n$ from $I_\mathrm{X}$ one should first determine the temperature and metallicity profiles by fitting X-ray spectra in annuli at different distances from the galaxy centre. The deprojected temperature and metallicity profiles can then be used to deproject the observed X-ray surface brightness profile and obtain the intrinsic gas number density profile. However, the typical temperature and metallicity variations within the hot gas of ETGs are relatively small and the strongest dependence of surface brightness is on density, so in practice it is possible to get an estimate of the gas density distribution directly from the X-ray surface brightness map even in the absence of detailed spectral information. From the gas density distribution $\rho_{\text{gas}} = \mu m_{\text{p}} n$ it is possible to infer the total mass of hot gas $\mathcal{M}_{\text{gas}} = \int \rho_{\text{gas}}\,\text{d}V$ by integrating over the volume. The detected gas mass is typically a small fraction of the stellar mass within $R_{\text{e}}$, but extrapolations of $I_{\text{X}}$ to large radii (where X-rays are hard to detect, because of the low $n$) suggest that, at least in the brightest ellipticals, the total mass of hot gas can be of the order of or higher than the total stellar mass.

When observing elliptical galaxies in X-rays, it is important to separate contributions from the hot ISM and from stellar X-ray binaries. In lower-luminosity ellipticals the X-ray emission is typically dominated by such discrete sources, so it is not easy to estimate the amount of hot gas. Examples of X-ray images of ellipticals are shown in Fig. 5.10,

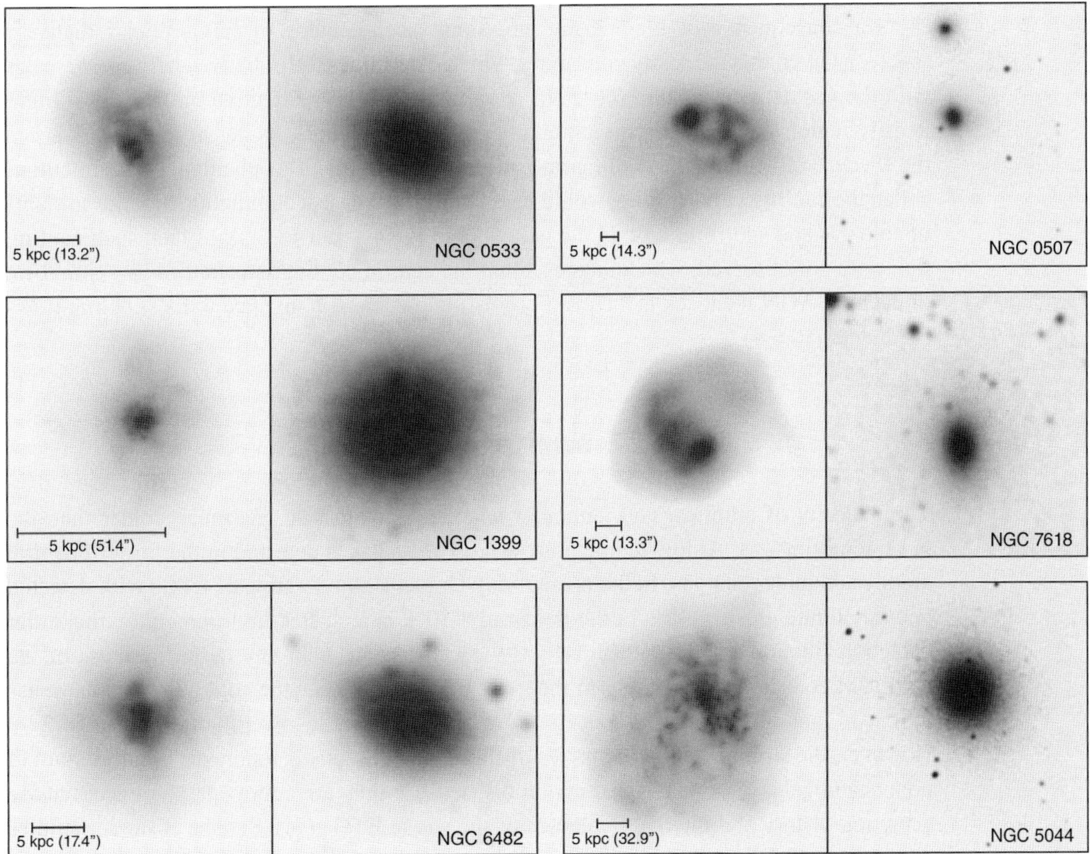


Fig. 5.10 X-ray and optical images of six elliptical galaxies. In each frame the X-ray image is in the *left panel* and the corresponding optical image is in the *right panel*. From the *Chandra X-ray Observatory*. Credit: T. Statler and S. Diehl (NASA/CXC/Ohio University; X-ray), and DSS (optical).

where they can be compared with the corresponding optical images. The distribution of the hot gas is often more extended than the stellar distribution. When the diffuse X-ray emission is observed spectroscopically, the chemical composition of the gas can be inferred by measuring the metal emission lines. The metallicity of the hot ISM of ellipticals is $Z \gtrsim 0.5Z_{\odot}$: the hot ISM is substantially enriched in metals, which suggests that it originated at least partly from gas lost by evolved stars. Though the gas ejected by stars is cold, it is expected to be rapidly thermalised to the temperature corresponding to the stellar velocity dispersion (of the order of the virial temperature), via shocks (§D.2.4) in the interactions with ejecta from other stars and with the hot ISM, and therefore to join the X-ray-emitting component.

The diffuse X-ray luminosity from hot gas in ellipticals correlates with other galaxy properties (see Tab. 5.2). In massive slow rotators the mass of the observed interstellar hot gas is comparable to the expected mass of gas lost by stars over the galaxy lifetime, while in lower-mass fast rotators the amount of interstellar hot gas falls short of the expectation. This suggests that only in the massive slow rotators are the physical conditions such that the gas lost by stars can be effectively retained and thermalised.

As a consequence of their X-ray emission, the hot halos of ETGs cool radiatively (§8.1.1). Given that the X-ray emissivity, which determines the radiative energy loss, scales with the gas density squared (eq. 5.9), cooling is especially effective in the central high-density regions. When this gravitationally confined gas cools, its density increases, so, in the absence of heating, the cooling rate increases with time. This phenomenon, sometimes called the **cooling catastrophe**, would lead to inflow of gas, accumulation of large amounts of cold gas and star formation in the galaxy core (§6.4.3 and §8.2.3). In fact, such **cooling flows** are not observed, which suggests that the central cooling is balanced by some form of heating. AGN feedback is the most promising mechanism invoked to halt cooling flows (§8.8 and §10.6.4).

### 5.2.2 Warm Gas, Cold Gas and Dust

The majority of elliptical and lenticular galaxies contain also gas much colder than the X-ray-emitting gas. At low temperatures ($T < 10^4$ K) gas is detected in the form of neutral atomic and molecular gas. In the present-day Universe, the fraction of ETGs with detected neutral atomic gas is $\approx$40% in the field and $\approx$ 10% in a cluster environment, with similar detection fraction for the molecular gas (see also §6.4.1). In any case, the mass of the cold ISM contributes negligibly to the mass budget of ETGs: the cold gas-to-stellar-mass ratios are found in the range $10^{-4}$–$10^{-2}$, orders of magnitude lower than in SFGs (§4.2.6). However, the structure and kinematics of these cold gaseous components are important to understand galaxy formation and evolution, because they are believed to be tracers of the accretion history and internal dynamics of the gas in ETGs (§10.8). The H I is sometimes observed in the form of discs or rings. In some cases the H I discs are small, well within the body of the stellar component of the galaxy, and are kinematically coupled to the stars. In a few cases, larger H I discs are observed (more extended than the stellar distribution), which are often kinematically decoupled from the stars. Molecular gas is found mainly in the central regions, being confined within $R_e$. Various molecular gas structures are observed, including central discs, rings and bars. The kinematic behaviour of such structures differs from galaxy to galaxy, ranging from corotating to counter-rotating with respect to the stars. In some cases the cold ISM structure is less ordered and the gas distribution appears filamentary and disturbed.

At higher temperatures, warm ($T \sim 10^4$ K) ionised gas is detected via optical emission lines (for instance H$\alpha$, [NII] and [OIII] lines; §4.2.3). The mass of ionised gas is of the order of $10^3$–$10^5 \mathcal{M}_\odot$ in normal ellipticals, so it is much less than the mass in stars and in hot gas. The origin of this warm gas is unclear (§8.1.4 and §10.6.4), but the observed line strengths suggest that emission is not powered only by collisional excitation. Different ionisation sources are believed to be responsible for ionising this warm gas: AGN emission (effective in the central regions), UV light from old stars (§5.1.3) and, in the most massive systems, interaction with the hot X-ray-emitting gas.

Another component of the ISM of ETGs is the interstellar dust, which is typically in the form of lanes that are easily detected, because they obscure optical light from background stars. Beside this interstellar dust, diffuse thermal emission from dust grains is also observed from circumstellar regions in ETGs (§3.2).

## 5.3 Mass Distribution

The mass distribution of ETGs can be estimated using different methods: the kinematics of stars, globular clusters and satellite galaxies, the properties of X-ray-emitting gas, and gravitational lensing. The results of these methods lead to the conclusion that, as found for SFGs (§4.3), most of the mass of ETGs is in the form of dark matter. The total mass profiles of ellipticals are harder to estimate than those of disc galaxies, because, especially at large radii, there are not kinematic tracers as good as the thin rotation-supported discs observed in SFGs. In massive ellipticals the density distribution is dominated by the stars in the centre and by the dark matter in the outskirts, similar to what happens in massive disc galaxies (§4.3.4). Dwarf ellipticals have dark-to-luminous mass ratios similar to normal ellipticals, so, also in this respect, they are distinct from less concentrated dwarfs such as dSphs (§6.3.1 and §6.3.2), which are much more dark matter-dominated. The methods described in this section allow one to estimate the total galaxy mass distribution, given by the sum of the stellar, gas and dark matter mass distributions.

### 5.3.1 The Virial Theorem Method

An order-of-magnitude estimate of the crossing time of stars in ETGs is $R_e/\sigma_0 \sim 10^7$ yr for $R_e \sim 1$ kpc and $\sigma_0 \sim 100\,\mathrm{km\,s^{-1}}$ (Tab. 5.1). The fact that this timescale is much shorter than the age of the stars ($\sim 10^9$–$10^{10}$ yr; §5.1.3) suggests that the system has had time to reach equilibrium (§8.9.2), because the stars have been orbiting for many periods.[4] A galaxy in equilibrium satisfies the **virial theorem**, which can be used to obtain an estimate of its total mass. Let us model the galaxy as a self-gravitating system with total (virial) mass $\mathcal{M}_{\rm vir}$ (§5.3.5), consisting of stars, gas and dark matter. In particular, the **kinetic energy** $K$ and the **gravitational potential energy** $W$ of such a system are related by the scalar virial theorem

$$2K + W = 0. \tag{5.12}$$

By definition, the **virial velocity dispersion** $\sigma_{\rm vir}$ and the **gravitational radius** $r_{\rm g}$ of the system are such that

$$K = \frac{1}{2}\mathcal{M}_{\rm vir}\sigma^2_{\rm vir} \tag{5.13}$$

and

$$W = -\frac{G\mathcal{M}^2_{\rm vir}}{r_{\rm g}}. \tag{5.14}$$

From the virial theorem (eq. 5.12) and the above two equations it follows that the total mass can be written as

$$\mathcal{M}_{\rm vir} = \frac{r_{\rm g}\sigma^2_{\rm vir}}{G}. \tag{5.15}$$

The quantities $\sigma_{\rm vir}$ and $r_{\rm g}$ are not directly observable, but, under reasonable assumptions, they can be related to the observed galaxy size and velocity dispersion. The relation between the central velocity dispersion of stars $\sigma_0$ (eq. 5.3) and $\sigma_{\rm vir}$ can be written as $\sigma_0^2 = a\sigma^2_{\rm vir}$, where $a$ depends on the spatial distribution of stars and dark matter, and on the velocity distribution of stars. The observable used as a proxy for $r_{\rm g}$ is the effective radius, which can be written as $R_{\rm e} = br_{\rm g}$ (where $b$ depends on the spatial distribution of stars and dark matter). In terms of $\sigma_0$ and $R_{\rm e}$ eq. (5.15) becomes

$$\mathcal{M}_{\rm vir} = \frac{k_{\rm vir}R_{\rm e}\sigma_0^2}{G}, \tag{5.16}$$

where $k_{\rm vir} \equiv 1/(ab)$ is the so-called **virial coefficient**. The value of $k_{\rm vir}$ can vary significantly from galaxy to galaxy, but a reference[5] number is $k_{\rm vir} \approx 5$. For this reason it is customary to define the **dynamical mass** of a spheroid as

$$\mathcal{M}_{\rm dyn} \equiv \frac{5R_{\rm e}\sigma_0^2}{G} \simeq 4.65 \times 10^{10}\left(\frac{R_{\rm e}}{\rm kpc}\right)\left(\frac{\sigma_0}{200\,{\rm km\,s^{-1}}}\right)^2 \mathcal{M}_\odot, \tag{5.17}$$

which is often used as a proxy for $\mathcal{M}_{\rm vir}$. In fact, though the estimate of the total mass using the virial theorem gives a useful reference measure, the uncertainty on $k_{\rm vir}$ (which depends

[4] In principle, a substantial fraction of the stars, though old, may have been accreted recently in a merger (§8.9), but in this case the morphology of the galaxy would be disturbed. Galaxies with no major sign of ongoing interaction can be safely assumed to be close to equilibrium.

[5] For instance, $k_{\rm vir} \simeq 4.7$ for a spherical, non-rotating, isotropic, one-component (star-only) galaxy with $R^{1/4}$ surface brightness profile (eq. 5.2) and position-independent mass-to-light ratio, when $\sigma_0$ is measured within a circular aperture of radius $R_{\rm e}/8$. The value of $k_{\rm vir}$ for spherical, non-rotating, isotropic, one-component stellar systems with Sérsic profiles (eq. 3.1) increases for decreasing Sérsic index $n$: $1.5 \lesssim k_{\rm vir} \lesssim 7.5$ for $10 \gtrsim n \gtrsim 2$.

on stellar orbits, dark and luminous matter distribution, galaxy shape and rotation support) prevents the accurate measurement of $\mathcal{M}_{\rm vir}$ with this method. However, when spatially resolved kinematic information is not available, eq. (5.16) is the only estimate of the total mass that can be obtained from observations of the galaxy stellar light. Therefore, mass estimates based on the virial theorem can be especially useful when comparing high- and low-redshift galaxies (see §11.1.7 and Tab. 11.2).

### 5.3.2 Dynamical Modelling

More accurate mass determinations using spatially resolved stellar kinematics are those based on dynamical modelling, in which galactic models are built and compared with the observed galaxy properties. Dynamical modelling allows us to estimate not only the galaxy virial mass $\mathcal{M}_{\rm vir}$, but also the total mass density distribution $\rho_{\rm tot}(\boldsymbol{x})$, which is the sum of the mass density distributions of all components (stars, dark matter and gas).

The simplest of these methods is **Jeans modelling**, based on the solution of the **Jeans equations** (Jeans, 1915) for a **tracer population**, that is, any component of the system in equilibrium in the total gravitational potential (for instance, stars, globular clusters, planetary nebulae or satellite galaxies). Let us consider a spherical galaxy in equilibrium and take a non-rotating spherical tracer population with density distribution $\rho(r)$. For this population the velocity dispersion tensor (eq. C.40) in spherical coordinates $(r, \vartheta, \varphi)$ is diagonal with components $\sigma_r^2 \equiv \sigma_{11}^2$, $\sigma_\vartheta^2 \equiv \sigma_{22}^2$ and $\sigma_\varphi^2 \equiv \sigma_{33}^2$, when $v_1 = v_r$, $v_2 = v_\vartheta$ and $v_3 = v_\varphi$. The radial velocity dispersion $\sigma_r(r)$ is related to $\rho(r)$ via the spherically symmetric Jeans equation

$$\frac{\mathrm{d}(\rho\sigma_r^2)}{\mathrm{d}r} + \frac{2\beta\rho\sigma_r^2}{r} = -\rho\frac{G\mathcal{M}}{r^2}, \tag{5.18}$$

where $\mathcal{M}(r)$ is the *total* mass within a radius $r$, obtained by integrating $\rho_{\rm tot}(r)$ as in eq. (4.26, first equality), and

$$\beta(r) \equiv 1 - \frac{\sigma_\vartheta^2 + \sigma_\varphi^2}{2\sigma_r^2} \tag{5.19}$$

is the **anisotropy parameter**. The definition of $\beta$ is such that $-\infty \leq \beta \leq 1$ and the velocity distribution is isotropic for $\beta = 0$, radially anisotropic for $\beta > 0$ (when $\beta = 1$ all orbits are radial) and tangentially anisotropic for $\beta < 0$ (when $\beta = -\infty$ all orbits are circular).

We want to calculate the tracer's line-of-sight velocity dispersion $\sigma_{\rm los}(R)$, where $R$ is the projected radial coordinate in the plane of the sky (§C.8). We first compute the surface density of the tracer population (eq. C.42), which in spherical symmetry can be written as

$$\Sigma(R) = 2\int_R^\infty \frac{\rho(r) r\,\mathrm{d}r}{\sqrt{r^2 - R^2}} \tag{5.20}$$

(see eq. 5.10). In the absence of rotation, the line-of-sight velocity dispersion at a distance $R$ from the centre is given by eq. (C.46). Combining this equation with the definition of $\beta$

(eq. 5.19) and considering the geometry of the system, it can be shown that $\sigma_{\rm los}(R)$ is given by

$$\sigma^2_{\rm los}(R)=\frac{2}{\Sigma(R)}\int_R^\infty\left[1-\beta(r)\frac{R^2}{r^2}\right]\frac{\rho(r)\sigma_r^2(r)r\,{\rm d}r}{\sqrt{r^2-R^2}}. \tag{5.21}$$

From the Jeans equation (eq. 5.18), assuming the anisotropy profile $\beta(r)$, it is possible to derive the total mass profile $\mathcal{M}(r)$ from the tracer's density distribution $\rho(r)$ and radial velocity dispersion profile $\sigma_r(r)$. However, $\rho(r)$ and $\sigma_r(r)$ are intrinsic quantities, which are not observable. The quantities that can be directly estimated from the observations are $\Sigma(R)$ and $\sigma_{\rm los}(R)$. By deprojecting $\Sigma(R)$ (eq. 5.20) via the **Abel inversion**,[6] one obtains the tracer's intrinsic density distribution

$$\rho(r)=-\frac{1}{\pi}\int_r^\infty\frac{{\rm d}\Sigma}{{\rm d}R}\frac{{\rm d}R}{\sqrt{R^2-r^2}}. \tag{5.26}$$

Similarly, by deprojecting $\sigma_{\rm los}(R)$ one can obtain $\sigma_r(r)$. However, deprojection of the observed profiles can be inconvenient, especially when the data are noisy. An alternative strategy is to build sets of models of given $\rho(r)$, $\beta(r)$ and $\mathcal{M}(r)$, compute $\Sigma(R)$ and $\sigma_{\rm los}(R)$ predicted by these models, compare these with the observed profiles of $\Sigma$ and $\sigma_{\rm los}$, and find the model that best represents the data.

The result of the above Jeans modelling depends crucially on the assumption on the anisotropy of the velocity distribution of the tracer population, which is parameterised by $\beta(r)$. Different choices of $\beta$ can lead to different estimates of the total mass, a phenomenon which is usually referred to as **mass–anisotropy degeneracy**. To break this degeneracy it is possible to use information on the line profiles (§5.1.2) or to combine Jeans modelling with other probes such as X-ray (§5.3.3) or gravitational lensing (§5.3.4) data.

When stars representative of the bulk of the stellar component of the galaxy are used as tracers (typically measuring integrated light profiles and spectra at optical wavelengths), Jeans modelling can probe the mass distribution within $\approx 2R_{\rm e}$, where the galaxy surface brightness is higher. The outer mass distribution is better probed by other tracers such as globular clusters, planetary nebulae and satellite galaxies. A Jeans analysis similar to

[6] The Abel inversion relates two functions $f$ and $g$ as follows. Given

$$f(x)=\int_x^\infty\frac{g(t){\rm d}t}{(t-x)^\alpha}, \tag{5.22}$$

with $0<\alpha<1$, the function $g$ can be written as

$$g(t)=-\frac{\sin(\pi\alpha)}{\pi}\int_t^\infty\frac{{\rm d}f}{{\rm d}x}\frac{{\rm d}x}{(x-t)^{1-\alpha}}. \tag{5.23}$$

When $\alpha=1/2$, substituting $t=r^2$ and $x=R^2$, we have

$$f(R^2)=2\int_R^\infty\frac{g(r^2)r\,{\rm d}r}{\sqrt{r^2-R^2}} \tag{5.24}$$

and

$$g(r^2)=-\frac{1}{\pi}\int_r^\infty\frac{{\rm d}f(R^2)}{{\rm d}R}\frac{{\rm d}R}{\sqrt{R^2-r^2}}. \tag{5.25}$$

Rewriting $f(R^2)$ as $\Sigma(R)$ and $g(r^2)$ as $\rho(r)$ in the above two equations, we get eqs. (5.20) and (5.26).

the one described above can also be performed by relaxing the assumption of spherical symmetry and considering axisymmetric systems. However, the Jeans analysis has intrinsic limitations due to the fact that in general it is not guaranteed that the model is consistent (i.e. having an everywhere positive distribution function; §C.8). To overcome this problem it is possible to resort to more sophisticated stellar dynamical models. For instance, in the so-called **Schwarzschild method** (Schwarzschild, 1979), based on orbit superposition, a library of numerically integrated orbits in a given gravitational potential is created and these orbits, opportunely weighted, are combined to reproduce the observed kinematic and photometric properties of the galaxy. Alternatively, one can use **distribution function-based methods**, in which consistent models are built starting from an analytic distribution function for the tracer population, depending either on the classical integrals of motion (such as energy and angular momentum) or on the action integrals (see Binney and Tremaine, 2008).

### 5.3.3 Measuring Mass with X-Ray-Emitting Gas

Let us consider a spherically symmetric galaxy in which the hot gas is in equilibrium in the total gravitational potential only thanks to the presence of pressure gradients. In this case, the Euler equation of hydrodynamics (eq. D.25) reduces to the equation of hydrostatic equilibrium (§4.6.2),

$$\frac{\mathrm{d}P}{\mathrm{d}r} = -\rho_{\mathrm{gas}}\frac{G\mathcal{M}}{r^2}, \tag{5.27}$$

where $P(r)$ is the gas pressure, $\rho_{\mathrm{gas}}(r)$ is the gas density and $\mathcal{M}(r)$ is the total mass within $r$. Using the ideal gas equation $P = \rho_{\mathrm{gas}} k_{\mathrm{B}} T/(\mu m_{\mathrm{p}})$ (§D.2.2), where $T(r)$ is the gas temperature at $r$, we get

$$\mathcal{M}(r) = -\frac{k_{\mathrm{B}} T r}{\mu m_{\mathrm{p}} G}\left(\frac{\mathrm{d}\ln\rho_{\mathrm{gas}}}{\mathrm{d}\ln r} + \frac{\mathrm{d}\ln T}{\mathrm{d}\ln r}\right). \tag{5.28}$$

We have seen in §5.2.1 that, using X-ray observations, it is possible to estimate $T(r)$ and $\rho_{\mathrm{gas}}(r)$, so the above equation can be used to recover the total mass profile of massive ellipticals. Usually, the radial variations of $T$ are not strong (typically smaller than a factor of 2), so the gaseous halos of ellipticals are quasi-isothermal. When the observational data are not good enough to constrain the temperature profile, the gas is assumed isothermal at temperature $T_0$, so $\mathcal{M}(r)$ is given by eq. (5.28) with $T = T_0$ and $\mathrm{d}\ln T/\mathrm{d}\ln r = 0$. An advantage of using the hot gas to measure $\mathcal{M}(r)$ is that this method, at variance with Jeans modelling, does not suffer from the mass–anisotropy degeneracy (§5.3.2), because the hot ISM is collisional and thus well described as a fluid with isotropic pressure (§D.2.2). However, a potential drawback of the X-ray-emitting gas method is that it is not guaranteed that the gas is actually in hydrostatic equilibrium: significant bulk motions such as inflows, outflows and rotation might be present and very difficult to constrain observationally, due to the relatively poor energy resolution of X-ray instruments (typically corresponding to gas speeds of at least a few hundred $\mathrm{km\,s^{-1}}$).

### 5.3.4 Measuring Mass with Gravitational Lensing

**Gravitational lensing** is an effect predicted by general relativity. Photons travel in a curved space-time, so the presence of mass deflects the light path similarly to an optical lens. When a galaxy acts as a gravitational lens, useful constraints on its mass distribution come from the analysis of the deflected images of background sources. When multiple images of the same source are produced, we speak of **strong gravitational lensing**. Strong lensing produces dramatic distortions of the sources, often in the form of beautiful **gravitational arcs** and **rings** (the so-called **Einstein rings**). The arcs are typically tangential, but sometimes it is possible to observe an image elongated towards the lens centre (which is called a radial arc). A spectacular example of double Einstein rings around a massive elliptical galaxy acting as a lens is shown in Fig. 5.11 (left). In order to have strong lensing features, high surface mass density is required, so strong lensing is typically caused by the central regions of galaxies or of galaxy clusters.

When the distortion is a slight deformation of the source (for instance a circular source is transformed into an elliptical image) we speak of **weak gravitational lensing**. Weak lensing produces small distortions of the shape and orientation of background galaxies, which typically appear to be tangentially stretched with respect to the lens mass distribution. Weak lensing does not require high surface density, so it is sensitive also to the mass distribution in the outer regions of galaxies. In fact, any line of sight in the Universe is weakly lensed, a phenomenon known as **cosmic shear**.

In configurations relevant to lens galaxies, the size of the lens is much smaller than its distances to the sources and to the observer. Therefore, sufficiently accurate results can be obtained in the so-called **thin lens approximation**, in which it is assumed that all the mass of the lens is distributed within a plane (called the **lens plane**) passing through the centre of the galaxy and perpendicular to the line of sight. In the following sections and in §C.9 we report some of the basic equations[7] describing the gravitational lensing effect in the limit of small deflection angle and thin lens (which is a good approximation in most applications to lens galaxies). We start from the case of a circularly symmetric lens before discussing the general case of a lens of arbitrary geometry.

#### Circularly Symmetric Lens

Let us consider the case in which the lens is, in the thin lens approximation, a circularly symmetric mass distribution (for instance, a spherical galaxy). The symmetry of the configuration implies that the path of the photon is confined to the plane defined by the initial direction of the photon and by the centre of the lens. The trajectory of the photon can be thought of as composed of two straight lines intersecting in the lens plane. The geometric configuration of this lens system is represented in Fig. 5.11 (right), where $D_{\mathrm{OS}}$, $D_{\mathrm{OL}}$ and $D_{\mathrm{LS}}$ are the angular diameter distances (§2.1.4) from the observer to the source,

[7] For a more detailed treatment of gravitational lensing we refer the reader to Wambsganss (1998) and Mollerach and Roulet (2002).

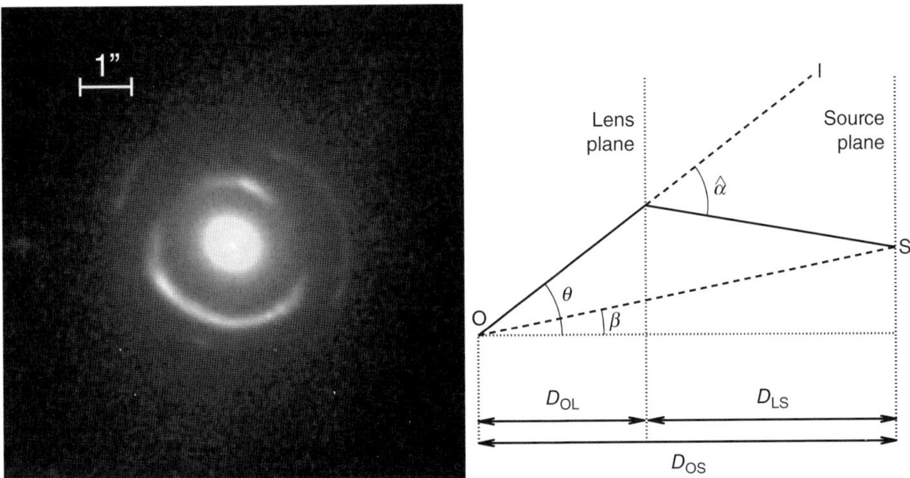


Fig. 5.11

*Left panel. HST* image of the gravitational lens system SDSSJ0946+1006. The lens galaxy (the central ellipse in the image) is a massive ($\mathcal{M}_\star \simeq 5.5 \times 10^{11} \mathcal{M}_\odot$) ETG at redshift $z \simeq 0.222$. Two structures of gravitational arcs are observed: the inner Einstein ring is produced by a source galaxy at $z \simeq 0.609$, while the outer ring is produced by a more distant source galaxy, at $z \simeq 2.41^{+0.04}_{-0.21}$ (no spectroscopic redshift measurement is available for this second source, so the redshift is estimated photometrically; §11.1.1). At the distance of the lens galaxy 1 arcsec $\simeq$ 3.6 kpc. Adapted from Sonnenfeld et al. (2012). © AAS, reproduced with permission. Courtesy of A. Sonnenfeld. *Right panel.* Geometry of the gravitational lensing in the thin lens approximation. The solid line indicates the path of photons emitted by the source (S), deflected in the lens plane (distant $D_{LS}$ from the source plane) by an angle $\hat{\alpha}$, and reaching the observer (O; distant $D_{OS}$ and $D_{OL}$ from the source and the lens planes, respectively). Taking as reference the optical axis (horizontal dotted line), the image (I) is observed at an angle ($\theta$) different from the angle ($\beta$) at which the source would be observed if undeflected.

from the observer to the lens and from the lens to the source, respectively. We define the **optical axis** of the lens system as the line from the observer to the centre of the lens.

Consider a source that, in the absence of lensing effects, would be observed at an angle $\beta$ with respect to the optical axis. A photon emitted by the source and passing at closest approach at a distance $R$ from the lens centre is deflected by an angle $\hat{\alpha}$ (the so-called **deflection angle**) and is therefore observed at an angle $\theta$. From simple geometric relations (see Fig. 5.11, right), in the limit of very small angles $\hat{\alpha}$, $\beta$ and $\theta$, we get

$$\theta D_{OS} = \beta D_{OS} + \hat{\alpha} D_{LS}. \tag{5.29}$$

This equation can be written in the following form, known as the **lens equation**:

$$\beta = \theta - \alpha, \tag{5.30}$$

where $\alpha \equiv \hat{\alpha} D_{LS}/D_{OS}$ is the **reduced deflection angle**.

It can be shown that the deflection angle of a circularly symmetric lens is

$$\hat{\alpha} = \frac{4G\mathcal{M}_{proj}(R)}{Rc^2}, \tag{5.31}$$

where $R = \theta D_{\rm OL}$ is the distance from the centre of the lens in the lens plane, and

$$\mathcal{M}_{\rm proj}(R) = 2\pi \int_0^R R'\Sigma(R')\mathrm{d}R' \tag{5.32}$$

is the projected mass within a cylinder of radius $R$. Here $\Sigma$ is the surface mass density of the lens. It follows that the reduced deflection angle is

$$\alpha = \frac{4G\mathcal{M}_{\rm proj}}{c^2\theta}\frac{D_{\rm LS}}{D_{\rm OL}D_{\rm OS}}, \tag{5.33}$$

where $\mathcal{M}_{\rm proj} = \mathcal{M}_{\rm proj}(\theta D_{\rm OL})$ and eq. (5.30) can be written as

$$\theta^2 - \beta\theta - \theta_{\rm E}^2 = 0, \tag{5.34}$$

where

$$\theta_{\rm E} \equiv \frac{R_{\rm E}}{D_{\rm OL}} \tag{5.35}$$

is the **Einstein angle** and the **Einstein radius** $R_{\rm E}$ is such hat

$$R_{\rm E} = \sqrt{\frac{4G\mathcal{M}_{\rm proj}(R_{\rm E})}{c^2}\frac{D_{\rm LS}D_{\rm OL}}{D_{\rm OS}}}. \tag{5.36}$$

When $\beta = 0$ (observer, lens and source lie on a straight line) the image is a ring of angular radius $\theta_{\rm E}$ (Einstein ring). The Einstein radius $R_{\rm E}$ is the characteristic radius of strong gravitational lensing.

One of the simplest possible models for a lens galaxy is a **singular isothermal sphere** with mass density distribution (eq. 4.25 with $\delta = 2$)

$$\rho(r) = \rho_0\left(\frac{r}{r_0}\right)^{-2} = \frac{v_{\rm c}^2}{4\pi G r^2}, \tag{5.37}$$

where $r_0$ is a reference radius, $\rho_0 \equiv \rho(r_0)$, and the constant $v_{\rm c}$ is the circular speed (eq. 4.23), which in this case is independent of the spherical radius $r$ (eq. 4.27 with $\delta = 2$). The singular isothermal sphere has mass profile

$$\mathcal{M}(r) = \frac{v_{\rm c}^2}{G}r, \tag{5.38}$$

which diverges at large radii. The gravitational potential generated by the density distribution (5.37) is

$$\Phi(r) = v_{\rm c}^2 \ln\left(\frac{r}{r_0}\right) + \Phi_0, \tag{5.39}$$

where $\Phi_0 \equiv \Phi(r_0)$. If a self-gravitating singular isothermal sphere is realised with particles in equilibrium with isotropic velocity distribution, the one-dimensional (1D) velocity dispersion $\sigma = v_{\rm c}/\sqrt{2}$ is independent of radius[8] (hence the term isothermal).

[8] For an isotropic system $\sigma = \sigma_r = \sigma_\vartheta = \sigma_\varphi$. The reader can verify that for a self-gravitating singular isothermal sphere $\sigma_r = v_{\rm c}/\sqrt{2}$ satisfies eq. (5.18) with $\beta = 0$.

When projected along a line of sight, the singular isothermal sphere has projected mass

$$\mathcal{M}_{\rm proj}(R) = \frac{\pi v_{\rm c}^2}{2G} R, \tag{5.40}$$

where we have used eqs. (5.20), (5.32) and (5.37). The deflection angle of a singular isothermal sphere is

$$\hat{\alpha} = \frac{4\pi\sigma^2}{c^2}, \tag{5.41}$$

independent of radius: from this equation it is apparent that the deflection angle is typically small (because $\sigma \ll c$) and that the strongest lensing effects are expected for galaxies with high $\sigma$. Combining eqs. (5.35), (5.36) and (5.40), we can write the Einstein angle of a singular isothermal sphere lens as

$$\theta_{\rm E} = \frac{4\pi\sigma^2}{c^2}\frac{D_{\rm LS}}{D_{\rm OS}} \simeq 2.6 \left(\frac{D_{\rm LS}}{D_{\rm OS}}\right)\left(\frac{\sigma}{300\,{\rm km\,s^{-1}}}\right)^2 \text{ arcsec}. \tag{5.42}$$

Eq. (5.42) shows that massive ellipticals can produce gravitational arcs on angular scales of the order of an arcsec (see Tab. 5.1).

## Effect of Convergence and Shear in Lens Galaxies

In the more general case of a non-circularly symmetric lens, the mapping between the sources and the images can be described, to a first approximation, in terms of the **convergence** $\kappa$ and the **shear** $\gamma = (\gamma_1, \gamma_2)$, which are functions of the angular position in the plane of the sky $\boldsymbol{\theta} = (\theta_1, \theta_2)$, where $\theta_1 \equiv x_1/D_{\rm OL}$ and $\theta_2 \equiv x_2/D_{\rm OL}$, with $(x_1, x_2)$ Cartesian coordinates labelling the lens plane. Here we briefly describe the effect of convergence and shear in lens galaxies: a few more details on gravitational lensing by non-circularly symmetric lenses are given in §C.9.

For a lens of surface mass density $\Sigma(\boldsymbol{R})$, we define the line-of-sight projected gravitational potential

$$\Phi_{\rm proj}(\boldsymbol{R}) = \int \Phi(\boldsymbol{r}) {\rm d}x_{\rm los}, \tag{5.43}$$

related to $\Sigma$ by the 2D Poisson equation

$$\nabla^2 \Phi_{\rm proj} = \frac{1}{D_{\rm OL}^2}\left(\frac{\partial^2 \Phi_{\rm proj}}{\partial \theta_1^2} + \frac{\partial^2 \Phi_{\rm proj}}{\partial \theta_2^2}\right) = 4\pi G \Sigma. \tag{5.44}$$

The convergence is defined as

$$\kappa(\boldsymbol{\theta}) \equiv \frac{1}{2}\left(\psi_{11} + \psi_{22}\right), \tag{5.45}$$

where

$$\psi_{ij} \equiv \frac{\partial^2 \psi}{\partial \theta_i \partial \theta_j} \tag{5.46}$$

and

$$\psi \equiv \frac{2}{c^2} \frac{D_{\rm LS}}{D_{\rm OS} D_{\rm OL}} \Phi_{\rm proj} \tag{5.47}$$

is the normalised projected gravitational potential. Using eq. (5.44) the convergence can be expressed in terms of the surface mass density as

$$\kappa(\boldsymbol{\theta}) = \frac{\Sigma(\boldsymbol{\theta})}{\Sigma_{\rm cr}}, \tag{5.48}$$

where

$$\Sigma_{\rm cr} \equiv \frac{c^2 D_{\rm OS}}{4\pi G D_{\rm OL} D_{\rm LS}} \tag{5.49}$$

is the so-called **critical surface density**. The two components of the shear are

$$\gamma_1 \equiv \frac{1}{2}\left(\psi_{11} - \psi_{22}\right) \tag{5.50}$$

and

$$\gamma_2 \equiv \psi_{12}. \tag{5.51}$$

The convergence $\kappa$ is responsible for changing the size of the image, but does not affect the shape. The shear $\gamma$ is responsible for the distortion of the image. A sufficient condition to produce multiple images is that at some point in the lens plane $\kappa > 1$, i.e. $\Sigma(\boldsymbol{\theta}) > \Sigma_{\rm cr}$, that is, the surface density exceeds the critical value defined by eq. (5.49). In the presence of shear, multiple images are produced even for lower surface densities, that is when $\kappa > 1 - \gamma$. In the regime of weak gravitational lensing ($\Sigma < \Sigma_{\rm cr}$), the effect of the shear is to systematically modify the ellipticity of lensed galaxies. For instance, a circular source is deformed into an ellipse with axis ratio

$$\frac{b}{a} = \frac{1-\kappa-\gamma}{1-\kappa+\gamma}. \tag{5.52}$$

When a sufficient number of weakly lensed images are available it is possible to calculate, in a given patch of the image, the average ellipticity and position angle of the distorted sources. When such an ellipticity map is constructed, it is possible to derive shear and convergence maps, i.e. to measure $\kappa(\boldsymbol{\theta})$ and $\gamma(\boldsymbol{\theta})$. An estimate of the mass surface density of the lens $\Sigma(\boldsymbol{R}) = \kappa(\boldsymbol{\theta})\Sigma_{\rm cr}$, where $\boldsymbol{R} = D_{\rm OL}\boldsymbol{\theta}$, is therefore obtained, because for each source $\Sigma_{\rm cr}$ (eq. 5.49) depends only on $D_{\rm LS}$, $D_{\rm OL}$ and $D_{\rm OS}$, which are known if the redshifts of the source and of the lens are known.

## Combining Weak and Strong Gravitational Lensing

From the above discussion it is clear that by measuring the properties of the lensed images, we can estimate the mass distribution of a galaxy that acts as a lens (see Treu, 2010, for a review). When we have a strong lensing arc, measuring the angular distance of the arc from the centre ($\theta_{\rm E}$) and the redshifts of the source and of the lens (hence $D_{\rm LS}$, $D_{\rm OL}$ and $D_{\rm OS}$ for a given cosmological model; §2.1.4), we get a measurement of the projected mass within $R_{\rm E}$ (eq. 5.36), related to $\theta_{\rm E}$ by eq. (5.35). The combination of

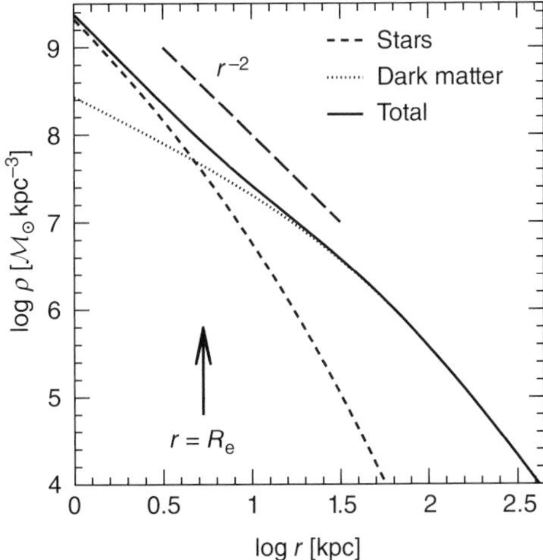


Fig. 5.12 Average stellar (dashed curve), dark matter (dotted curve) and total (solid curve) mass density profiles of a sample of 22 lens ETGs, based on a combination of strong and weak lensing measurements. The arrow indicates the average effective radius. For comparison the power law $r^{-2}$ is also plotted (long-dashed line). Data from Gavazzi et al. (2007).

strong lensing and kinematic data can be used to recover the total mass density profile of a lens galaxy in the central regions, because strong lensing measurements break the mass–anisotropy degeneracy (§5.3.2). When weak lensing measurements are available, from the convergence map it is possible to estimate the surface mass density of the lens galaxy also in the galaxy outskirts. However, the weak lensing signal produced by an individual galaxy is in general so low that useful measurements are obtained only by stacking (i.e. combining by superposition) images of several lens galaxies with similar properties. An example of the result of the combination of weak and strong gravitational lensing analyses, in addition to measurements of central velocity dispersion, is given in Fig. 5.12, showing the average stellar, dark matter and total mass density profiles of a sample of lens ETGs.

### 5.3.5 Baryonic, Dark and Total Mass Distributions

Thanks to the application of the methods described above to observational data, we have now a census of the luminous and dark mass density distributions of present-day ETGs. Provided that a 'galaxy boundary' is defined, from the estimate of the density distribution it is possible to compute the integrated total mass of a galaxy. Assuming for simplicity a spherical mass distribution, it is convenient to take as boundary the virial radius[9] $r_{vir}$ (§7.3.2), so the integrated mass of the galaxy is the virial mass $\mathcal{M}_{vir} \equiv \mathcal{M}(r_{vir})$, where $\mathcal{M}(r)$ is the total mass contained within a sphere of radius $r$ with origin at the galaxy centre. Similar to the total luminosities, also the virial masses of ETGs span a large interval ($10^8\mathcal{M}_\odot \lesssim \mathcal{M}_{vir} \lesssim 10^{13}\mathcal{M}_\odot$; Tab. 5.1). We have seen above that it is possible

[9] The virial radius must not be confused with the gravitational radius $r_g$ defined by eq. (5.14).

to estimate $\mathcal{M}_\star$ (from SED fitting, under some assumption on the IMF; §3.3, §8.6 and §11.1.3) and $\mathcal{M}_{gas}$ (from X-ray observations; §5.2.1). The total mass is the sum of the mass in stars, gas and dark matter ($\mathcal{M}_{vir} = \mathcal{M}_\star + \mathcal{M}_{gas} + \mathcal{M}_{DM}$), so a measure of $\mathcal{M}_{vir}$ gives an estimate of the dark matter mass $\mathcal{M}_{DM} = \mathcal{M}_{vir} - \mathcal{M}_\star - \mathcal{M}_{gas}$. The characteristic mass components of different classes of ETGs are summarised in Tab. 5.1. The baryon fraction $(\mathcal{M}_\star + \mathcal{M}_{gas})/\mathcal{M}_{vir}$ is of the order of $10^{-2}$ in ellipticals (see also §10.10). However, similar to what is found in spirals (§4.3.4), the baryons are more concentrated than the dark matter: in ellipticals the baryon fraction measured within $R_e$ is 60–90%.

The total mass density distribution $\rho_{tot}(r)$ of massive ellipticals is well approximated, out to $\approx 4R_e$, by a power law $\rho_{tot} \propto r^{-\gamma}$, with $\gamma \approx 2$ (similar to the density profile of a singular isothermal sphere; eq. 5.37). The density distribution is typically dominated by stars for $R \lesssim R_e$ and by dark matter for $R \gtrsim R_e$ (Fig. 5.12). Unfortunately, the exact radial dependence of the dark matter density profile (§7.5.2) is hard to determine observationally, especially in the inner regions, where the stellar mass density is dominant, mainly because of uncertainties in the stellar IMF. In the absence of observational constraints on the IMF, different assumptions on the IMF (§8.3.9) lead to differences in the stellar mass of a factor of $\approx 2$.

## 5.4 Structural and Kinematic Scaling Relations

ETGs are observed to follow several scaling relations among global parameters, analogous to those of SFGs (§4.4). In §5.1.3 we have presented scaling relations involving the properties of the stellar populations. Here we focus on structural and kinematic scaling laws.

### 5.4.1 The Fundamental Plane

One of the most studied scaling relations of ETGs is the so-called **fundamental plane** (Djorgovski and Davis, 1987; Dressler et al., 1987), which relates the size, velocity dispersion and surface brightness of elliptical galaxies. The fundamental plane is an empirical correlation that defines a plane in the 3D parameter space ($\log R_e$, $\log \sigma_0$, $\log\langle I\rangle_e$), where $R_e$ is the effective radius, $\sigma_0$ is the central velocity dispersion and

$$\langle I\rangle_e \equiv \frac{L}{2\pi R_e^2} \tag{5.53}$$

is the average surface brightness within $R_e$ ($L$ is the galaxy luminosity). The fundamental plane can be written as

$$\log R_e = \alpha \log \sigma_0 + \beta \log\langle I\rangle_e + \gamma, \tag{5.54}$$

where the values of the coefficients $\alpha$ and $\beta$ depend on the photometric band of the observations, with reference ranges $1 \lesssim \alpha \lesssim 1.4$ and $-0.9 \lesssim \beta \lesssim -0.75$ in the optical, and $\gamma$ is a normalisation whose value depends on the adopted units (see Fig. 5.13, left). In a

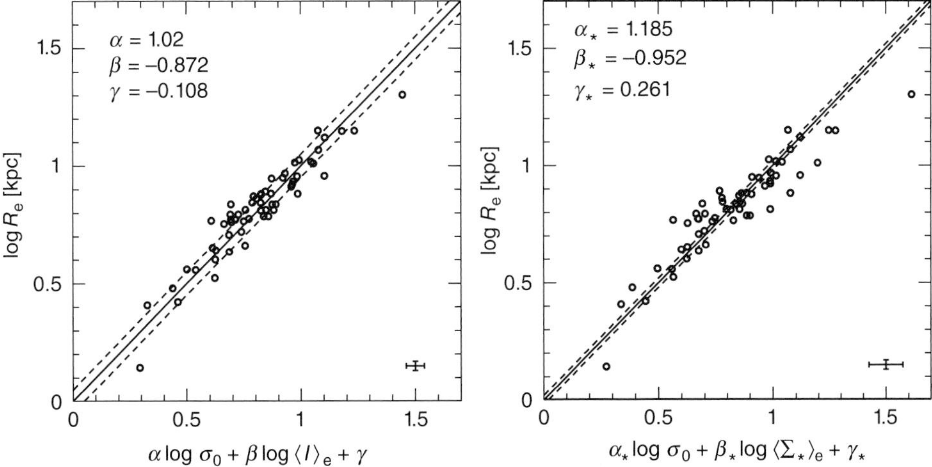


Fig. 5.13 Edge-on views of the fundamental plane (*left panel*) and of the stellar mass fundamental plane (*right panel*) for the Sloan Lens ACS Survey sample of present-day ETGs (circles). Here the central velocity dispersion $\sigma_0$ (in units of 100 km s$^{-1}$) is measured within $R_e/2$, the surface brightness $\langle I \rangle_e$ (in units of $10^9 L_\odot$ kpc$^{-2}$) and the effective radius $R_e$ (in kpc) are measured in the $V$ band, and the stellar mass surface density (in units of $10^9 \mathcal{M}_\odot$ kpc$^{-2}$) is derived assuming a Chabrier IMF (§8.3.9). The solid lines are the best fits (with values of the parameters indicated in each panel) and the dashed lines give a measure of the estimated intrinsic scatter. In both panels the cross in the bottom right corner indicates typical error bars. Data from Auger et al. (2010).

given observation band, ellipticals are found to lie close to the best-fitting fundamental plane relation, with small scatter ($\approx 15\%$ in $R_e$). Given this observed almost planar distribution in the 3D parameter space, it is useful to define the concept of the **edge-on fundamental plane**, which is the distribution of data in the 2D space in which the best-fitting plane is seen edge-on. The edge-on fundamental plane is shown in the left panel of Fig. 5.13: in this specific case the axes of the 2D space are $x = 1.02 \log \sigma_0 - 0.872 \log\langle I \rangle_e + \text{const}$ and $y = \log R_e$. Galaxies are not found everywhere within the plane defined by eq. (5.54): though the distribution within the plane is relatively broad, there are avoided zones. The presence of these zones in the **face-on fundamental plane** gives independent and additional information with respect to the edge-on fundamental plane, as we see in §5.4.2.

Once calibrated on galaxies with independently measured distance, the fundamental plane can be used, at low redshift,[10] to estimate the distance of elliptical galaxies, because it relates the distance-independent observables $\sigma_0$ and $\langle I \rangle_e$ with $R_e$ measured in physical units (usually kpc), which, combined with the angular measure of $R_e$, gives the angular diameter distance (§2.1.4). However, the importance of the fundamental plane relation is more due to the fact that its very existence gives us crucial information on structural

[10] Higher-redshift ellipticals, in general, do not lie on the present-day fundamental plane, because of the evolution in luminosity, size and velocity dispersion (§11.3.1).

and kinematic properties of present-day ellipticals and therefore on their formation and evolution histories.

The existence of the fundamental plane can be interpreted as follows. Defining the **total** (dark matter plus stars) **mass-to-light ratio** $\Upsilon \equiv \mathcal{M}_{\rm vir}/L$, the virial theorem (eq. 5.16) can be written in terms of $\sigma_0$, $R_e$ and $L$ as

$$L = \frac{k_{\rm vir} R_e \sigma_0^2}{G\Upsilon}. \tag{5.55}$$

Replacing $L$ with $\langle I \rangle_e$ and $R_e$ via eq. (5.53), the above relation takes a form easily comparable with the fundamental plane (eq. 5.54):

$$\log R_e = 2 \log \sigma_0 - \log\langle I \rangle_e + \log\left(\frac{k_{\rm vir}}{\Upsilon}\right) + \text{const.} \tag{5.56}$$

If all ellipticals were structurally and kinematically **homologous**[11] and had the same total mass-to-light ratio, the ratio $k_{\rm vir}/\Upsilon$ would be a constant and we should observe a fundamental plane-like relation with $\alpha = 2$ and $\beta = -1$. In fact, there is no reason why the ratio $k_{\rm vir}/\Upsilon$ should be the same for different galaxies. $\Upsilon$ depends on the stellar mass-to-light ratio and on the dark-to-stellar mass fraction, while $k_{\rm vir}$ depends on the light profile, on the relative distribution of dark and luminous matter and on the orbital distribution of stars (§5.3.1). Formally, virialised systems could lie everywhere in the space of parameters $(\langle I \rangle_e, \sigma_0, R_e)$. The fact that ellipticals are confined close to the fundamental plane has to do with the galaxy formation process and not with equilibrium. The observed values of $\alpha$ and $\beta$ deviate significantly from 2 and $-1$: this **tilt of the fundamental plane** implies that the ratio $k_{\rm vir}/\Upsilon$ varies systematically with $\sigma_0$ and $\langle I \rangle_e$. Several pieces of evidence suggest that the tilt of the fundamental plane is mainly due to a systematic variation of $\Upsilon$: more luminous (higher $\sigma_0$; §5.4.2) galaxies tend to have higher stellar mass-to-light ratios and dark matter fractions.

### 5.4.2 The Kormendy and Faber–Jackson Relations

Before the discovery of the fundamental plane relation, elliptical galaxies were found to obey other important two-variable scaling laws. The **Kormendy relation**, originally discovered by Kormendy (1977) as a correlation between $R_e$ and $I_e$ (surface brightness at $R_e$), is also found as a correlation between $R_e$ and $\langle I \rangle_e$ (average surface brightness within $R_e$; eq. 5.53), which can be recast in the form of a correlation between $R_e$ and $L$:

$$\log\left(\frac{R_e}{\rm kpc}\right) = A + a \log\left(\frac{L}{10^{11} L_\odot}\right). \tag{5.57}$$

[11] Two stellar systems are said to be 'structurally homologous' if their normalised density distributions are the same. For instance, for one-component (star-only) spherical stellar systems the normalised density can be obtained by considering radii in units of the **half-mass radius** $r_{\rm half}$ (the spherical radius containing 50% of the mass) and densities in units of the density at $r_{\rm half}$. Two structurally homologous stellar systems are also kinematically homologous if their velocity distributions are the same: for example, when both systems have isotropic velocity distributions.

The normalisation and slope depend on the band of the observations, but reference values are $A \approx 0.8$ (for instance in the $r$ band) and $0.6 \lesssim a \lesssim 0.7$. Combining eqs. (5.53) and (5.57), the effective surface brightness can be written as a function of luminosity as

$$\langle I \rangle_{\rm e} = \frac{L}{2\pi R_{\rm e}^2} \propto L^{1-2a}. \tag{5.58}$$

As in general we find $a > 1/2$, the observed size–luminosity relation is such that (at least among bright ellipticals) more luminous galaxies tend to have lower effective surface brightness. We note that this anticorrelation between luminosity and effective surface brightness found for bright ETGs extends to dEs, but not to other types of dwarf pressure-supported systems, such as dSphs and ultra-faint dwarfs (UFDs), which are among the lowest surface brightness galaxies known (Fig. 3.4).

The **Faber–Jackson relation** (Faber and Jackson, 1976) is an empirical correlation between central velocity dispersion $\sigma_0$ and luminosity, traditionally expressed in the form

$$\log\left(\frac{L}{10^{11}L_\odot}\right) = B + b\log\left(\frac{\sigma_0}{200\,{\rm km\,s^{-1}}}\right), \tag{5.59}$$

with best-fitting power-law index $b \approx 4$ and normalisation (for instance, in the $r$ band) $B \approx -0.2$. There are indications that a single power law is not the best representation of the observed distribution of galaxies in the $L$–$\sigma_0$ space. The trend is that the correlation is shallower at lower luminosities ($b \lesssim 3$) and steeper at the highest luminosities ($b \gtrsim 5$). The Faber–Jackson relation is conceptually similar to the Tully–Fisher relation of spirals (§4.4.1): both suggest that galaxy luminosity and kinematics are strictly linked by the process of galaxy formation (see §10.10.3).

By definition, the $R_{\rm e}$–$L$ and $\sigma_0$–$L$ relations can be thought of as non-edge-on projections of the fundamental plane. In fact, these relations are not as tight as the fundamental plane and therefore might be expected to put less stringent constraints on galaxy formation and evolution models. However, it must be stressed that the $R_{\rm e}$–$L$ and $\sigma_0$–$L$ relations contain additional information with respect to the edge-on fundamental plane (eq. 5.54), because galaxies with the same position in the edge-on fundamental plane can be located in very different places in the $L$–$R_{\rm e}$ and $L$–$\sigma_0$ spaces (and in the face-on fundamental plane; §5.4.1).

### 5.4.3 Scaling Relations Involving Stellar Mass

Estimates of the total stellar mass $\mathcal{M}_\star$ (based on SED fitting; §3.3, §8.6 and §11.1.3) are now available for large samples of present-day ETGs. It is therefore natural to study correlations between $\mathcal{M}_\star$, $R_{\rm e}$ and $\sigma_0$. Stellar mass is believed to be a more fundamental quantity than luminosity, being less affected by stellar evolution effects, so the scaling relations involving $\mathcal{M}_\star$ could be even more useful than those involving $L$. Provided the stellar mass is robustly estimated, the study of the redshift evolution of scaling laws involving stellar mass is a powerful tool in the attempt to understand galaxy formation and evolution.

ETGs in the present-day Universe are observed to follow a **stellar mass fundamental plane** (Fig. 5.13, right), which, in analogy to the classical fundamental plane (eq. 5.54), can be written in the form

$$\log R_{\rm e} = \alpha_\star \log \sigma_0 + \beta_\star \log\langle\Sigma_\star\rangle_{\rm e} + \gamma_\star, \tag{5.60}$$

where $\langle\Sigma_\star\rangle_{\rm e} \equiv \mathcal{M}_\star/(2\pi R_{\rm e}^2)$ is the mean stellar mass surface density within $R_{\rm e}$. Depending on the sample and on the details of the derivation of the involved quantities, the best-fitting coefficients are found in the ranges $1.2 \lesssim \alpha_\star \lesssim 1.5$ and $-1 \lesssim \beta_\star \lesssim -0.9$. In analogy to eq. (5.56), the virial theorem (eq. 5.16) can be rewritten as

$$\log R_{\rm e} = 2\log \sigma_0 - \log\langle\Sigma_\star\rangle_{\rm e} + \log\,(k_{\rm vir} f_{\rm m,\star}) + {\rm const}, \tag{5.61}$$

where $f_{\rm m,\star} \equiv \mathcal{M}_\star/\mathcal{M}_{\rm vir}$ (see §10.1). The stellar mass fundamental plane is less tilted (§5.4.1) with respect to eq. (5.61) than the traditional fundamental plane with respect to eq. (5.56). As the tilt means a systematic variation of either $k_{\rm vir}/\Upsilon$ or $k_{\rm vir} f_{\rm m,\star}$, this finding suggests that the tilt of the traditional fundamental plane is partly due to a systematic variation of the stellar mass-to-light ratio $\mathcal{M}_\star/L$, because $\Upsilon = (\mathcal{M}_\star/L) f_{\rm m,\star}^{-1}$. However, the fact that the stellar mass fundamental plane has non-negligible tilt means that also the dynamical-to-stellar mass ratio $\mathcal{M}_{\rm dyn}/\mathcal{M}_\star$ varies systematically ($\mathcal{M}_{\rm dyn}$ is defined in eq. 5.17). A direct interpretation of the tilt of the stellar mass fundamental plane is obtained by measuring the correlation

$$\frac{\mathcal{M}_{\rm dyn}}{\mathcal{M}_\star} \propto \mathcal{M}_\star^{\eta}, \tag{5.62}$$

which is found to have best-fitting index in the range $0.2 \lesssim \eta \lesssim 0.3$, meaning that either the dark matter content (parameterised by $f_{\rm m,\star}$) or the structural and/or kinematic properties of the stars (parameterised by $k_{\rm vir}$) vary systematically with $\mathcal{M}_\star$. When direct measurements of total mass (stars plus dark matter) are available (for instance, the projected mass $\mathcal{M}_{\rm proj}$ within $\approx R_{\rm e}$ from strong gravitational lensing; §5.3.4), we find that $\mathcal{M}_{\rm proj} \propto \mathcal{M}_{\rm dyn}$, meaning that the systematic variation of the dynamical mass with stellar mass is mainly driven by the dark matter fraction and not much by **non-homology**. To understand this, let us consider two systems with the same values of $\mathcal{M}_\star$ and $R_{\rm e}$, and the same stellar density distribution, but different values of $\sigma_0$ (and therefore of $\mathcal{M}_{\rm dyn}$; eq. 5.17). If also the dark matter distribution (and thus the total gravitational potential) were the same, the difference in $\mathcal{M}_{\rm dyn}$ would be due to different stellar velocity distributions (kinematic non-homology), but the two systems would have different values of $\mathcal{M}_{\rm proj}/\mathcal{M}_{\rm dyn}$.

The distribution of present-day ETGs in the stellar mass–size plane (see also §3.1.4) is well represented by a power law

$$\log\left(\frac{R_{\rm e}}{\rm kpc}\right) = A_\star + a_\star \log\left(\frac{\mathcal{M}_\star}{10^{11}\mathcal{M}_\odot}\right), \tag{5.63}$$

with index in the range $0.5 \lesssim a_\star \lesssim 0.8$ (similar to the Kormendy relation; eq. 5.57) and normalisation $0.5 \lesssim A_\star \lesssim 0.6$ (assuming Chabrier IMF; §8.3.9). The stellar mass analogue of the Faber–Jackson relation (eq. 5.59) is

$$\log\left(\frac{\mathcal{M}_\star}{10^{11}\mathcal{M}_\odot}\right) = B_\star + b_\star \log\left(\frac{\sigma_0}{200\,{\rm km\,s^{-1}}}\right), \tag{5.64}$$

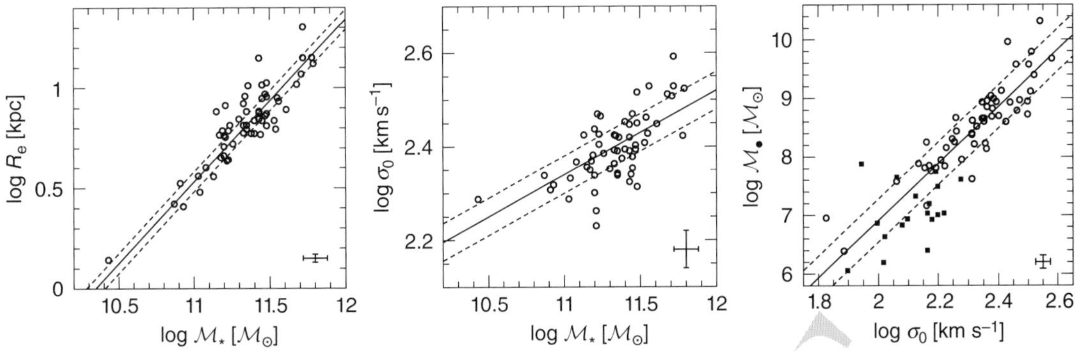


Fig. 5.14 Effective radius $R_e$ (*left panel*) and central velocity dispersion $\sigma_0$ (*middle panel*) as functions of stellar mass for the Sloan Lens ACS Survey sample of massive present-day ETGs (circles). Here $R_e$ is measured in the $V$ band, $\sigma_0$ is measured within $R_e/2$ and $\mathcal{M}_\star$ is derived assuming a Chabrier IMF (§8.3.9). Data from Auger et al. (2010). *Right panel*. Mass of the central SMBH as a function of the central stellar velocity dispersion for a sample of both ETGs (circles) and bulges of spiral galaxies (squares). Data from Saglia et al. (2016). In each panel the solid line is the best fit ($R_e \propto \mathcal{M}_\star^{0.81}$, $\sigma_0 \propto \mathcal{M}_\star^{0.18}$ and $\mathcal{M}_\bullet \propto \sigma^{4.87}$), the dashed lines indicate the intrinsic scatter and the cross in the bottom right corner represents typical error bars.

with $4 \lesssim b_\star \lesssim 5$ and normalisation $-0.2 \lesssim B_\star \lesssim 0$ (assuming Chabrier IMF; §8.3.9). Examples of the $R_e$–$\mathcal{M}_\star$ and $\sigma_0$–$\mathcal{M}_\star$ relations observed for a sample of massive ($\log \mathcal{M}_\star/\mathcal{M}_\odot \gtrsim 10.5$) present-day ETGs are shown in Fig. 5.14 (left and middle panels). For this sample, the intrinsic scatter of both relations is $\approx 0.1$ dex at fixed $\mathcal{M}_\star$. We note that the $R_e$–$\mathcal{M}_\star$ relation shown in Fig. 5.14 corresponds to the high-mass end of the relation shown in the right panel of Fig. 3.5. In §10.10.3 we discuss the origin of the stellar mass scaling relations in a cosmological framework.

## 5.4.4 Scaling Relations Involving Black Hole Mass

SMBHs are believed to be ubiquitous in the centre of galaxies with a spheroidal stellar component, such as ETGs and early spirals. The presence and mass of central SMBHs have been inferred in nearby galaxies by studying the kinematics of stars and ionised gas as dynamical tracers. When high-resolution kinematic data are available, the central kinematics of the tracers reveals that the gravitational potential at the galaxy centre cannot be explained by an extended mass distribution, but requires a central massive object. The mass of the central SMBH $\mathcal{M}_\bullet$, which is found in the range $10^6$–$10^{10}\mathcal{M}_\odot$ (see also §3.6.1) has been estimated in sufficiently large samples of galaxies to study correlations between $\mathcal{M}_\bullet$ and other galaxy properties. The black hole mass is found to scale linearly with the spheroid stellar mass $\mathcal{M}_\bullet \propto \mathcal{M}_\star$ (this scaling law is known as the **Magorrian relation** after Magorrian et al., 1998): the normalisation is such that $\mathcal{M}_\bullet/\mathcal{M}_\star \approx 0.001$–$0.005$ (depending on the sample) and the intrinsic scatter is $\approx 0.3$ dex. The black hole mass correlates also with the stellar central velocity dispersion:

$$\log\left(\frac{\mathcal{M}_\bullet}{10^8\mathcal{M}_\odot}\right) = B_\bullet + b_\bullet \log\left(\frac{\sigma_0}{200\,\mathrm{km\,s^{-1}}}\right), \tag{5.65}$$

with $4 \lesssim b_\bullet \lesssim 5$ and $B_\bullet \approx 0.5$. The intrinsic scatter of this **black hole mass–velocity dispersion relation** is $\approx 0.3$ in $\log \mathcal{M}_\bullet$ at fixed $\sigma_0$. The observed $\mathcal{M}_\bullet$–$\sigma_0$ relation is shown in Fig. 5.14 (right) for a sample of both early-type and star-forming galaxies.

It is important to note that $\sigma_0$ is not significantly influenced by the presence of the central black hole. The characteristic scale over which the black hole is dynamically important is the radius of its **sphere of influence**,

$$r_{\rm infl} \equiv \frac{G\mathcal{M}_\bullet}{\sigma_0^2} \simeq 0.03 \left( \frac{\sigma_0}{200\,{\rm km\,s^{-1}}} \right)^{b_\bullet - 2} {\rm kpc}, \tag{5.66}$$

where we have eliminated $\mathcal{M}_\bullet$ using eq. (5.65) with $\mathcal{M}_\bullet = 3 \times 10^8 \mathcal{M}_\odot$ when $\sigma_0 = 200\,{\rm km\,s^{-1}}$. It is clear that $r_{\rm infl}$ is much smaller than the characteristic aperture radius $R_{\rm ap}$ (§5.1.2) over which $\sigma_0$ is measured. For instance for a nearby ETG with $\mathcal{M}_\star = 10^{11} \mathcal{M}_\odot$ and $\sigma_0 = 200\,{\rm km\,s^{-1}}$, $r_{\rm infl} \approx 0.03$ kpc (eq. 5.66), while $R_{\rm ap} \sim 1$ kpc.

The mass of the central black hole is found to correlate also with other galaxy properties, such as the number of hosted globular clusters and the Sérsic index that best fits the observed surface brightness profile. For galaxies with both spheroidal and disc components $\mathcal{M}_\bullet$ appears to correlate somewhat better with the properties of the bulge (mass and luminosity), while, at least in the present-day Universe, $\mathcal{M}_\bullet$ and the global properties of the galaxy, when the disc component is included, are more weakly related. The scaling between SMBH mass and stellar velocity dispersion seems to hold for different galaxy types.

# 6 The Environment of Present-Day Galaxies

In Chapters 3–5 we have described present-day galaxies, focusing on their internal properties. However, galaxies are by no means closed systems, so a fundamental ingredient to understand the formation and evolution of galaxies is their **environment**, that is, the ensemble of properties of the region of the Universe in which each galaxy is located. In this chapter we describe the environment of present-day galaxies from the smallest scale (pairs of interacting galaxies) to the very large scale (large-scale structure of the Universe).

## 6.1 Interacting Galaxies

A significant fraction of the observed galaxies are found to be interacting. In the simplest case a pair of **interacting galaxies** appears as two nearby galaxies, with clearly distinct bodies, which are characterised in the outskirts by tidal features (§8.9.4), are connected by stellar and gaseous **bridges**, or have, in general, very disturbed morphology. These features reveal not only that the two systems are physically close to each other, but also that their structures are mutually affected by each other's gravitational fields. These kinds of systems are often classified as **peculiar galaxies**. Spectacular and prototypical examples of peculiar galaxies are NGC 4038/4039 (known as the Antennae) and NGC 4676 (the Mice), which are believed to be SFGs caught in the early stage of a merger process (Toomre and Toomre, 1972; §8.9). When, as in these cases, the involved galaxies are gas-rich, the interaction is associated with important star formation, often a starburst (§4.5). These processes are called dissipative or 'wet' mergers (§8.9). When the interacting galaxies are gas-poor, there is no substantial associated star formation: these are dissipationless or 'dry' mergers (§8.9). Not all peculiar galaxies are ongoing mergers: when the disturbance is milder, the two galaxies might be undergoing only a temporary interaction (fly-by; §8.9).

**Ring galaxies**, such as the so-called Cartwheel galaxy (ESO 350-40) or the peculiar galaxy Arp 148 (leftmost top panel of Fig. 6.1), show beautiful large-scale rings of star-forming gas, which likely originate from the perturbation due to the interaction with a smaller companion galaxy. In some cases the signatures of interaction are less evident. Examples are **shell galaxies** such as NGC 474 (Fig. 5.4), NGC 1344 and NCG 3923, ETGs in which deep photometry reveals the presence of shells in the outer surface brightness distributions. These shells are believed to be the remnant of the accretion of a satellite disc galaxy. Finally, there are cases of galaxies, such as NGC 5907 (Fig. 4.4) and NGC 474 itself, that show in deep images extended star streams strikingly similar to those expected along the orbit of an accreted companion in a minor merger (§10.4; Fig. 10.9).

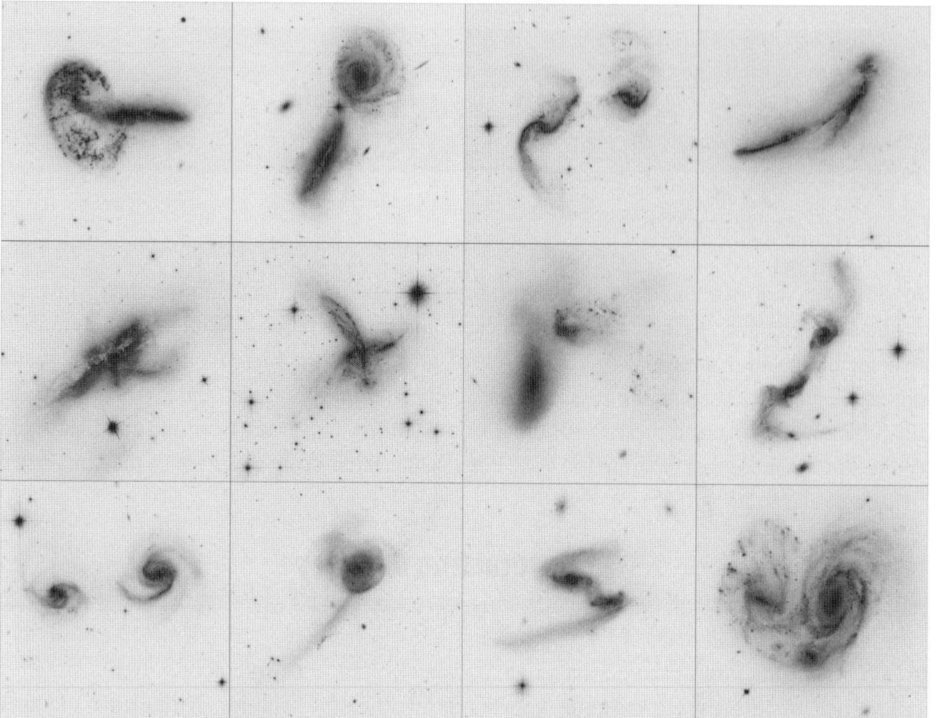

Fig. 6.1 A gallery of images of interacting galaxies. From top to bottom and from left to right: Arp 148, UGC 9618, Arp 256, NGC 6670 (*first row*), NGC 6240, ESO 593-8, NGC 454, UGC 8335 (*second row*), NGC 6786, NGC 17, ESO 77-14, NGC 6050 (*third row*). The frames have different sizes, ranging from $\approx$ 30 kpc (Arp 148) to $\approx$ 130 kpc (ESO 77-14). Credit: NASA, ESA, the Hubble Heritage (STScI/AURA)-ESA/Hubble Collaboration, and A. Evans (University of Virginia, Charlottesville/NRAO/Stony Brook University).

An idea of the variety of the appearance of colliding galaxies in different phases of interaction is given by Fig. 6.1, showing 12 spectacular examples of such systems. Interacting galaxies are not only aesthetically beautiful, but they are important for our understanding of galaxy evolution, because galaxy interactions, collisions and mergers are believed to be among the fundamental mechanisms characterising the assembly and the shaping of galaxies (§8.9). It is therefore interesting to study the properties of interacting galaxies in a quantitative way.

A particularly important quantity for comparison with both high-redshift observations and theoretical models is the **merger rate** of galaxies in the present-day Universe. The merger rate, the number of mergers per unit time per unit volume, is

$$\Gamma_{\rm merg} \equiv \frac{n_{\rm merg}}{\tau_{\rm merg}}, \tag{6.1}$$

where $n_{\rm merg}$ is the number of galaxies, per unit volume, that are undergoing a merger and $\tau_{\rm merg}$ is the merging timescale (§8.9.5). It is useful to define the **fractional merger rate** (number of mergers per galaxy per unit time)

$$\mathcal{R}_{\text{merg}} \equiv \frac{f_{\text{merg}}}{\tau_{\text{merg}}}, \tag{6.2}$$

where $f_{\text{merg}} \equiv n_{\text{merg}}/n_{\text{gal}}$ is the fraction of galaxies that are merging and $n_{\text{gal}}$ is the total galaxy number density. Measurements of $n_{\text{merg}}$ can be obtained either by identifying as mergers objects with disturbed morphology or by counting the number of close pairs of galaxies per unit volume, which are expected to end up merging (§11.3.2). The observationally determined values of the fraction of mergers and of the merger rate depend on the details of the definition of the involved quantities, but reference numbers in the present-day Universe are $f_{\text{merg}} \approx 0.01$ and $\mathcal{R}_{\text{merg}} \approx 0.02\,\text{Gyr}^{-1}$ (for $\tau_{\text{merg}} \approx 0.5\,\text{Gyr}$).

## 6.2 Groups of Galaxies

Most galaxies are not isolated: they tend to cluster in gravitationally bound aggregations. These systems are usually characterised by the number of member galaxies, by the size of the spatial distribution of galaxies and by the dispersion of the galaxy velocity distribution. When these aggregations consist of less than about 50 members, they are called **groups of galaxies**, having characteristic scales $\lesssim 1$ Mpc, and velocity dispersions $\lesssim 500\,\text{km s}^{-1}$ (Tab. 6.1; Fig. 6.2). More numerous aggregations of galaxies are classified as clusters of galaxies (§6.4). However, there is not a clear-cut distinction between groups and clusters. When an estimate of the virial mass $\mathcal{M}_{\text{vir}}$ is available, it is often taken as mass threshold $\mathcal{M}_{\text{vir}} \approx 10^{14}\mathcal{M}_{\odot}$: systems with lower mass are groups and systems with higher mass are clusters.

Our own Milky Way belongs to a group of galaxies named the Local Group, which we describe in detail in §6.3. The Local Group is not representative of the whole population of galaxy groups, which are found to show a great deal of variety in terms of their galaxy

**Table 6.1** Main properties of groups and clusters of galaxies in the present-day Universe

| | Local Group | Groups | Clusters |
|---|---|---|---|
| Number of galaxies | $7^a$ | 4–50 | 50–1000 |
| Stellar mass ($\mathcal{M}_{\odot}$) | $1\text{–}1.5 \times 10^{11}$ | $\lesssim 10^{12}$ | $10^{12}\text{–}10^{13}$ |
| Velocity dispersion (km s$^{-1}$) | — | 100–500 | 500–2000 |
| X-ray luminosity (erg s$^{-1}$) | — | $10^{41}\text{–}10^{43}$ | $10^{43}\text{–}10^{45}$ |
| Gas mass ($\mathcal{M}_{\odot}$) | — | $\lesssim 10^{13}$ | $10^{13}\text{–}10^{14}$ |
| Gas temperature ($10^7$ K) | — | 0.5–2 | 2–10 |
| Virial mass ($\mathcal{M}_{\odot}$) | $2\text{–}5 \times 10^{12}$ | $\lesssim 10^{14}$ | $10^{14}\text{–}10^{15}$ |
| Virial radius (Mpc) | $\approx 1$ | 0.4–1 | 1–3 |
| Baryon fraction$^b$ | — | 0.5–0.8 | 0.7–1 |

$^a$ Only galaxies with $M_V < -16$.
$^b$ In units of the cosmic baryon fraction $f_b \simeq 0.16$ (§2.5).

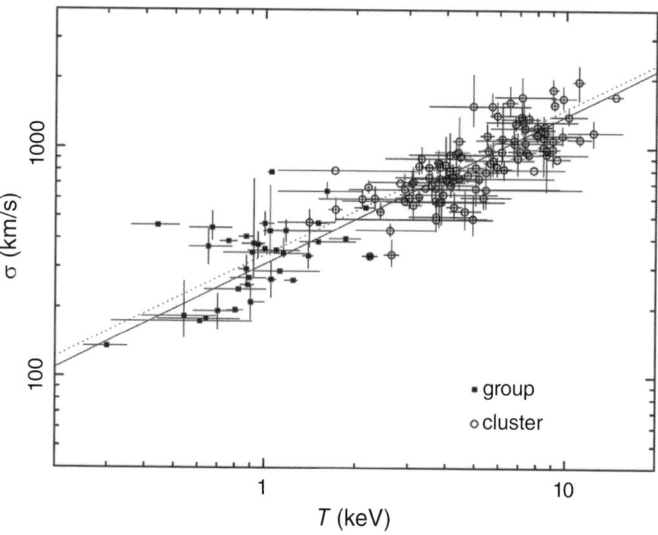


Fig. 6.2 Line-of-sight galaxy velocity dispersion $\sigma$ as a function of the plasma temperature $T$ for galaxy groups and clusters. The quantity on the abscissa, here indicated with $T$ as is standard in X-ray astronomy, is in fact the corresponding energy $k_B T$ (1 keV $\simeq 1.6 \times 10^{-9}$ erg; 1 keV/$k_B \simeq 1.16 \times 10^7$ K). The dotted and solid lines are, respectively, the best fits to the distributions of the group and cluster samples. From Xue and Wu (2000). © AAS, reproduced with permission.

content. Groups are usually classified depending on whether most of the luminous galaxies are early-type or star-forming. Besides systems dominated by spiral galaxies such as the Local Group (**spiral groups**), there are groups dominated by ellipticals or S0s (**evolved groups**), and groups in which there are not massive dominant galaxies, but only dwarf galaxies (**dwarf associations**).

Groups are environments favourable to galaxy interactions and mergers, because the galaxy density is higher than in the field and the relative speeds between the group's members are of the order of the galaxy internal stellar velocity dispersions (§8.9.5). Therefore we expect that galaxies that are members of groups undergo substantial transformation as a consequence of close encounters with other galaxies and that two or more group members can merge to form a more massive galaxy (§8.9). Evolved groups are so named because they are considered dynamically evolved, in the sense that the group's population has been already transformed by galaxy mergers, while spiral groups are considered to be at an earlier stage.

Similar to massive ETGs (§5.2.1), galaxy groups have hot X-ray-emitting gas halos at the system's virial temperature (§8.2.1), in the range $5 \times 10^6\,\mathrm{K} \lesssim T \lesssim 2 \times 10^7\,\mathrm{K}$ (Fig. 6.2). The temperature of the hot **intragroup medium** scales roughly with $\sigma_{\rm los}^2$, where $\sigma_{\rm los}$ is the galaxy–galaxy line-of-sight velocity dispersion (Fig. 6.2), as expected if the galaxies and the hot gas are both in equilibrium in the same gravitational potential (§6.4.2). The virial mass of a group $\mathcal{M}_{\rm vir}$ is traditionally estimated with the virial theorem method (§5.3.1) based on measurements of $\sigma_{\rm los}$ and of the characteristic group size. Alternatively, $\mathcal{M}_{\rm vir}$ is measured using X-ray (§5.3.3) or weak gravitational lensing (§5.3.4) data. Typical values of the total mass-to-light ratio of galaxy groups are $\mathcal{M}_{vir}/L = 100–400$ in units of $(\mathcal{M}/L)_\odot$; thus stars contribute very little to the group mass. Note that most of the baryonic mass in groups is in the intragroup medium, but, even accounting for the gas, the mass baryon fraction is $\approx 10\%$ (Tab. 6.1), so groups are dark matter-dominated systems.

**Compact groups**, sometimes called **Hickson compact groups** from the pioneering work of Hickson (1982), are very dense associations of typically four or five bright galaxies distributed over a few tens of kpc. Compact groups are relatively rare and deserve a special classification among the population of galaxy groups because they are expected to have peculiar dynamical evolution. Due to the high density, galaxy interactions and mergers are expected to be very likely in compact groups: therefore, these systems are believed to be short-lived associations that will be radically transformed by mergers with the formation of a dominant central massive galaxy. If all the bright galaxies in a small group merge, the end product will be a **fossil group**: a luminous elliptical galaxy with no luminous companions, but only small satellites, and surrounded by an extended (group-size) X-ray luminous gaseous halo, which is the remnant of the intragroup medium.

## 6.3 The Local Group

The Milky Way is part of a small group of galaxies called the **Local Group**. It is a rather loose group of $\sim 1$ Mpc in size that contains two large spiral galaxies (the Milky Way and M 31), a smaller disc galaxy (M 33) and many tens of dwarf galaxies of different sizes and properties. The proximity of these galaxies to us makes this an ideal place to study their detailed evolution and mutual interactions. The excellent spatial resolution that can be achieved for many of these dwarf galaxies allows us to trace the distribution of luminous and dark mass in great detail. Moreover, the possibility of resolving single stars down to faint magnitudes allows a careful reconstruction of the SFHs in a way that is not achievable anywhere else in the Universe. Finally, the determination of proper motions and the 3D velocity vectors, greatly refined thanks to *Gaia*, opens the possibility to accurately determine galaxy orbits and study their infall history into the Local Group.

### 6.3.1 Galaxies in the Local Group

Other than the Milky Way (§4.6), the Local Group hosts another large spiral: Andromeda (M 31), a non-barred Sb spiral seen at a high inclination angle (78°) with respect to our line of sight. M 31 has a stellar mass (and a virial mass; §7.3.2) a factor of 1.5–2 higher than the Milky Way. The H I mass is also slightly higher, but the current SFR appears lower (SFR $\approx 0.7\,\mathcal{M}_\odot\,\mathrm{yr}^{-1}$). M 31 has around twice as many globular clusters as the Milky Way and also more dwarf satellites. The Milky Way and M 31 are approaching each other (§10.3.2) but they are still quite far apart. The distance between them ($\approx 780$ kpc) is larger than the sum of their virial radii (both in the range 250–350 kpc; §7.3.2). In this sense, the Local Group can be seen as formed by the two subgroups of the two main spirals.

The outer stellar halo of M 31 appears crisscrossed by a number of stellar streams (Fig. 4.4), fossils of past encounters with its satellites and the partial or total disruption of some of them. An interaction (fly-by) has also possibly occurred between M 31 and the third largest SFG of the Local Group, Triangulum (M 33). The signature of this could be the presence of a plume of stars coming out of the M 33 disc in the direction of M 31 (Fig. 4.4) and also a significant warp of its outer H I disc. M 33 is a relatively low-mass disc galaxy of type Sc with a total luminosity that is one-fifth of that of M 31 and a high star formation rate considering its mass (SFR $\approx 0.5\ \mathcal{M}_\odot\ \mathrm{yr}^{-1}$). The general physical properties of M 33 and Local Group dwarf galaxies are presented in Tab. 6.2. Fig. 6.3 gives an overview of the locations and the types of galaxies present in the Local Group.

The satellite galaxies in the Local Group cover a wide range of mass and luminosity. Broadly speaking, they fall into two categories: the gas-rich and the gas-poor dwarfs, which are clearly segregated as we discuss in §6.3.2. The gas-poor satellites are called **dwarf spheroidals** (dSphs) with one exception: M 32, the only dwarf elliptical of the Local Group and the closest elliptical galaxy to us. Despite having a total luminosity similar to the brightest dSphs, the central surface brightness of M 32 is about 10 magnitudes (i.e. a factor of 10 000) brighter; thus it is remarkably more centrally concentrated than the dSph systems (Fig. 3.4). Gas-rich satellites are generically called dwarf irregulars (dIrrs) as we have seen in Chapter 4. They are more disc-like but have global properties similar to dSphs (Tab. 6.2). Some satellites of M 31 (e.g. NGC 147) appear disc-like but are also devoid of gas and, for this reason, are classified as dSphs. Finally, a few dwarfs are considered **dSph/dIrr transition types** sharing characteristics of the two classes: in particular they have a small amount of gas and low SFRs.

**Table 6.2** Properties of Local Group galaxies

| | M 33 | M 32 | dIrrs | dSphs | UFDs |
|---|---|---|---|---|---|
| Magnitude ($M_V$) | −18.8 | −16.4 | −17 to −10 | −16.5 to −9 | −8 to −1.5 |
| Luminosity $\left(L_V/10^6\ L_{V,\odot}\right)$ | 2800 | 300 | 1–500 | 0.3–300 | 0.0003–0.1 |
| $\mu_0$ (mag arcsec$^{-2}$)[a] | 21.4 | 11.1 | 20–25 | 21–26 | >26 |
| Effective radius (kpc) | 2.0 | 0.11 | 0.4–2 | 0.2–1 | 0.02–0.5 |
| Stellar mass ($10^6\ \mathcal{M}_\odot$) | 2900 | 320 | 1–500 | 0.3–300 | 0.0003–0.2 |
| H I mass ($10^6\ \mathcal{M}_\odot$) | 1400 | — | 0.4–500 | — | $\lesssim 0.3$ |
| $\mathcal{M}_{\mathrm{dyn}}$ ($10^6\ \mathcal{M}_\odot$)[b] | $5 \times 10^4$ | 1082 | 20–3000 | 10–800 | 0.5–10 |
| $\mathcal{M}_{\mathrm{dyn}}/L_V$ ($\mathcal{M}_\odot/L_{V,\odot}$) | 17 | 3 | ~10 | 2–100 | 50–5000 |
| SFR ($\mathcal{M}_\odot\ \mathrm{yr}^{-1}$) | ≈0.5 | — | 0.001–0.1 | — | $<2 \times 10^{-5}$ |
| Metallicity ([Fe/H]) | −0.6 | −0.25 | −1 to −2 | −0.5 to −2 | −1.8 to −2.7 |

Most values in this table from McConnachie (2012).

[a] Central surface brightness in the $V$ band, ranges are indicative.

[b] Calculated with eq. (5.17) for dispersion-dominated systems and as $(R_{\mathrm{max}} v_{\mathrm{rot}}^2/G)^{1/2}$ for M 33 and dIrrs, $R_{\mathrm{max}}$ being the outermost radius of the H I rotation curve and $v_{\mathrm{rot}} = v_{\mathrm{rot}}(R_{\mathrm{max}})$.

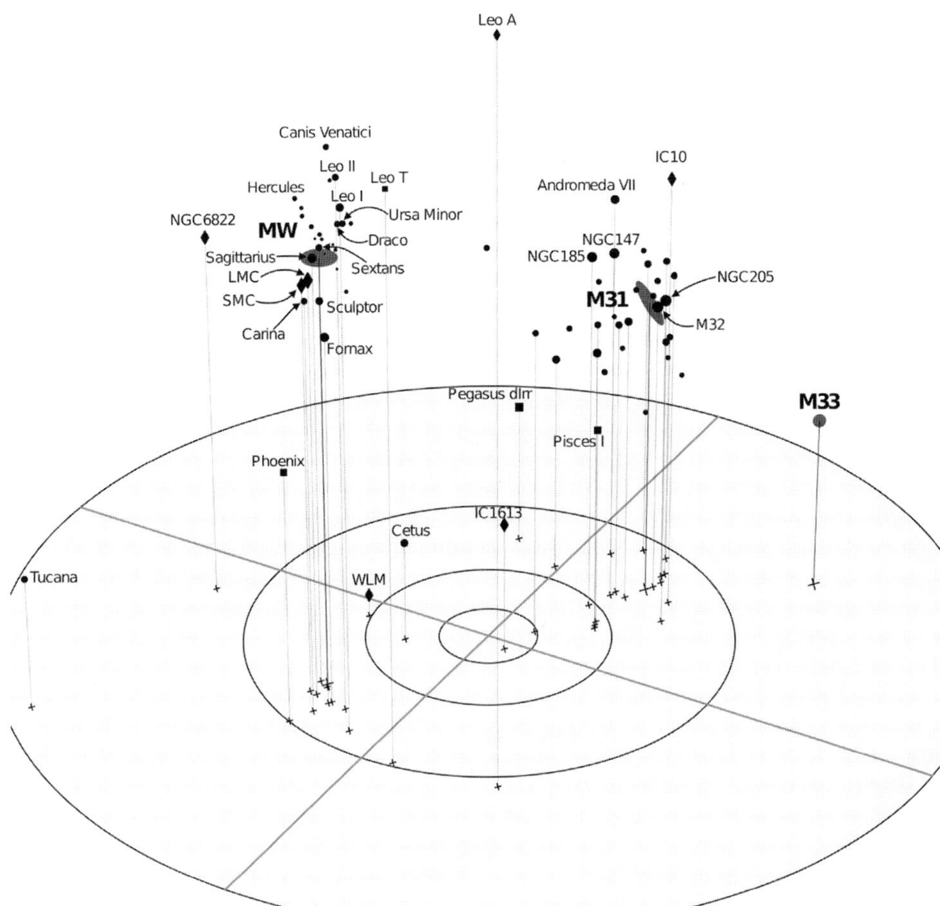


Fig. 6.3

A 3D view of the Local Group of galaxies showing the Milky Way (MW), M 31, M 33 and the dwarf satellites. The bottom plane is parallel to the plane of the disc of the Milky Way and the rings are at $R = 100$, 250, 500 and 1000 kpc from the centre, defined as the midpoint of the projected separation between the Milky Way and M 31; dSphs are indicated with circles, dIrrs with diamonds and transition types with squares. The size of the symbols is a measure of the galaxy luminosity in the $V$ band. Only galaxies with absolute magnitude $M_V < -9$ are projected onto the plane (crosses). All data from McConnachie (2012). Figure courtesy of G. Iorio.

The Milky Way has its population of satellites, most of which are dwarf spheroidal galaxies with no gas. Two relatively large SFGs (the Large Magellanic Cloud, LMC, and the Small Magellanic Cloud, SMC) are currently passing close ($\approx 50$ kpc) to the Milky Way. The LMC is considered a low-mass disc galaxy having a mass a factor of 2 smaller than M 33, while the SMC can be considered the largest dIrr in the Local Group. The Magellanic Clouds are interacting very strongly with each other and with the Milky Way. The effects of these interactions are very clear: a large bridge (§6.1) of gas (mostly H I) and stars that have been stripped by the LMC and the SMC is connecting the pair. In addition, a very extended stream of gas (the **Magellanic Stream**) appears to be escaping from them.

This stream extends for $\approx 200$ degrees (mostly across the southern Galactic sky), which correspond to several tens of kpc, and contains $\sim 10^9\ \mathcal{M}_\odot$ of gas. The Magellanic Stream originates from the interaction between the fast-moving[1] Magellanic Clouds and the Milky Way. An important effect is thought to be ram-pressure stripping (§8.9.4) by the Galactic corona (hot gas surrounding the Milky Way; §6.7) but tidal stripping may also have a significant role (§8.9.4).

## 6.3.2 Distribution and Masses of Dwarf Galaxies

The two main types of dwarf galaxies in the Local Group are not randomly distributed with respect to the main galaxies. There is a clear segregation between the gas-poor galaxies, which are preferentially located close to M 31 and the Milky Way, and the gas-rich galaxies that lie in the outskirts of the Local Group. Fig. 6.4 shows the H I mass of the satellite galaxies of the Milky Way as a function of their Galactocentric distance. All galaxies within the virial radius of the Milky Way ($r_{\rm vir} \approx 280$ kpc for $\mathcal{M}_{\rm vir} \simeq 1.3 \times 10^{12}\ \mathcal{M}_\odot$; see Tab. 4.1), with the exception of the LMC and the SMC, are devoid of

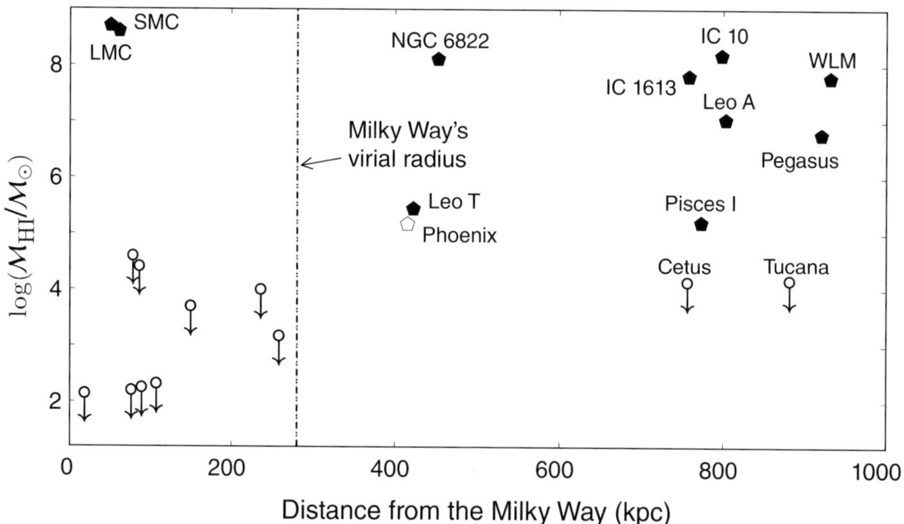


Fig. 6.4 The H I gas content of dwarf galaxies (dSphs, circles; dIrrs, pentagons) as a function of their Galactocentric distance. All the downward-pointing arrows are upper limits and are dSphs (where essentially no gas is detected); the H I detection of Phoenix dwarf (open pentagon) is debatable due to potential contamination from the Galactic emission. For simplicity, with the exception of the gas-rich Leo T, we do not include the UFDs, which would appear as upper limits within the virial radius of the Milky Way. Note the clear segregation between gas-poor and gas-rich systems within and outside the virial radius of the Milky Way, respectively. All the obvious satellites of M 31 (within its virial radius) have been left out of this plot. Data from McConnachie (2012) and Grcevich and Putman (2009).

[1] The measurements of proper motions indicate that they are on highly elongated orbits and currently close to pericentre.

gas. All galaxies beyond the virial radius, with two exceptions (Cetus and Tucana), contain cold gas.

The segregation between gas-rich and gas-poor dwarfs shows that the evolutionary paths of these galaxies are determined by the environment and in particular by their distance from the main galaxies (dwarf satellites of M 31 display a similar behaviour). This is a rescaled version of the morphology–density relation that we discuss in §6.5.1. The main effect of the proximity to a large galaxy is the removal of the cold ISM from the dwarf due to the combined effect of ram-pressure stripping (§8.9.4) and tidal stripping (§8.9.4).

The internal dynamics of dwarfs in the Local Group can be studied with great accuracy using velocity measurements of individual stars (especially for dSphs) and of the H I (for dIrrs). The kinematics of the gas in dwarf irregulars tends to be dominated by rotation except for the smallest galaxies like Leo A ($M_V \simeq -12.1$) where $V/\sigma \lesssim 1$. In contrast, the kinematics of dSphs, seen through thousands of stars in some cases, is dominated by velocity dispersion although some rotation has been detected in a few cases. The application of the virial theorem returns dynamical masses that are much higher than the stellar mass, with typical $\mathcal{M}_{\rm dyn}/L$ ratios of up to 100 $\mathcal{M}_\odot/L_\odot$. More careful dynamical modelling (see §5.3.2) indicates that these systems are dark matter-dominated down to their central regions. An example is shown in Fig. 6.5, where we see that the radial profile of the velocity dispersion in the Fornax dSph clearly requires a large amount of dark matter at every radius. This result is common to other dSphs.

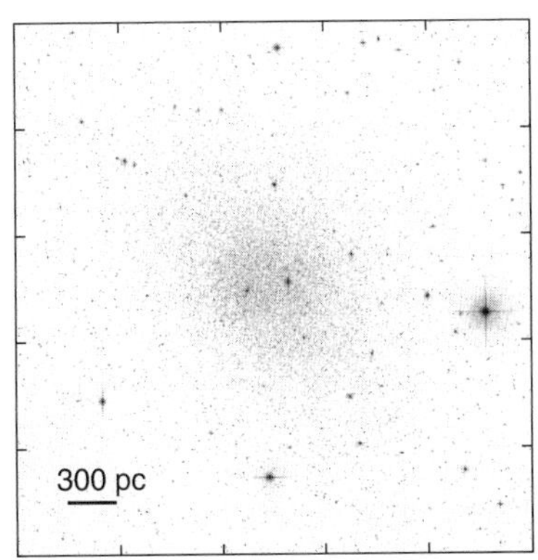


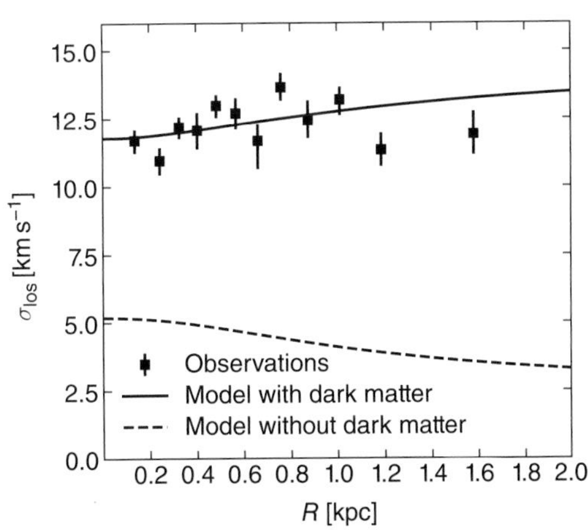


**Fig. 6.5** *Left panel.* Optical image of the Fornax dSph ($M_V \simeq -13.3$), satellite of the Milky Way located at a distance of $\approx$140 kpc. *Right panel.* Radial profile of the line-of-sight stellar velocity dispersion (squares) obtained from spectra of individual stars in the Fornax dSph compared with the results of two dynamical models. The model without dark matter (dashed curve) is obtained assuming a stellar mass-to-light ratio $\Upsilon_\star = 2$ in the $V$ band (corresponding to $\mathcal{M}_\star \simeq 2.8 \times 10^7\ \mathcal{M}_\odot$) and shows the maximum contribution that we should expect by the stellar component alone. Clearly, this model fails to reproduce the observed velocity dispersion profile, which, instead, shows the need for large amounts of dark matter at every radius. The solid curve is the best-fitting model with a dark matter halo (Pascale et al., 2018). Optical image credit: ESO/DSS2. Figure courtesy of R. Pascale.

The smallest dwarf galaxies in the Local Group are called **ultra-faint dwarfs** (UFDs) and were first discovered using the SDSS. When compared to UFDs, the previously known dwarf galaxies are often referred to as **classical dwarfs**. Some of the UFDs are so small that they contain only a few hundred stars (Tab. 6.2). They are close to either the Milky Way or M 31, and are several magnitudes fainter than classical dwarfs and they all lack gas, except Leo T ($M_V \simeq -8.0$). Their total mass-to-light ratios are estimated to be as high as a few thousands in some cases. However, the few stars available and the potential contamination of stars belonging to the Milky Way's halo (§4.6.1) make these determinations uncertain. The discovery of the UFDs provides a partial solution to the problem of the missing satellites that we discuss in §10.9.2.

### 6.3.3 Star Formation Histories and Chemical Pattern

The study of resolved stellar populations in dwarf galaxies allows us to reconstruct precise SFHs for most dwarf galaxies in the Local Group. In fact, the Local Group is the only environment where such a study can be carried out with the current instrumentation. As the dwarf galaxies are close by and relatively diffuse, we can separate individual stars and build **colour–magnitude diagrams** (CMDs) that go some magnitudes below the main-sequence turn-off of the oldest stellar populations (§C.5). Two examples are shown in the top panels of Fig. 6.6 for the Cetus dSph ($M_V \simeq -11.2$) and the Leo A dIrr. These CMDs are different from those of single star clusters, which appear narrower and closely follow a specific isochrone (§8.6.2) that identifies the age of the cluster. The CMDs of dwarfs are instead very broad as they contain a range of stellar populations that were born in the dwarfs at different times (i.e. they have different ages) and we can envision them as the sum of several CMDs of single populations. Making use of stellar evolution models (§C.5) allows us to derive very accurate SFHs (§4.1.5), shown in the bottom panels of Fig. 6.6. The SFHs of dwarf galaxies have various shapes and are sometimes intermittent, in the sense that there are periods when the star formation is low or switched off. The clearest example is the Carina dSph ($M_V \simeq -9.1$) with at least three distinct episodes of star formation.

In general terms, one can distinguish between early star formation (like Cetus) and prolonged/ongoing star formation (like Leo A). Several dSphs have formed all their stars within a few gigayears at early times and then stopped probably because the interaction with the Milky Way or M 31 drove away their gas (Fig. 6.4). Dwarf irregulars are currently forming stars. An interesting result of these resolved stellar population analyses of Local Group dwarfs is that they all started forming stars at the earliest epochs. In fact, there is no known galaxy in our vicinity without an old ($\gtrsim 10$ Gyr) stellar population. The example of Leo A is particularly instructive: a few hundred stars formed early and then a shutdown of the star formation occurred, to be restarted 4 Gyr later, when $\approx 95\%$ of its stellar mass has been formed (Fig. 6.6, bottom right). It is only thanks to the careful study of resolved stars that this tiny level of early star formation could be revealed in this galaxy.

In addition to kinematic and dynamical information, the spectra of individual stars in dwarf galaxies return key information about their chemistry and the evolution of their

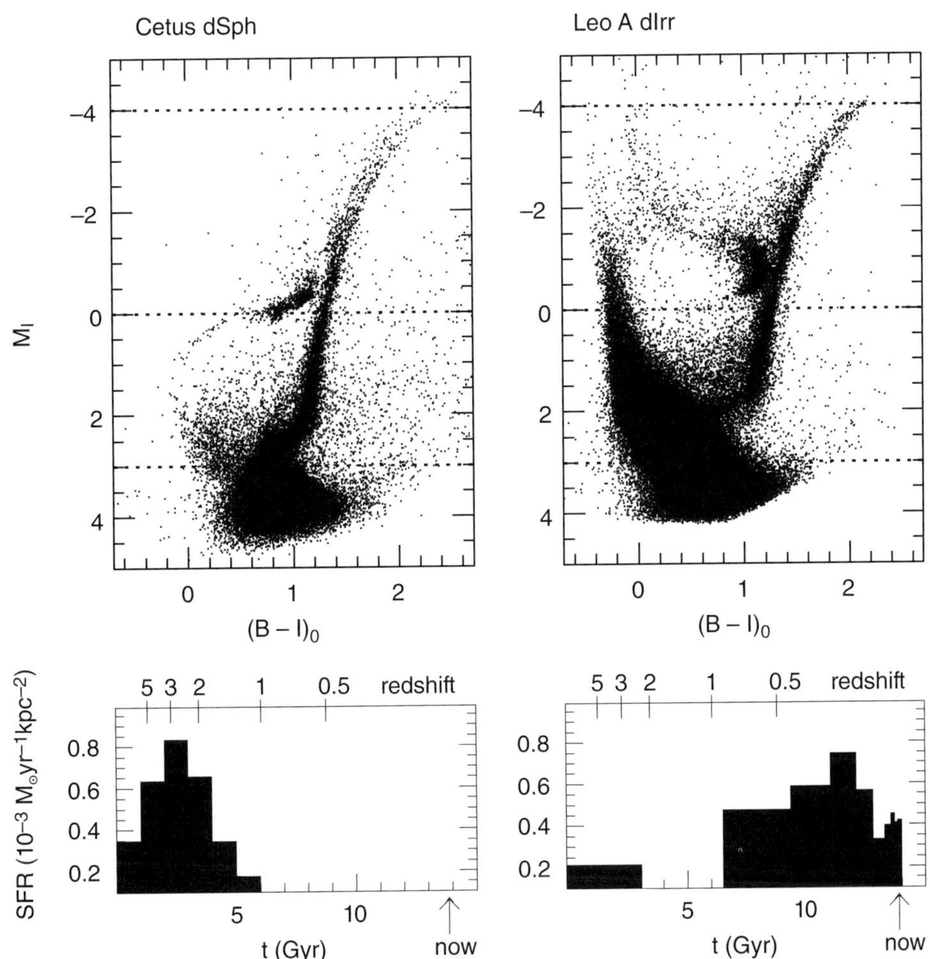


Fig. 6.6 CMDs (*top*) and SFHs (SFR surface density as a function of time; *bottom*) for two representative dwarf galaxies in the Local Group. Cetus (*left*) is a typical dSph with an early star formation activity that stopped about 8 Gyr ago. Leo A (*right*) is a dIrr that formed the majority of its stars at relatively recent times. Data from Monelli et al. (2010) and Cole et al. (2007), for Cetus and Leo A, respectively. Figure courtesy of E. Tolstoy.

ISM. In general, dwarf galaxies are metal-poor, with metallicities that overlap with some of the halo stars of the Milky Way (Fig. 4.24). However, the [$\alpha$/Fe] ratio of stars in dSphs reveals a pattern rather distinct from that of the Milky Way at the same [Fe/H]. Fig. 6.7 shows the location in the [Fe/H]–[Mg/Fe] space of stars in the Sculptor dSph galaxy, which we can take as a representative small dSph. We have already discussed that the 'knee' visible in this diagram is caused by the onset of Type Ia SNe and their release of large quantities of iron (with respect to $\alpha$-elements) in the ISM (§4.6.1). This enrichment occurs on timescales of order of 1 Gyr from the start of the star formation in a galaxy. The knee in Sculptor's stars is clearly at lower metallicity ([Fe/H] $\sim -2$) with respect to that of the Milky Way ([Fe/H] $> -1$). This means that the enrichment of iron in Sculptor has occurred

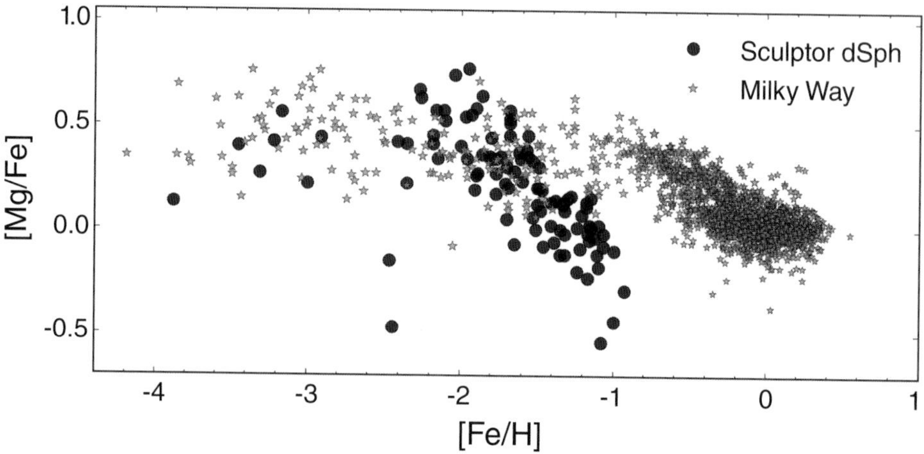


Fig. 6.7 Ratio between the abundances of the $\alpha$-element Mg and Fe versus metallicity ([Fe/H]) in stellar photospheres. The grey stars show the abundances of stars in the Milky Way (halo, thick and thin disc; same as Fig. 4.24), while the black circles show stars of the Sculptor dSph galaxy ($M_V \simeq -11.1$). The typical errors are $\approx 0.1$ dex increasing towards lower metallicities. Data for Sculptor from Tolstoy et al. (2009), Starkenburg et al. (2013) and Jablonka et al. (2015).

less efficiently and, when Type Ia SNe started to explode, the Fe abundance in its ISM was still rather low. This result is general for other dwarf galaxies, including dIrrs. However, the location of the knee in the metallicity axis is not always in the same position as it seems to roughly scale with the galaxy total mass (the higher the mass, the higher the [Fe/H] of the knee).

A diagram like Fig. 6.7 also provides a way to 'recognise' stars that have formed in a dwarf galaxy and are now incorporated in the halo of the Milky Way because the satellite has been disrupted. This information can be combined with kinematic measurements, for instance from *Gaia* (Fig. 4.23), to build a complete picture of the formation of the Galactic stellar halo (§10.4).

## 6.4 Clusters of Galaxies

**Clusters of galaxies** are gravitationally bound aggregations of galaxies containing from $\approx 50$ up to a few $\times\ 10^3$ galaxies with velocity dispersion of the order of a thousand $\mathrm{km\,s^{-1}}$. The most regular clusters are believed to be almost virialised, i.e. close to dynamical equilibrium. Clusters have virial masses $\gtrsim 10^{14} \mathcal{M}_\odot$ and virial radii $\gtrsim 1$ Mpc. Though clusters of galaxies were originally identified as overdensities of galaxies in optical surveys, we now know that the mass in stars is by no means the dominant mass component of clusters. Most of the mass in clusters (84–90%) is in the form of dark matter, while only 10–16% of the mass is in the form of baryons, mostly hot gas (stars represent only 1–5% of the total mass). The baryon fraction of clusters of galaxies increases for increasing cluster virial mass, reaching for the most massive clusters ($\mathcal{M}_{\rm vir} \sim 10^{15}\mathcal{M}_\odot$) values close to the cosmological baryon fraction $f_{\rm b} = \Omega_{\rm b,0}/\Omega_{\rm m,0} \simeq 0.16$ (Tab. 6.1; §2.5). This suggests that clusters of galaxies, especially the most massive ones, have efficiently retained their baryons during their formation (which is instead not the case for galaxies; §6.7 and §10.10.1). Tab. 6.1 reports reference values of a few quantities characterising clusters of galaxies. Here below we review the properties of galaxy clusters that are most relevant to the question of galaxy formation and evolution.

### 6.4.1 Galaxies in Clusters

Most of the optical emission of galaxy clusters is due to light from stars in galaxies (Fig. 6.8, left). Optical catalogues of clusters of galaxies are constructed by first identifying overdensities in galaxy surveys. Then measurements of galaxy redshifts are used to select *bona fide* members and exclude foreground and background objects.[2] Following Abell (1958), who constructed one of the first and most important catalogues of clusters of galaxies, clusters are classified in classes of **richness**, defined as the number of galaxies, no more than two magnitudes fainter than the third brightest galaxy, within a fixed radius (usually 2 Mpc), known as the **Abell radius**. The luminosity distribution of galaxies within

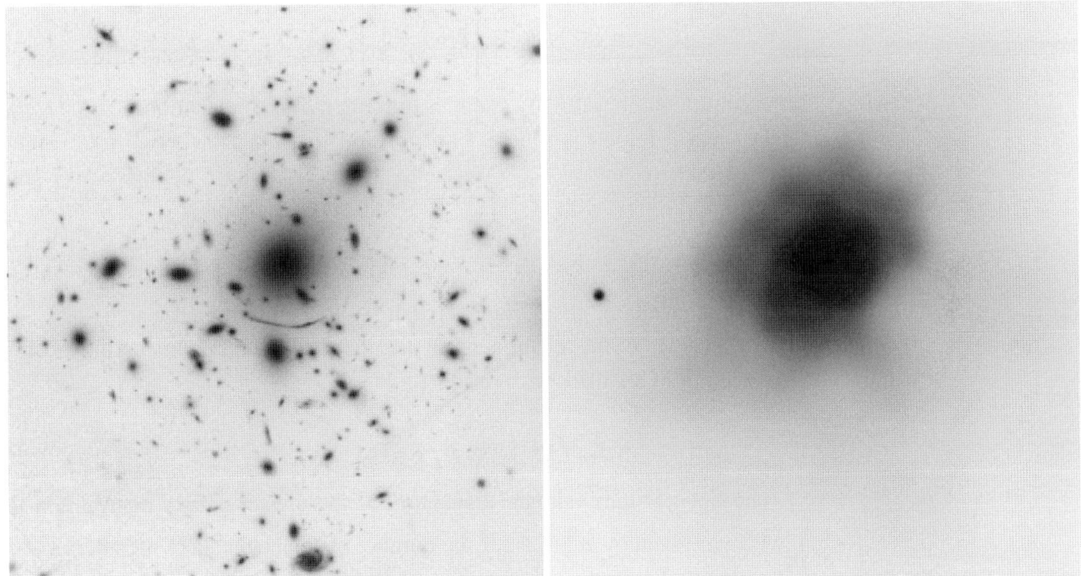

Fig. 6.8 Optical (*left panel*) and X-ray (*right panel*) images of the cluster of galaxies Abell 383 on the same scale (the width of each image corresponds to about 0.5 Mpc). See Newman et al. (2011) and Morandi and Limousin (2012) for details. Credit: NASA/STScI, ESO/VLT and SDSS (Optical); NASA and *Chandra* X-ray Center (X-rays).

[2] Galaxies that appear close in projection on the plane of the sky might be very far from each other along the line of sight: members of the same cluster must also have very similar redshifts.

a cluster is well described by a Schechter LF (§3.5.1), with values of the parameters $\alpha$ and $L^*$ similar to those found for field galaxies (§3.5.1, Tab. 3.1). Based on the LF, the richness of a cluster can be alternatively defined as the number of galaxies more luminous than $L^*$.

Galaxy clusters are classified on the basis of the spatial distribution of galaxies as regular or irregular. **Regular clusters** (such as the Coma cluster) appear close to circular symmetry and do not show important substructures, which instead characterise **irregular clusters** (such as Virgo, the cluster of galaxies closest to us). In general, ellipticals and S0s are more common than spirals in clusters (§6.5.1), and the fraction of spirals is higher in irregular than in regular clusters. Moreover, the fraction of ellipticals is higher in the central parts than in the outskirts of the clusters. Many spirals in clusters have less cold neutral atomic gas than field spirals: these gas-poor galaxies are sometimes called **anaemic spirals**.

Quantitatively, the distribution of galaxies in a cluster can be characterised by measuring the surface number density of galaxies $\Sigma_{\rm gal}$ as a function of the projected distance $R$ from the galaxy centre. For regular clusters, a reasonably good model is a spherically symmetric distribution, for which (eq. 5.20)

$$\Sigma_{\rm gal}(R)=2\int_R^{\infty}\frac{n_{\rm gal}(r)r\,{\rm d}r}{\sqrt{r^2-R^2}}, \tag{6.3}$$

where $n_{\rm gal}(r)$ is the intrinsic number density of galaxies. Based on early studies of the Coma cluster, $n_{\rm gal}(r)$ has often been modelled with a **King analytic profile** (King, 1972)

$$n_{\rm gal}(r)=n_{\rm gal,0}\left[1+\left(\frac{r}{r_{\rm c}}\right)^2\right]^{-3/2}, \tag{6.4}$$

where $n_{\rm gal,0}$ is the central number density and $r_{\rm c}$ is the core radius. In this model, $n_{\rm gal}\approx n_{\rm gal,0}$ for $r\ll r_{\rm c}$, so the central region is a core with uniform density. However, more recent studies have shown that in general $n_{\rm gal}(r)$ is better represented by a double power-law model: in the centre $n_{\rm gal}\propto r^{-\alpha_{\rm in}}$ with $1\lesssim\alpha_{\rm in}\lesssim 1.5$ and in the outer parts $n_{\rm gal}\propto r^{-\alpha_{\rm out}}$ with $\alpha_{\rm out}\approx 3$–$4$.

Let us consider a cluster for which we have measurements of galaxy redshifts. Once member galaxies are selected, the cosmological redshift of the cluster is estimated as $z_{\rm clust}\equiv\langle z\rangle$, where $\langle z\rangle$ is the average of the redshifts of all members. In the rest frame of the cluster the line-of-sight velocity of a galaxy with redshift $z$ is

$$v_{\rm los}=\frac{c\,(z-z_{\rm clust})}{1+z_{\rm clust}} \tag{6.5}$$

(see eq. C.35). The line-of-sight velocity dispersion of the cluster galaxy distribution is $\sigma_{\rm los}=\sqrt{\langle v_{\rm los}^2\rangle}$, with $v_{\rm los}$ given by eq. (6.5). If the number of selected galaxy members is $N$, the line-of-sight velocity dispersion can be written as

$$\sigma_{\rm los}=\left[\frac{1}{(N-1)}\sum_{i=1}^{N}\frac{c^2\,(z_i-z_{\rm clust})^2}{(1+z_{\rm clust})^2}\right]^{1/2}. \tag{6.6}$$

As happens for galaxy groups, $\sigma_{\rm los}^2$ is found to be proportional to the temperature of the intracluster medium (Fig. 6.2; see also §6.4.2) and can be used to estimate the cluster mass using the virial theorem (§5.3.1 and §6.4.4).

In analogy to the line-of-sight velocity distribution of stars in galaxies (§5.1.2 and §C.8), it is useful to consider the distribution of the line-of-sight velocities of galaxies in a cluster as a function of distance from the cluster centre. In the case of clusters, the procedure is very similar to the one used to measure the line-of-sight velocity dispersion profiles of nearby dSphs, for which we have measurements of the velocities of individual stars (Fig. 6.5). In practice, one usually constructs a line-of-sight velocity dispersion profile by measuring

$$\sigma_{\rm los} = \sqrt{\langle (v_{\rm los} - \langle v_{\rm los} \rangle)^2 \rangle} \tag{6.7}$$

in annuli of different radii around the cluster centre, where the averages are computed only over galaxies within each annulus. We typically find that $\sigma_{\rm los}$ depends on the position in the cluster, slightly decreasing for increasing projected distance from the centre.

The most luminous galaxy in a cluster, called the **brightest cluster galaxy** (BCG), is invariably a giant elliptical (Tab. 5.1). The BCGs, which typically sit at the centre of their clusters, are the most luminous galaxies in the Universe, with luminosities up to $10^{12} L_\odot$. The estimated total mass of the most massive BCGs is $\gtrsim 10^{13} \mathcal{M}_\odot$, although, given the special location of these systems, it is not straightforward to separate the mass associated with the galaxy from that associated with the cluster. BCGs are not statistically drawn from the LF of cluster galaxies, in the sense that, given their luminosity, they would be expected to be extremely rare on the basis of the LF of normal galaxies in clusters. Moreover, many BCGs have double or multiple nuclei: two or more surface brightness peaks are observed in the central regions. The above peculiar properties strongly suggest that BCGs formed via accretion of companion cluster members, due to dynamical friction (§8.9.3), a phenomenon known by the picturesque name of **galactic cannibalism**.

In some BCGs an extended low surface brightness envelope of stars is detected: for historical reasons,[3] the BCGs in which this feature is present are often called **cD galaxies**. In cD galaxies the central surface brightness profile is well fitted by a Sérsic law (eq. 3.1) with $n \gtrsim 4$, but the extended halo deviates from the outward extrapolation of the Sérsic fit (an excess of light is observed in the outer regions). The luminous halo can be considered as associated with the BCG or with the cluster. In the latter case the luminous halo is called **intracluster light** (stars in clusters of galaxies that are not in galaxies). The intracluster light is generally less than $\approx 10\%$ of the total luminosity of stars in the galaxies of the cluster.

In regular clusters, the central BCG is almost always a radio galaxy (§3.6 and §8.8), but radio sources associated with AGNs are found also in non-central cluster galaxies. These non-central radio galaxies are often characterised by a distorted **head–tail** shape due to interaction with the intracluster medium (§6.4.2). In a small fraction of galaxy clusters, typically irregular clusters undergoing cluster–cluster mergers, radio observations have revealed also extended emission that is not associated with galaxies (**radio relics** and **radio**

[3] In classification schemes proposed in the 1950s, galaxies like ellipticals, but with distinct outer halos with shallow surface brightness profiles, were named D galaxies. The most luminous D galaxies were given the name cD, where the prefix 'c' was used in a manner similar to the (no longer used) notation for supergiant stars in stellar spectroscopy.

**halos**), which are interpreted as being due to synchrotron radiation (§D.1.7) produced by non-thermal (relativistic) electrons moving in the cluster magnetic field.

### 6.4.2 The Intracluster Medium

The dominant baryonic component of galaxy clusters is the **intracluster medium** (ICM): almost fully ionised plasma at about the system's virial temperature (§8.2.1), $10^7$–$10^8$ K, permeating the entire cluster volume. Here we briefly describe how this hot ICM is detected observationally and report its main physical properties.

#### X-Ray Emission

The ICM is responsible for the diffuse X-ray emission of clusters of galaxies (Fig. 6.8, right panel), which are extremely luminous X-ray sources, with luminosities in the range $10^{43}\,\mathrm{erg\,s^{-1}} \lesssim L_X \lesssim 10^{45}\,\mathrm{erg\,s^{-1}}$. Given the high temperatures of the ICM, its X-ray emission is dominated by thermal bremsstrahlung (§D.1.6). Compared to the selection of clusters in optical galaxy surveys, which can be contaminated by overdensities due to chance alignment of galaxies, the X-ray selection has the advantage of automatically including only real clusters. The presence of the diffuse X-ray emission guarantees that there is a plasma confined by a deep gravitational potential well.

Galaxy clusters present a variety of morphologies of the X-ray emission and are classified from regular (almost circularly symmetric) to irregular (asymmetric, with much substructure), in a way similar to the optical classification (§6.4.1). For sufficiently regular clusters, as in the case of the X-ray gas of massive ellipticals (§5.2.1), deprojection of spatially resolved X-ray spectroscopic measurements allows us to reconstruct the intrinsic density and temperature distributions, typically under the assumption of spherical symmetry (Figs. 6.11a and 6.11b). For massive clusters, the soft (0.5–2 keV) X-ray emissivity is almost independent of temperature (which mainly affects the high-energy cut-off of the spectrum; eq. D.18), so the gas density profile can be obtained directly from the X-ray surface brightness profile (§5.2.1). X-ray spectra of galaxy clusters reveal the presence of emission lines of highly ionised metals. The most important feature of the X-ray spectrum of a galaxy cluster is the **7 keV Fe line complex**, which is the combination of several K-shell[4] emission lines of iron in various ionisation states, but, at typical cluster temperatures, predominantly of the (He-like) Fe XXV and (H-like) Fe XXVI ions. Except for the highest-temperature systems, in which lighter elements are fully ionised, the X-ray spectra of galaxy clusters, an example of which is shown in Fig. 6.9, are also characterised by many lower-energy lines from highly ionised C, N, O, Ne, Mg, Si, S and Ni, as well as lower-energy Fe lines. The element abundance in the ICM can be estimated from measurements of the emission line equivalent width (eq. 3.6). A reference value of the iron abundance of the ICM is $[\mathrm{Fe/H}] \approx -0.5$, i.e. about one-third of the solar value (§C.6.1).

[4] K-shell emission lines are produced by electron transitions to the innermost energy level (with principal quantum number $n = 1$). L-shell lines correspond to transitions to levels with $n = 2$.

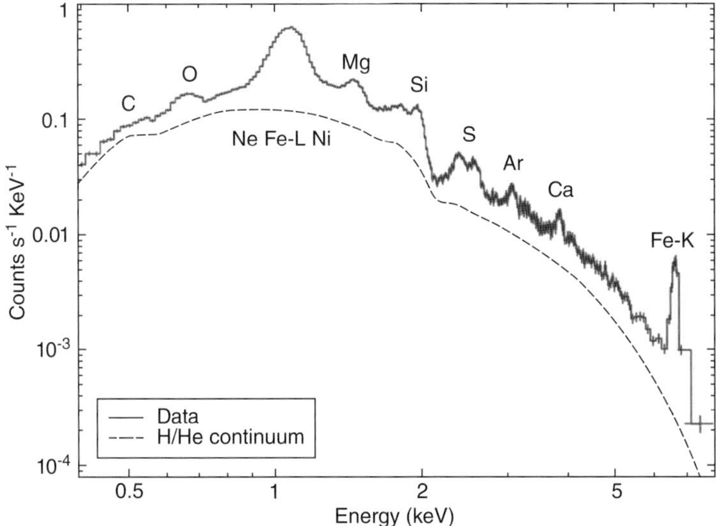


Fig. 6.9

X-ray spectrum of the Centaurus cluster of galaxies, which has an ICM temperature of $\approx 2 \times 10^7$ K. The strongest lines of each element are indicated ('Ne Fe-L Ni' refers to the peak at $\approx 1$ keV, which is due to a combination of lines). In the case of iron, 'Fe-L' and 'Fe-K' refer to recombination of electrons to the L and K shells, respectively. The dashed curve shows the estimated contribution to the spectrum due to hydrogen and helium. From Sanders and Fabian (2006).

Spatially resolved spectral observations reveal the presence of **ICM metallicity gradients**, with higher metallicity in the central regions.

## Sunyaev–Zeldovich Effect

Clusters of galaxies can be detected in the microwaves because of an effect on the CMB (§2.4) known as the **Sunyaev–Zeldovich effect** (SZE; Sunyaev and Zeldovich, 1970). The CMB is characterised by a black-body spectrum (§D.1.3) at temperature $T_{\rm CMB} \simeq 2.726$ K. The free electrons of the ICM scatter the CMB photons and produce a distortion of the CMB spectrum. The importance of the photon–electron interaction is quantified by the optical depth (§D.1.1)

$$\tau = \sigma_{\rm T} \int n_{\rm e}(x_{\rm los}) {\rm d}x_{\rm los}, \tag{6.8}$$

where the integral is through the entire cluster along the line of sight (labelled by $x_{\rm los}$), $n_{\rm e}$ is the ICM electron number density and

$$\sigma_{\rm T} = \frac{8\pi}{3}\left(\frac{e^2}{m_{\rm e}c^2}\right)^2 \simeq 6.65 \times 10^{-25}\ {\rm cm}^2 \tag{6.9}$$

is the **Thomson cross section** ($e$ is the electron charge and $m_{\rm e}$ is the electron mass). For typical electron number density $n_{\rm e} \sim 10^{-2}$ cm$^{-3}$ and cluster size $\sim 1$ Mpc, the optical depth is of the order of $\tau \sim 10^{-2}$, meaning that there is $\sim 1\%$ probability of scattering.

As the characteristic energy of the CMB photons is much lower than the energy of the cluster electrons, the result of the interaction is an inverse Compton scattering (§8.1.3), with a consequent distortion of the CMB spectrum at the position of the cluster in the plane of the sky. Here we report the main equations describing the SZE in the case of non-relativistic electrons (relativistic corrections become non-negligible only for the most massive clusters). We refer the reader to Birkinshaw (1999) for the details of the derivation of these equations.

A photon of original frequency $\nu$ after the scattering has frequency $\nu + \Delta\nu$, with $\Delta\nu/\nu \approx k_B T_e/(m_e c^2)$, where $T_e$ is the temperature of the electrons (i.e. of the ICM). Thus, the distorted spectrum is not a black-body spectrum because $\Delta\nu$ depends on $\nu$ at given $T_e$. At given frequency $\nu$, the spectrum of the emerging radiation can be characterised by the brightness temperature $T_b$ (§C.2). For the unperturbed spectrum, which is a black body, $T_b = T_{CMB}$ by definition. The spectral distortion can be written in terms of $T_b$ as

$$\frac{\Delta T_b}{T_b} = \frac{\Delta I_\nu}{I_\nu}\frac{d\ln T_b}{d\ln I_\nu} = f(x)y, \tag{6.10}$$

where $I_\nu$ is the intensity (eq. C.1) at frequency $\nu$, $x \equiv h\nu/(k_B T_b)$, $f(x) = x\coth(x/2) - 4$ and

$$y \equiv \int \frac{k_B T_e(\tau)}{m_e c^2} d\tau = \int \frac{k_B T_e(x_{los})}{m_e c^2} n_e(x_{los})\sigma_T \, dx_{los} \tag{6.11}$$

is the **Compton $y$-parameter**, which is a dimensionless quantity.

As illustrated by Fig. 6.10, the distortion of the spectrum is such that the intensity (left panel) and the brightness temperature (right panel) decrease at $\nu < 218$ GHz and increase

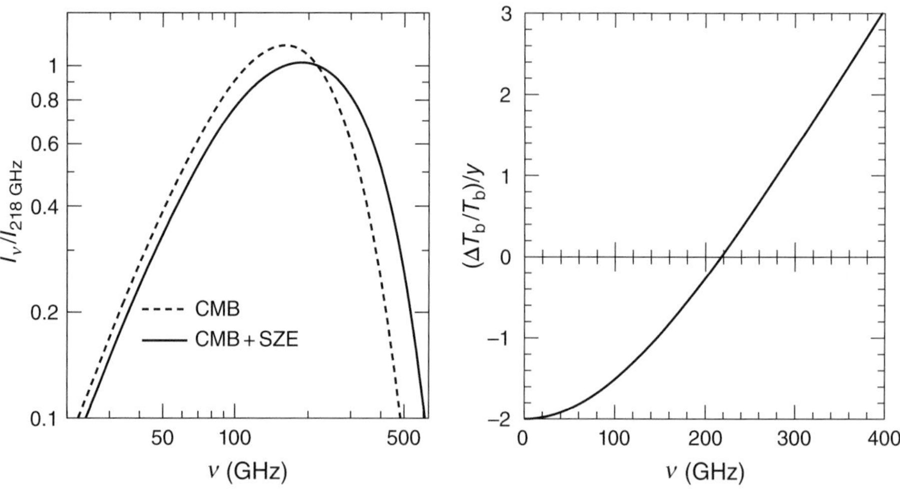


Fig. 6.10 *Left panel.* Unperturbed black-body CMB spectrum (dashed curve) and CMB spectrum distorted by the thermal SZE (solid curve). For the purpose of illustrating the effect, here we have assumed an unrealistically high value of the Compton $y$-parameter, $y = 0.05$ (realistic values for a massive cluster of galaxies are $y \sim 10^{-4}$). The curves are normalised at 218 GHz, where the thermal SZE is null. *Right panel.* Relative variations of the CMB brightness temperature $\Delta T_b/T_b$ (in units of $y$) as a function of frequency due to the SZE: $\Delta T_b/T_b$ is of the order of $y$.

at $\nu > 218$ GHz. As the measurements are often performed at low frequency ($\nu < 218$ GHz), the SZE is sometimes referred to as **microwave diminution**. In the Rayleigh–Jeans regime (low frequency, $h\nu \ll k_B T_b$; eq. C.10) $f(x) \to -2$ and the variation of brightness temperature is

$$\frac{\Delta T_b}{T_b} = -2y = -\frac{2k_B\sigma_T}{m_e c^2}\int n_e(x_{los})T_e(x_{los})\mathrm{d}x_{los}. \tag{6.12}$$

Therefore, in low-frequency data, galaxy clusters appear as lower-temperature regions in CMB maps. The SZE is small ($\Delta T_b/T_b < 10^{-3}$), but independent of the cluster distance, so it is extremely important especially for studying high-redshift clusters (§11.2.13).

With spatially resolved SZE observations it is possible, for a given cluster, to obtain a map of $\Delta T_b/T_b$ and therefore a map of the Compton $y$-parameter (eq. 6.11), which can be used to constrain the gas pressure distribution within the cluster ($P \propto n_e T_e \propto y$). When the SZE observation of the cluster is not spatially resolved, the only available measure is the **Compton integrated $Y$-parameter**

$$Y = \int y\,\mathrm{d}A = \frac{k_B\sigma_T}{m_e c^2}\int n_e T_e\,\mathrm{d}V, \tag{6.13}$$

where $\mathrm{d}A$ is the projected area element, $\mathrm{d}V = \mathrm{d}A\,\mathrm{d}x_{los}$ is the volume element and the integrals are performed over the entire cluster. Under the assumption that the temperature is not strongly dependent on position, we can substitute $T_e$ in eq. (6.13) with its average $\langle T_e\rangle$ to obtain

$$M_{gas} = \int \rho\,\mathrm{d}V = \mu m_p \frac{n}{n_e}\int n_e\,\mathrm{d}V \approx \frac{n}{n_e}\frac{\mu m_p m_e c^2}{k_B\sigma_T\langle T_e\rangle}Y, \tag{6.14}$$

where $\rho$ is the gas density, $n = \rho/(\mu m_p)$, we have assumed that the ratio $n/n_e$ is position-independent ($n/n_e \approx 2$ for a fully ionised plasma) and we have used eq. (6.13). Thus, if $\langle T_e\rangle$ is estimated from X-ray spectra, $Y$ can be used to estimate the total mass of gas.

The effect described above is due to the thermal motion and is therefore sometimes called the **thermal SZE**. If a cluster has non-negligible peculiar velocity (i.e. it is moving with respect to the CMB), there is an additional spectral distortion of the CMB, known as the **kinetic SZE**, which is typically at least an order of magnitude weaker than the thermal SZE and therefore can only be observed at frequencies close to 218 GHz, at which the thermal SZE is negligible.

## Density and Temperature Distributions

The ICM shares several properties with the hot gaseous halos of elliptical galaxies (§5.2.1). The gas distribution is such that the gas density spans a few orders of magnitude from the centre to the outskirts. The value of the central gas number density varies from cluster to cluster, in the reference range $10^{-3}$–$10^{-1}$ cm$^{-3}$ (Fig. 6.11a). The ICM temperature varies less with distance from the cluster centre, so that, as a zeroth-order approximation, the gas can be modelled as isothermal. A simple but successful model of the density profile of the ICM in clusters is the so-called **$\beta$-model**. In this model the distributions of both gas and galaxies are assumed to be isothermal, in equilibrium in the cluster gravitational potential.

Let us consider a spherically symmetric cluster: if the velocity distribution of the galaxies is isotropic and isothermal ($\sigma_r$ independent of radius; §5.3.2), then $\sigma_{\rm los} = \sigma_r$ (eq. 5.21) and the Jeans equation (eq. 5.18) can be written as

$$\sigma_{\rm los}^2 \frac{{\rm d}\ln n_{\rm gal}}{{\rm d}\ln r} = -\frac{GM(r)}{r}, \tag{6.15}$$

where $n_{\rm gal}(r)$ is the galaxy number density. The hydrostatic equilibrium equation (eq. 5.27) for an isothermal gas reads

$$\frac{k_{\rm B}T}{\mu m_{\rm p}} \frac{{\rm d}\ln n}{{\rm d}\ln r} = -\frac{GM(r)}{r}, \tag{6.16}$$

where $n$ is the gas number density. Combining the two equations above, we get

$$n \propto n_{\rm gal}^{\beta}, \tag{6.17}$$

where the dimensionless parameter $\beta \equiv \mu m_{\rm p}\sigma_{\rm los}^2/(k_{\rm B}T)$ is the ratio between galaxy 'temperature' $\mu m_{\rm p}\sigma_{\rm los}^2/k_{\rm B}$ and gas temperature $T$. For instance, if the galaxy distribution is a King analytic model (eq. 6.4), the gas density profile is

$$n(r) = n_0 \left[1 + \left(\frac{r}{r_{\rm c}}\right)^2\right]^{-3\beta/2}. \tag{6.18}$$

The X-ray surface brightness $I_{\rm X}$ is given by eq. (5.10), where $j_{\rm X}$ is the X-ray emissivity (eq. 5.9): using eq. (6.18) we get

$$I_{\rm X}(R) \propto \left[1 + \left(\frac{R}{r_{\rm c}}\right)^2\right]^{-3\beta+\frac{1}{2}}. \tag{6.19}$$

Fitting this function to the X-ray surface brightness profiles of observed clusters returns $1/2 \lesssim \beta \lesssim 1$, with a reference value $\beta \approx 2/3$. It follows that the gas distribution is more extended than the galaxy distribution: for instance, for $\beta = 2/3$, at large radii ($r \gg r_{\rm c}$) $n \propto r^{-2}$ for the gas, while $n_{\rm gal} \propto r^{-3}$.

Though very useful as a reference model, the $\beta$-model does not capture the variety of gas properties observed in clusters. In fact, when X-ray spectroscopic observations allow us to measure the temperature profile, measurable deviations from isothermality are found (Fig. 6.11b). Moreover, we have seen in §6.4.1 that the galaxy density distribution might deviate significantly from an isotropic analytic King model.

### 6.4.3 Cool-Core and Non-Cool-Core Clusters

Though the ICM is close to isothermal, measurements of the ICM temperature profiles reveal the presence of temperature gradients (Fig. 6.11b). When in the inner parts the gas temperature increases outwards, the central region of the cluster is called a **cool core** and the cluster is called a **cool-core cluster**. The existence of cool cores does not come as a surprise, because the ICM loses energy via radiative cooling (§8.1.1) and, given the dependence of the cooling time $t_{\rm cool}$ on gas temperature and density (eq. 8.4), $t_{\rm cool}$ is relatively short in the cluster centre (where the gas density is higher) and increases outwards (Fig. 6.11d). In cool-core clusters the cooling time in the cluster core is shorter

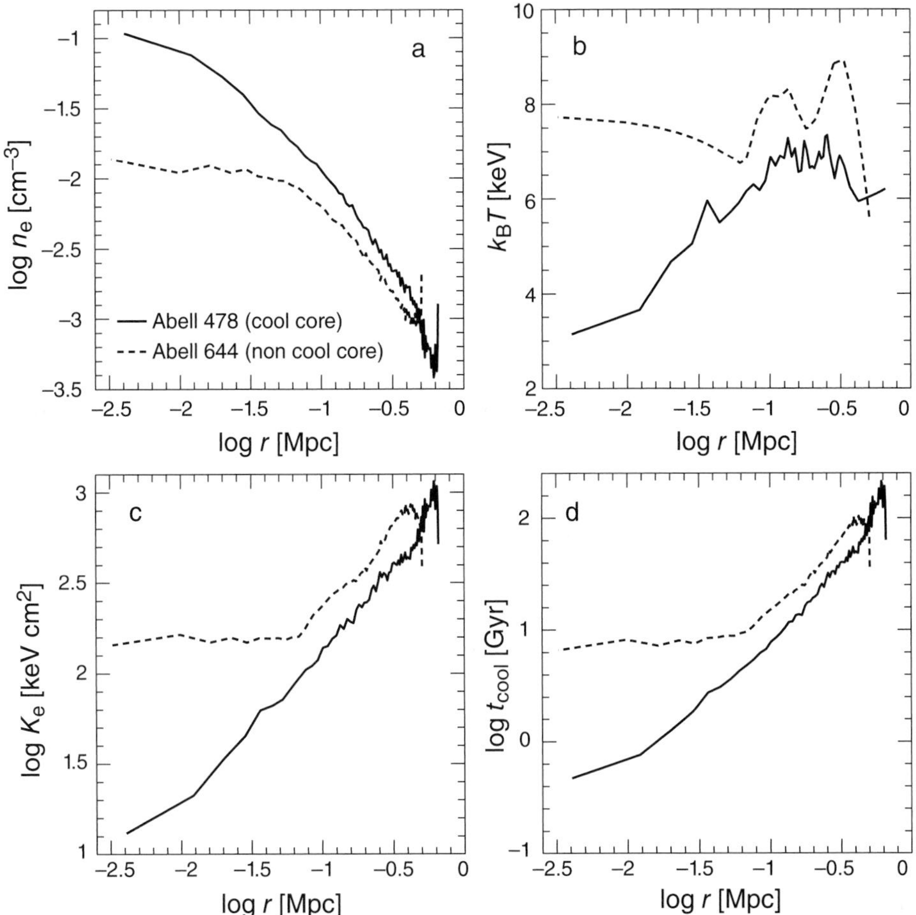


Fig. 6.11 Electron number density (*panel a*), temperature (*panel b*), 'entropy' (*panel c*) and cooling time (*panel d*) profiles of the ICM of the galaxy clusters Abell 478 (a cool-core cluster) and Abell 644 (a non-cool-core cluster). Data from Cavagnolo et al. (2009).

than the cluster age $t_{age}$ ($\sim 10^{10}$ yr for low-redshift clusters). The size of the cool core is usually defined, under the assumption of spherical symmetry, by the **cooling radius** $r_{cool}$, which is such that $t_{cool} < t_{age}$ if $r < r_{cool}$. The existence of cool cores is related to the question of cooling flows (§5.2.1), which has important implications for the theory of galaxy formation (§8.2.3 and §10.6.4).

A quantity often used in the study of the ICM, which is a proxy for the gas thermodynamic entropy, is the **entropy index**

$$K \equiv \frac{P}{\rho^{\gamma}}, \tag{6.20}$$

where $\gamma$ is the adiabatic index. For a monoatomic gas ($\gamma = 5/3$)

$$K = \frac{k_B T}{\mu m_p \rho^{2/3}}, \tag{6.21}$$

**Table 6.3** Distinctive properties of prototypical cool-core and non-cool-core clusters

| Cool-core clusters | Non-cool-core clusters |
|---|---|
| Outward-increasing central temperature profile | Flatter temperature profile |
| Higher central gas density | Lower central gas density |
| Steep central entropy profile | Shallow central entropy profile |
| Regular | Irregular |
| Central dominant galaxy | No central dominant galaxy |
| Circularly symmetric | Asymmetric |
| Little substructure | Substructures |
| No diffuse radio halo | Diffuse radio halo |
| Central radio galaxy | No central radio galaxy |

in units of $\mathrm{erg\,cm^2\,g^{-5/3}}$. The entropy index $K$, which is often shortened to 'entropy' in the literature, is related to the thermodynamic entropy per unit mass $s$ by

$$s = \frac{k_\mathrm{B}}{\mu m_\mathrm{p}} \ln K^{3/2} + \mathrm{const.} \tag{6.22}$$

Sometimes also the quantity $K_\mathrm{e} \equiv k_\mathrm{B} T n_\mathrm{e}^{-2/3}$ is called 'entropy' (in units of $\mathrm{keV\,cm^2}$). When measuring ICM entropy profiles of clusters, we find a variety of shapes, but the common trend is that the entropy increases outwards (Fig. 6.11c). Thus, the gas distribution is convectively stable according to the **Schwarzschild criterion** (see e.g. Shu, 1992), which requires $\nabla s \cdot \nabla P < 0$ for stability (i.e. $\mathrm{d}s/\mathrm{d}r > 0$ in spherical symmetry, because $\mathrm{d}P/\mathrm{d}r < 0$ for a gas in hydrostatic equilibrium; eq. 5.27).

Besides the relatively short cooling times and the outward-increasing gas temperature, distinctive features of cool cores are a central spike in the X-ray surface brightness distribution (produced by the high central gas density and thus emissivity; Fig. 6.11a) and the steep entropy profile (Fig. 6.11c). Clusters of galaxies that do not possess a cool core are classified as **non-cool-core clusters**. The presence or absence of the cool core is found to correlate with other observed cluster properties, so, broadly speaking, galaxy clusters appear to belong to two distinct classes: cool-core clusters (more regular and symmetric) and non-cool-core clusters (more irregular and rich in substructures). The origin of this dichotomy is believed to be related to the merger history of the cluster: non-cool-core clusters are (or have been recently) undergoing a merger, while cool-core clusters have not had any recent merger. The main properties of these two families of galaxy clusters are reported in Tab. 6.3.

### 6.4.4 Cluster Mass and Scaling Relations

The determination of the masses of galaxy clusters is important not only to estimate the relative contribution of baryons and dark matter, but also because clusters trace the high-mass end of the halo mass function (§7.4.3), which is an important test for cosmological models. The mass distribution of a galaxy cluster can be estimated with different methods, similar to those described to estimate the mass of elliptical galaxies (§5.3): (1) virial theorem and dynamical modelling, based on the kinematics of the cluster's galaxies (i.e. considering the motion of the galaxies within the cluster; §5.3.1 and §5.3.2), (2) hydrostatic equilibrium, based on X-ray and SZE observations of the ICM (§5.3.3), and (3) gravitational lensing (both weak and strong; §5.3.4). When a central BCG is present, the dynamical modelling method can be applied also using as tracer the stellar component of the BCG, i.e. using measurements of the stellar velocity dispersion of the BCG. In some cases the combination of different methods allows us to robustly measure the mass profile of a cluster from kpc up to Mpc scales, as illustrated in Fig. 6.12 for seven massive clusters. Apart from the very central regions, where the stellar density of the BCG dominates, these mass profiles cannot be accounted for by the baryons and are then tracing the distribution of the dark matter, which largely dominates the mass budget (Tab. 6.1). Thanks to this kind of measurement, the virial mass $\mathcal{M}_{\rm vir}$ of galaxy clusters is a quantity reasonably well estimated from observations, with uncertainties of the order of 10–20%.

A few global properties of observed galaxy clusters, such as X-ray luminosity $L_{\rm X}$, virial mass $\mathcal{M}_{\rm vir}$, SZE $Y$-parameter and ICM average entropy $K$, are found to be significantly correlated with the ICM average temperature $T$. These empirical scaling relations are very useful tools for our understanding of cosmological structure formation. For instance, low-redshift clusters of galaxies are found to obey approximately the power-law scaling relations $\mathcal{M}_{\rm vir} \propto T^{1.5}$ and $L_{\rm X} \propto T^3$. Power-law correlations are also found by considering $K$ or $Y$ as functions of $T$, $L_{\rm X}$ or $\mathcal{M}_{\rm vir}$. The analysis of these scaling relations suggests a lack of low-entropy gas in lower-mass systems, likely due to the fact that the gas, before falling into the clusters, was preventively heated by non-gravitational processes (§10.7.3).

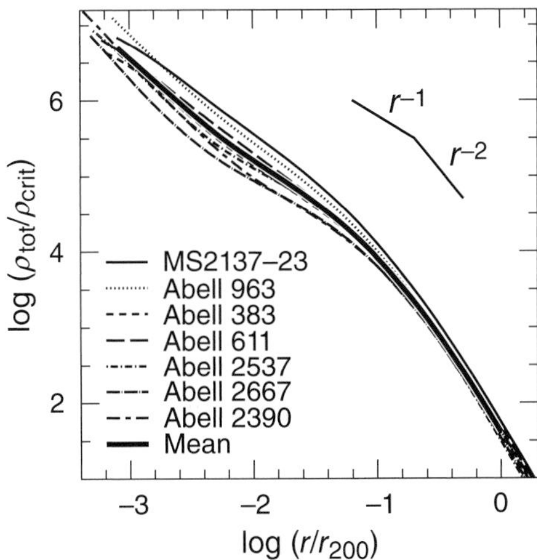


Fig. 6.12 Spherically averaged profiles of the total density, normalised by the virial radius $r_{200}$ and the critical density $\rho_{\rm crit}$ (eq. 2.23), of seven massive (virial mass $4\times10^{14}\mathcal{M}_\odot \lesssim \mathcal{M}_{200} \lesssim 2\times10^{15}\mathcal{M}_\odot$) clusters of galaxies ($r_{200}$ and $\mathcal{M}_{200}$ are defined in §7.3.2). The mass profiles are obtained by combining different probes: strong and weak gravitational lensing, X-ray data and stellar kinematics of the BCG. The mean density profile is represented by the thick curve. Power laws with logarithmic slopes −1 and −2 are shown for comparison. Adapted from Newman et al. (2013). © AAS, reproduced with permission.

The existence of correlations involving $\mathcal{M}_{\rm vir}$ allows us to infer $\mathcal{M}_{\rm vir}$ for clusters with no direct measurement of mass, which can be used to measure the cluster mass function to be compared with the halo mass function predicted by different cosmological models (§7.4.3).

## 6.5 The Influence of the Environment on Galaxy Properties

Present-day galaxies are located in a variety of environments with a wide range of densities and shapes, from high-density clumps with typical sizes of 1–2 Mpc (which correspond to the size of galaxy clusters), to regions resembling sheets, walls or bubbles. Long chains of galaxies, called **filaments**, with lengths up to 50–100 Mpc are also observed. These filaments intersect high-density nodes populated by the densest clusters of galaxies, where the density of galaxies can be as high as $\sim 1000$ times the average (§6.4.1). Within this complex network of galaxy distributions (the so-called **cosmic web**), vast regions with very few (or no) galaxies are also present. These lowest-density regions have been named **voids** and are extended on scales larger than 10–50 Mpc. Galaxies located in regions with intermediate densities between voids and clusters are usually called **field galaxies**. An important question is therefore whether and how the environment (i.e. the average density of galaxies in a given volume) influences galaxy properties.

### 6.5.1 The Morphology–Density Relation

One of the most important environmental effects is known as the **morphology–density relation** (Dressler, 1980). This consists in a strong dependence of the fraction of galaxy morphological types on the environmental density. The underlying environmental density around a given galaxy can be estimated with several methods (§11.1.9). A frequently used estimator is the sky surface density ($\Sigma_N$; Fig. 6.13) of galaxies around a central galaxy obtained by counting the $N$ nearest neighbours within a given redshift interval. The observational results show that the percentage of ETGs increases from $\approx 10$–20% in underdense regions to $\approx 80$–90% in the densest clusters, whereas the trend of SFGs (spirals and irregulars) is the opposite. This implies that ETGs are preferentially located in high-density environments, while SFGs are more common in the field. The influence of the environment is evident when the bimodal distribution of galaxy colours is displayed as a function of the environmental density. This is visible in Fig. 6.13, as the red sequence becomes progressively dominant in the densest environment whereas the blue cloud shows an opposite trend (§3.4.1; Fig. 3.9). The morphology–density relation is reflected also in the distribution functions of luminosity and stellar mass. Denser environments contain more luminous systems (i.e. the ETGs), whereas galaxies with $L < L^*$ (eq. 3.7) are preferentially located in relatively underdense environments. Similarly, the SMF shows that massive (i.e. red early-type) galaxies are preferentially located in high-density regions. This is shown by the SMF of ETGs, where the characteristic stellar mass ($\mathcal{M}^*$; §3.5.2) increases with the density of the environment where ETGs are located. Also within galaxy

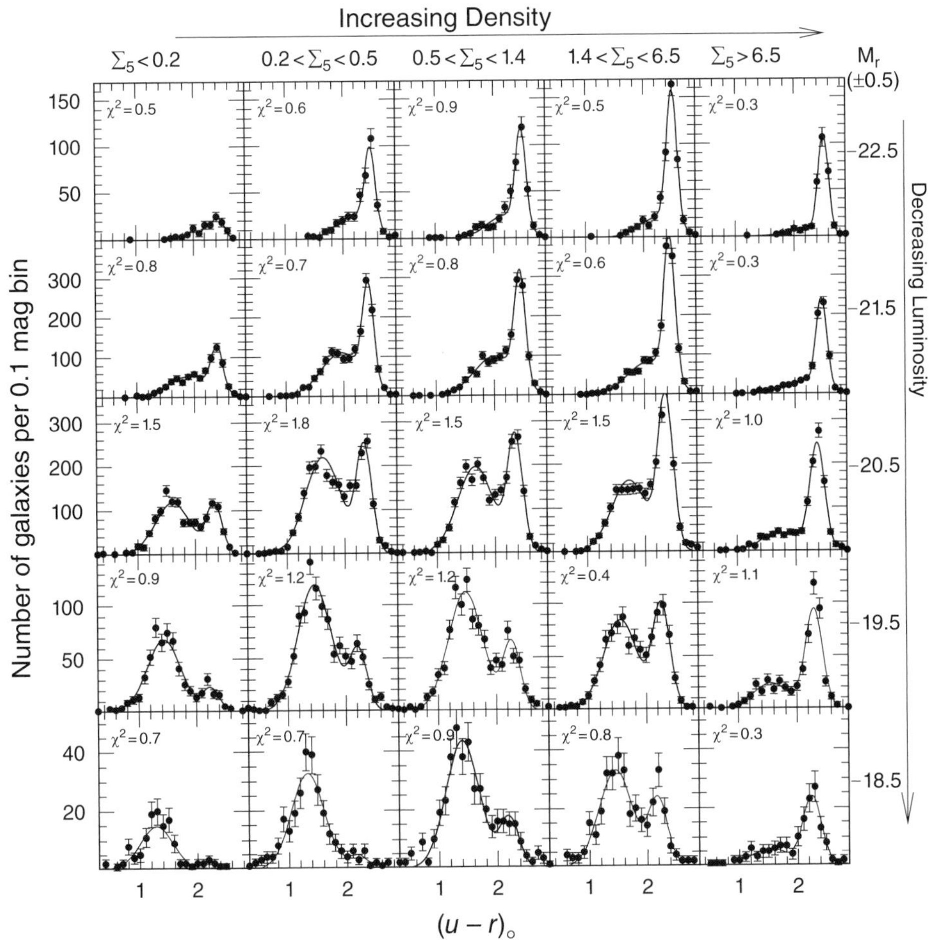


Fig. 6.13 The rest-frame $u-r$ colour distribution as a function of absolute magnitude ($M_r$; right $y$-axis) and environmental density $\Sigma_5$ (top $x$-axis) measured as the surface density of galaxies in units of $\mathrm{Mpc}^{-2}$ within a projected radius of 5 Mpc. The red sequence is centred at $(u-r)_0 \approx 2.3$, whereas the blue cloud is at $(u-r)_0 \approx 1.5$. The progressive depopulation of the red sequence for decreasing $\Sigma_5$ and luminosity, and the opposite trend of the blue cloud, are clearly visible. From Balogh et al. (2004). © AAS, reproduced with permission.

clusters, a radial segregation is often present, with ETGs and SFGs/gas-rich systems preferentially located in the inner and outer regions, respectively.

Other clear environmental effects have been found. For example, galaxies with the highest Sérsic indices ($n$), the largest D4000 continuum break or the lowest SFR are preferentially located in the densest environments. This is not surprising because it simply reflects the properties of ETGs having the largest values of $n$ and D4000, and the weakest or absent star formation (§3.3).

At the lowest end of environmental densities, galaxies in voids provide an important possibility to investigate the effects of evolving in isolation. Void galaxies do not look anomalous, and are mostly small discs with star formation and gas content rather similar to those of galaxies located in moderate-density environments. In contrast, the densest environments (i.e. clusters of galaxies) show a wide range of phenomena due to the high density of galaxies and the presence of diffuse hot gas (§6.4.2).

The general emerging picture is that the environmental effects become especially evident when galaxies are classified based on their structure (e.g. morphology or Sérsic index). The morphology–density relation is a clear example of this. However, for the scaling relations between stellar population properties (e.g. colours and ages) and mass (Fig. 5.9), the environmental effects seem to remain subdominant with respect to those of galaxy stellar or dynamical mass.

An important question is on what scales the environment is influential. Some results suggest that the local environment (i.e. on scales of 50–500 kpc, typical of galaxy groups) plays a more important role on galaxy properties than the global one on larger scales.

## 6.6 Large-Scale Structure and Galaxy Clustering

The average distance between galaxies is $\sim 1$ Mpc, about two orders of magnitude times the size of the stellar disc of the Milky Way. This means that it is necessary to observe large volumes of the Universe to study the so-called **large-scale structure** (LSS), i.e. how galaxies are spatially distributed on scales much larger than 1 Mpc. The knowledge of the LSS is based on the measurements of positions and redshifts of a large number of galaxies over wide sky areas. For example, the SDSS-III sample contains nearly one million galaxies at $0.2 < z < 0.7$ selected from about 8500 square degrees, corresponding to a probed volume of the Universe of $\approx 13$ Gpc$^3$.

Overall, the LSS looks like a complicated 'spider web' extended over scales of several hundred Mpc (Fig. 6.14). There are several approaches to characterise statistically and quantitatively the distribution of galaxies. The way galaxies clump together is called **clustering**, and the most widely used approach relies on the measurement of the correlation functions. The basic principle behind this method is that the distribution of galaxies can be seen as the superposition of high-density (e.g. clusters) or low-density (e.g. voids) fluctuations on the mean density field.

In 3D space, the simplest approach is to derive the **two-point correlation function** $\xi_{\rm gal}(r)$, which is a measure of the excess probability, relative to an unclustered Poissonian distribution, of finding a galaxy located at distance $r$ from a galaxy selected at random within a volume $dV$. The probability can be written as

$$d\mathcal{P} = n[1 + \xi_{\rm gal}(r)]dV, \tag{6.23}$$

where $r$ is a comoving distance and $n$ is the average number density of galaxies within the whole sampled volume. For a random distribution, $\xi_{\rm gal}(r) = 0$ and the probability is $d\mathcal{P} = n\,dV$. In practice, $\xi_{\rm gal}(r)$ is a measure of the amplitude of galaxy clustering as a function of the spatial scale. If $\xi_{\rm gal}(r) > 0$, the clustering is in excess with respect to random, and this corresponds to a case where galaxies tend to cluster. Instead, $\xi_{\rm gal}(r) < 0$ indicates a

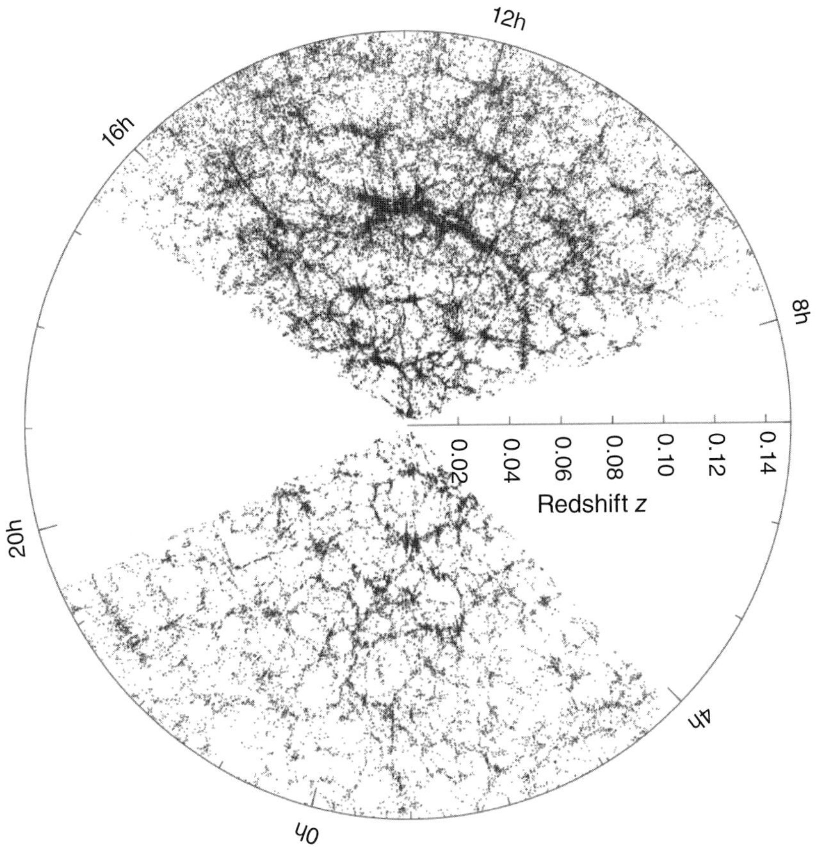


Fig. 6.14 The large-scale structure displayed in the plane of right ascension and redshift ($z < 0.15$) based on SDSS data. Each dot is a galaxy with measured spectroscopic redshift. The two conical regions show the sky coverage as seen from an observer located at the centre of the circle and within 6° of the equator. From Blanton et al. (2003). © AAS, reproduced with permission.

deficit of clustering compared to a Poissonian distribution, and it is relative to a case where galaxies tend to avoid each other. The observations show that $\xi_{\rm gal}(r)$ is positive and can be described as a power-law function of the spatial scale $r$:

$$\xi_{\rm gal}(r) = \left(\frac{r}{r_0}\right)^{-\gamma}, \tag{6.24}$$

where $r_0$ is the **clustering scalelength** at which $\xi_{\rm gal} = 1$. This power-law shape is valid typically on scales $< 10$ Mpc, with $\gamma \approx 1.8$ and $r_0 \approx 5$ Mpc. The function $\xi_{\rm gal}(r)$ remains positive up to $r \approx 30$–$50$ Mpc (the typical size of voids). On larger scales, $\xi_{\rm gal}(r)$ starts to oscillate around zero and the spatial distribution of galaxies becomes rather uniform due to the properties of homogeneity and isotropy of the Universe (§2.1). On scales of $\approx 100h^{-1}$ Mpc, $\xi_{\rm gal}(r)$ shows an excess above zero due to the BAOs (§2.4). The two-point correlation function is useful to describe the general clustering properties of galaxy populations. Other methods, such as the higher-order correlation functions (e.g. the three-point correlation function), have been developed to describe the shape of large-scale structure (e.g. the filaments, sheets and voids within the cosmic web). However, the details of these techniques are beyond the scope of this book, and we refer the interested reader to Peebles (2001) for a review.

If the observations are limited to two dimensions on the sky plane (i.e. without knowing the redshift of each galaxy), it is still possible to have information on the clustering through the **two-point angular correlation function** $w(\theta)$:

$$d\mathcal{P} = N[1 + w(\theta)]d\Omega, \tag{6.25}$$

where $d\mathcal{P}$ is the probability to find a galaxy at an angular distance $\theta$ from a galaxy selected at random within a solid angle $d\Omega$ on a sky area where the mean surface density of galaxies is $N$. In practice, $w(\theta)$ is the projection of $\xi_{\rm gal}(r)$ on the sky plane, and its shape is described by a power law with a slope $\delta = 1 - \gamma$. There exists a relation between $w(\theta)$ and $\xi_{\rm gal}(r)$ called the Limber equation (Limber, 1953) which allows us to derive $\xi_{\rm gal}(r)$ from the measured $w(\theta)$, provided that the galaxy redshift distribution is known (e.g. from photometric redshifts; §11.1.1).

The amount of clustering depends on galaxy properties. In general, ETGs are more clustered than SFGs. This is due to the tendency of a galaxy with high mass, red colours and E/S0 morphology to be preferentially located in a high-density environment, i.e. to be more clustered. Instead, SFGs are less clustered because they populate lower-density environments. The clustering scalelength depends on luminosity, increasing from $r_0 \approx 3$ Mpc for $M_r \approx -17$ to $r_0 \approx 10$ Mpc for $M_r \approx -23$, whereas the slope $\gamma$ remains almost constant. If colour bimodality is considered (§3.4.1), red sequence galaxies have on average $r_0 \approx 5$–6 Mpc and $\gamma \approx 2$, whereas the blue cloud systems have $r_0 \approx 3$–4 Mpc with $\gamma \approx 1.7$.

The LSS has profound links with cosmology because it allows us to reconstruct the 3D map of the matter in the Universe. This map is rich in information on how structures succeeded in collapsing influenced by gravity on different scales. The formation of galaxies takes place at the centre of dark matter halos formed through the collapse of overdensities of the dark matter distribution (§7.2). The mass distribution, clustering and evolution of the halos depend only on the cosmological model and the properties of dark matter particles. Instead, the efficiency of galaxy formation within the halos depends on a variety of baryonic matter processes (e.g. gas cooling, gas heating, star formation, feedback; Chapter 8). This implies that the observed galaxy clustering at a given redshift depends not only on the underlying spatial distribution of dark matter halos, but also on how the halos are actually populated by galaxies that formed within them. The evolution of clustering depends on gravity, on the expansion rate of the Universe and, in turn, on cosmological parameters such as $\Omega_\Lambda$ and $\Omega_{\rm m}$ (§2.2.2). For instance, the voids and the density contrast between low- and high-density regions of the cosmic web are higher for larger $\Omega_\Lambda$. As a consequence, the LSS can also be exploited to constrain the cosmological parameters and test alternative models in which general relativity is modified. However, most matter ($\approx 84\%$) is dark, and hence galaxies are luminous tracers which do not necessarily reflect the spatial distribution of the underlying dark matter. For this reason, galaxies are biased tracers of the LSS. It is therefore fundamental to estimate the so-called **galaxy bias** factor which allows us to establish the relationship between the distribution of luminous and dark matter. The bias is defined as

$$b = \left( \frac{\xi_{\rm gal}}{\xi} \right)^{1/2}, \tag{6.26}$$

where $\xi$ is the correlation function of the underlying total mass field (eq. 7.31), dominated by dark matter. If $b = 1$, this means that the light traces the mass exactly, whereas $b > 1$ indicates that light is a biased tracer of mass. The bias of a given galaxy population is called absolute or relative when it is relative to dark matter or to another galaxy population, respectively. The bias is a function of spatial scale, and it can be derived by comparing the observed clustering of galaxies ($\xi_{\rm gal}$) with that of dark matter ($\xi$) measured in an $N$-body cosmological simulation with dark matter only (§7.3.2). Another methodology is to estimate $b$ using the ratio of the two-point and three-point correlation functions, because they have different dependencies on the bias. The bias of galaxies depends on galaxy luminosity and mass. Galaxies with higher masses and luminosities (i.e. ETGs) are more biased. The bias normalised to the value $b^*$ of galaxies with characteristic luminosity $L^*$ (§3.5.1) increases as $b/b^* \approx A + B(L/L^*)$, where $L$ is the galaxy luminosity, and $A$ and $B$ are constants such that $b = b^*$ when $L = L^*$ and which depend on the filter used for the observations (for instance, in the $r$ band, $A \approx 0.85$ and $B \approx 0.15$). Galaxy clusters (§6.4) have biases larger than galaxies because they are located in the highest-density peaks of the LSS.

## 6.7 Baryon Budget

One of the most accurately determined cosmological parameters is the baryon fraction $f_{\rm b}$ (§2.5). This is the ratio between the baryonic mass and the total (baryon and dark matter) mass in the Universe. If we naïvely apply this to structure formation we could expect to observe this fraction in every structure that formed in the Universe, from galaxy clusters to dwarf galaxies. This is not at all what is observed. In fact, the amount of baryons that are *observed* in cosmic structures is typically much below the cosmological fraction (see also §10.10.1). This lack of baryons (gas and stars) in structures goes under the name of the **missing baryon problem** and it has a number of possible solutions that we outline here.

First, let us quantitatively state the problem for the Milky Way. The virial (total) mass of our Galaxy has been estimated using the kinematics of planetary nebulae, globular clusters, dSphs and stellar streams to be $\mathcal{M}_{\rm vir,MW} \approx 1.3 \times 10^{12}\, \mathcal{M}_\odot$ (Tab. 4.1). Multiplying this by the universal baryon fraction we obtain the baryonic mass that one would expect the Milky Way to contain:

$$\mathcal{M}_{\rm b,MW} \approx 2.1 \times 10^{11} \left( \frac{f_{\rm b}}{0.16} \right) \left( \frac{\mathcal{M}_{\rm vir,MW}}{1.3 \times 10^{12}\, \mathcal{M}_\odot} \right) \mathcal{M}_\odot. \tag{6.27}$$

As we have seen in §4.6, the stellar disc of our Galaxy has a total mass of $\approx 4 \times 10^{10}\, \mathcal{M}_\odot$ to which we can add $\approx 1 \times 10^{10}\, \mathcal{M}_\odot$ for the bulge. The stellar halo does not appreciably contribute and thus the stellar mass of the Milky Way is $\mathcal{M}_{\star,\mathrm{MW}} \approx 5 \times 10^{10}\, \mathcal{M}_\odot$. The ISM also does not add a significant contribution with $\approx 4 \times 10^9\, \mathcal{M}_\odot$ of H I and a somewhat smaller amount of molecular and ionised gas. In the end, we can safely take the total baryonic mass of the Milky Way (in stars and ISM) to be $\mathcal{M}_{\mathrm{b,MW}} \approx 6 \times 10^{10}\, \mathcal{M}_\odot$. Thus, the Milky Way appears to be missing about 70% of the baryons that should be associated with its dark matter halo.

The above result is general and it has been obtained for galaxies and galaxy groups covering a wide range in masses. It appears that the only structures in the Universe where we observe the predicted baryon fractions are the most massive galaxy clusters, in which most of the mass is in the hot ICM (§6.4.2). Smaller clusters and galaxy groups are missing a few tens of per cent of the baryonic mass (Tab. 6.1), spiral galaxies around 50–80%, and more and more for galaxies of lower masses. Dwarf galaxies can contain ~ 1% of the baryons that we should associate with their potential wells using the cosmological fraction $f_\mathrm{b}$. Tab. 6.4 gives an account of the baryon budget in structures (galaxies and galaxy clusters) in the different components. From this we see that $\approx 11\%$ of the universal baryons are in stars, in the cold ISM of galaxies and in the ICM of clusters. The rest is elsewhere.

We identify two 'media' external to galaxies where the baryons can reside: the **intergalactic medium** (IGM) and the **circumgalactic medium** (CGM). We can separate

**Table 6.4** Baryon budget in the present-day Universe

| Component | Contribution to $\Omega_0$[a] | Fraction of $\Omega_{\mathrm{b},0}$[b] |
|---|---|---|
| In galaxies | | |
| Main-sequence stars | $0.00205 \pm 0.00042$ | 4.2% |
| White dwarfs | $0.00036 \pm 0.00008$ | 0.7% |
| Stellar remnants | $0.00012 \pm 0.00003$ | 0.2% |
| Substellar objects[c] | $0.00014 \pm 0.00007$ | 0.3% |
| Neutral atomic gas | $0.00062 \pm 0.00010$ | 1.3% |
| Molecular gas | $0.00016 \pm 0.00006$ | 0.3% |
| Intracluster medium | $0.0018 \pm 0.0007$ | 3.7% |
| Intergalactic medium | | |
| Ly$\alpha$ forest[d] | | $\approx 29\%$ |
| WHIM[e] | | $\gtrsim 20\%$ |
| Circumgalactic medium (all phases) | | rest |

[a] The values for galaxies and intracluster gas are from Fukugita and Peebles (2004) and are calculated assuming $\Omega_0 = 1$.

[b] The percentage of baryons in a certain component with respect to the total amount of cosmological baryonic matter, calculated assuming $\Omega_{\mathrm{b},0} h^2 = 0.0224$ (Tab. 2.1).

[c] Brown dwarfs with stellar masses below $\mathcal{M} = 0.08\, \mathcal{M}_\odot$ (§8.3.9).

[d] Local Ly$\alpha$ forest from Penton et al. (2004).

[e] Warm–hot intergalactic medium from Nicastro et al. (2018).

these two considering as a demarcation the virial radii of galaxies. Material within the virial radius of a galaxy constitutes its circumgalactic medium. The potential wells of galaxies and other universal structures grow thanks to the continuous accretion of dark matter and gas (§7.4.4 and §8.2). If the accreted material has the universal baryon fraction and the baryons remain in the potential wells once they have been accreted, we should expect to observe *exactly* the cosmological baryon fraction in the CGM of galaxies. Conversely, if gas accretion occurs at values lower than the universal baryon fraction or there is efficient ejection of baryons from the potential wells (see §8.7), then a fraction of these baryons will be still in (or back to) the IGM.

It is important to realise, however, that the space beyond the virial radii of galaxies (IGM) at $z = 0$ must necessarily contain *some* baryons as these are associated with the dark matter that has not yet collapsed. Cosmological simulations (§10.11.1) show that this fraction should amount to almost half of the whole universal dark matter. Thus it is natural that $\approx 50\%$ of the present-day baryons are still in the IGM. These IGM inhabitants are rightfully there and we should not find them inside the virial radii of any galaxy, simply because gravity has not yet reached them.

Establishing the amount of baryons that the IGM and CGM contain is challenging because the gas in both media is difficult to observe as it is at very low densities. Observations are typically carried out in absorption towards distant QSOs as they are sensitive to low column densities (we discuss this in §9.8). Lyman-$\alpha$ (Ly$\alpha$) absorptions in QSO spectra have revealed that about 30% of the local baryons could be in an IGM component at relatively low temperatures ($T \sim 10^4$ K). We call this feature the **local Lyman-$\alpha$ forest** and we interpret it as the baryon content of cosmic filaments (§9.8.1). At high redshifts ($z > 2$), most of the baryons are in the Ly$\alpha$ forest and not yet collapsed as is most of the dark matter (§9.8.2).

Gas in the IGM could also be at higher temperatures between $T \sim 10^5$ K and a few $\times 10^6$ K. This material is called **warm–hot intergalactic medium** (WHIM). Rarefied gas at these temperatures is extremely difficult to detect. Observations have been carried out looking for absorption features of high-excitation species like O VII, O VIII and Ne IX in the X-ray spectra of distant quasars. They have revealed that a few tens of per cent of the baryons in the IGM could be in the form of WHIM but the precise value remains uncertain. Given that 50% of the baryons should be still in the IGM, and if we believe the results of the local Ly$\alpha$ forest, then an amount of WHIM of at least 20% of the total is expected.

The rest of the baryons, roughly 40% of the total, could then be in the IGM or hidden within the CGM of galaxies. The CGM has been probed in emission using X-ray observations that have revealed the presence of hot gas at nearly the virial temperature (§8.2.1). This component is called the hot halo or **corona**[5] and constitutes the hot CGM. The coronae are expected to extend out to the virial radius (at least in massive galaxies) and to contain a fraction of missing baryons that is not well constrained and could be between

[5] Note that the term corona is often used for disc galaxies, while the medium around ETGs is usually referred to as hot halo (§5.3.3).

10% and 50% of $f_b M_{vir}$ (eq. 6.27). Other than this, the CGM of galaxies also hosts gas at lower temperatures. This material, again detected using absorption features in the spectra of QSOs mostly with UV spectrographs (§9.8.3), has temperatures from a few times $10^5$ K down to $T \sim 10^4$ K. The gas at $T \sim 10^4$ K is seen through Ly$\alpha$ absorption and low-excitation ions, for instance the metal species of Si II, Si III, Mg II and C II. The warmer material is often observed in O VI absorption. As a consequence of this spread in temperatures, the CGM of galaxies is said to be a *multi-phase* medium. The total mass of the various phases is unfortunately uncertain and a definitive answer to the question of the missing baryons has not been reached. For reviews on this topic see Bregman (2007) and Tumlinson et al. (2017).

# 7 Formation, Evolution and Properties of Dark Matter Halos

Galaxies are believed to form from collapse and cooling of gas in dark matter halos. Dark matter halos form as a consequence of the gravitational instability and the growth of primordial perturbations in the matter density distribution of the Universe. In this chapter we describe the main steps that lead from the linear evolution of density perturbations to the virialisation of dark matter halos. We then describe the main structural and kinematic properties of dark halos. We note that this is only a brief overview of the complex process of structure formation in the Universe. We refer the reader to Coles and Lucchin (2002), Mo et al. (2010) and Schneider (2015) for more details.

## 7.1 Observational Evidence for Dark Matter Halos

The proposal that a substantial part of the matter in the Universe is in the form of dark matter dates back to much earlier than the study of structure formation in cosmological models. Back in the 1930s, Zwicky (1933) suggested that most of the mass in the Coma cluster of galaxies is in the form of dark matter. Nowadays, thanks to the combination of different methods to measure the mass of groups and clusters of galaxies, we have detailed information not only on the total mass of their dark matter halos, but also on their mass distribution (§6.2 and §6.4.4), which can be compared with the cosmological predictions that we discuss in the following sections.

Moving to the smaller scales of galaxies, Ostriker and Peebles (1973) envisaged the presence of dark matter in galaxies, based on their conclusion that massive spherical halos are necessary to stabilise galactic discs against bar instability (§10.2.1). In the 1970s it was also realised that extended halos of dark matter are necessary to explain the observed rotation curves of disc galaxies, and in particular the flat outer parts traced by H I (§4.3.3). In general these rotation curves can be reproduced assuming cosmologically motivated halos on large scales, while the theoretical interpretation of the observationally inferred central distribution of dark matter is not always straightforward (§4.3.5 and §7.5.5). While HSB galaxies are baryon-dominated in the inner regions, dark matter is dominant down to the centre in LSB galaxies, so among SFGs there is a diversity in the dark matter distribution.

Disentangling the distributions of luminous and dark matter in ETGs is more difficult than in disc galaxies, but, mainly thanks to gravitational lensing measurements (§5.3.4), it is now well established that dark matter halos are required also in ETGs (§5.3.5): stars

**Table 7.1** Main methods used to infer the properties of dark matter halos in different astrophysical systems

| Method | dSphs | Disc galaxies | Early-type galaxies | Groups of galaxies | Clusters of galaxies |
|---|---|---|---|---|---|
| Virial theorem[a] (§5.3.1) | ✓ | | ✓ | ✓ | ✓ |
| Rotation curve (§4.3) | | ✓ | | | |
| Dynamical modelling (§5.3.2) | ✓ | ✓ | ✓ | | ✓ |
| X-ray-emitting gas (§5.3.3) | | | ✓ | ✓ | ✓ |
| Gravitational lensing (§5.3.4) | | ✓ | ✓ | ✓ | ✓ |

[a] In the case of galaxies, using the virial theorem means in practice estimating the dynamical mass $\mathcal{M}_{\rm dyn}$ (eq. 5.17).

typically dominate in the central regions, while dark matter is dominant at radii of the order of or larger than $R_{\rm e}$ (Fig. 5.12).

On the very small scales we find extreme behaviours in terms of dark matter content: while, on the one hand, there is no evidence for the presence of dark matter in globular clusters (§4.6.1 and §4.1.3), on the other hand the much more diffuse dSphs and UFDs are the most dark matter-dominated systems we know of, with mass-to-light ratios as high as $10^2$–$10^3$ (§6.3.2). A summary of the methods used to measure the dark matter mass in different astrophysical systems is given in Tab. 7.1.

## 7.2 Dark Matter and Structure Formation

The observational evidence summarised in §7.1 suggests that dark matter must have a fundamental role in structure formation. In fact, the following independent argument demonstrates that galaxies could not have formed in a Universe made only of baryons.

Before photon decoupling ($z > z_{\rm rec}$, where $z_{\rm rec} \approx 1000$ is the redshift of recombination; §2.4 and §9.1) baryons are tightly coupled with photons through Thomson scattering (§2.4). The Jeans mass (§8.3.2) of the coupled baryon–photon fluid is very high ($\sim 10^{16}\mathcal{M}_\odot$), because the characteristic velocity of the fluid is the speed of light. As a consequence, no significant growth of photon–baryon perturbations on galaxy scales ($\ll 10^{16}\mathcal{M}_\odot$) is possible. Around $z = z_{\rm rec}$ baryons decouple from photons, because the mean free path of Thomson scattering becomes very large as a consequence of recombination. As soon as the baryons are decoupled, the Jeans mass drops to $\sim 10^6\mathcal{M}_\odot$, because the characteristic velocity of the baryonic fluid is the sound speed (§D.2.4), so the baryonic density perturbations are able to grow. However, we know from observations of the CMB (§2.4) that fluctuations in baryons and radiation at $z \approx z_{\rm rec}$ were as small as $\delta \sim 10^{-5}$, where $\delta$ is the density contrast (eq. 7.1). A quantitative analysis of the growth of density perturbations shows that, to form the observed structures (galaxies and clusters of galaxies) as quickly as observed, there must have been non-baryonic matter fluctuations that at $z \sim 10^3$ were about $\delta \sim 10^{-3}$, two orders of magnitude larger than photon–baryon fluctuations, and thus necessarily already decoupled from photons. The above argument implies[1] that structure formation was driven by non-baryonic dark matter (see below). As soon as the baryons decouple from photons, they quickly catch up with dark matter perturbations by falling into their gravitational potential wells.

All the matter in the Universe that is not detectable by emitted or absorbed radiation is called **dark matter**. An important distinction is between baryonic and non-baryonic dark matter. **Baryonic dark matter** is ordinary matter composed of **baryons**,[2] which does not emit detectable radiation (for instance, collapsed objects such as black holes and neutron stars). Any kind of dark matter that is not baryonic is called **non-baryonic dark matter**, and is characterised by having no electromagnetic interaction. The analysis of the CMB (§2.4) and the study of the Big Bang nucleosynthesis (§2.5) suggest that most of the dark matter in the Universe must be non-baryonic.[3] Therefore, nowadays it is common practice to refer to non-baryonic dark matter as dark matter *tout court* (this is the convention that we adopt in this textbook, when not specified otherwise).

Most candidate dark matter particles are **thermal relics**, i.e. particles that in the early Universe were in thermal equilibrium with radiation, from which they decoupled at some time. Depending on their thermal speed (eq. 4.9) at the time of decoupling, candidate dark matter particles are usually classified as hot, cold and warm. **Hot dark matter** (HDM) is made of very low-mass particles: prototypes are neutrinos with $m_{\rm DM}c^2 \lesssim 0.2\,{\rm eV}$ (we note, for comparison, that the proton mass is $m_{\rm p}c^2 \simeq 0.9\,{\rm GeV}$). Neutrinos decouple from baryons (and therefore from photons, then coupled to baryons) at very high redshift ($z \sim 10^{10}$), when they are still relativistic (§2.2.2). At later times massive neutrinos become non-relativistic and their thermal speed decreases with cosmic time. However, even at the present time, the thermal speed of neutrinos is $\gtrsim 100\,{\rm km\,s^{-1}}$, which prevents them from being gravitationally bound on galaxy scales (a phenomenon called **free streaming**), because the characteristic speed corresponding to the potential wells of galaxies is of the order of $10^2\,{\rm km\,s^{-1}}$ (Tabs. 4.1 and 5.1). This effectively excludes that galactic dark halos are made of HDM. **Cold dark matter** (CDM) is made of much more massive particles (with masses $m_{\rm DM}c^2 \gtrsim 1\,{\rm GeV}$), generally referred to as **weakly interacting massive particles** (WIMPs). WIMPs decouple very early, but, when they decouple, they are already non-relativistic. Their thermal speed is negligible (effectively it can be taken to be zero) and therefore they can form structures on all scales. An alternative to WIMPs is represented by the **axions**, dark matter particle candidates that, though very light, would

1 This conclusion holds under the assumption that general relativity is the correct theory of gravity. In principle, non-baryonic dark matter is not necessarily required to explain structure formation in alternative gravity theories.

2 It is customary in astrophysics and cosmology to use the term 'baryons' to indicate all matter made of ordinary particles (essentially quarks and electrons), independent of whether they are actually baryons or leptons. This is justified because electrons (leptons) contribute negligibly to the mass budget compared to protons and neutrons (baryons).

3 Exceptions are the **primordial black holes**, which are dark matter candidates that would have formed in the very early Universe (before Big Bang nucleosynthesis; §2.5). Primordial black holes, though made of baryons, would effectively behave in many respects as non-baryonic dark matter.

in fact behave as CDM because they would have very low speeds (axions are not thermal relics). **Warm dark matter** (WDM) particles have masses intermediate between HDM particles and WIMPs (typically $m_{\rm DM}c^2 \sim$ keV). They are non-relativistic at decoupling and their thermal speed, though lower than those of HDM, is non-negligible and such that gravitational binding of WDM is prevented on small enough scales, comparable to those of dwarf galaxies.

Independent of whether it is cold, warm or hot, dark matter is collisionless (§C.8), because dark matter particles are believed to interact significantly among themselves only through gravity[4] and their mass, even in the case of CDM, is so small that two-body gravitational interactions between individual particles are negligible in comparison with the mean gravitational field. Gravitational lensing measurements (§5.3.4) of merging clusters of galaxies (§6.4.1), such as the so-called Bullet cluster (1E 0657–558), provide observational evidence that the dark matter is largely collisionless, because during the merging it follows the collisionless distribution of galaxies, while it departs from the collisional ICM. Based on measurements of the power spectrum of the cosmological density fluctuations (§7.4.1), CDM is preferred in the standard cosmological model, so hereafter we focus on structure formation driven by collisionless CDM.

## 7.3 Evolution of a Density Perturbation

It follows from the discussion in §7.2 that structure formation is driven by the growth of perturbations of the (non-relativistic) dark matter component. Here we thus consider the evolution of density perturbations, assuming for simplicity a Universe composed only of non-relativistic matter, which represents dark matter before decoupling, but the entire matter component (dark matter and baryons) after decoupling. Given that dark matter is dominant, in the present description of structure formation we neglect the complications related to the physics of baryons, which we defer to Chapter 8.

The matter density $\rho(\boldsymbol{r},t)$ at position $\boldsymbol{r}$ and time $t$ can be written as $\rho(\boldsymbol{r},t) = \overline{\rho}(t)[1 + \delta(\boldsymbol{r},t)]$, where $\overline{\rho}(t)$ is the mean density of the Universe at time $t$ and the dimensionless quantity

$$\delta(\boldsymbol{r},t) \equiv \frac{\rho(\boldsymbol{r},t) - \overline{\rho}(t)}{\overline{\rho}(t)} \tag{7.1}$$

is called the **density contrast**. At a sufficiently early time $t_{\rm i}$, the density perturbation must have been small ($|\delta| \ll 1$), because the early Universe is almost homogeneous (§2.1 and §2.4). Let us consider a region of the Universe that at $t_{\rm i}$ has density $\rho = (1+\delta)\overline{\rho}$ with $\delta > 0$. Because of the effect of its self-gravity, this slightly overdense region expands at a rate slower than the expansion rate of the Universe. As a consequence, the density contrast of the overdensity increases with time. As long as $\delta \ll 1$, the overdensity is called a **linear density perturbation** and the associated growth is called **linear evolution of the density**

[4] Alternative approaches postulate that dark matter particles have non-negligible non-gravitational interactions among themselves. If this is the case the dark matter is called **self-interacting**.

**perturbation**.[5] If the initial overdensity is sufficiently large, at some time $t > t_\mathrm{i}$ it will become non-linear ($\delta \gtrsim 1$), it will cease to expand and it will start to recollapse due to the effect of its own gravity to form a virialised system (that is, a system in equilibrium, thus satisfying the virial theorem; eq. 5.12).

### 7.3.1 Linear Evolution of a Perturbation

Let us assume that the Universe is dominated by collisionless CDM and that the evolution of the density perturbation can be described by Newtonian gravity.[6] The linear evolution of a density perturbation should be followed using the collisionless Boltzmann equation (eq. C.37), which is appropriate for a collisionless component such as dark matter. However, under the assumption that the dark matter is cold (i.e. with negligible velocity dispersion), the collisionless Boltzmann equation behaves as the hydrodynamic equations (§D.2) for a fluid with zero pressure. The relevant hydrodynamic equations are the mass conservation equation (eq. D.22) and the Euler equation (eq. D.25) that, neglecting the pressure gradient, reads

$$\rho\frac{\partial \boldsymbol{u}}{\partial t} + \rho\boldsymbol{u}\cdot\boldsymbol{\nabla}\boldsymbol{u} = -\rho\boldsymbol{\nabla}\Phi, \tag{7.2}$$

where $\boldsymbol{u}$ is the velocity of the fluid, coupled with the **Poisson equation**

$$\nabla^2\Phi = 4\pi G\rho, \tag{7.3}$$

which relates the gravitational potential $\Phi$ to the density distribution $\rho$. It is convenient to rewrite the above equations in terms of the comoving coordinates $\boldsymbol{x}$ related to the physical coordinates $\boldsymbol{r}$ by $\boldsymbol{r} = a\boldsymbol{x}$, where $a(t)$ is the scale factor (§2.1.1). The gravitational potential of the perturbation is

$$\Phi_\mathrm{pert} = \Phi - \Phi_0, \tag{7.4}$$

where $\Phi_0$ is the unperturbed gravitational potential. From eqs. (D.22) and (7.2) we get

$$\frac{\partial\rho}{\partial t} + \frac{3\dot{a}}{a}\rho + \frac{1}{a}\boldsymbol{\nabla}_{\boldsymbol{x}}\cdot(\rho\boldsymbol{v}) = 0 \tag{7.5}$$

and

$$\frac{\partial\boldsymbol{v}}{\partial t} + \frac{1}{a}\boldsymbol{v}\cdot\boldsymbol{\nabla}_{\boldsymbol{x}}\boldsymbol{v} + \frac{\dot{a}}{a}\boldsymbol{v} = -\frac{1}{a}\boldsymbol{\nabla}_{\boldsymbol{x}}\Phi_\mathrm{pert}, \tag{7.6}$$

where $\partial/\partial t$ is computed at fixed $\boldsymbol{x}$, $\boldsymbol{\nabla}_{\boldsymbol{x}} = \partial/\partial\boldsymbol{x}$ is the gradient in comoving coordinates and $\boldsymbol{v} = \boldsymbol{u} - \dot{a}\boldsymbol{x}$ is the peculiar velocity (that is, the velocity perturbation) in comoving coordinates.[7] Combining the time derivative of eq. (7.5) with eq. (7.6) and linearising, that

[5] A small perturbation is called linear, because the effect of the perturbation can be studied by linearising the equations, that is, keeping only first-order terms in the perturbation and neglecting higher-order terms.

[6] To be rigorous one should use general relativity, but the Newtonian approximation is sufficient as long as the perturbations are small.

[7] The gradient in comoving coordinates $\boldsymbol{\nabla}_{\boldsymbol{x}} = \partial/\partial\boldsymbol{x}$ is related to the physical gradient $\boldsymbol{\nabla} = \partial/\partial\boldsymbol{r}$ by $\boldsymbol{\nabla} = (1/a)\boldsymbol{\nabla}_{\boldsymbol{x}}$. The partial time derivatives at fixed $\boldsymbol{r}$ and at fixed $\boldsymbol{x}$ are related by

$$\left(\frac{\partial}{\partial t}\right)_{\boldsymbol{r}} = \left(\frac{\partial}{\partial t}\right)_{\boldsymbol{x}} - \frac{\dot{a}}{a}\boldsymbol{x}\cdot\boldsymbol{\nabla}_{\boldsymbol{x}}. \tag{7.7}$$

In the derivation of eq. (7.6) we have used $\ddot{a}\boldsymbol{x} = -(1/a)\boldsymbol{\nabla}_{\boldsymbol{x}}\Phi_0$.

is, replacing $\rho$ with $\overline{\rho}(1+\delta)$ and considering the limit $\delta \ll 1$ (thus keeping only terms of first order in $\delta$ and neglecting higher-order terms), we get

$$\frac{\partial^2 \delta}{\partial t^2} + \frac{2\dot{a}}{a}\frac{\partial \delta}{\partial t} = 4\pi G\overline{\rho}\delta, \tag{7.8}$$

where we have used

$$\nabla_{\boldsymbol{x}}^2 \Phi_{\rm pert} = 4\pi G\overline{\rho}\delta, \tag{7.9}$$

where $\nabla_{\boldsymbol{x}}^2$ is the Laplacian in comoving coordinates. The linearity of eq. (7.8) suggests to factorise the density perturbation as

$$\delta(\boldsymbol{x}, t) = D(t)\delta(\boldsymbol{x}, t_0), \tag{7.10}$$

where $D(t)$ is the **growth factor** of cosmological perturbations, which is normalised so that $D(t_0)=1$, where $t_0$ is the present time ($z=0$), such that $a(t_0)=1$ by definition (§2.1.1). Here $\delta(\boldsymbol{x}, t_0)$ is the overdensity *linearly extrapolated* to the present time.[8] Eqs. (7.8) and (7.10) imply that the *shape* of the perturbation (in comoving space) does not vary with time, while its *amplitude* is determined by $D(t)$, which must satisfy

$$\frac{\mathrm{d}^2 D}{\mathrm{d}t^2} + \frac{2\dot{a}}{a}\frac{\mathrm{d}D}{\mathrm{d}t} = 4\pi G\overline{\rho}D. \tag{7.11}$$

For instance, in an EdS cosmological model (§2.3.1), in which $\overline{\rho} = 1/(6\pi G t^2)$ (eq. 2.34), we get

$$\frac{\mathrm{d}^2 D}{\mathrm{d}t^2} + \frac{4}{3t}\frac{\mathrm{d}D}{\mathrm{d}t} - \frac{2}{3t^2}D = 0, \tag{7.12}$$

where we have used eq. (2.33) and its time derivative. Eq. (7.12) has solution $D(t)=a(t)$, so, in the linear regime, the perturbation grows in time as the scale factor in an EdS Universe. For different cosmological parameters the solution of eq. (7.11) is generally different. However, for a standard $\Lambda$CDM model with $\Omega_{\rm m,0} \approx 0.3$ and $\Omega_{\Lambda,0} \approx 0.7$ (§2.3), $D(t)$ does not differ from the function computed for an EdS model by more than 30% over the large redshift range $0 \leq z \leq 100$, as we see in Fig. 7.1 (left).

The function $D(t)$ encapsulates all the information about the linear growth of perturbations in a given cosmological model. In §7.4 we make use of $D(t)$ to study the evolution of the statistical properties of perturbations across cosmic time.

### 7.3.2 Non-Linear Evolution of an Overdensity

When a matter overdensity has become non-linear its evolution cannot be rigorously described analytically. It is possible to study the problem numerically, with the help of cosmological $N$-body simulations (§7.5.1) or to resort to analytic approximations. The simplest analytic description of the non-linear evolution of overdensities is the **spherical collapse model**, which we consider in this section.

[8] In general $\delta(\boldsymbol{x}, t_0)$ is *not* the present-day overdensity, which might well have entered the non-linear regime. Clearly, if this is the case, the present-day overdensity cannot be described by the linearised equations.

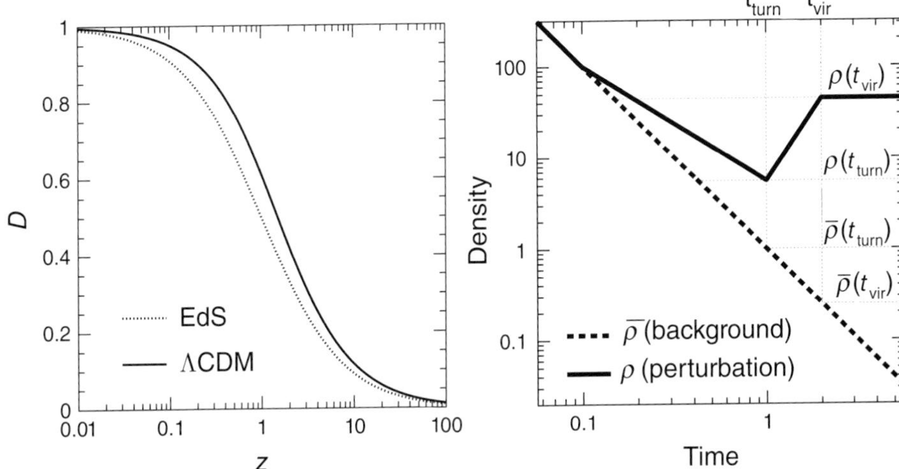


Fig. 7.1

*Left panel*. Growth factor $D$ for a $\Lambda$CDM ($\Omega_{m,0}=0.3, \Omega_{\Lambda,0}=0.7$) model and for an EdS ($\Omega_m=1, \Omega_\Lambda=0$) model as a function of redshift. For an EdS Universe $D=a=(1+z)^{-1}$, while for $\Lambda$CDM $D$ is computed by integrating numerically equation 28 of Carroll et al. (1992). *Right panel*. Schematic representation of the evolution of the perturbation density $\rho$ and of the background density $\overline{\rho}$ in the spherical collapse model. At the time of turnaround $t_{turn}$, $\rho \simeq 5.55\overline{\rho}$ (eq. 7.13). At the time of virialisation $t_{vir}=2t_{turn}$, $\rho \simeq 178\overline{\rho}$ (eq. 7.20).

## Spherical Collapse Model

The right panel of Fig. 7.1 illustrates schematically the evolution of a density perturbation in the spherical collapse model. Let us consider at an initial time $t_i \approx t_{rec}$ (relatively close to recombination; §2.4) a spherical region with uniform density with density contrast (eq. 7.1) $0<\delta \ll 1$. This overdensity is expected to grow in the linear regime, as described in §7.3.1, up to reaching $\delta \sim 1$. Initially the spherical region expands, but more slowly than the background, so $\delta$ increases. At the **turnaround time** $t_{turn}$ the expansion of the sphere stops and the contained matter starts to collapse. Assuming an EdS model of the Universe, the background density at time $t_{turn}$ is $\overline{\rho}(t_{turn})=1/(6\pi G t_{turn}^2)$ (eq. 2.34). It can be shown that the density of the spherical region at time $t_{turn}$ is

$$\rho(t_{turn})=\left(\frac{3\pi}{4}\right)^2 \overline{\rho}(t_{turn}) \simeq 5.55\overline{\rho}(t_{turn}). \tag{7.13}$$

In this idealised model, at time $t_{coll}=2t_{turn}$ the homogeneous sphere collapses into a singularity, so $t_{coll}$ can be formally taken as the **collapse time** of the overdensity. For a mass shell to collapse at time $t_{coll}$, the contained overdensity, *linearly extrapolated*[9] to $t_{coll}$, must be $\delta(t_{coll})=\delta_{coll}\simeq 1.686$. Though this result has been obtained for an EdS model, the value of $\delta_{coll}$ has only weak dependence on the cosmological parameters. For a $\Lambda$CDM model $\delta_{coll}$ varies weakly with the collapse time, but it deviates from 1.686 by less than 1%.

[9] This is not the actual overdensity at $t_{coll}$, which formally diverges.

As denser regions take less time to collapse (because higher densities counteract the expansion more efficiently), we expect systems that collapse at higher redshift to originate from higher overdensities. It is useful to state this result in terms of the overdensity linearly extrapolated to the present time $\delta_0 \equiv \delta(t_0)$. An overdensity with $\delta_0 = \delta_{\rm coll}$ collapses at $z = 0$ $(t = t_0)$, while an overdensity with $\delta_0 > \delta_{\rm coll}$ collapses at higher redshift $(t < t_0)$. This translates into the following criterion for collapse as a function of redshift: at any given redshift $z$, only overdensities with $\delta_0 > \delta_{\rm c}(z)$ have collapsed, where the **critical overdensity for collapse** $\delta_{\rm c}(z)$ is an increasing function of redshift such that $\delta_{\rm c}(0) = \delta_{\rm coll} \simeq 1.686$.

## Virialisation

Though the spherical collapse model is a very useful approximation, in fact the collapse of a matter overdensity is not expected to be spherical and the density distribution is not expected to be uniform, so, during the collapse, dark matter particles experience phase mixing and violent relaxation (§8.9.2). As a consequence, instead of collapsing into a singularity, the infalling particles form an equilibrium system that satisfies the virial theorem (eq. 5.12). The overdensity will be virialised at a time $t_{\rm vir}$ of the order of $t_{\rm coll} = 2t_{\rm turn}$; here, for simplicity, we assume

$$t_{\rm vir} = 2t_{\rm turn}. \tag{7.14}$$

A first quantitative description of the process of virialisation can be obtained by assuming that at $t_{\rm turn}$ the system is a homogeneous sphere of radius $r_{\rm turn}$ with zero kinetic energy (because it is turning over) and that at $t_{\rm vir}$ the system is a virialised spherical system with uniform density and radius $r_{\rm vir}$. Though crude, this approximation allows us to estimate the characteristic overdensity of virialised cosmological halos. Let $K$, $W$ and $E = K + W$ denote, respectively, the kinetic, gravitational potential and total energy of the system. At time $t_{\rm turn}$, $E = W$ (because $K = 0$), so, in the homogeneous sphere approximation,

$$E = -\frac{3GM^2}{5r_{\rm turn}}, \tag{7.15}$$

where

$$M = \frac{4}{3}\pi\rho(t_{\rm turn})r_{\rm turn}^3 \tag{7.16}$$

is the mass of the perturbation, which does not vary with time. At time $t_{\rm vir}$ the system satisfies the virial theorem (eq. 5.12) $K = -W/2$, so its total energy is $E = W/2$, which, in the homogeneous sphere approximation, is

$$E = -\frac{3GM^2}{10r_{\rm vir}}, \tag{7.17}$$

where $r_{\rm vir}$ is the radius of the virialised system. As the total mass energy $E$ is conserved, combining eqs. (7.15) and (7.17) we get $r_{\rm vir} = r_{\rm turn}/2$. The final density of the perturbation is therefore

$$\rho(t_{\rm vir}) = \frac{3\mathcal{M}}{4\pi r_{\rm vir}^3} = \frac{3\mathcal{M}}{4\pi r_{\rm turn}^3}\left(\frac{r_{\rm turn}}{r_{\rm vir}}\right)^3 = 8\rho(t_{\rm turn}). \tag{7.18}$$

The overdensity at $t_{\rm vir}$ is

$$\frac{\rho(t_{\rm vir})}{\overline{\rho}(t_{\rm vir})} = \frac{8\rho(t_{\rm turn})}{\overline{\rho}(t_{\rm turn})}\frac{\overline{\rho}(t_{\rm turn})}{\overline{\rho}(t_{\rm vir})}, \tag{7.19}$$

which, using eqs. (2.34), (7.13) and (7.14), gives

$$\frac{\rho(t_{\rm vir})}{\overline{\rho}(t_{\rm vir})} = 18\pi^2. \tag{7.20}$$

Therefore, in the above approximated calculation of spherical collapse, assuming EdS cosmology, we expect virialised halos to be characterised by an overdensity $\rho/\overline{\rho} \simeq 178$ (Fig. 7.1, right).

In fact, virialised dark matter halos do not have uniform density distribution and do not have a well defined edge. It is however useful to define the *size* and *mass* of dark matter halos. The average matter density of the Universe appearing in eq. (7.20) can be written as $\overline{\rho} = \Omega_{\rm m}\rho_{\rm crit}$, where $\Omega_{\rm m}(z)$ is the matter density parameter (§2.2.2) and $\rho_{\rm crit}(z)$ is the critical density (eq. 2.23), related to the present-day critical density $\rho_{\rm crit,0}$ (eq. 2.24) by $\rho_{\rm crit}(z)/\rho_{\rm crit,0} = H^2(z)/H_0^2$ (eq. 2.37). Given that $\Omega_{\rm m} = 1$ for an EdS Universe, the above calculation suggests that one defines the **virial radius** $r_\Delta$ of a virialised halo at redshift $z$ as the radius of a sphere containing an overdensity with average density[10] $\Delta_{\rm c}\rho_{\rm crit}$, where $\Delta_{\rm c}$ is the **critical overdensity for virialisation**. For an EdS model $\Delta_{\rm c} = 18\pi^2$, independent of redshift, but $\Delta_{\rm c}$ depends on $z$ in general. An analytic approximation of $\Delta_{\rm c}$ for $\Omega_{\rm m} \neq 1$ and $\Omega_{\rm m} + \Omega_\Lambda = 1$ is

$$\Delta_{\rm c}(z) = 18\pi^2 + 82y - 39y^2, \tag{7.21}$$

where $y = \Omega_{\rm m}(z) - 1$ and

$$\Omega_{\rm m}(z) = \frac{\Omega_{\rm m,0}(1+z)^3}{\Omega_{\rm m,0}(1+z)^3 + \Omega_{\Lambda,0}}. \tag{7.22}$$

Eq. (7.22), where $\Omega_{\rm m} = \rho_{\rm m}/\rho_{\rm crit}$, is obtained from eqs. (2.30) and (2.37), without the radiation term, which is negligible at the redshifts relevant for the virialisation of halos. The value of $\Delta_{\rm c}$ as defined in eq. (7.21) depends on redshift, varying from $\Delta_{\rm c} \simeq 101$ at $z = 0$ to $\Delta_{\rm c} \simeq 178$ at high redshift for the standard $\Lambda$CDM cosmology (Fig. 7.2, left). However, different choices of $\Delta_{\rm c}$ are also possible. A simple and widely adopted choice is $\Delta_{\rm c} = 200$, independent of redshift.

Once the reference overdensity for virialisation $\Delta_{\rm c}$ is chosen, it is possible to define the **virial mass**

$$\mathcal{M}_\Delta \equiv \frac{4\pi}{3}\Delta_{\rm c}\rho_{\rm crit}r_\Delta^3, \tag{7.23}$$

[10] We warn that this definition of $r_\Delta$ is not always adopted in the literature: in some works $r_\Delta$ is defined as containing an overdensity $\Delta_{\rm c}\overline{\rho} = \Delta_{\rm c}\Omega_{\rm m}\rho_{\rm crit}$.

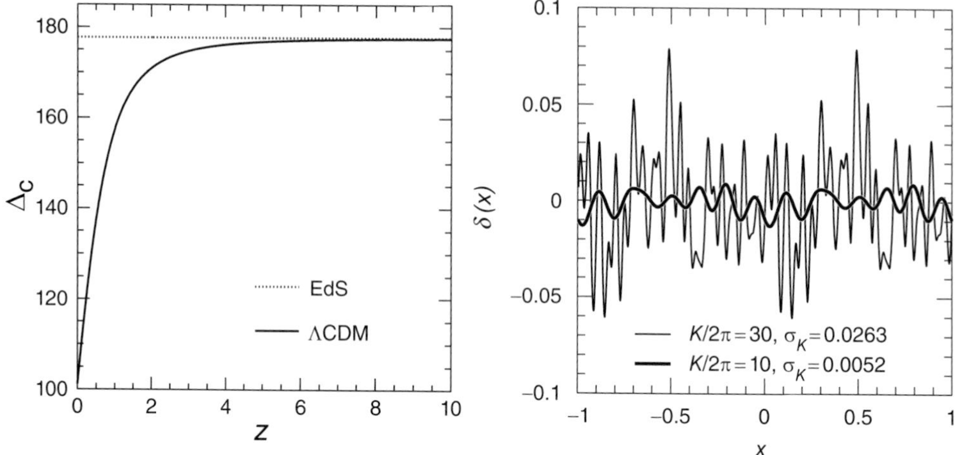


**Fig. 7.2** *Left panel*. Critical overdensity for virialisation $\Delta_c$ as a function of redshift $z$ for EdS ($\Omega_m = 1$) and $\Lambda$CDM ($\Omega_{m,0} = 0.3$, $\Omega_{\Lambda,0} = 0.7$) cosmological models. For an EdS Universe $\Delta_c = 18\pi^2$, independent of redshift, while for $\Lambda$CDM $\Delta_c(z)$ is given by the analytic approximation of Bryan and Norman (1998) (eq. 7.21). *Right panel.* Illustration of the effect of filtering a fluctuation field by eliminating fluctuations on scales smaller than $2\pi/K$. The thin curve is a 1D Gaussian random field (§7.4.3) $\delta(x)$ including short-wavelength fluctuations down to $2\pi/K \simeq 0.03$. The thick curve is the same Gaussian random field, but filtered eliminating fluctuations with wavelengths shorter than $2\pi/K = 0.1$. The variance $\sigma_K^2$ of the field is much lower in the latter case.

which is the mass contained within a sphere of radius $r_\Delta$, centred at the peak of the halo density distribution. A characteristic speed of a halo is the circular speed (eq. 4.23) at $r_\Delta$:

$$v_\Delta = \sqrt{\frac{GM_\Delta}{r_\Delta}}. \tag{7.24}$$

For instance, when $\Delta_c = 200$ is assumed, the virial radius and mass are indicated with $r_{200}$ and $M_{200}$, respectively, and the corresponding circular speed as $v_{200}$. When $\Delta_c$ depends on redshift, as in eq. (7.21), we indicate the virial radius with $r_{vir}$ and the virial mass with $M_{vir}$. Combining eqs. (2.23) and (7.23) we get

$$r_{vir} = \left[\frac{2GM_{vir}}{\Delta_c(z)H^2(z)}\right]^{1/3}. \tag{7.25}$$

The circular speed at $r_{vir}$ is

$$v_{vir} = \sqrt{\frac{GM_{vir}}{r_{vir}}} = [GM_{vir}H(z)]^{1/3}\left[\frac{\Delta_c(z)}{2}\right]^{1/6}. \tag{7.26}$$

At $z = 0$, for $\Delta_c \simeq 101$ (eq. 7.21; Fig. 7.2)

$$r_{vir} \simeq 259.0\left(\frac{M_{vir}}{10^{12}M_\odot}\right)^{1/3}\left(\frac{h}{0.7}\right)^{-2/3} \text{kpc} \tag{7.27}$$

and

$$v_{\rm vir} \simeq 128.9 \left(\frac{M_{\rm vir}}{10^{12} M_\odot}\right)^{1/3} \left(\frac{h}{0.7}\right)^{1/3} \ {\rm km\,s^{-1}}. \tag{7.28}$$

At $z=0$, for $\Delta_c = 200$

$$r_{200} \simeq 206.3 \left(\frac{M_{200}}{10^{12} M_\odot}\right)^{1/3} \left(\frac{h}{0.7}\right)^{-2/3} \ {\rm kpc} \tag{7.29}$$

and

$$v_{200} \simeq 144.4 \left(\frac{M_{200}}{10^{12} M_\odot}\right)^{1/3} \left(\frac{h}{0.7}\right)^{1/3} \ {\rm km\,s^{-1}}. \tag{7.30}$$

## 7.4 Statistical Properties of Cosmological Perturbations and Dark Matter Halos

In §7.3 we have focused on an individual overdensity, studying its linear and non-linear evolution, up to the formation of a virialised dark matter halo. Here we take a different perspective and we consider the statistical properties of the cosmological perturbations and their evolution. We then use these results to study the cosmological dark matter halos as an evolving population.

### 7.4.1 The Power Spectrum of Cosmological Perturbations

At a given time, the perturbation field $\delta(\boldsymbol{r})$ (eq. 7.1) can be considered a random field: the actual perturbation field $\delta(\boldsymbol{r})$ is just a realisation of an underlying probability distribution. The statistical properties of this distribution can be described by defining the **correlation function** $\xi$ and the **power spectrum** $P$. In comoving coordinates $\boldsymbol{x}$ (§2.1.1), the correlation function[11] is

$$\xi(x) \equiv \langle \delta(\boldsymbol{x}')\delta(\boldsymbol{x}' + \boldsymbol{x})\rangle, \tag{7.31}$$

where $x = |\boldsymbol{x}|$ and $\langle \cdots \rangle$ represents the average over $\boldsymbol{x}'$. The correlation function $\xi$ depends only on the magnitude of $\boldsymbol{x}$ because of the expected isotropy of the fluctuation density field. To estimate the characteristic spatial scales of the density fluctuations, it is useful to analyse $\delta(\boldsymbol{x})$ in terms of Fourier series. Given that the Universe is homogeneous and isotropic on large scales (§2.1), it is possible to take a cubic volume of the Universe $V$ sufficiently large that $\delta(\boldsymbol{x})$ can be considered periodic (because all fluctuations are on scales much smaller than the box size) and write

$$\delta(\boldsymbol{x}) = \sum_{\boldsymbol{k}} \delta_{\boldsymbol{k}} e^{i\boldsymbol{k}\cdot\boldsymbol{x}}, \tag{7.32}$$

[11] The matter correlation function defined here would coincide with the galaxy correlation function defined in §6.6 if galaxies perfectly traced mass (i.e. if the galaxy bias were $b = 1$).

where

$$\delta_{\boldsymbol{k}} = \frac{1}{V}\int_V \delta(\boldsymbol{x})e^{-\mathrm{i}\boldsymbol{k}\cdot\boldsymbol{x}}\mathrm{d}^3\boldsymbol{x} \tag{7.33}$$

are the Fourier coefficients of $\delta(\boldsymbol{x})$ and $\boldsymbol{k}$ is a vector with Cartesian components $k_i = 2\pi N_i/l$, where $N_i$ are integers ($i = 1,2,3$) and $l = V^{1/3}$ is the size of the cube's edge. We define the power spectrum of density fluctuations as

$$P(k) \equiv V\langle|\delta_{\boldsymbol{k}}|^2\rangle, \tag{7.34}$$

where $k = |\boldsymbol{k}|$ is the wavenumber and $\langle\cdots\rangle$ represents now the average over different realisations of the perturbation field. Similarly to the correlation function, the power spectrum $P$ depends only on the magnitude of $\boldsymbol{k}$ because of the assumption of isotropy. $P(k)$ measures the amount of fluctuations on scale $\lambda = 2\pi/k$. By definition (eqs. 7.31 and 7.34), the correlation function and the power spectrum are the Fourier transforms of one another, so they are related by

$$\xi(x) = \frac{1}{(2\pi)^3}\int P(k)e^{\mathrm{i}\boldsymbol{k}\cdot\boldsymbol{x}}\mathrm{d}^3\boldsymbol{k} = \frac{1}{2\pi^2}\int_0^\infty P(k)\frac{\sin(kx)}{kx}k^2\mathrm{d}k \tag{7.35}$$

and

$$P(k) = \int \xi(x)e^{-\mathrm{i}\boldsymbol{k}\cdot\boldsymbol{x}}\mathrm{d}^3\boldsymbol{x} = 4\pi\int_0^\infty \xi(x)\frac{\sin(kx)}{kx}x^2\mathrm{d}x. \tag{7.36}$$

A global measure of the amount of fluctuations is the variance of the perturbation field

$$\sigma^2 = \frac{1}{2\pi^2}\int_0^\infty P(k)k^2\mathrm{d}k, \tag{7.37}$$

which is due to the contribution of modes with all possible wavenumbers. It is useful to define the variance of the field due only to fluctuations on scales larger than a given threshold, which is a key quantity, for instance, to estimate the mass function of dark matter halos (§7.4.3). Filtering the power spectrum by smoothing the field on scales smaller than $2\pi/K$ we obtain the **mass variance**

$$\sigma_K^2(\mathcal{M}) = \frac{1}{2\pi^2}\int_0^\infty W_K(k)P(k)k^2\mathrm{d}k, \tag{7.38}$$

where $W_K$ is a $k$-space filter function (such that $W_K \approx 1$ for $k \lesssim K$ and $W_K \approx 0$ for $k \gtrsim K$) and the mass $\mathcal{M}$ associated with the wavenumber $K$ (i.e. spatial scale $2\pi/K$) is

$$\mathcal{M} \propto K^{-3} \tag{7.39}$$

(the coefficient of proportionality, which has the dimensions of density, depends on the specific form of the function $W_K$). As illustrated with a simple 1D example in Fig. 7.2 (right), if the field is filtered by excluding modes with wavenumbers larger than a given value $K$, the variance is expected to decrease for decreasing $K$, because the effect of the filter is to average out small-scale fluctuations. When the power spectrum is a power law $P(k) \propto k^{n_s}$, where $n_s$ is the **spectral index**, the mass variance scales as

$$\sigma_K^2 \propto \int_0^K k^{n_s+2}\mathrm{d}k \propto K^{n_s+3} \propto \mathcal{M}^{-(3+n_s)/3}. \tag{7.40}$$

### 7.4.2 Evolution of the Linear Power Spectrum

The primordial power spectrum is believed to be a power law

$$P_{\rm prim}(k) \propto k^{n_s}, \tag{7.41}$$

where $n_s$ is the spectral index. When $n_s = 1$ the spectrum is called a **Harrison–Zeldovich power spectrum** (Harrison, 1970; Zeldovich, 1972), which is a good approximation of the power spectrum expected at very early times on the basis of the simplest inflation models (§2.3.2). At the time of recombination the perturbation spectrum of the CDM has been modified with respect to the primordial power spectrum $P_{\rm prim}$. The power spectrum at a time $t$ after recombination can be parameterised as

$$\frac{P(k,t)}{P_0(k_0)} = D^2(t)T^2(k)\left(\frac{k}{k_0}\right)^{n_s}, \tag{7.42}$$

where $T(k)$ is the **transfer function**, $k_0$ is a reference wavenumber and $P_0(k) \equiv P(k,t_0)$, where $t_0$ is the present time. The transfer function is a measure of the fact that perturbations with different characteristic scales (thus different $k$) evolve differently in the lapse of time between the end of inflation, when the power spectrum is the primordial power law, and recombination, when the power spectrum contains information on the initial conditions for structure formation. Eq. (7.42) gives at any later time the **linearly extrapolated power spectrum**, which is the spectrum of the perturbations computed in the hypothesis that all the perturbations remain linear.[12] It is apparent that the shape of the linearly extrapolated power spectrum does not change with time, while its amplitude scales with time as the growth factor squared, $D^2(t)$ (eq. 7.10). It follows that the present-day ($t = t_0$) linearly extrapolated power spectrum $P_0(k)$ is given by

$$\frac{P_0(k)}{P_0(k_0)} = T^2(k)\left(\frac{k}{k_0}\right)^{n_s}, \tag{7.43}$$

because $D(t_0) = 1$ by definition.

The CDM transfer function is shown in Fig. 7.3 (left). As apparent, in CDM cosmology the effect of the transfer function is to partially suppress fluctuations on small scales (large $k$). The shape of the transfer function is strictly related to the presence of a finite horizon distance (§2.1.4), that is, to the fact that, at any time $t$, the size of the observable (and causally connected) Universe is of the order of $ct$. While perturbations on scales larger than the size of the horizon can always grow (because gravity is the only force at work), the growth of perturbations on smaller scales can be damped by radiation pressure, as long as matter is coupled with radiation (i.e. before $z_{\rm rec} \approx 1000$; §2.4). In particular, for CDM density fluctuations, perturbations on scales that enter the horizon during the radiation-dominated era ($z > z_{\rm eq} \approx 3400$; §2.3.1) grow more slowly than those that enter during the matter era ($z < z_{\rm eq}$), because the radiation-driven expansion timescale is shorter than the collapse timescale. As a consequence, the size of the horizon ($2\pi/k_{\rm eq} \approx 125$ Mpc) at

[12] This hypothesis is clearly not justified at sufficiently late times and small scales, but the linearly extrapolated power spectrum, even at the present time, is a well defined quantity and a useful tool to compare theory and observations (taking care to correct for non-linear effects on small scales).

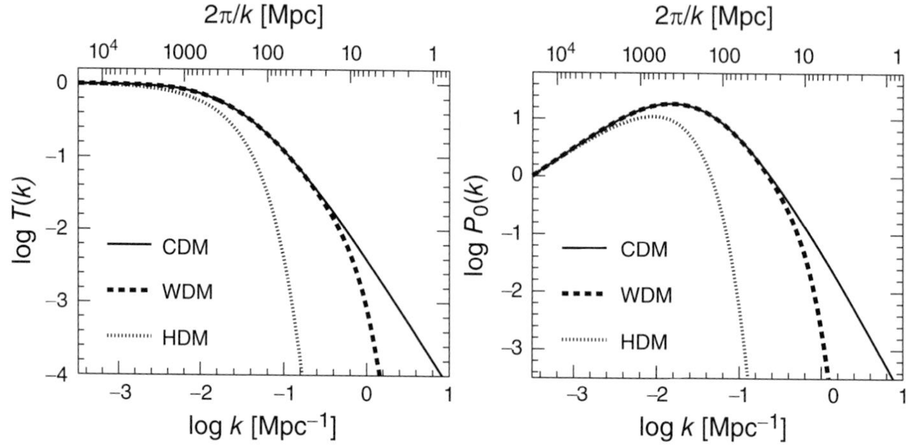


Fig. 7.3 *Left panel.* Transfer function $T$ of cosmological density fluctuations (eq. 7.42) as a function of the wavenumber $k$ for CDM, WDM and HDM. The curves are obtained using the fitting functions of Bardeen et al. (1986) and Bode et al. (2001), assuming present-day matter density parameter $\Omega_{\mathrm{m},0} = 0.3$ and dimensionless Hubble constant $h = 0.7$. In the case of WDM the dark matter particle mass is $m_{\mathrm{DM}}c^2 = 0.1$ keV. *Right panel.* Present-day linearly extrapolated power spectrum $P_0$ of cosmological density fluctuations as a function of the wavenumber $k$, for CDM, WDM and HDM. The power spectrum, given by eq. (7.43), is normalised to $P_0(k_0)$, with $\log(k_0/\,\mathrm{Mpc}^{-1}) = -3.5$. The transfer functions are computed as in the left panel.

the time of matter–radiation equivalence $z_{\mathrm{eq}}$ is a characteristic scale in the CDM transfer function: $T \to 1$ for $k \ll k_{\mathrm{eq}}$, while $T \to 0$ for $k \gg k_{\mathrm{eq}}$. The present-day linearly extrapolated power spectrum (eq. 7.43) in a ΛCDM Universe, shown in Fig. 7.3 (right), is obtained by combining the initial Harrison–Zeldovich power-law power spectrum with the CDM transfer function shown in the left panel of Fig. 7.3.

In addition to the CDM transfer function and power spectrum, Fig. 7.3 plots also the same quantities, but in the case of WDM (with dark matter particle mass $m_{\mathrm{DM}}c^2 = 0.1$ keV) and HDM (for instance, neutrinos; §7.2). When the dark matter is not cold, the transfer function accounts also for the effect of free streaming (§7.2): particles with non-negligible kinetic energy are not confined by the potential wells of small-scale fluctuations, which are therefore further damped (similar to what happens to perturbations smaller than the Jeans length in a fluid; §8.3.2). As a consequence, the WDM power spectrum differs from the CDM power spectrum for the presence of a cut-off at high $k$. The cut-off occurs at lower values of $k$ for smaller $m_{\mathrm{DM}}$ and is dramatic in the case of very light (HDM) dark matter particles, which are relativistic at decoupling (§7.2). In other words, compared to a CDM Universe, the linear density field of a WDM Universe is smoother on small scales. In an HDM Universe significant linear fluctuations are present only on relatively large scales.

The shape of the cosmological power spectrum that we discussed in this section is a key ingredient for cosmological structure formation. For instance, in the next section we show how the mass function of dark matter halos can be computed for a given cosmological model by combining the mass variance (§7.4.1), which is a function of the power spectrum, with the results of the spherical collapse model (§7.3.2).

### 7.4.3 Halo Mass Function

Similar to the distribution functions of galaxies, such as the LFs and SMFs (§3.5), a fundamental quantity to describe statistically the population of dark matter halos at a given epoch is the **halo mass function** (HMF), which can be estimated analytically using the **Press–Schechter formalism** (Press and Schechter, 1974). With the aim to evaluate the number of halos more massive than a given mass $\mathcal{M}$, we start by defining $\delta_K$, the density fluctuation field filtered on a mass scale $\mathcal{M} \propto K^{-3}$, which is obtained from $\delta$ (eq. 7.1) by averaging out all perturbations on scales smaller than $\approx K^{-1}$ (§7.4.1). If the fluctuation field is a **Gaussian random field**,[13] the probability that the overdensity has a value between $\delta_K$ and $\delta_K + \mathrm{d}\delta_K$ is

$$p(\delta_K)\mathrm{d}\delta_K = \frac{1}{\sqrt{2\pi}\sigma_K} \exp\left(-\frac{\delta_K^2}{2\sigma_K^2}\right)\mathrm{d}\delta_K, \tag{7.44}$$

where $\sigma_K^2(\mathcal{M})$ is the mass variance (eq. 7.38). Given a critical overdensity for collapse $\delta_\mathrm{c}$ (§7.3.2), the probability of finding a fluctuation larger than $\delta_\mathrm{c}$ is

$$\mathcal{P}(\mathcal{M}) = \int_{\delta_\mathrm{c}}^{\infty} p(\delta_K)\mathrm{d}\delta_K, \tag{7.45}$$

where we are considering the fluctuation field averaged on mass scale $\mathcal{M}$ (i.e. filtered by eliminating fluctuations on scales smaller than $2\pi/K$; §7.4.1). The fraction of collapsed halos with mass between $\mathcal{M}$ and $\mathcal{M} + \mathrm{d}\mathcal{M}$ is

$$\left|\frac{\mathrm{d}\mathcal{P}}{\mathrm{d}\mathcal{M}}\right| = \left|\frac{\mathrm{d}\mathcal{P}}{\mathrm{d}\sigma_K}\frac{\mathrm{d}\sigma_K}{\mathrm{d}\mathcal{M}}\right|. \tag{7.46}$$

From eqs. (7.44) and (7.45) we get

$$\frac{\mathrm{d}\mathcal{P}}{\mathrm{d}\sigma_K} = \int_{\delta_\mathrm{c}}^{\infty} \frac{\mathrm{d}p}{\mathrm{d}\sigma_K}\mathrm{d}\delta_K = \frac{1}{\sqrt{2\pi}}\frac{\delta_\mathrm{c}}{\sigma_K^2} \exp\left(-\frac{\delta_\mathrm{c}^2}{2\sigma_K^2}\right), \tag{7.47}$$

so

$$\left|\frac{\mathrm{d}\mathcal{P}}{\mathrm{d}\mathcal{M}}\right| = \frac{1}{\sqrt{2\pi}}\frac{\delta_\mathrm{c}}{\sigma_K^2} \exp\left(-\frac{\delta_\mathrm{c}^2}{2\sigma_K^2}\right)\left|\frac{\mathrm{d}\sigma_K}{\mathrm{d}\mathcal{M}}\right|. \tag{7.48}$$

The number density of halos with mass between $\mathcal{M}$ and $\mathcal{M} + \mathrm{d}\mathcal{M}$ is then given by

$$\frac{\mathrm{d}n}{\mathrm{d}\mathcal{M}}(\mathcal{M}) = 2\frac{\overline{\rho}}{\mathcal{M}}\left|\frac{\mathrm{d}\mathcal{P}}{\mathrm{d}\mathcal{M}}\right| = \sqrt{\frac{2}{\pi}}\frac{\overline{\rho}}{\mathcal{M}^2}\frac{\delta_\mathrm{c}}{\sigma_K} \exp\left(-\frac{\delta_\mathrm{c}^2}{2\sigma_K^2}\right)\left|\frac{\mathrm{d}\ln\sigma_K}{\mathrm{d}\ln\mathcal{M}}\right|, \tag{7.49}$$

where the factor of 2 is introduced to approximately account for the mass in underdense regions, which is formally not allowed to collapse. The HMF given by eq. (7.49) depends on the power spectrum $P(k)$, because the function $\sigma_K(\mathcal{M})$ is determined by the shape of the power spectrum (§7.4.1). At given $P(k)$, the HMF depends on redshift through the critical overdensity for collapse $\delta_\mathrm{c}(z)$, which is an increasing function of $z$ (§7.3.2).

[13] A random field $\delta(\boldsymbol{x})$ is Gaussian if its modes $\delta_{\boldsymbol{k}}$ (eq. 7.33), which can be written as $\delta_{\boldsymbol{k}} = |\delta_{\boldsymbol{k}}|\exp(\mathrm{i}\theta_k)$, where $|\delta_{\boldsymbol{k}}|$ is the amplitude and $\theta_k$ is the phase, are independent and have random phases. The statistical properties of a Gaussian random field are fully determined by its power spectrum.

The Press–Schechter formalism has been taken as the starting point to build more sophisticated analytic models to compute the statistical properties of dark matter halos, such as the **extended Press–Schechter formalism** (Bond et al., 1991). The extended Press–Schechter theory, accounting for the fact that dark matter clumps can contain subclumps and taking care of the normalisation factor self-consistently (allowing also for matter in initially underdense regions to end up in collapsed halos), predicts that the HMF $\mathrm{d}n/\mathrm{d}\mathcal{M}$ is given by eq. (7.49), so the normalisation used by Press and Schechter (1974) turns out to be correct. The extended Press–Schechter theory is found to compare well with the results of cosmological $N$-body simulations (§7.5.1) and is nowadays at the basis of most analytic models of structure formation. The extended Press–Schechter formalism also connects the properties of progenitor and descendant halos, thus allowing us to build halo merger trees (§7.4.4), which is not possible within the framework of the original Press–Schechter formalism.

The HMF in a $\Lambda$CDM Universe, computed using the extended Press–Schechter formalism, is plotted for different redshifts in Fig. 7.4. The shape of the HMF is qualitatively similar to the SMF of galaxies (§3.5.2), with a power-law regime at low masses and an exponential cut-off at high masses, but with important quantitative differences, as discussed in §10.10.1, where we briefly describe how we can connect the dark and luminous matter distributions in the Universe. From Fig. 7.4 it is apparent that, at fixed redshift, the number density of halos decreases for increasing mass, because the mass variance decreases for increasing mass (eq. 7.40). At fixed mass, the number density is higher at lower redshift, especially at the higher masses, because the critical overdensity for collapse $\delta_c$ decreases with cosmic time (§7.3.2).

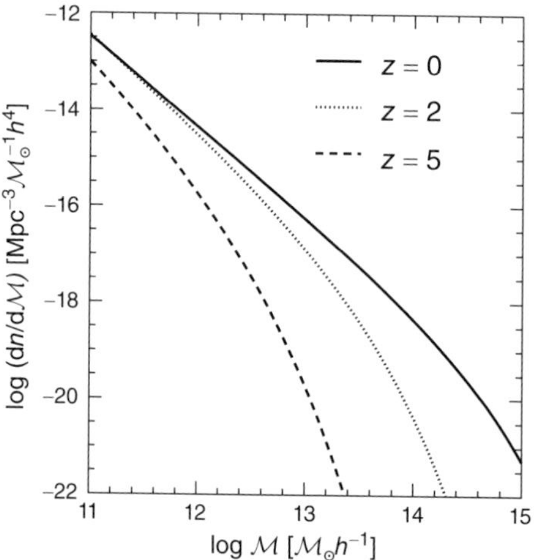


Fig. 7.4 HMFs at different redshifts in a $\Lambda$CDM Universe, computed by Angrick and Bartelmann (2010) with the extended Press–Schechter formalism combined with a refinement of the ellipsoidal collapse model of Sheth and Tormen (1999). Figure adapted from Angrick and Bartelmann (2010). Data courtesy of C. Angrick.

### 7.4.4 Hierarchical Merging of Cold Dark Matter Halos: Merger Trees and Merger Rates

In the CDM cosmology (§2.3) the growth of structure occurs bottom-up. Smaller systems virialise first and larger systems form later (see Fig. 7.4), mainly by incorporating smaller virialised systems, but also accreting diffuse matter (that is, matter that does not belong to a virialised halo). The process in which two or more virialised systems coalesce and form a new virialised system is known as merging (§8.9). As dark matter halos grow mainly through subsequent mergers of smaller systems, the assembly process of dark matter halos is known as **hierarchical merging**. Neglecting accretion of diffuse matter, the growth history of a dark matter halo can be characterised by its **merger tree**, which is a tree-shaped diagram, similar to a genealogical tree, plotting the progenitors of a halo back in cosmic time (Fig. 7.5, left). In a merger tree a merger occurs at each branch attachment. At any given time, we define the main progenitor halo as the most massive among the progenitors at that time. The **merger history** of a dark matter halo is characterised by both mergers with halos of similar mass (**major mergers**) and accretion of much smaller halos (**minor mergers**). The fundamental parameter characterising a merger between two

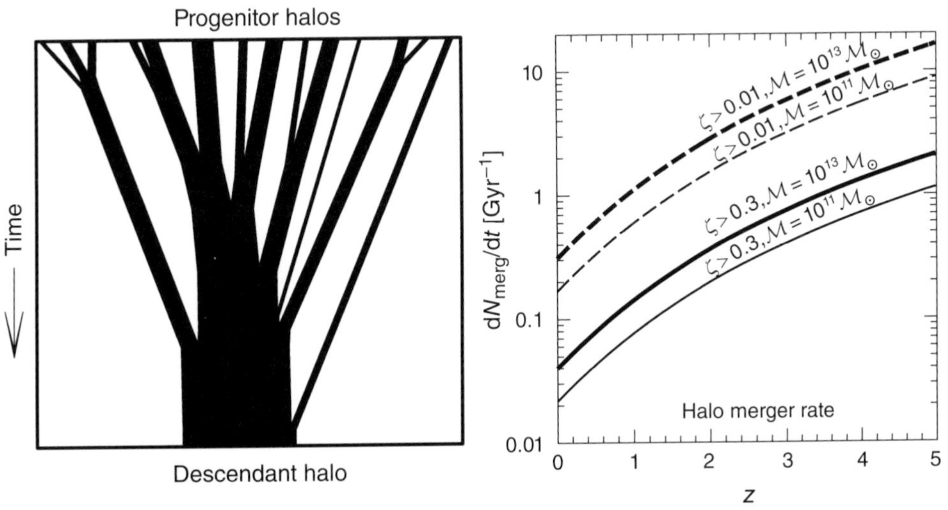


Fig. 7.5 *Left panel*. Schematic representation of a merger tree. The descendant halo (bottom of the figure, late times) is represented by the trunk of the tree. The initial progenitors (top of the figure, early times) are the thinnest branches of the tree. At any given time each halo is represented in black by a horizontal line with length proportional to its mass. The sum of the line lengths is the same at all times, because in this scheme it is assumed that a halo grows only through mergers and there is no mass loss. Halo mergers correspond to branch attachments. *Right panel*. Number of mergers per halo per unit time as a function of redshift as measured in dark matter-only cosmological simulations. The solid curves represent the merger rate considering only major mergers (merger mass ratio $\zeta > 0.3$). The dashed curves represent the merger rate including both major and minor mergers with $\zeta > 0.01$. At given redshift, the merger rate is higher for halos with virial mass $10^{13}\mathcal{M}_\odot$ (thick curves) than for halos with virial mass $10^{11}\mathcal{M}_\odot$ (thin curves). The curves are computed using the fits to the average merger rates by Fakhouri et al. (2010).

systems of masses, respectively, $\mathcal{M}$ and $\mathcal{M}'$ is the **merger mass ratio** $\zeta \equiv \mathcal{M}'/\mathcal{M}$ (usually defined taking $\mathcal{M}' \leq \mathcal{M}$, so $\zeta \leq 1$). Conventionally, the value of $\zeta$ discriminating between minor and major mergers is taken in the range $1/4 \lesssim \zeta \lesssim 1/3$.

Statistically, the merger history of dark matter halos can be quantified by calculating the halo merger rate, that is, the probability that, at a given time, a halo of mass $\mathcal{M}$ merges with a halo of mass $\mathcal{M}'$. The halo merger rate can be estimated analytically using the extended Press–Schechter formalism (§7.4.3). Alternatively, merger rates can be measured numerically in cosmological $N$-body simulations (§7.5.1). The merger rate can be seen as a function of halo mass $\mathcal{M}$, mass ratio $\zeta$ and redshift $z$. The number of mergers per halo per unit time ($dN_{\rm merg}/dt$) increases for increasing $z$, is weakly dependent on halo mass (slightly increasing for increasing $\mathcal{M}$) and is strongly decreasing ($\propto \zeta^{-2}$) for increasing mass ratio $\zeta$. Examples of merger rates measured in cosmological $N$-body simulations are shown in the right panel of Fig. 7.5 as functions of redshift, for different halo masses and ranges of mass ratios. The quantity $dN_{\rm merg}/dt$, here defined for halos, is analogous to the fractional merger rate of galaxies $\mathcal{R}_{\rm merg} = f_{\rm merg}/\tau_{\rm merg}$ (eq. 6.2), for which observational estimates are given in §6.1 and §11.3.2.

Fig. 7.6 plots the mean mass assembly history of dark matter halos, due to both mergers and diffuse accretion, as found in cosmological $N$-body simulations: in the left panel the mass of representative halos is plotted as a function of $z$, while the right panel shows the corresponding dark matter mass accretion rate. This kind of information can be used

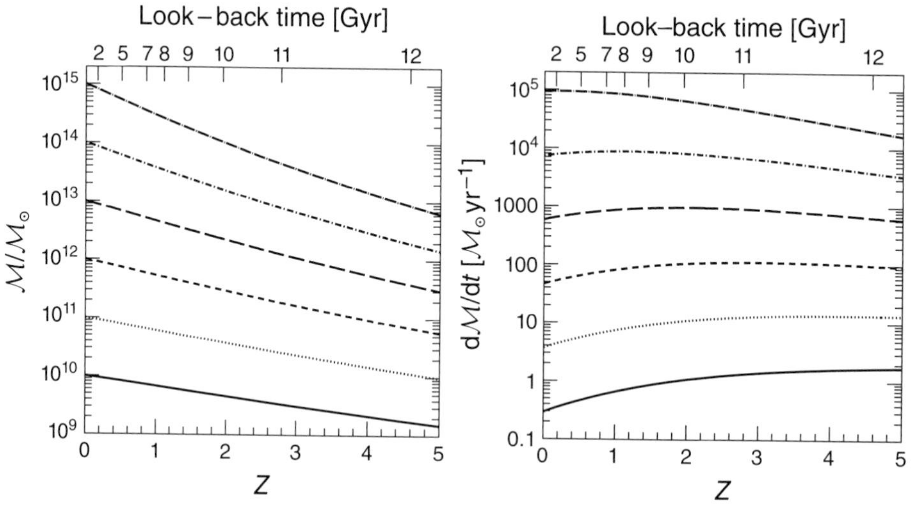


Fig. 7.6 *Left panel.* Mean halo virial mass as a function of redshift (or look-back time), for halos of different $z = 0$ mass, as measured in dark matter-only cosmological simulations. The curves are computed using the fits to the mean halo mass assembly histories by Fakhouri et al. (2010). *Right panel.* Mean mass accretion rate of dark matter onto halos as a function of redshift (or look-back time) for halos of $z = 0$ mass $10^{10}$, $10^{11}$, $10^{12}$, $10^{13}$, $10^{14}$ and $10^{15}\,\mathcal{M}_\odot$, as measured by Fakhouri et al. (2010) in the same cosmological simulations as in the left panel. The line styles refer to the same $z = 0$ halo masses as in the left panel.

to predict the mass of the progenitors of present-day dark matter halos and their mass accretion history.

## 7.5 Structural and Kinematic Properties of Dark Matter Halos

So far we have characterised dark matter halos only with global quantities, such as the virial mass and the virial radius. Here we look at halos in more detail, by focusing on their internal properties.

### 7.5.1 Cosmological Numerical Simulations

While the statistical properties of the dark matter halo population can be sufficiently well described with analytic methods, such as the extended Press–Schechter formalism (§7.4.3), our understanding of the internal structure and kinematics of dark matter halos relies on the results of **cosmological numerical simulations**. In these simulations the evolution of a representative cubic volume (simulation box) of the Universe is followed from high to low redshift under the assumption of periodic boundary conditions. In practice, the Universe is modelled by formally replicating infinite times the simulation box and placing each box adjacent to another identical box (see also §7.4.1), which is justified, for sufficiently large boxes, because the Universe is homogeneous on large scales (§2.1). The simplest cosmological simulations are **dark matter-only cosmological $N$-body simulations**, in which the matter distribution of the Universe is represented by a single dissipationless component, representing the dark matter, while the contribution of baryons is not explicitly accounted for. In these simulations gravity is the only force considered. Though simplified, these dark matter-only simulations are very useful for several purposes, because dark matter dominates the matter component of the Universe (§2.5). The initial conditions are represented by a homogeneous distribution of particles,[14] perturbed according to the power spectrum as given in eq. (7.42), with parameters and transfer function determined by the adopted cosmological model. Snapshots of the simulations are produced, containing information on masses, positions and velocities of dark matter particles at given redshift. **Halo finding algorithms**, often referred to as **halo finders**, are then applied to each snapshot to identify individual halos and subhalos (§7.5.3). The halo finders therefore provide halo merger trees and mass functions at different redshifts. Moreover, once the particles belonging to a halo are identified, the internal structure and kinematics of the halo can be studied.

Cosmological $N$-body simulations are also a fundamental tool to study the process of structure formation and the LSS predicted by a given cosmological model. The dark matter density distribution found in these simulations is characterised by elongated structures (filaments), almost empty regions (voids) and high-density regions (halos) typically located

[14] In these simulations the particles (with typical masses $\gtrsim 10^6\,\mathcal{M}_\odot$) do not represent individual dark matter particles, but are just tracers of the collisionless dark matter distribution.

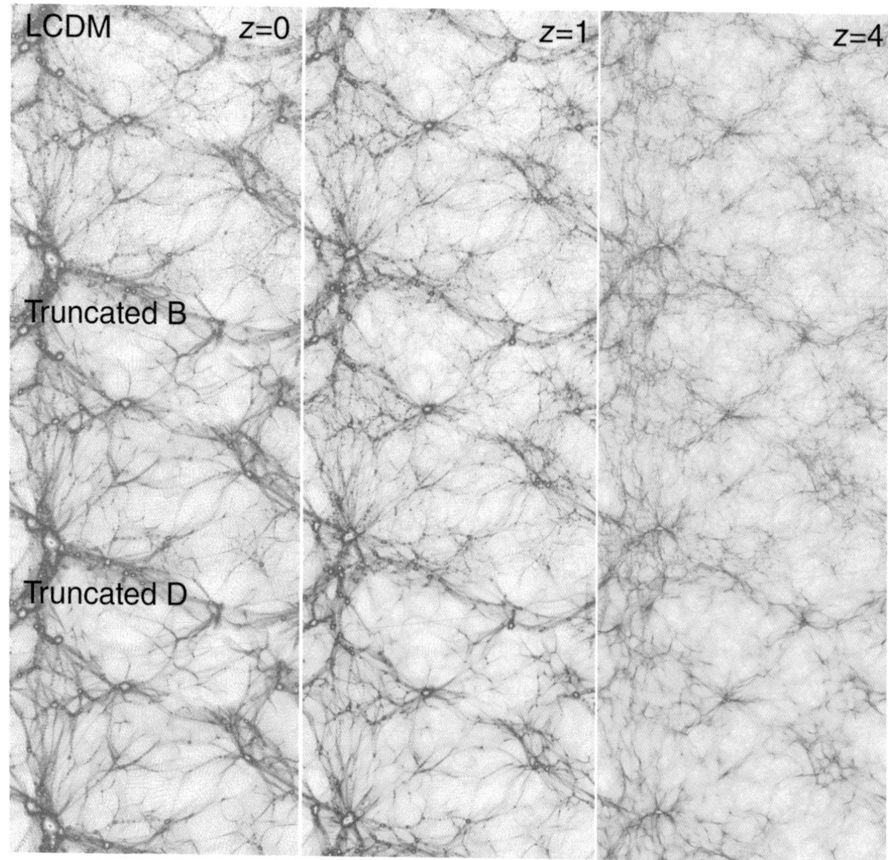


Fig. 7.7 Structure formation in CDM and WDM cosmologies. The top row of panels (model 'LCDM') show the projected dark matter density distribution in thin slices ($20 \times 20 \times 2h^{-3}\,\mathrm{Mpc}^3$) in a dark matter-only $\Lambda$CDM cosmological simulation at redshifts $z = 0$, $z = 1$ and $z = 4$ (from left to right; darker regions have higher density). The panels below show, at the same redshifts, the results of similar simulations in which the initial fluctuation power spectrum is truncated on small scales ($k > k_{\mathrm{cut}}$), mimicking the effect of WDM (Fig. 7.3). In model 'Truncated B' (middle row of panels) the power spectrum is truncated at smaller spatial scales ($k_{\mathrm{cut}} = 19.06h\,\mathrm{Mpc}^{-1}$) than in model 'Truncated D' ($k_{\mathrm{cut}} = 8.85h\,\mathrm{Mpc}^{-1}$; bottom row of panels). Qualitatively, the figure can be interpreted in the sense that the mass of the dark matter particle decreases from top ('LCDM') to bottom ('Truncated D'). From Power (2013).

at the intersection of filaments (Fig. 7.7). This simulated cosmic web is clearly reminiscent of the observed large-scale distribution of galaxies (Fig. 6.14). In detail, the properties of the predicted cosmic web depend on the specific cosmological model. For instance, in Fig. 7.7 the dark matter density distribution predicted by a CDM model is compared with two WDM-like models. It is apparent that WDM models predict less small-scale structure than the CDM model (in which filaments are made of small substructures) and that the effect is stronger for lower-mass dark matter particles, for which the fluctuation power spectrum is truncated at larger scales (§7.2 and §7.4.2; Fig. 7.3).

The evolution of the dark matter, which is dissipationless and collisionless, is determined only by gravity, so dark matter-only simulations are basically limited only by numerical resolution, that is by the size of the simulated box and by the number of particles (or, equivalently, by the masses of the simulation particles). In §10.11.1 we consider hydrodynamic cosmological simulations, in which, in addition to the dark matter, the evolution of the baryonic component (gas and stars) is studied.

## 7.5.2 Density Profiles of Dark Matter Halos

Cosmological simulations of structure formation using only dark matter (§7.5.1) show that the shape of the density profile of dark matter halos does not vary dramatically either as a function of mass (from dwarf galaxy to galaxy cluster-size halos) or from halo to halo, at a given mass. The density profile of a simulated halo is usually defined by fixing the halo centre (for instance as the location of the particle of the halo with the minimum gravitational potential energy) and then measuring the mass contained in spherical shells centred on the halo centre.

### Navarro–Frenk–White and Einasto Profiles

A two-parameter analytic fitting function that gives a good representation of the density profiles of the halos obtained in dark matter-only cosmological simulations is the **Navarro–Frenk–White (NFW) profile** (Navarro et al., 1996)

$$\rho(r) = \frac{4\rho_s}{(r/r_s)\,(1 + r/r_s)^2}, \tag{7.50}$$

where $r_s$ is the **scale radius** and $\rho_s$ is the density at $r_s$. The corresponding mass profile (first equality in eq. 4.26) is

$$\mathcal{M}(r) = 16\pi\rho_s r_s^3 \left[\ln\left(1 + \frac{r}{r_s}\right) - \frac{r/r_s}{1 + r/r_s}\right]. \tag{7.51}$$

The density at $r_s$ can be written as

$$\rho_s = \frac{\mathcal{M}_\Delta}{16\pi r_s^3\,[\ln(1 + c_\Delta) - c_\Delta/(1 + c_\Delta)]}, \tag{7.52}$$

where $\mathcal{M}_\Delta = \mathcal{M}(r_\Delta)$ is the virial mass, $r_\Delta$ is the virial radius (§7.3.2) and $c_\Delta \equiv r_\Delta/r_s$ is the **concentration**. The gravitational potential of the NFW model, related to the NFW density (eq. 7.50) by the Poisson equation (eq. 7.3), has the simple analytic form

$$\Phi(r) = -16\pi G\rho_s r_s^2 \frac{\ln(1 + r/r_s)}{r/r_s}. \tag{7.53}$$

An important quantity to characterise the density profile of a dark matter halo is the logarithmic density slope

$$\gamma(r) \equiv \frac{\mathrm{d}\ln\rho}{\mathrm{d}\ln r} = \frac{r}{\rho}\frac{\mathrm{d}\rho}{\mathrm{d}r}. \tag{7.54}$$

For an NFW profile

$$\gamma(r) = -\frac{1 + 3r/r_s}{1 + r/r_s}, \tag{7.55}$$

so $\gamma = -1$ at $r = 0$ and $\gamma \to -3$ for $r \to \infty$. The scale radius $r_s$ is such that $\gamma(r_s) = -2$, corresponding to the slope of the singular isothermal sphere (§5.3.4).

Another functional form that represents well the density profiles of dark matter halos produced in dark matter-only cosmological simulations is the **Einasto profile**[15] (Einasto, 1965), originally proposed to describe stellar components of galaxies, which can be written as

$$\rho(r) = \rho_s \exp\left\{-\frac{2}{\alpha}\left[\left(\frac{r}{r_s}\right)^{\alpha} - 1\right]\right\}, \tag{7.56}$$

where $\alpha$ is the **Einasto index** and, as in the NFW model, $r_s$ is the scale radius and $\rho_s \equiv \rho(r_s)$. For the Einasto model the logarithmic density slope is

$$\gamma(r) = -2\left(\frac{r}{r_s}\right)^{\alpha}, \tag{7.57}$$

so, as for the NFW model, $r_s$ is such that $\gamma(r_s) = -2$. The mass profile of the Einasto model is

$$\mathcal{M}(r) = \frac{4\pi}{\alpha}\left(\frac{\alpha}{2}\right)^{3/\alpha} e^{2/\alpha}\rho_s r_s^3 \left[\Gamma\left(\frac{3}{\alpha}\right) - \Gamma\left(\frac{3}{\alpha}, \eta^{\alpha}\right)\right], \tag{7.58}$$

where

$$\eta \equiv \left(\frac{2}{\alpha}\right)^{1/\alpha} \frac{r}{r_s}, \tag{7.59}$$

$\Gamma(z)$ is the Gamma function and $\Gamma(z, x)$ is the upper incomplete Gamma function (§3.5.1). As for the NFW model, the concentration of an Einasto model is defined as $c_\Delta = r_\Delta/r_s$, where $r_\Delta$ is the virial radius.

Note that, while the NFW profile has two parameters ($\rho_s$ and $r_s$), the Einasto profile has three parameters ($\rho_s$, $r_s$ and $\alpha$). The shape of the Einasto density profile varies for varying $\alpha$. The density profiles and logarithmic density slopes of the NFW profile and of Einasto models with different values of the Einasto index $\alpha$ are shown in Fig. 7.8. We note that, over the radial range $0.1 \lesssim r/r_s \lesssim 10$, an Einasto profile with $\alpha \simeq 0.2$ is very similar to an NFW profile, consistent with the fact that both models represent well the numerical profiles found in cosmological $N$-body simulations. As we see in Fig. 7.10 (right), at least at $z = 0$, $\alpha$ measured in simulated halos spans a relatively small range around 0.2.

## Concentration–Mass–Redshift Relation

Above we have defined the concentration for NFW and Einasto profiles. More generally, the concentration of a halo is defined as $c_\Delta = r_\Delta/r_{-2}$, where $r_{-2}$ is the radius such that

[15] The **Einasto law** (eq. 7.56) is, from a mathematical point of view, the same function as the Sérsic law (eq. 3.1), with the Einasto index $\alpha$ corresponding to $1/n$, where $n$ is the Sérsic index. However, the Sérsic law, usually applied to the surface brightness profiles of galaxies, is a function of the projected radius $R$, while the Einasto law, usually applied to the 3D density profiles of halos, is a function of the intrinsic radius $r$.

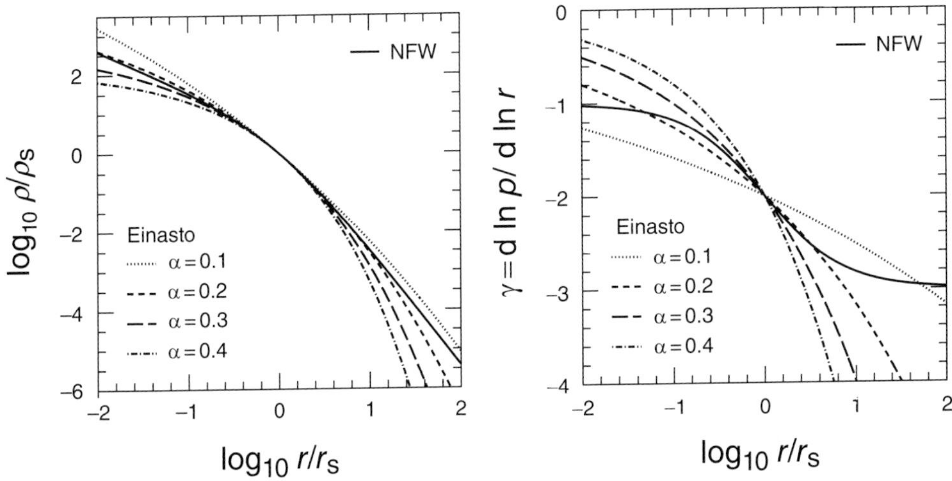


Fig. 7.8 Density (*left panel*) and logarithmic density slope (*right panel*) as functions of radius for Einasto and NFW models.

the logarithmic density slope $\gamma(r_{-2}) = -2$ (for both NFW and Einasto models $r_s = r_{-2}$). Consistent with the notation introduced in §7.3.2, when the critical overdensity $\Delta_c$ depends on redshift, as in eq. (7.21), the concentration is $c_{vir} \equiv r_{vir}/r_{-2}$. In dark matter-only cosmological simulations (§7.5.1) the concentration is found to correlate with both halo mass and redshift. At fixed $z$, $c_\Delta$ decreases for increasing mass; at fixed mass, $c_\Delta$ decreases for increasing $z$ (Fig. 7.9, left). As individual halos grow in mass with time, the concentration–mass–redshift relation indicates that individual halos tend to increase their concentration with cosmic time (Fig. 7.9, right) by accreting mass mainly in the outskirts, while leaving almost unaltered the central density distribution formed at high redshift. The correlation between concentration and virial mass is well approximated by a power law

$$\log c_\Delta = a + b \log\left(\frac{\mathcal{M}_\Delta}{10^{12}h^{-1}\mathcal{M}_\odot}\right), \tag{7.60}$$

where $a = a(z)$ and $b = b(z)$ are redshift-dependent parameters. For $\Delta_c = 200$, at $z = 0$, $a \approx 0.9$ and $b \approx -0.1$, so $c_{200} \approx 8$ for $\mathcal{M}_{200} = 10^{12}h^{-1}\mathcal{M}_\odot$ (Fig. 7.9, left). For $\Delta_c(z)$ (eq. 7.21), at $z = 0$, $a \approx 1$ and $b \approx -0.1$, so $c_{vir} \approx 10$ for $\mathcal{M}_{vir} = 10^{12}h^{-1}\mathcal{M}_\odot$. The scatter in concentration at given mass and redshift is $\approx 0.1$–$0.2$ dex.

## Cusps and Cores

An important property of dark matter halos is the central logarithmic density slope $\gamma_0 \equiv \lim_{r\to 0}\gamma(r)$, where $\gamma$ is given by eq. (7.54). We speak of density **cusps** when $\gamma_0 < 0$ (the density diverges in the centre) and of density **cores** when $\gamma_0 \approx 0$, as illustrated by Fig. 7.10 (left). In fact models with $\gamma > -0.5$ in the central regions, though formally cuspy, can be considered to have a central core (the density increases mildly towards the centre), so a quantitative distinction between cusps and cores can be obtained by fixing $\gamma_0 = -0.5$ as a threshold value. A general result of dark matter-only cosmological simulations is that dark

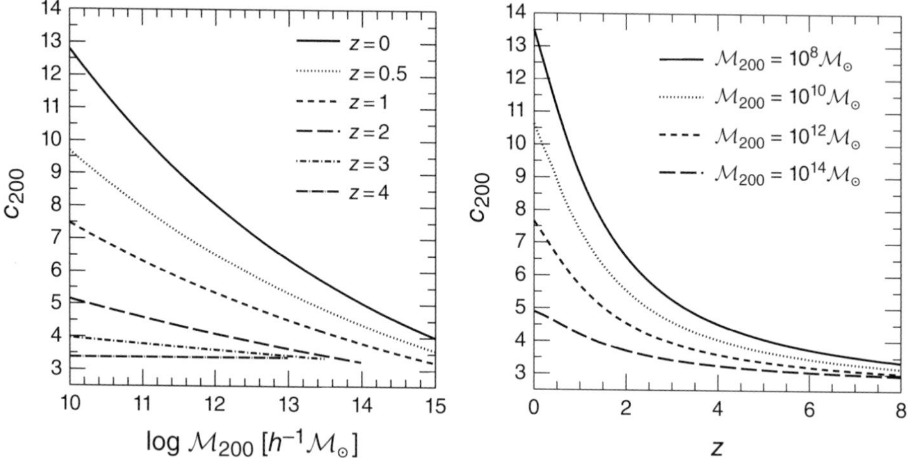


**Fig. 7.9** *Left panel*. Halo concentration $c_{200}$ as a function of halo mass $\mathcal{M}_{200}$ at different redshifts as found in dark matter-only cosmological simulations. The curves represent the best fits given by Dutton and Macciò (2014). Not all curves span the full mass range, because high-mass halos are rare at high redshift (Fig. 7.4). *Right panel*. Redshift evolution of the concentration of individual dark matter halos of different present-day ($z = 0$) virial masses, as indicated by the labels. The curves are computed using a model based on the extended Press–Schechter formalism (§7.4.3). Figure adapted from Correa et al. (2015). Data courtesy of C. Correa.

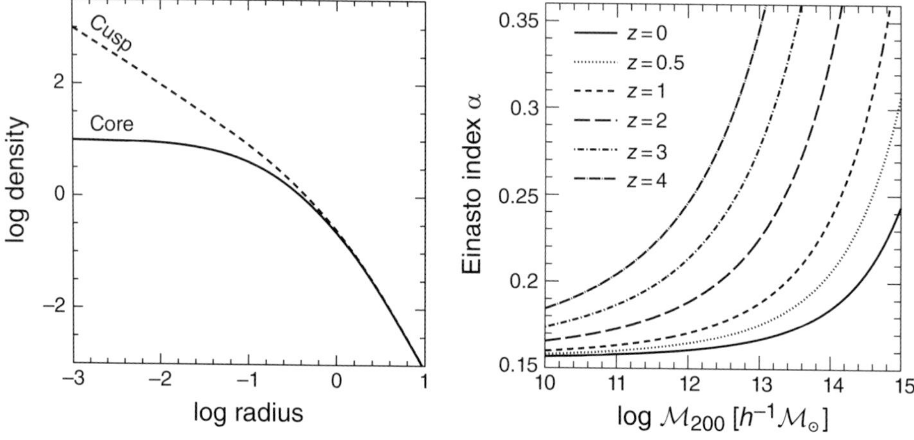


**Fig. 7.10** *Left panel*. Illustrative examples of cuspy (divergent at small radii, with a central cusp; dashed curve) and cored (constant at small radii, with a central core; solid curve) density profiles (in arbitrary units). *Right panel*. Median best-fitting Einasto index $\alpha$ as a function of halo mass and redshift as found in dark matter-only cosmological simulations. Figure adapted from Dutton and Macciò (2014).

matter halos are cuspy. The NFW profile is characterised by a strong central density cusp, because $\gamma \leq -1$ at all radii and $\gamma_0 = -1$ (eq. 7.55). Though formally the Einasto profile is not cuspy ($\gamma_0 = 0$, eq. 7.57), in fact for values of $\alpha$ representative of cosmological halos ($\alpha \lesssim 0.3$), also the Einasto fit has $\gamma < -0.5$ down to $r \approx r_s/100$ (Fig. 7.8, right), so these profiles are effectively cuspy.

Based on what can be inferred from the observational data, there is not strong evidence for the presence of central cusps in the dark halos of galaxies (§7.1). However, this is not necessarily in contrast with the results of dark matter-only cosmological simulations, because these simulations neglect the effect of baryon physics, which can influence substantially the density distribution of dark matter in galaxies, possibly turning cusps into cores during the process of galaxy formation and evolution (§7.5.5 and §10.9.2).

### Structural Non-Homology of Dark Matter Halos

The NFW law (eq. 7.50) represents well the density profiles of dark matter halos of dark matter-only cosmological simulations over a wide mass interval from the galaxy cluster scale down to the dwarf galaxy scale, independently of redshift. In this sense, the density profile of dark matter halos is sometimes said to be *universal*. In fact, in high-resolution dark matter-only cosmological simulations, there is evidence for a weak systematic variation of the shape of dark matter halo profiles with mass and redshift, known as structural non-homology (§5.4.1 and §5.4.3), which can be quantified by fitting the density profiles using the Einasto law (eq. 7.56) with free shape parameter $\alpha$. The best-fitting values of $\alpha$ depend on mass and redshift (Fig. 7.10, right). At $z = 0$ the median best-fitting value of $\alpha$ tends to increase with halo mass. At fixed halo mass $\alpha$ is larger at higher redshift. There are indications that the parameter $\alpha$ is sensitive to the mass accretion history of the halo: for instance, late infall of satellite halos is believed to produce an extended envelope, which is a feature of low-$\alpha$ Einasto profiles. It must be noted that, though clear trends are evident when the average $\alpha$ of simulated halos is considered, the distribution of $\alpha$ of individual halos of similar mass at similar redshift has significant scatter ($\approx 0.2$ dex).

## 7.5.3 Shape of Dark Matter Halos

The density profiles of halos in cosmological $N$-body simulations are often computed by considering the mass distribution in spherical shells (§7.5.2). However, it is now well established that the **shape** of simulated dark matter halos is significantly different from a spherically symmetric distribution. Deviations from spherical symmetry are due both to **triaxiality** of the large-scale density distribution and to the fact that the dark matter distribution is not smooth on small scales, because of the presence of **substructures**. The deviation from spherical symmetry and the abundance of substructures is apparent in snapshots of dark matter halos obtained in high-resolution cosmological $N$-body simulations, such as that shown in Fig. 7.11.

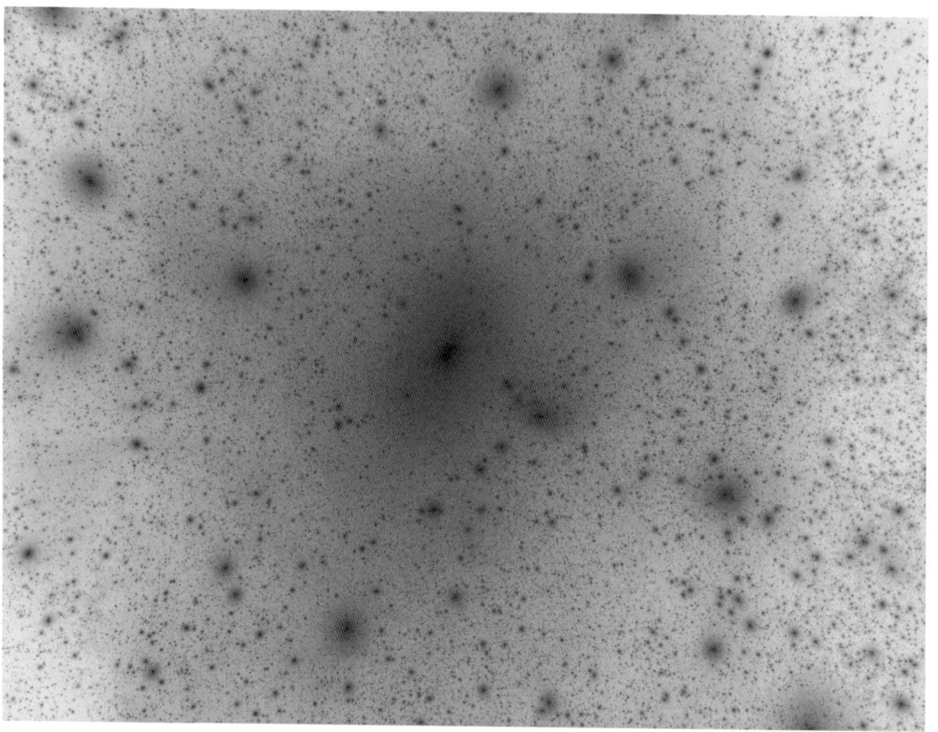

Fig. 7.11 Snapshot at $z = 0$ of a Milky Way-size dark matter halo in the dark matter-only cosmological simulation Aquarius (Springel et al., 2008). The halo, dubbed Aq-A-1, has virial mass $\mathcal{M}_{200} \simeq 1.8 \times 10^{12} \mathcal{M}_\odot$ and virial radius $r_{200} \simeq 246$ kpc at $z = 0$. The greyscale is proportional to the logarithm of the projected dark matter density (the density is higher in darker regions). The extended dark region at the centre of the snapshot is the central density peak of the halo, while the smaller dark regions are subhalos. The horizontal dimension of the plot corresponds to about 1 Mpc. Courtesy of V. Springel.

## Triaxiality

The isodensity surfaces of dark matter halos in cosmological simulations can be described by concentric ellipsoids. Ellipsoids are characterised by their intrinsic principal semi-axes $c_{\rm int} \leq b_{\rm int} \leq a_{\rm int}$ and are classified as prolate, oblate or triaxial (§5.1.1). A quantity often used to classify halos on the basis of their shape is the **triaxiality parameter**

$$T = \frac{a_{\rm int}^2 - b_{\rm int}^2}{a_{\rm int}^2 - c_{\rm int}^2}, \tag{7.61}$$

such that an oblate spheroid has $T = 0$ and a prolate spheroid has $T = 1$. In general, halos are not spheroids ($T \neq 0, T \neq 1$), but, broadly speaking, triaxial halos are classified as oblate when $0 < T < 1/3$, prolate when $2/3 < T < 1$ and triaxial for intermediate values of $T$. In this sense, most halos in dark matter-only cosmological simulations are found to be prolate, a significant fraction of halos are triaxial, while oblate halos are very rare.

A simple quantitative measure of the deviation from spherical symmetry of halos is the smallest-to-largest axis ratio $s \equiv c_{\rm int}/a_{\rm int}$. The trend is that the **shape parameter** $s$

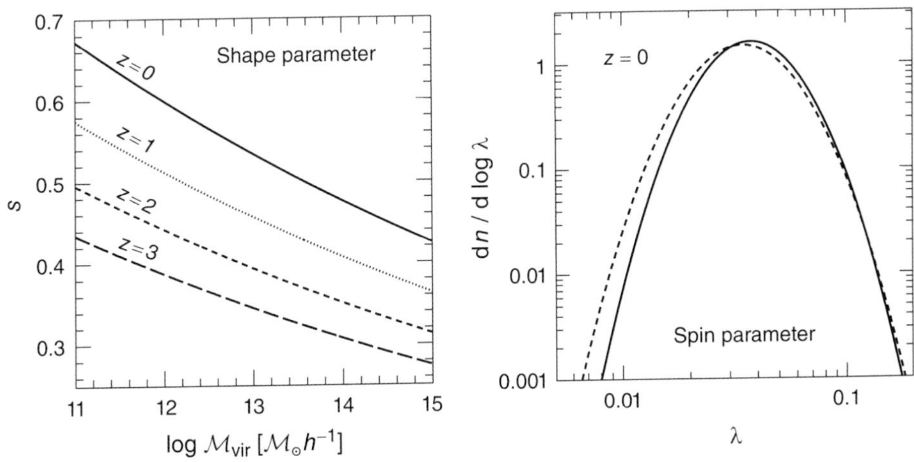


Fig. 7.12 *Left panel*. Average shape parameter (i.e. smallest-to-largest axis ratio $s \equiv c_{\rm int}/a_{\rm int}$) of dark matter halos as a function of virial mass and redshift, as found in dark matter-only cosmological simulations. Data from Allgood et al. (2006). *Right panel*. Lognormal (eq. 7.65) fits to the distribution of the spin parameter $\lambda$ of dark matter halos at $z=0$, as measured in a dark matter-only cosmological simulation for virial masses in the range $(10^{11}$–$10^{14})h^{-1}\mathcal{M}_\odot$. The solid curve, with $\log\lambda_0=-1.423$ and $\sigma_\lambda=0.248$, refers to $\lambda$ defined as in eq. (7.63). The dashed curve, with $\log\lambda_0=-1.459$ and $\sigma_\lambda=0.268$, refers to $\lambda$ defined as in eq. (7.64). The virial quantities are computed assuming $\Delta_c\simeq 101$ (eq. 7.21). Data from Rodríguez-Puebla et al. (2016).

decreases for increasing mass and is on average smaller at higher redshift (Fig. 7.12, left). Hydrodynamic cosmological simulations (§10.11.1) indicate that, in the presence of baryons, gas cooling (§8.1.1) tends to make the halos more spherical.

## Substructures

Dark matter halos are characterised by significant substructure, due to the presence of **subhalos**, i.e. dark matter halos that orbit within larger halos (Fig. 7.11). Subhalos of a massive galactic halo are believed to host satellite galaxies; therefore it is important to characterise the properties of the subhalo population in order to compare the predictions of cosmological models with observations (§10.9.2 and §10.10.1). It is possible that some subhalos do not host any galaxy: if these systems have also lost their gas, they would be completely *dark* and manifest themselves only through the effect of their gravitational potential.

The properties of the subhalos of a given dark matter halo are determined by the combination of the characteristics of the subhalo population at the time of accretion and of the subsequent evolution. In particular, the subhalos, after infalling into their host halos, evolve via dynamical friction (§8.9.3), tidal stripping (§8.9.4) and interactions with other subhalos.

The fundamental property of the dark matter subhalo population is their mass function. The **subhalo mass function** can be estimated using cosmological $N$-body simulations and

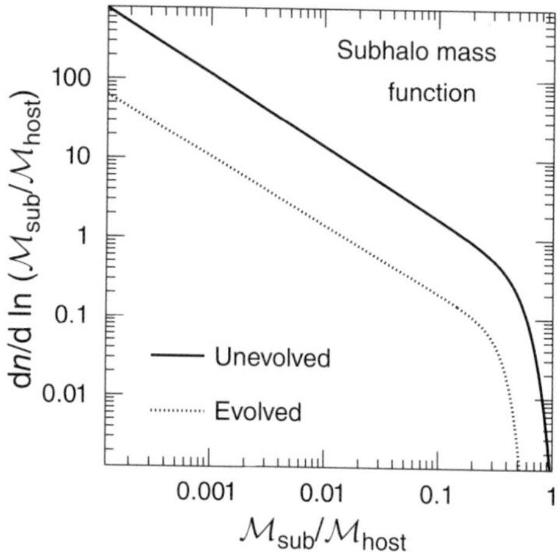


Fig. 7.13 Unevolved (solid curve) and evolved (dotted curve) subhalo mass functions estimated with a model calibrated on cosmological $N$-body simulations. These functions are weakly dependent on host halo mass and redshift. The curves are given by eq. (7.62) with parameters taken from table A1 of Jiang and van den Bosch (2016), assuming subhalo mass fraction 0.1 in the case of the evolved subhalo mass function.

analytic methods based on the extended Press–Schechter formalism (§7.4.3). For a given halo and at a given time $t$, we distinguish the **unevolved subhalo mass function**, in which we consider the masses of the subhalos at the time of accretion, from the **evolved subhalo mass function**, in which we consider the masses of the subhalos at the time $t$. The evolved function differs from the unevolved function because, as a consequence of tidal stripping, some subhalos are completely disrupted and others suffer substantial mass loss.

Examples of evolved and unevolved subhalo mass functions are given in Fig. 7.13. The results of cosmological $N$-body simulations indicate that the unevolved subhalo mass function is universal, i.e. independent of host halo mass and redshift. The number of subhalos per unit logarithmic mass can be parameterised as

$$\frac{\mathrm{d}n}{\mathrm{d}\ln\mu} = \gamma\mu^{\alpha}e^{-\beta\mu^{\omega}}, \tag{7.62}$$

where $\mu \equiv \mathcal{M}_{\rm sub}/\mathcal{M}_{\rm host}$ is the ratio between the subhalo mass and the host halo mass, and $\alpha$, $\beta$, $\gamma$ and $\omega$ are dimensionless parameters, with reference values $\alpha \approx -0.9$, $\beta \approx 6$, $\gamma \approx 0.2$ and $\omega \approx 3$. The evolved subhalo mass function can be described by the same functional form as the unevolved function (eq. 7.62), with normalisation $\gamma$ decreasing with time since the formation time of the host halo. Consistently, also the fraction of the total halo mass locked in subhalos, which is typically of the order of 10%, tends to decrease as the system evolves.

### 7.5.4 Spin of Dark Matter Halos

Though pressure-supported, dark matter halos are characterised by non-negligible angular momentum. This angular momentum is believed to be acquired in the early phases of halo formation, due to **tidal torques** from neighbouring overdensities. Tidal torques occur because dark matter halos are not spherical (§7.5.3) and are surrounded by an inhomogeneous matter distribution that produces tidal forces, because its gravitational field is not uniform. The effect of tidal forces onto a non-spherical halo is to induce a torque and then rotation. A dimensionless parameter often used to quantify the rotation support of halos is the **spin parameter**

$$\lambda \equiv \frac{J|E|^{1/2}}{G\mathcal{M}_{\Delta}^{5/2}}, \tag{7.63}$$

where $J$ is the magnitude of the total angular momentum, $E$ is the total energy and $\mathcal{M}_{\Delta}$ is the virial mass of the halo. A non-rotating system has $\lambda = 0$, while a fully rotation-supported, razor-thin, self-gravitating exponential disc with mass $\mathcal{M}_{\Delta}$ has $\lambda \simeq 0.4255$. An alternative, but almost equivalent, definition of the spin parameter of dark halos is

$$\lambda \equiv \frac{J}{\sqrt{2}\mathcal{M}_{\Delta} r_{\Delta} v_{\Delta}}, \tag{7.64}$$

where $v_{\Delta}$ is the circular speed at the virial radius $r_{\Delta}$ (eq. 7.24).

In dark matter-only cosmological simulations, the shape of the probability distribution of the spin parameter is found to be almost independent of redshift and halo mass. This empirically determined probability distribution of the halo spin parameter is usually fitted with a lognormal distribution,

$$\frac{\mathrm{d}n}{\mathrm{d}\log\lambda} = \frac{1}{\sqrt{2\pi}\sigma_{\lambda}} \exp\left[-\frac{(\log\lambda - \log\lambda_0)^2}{2\sigma_{\lambda}^2}\right], \tag{7.65}$$

with average $\log\lambda_0$ and variance $\sigma_{\lambda}^2$. The mean value of $\lambda$ is 0.03–0.04 at $z = 0$ and has only a slight decrease for increasing redshift. As an example, in Fig. 7.12 (right) we plot the lognormal probability density function that best fits the distribution of $\lambda$ at $z = 0$ measured in a dark matter-only cosmological simulation. The fact that the distribution of $\lambda$ peaks at $\sim 10^{-2}$ means that rotation is dynamically unimportant for dark matter halos. However, the presence of halo spin is fundamental in the context of galaxy formation, because it is at the basis of the formation of galactic discs (§10.1). In the presence of baryons, the rotation of halos tends to increase, mainly due to transfer of angular momentum from baryons to dark matter particles (§7.5.5).

### 7.5.5 Effect of Baryons on Dark Matter Halos

So far in this chapter we have mainly described the properties of dark halos expected in a dark matter-only Universe. In §7.1 we have seen that the observational evidence in favour of dark matter halos is abundant. However, the observationally determined properties of these halos are not always consistent with the properties predicted by dark matter-only cosmological simulations (§7.5.2), especially on the small scales. This is interpreted as an indication that the physics of baryons has important effects on the dark matter halos. Though the cosmic baryon fraction is just 16% (§2.5), the influence of baryons on the structure and kinematics of dark halos can be significant.

A first important effect is that, due to dissipation (that is, the loss of energy by emission of radiation; §8.1.1), baryons end up occupying the central regions of dark halos, producing a gravitational influence on the dark matter, which thus readjusts its density distribution, becoming more concentrated (a phenomenon known as **contraction**). If the redistribution of dark matter occurs *slowly*, that is on timescales much longer than the orbital time of dark matter particles, the process takes the name of **adiabatic contraction** and can be treated analytically (see Blumenthal et al., 1986). The net effect of contraction is that the dark matter density profile is steeper in the central parts than in the absence of baryons.

Another process that can modify the shape of the dark matter halo is dynamical friction (§8.9.3) between the baryons and the dark matter. For instance, in a barred galaxy the bar experiences dynamical friction against the dark matter, thus heating and transferring angular momentum to the dark halo, with consequent slowing down of the bar (§10.2) and modification of the halo density distribution. Similarly, massive gaseous clumps, expected in the early phases of the formation of galaxies (§10.3.3 and §10.9.2), are decelerated by dynamical friction against the dark matter, which is then heated, making the halo less dense in the centre. The net effect of dynamical friction heating is opposite with respect to contraction: as a result the halo density profile is shallower in the central parts than in the absence of baryons.

The dark matter distribution can be also affected by the removal of a large fraction of the gas through feedback (§8.7 and §8.8) or, if the galaxy is in a group or a cluster, ram-pressure stripping (§8.9.4). If the gas contributes non-negligibly to the mass in the centre of the galaxy, its removal has consequences for the overall gravitational potential, possibly flattening central dark matter cusps.

Overall, the effect of baryons on the dark matter distribution is not easy to predict: as a general rule, galaxies with low stellar-to-halo mass ratio (such as dwarfs; §10.10.1) are more subject to expansion than to contraction (§10.9.2). Contraction is dominant in galaxies that are more baryon-dominated in the centre, with the possible exception of disc galaxies with massive bars, in which dynamical friction heating effectively counteracts contraction.

# 8 Main Ingredients of Galaxy Formation Theory

Galaxy formation is the outcome of the complex interplay between dark matter, gas and stars in the cosmological framework. In Chapter 7 we have described the formation and evolution of dark matter halos. In the present chapter we focus on the physics of baryons, introducing some of the processes that are considered the main ingredients of the theory of galaxy formation (see also Fig. 1.3): gas cooling and heating (§8.1), gas infall and accretion onto dark halos (§8.2), star formation (§8.3), feedback from stars (§8.7) and AGNs (§8.8), and galaxy mergers (§8.9). Other fundamental tools presented in this chapter are models of the evolution of the ISM (§8.4), of chemical evolution (§8.5) and of galaxy spectra (§8.6). These ingredients are used in Chapters 9 and 10, where we describe the essential properties of the currently favoured theories of galaxy formation.

## 8.1 Thermal Properties of Astrophysical Gases

Generally speaking, an astrophysical gas can be subject to different forms of both cooling and heating. For instance, it can cool or be heated adiabatically (because of expansion or contraction), or be heated gravitationally via shocks (§8.2.2). In this section we focus on cooling and heating mechanisms that are due not to adiabatic or gravitational processes, but to emission (§8.1.1), absorption (§8.1.2) or scattering (§8.1.3) of radiation, or interaction with cosmic rays (§8.1.2). In §8.1.4 we introduce the concept of thermal instability.

### 8.1.1 Radiative Cooling

When a cosmic gas emits radiation it cools because it loses the energy carried by the emitted photons, which ultimately is subtracted from the kinetic energy of the gas particles. This fundamental process is called **radiative cooling**. We consider here the case, relevant in the context of galaxy formation, of an optically thin gas, that is gas in which the emitted photons are not absorbed by the gas itself and thus escape from the system. Depending on the gas density, temperature and ionisation state, there are several mechanisms responsible for radiative cooling. In this respect a discriminant temperature is $T \approx 10^4$ K, which roughly separates, for hydrogen, regimes of ionised and neutral gas.

## Gas at Temperature Higher than $10^4$ K

For $T \gtrsim 10^4$ K the main mechanisms determining radiative cooling of the gas are the following processes involving electrons and ions (or atoms), which can be classified on the basis of whether the electron is free or bound before and after the emission of radiation.

1. Bremsstrahlung (free–free; §D.1.6): interacting with an ion, a free electron emits radiation because its velocity changes when it is deflected.
2. Recombination (free–bound; §D.1.5): an electron emits a photon when it recombines with an ion to form a neutral atom or a less ionised ion.
3. Radiative de-excitation (bound–bound; §D.1.5): an electron bound to a neutral atom or to an ion emits radiation when moving from a higher to a lower energy level.

Together with these radiative processes, the following collisional processes are fundamental in determining the conditions for radiative cooling.

1. Collisional ionisation (bound–free): an electron bound to a neutral atom or to an ion is freed by a collision (typically with a free electron; §4.2.4).
2. Collisional excitation (bound–bound): a bound electron is excited to a higher energy level by a collision (typically with a free electron; §4.2.3).

The combined effect of the above mechanisms can be computed, provided we have information on the rate of ionisation and recombination of the various species involved. A useful approximation is that of the **collisional ionisation equilibrium**,[1] which assumes that there is perfect balance between the rates of collisional ionisation and recombination (photoionisation from any external radiation field is neglected; see §8.1.2), and that ions and neutral atoms are always in the ground state (there is no thermodynamic equilibrium; §D.1.4), a situation that occurs when the gas number density is much lower than the critical density $n_{\rm crit}$ (§4.2.3). Moreover, in the collisional ionisation equilibrium approximation, when ions and neutral atoms are excited they immediately emit radiation (radiative de-excitation; §4.2.3).

We now describe quantitatively the cooling rate of an optically thin astrophysical gas under the hypotheses of the collisional ionisation equilibrium model. Let us consider a gas with internal energy per unit volume

$$\mathcal{E} = \frac{3}{2} n k_{\rm B} T, \tag{8.1}$$

where $n$ is the (total) gas number density and $T$ is the gas temperature ($\mathcal{E}$ has units of erg cm$^{-3}$; §D.2.3). Assuming that the fluid is static ($\boldsymbol{u} = 0$), and neglecting gravity ($\Phi = 0$), heating ($\mathcal{H} = 0$) and thermal conduction ($\boldsymbol{F}_{\rm cond} = 0$), the energy equation of fluids (eq. D.29) reduces to $\partial\mathcal{E}/\partial t = -\rho\mathcal{C}$, where $\rho$ is the gas density and $\mathcal{C}$ is the cooling rate. For $T \gtrsim 10^4$ K, $\mathcal{C}$ scales linearly with the gas density, so we can write $\rho\mathcal{C} = n_{\rm t} n_{\rm e} \Lambda(T)$, where $n_{\rm t}$ is the atomic number density (including all neutral atoms and ions), $n_{\rm e}$ is the

[1] The simple assumptions of the collisional ionisation equilibrium model are not always satisfied in astrophysical systems. In some cases a more realistic treatment, based on complex non-equilibrium models, is required (see Sutherland and Dopita, 1993).

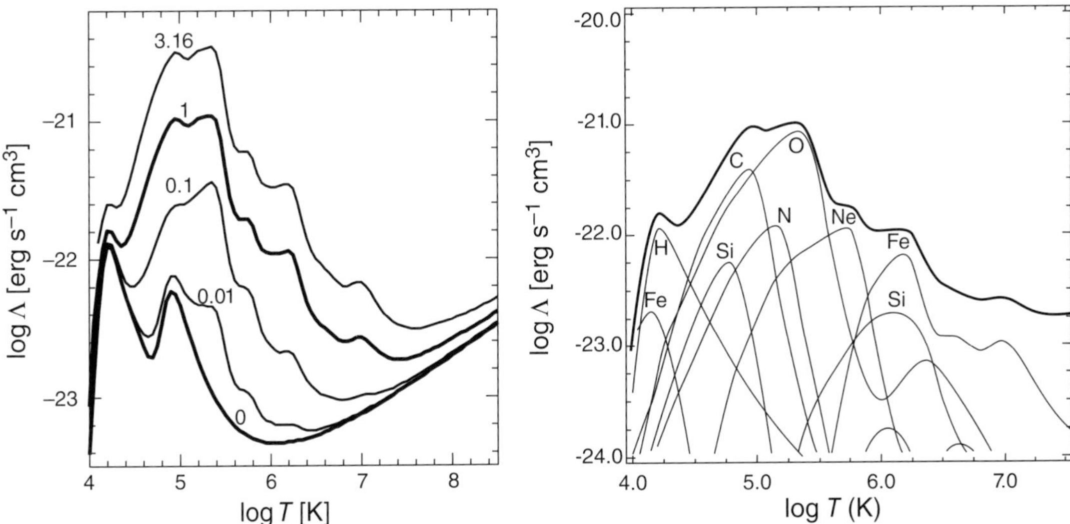


Fig. 8.1 *Left panel.* Collisional ionisation equilibrium cooling functions in the temperature range $10^4$ K < $T$ < $10^{8.5}$ K for different values of the gas metallicity (the labels indicate the values of $Z/Z_\odot$; thicker lines are used for $Z = 0$ and $Z = Z_\odot$). Data from Sutherland and Dopita (1993). *Right panel.* Dominant contribution of different species to the solar-metallicity collisional ionisation equilibrium cooling function shown in the left panel. Adapted from Sutherland and Dopita (1993). © AAS, reproduced with permission.

electron number density (thus $n = n_\mathrm{t} + n_\mathrm{e}$) and $\Lambda$ (in units of $\mathrm{erg\,s^{-1}\,cm^3}$) is the **cooling function**,[2] depending only on the gas temperature $T$ and metallicity $Z$. As is standard in the astrophysical literature, in this textbook we indicate explicitly only the dependence of $\Lambda$ on $T$, but it is understood that for any given value of $Z$ the function $\Lambda(T)$ is different. At a given position the energy loss per unit time per unit volume due to radiative cooling is

$$\frac{\mathrm{d}\mathcal{E}}{\mathrm{d}t} = -n_\mathrm{t} n_\mathrm{e} \Lambda(T). \tag{8.2}$$

The energy lost in the cooling process is emitted as radiation with bolometric emissivity $j = |\mathrm{d}\mathcal{E}/\mathrm{d}t|$.

The cooling function $\Lambda(T)$ is plotted in Fig. 8.1 (left panel) in the temperature interval $4 < \log(T/\mathrm{K}) < 8.5$ for plasmas with different metallicities. Let us focus first on a zero-metallicity ($Z = 0$) gas made only of hydrogen and helium with primordial cosmological composition (hydrogen mass fraction $X = 0.75$ and helium mass fraction $Y = 0.25$; §2.5). The cooling function has a peak at $T \approx 2 \times 10^4$ K, at which collisional excitation (and consequent radiative de-excitation) and ionisation (and consequent recombination) of neutral hydrogen are most effective. At higher temperatures hydrogen is completely ionised and its contribution to radiative cooling drops. The second peak in the metal-free cooling function, at $T \approx 10^5$ K, is due to collisional excitation and ionisation (and radiative de-excitation and recombination) of helium. At even higher temperatures the contribution

[2] The definition of $\Lambda$ is not unique in the literature. For instance, $\Lambda$ can be defined by $\rho C = n_\mathrm{e}^2 \Lambda$ or $\rho C = n_\mathrm{H}^2 \Lambda$, where $n_\mathrm{H}$ is the hydrogen number density.

of helium drops (it is fully ionised) and bremsstrahlung becomes dominant, asymptotically reaching $\Lambda(T) \propto T^{1/2}$ for high temperatures.

At a given temperature the cooling rate is higher for higher metallicity (Fig. 8.1, left). This is due to the fact that atoms and ions more massive than helium (C, O, Fe, Si, N and Ne) have plenty of line transitions especially at $T \approx 10^5$ K (Fig. 8.1, right), which are the dominant coolants at these temperatures even for relatively low abundances.

In order to quantify the characteristic timescales over which radiative cooling is important, we define the **cooling time**

$$t_{\rm cool} \equiv \frac{\mathcal{E}}{|\mathrm{d}\mathcal{E}/\mathrm{d}t|} = \frac{3nk_{\rm B}T}{2n_{\rm t}n_{\rm e}\Lambda(T)}, \tag{8.3}$$

where we have used eqs. (8.1) and (8.2). The timescale $t_{\rm cool}$, as defined in eq. (8.3), is the time in which the gas would radiate away all its thermal energy if the cooling rate were constant. In fact $n$ and $T$ change during the cooling process, but in most cases $t_{\rm cool}$ is a reliable order-of-magnitude estimate of the actual cooling time. It is convenient to rewrite eq. (8.3) as

$$t_{\rm cool} = \frac{3}{2}\left(\frac{n}{n_{\rm t}}\right)\left(\frac{n}{n_{\rm e}}\right)\frac{k_{\rm B}T}{n\Lambda(T)} \approx \frac{6k_{\rm B}T}{n\Lambda(T)}, \tag{8.4}$$

where the last approximate equality holds for a fully ionised gas. The ratios $n_{\rm t}/n$ and $n_{\rm e}/n$ depend on both $Z$ (the average number of electrons per nucleus varies with $Z$) and $T$ (the degree of ionisation of the plasma depends on temperature). For $T \gtrsim 3 \times 10^4$ K and $0 \leq Z \leq Z_\odot$, $n/n_{\rm t} \approx n/n_{\rm e} \approx 2$, with only weak dependence on $T$ and $Z$. In the end, at given $T$ and $Z$, $t_{\rm cool} \propto 1/n$; thus higher-density plasmas have shorter cooling times.

## Gas at Temperature Lower than $10^4$ K

At temperatures lower than $\approx 10^4$ K virtually all atoms are neutral (including H) and there are very few free electrons, so it is more convenient to define the cooling function $\Lambda$ to be such that $\rho C = n_{\rm H}^2\Lambda$. Note that, at $T < 10^4$ K, $\Lambda$ depends not only on $T$ and $Z$, but also on the gas density: cooling is less efficient at higher densities, because collisional de-excitation takes the place of radiative de-excitation and thus suppresses metal cooling. The few free electrons are not sufficiently energetic to excite the inner energy levels of hydrogen. As a consequence, the atomic cooling of a metal-free gas drops rapidly to zero for $T < 10^4$ K, as illustrated by Fig. 8.2, showing the collisional ionisation equilibrium atomic cooling function down to $T \approx 50$ K for a representative selection of metal abundances, assuming $n_{\rm H} = 1\ {\rm cm}^{-3}$. Metal-enriched gas, also at these lower temperatures, cools more efficiently than metal-free gas, because some heavier elements have low-energy transitions that can be excited at $T < 10^4$ K (for instance the singly ionised carbon; §8.3). Note however that, even in the presence of metals, the atomic cooling rate at $T < 10^4$ K is orders of magnitude lower than at $T > 10^4$ K.

An important question, especially in the context of the formation of the first stars and galaxies, is whether it is possible for gas with primordial composition to cool significantly below $10^4$ K (§9.4). While atomic cooling is completely inefficient in the absence of metals, this low-temperature gas can radiate its energy through molecules. Quantitatively,

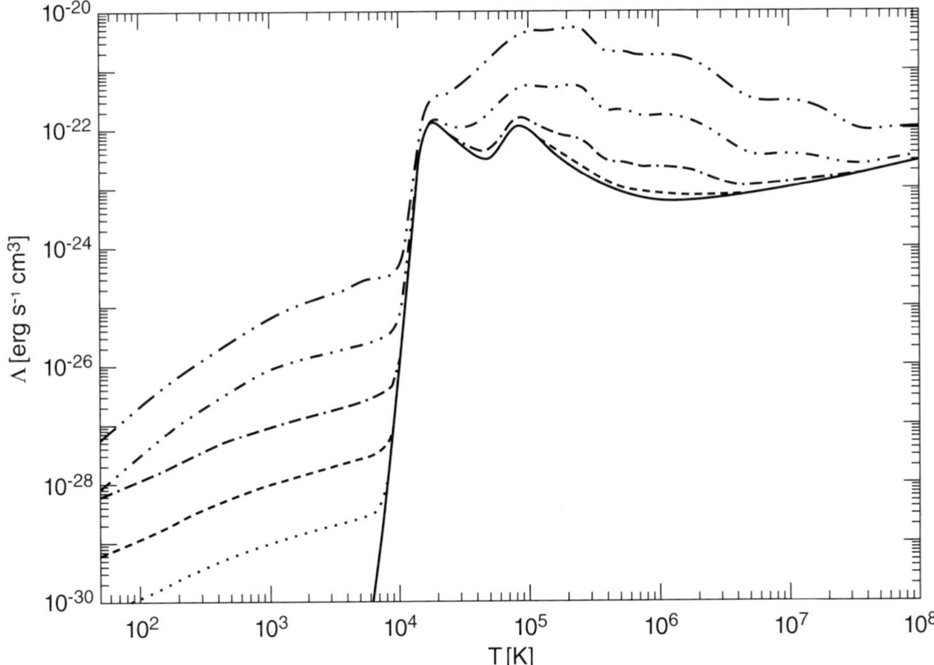


Fig. 8.2 Collisional ionisation equilibrium atomic cooling functions ($\Lambda = \rho C/n_{\rm H}^2$) for gases with metallicities $Z = 0$ (solid), $10^{-3}Z_\odot$ (dotted), $10^{-2}Z_\odot$ (dashed), $10^{-1}Z_\odot$ (dot-dash-dashed), $1Z_\odot$ (dot-dot-short dashed) and $10Z_\odot$ (dot-dot-long dashed) in the temperature range $50\,{\rm K} < T < 10^8\,{\rm K}$. It is assumed that the hydrogen number density is $n_{\rm H} = 1\,{\rm cm}^{-3}$. From Smith et al. (2008).

the molecular contribution to the cooling function depends on the density of the gas. An example of a molecular cooling function is shown in Fig. 9.2 (right), where the contribution of $H_2$ is compared with atomic cooling of a metal-free gas. It is apparent that molecular cooling, though having cooling rates that are low in absolute terms, can be very important, because it peaks right at the drop of atomic hydrogen cooling. The characteristics of molecular cooling are discussed in §4.2.5 and §9.4.2.

## 8.1.2 Photoionisation and Cosmic-Ray Heating

In the presence of an ionising background radiation field, the cosmic gas is heated by a mechanism called **photoionisation heating** or **photoheating**. For a radiation field to be ionising, its photons must be sufficiently energetic to strip electrons from the atoms of the gas. For instance, the binding energy of a ground-state electron of hydrogen is 13.6 eV, so in order to ionise hydrogen the photons must have frequencies in the far-UV or more energetic bands (§4.2.3). The ionising radiation most relevant in the context of galaxy formation is the extragalactic **ultraviolet background** (UVB) radiation produced by massive stars and QSOs (§9.5.3 and §9.8.2). The intensity of the ionising UVB depends on redshift: it increases with cosmic time from reionisation till $z \approx 2$, at the peak of the cosmic star formation density (Fig. 11.33) and AGN emission (Fig. 11.44), and then declines till $z = 0$.

In the ionisation process the gas is heated, because the energy $\Delta E = h\nu - \chi$, where $h\nu$ is the energy of the ionising photon and $\chi$ is the ionisation energy, is acquired by the freed electron as kinetic energy, which is transferred to other particles via collisions. In addition to directly heating the gas, photoionisation has also the effect of inhibiting radiative cooling, because, by ionising the gas, it eliminates the cooling processes involving bound electrons (§8.1.1). Of course, for photoionisation heating to be effective there must be a reservoir of neutral atoms to ionise: for instance, photoionisation of hydrogen is important only for relatively low-temperature gas ($T < 10^5$ K). But at these temperatures the impact of the photoionisation is strong: the hydrogen peak of the cooling function at $T \approx 2 \times 10^4$ K (Fig. 8.1) can be almost completely suppressed. This is apparent from Fig. 8.3 showing, as a function of temperature, the net cooling function $\Lambda - \Gamma$, where $\Lambda$ and $\Gamma$ are, respectively, the cooling (eq. 8.2) and heating functions, in the presence of UVB. The heating function $\Gamma$ (in units of $\mathrm{erg\,s^{-1}\,cm^3}$) is related to the heating rate $\mathcal{H}$ (eq. D.29) by $\rho\mathcal{H} = n_t n_e \Gamma$. The overall effect of photoionisation heating depends on both the gas number density and the properties of the radiation field, which, in the case of the extragalactic

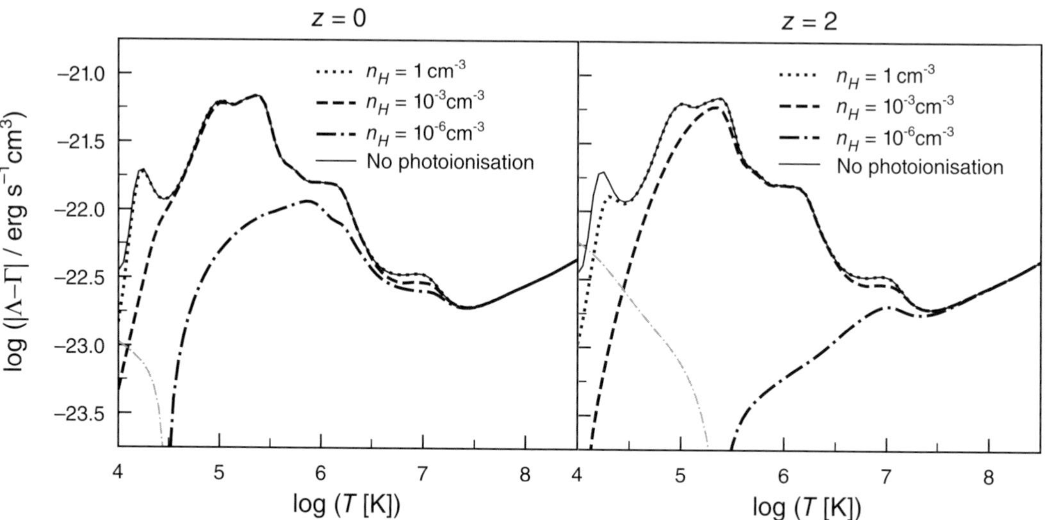


Fig. 8.3 *Left panel.* Net cooling function $\Lambda - \Gamma$ in the temperature range $10^4\ \mathrm{K} \leq T \leq 10^{8.5}$ K for a solar-metallicity gas at redshift $z = 0$, for three values of the hydrogen number density. Here $\Lambda$ and $\Gamma$ are, respectively, the cooling and heating functions. The curves have been obtained accounting for both collisional processes (ionisation, excitation and recombination) and photoionisation by a uniform extragalactic UVB (Haardt and Madau, 2012) using the software CLOUDY (Ferland et al., 2013). The solid curve indicates the collisional ionisation equilibrium cooling function in the absence of UVB (§8.1.1; slight differences with respect to the solar-metallicity curve shown in Fig. 8.1 just reflect different prescriptions adopted in the calculations). The thin grey dash-dotted curve indicates the net cooling function in the region where heating dominates over cooling. *Right panel.* Same as left panel, but for gas at $z = 2$, when the ionising UVB is more intense. Courtesy of L. Armillotta.

UVB, vary with redshift. For instance, Fig. 8.3 shows that the effect of photoionisation heating is, for given hydrogen number density $n_{\rm H}$, weaker at $z=0$ (left panel) than at $z=2$ (right panel) and, at given $z$, stronger for lower values of $n_{\rm H}$. In addition to the extragalactic UVB, the gas within a galaxy is photoionised also by internal sources hosted by the same galaxy, such as young massive stars (§4.2.3) and the AGN (§8.8).

Not only ionising photons, but also cosmic rays (relativistic particles, mainly protons and electrons, produced by SNRs and AGNs; §4.6.2) can contribute to heat astrophysical gases. Even relatively low-energy cosmic rays are effective in ionising atoms and dissociating molecules, and thus injecting kinetic energy into atomic and molecular gases. Such **cosmic-ray heating** is especially important for molecular clouds, which, due to their high densities, are self-shielded from UV photons (§4.2.5 and §8.3.1). Cosmic rays can also heat relatively high-temperature plasmas by exciting Alfvén waves (§8.3.6) that are then dissipated: this flavour of cosmic-ray heating may play a role in the thermal balance of the ICM (§6.4.2).

### 8.1.3 Compton Heating and Cooling

Important cooling and heating effects in astrophysical plasma are due to **Compton scattering**, that is the process of interaction between a photon and a free electron. Compton scattering is a quantum-mechanical process, but, when the photon energy is much lower than the rest-mass energy of the electron ($h\nu \ll m_{\rm e}c^2$), quantum effects can be neglected (this low-energy limit is the classical Thomson scattering, with cross section $\sigma_{\rm T}$ given by eq. 6.9).

Let us consider a population of thermal electrons at temperature $T$ in the presence of background radiation. If, on average, the energies of the photons are high compared to the kinetic energies of the electrons ($h\nu \gg k_{\rm B}T$), energy is transferred from the photons to the electrons. In this process, usually referred to as **Compton heating**, the plasma gains energy in the interaction with the background radiation. Compton heating of the interstellar and intergalactic medium is believed to occur in the presence of X-ray and gamma-ray emission from AGNs (§3.6). If instead the energies of the electrons are higher than the energies of the photons ($h\nu \ll k_{\rm B}T$), energy is transferred from the electrons to the photons. This mechanism, called **Compton cooling** or **inverse Compton scattering**, represents a cooling process for the plasma. An important application of Compton cooling is inverse Compton scattering of hot gas against the low-energy photons of the CMB (§6.4.2), but at high redshifts the CMB photons can also produce Compton heating (§9.2).

### 8.1.4 Thermal Instability

The phenomenon of **thermal instability** is believed to be at the origin of the multi-phase ISM (§4.6.2) and is in general important for our understanding of the process of galaxy formation (§10.6.4). Here we outline the essence of the thermal stability analysis. Let us consider an astrophysical gas in which $\mathcal{C}$ and $\mathcal{H}$ are the rates of cooling and heating not due to adiabatic or gravitational processes (§D.2.3). In the context of galaxies, $\mathcal{C}$ is usually due to radiative cooling (§8.1.1), while $\mathcal{H}$ is due to either photoionisation or heating from cosmic rays (§8.1.2). If either cooling or heating is dominant, the gas is out of equilibrium and is thus expected to be fast evolving. For instance, if radiative cooling is dominant and unbalanced by any form of heating, a cooling catastrophe (§5.2.1) will occur, in which the gas will become colder and denser, it will collapse and it will likely end up forming stars (§8.3). An interesting regime is one in which cooling is balanced by heating ($\mathcal{C}=\mathcal{H}$). A crucial question is whether such an equilibrium is stable: what happens if a parcel of gas is perturbed so that, for instance, it is colder than the surrounding medium? If the perturbation does grow (the temperature of the parcel of gas becomes lower and lower) the gas is said to be thermally *unstable*, otherwise the gas is thermally *stable*.

For an optically thin gas subject to both cooling and heating, we define the **energy loss function** $\mathcal{L}\equiv\mathcal{C}-\mathcal{H}$ (energy loss per unit time per unit mass, in units of $\mathrm{erg\,s^{-1}\,g^{-1}}$). For the second law of thermodynamics, the variation of the specific (i.e. per unit mass) entropy $s$ (eq. 6.22) in the time interval $\mathrm{d}t$ is given by

$$T\frac{\mathrm{d}s}{\mathrm{d}t}=-\mathcal{L}. \tag{8.5}$$

If we start from an unperturbed medium with specific entropy $s_0$, temperature $T_0$ and energy loss function $\mathcal{L}_0$, and we perturb it with a *small* (linear; §7.3) perturbation $s_0+\delta s$ and $T_0+\delta T$, the above equation gives

$$(T_0+\delta T)\frac{\mathrm{d}(s_0+\delta s)}{\mathrm{d}t}=-(\mathcal{L}_0+\delta\mathcal{L}). \tag{8.6}$$

For an unperturbed medium in equilibrium ($\mathcal{L}_0=0$, $\mathrm{d}s_0/\mathrm{d}t=0$), this equation gives

$$T_0\frac{\mathrm{d}\delta s}{\mathrm{d}t}=-\delta\mathcal{L}, \tag{8.7}$$

where we have neglected the second-order term proportional to $\delta s\delta T$. In the case of isobaric perturbations ($\delta P=0$), from eq. (D.26) we get $\delta T/T_0=-\delta\rho/\rho_0$, so, using eq. (6.22), the entropy perturbation can be written as

$$\delta s=\frac{5}{2}\frac{k_\mathrm{B}}{\mu m_\mathrm{p}}\frac{\delta T}{T_0}. \tag{8.8}$$

Combining this expression with eq. (8.7) gives

$$\frac{5}{2}\frac{k_\mathrm{B}}{\mu m_\mathrm{p}}\frac{\mathrm{d}\delta T}{\mathrm{d}t}=-\delta\mathcal{L}. \tag{8.9}$$

It follows from this equation that if $\delta T$ and $\delta\mathcal{L}$ have the same sign, $|\delta T|$ decreases with time, so the system is thermally stable (if the perturbed parcel of gas is colder than the surrounding gas, it is heated; if it is warmer, it cools). Otherwise we have thermal instability ($|\delta T|$ increases with time). Therefore, there is thermal *instability* if

$$\left(\frac{\partial\mathcal{L}}{\partial T}\right)_P=\left(\frac{\partial\mathcal{L}}{\partial T}\right)_\rho-\frac{\rho_0}{T_0}\left(\frac{\partial\mathcal{L}}{\partial\rho}\right)_T<0, \tag{8.10}$$

where the subscripts indicate the thermodynamic variable that is kept constant. The above instability criterion is often referred to as the **Field instability criterion** (Field, 1965).

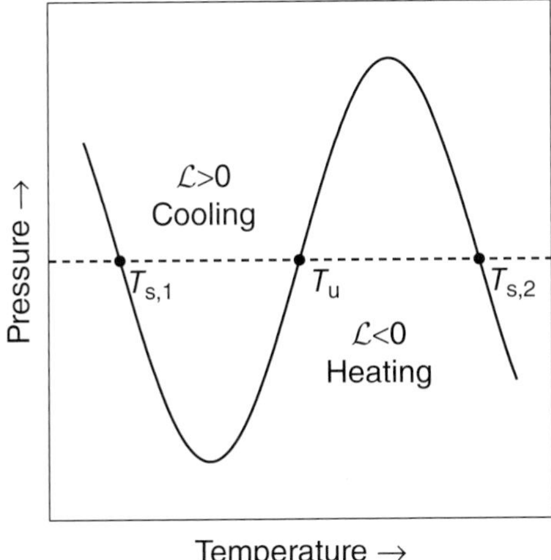


Fig. 8.4 The solid line represents qualitatively the locus of vanishing energy loss function ($\mathcal{L} = 0$) in the gas temperature–pressure plane. The gas cools above the curve ($\mathcal{L} > 0$) and is heated below the curve ($\mathcal{L} < 0$). The dashed line is the locus of isobaric perturbations. Points $T_{s,1}$, $T_u$ and $T_{s,2}$ are in pressure equilibrium and such that heating balances cooling. Points $T_{s,1}$ and $T_{s,2}$ are thermally stable; point $T_u$ is thermally unstable. Figure inspired by Field et al. (1969) and Balbus (1995).

Physically, this result can be interpreted as follows. The energy loss function $\mathcal{L}$ can be seen as a function of $P$ and $T$: in Fig. 8.4 the solid line represents the locus $\mathcal{L}(T, P) = 0$, that is, the fact that heating balances cooling in the unperturbed medium. Above the curve (at given $T$, higher $P$, thus higher $\rho$ and higher cooling rate; §8.1.1) the medium is cooling ($\mathcal{L} > 0$); below the curve (at given $T$, lower $P$, thus lower $\rho$ and lower cooling rate) the medium is being heated ($\mathcal{L} < 0$). Let us consider a given value of the pressure (horizontal line in the plot). Unperturbed gas at this pressure can be found at three temperatures (these points, labelled $T_{s,1}$, $T_u$ and $T_{s,2}$, are points of pressure equilibrium and such that heating balances cooling). When a parcel of gas is perturbed isobarically, it moves horizontally. Points $T_{s,1}$ and $T_{s,2}$ are thermally *stable*, because the parcel of gas moves into the heating region if $\delta T < 0$ and into the cooling region if $\delta T > 0$: the instability criterion (eq. 8.10) is not satisfied. Point $T_u$ is thermally *unstable* because the parcel of gas moves into the cooling region if $\delta T < 0$ and into the heating region if $\delta T > 0$: the instability criterion (eq. 8.10) is satisfied and the perturbation grows. For instance, in the case of the neutral atomic gas of the ISM of the Milky Way (§4.6.2), $T_{s,1}$ can be identified with the temperature of the CNM and $T_{s,2}$ with that of the WNM.

Though the instability criterion (eq. 8.10) has been derived for isobaric perturbations, more general thermal stability analyses confirm that the sign of $(\mathrm{d}\mathcal{L}/\mathrm{d}T)_P$ is the crucial quantity to determine the thermal stability properties of a medium. Assuming that the heating rate is independent of temperature and density, the thermal stability of an astrophysical gas is determined by the cooling rate $C \propto \rho\Lambda(T)$ (§8.1.1), which is proportional to the gas density and depends on the temperature through the cooling function $\Lambda$. The temperature dependence of the cooling function (Figs. 8.1 and 8.2) implies that the thermal instability criterion of eq. (8.10) is satisfied in astrophysical gases over wide temperature ranges. Therefore, we might expect the growth of thermal perturbations to be a common phenomenon in the processes of galaxy formation and evolution. However, we stress that in the above simplified analysis we have neglected several potentially important factors: in particular, **thermal conduction** (i.e. the flow of heat within a medium due to the presence of temperature gradients; §D.2.3), stratification of the gas in a gravitational field, rotation, coherent gas flows, turbulence and magnetic fields can affect dramatically the thermal stability properties of astrophysical gases.

## 8.2 Gas Accretion and Cooling in Dark Matter Halos

Having introduced some general thermal properties of astrophysical gases, we now specialise to the process of gas accretion and cooling in dark matter halos, which is a fundamental step in the formation of galaxies.

### 8.2.1 Virial Temperature

A reference temperature for the gas in a dark matter halo is the halo **virial temperature** $T_{\rm vir}$, which is defined in analogy to the virial temperature of a self-gravitating unmagnetised gas distribution in hydrostatic equilibrium. For such a gaseous system, the internal energy is

$$U = \frac{3}{2}\frac{k_{\rm B}T_{\rm vir}M_{\rm gas}}{\mu m_{\rm p}} \tag{8.11}$$

and the gravitational potential energy is

$$W = -\frac{GM_{\rm gas}^2}{r_{\rm g}}, \tag{8.12}$$

where $M_{\rm gas}$ is the total gas mass and $r_{\rm g}$ is the gravitational radius (§5.3.1). In this case the virial theorem (eqs. 5.12 and 8.71) can be written as

$$3\frac{k_{\rm B}T_{\rm vir}}{\mu m_{\rm p}} - \frac{GM_{\rm gas}}{r_{\rm g}} = 0, \tag{8.13}$$

so the virial temperature of a self-gravitating gaseous system is

$$T_{\rm vir} = \frac{G\mu m_{\rm p}M_{\rm gas}}{3k_{\rm B}r_{\rm g}}. \tag{8.14}$$

Let us now consider an isothermal gas distribution (at temperature $T_{\rm gas}$) in hydrostatic equilibrium in a dark matter halo modelled, for simplicity, as a singular isothermal sphere (§5.3.4) with mass profile given by eq. (5.38) and gravitational potential given by eq. (5.39). Neglecting the self-gravity of the gas (i.e. assuming that the gravitational potential is due only to the dark matter), the hydrostatic equilibrium equation (eq. 6.16) gives

$$\frac{\mathrm{d}\ln n}{\mathrm{d}\ln r} = -\frac{\mu m_{\mathrm{p}} v_{\mathrm{c}}^2}{k_{\mathrm{B}} T_{\mathrm{gas}}}, \tag{8.15}$$

where $n(r)$ is the gas number density and $v_{\mathrm{c}}$ is the circular speed, which in this case is independent of radius. Thus, the gas density distribution is the power law

$$n(r) = n_0 \left(\frac{r}{r_0}\right)^{-2T_{\mathrm{vir}}/T_{\mathrm{gas}}}, \tag{8.16}$$

where $r_0$ is a reference radius, $n_0 \equiv n(r_0)$ and

$$T_{\mathrm{vir}} \equiv \frac{\mu m_{\mathrm{p}}}{2k_{\mathrm{B}}} v_{\mathrm{c}}^2 \tag{8.17}$$

is the virial temperature of the potential well. When $T_{\mathrm{gas}} = T_{\mathrm{vir}}$ the gas density profile has logarithmic slope $-2$, so in this case the gas distribution follows that of the host dark matter halo. The gas density profile is steeper than $r^{-2}$ when $T_{\mathrm{gas}} < T_{\mathrm{vir}}$ and shallower when $T_{\mathrm{gas}} > T_{\mathrm{vir}}$ (eq. 8.16).

Though dark matter halos are not singular isothermal spheres, it is useful, based on eq. (8.17), to define the virial temperature of a dark matter halo as

$$T_{\mathrm{vir}} \equiv \frac{\mu m_{\mathrm{p}}}{2k_{\mathrm{B}}} v_{\mathrm{vir}}^2 \simeq 1.45 \times 10^6 \left(\frac{v_{\mathrm{vir}}}{200\,\mathrm{km\,s^{-1}}}\right)^2 \mathrm{K}, \tag{8.18}$$

where $v_{\mathrm{vir}}$ is the circular speed at the virial radius $r_{\mathrm{vir}}$ (§7.3.2) and in the second rough equality we have assumed $\mu = 0.6$. Using eq. (7.26) the virial temperature can be written as

$$T_{\mathrm{vir}} = \frac{\mu m_{\mathrm{p}}}{2^{4/3} k_{\mathrm{B}}} (G\mathcal{M}_{\mathrm{vir}})^{2/3} \left[\Delta_{\mathrm{c}}(z) H^2(z)\right]^{1/3}, \tag{8.19}$$

where $\mathcal{M}_{\mathrm{vir}}$ is the virial mass (§7.3.2), $\Delta_{\mathrm{c}}(z)$ (eq. 7.21) is the redshift-dependent cosmological overdensity for virialisation and $H(z)$ is the Hubble parameter as a function of redshift, which is given by eq. (2.37) for the standard cosmological model. At $z=0$

$$T_{\mathrm{vir}} \simeq 6.0 \times 10^5 \left(\frac{\mathcal{M}_{\mathrm{vir}}}{10^{12}\mathcal{M}_\odot}\right)^{2/3} \left(\frac{h}{0.7}\right)^{2/3} \mathrm{K}, \tag{8.20}$$

for $\mu = 0.6$. As we see in the following sections, $T_{\mathrm{vir}}$ is a fundamental quantity for the process of accretion of gas onto dark matter halos and thus for galaxy formation.

### 8.2.2 Shock Heating

In this section and in §8.2.4, we briefly describe the process of cosmological gas accretion onto dark matter halos, referring the reader to Birnboim and Dekel (2003) for an in-depth analysis. We consider the case in which, before accretion, the gas temperature $T_{\mathrm{pre}}$ is much lower than the virial temperature $T_{\mathrm{vir}}$ of the halo (this is always the case for sufficiently massive halos; §9.3). Here we further assume that radiative cooling is negligible (we will discuss the effects of cooling in §8.2.3 and §8.2.4). When the accreting gas crosses the virial radius its infall speed is of the order of the circular speed $v_{\mathrm{vir}} \sim \sqrt{k_{\mathrm{B}} T_{\mathrm{vir}}/(\mu m_{\mathrm{p}})}$ (eq. 8.18). The sound speed of the infalling gas is $c_{\mathrm{s}} \sim \sqrt{k_{\mathrm{B}} T_{\mathrm{pre}}/(\mu m_{\mathrm{p}})}$ (§D.2.4) so the condition $T_{\mathrm{pre}} \ll T_{\mathrm{vir}}$ implies that the motion of the infalling gas is highly supersonic.

Due to its collisional nature, the infalling gas is expected to produce shocks (§D.2.4) either at the centre of the halo or when it impacts previously accreted gas. In the absence of cooling, the **shock radius** $r_{\rm sh}$, which is the radius at which this discontinuity occurs, is estimated to be close to the virial radius $r_{\rm vir}$. The effect of the shock is described in detail by the equations that link the properties of the upstream (pre-shock) and downstream (post-shock) fluid (§D.2.4). However, a useful simplified description of the process is the following.

Assuming that the infalling gas is at rest when at large distance from the halo, energy conservation implies that its infalling speed at the shock radius is

$$v_{\rm sh} = \sqrt{-2\Phi(r_{\rm sh})}, \tag{8.21}$$

where $\Phi$ is the gravitational potential of the halo, which we assume to be spherical ($v_{\rm sh}$ as given by eq. 8.21 is just the escape speed at $r_{\rm sh}$; eq. 4.21). Under the assumption $r_{\rm sh} \approx r_{\rm vir}$ and that the halo is truncated at $r_{\rm vir}$, so $\Phi(r_{\rm vir}) = -G\mathcal{M}_{\rm vir}/r_{\rm vir}$, we get $\Phi(r_{\rm sh}) \approx -v^2_{\rm vir}$ and, from eq. (8.21), $v^2_{\rm sh} \approx 2v^2_{\rm vir}$. Assuming that all the gas kinetic energy ($\mathcal{M}_{\rm gas}v^2_{\rm sh}/2$) is converted into internal energy (eq. 8.11) at $r_{\rm sh}$,

$$\frac{3}{2}\frac{k_{\rm B}T_{\rm post}}{\mu m_{\rm p}} = \frac{1}{2}v^2_{\rm sh} \approx v^2_{\rm vir}, \tag{8.22}$$

we find that the temperature of the shocked gas is

$$T_{\rm post} \approx \frac{2\mu m_{\rm p}}{3k_{\rm B}}v^2_{\rm vir} = \frac{4}{3}T_{\rm vir}, \tag{8.23}$$

where we have used eq. (8.18). This simple analysis shows that, if cooling is inefficient, the gas infalling into a dark matter halo is shock heated at the virial radius to a temperature close to the virial temperature of the halo. Similar results are obtained with more rigorous analytic calculations of the pre-shock and post-shock gas properties (§D.2.4), and with numerical hydrodynamic simulations (§10.11.1).

### 8.2.3 Cooling of Virial-Temperature Gas

Based on arguments similar to those presented in §8.2.2, classical galaxy formation models relied on the assumption that all gas falling into dark matter halos is heated to the virial temperature. Though this assumption is simplistic (see §8.2.4), before considering more sophisticated models, it is very useful to understand the evolution of virial-temperature gas in dark matter halos. In this section we present a simple, but quantitative, description of this process.

A cornerstone of the theory of galaxy formation, laid in the late 1970s (Binney, 1977; Rees and Ostriker, 1977; Silk, 1977; White and Rees, 1978), is that the characteristic masses of galaxies are determined by a few fundamental timescales: the cooling time $t_{\rm cool}$, the dynamical time $t_{\rm dyn}$ and the Hubble time $t_{\rm H}$. In order to show how these timescales affect the evolution of cosmological gaseous halos, it is instructive to focus on a simplified case. Let us consider a spherical virialised dark matter halo, with mass $\mathcal{M}_{\rm vir}$ and radius $r_{\rm vir}$, hosting an isothermal distribution of gas with average number density $n$ and temperature $T$ of the order of the halo virial temperature $T_{\rm vir}$ (§8.2.1). For given metallicity $Z$, the cooling time of the gas $t_{\rm cool}(n,T)$ is given by eq. (8.3). The **dynamical time** of the system is

$$t_{\rm dyn} \equiv \frac{1}{\sqrt{G\langle\rho\rangle}}, \tag{8.24}$$

where $\langle\rho\rangle = 3M_{\rm vir}/(4\pi r_{\rm vir}^3)$ is the average mass density of the halo. Assuming $n = f_{\rm gas}\langle\rho\rangle/(\mu m_{\rm p})$, where $f_{\rm gas} \equiv M_{\rm gas}/M_{\rm vir}$ is the gas fraction, the dynamical time reads

$$t_{\rm dyn}(n) = \sqrt{\frac{f_{\rm gas}}{G\mu m_{\rm p} n}}. \tag{8.25}$$

Let us focus first on $t_{\rm cool}$ and $t_{\rm dyn}$. If $t_{\rm cool} \ll t_{\rm dyn}$ the gas will cool, lose pressure support and then collapse on timescales of the order of $t_{\rm dyn}$. This catastrophic cooling is expected to lead to physical conditions favourable for the gas to form stars. If $t_{\rm cool} \gg t_{\rm dyn}$ the gas is expected to remain hot, approximately in equilibrium in the halo potential well. Therefore, the condition $t_{\rm cool} = t_{\rm dyn}$ can be taken as a discriminant, in the $n$–$T$ space, to predict whether a halo will form a galaxy or not. Fig. 8.5 shows, as solid curves, loci of $t_{\rm cool} = t_{\rm dyn}$, in the gas density–temperature plane, for two values of the gas metallicity ($Z = 0$ and $Z = Z_\odot$). For fixed metallicity and temperature, $t_{\rm cool} \propto n^{-1}$ (eq. 8.4) and $t_{\rm dyn} \propto n^{-1/2}$ (eq. 8.25), therefore, at a given $T$, efficient cooling and star formation are only expected in sufficiently high-density systems.

Let us now consider the role of the Hubble time $t_{\rm H}$ (§2.1.2). Loci of $t_{\rm cool} = t_{\rm H}$ are shown as dashed curves in the diagrams in Fig. 8.5. Given that $t_{\rm dyn} < t_{\rm H}$ is a necessary condition for a halo to be virialised (the halo must have had time to reach equilibrium; §7.3.2), we have the following three possible regimes.

1. $t_{\rm cool} < t_{\rm dyn}$ (rapid cooling): gas cooling is efficient and the conditions are favourable for rapid galaxy formation.
2. $t_{\rm dyn} < t_{\rm cool} < t_{\rm H}$ (slow cooling): cooling is important on relatively long timescales, but shorter than the age of the system. For simplicity, here we identify the age of the system with the present-day age of the Universe. Of course this is an overestimate for systems at $z > 0$, which however does not affect the main results of the present analysis.
3. $t_{\rm cool} > t_{\rm H}$ (no cooling): cooling is unimportant, the virial-temperature gas remains hot and star formation does not occur.

The plots in Fig. 8.5 also show, as horizontal lines, the values of $n$ expected at different redshifts for halos with overdensity $\Delta_{\rm c}(z)$ (eq. 7.21) and gas fraction $f_{\rm gas} = \Omega_{\rm b}/\Omega_{\rm m} = 0.16$ (§2.5). The average gas number density in halos with overdensity $\Delta_{\rm c}$, and thus average mass density $\langle\rho\rangle = \Delta_{\rm c}\rho_{\rm crit}$ (eq. 7.23), is

$$n = f_{\rm gas}\frac{\Delta_{\rm c}\rho_{\rm crit}(z)}{\mu m_{\rm p}} = \frac{3\Delta_{\rm c} f_{\rm gas}}{8\pi G\mu m_{\rm p}} H^2(z), \tag{8.26}$$

where $H(z)$ is given by eq. (2.37), so $n$ increases for increasing $z$. From the left panel of Fig. 8.5 it is apparent that galaxy formation is favoured in halos with intermediate virial temperatures $10^4\,{\rm K} \lesssim T_{\rm vir} \lesssim 10^6\,{\rm K}$ and disfavoured in high-mass ($T_{\rm vir} \gtrsim 10^6$ K) and in low-mass ($T_{\rm vir} \lesssim 10^4$ K) halos. The behaviour at $T_{\rm vir} < 10^4$ K, which is not shown in Fig. 8.5,

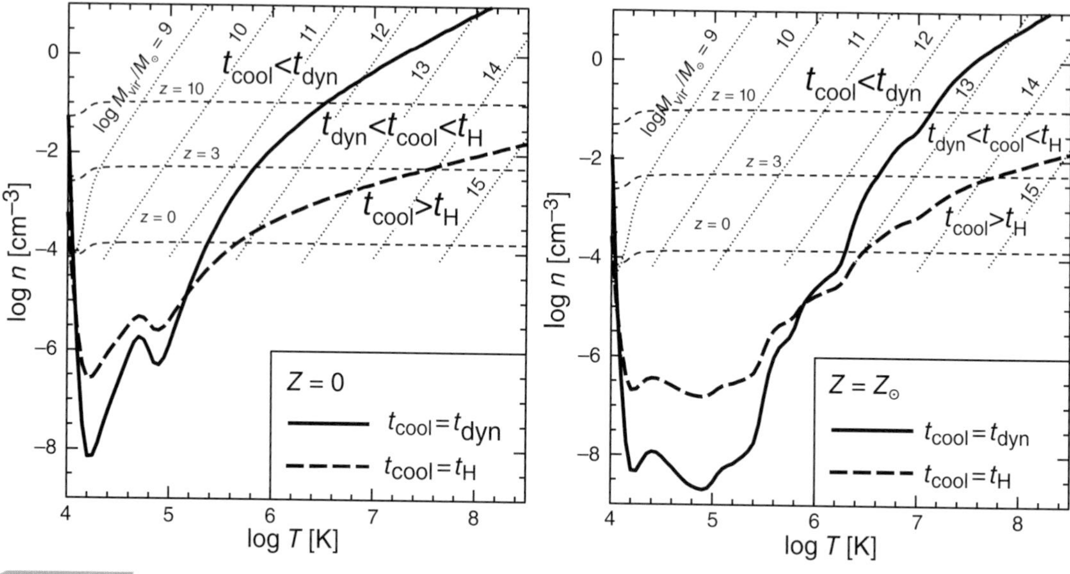


**Fig. 8.5** *Left panel.* The thick solid and dashed curves are, respectively, the loci of $t_{cool} = t_{dyn}$ and $t_{cool} = t_H$, in the gas density–temperature plane $n$–$T$, for a metal-free gas ($Z = 0$) gravitationally supported by a dark matter halo with virial temperature $T$. Here $t_{cool}$ is the cooling time (eq. 8.4), $t_{dyn}$ is the halo dynamical time (eq. 8.25) and $t_H = 13.8$ Gyr is the present-day age of the Universe. The dotted curves indicate the loci of dark matter halos of different masses with overdensity $\Delta_c(z)$ (eq. 7.21) and gas fraction $f_{gas} = 0.16$. The horizontal thin dashed curves indicate, under the same assumptions, the average values of $n$ expected for virialised halos at different redshifts. There are three regimes, $t_{cool} < t_{dyn}$ (rapid cooling), $t_{dyn} < t_{cool} < t_H$ (slow cooling) and $t_{cool} > t_H$ (no cooling), which regulate galaxy formation (see text for a discussion). The curves are computed by taking from Sutherland and Dopita (1993) the tabulated values of $\Lambda$, $n_e/n$, $n_t/n$ and $\mu$ as functions of $T$ for given $Z$. The shape of the thin dotted and dashed curves for $T \to 10^4$ K reflects the dependence of $\mu$ on $T$. *Right panel.* Same as left panel, but for gas with solar metallicity ($Z = Z_\odot$). Due to the contribution of metals, the cooling time is shorter than for $Z = 0$ and efficient galaxy formation is possible over larger regions of the $n$–$T$ space. Figures inspired by Rees and Ostriker (1977).

can be inferred by noting that the efficiency of cooling decreases dramatically at $T < 10^4$ K; see Figs. 8.2 and 9.2 (right).

The effect of increasing metallicity is to make cooling more efficient, so when $Z = Z_\odot$ (Fig. 8.5, right) galaxy formation is allowed over a larger range of virial temperatures than when $Z = 0$. In other words, in the presence of metals the critical gas density for collapse, at given $T$, is lower. For given overdensity $\Delta_c(z)$ and gas fraction $f_{gas}$, it is also possible to draw in the $n$–$T$ diagram loci of constant halo mass $\mathcal{M}_{vir}$ (dotted curves in Fig. 8.5), given by

$$n = \frac{6k_B^3}{\pi G^3 m_p^4} \frac{f_{gas} T^3}{\mu^4 \mathcal{M}_{vir}^2}, \tag{8.27}$$

which is obtained from eq. (8.26), using eq. (8.19) to eliminate $\Delta_c H^2$, and assuming $T = T_{vir}$.

Typical values of $n$ and $T$ for cosmological halos are given by the intersections of curves of constant $z$ and curves of constant $\mathcal{M}_{\rm vir}$. For instance, it is interesting to note that, over the entire redshift range $0 \lesssim z \lesssim 10$, the curve $\mathcal{M}_{\rm vir} = 10^{12}\mathcal{M}_\odot$ lies below the curve $t_{\rm cool} = t_{\rm dyn}$ when $Z = 0$, while it lies above it when $Z = Z_\odot$. In other words, a halo with virial mass $10^{12}\mathcal{M}_\odot$ is expected to host efficient galaxy formation only if the gas is enriched in metals. We note that in Fig. 8.5 the cooling time has been computed using the collisional ionisation equilibrium cooling function (§8.1.1). The effect of the UVB (§8.1.2) is to make the cooling rate lower, especially at $T \lesssim 10^5$ K (Fig. 8.3), and thus to move the $t_{\rm cool} = t_{\rm H}$ and $t_{\rm cool} = t_{\rm dyn}$ curves upwards in Fig. 8.5.

Summarising, based on the simple model represented in Fig. 8.5, we should expect to have efficient galaxy formation only in halos with $10^9\mathcal{M}_\odot \lesssim \mathcal{M}_{\rm vir} < 10^{13}\mathcal{M}_\odot$, consistent with the fact that the most massive galaxies have $\mathcal{M}_\star \sim 10^{12}\mathcal{M}_\odot$. In higher-mass halos the model predicts, in agreement with observations, that the gas remains hot and does not efficiently form stars: in this regime we have groups and clusters of galaxies, where galaxy formation is not favoured. Halos less massive than $10^9\mathcal{M}_\odot$ are not expected to form stars efficiently, but a detailed treatment of cooling at $T < 10^4$ K is necessary to make quantitative predictions (§9.4.2).

It must be stressed that the picture presented in this section, though capturing the main properties of cooling and collapse in dark matter halos, must not be taken as a detailed quantitative description of these processes, because it is based on several simplifying assumptions. Specifically, as the dark matter halos are not homogeneous systems and their gas is expected neither to be isothermal nor to have uniform density, even in spherical symmetry the ratio $t_{\rm dyn}/t_{\rm cool}$ varies with radius. Moreover, a dark matter halo is not a stationary system, but it is continuously assembling dark matter and gas over timescales often comparable with the cooling time. Finally, and most importantly, it is not necessarily the case that the cooling gas is 'initially' at the virial temperature. In the next section we discuss the more realistic case in which this last assumption is relaxed.

### 8.2.4 Hot and Cold Accretion Modes

In §8.2.3 we have studied the case in which the gas infalling into dark matter halos is first shock heated to the virial temperature, close to the virial radius, and later cools radiatively, condensing at the centre of the halo. In fact, this scenario is not, in general, a realistic representation of the accretion mode of dark matter halos, because heating and cooling are at work at the same time and it is not necessarily the case that, before cooling, the gas is heated to $T_{\rm vir}$. Both analytic calculations and numerical hydrodynamic simulations indicate that two modes of gas accretion are expected to occur: the **hot mode**, in which the gas is shock heated to the virial temperature, and the **cold mode**, in which the gas is still cold when it reaches the centre of the halo.

Qualitatively, we expect the cold mode to be dominant in systems in which cooling is more efficient, that is in systems with $T_{\rm vir}$ similar to the temperature at which the cooling function peaks (§8.1.1). In practice, hydrodynamic models indicate that virial shocks occur only in halos more massive than a critical virial mass $\mathcal{M}_{sh}$. The value of $\mathcal{M}_{sh}$ depends mainly on metallicity $Z$: $\mathcal{M}_{sh}$ is higher for higher $Z$, because cooling is more effective in a more metal-rich gas (§8.1.1). The value of $\mathcal{M}_{sh}$ depends also on the shock-to-virial radius ratio $r_{sh}/r_{vir}$, which is somewhat uncertain, but, according to numerical simulations, can be as low as $\approx 0.1$ in the presence of cooling. As a rule, higher halo masses are needed to have virial shocks at larger radii. The results of analytic models and numerical simulations have provided estimates of $\mathcal{M}_{sh}$ as a function of $Z$ and $r_{sh}$. In principle, $\mathcal{M}_{sh}$ is expected to depend also on redshift $z$ (because the gas density is higher and the metallicity lower at higher $z$), but it turns out that the two effects compensate and the dependence of $\mathcal{M}_{sh}$ on $z$ is weak.

A plot of $\mathcal{M}_{sh}$ as a function of $z$ is shown in Fig. 8.6, based on models of isolated spherically symmetric halos, assuming $r_{sh} = 0.1 r_{vir}$ and mean present-day metallicity $Z = 0.1 Z_\odot$ (the metallicity is assumed to decrease with increasing $z$). For $0 \leq z \leq 5$, $\mathcal{M}_{sh}$ is in the range $5\text{–}7 \times 10^{11} \mathcal{M}_\odot$. At all redshifts, halos with $\mathcal{M}_{vir} < \mathcal{M}_{sh}$ are not effective in shock heating the gas, so the gas infalling into these dark matter halos is not heated to $T_{vir}$ before reaching the halo centre: in these halos the gas is accreted in the cold mode. Virial shocks occur for halo masses above the critical mass, so, on the basis of simple spherical models of isolated halos, one would expect the hot mode to be always dominant in these cases. In fact, accretion is not spherically symmetric and cold streams of gas can

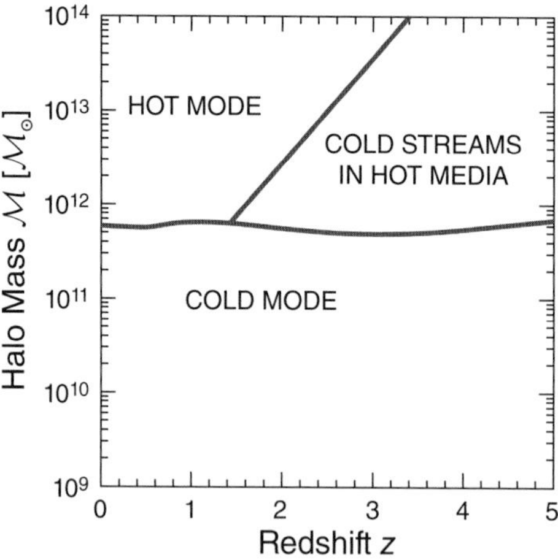


**Fig. 8.6** The horizontal curve indicates the critical dark matter halo virial mass $\mathcal{M}_{sh}$ for stable shocks as a function of redshift for shock radius $r_{sh} = 0.1 r_{vir}$ and mean present-day metallicity $Z = 0.1 Z_\odot$, assuming that the halo is spherical and isolated. Below the critical mass the gas is accreted in the cold mode: it reaches the halo centre, without being heated to the virial temperature. For masses above $\mathcal{M}_{sh}$, virial shocks occur, but the accretion mode depends on redshift. At higher redshift (to the right of the diagonal solid line) the cold mode prevails, because cold gas is accreted mainly in narrow streams that penetrate down to the halo centre without being heated. At lower redshift (to the left of the diagonal solid line) a higher fraction of the accreting gas is heated to the virial temperature. Figure adapted from Dekel et al. (2013).

reach the system's centre even in halos with $\mathcal{M}_{vir} > \mathcal{M}_{sh}$. Cosmological hydrodynamic simulations suggest that the accretion mode for $\mathcal{M}_{vir} > \mathcal{M}_{sh}$ halos depends on redshift. At lower redshift ($z \lesssim 1.5$, for $\mathcal{M}_{vir} \gtrsim \mathcal{M}_{sh}$) a significant fraction of the infalling gas is heated. In massive halos at higher redshift ($z \gtrsim 1.5$, for $\mathcal{M}_{vir} \gtrsim \mathcal{M}_{sh}$) filamentary cold accretion is dominant. In this last case the cold mode can operate, despite the high mass of the halo, because cold gas is accreted mainly in narrow **gas filaments** (or **streams**) that penetrate down to the halo centre without being heated. Shocks are not effective along the directions of the filaments, because there the density is higher and the cooling time is shorter. Far from the filaments the quasi-spherical hot mode accretion is at work. We further discuss the accretion modes in the context of galaxy formation in §10.5 and §10.7.2.

## 8.3 Star Formation

Star formation is the transformation of gas into stars. This is a fundamental step in galaxy formation and evolution that, after the collapse of gas into the dark matter halo described in the previous sections, eventually leads to the build-up of the stellar component of galaxies. The main ingredient for star formation is very cold ($T \lesssim 30$ K) gas at high densities ($n \gtrsim 10^2$ cm$^{-3}$). Typically, to achieve these conditions the gas has first to form molecules that allow a very efficient cooling to low temperatures. For instance, molecular gas can cool very effectively through the CO rotational transitions that we described in §4.2.5. The formation of molecular clouds with a complex and somewhat fractal internal structure (Fig. 4.25) is the next key step. Conditions such that gravity overcomes other competing forces are met in the densest regions of these clouds. There, the gravitational collapse can take place and a fraction of the gas is finally turned into stars. In the end, star formation is a very complex phenomenon that involves a variety of physical and chemical processes occurring in an extremely large range of scales going from the size of a galaxy (tens of kpc, ~ $10^{23}$ cm) to the size of a star (~ $10^{10}$ cm)! Several aspects of star formation are still not fully understood and are the subject of active investigation.

In this book, we follow the line of thought just described that has been historically regarded as the main mechanism for star formation in present-day galaxies. Some theoretical models predict that stars can also form directly from atomic gas (without molecules), as the fine-structure transitions (§4.2.1) of some elements, in particular singly ionised carbon ($C^+$), can cool the gas down to $T \approx 30$ K (Fig. 8.2). The formation of molecular gas, in this scenario, is not needed, as the only requirement for star formation is the presence of cold and high-density gas, which can be achieved by atomic gas cooling (although molecules are still needed to go below 30 K). If this is the case, the fact that molecules are found alongside star-forming regions is *coincidental* as both phenomena have the same common cause: molecules also happen to require high-density environments to form (§8.3.1). We do not enter into the details of this theory and instead point the interested reader to Krumholz (2014). It is however important to stress that, although in the following we often mention molecular clouds, most of the calculations presented would remain valid for star formation occurring directly from atomic gas. Moreover the two scenarios are also not necessarily mutually exclusive, as stars can form from molecular or atomic gas depending on the environment. The reader may also notice that, in the following sections, we often refer to star formation in local galaxies and the Milky Way in particular, as these are the sites where new stars can truly be seen *emerging* from gas clouds. However, most of the theoretical concepts that we describe are, in fact, general and there is no compelling reason why they should not apply to star formation in more distant galaxies.

### 8.3.1 Molecule Formation

We consider, as the first step towards efficient star formation in the ISM of a galaxy, the formation of molecules (see also §9.4.1). The main reason that makes molecular clouds the sites of star formation is that in them radiative cooling can be more efficient than in the diffuse ISM. This is mostly due to the numerous rotational levels possessed by molecules. The consequence of this very efficient molecular cooling and the critical self-shielding (§4.2.5) that occurs at densities $n \gtrsim 10^2\,\mathrm{cm}^{-3}$ is that the gas in molecular clouds reaches very low temperatures ($T \approx 10$–$15\,\mathrm{K}$). This, in turn, implies low internal energy (low pressure) and that they are potentially prone to gravitational collapse.

The general chemical reaction that leads to the formation of a molecule $M$ can be written

$$A + B \rightarrow M + R, \tag{8.28}$$

where $A$ and $B$ are atoms or molecules and $R$ is a remnant (atom, molecule, electron or photon). Any term in this reaction can be neutral or have some electric charge. Often, to lead to the formation of a molecule, a number of reactions similar to eq. (8.28) must take place. The molecule $M$ can be destroyed following the reaction

$$M + X \rightarrow C + D, \tag{8.29}$$

where $X$ is often a photon (photodissociation), while $C$ and $D$ are product particles. From the above we can write the time evolution of the molecular species $M$ as

$$\frac{\mathrm{d}n(M)}{\mathrm{d}t} = k_{\mathrm{f}}n(A)n(B) - n(M)\left[\beta + k_{\mathrm{d}}n(X)\right], \tag{8.30}$$

where $n(i)$ is the density of the species $i$, $k_{\mathrm{f}}$ and $k_{\mathrm{d}}$ are the formation and destruction coefficients of $M$ and $\beta$ is its photodissociation rate. The formation and destruction coefficients are often given in $\mathrm{cm}^3\,\mathrm{s}^{-1}$, while $\beta$ is in $\mathrm{s}^{-1}$. In equilibrium conditions, eq. (8.30) equals zero as molecule formation is fully balanced by destruction processes. The assumption of equilibrium is sometimes used to obtain a rough estimate of $n(M)$.

As an example, we give the sequence of reactions for the formation of CO in Galactic molecular clouds:[3]

$$
\begin{aligned}
C^+ + H_2 &\rightarrow CH_2^+ + \gamma \\
CH_2^+ + e^- &\rightarrow \begin{cases} CH + H & (25\%) \\ C + H_2 & (12\%) \\ C + H + H & (63\%) \end{cases} \\
CH + O &\rightarrow CO + H,
\end{aligned}
\tag{8.31}
$$

where $\gamma$ is a photon and the percentages indicate the probability of obtaining the various products in the intermediate reaction. The formation coefficients vary greatly from $k_f \simeq 6 \times 10^{-16}\,\mathrm{cm^3\,s^{-1}}$ to $k_f \simeq 2 \times 10^{-6}\,\mathrm{cm^3\,s^{-1}}$ for the first and the second reaction, respectively, at $T = 40\,\mathrm{K}$. Once formed, CO can be photodissociated by a UV photon following the reaction

$$CO + \gamma \rightarrow C + O \tag{8.32}$$

$$C + \gamma \rightarrow C^+ + e^-, \tag{8.33}$$

at a rate $\beta \simeq 2 \times 10^{-10}\mathrm{s^{-1}}$ in the absence of self-shielding. Note that the last end product can, in principle, restart the cycle all over again thanks to high abundance of $H_2$ in molecular clouds.

Concerning $H_2$ molecules, in §9.4.1 we see the formation mechanisms that take place in primordial galaxies for a gas of zero metallicity. However, in the ISM of galaxies at a further stage of their evolution, the main mechanism is very different, as their formation takes place *on the surface of dust grains*. This process requires the following steps. First, through collisions between H atoms and dust grains, the H atoms *stick* on the surface of the grains: this process is called **adsorption**. Second, an H atom migrates on the surface having therefore a non-negligible probability to encounter another H atom. This encounter can lead to the formation of an $H_2$ molecule, where the energy released by the transition is absorbed by the dust grain (excitation of its vibrational levels). Finally, the newly formed molecule escapes from the surface of the grain: this process is called **desorption**.

The rate of $H_2$ formation on the surface of dust grains is proportional to the density of atoms and grains, the cross section of the grain, the thermal speed of the atoms and other factors that describe the probability that the atoms will stick to the grain and will migrate across it. These latter depend on the temperature of the dust: if the dust is warmer this probability is reduced. Note that the dust grain can also shield the new molecules from UV photons, protecting them from photodissociation.

As described in §4.2.5, in a galaxy like the Milky Way and in nearby SFGs (with the possible exception of dwarf galaxies), star formation largely occurs in large molecular

[3] This is not the only route that leads to CO formation. Another channel is triggered by the $H_2$ excitation due to the impact of a cosmic ray. This route passes through the formation of an OH molecule, which eventually leads to an $HCO^+$ molecule that, reacting with an $e^-$, dissociates into CO and H. Which of the two channels dominates is regulated by the local cosmic-ray flux.

clouds, usually referred to as giant molecular clouds (GMCs). These structures have masses of $10^5$ to a few$\times 10^6\ \mathcal{M}_\odot$ and host a hierarchy of smaller and smaller structures down to the dense cores where star formation is directly observed (Fig. 4.25). For the typical properties of these structures see Tab. 4.4. We now proceed to investigate the dynamical formation mechanisms of these clouds.

### 8.3.2 Jeans Criterion for Gravitational Instability

The classical criterion that describes under what conditions a portion of gas undergoes gravitational collapse is the **Jeans criterion** (Jeans, 1902). Let us consider a homogeneous, infinite and isothermal medium with initial constant pressure ($P_0$), density ($\rho_0$) and a constant gravitational potential ($\Phi_0$). We assume the medium to be initially static ($\boldsymbol{u}_0=0$) and introduce small (linear; §7.3) perturbations of all these quantities such that

$$P=P_0+\delta P, \quad \rho=\rho_0+\delta\rho, \quad \boldsymbol{u}=\delta\boldsymbol{u} \quad \text{and} \quad \Phi=\Phi_0+\delta\Phi. \tag{8.34}$$

We substitute these perturbations into the continuity (eq. D.22), Euler (eq. D.25) and Poisson (eq. 7.3) equations to obtain

$$\frac{\partial\delta\rho}{\partial t}+\rho_0\boldsymbol{\nabla}\cdot\delta\boldsymbol{u}=0, \tag{8.35}$$

$$\frac{\partial\delta\boldsymbol{u}}{\partial t}=-\frac{1}{\rho_0}\boldsymbol{\nabla}\delta P-\boldsymbol{\nabla}\delta\Phi, \tag{8.36}$$

$$\nabla^2\delta\Phi=4\pi G\delta\rho, \tag{8.37}$$

where we have neglected all second-order terms.[4] We write the pressure perturbation as $\delta P=c_{\rm s}^2\delta\rho$, where $c_{\rm s}$ is the sound speed (§D.2.4). We then take the time derivative of eq. (8.35) and the divergence of eq. (8.36). Combining these two equations with eq. (8.37), we obtain an expression that contains only density perturbations:

$$\frac{\partial^2\delta\rho}{\partial t^2}=c_{\rm s}^2\nabla^2\delta\rho+4\pi G\rho_0\delta\rho, \tag{8.38}$$

from which we aim to find a criterion for the development of the perturbation.

In the linear perturbation analysis, the sum of two solutions is still a solution to the problem (this is called the **superposition principle**). Given the perturbation $\delta Q$ of a generic quantity $Q$, let us look for simple solutions of the form

$$\delta Q=\widehat{\delta Q}e^{{\rm i}(\boldsymbol{k}\cdot\boldsymbol{x}-\omega t)}. \tag{8.39}$$

A generic solution $\delta Q(\boldsymbol{x},t)$ can then be constructed by summing together many solutions of this type, and the Fourier theorem tells us that in this way we can obtain *all* possible solutions:

$$\delta Q(\boldsymbol{x},t)=\frac{1}{(2\pi)^4}\int\widehat{\delta Q}(\boldsymbol{k},\omega)e^{{\rm i}(\boldsymbol{k}\cdot\boldsymbol{x}-\omega t)}{\rm d}^3\boldsymbol{k}\,{\rm d}\omega. \tag{8.40}$$

[4] Note that the unperturbed part of these equations leads to the contradiction that the density of the medium should be null. The results for the perturbed part are not affected by this inconsistency, the neglect of which has been referred to as the **Jeans swindle**.

It follows that, in practice, we can study the evolution of a generic perturbation by simply substituting eq. (8.39) in the equation describing the evolution of any perturbed quantity $\delta Q$.

The aim of the perturbation analysis then becomes to find the so-called **dispersion relation** $\omega(\boldsymbol{k})$, which describes all the solutions in a very compact form. One usually assumes the wavevector $\boldsymbol{k}$ to be real and the fate of the perturbation is determined by the value of the imaginary part of the frequency $\omega$. In particular, we can have the following cases (consider eq. 8.39 to understand).

1. $\mathrm{Im}(\omega)=0$ ($\omega$ real): the perturbation oscillates in time and there is no development of instability.
2. $\mathrm{Im}(\omega)<0$: the perturbation is damped in time and there is no instability.
3. $\mathrm{Im}(\omega)>0$: the perturbation grows exponentially with time; this is the condition for instability.

Coming back to the Jeans analysis, we can rewrite the density perturbation as

$$\delta\rho=\widehat{\delta\rho}\, e^{\mathrm{i}(\boldsymbol{k}\cdot\boldsymbol{x}-\omega t)} \tag{8.41}$$

and substitute it into eq. (8.38). We perform the derivatives (e.g. $\partial\delta\rho/\partial t=-\mathrm{i}\delta\rho\omega$ and $\boldsymbol{\nabla}\delta\rho=\mathrm{i}\delta\rho\boldsymbol{k}$) and divide everywhere by $\delta\rho$ to obtain

$$\omega^2=c_{\mathrm{s}}^2k^2-4\pi G\rho_0, \tag{8.42}$$

which is the dispersion relation for the Jeans instability ($k=|\boldsymbol{k}|$ is the wavenumber). The condition for instability is therefore that the r.h.s. of eq. (8.42) is negative. After the substitution of $k$ with the wavelength $\lambda=2\pi/k$ we find that the gravitational instability can develop if

$$\lambda>\lambda_{\mathrm{J}}\equiv\sqrt{\frac{\pi c_{\mathrm{s}}^2}{G\rho}}\simeq 171.9\left(\frac{c_{\mathrm{s}}}{1\ \mathrm{km\,s^{-1}}}\right)\left(\frac{\mu n}{1\ \mathrm{cm^{-3}}}\right)^{-1/2}\ \mathrm{pc}, \tag{8.43}$$

where $\lambda_{\mathrm{J}}$ is the **Jeans length** and we have generalised the expression by dropping the subscript 0 in the density. Thus, a gas cloud feels its own gravity overcoming its internal pressure whenever its size is larger than the Jeans length; these structures are expected to undergo gravitational collapse. For clouds much smaller than $\lambda_{\mathrm{J}}$, the self-gravity is negligible and they will tend to disperse. They can, however, survive if they are 'confined' by the pressure exerted by an external medium, in which case they are referred to as **pressure-confined clouds**. Note that, with the exception of the GMCs, most of the clouds in the ISM of a present-day SFG (for instance a cloud of the CNM; §4.6.2) can be considered pressure-confined.

The Jeans criterion is often given in terms of mass and it states that a portion of gas of density $\rho$ and sound speed $c_{\mathrm{s}}$ feels its own gravity overcoming the internal pressure whenever its mass is

$$\mathcal{M}>\mathcal{M}_{\mathrm{J}}\equiv\frac{\pi^{5/2}}{6G^{3/2}}\frac{c_{\mathrm{s}}^3}{\rho^{1/2}}\simeq 6.6\times10^4\left(\frac{c_{\mathrm{s}}}{1\ \mathrm{km\,s^{-1}}}\right)^3\left(\frac{\mu n}{1\ \mathrm{cm^{-3}}}\right)^{-1/2}\mathcal{M}_{\odot}. \tag{8.44}$$

$\mathcal{M}_J$ is called the **Jeans mass**[5] and this criterion implies that a gaseous structure more massive than $\mathcal{M}_J$ should undergo gravitational collapse.

In general, if the conditions for instability are fulfilled ($\mathrm{Im}(\omega)>0$), the timescale for the linear growth of the perturbation is $\tau \sim \mathrm{Im}(\omega)^{-1}$. In the case of the Jeans analysis, if in eq. (8.42) the gravity term largely overcomes the pressure gradient term $4\pi G\rho_0 \gg c_s^2 k^2$ (highly unstable), we obtain $\tau \sim (4\pi G\rho_0)^{-1/2}$. More precisely, the timescale for *unimpeded* gravitational collapse of a gas cloud is called the **free-fall time** and it can be estimated, for instance, for a spherical homogeneous sphere of density $\rho$ to be

$$t_{\rm ff} = \sqrt{\frac{3\pi}{32G\rho}} \simeq 1.1\times10^6 \left(\frac{\mu}{2.3}\right)^{-1/2} \left(\frac{n_{\rm H_2}}{10^3\,{\rm cm}^{-3}}\right)^{-1/2} \,{\rm yr}, \tag{8.45}$$

where we have indicated a molecular weight and a number density suitable for molecular clouds (Tab. 4.4).

It is interesting to note that the Jeans length can be expressed in terms of the free-fall time as

$$\lambda_J \approx c_s\, t_{\rm ff}. \tag{8.46}$$

To understand the relevance of eq. (8.46) we use the concept of **sound crossing time** of a gaseous system defined as

$$t_s \equiv \frac{\lambda}{c_s}, \tag{8.47}$$

where $\lambda$ is the size of the system. Sound waves are naturally produced in a fluid subject to a generic perturbation (for instance a compression) and the sound crossing time can be seen as the time that the system takes to *react* to such perturbations. The Jeans criterion for gravitational collapse can thus be approximately rewritten in terms of timescales as

$$t_{\rm ff} < t_s. \tag{8.48}$$

When this inequality is satisfied, the internal pressure of the system does not have enough time to react to the gravitational perturbation and the collapse cannot be halted.

The Jeans criterion is a useful estimate of the conditions under which we can expect gravitational collapse to occur when gravity in the gas is counteracted by internal energy (or thermal pressure) alone. We can express the Jeans mass in terms of gas temperature (using eq. D.30) and indicate typical numbers of molecular clouds in the present-day Milky Way (Tab. 4.4) to obtain

$$\begin{aligned} \mathcal{M}_J &= \frac{\pi^{5/2}}{6}\left(\frac{k_B}{\mu m_p G}\right)^{3/2} \frac{T^{3/2}}{\rho^{1/2}} \\ &\simeq 31.3 \left(\frac{\mu}{2.3}\right)^{-2} \left(\frac{n}{3\times10^2\,{\rm cm}^{-3}}\right)^{-1/2} \left(\frac{T}{15\,{\rm K}}\right)^{3/2} \mathcal{M}_\odot, \end{aligned} \tag{8.49}$$

implying that all GMCs should be gravitationally unstable and collapse in a time of the order of a free-fall time (eq. 8.45). However, in §4.6.2, we saw that the ages (lifetimes)

[5] Note that, in other texts, one may encounter slightly different coefficients for the Jeans mass depending on the adopted geometry. Here we have assumed a spherical cloud of radius $\lambda_J/2$ and thus $\mathcal{M}_J = (4\pi/3)\rho(\lambda_J/2)^3$.

of GMCs are $\sim 10^7$ yr, largely exceeding their free-fall time (eq. 8.45). This suggests that the GMCs are stabilised against self-gravity more than one would expect considering the thermal pressure alone, as we discuss in §8.3.5. The Jeans mass is, instead, of the order of the masses of the smallest structures inside GMCs: the dense cores, some of which are, indeed, experiencing gravitational collapse as we see stars forming inside them. In the following sections, we investigate the physical mechanisms that can play a role in the formation of GMCs and we study the dynamical state (equilibrium or collapse) of gas clouds.

### 8.3.3 Formation of Giant Molecular Clouds

In general, the ISM of SFGs does not show uniform gas densities but a variety of mass concentrations, the most prominent of which are the GMCs (§4.6.2). These structures have gas densities ($n \gtrsim 10^2\,\mathrm{cm}^{-3}$; Tab. 4.4) typically two or more orders of magnitude higher than that of the average (atomic) ISM and it is natural to imagine that they are formed by some kind of *instability*. Given that a large fraction of newly born stars are located inside GMCs, the formation of these latter seems to be an important step in the onset of star formation in galaxies. In this section, we investigate potential mechanisms that may make a gas disc unstable, leading to its *fragmentation* into clouds and eventually to star formation. We start from the simple application of the Jeans criterion to the ISM as a whole.

#### Jeans Instability in the ISM

As a starting point to investigate the formation of GMCs, we use the Jeans criterion (eq. 8.43) applied to the *average* atomic ISM of the present-day Milky Way. If we take $c_s \approx 3\,\mathrm{km\,s^{-1}}$ and $n \approx 1\,\mathrm{cm}^{-3}$ (roughly between the CNM and the WNM; Tab. 4.3) eqs. (8.43) and (8.44) give Jeans lengths of $\approx 500\,\mathrm{pc}$ and masses of $\sim 10^6\,\mathcal{M}_\odot$. These masses are quite close to those of GMCs (Tab. 4.4) and we could naïvely conclude that their formation is regulated by the Jeans instability. Instead, there are a number of caveats that we must consider. First, the strong dependence on $c_s$ can lead to large uncertainties. In fact, the characteristic speed of the ISM is not necessarily 3 $\mathrm{km\,s^{-1}}$, as it ranges from $\lesssim 1\,\mathrm{km\,s^{-1}}$ (CNM) to $\approx 8\,\mathrm{km\,s^{-1}}$ (WNM) (Tab. 4.3). Second, the size of a typical mass concentration predicted by eq. (8.43) is larger than the scaleheight of the H I disc (§4.6.2) beyond which the average gas density starts to decrease appreciably. However, one of the assumptions of the Jeans instability is that the unperturbed medium should be virtually infinite, which means homogeneous in 3D and on scales much larger than $\lambda_J$. Third, the ISM is turbulent and the energy that counteracts gravitational instabilities cannot just be that associated with the thermal pressure. To a crude approximation, we can consider an equivalent **turbulent pressure**

$$P_{\mathrm{turb}} = \rho\sigma_{\mathrm{turb}}^2, \tag{8.50}$$

with the turbulent velocity dispersion $\sigma_{\mathrm{turb}} = \sigma_{\mathrm{H\,I}} \approx 10\,\mathrm{km\,s^{-1}}$, with $\sigma_{\mathrm{H\,I}}$ the observed H I velocity dispersion (§4.2.2). If we use this pressure in the derivation described in §8.3.2, it would lead to the substitution of $\sigma_{\rm turb}$ in place of $c_{\rm s}$ in eqs. (8.43) and (8.44), thus to much larger values for the critical sizes and masses, incompatible with those of the GMCs. Finally, this simple application of the Jeans instability criterion to the galaxy ISM is unwarranted because of the potential importance of rotation. In the next section, we investigate the conditions for gravitational instability in a rotating galaxy disc.

## Local Gravitational Instabilities in Rotating Discs

The aim of this section is to find a criterion for gravitational instabilities in a rotating disc. In particular, we investigate the conditions for *local* instabilities, i.e. produced by perturbations that involve portions of the disc much smaller than its size. These are the kind of instabilities that we may expect to form *clumps* that then undergo gravitational collapse. Global disc instabilities are instead discussed in §10.2.1.

A proper account of the gravitational instability in a rotating disc is achieved by introducing linear perturbations in the Euler equation (§D.2.1) written for a rotating fluid (gas) in cylindrical coordinates together with the usual continuity (eq. D.22) and Poisson (eq. 7.3) equations. Assuming that the gas is self-gravitating (dominates the potential) and that the disc is razor-thin (zero density for $z \neq 0$), one can substitute volume densities $\rho$ with surface densities $\Sigma$. With this prescription, the pressure acts only in the disc plane. Moreover, we assume that the gas behaves adiabatically ($P \propto \Sigma^{\gamma}$, with $P$ and $\Sigma$ the pressure and surface density and $\gamma$ the adiabatic index), has sound speed $c_{\rm s}$ (that quantifies its random motions) and it is in pure circular motion, in the sense that it rotates at the circular speed (§4.3.1): $v_{R,0} = 0$ and $v_{\varphi,0}(R) = (R\,{\rm d}\Phi_0/{\rm d}R)^{1/2} \gg c_{\rm s}$, with $\Phi_0$ the unperturbed potential. The perturbed state of the surface density can be written as $\Sigma = \Sigma_{\rm gas} + \delta\Sigma$, where $\Sigma_{\rm gas}$ is the unperturbed density and $\delta\Sigma$ is the perturbation. We assume that the sizes of the density perturbations are small with respect to the radius $R$ and we write a generic perturbation in the conveni ent form

$$\delta\Sigma(R, \varphi, t) = A(R, t) e^{{\rm i}[m\varphi + s(R,t)]}, \tag{8.51}$$

where $A(R,t)$ is the amplitude, $\varphi$ is the azimuthal angle and $s(R,t)$ is a so-called **shape function**. The real part of eq. (8.51) represents a generic *density wave*, suitable to describe spiral arms in galaxies (§4.1.1 and §10.1.6). The curves defined by $m\varphi + s(R,t) = 2n\pi$ with $n = 0, \pm 1, ...$ correspond to the peak density in spiral arms whereas $m$ defines the **mode of the perturbation**, in practice, the number of arms. Fig. 8.7 gives a practical depiction of how these curves look to an external observer viewing the disc face on.

To proceed, we make the approximation that the arms are tightly wound (small pitch angle; §4.1.1). This is referred to as the **tight-winding approximation** and the spiral pattern that one might expect is shown in the right panel of Fig. 8.7. Having tightly wound arms allows us to make other simplifications, for which we refer the interested reader to other texts (e.g. Mo et al., 2010). In practice, one now introduces a simplified version of eq. (8.51) into the continuity, Euler and Poisson equations keeping only the first-order terms, in analogy to what was done for the Jeans instability (§8.3.2). At the end, we find

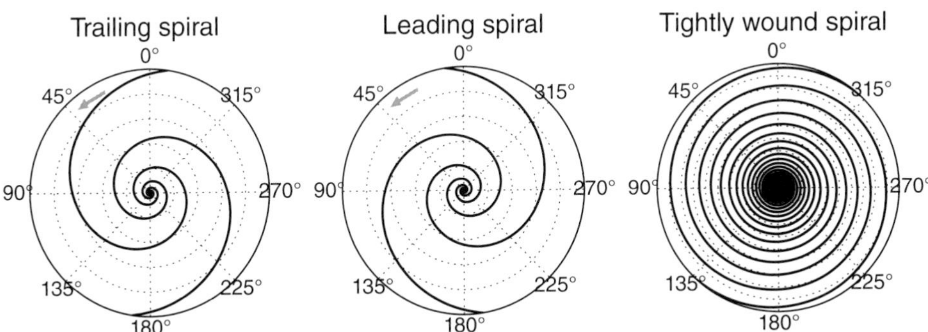


Fig. 8.7

Three examples of $m=2$ spiral perturbations produced by eq. (8.51): the solid curves represent the peak density. If the rotation of the disc is anticlockwise (grey arrow), then the *left panel* shows a trailing pattern (the tails of the spiral are left behind) and the *middle panel* shows a leading pattern. The pitch angle can be seen as the angle between the solid curve of the spiral and the dotted circles where they intersect. The *right panel* shows the same trailing pattern as the left panel, but tightly wound (small pitch angle). In this example we have used a logarithmic shape function $s(R,t)=\alpha \ln(R/R_0)$ with parameters $\alpha$ and $R_0$. The leading pattern is obtained with a negative $\alpha$, the tightly wound pattern with a large (positive) $\alpha$.

the following dispersion relation for gravitational instability in a rotating gaseous disc (Lin and Shu, 1964):

$$(m\Omega - \omega)^2 = \kappa^2 - 2\pi G\Sigma_{\rm gas}|k| + c_{\rm s}^2 k^2, \tag{8.52}$$

where $\Omega$ and $\kappa$ are the angular (§4.1.1) and epicycle frequencies. The **epicycle frequency** is the oscillation frequency that a *test* particle acquires if perturbed (displaced radially) from its circular motion in an axisymmetric potential. Its relation to the angular frequency reads

$$\kappa^2 \equiv \frac{1}{R^3}\frac{\mathrm{d}}{\mathrm{d}R}(R^2\Omega)^2 = R\frac{\mathrm{d}\Omega^2}{\mathrm{d}R} + 4\Omega^2. \tag{8.53}$$

Fig. 8.8 (top) shows the trend with radius of these frequencies for a typical galactic rotation curve. Finally, $\omega$ and $k$ in eq. (8.52) are the frequency and the radial wavenumber of the perturbation (this latter being analogous to the wavenumber of the Jeans instability analysis; §8.3.2).

Given the geometry of a tightly wound spiral (Fig. 8.7, right), we can further simplify the problem by considering axisymmetric perturbations ($m=0$, concentric rings) to obtain

$$\omega^2 = \kappa^2 - 2\pi G\Sigma_{\rm gas}|k| + c_{\rm s}^2 k^2, \tag{8.54}$$

which is an equation rather similar to the Jeans dispersion relation (eq. 8.42) except for the added term $\kappa^2$. The interpretation of this difference is that, in the presence of rotation, the gravity (second term on the r.h.s. of eq. 8.54) is counteracted not only by the gas pressure (third term) but also by the conservation of angular momentum in the disc (first term). However, note the other important difference that the Jeans analysis assumes a 3D geometry, while here we are considering a 2D flat disc.

We now define the two parameters

$$Q_{\rm gas} \equiv \frac{c_{\rm s}\kappa}{\pi G \Sigma_{\rm gas}} \qquad \text{and} \qquad \lambda_{\rm crit} \equiv \frac{4\pi^2 G \Sigma_{\rm gas}}{\kappa^2} \tag{8.55}$$

and rewrite the dispersion relation as

$$\omega^2 = \frac{4\pi^2 G \Sigma_{\rm gas}}{\lambda_{\rm crit}} \left[1 - \frac{\lambda_{\rm crit}}{\lambda} + \frac{Q_{\rm gas}^2}{4}\left(\frac{\lambda_{\rm crit}}{\lambda}\right)^2\right], \tag{8.56}$$

with $\lambda = 2\pi/|k|$ the wavelength of the perturbation, in practice the radial separation between the arms of the tightly wound spiral, which is by definition very small ($\lambda \ll R$). Eq. (8.56) shows that, for certain values of $Q_{\rm gas}$ and $\lambda$, the wave frequency can become imaginary and thus we can expect an instability to develop (§8.3.2). In particular, if $Q_{\rm gas} = 1$ and $\lambda = \lambda_{\rm crit}/2$ we have $\omega^2 = 0$, which is the demarcation between stability and instability. For $Q_{\rm gas} > 1$, $\omega^2 > 0$ for every $\lambda$. Instead, for $Q_{\rm gas} < 1$, the range of sizes ($\lambda$) of unstable perturbations becomes progressively larger as the positive term on the r.h.s. of eq. (8.56) decreases (for $Q \to 0$ any $\lambda < \lambda_{\rm crit}$ is unstable). In the end, the general criterion for gravitational instability is simply

$$Q_{\rm gas} < 1 \tag{8.57}$$

and $\lambda = \lambda_{\rm crit}/2$ represents the most unstable perturbation (the one that grows the fastest). The parameter $Q_{\rm gas}$, often indicated simply as $Q$, is known as the **Toomre *Q* parameter** (Toomre, 1964) and eq. (8.57) is the **Toomre criterion** for instability. Note that if the gas is turbulent, as observed in disc galaxies (§4.2.2), one would need to use the turbulent pressure (eq. 8.50) instead of the thermal pressure. In general, it is common to simply substitute the sound speed in the first of eqs. (8.55) with the observed velocity dispersion of the gas $\sigma_{\rm gas}$.

In Fig. 8.8 (bottom panel) we show the value of $Q_{\rm gas}$ in a simple model galaxy with properties similar to the present-day Milky Way. We see that $Q_{\rm gas}$ remains above one at every radius. Such a disc is said to be supercritical and, as a consequence of this, we should not expect any fragmentation and formation of GMCs. However, this straightforward application of the Toomre criterion to present-day spirals may not be justified given the initial assumption that the gas should dominate the mass in the disc. This assumption is not satisfied in most cases, with the exception of some low-mass SFGs (§4.2.6). The situation could be different in high-$z$ galaxies, where the stellar component has recently started to form and the discs can be gas-dominated (§11.1.4).

If the disc potential is completely dominated by the stars, we can use the results of a similar analysis to that shown above for collisionless stellar discs (Lin and Shu, 1966). In this case, the instability criterion is slightly different and reads

$$Q_\star \equiv \frac{\sigma_R \kappa}{3.36 G \Sigma_\star} < 1, \tag{8.58}$$

where $\sigma_R$ is the radial stellar velocity dispersion. Fig. 8.8 shows that, using $Q_\star$ instead of $Q_{\rm gas}$, we can obtain values much closer to one in the inner disc.

In intermediate situations (similar densities of gas and stars), a number of treatments have been proposed. As an example, we consider a formulation that also takes into account

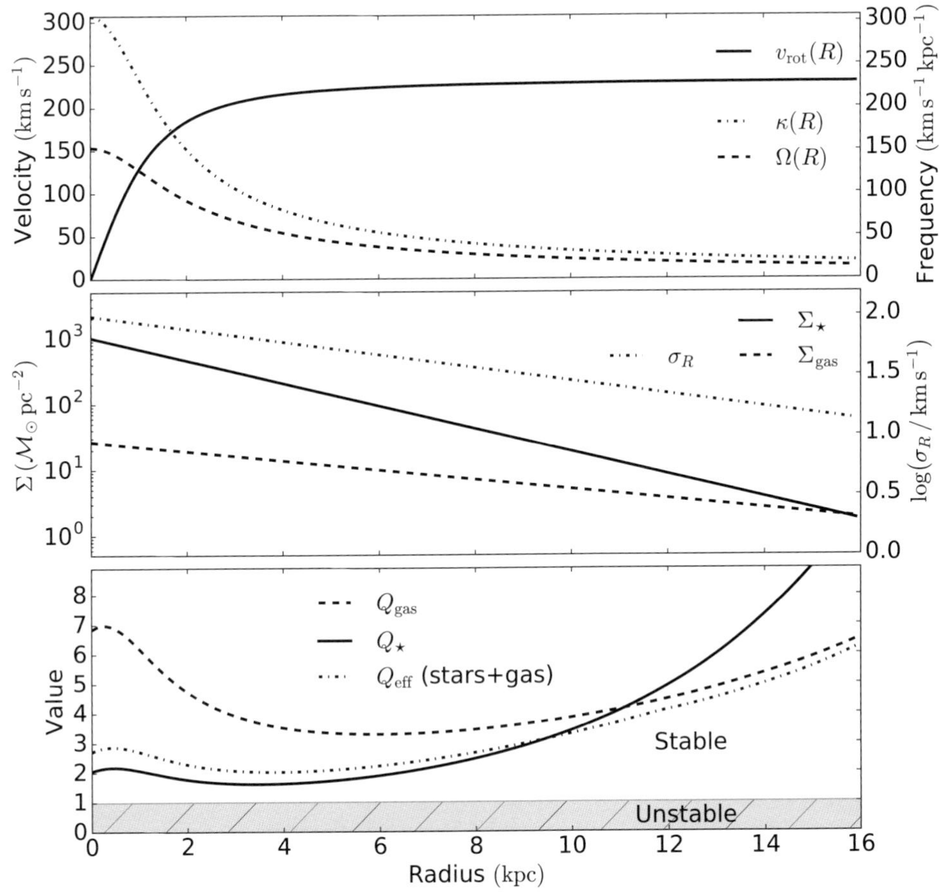


**Fig. 8.8** Radial profiles of various quantities in the equatorial plane of a model disc galaxy with properties similar to the present-day Milky Way. *Top panel*. Rotation curve, angular frequency $\Omega$ and epicycle frequency $\kappa$. Here we have used a gravitational potential with a circular speed that resembles real rotation curves (§4.3.3): $\Phi(R,0) = \frac{1}{2}v_0^2 \ln(R_c^2 + R^2)$ with $R_c = 1.5$ kpc and $v_0 = 230$ km s$^{-1}$. *Middle panel*. Assumed exponential gaseous and stellar disc surface densities ($R_{d,\star} = 2.5$ kpc, $\mathcal{M}_\star = 4 \times 10^{10}\,\mathcal{M}_\odot$, $R_{d,g} = 6$ kpc and $\mathcal{M}_{\rm gas} = 6 \times 10^9\,\mathcal{M}_\odot$) and exponentially declining stellar radial velocity dispersion $\sigma_R$. *Bottom panel*. The $Q$ parameters from the first of eqs. (8.55) and eq. (8.58) using $\sigma_{\rm gas} = 8$ km s$^{-1}$ in place of $c_s$. $Q_{\rm eff}$ is calculated using the prescription of Romeo and Falstad (2013).

the disc thickness and we show, in Fig. 8.8, the resulting *effective* parameter $Q_{\rm eff}$, which is a combination of the $Q$ parameters for gas (eq. 8.57) and stars (eq. 8.58) for our model galaxy. Note how it closely follows $Q_\star$ in the inner disc, while, as the stellar density falls off more rapidly than the gas density, it approaches $Q_{\rm gas}$ in the outer parts. With this formulation, we find that $Q_{\rm eff} \approx 2$–3 everywhere in the inner disc.

In the end, all the criteria presented in this section return, for our model galaxy, $Q > 1$ at every radius. Similar values have been found for a large number of star-forming present-day discs for which $\Sigma_{\rm gas}$, $\sigma_{\rm gas}$ and $\kappa$ can be measured with high precision. How can we interpret these results? If we apply the Toomre criterion *precisely* we should conclude that these discs are stable and GMCs should not form out of local gravitational instabilities. We would then deduce that some other instability is at work to generate GMCs and eventually lead to star formation. In reality though, there are a number of subtleties that could play a role. First, some of the assumptions of the above instability analysis are crude, for instance the assumptions of a razor-thin disc (although revised for $Q_{\rm eff}$), tightly wound spirals and axisymmetric perturbations. Second, values of $Q \approx 2$–3, although not indicative of a fully unstable disc, may represent a condition of only marginal stability. In particular, with these values of $Q$ the disc could be unstable against *non-axisymmetric* perturbations, more general than the ones that we have considered here. In general, apart from the exact threshold between stability and instability, one aspect that we must certainly value in the above analysis is the dependence of eqs. (8.57) and (8.58) on surface density and velocity dispersion (or sound speed). These tell us that, at a given $\kappa$, discs are more prone to local gravitational instabilities if they are 'heavy' (high $\Sigma_{\rm gas}$ or $\Sigma_\star$) and if they are dynamically 'cold' (low $\sigma_{\rm gas}$ or $\sigma_R$).

## The Role of Spiral Arms

In present-day disc galaxies, GMCs are seldom observed outside the regions of the spiral arms. This may indicate that the arms themselves are instrumental for the formation of the GMCs. In the classical density wave theory (§10.1.6), the spiral is a perturbation that rotates as a rigid body. This means that the angular speed of the spiral, called the **spiral pattern speed** $\Omega_{\rm sp}$, is a constant. Conversely, the gas is in differential rotation with a nearly flat rotation curve and thus its angular speed changes with radius nearly as $\Omega_{\rm gas}(R) \propto 1/R$ (Fig. 8.8, top). Thus, the angular speeds of the spiral and of the gas will be different from each other everywhere except at the **corotation radius** ($R_{\rm cr,sp}$) defined as the radius at which

$$\Omega_{\rm gas}(R_{\rm cr,sp}) = \Omega_{\rm sp}. \tag{8.59}$$

Inside this radius, the gas is faster than the spiral pattern ($\Omega_{\rm gas}(R) > \Omega_{\rm sp}$) and the opposite outside. Thus, in general, in most of the disc the gas enters the arms at high (often supersonic) speeds. This produces shocks and compressions which, in turn, can enhance gravitational and thermal instabilities (§8.1.4) leading to the formation of molecular clouds.

The determination of the pattern speed is quite challenging but it has been carried out in some nearby galaxies, returning typical values found for $\Omega_{\rm sp}$ of a few tens of km s$^{-1}$ kpc$^{-1}$. It is important to note that it appears that real disc galaxies have typically (likely always) **trailing spirals**. This means that the disc is rotating leaving the *tails* of the spiral behind (Fig. 8.7, left). A spiral rotating in the opposite way is called a **leading spiral** (Fig. 8.7, middle). This observation is relevant when we discuss the formation of the spiral arms (§10.1.6).

We end this section with a brief summary of what we have discussed so far about star formation in galaxies. Given that most star formation in present-day galaxies occurs in GMCs, we have investigated the mechanism of GMC formation.

- The general idea is that massive and dense clouds form out of a more homogeneous medium as a result of *some kind* of instability.
- We have considered as starting point the Jeans criterion that tells us the conditions for collapse in a homogeneous and infinite medium.
- In the case of a galaxy disc, however, rotation (angular momentum conservation) must be taken into account and this leads to other criteria for local gravitational instability that we wrote in terms of the $Q$ parameters.
- These criteria tell us, in general, that discs are more prone to fragment and form GMCs if the gas density is high and/or the velocity dispersion is low.
- Finally, we have seen that spiral arms, via compression of the ISM, can also facilitate GMC formation.

In the end, we still lack a unique theory that explains the formation of GMCs in galaxy discs and it is possible that more than one mechanism is at work. In the following we leave behind the problem of GMC formation and instead investigate their dynamical state and the conditions for gravitational collapse.

### 8.3.4 Equilibrium of an Isothermal Sphere

In this section and beyond we study the dynamical state of gas clouds (in particular molecular clouds) that have already formed, potentially as a consequence of the mechanisms described in §8.3.3. The general aim is to understand under what conditions we can expect a cloud to collapse and form stars. We start by outlining the theory of an isothermal sphere of gas in hydrostatic equilibrium, in which gravity is balanced by internal energy alone. For more details on this topic we refer the reader to Stahler and Palla (2005).

Let us consider an *isothermal* gaseous sphere in equilibrium with its own gravity. To find the density distribution $\rho(r)$ of such a sphere, we start by writing the hydrostatic equilibrium equation (§4.6.2) and the Poisson equation (eq. 7.3) in spherical coordinates:

$$\frac{1}{\rho}\frac{\mathrm{d}P}{\mathrm{d}r} = -\frac{\mathrm{d}\Phi}{\mathrm{d}r}, \tag{8.60}$$

$$\frac{1}{r^2}\frac{\mathrm{d}}{\mathrm{d}r}\left(r^2\frac{\mathrm{d}\Phi}{\mathrm{d}r}\right) = 4\pi G\rho, \tag{8.61}$$

where $\Phi$ is the gravitational potential. Given the assumption of isothermality, the pressure is $P = c_s^2\rho$, where the constant $c_s^2 = k_B T/(\mu m_p)$ is the isothermal sound speed (§D.2.4) squared. Then eq. (8.60) can be easily integrated to obtain

$$\rho = \rho_c \exp\left(-\frac{\Phi - \Phi_c}{c_s^2}\right), \tag{8.62}$$

where $\rho_c$ and $\Phi_c$ are the central ($r = 0$) density and gravitational potential, respectively. Introducing the two dimensionless variables

$$\Psi \equiv \frac{\Phi - \Phi_c}{c_s^2} \quad \text{and} \quad \xi \equiv \left(\frac{4\pi G\rho_c}{c_s^2}\right)^{1/2} r, \tag{8.63}$$

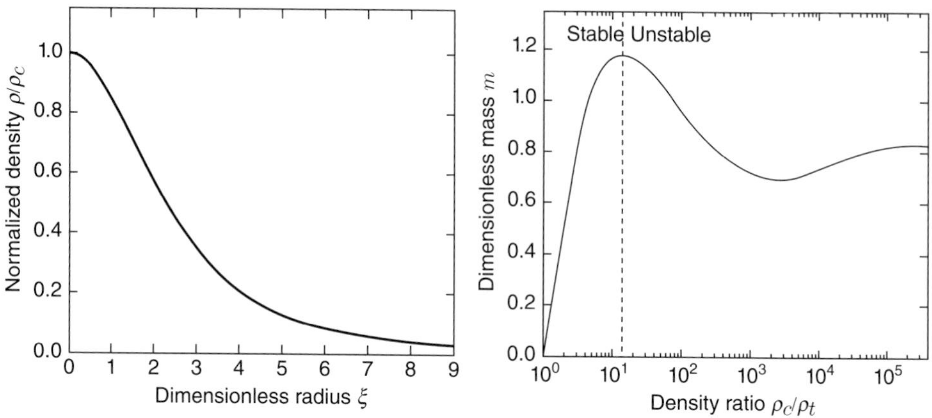


Fig. 8.9 *Left panel.* Normalised density of a Bonnor–Ebert sphere as a function of the dimensionless radius, solution of the isothermal Lane–Emden equation (eq. 8.64). *Right panel.* Dimensionless masses of Bonnor–Ebert spheres as a function of the ratio between their central and truncation densities.

eqs. (8.61) and (8.62) give the *isothermal* **Lane–Emden equation**:[6]

$$\frac{1}{\xi^2}\frac{\mathrm{d}}{\mathrm{d}\xi}\left(\xi^2\frac{\mathrm{d}\Psi}{\mathrm{d}\xi}\right)=e^{-\Psi}. \tag{8.64}$$

We look for solutions $\Psi(\xi)$ with boundary conditions $\Psi(0)=0$ (eq. 8.63) and $\Psi'(0)=0$, which avoids divergence of the density in the centre. Eq. (8.64) has no analytic solution with these boundary conditions and it has to be solved numerically. Once we have found $\Psi(\xi)$, eqs. (8.62) and (8.63) allow us to find the density $\rho(\xi)$. The numerical solution for the density is shown in Fig. 8.9 (left panel). The density has a central core and an asymptotic behaviour at large radii like a singular isothermal sphere (eq. 5.37): $\rho\propto\xi^{-2}$. The profile in Fig. 8.9 (left) is called a **Bonnor–Ebert sphere** from the two authors who found it independently (Ebert, 1955; Bonnor, 1956). It represents the generic density distribution of a non-singular isothermal sphere in hydrostatic equilibrium with its own gravity.

We can now derive the gas mass of a Bonnor–Ebert sphere by integrating the density from $r=0$ to an external truncation radius $r_\mathrm{t}$:

$$\mathcal{M}=4\pi\int_0^{r_\mathrm{t}}\rho(r)r^2\mathrm{d}r=4\pi\left(\frac{c_\mathrm{s}^2}{4\pi G\rho_\mathrm{c}}\right)^{3/2}\int_0^{\xi_\mathrm{t}}\rho(\xi)\xi^2\mathrm{d}\xi \tag{8.65}$$

$$=\frac{1}{\sqrt{4\pi\rho_\mathrm{c}}}\frac{c_\mathrm{s}^3}{G^{3/2}}\xi_\mathrm{t}^2\left(\frac{\mathrm{d}\Psi}{\mathrm{d}\xi}\right)_{\xi=\xi_\mathrm{t}}, \tag{8.66}$$

where, in the second equality, we have substituted a dimensionless truncation radius $\xi_\mathrm{t}$ related to $r_\mathrm{t}$ by the second of eqs. (8.63). Moreover, for the last equality, we have integrated the Lane–Emden equation (eq. 8.64) in $\xi$ after multiplying it by $\xi^2$ on both sides and we

[6] The *generic* Lane–Emden equation is obtained using a polytropic equation of state of the kind $P\propto\rho^{1+(1/n)}$, with $n$ the polytropic index, and it is typically employed for the study of stellar interiors. The isothermal case corresponds to $n\rightarrow\infty$.

have used the fact that, from eqs. (8.62) and (8.63), $\rho = \rho_c \exp(-\Psi)$. The truncation of the sphere at $r_t$ is both necessary in order to have convergence of the mass (given the asymptotic dependence of the density as $r^{-2}$) and justified by the fact that any gas cloud effectively *ends* when it reaches pressure equilibrium with the external medium. Thus, $r_t$ is simply defined as the radius at which the pressure $P$ of the gas in the cloud ($P$ decreases with radius, being proportional to $\rho$) equals the external pressure $P_{ext} = P(r_t) = c_s^2 \rho_t$, where $\rho_t \equiv \rho(r_t)$.

We multiply and divide eq. (8.66) by $c_s \rho_t^{1/2}$ and define a dimensionless mass

$$m \equiv \left( \frac{P_{ext}^{1/2} G^{3/2}}{c_s^4} \right) \mathcal{M}, \tag{8.67}$$

and the final expression for the dimensionless mass of a Bonnor–Ebert sphere becomes

$$m = \frac{1}{2\sqrt{\pi}} \left( \frac{\rho_c}{\rho_t} \right)^{-1/2} \xi_t^2 \left( \frac{d\Psi}{d\xi} \right)_{\xi=\xi_t}. \tag{8.68}$$

Eq. (8.68) shows that the mass of these spheres is a function of the density contrast $\rho_c/\rho_t$ and of the truncation radius $\xi_t$. However, given that the density of the cloud is a monotonic function of $\xi$ (Fig. 8.9, left), $\xi_t$ can also be written as a function of the density contrast $\rho(\xi_t)/\rho_c$. Thus, the dimensionless mass $m$ is a function of the density contrast $\rho_c/\rho_t$ only (Fig. 8.9, right). At low density contrasts, the mass shows a linear dependence on $\log(\rho_c/\rho_t)$; it then turns over and eventually reaches the asymptotic constant value of $\sqrt{2/\pi}$. The peak occurs at the critical value $(\rho_c/\rho_t)_{crit} \simeq 14$ corresponding to $m_{crit} \simeq 1.18$. The mass at which this maximum occurs is called the **Bonnor–Ebert mass**

$$\mathcal{M}_{BE} \equiv \frac{m_{crit}}{G^{3/2}} \frac{c_s^3}{\rho_t^{1/2}} \simeq 1.18 \left( \frac{k_B}{\mu m_p G} \right)^{3/2} \frac{T^{3/2}}{\rho_t^{1/2}}, \tag{8.69}$$

where we have used the conversion between isothermal sound speed and temperature (eq. D.30). Eq. (8.69) represents the maximum mass that an isothermal cloud can have to be in hydrostatic equilibrium and stable. Larger masses cannot be in hydrostatic equilibrium, meaning that it is not possible to find equilibrium configurations: the curve in Fig. 8.9 (right) represents *all* the possible clouds in equilibrium. As for the stability, it can be shown that clouds with a density contrast larger than 14 (beyond the peak in Fig. 8.9, right) are *not* stable under gravitational perturbation. Indeed, they react to perturbations by lowering their internal pressure, which then leads to a gravitational collapse. Note that eq. (8.69) has strong similarities to eqs. (8.44) and (8.49) for the Jeans mass and it is reassuring that different treatments of the problem lead to a similar mass threshold for the gravitational collapse.

We can test the applicability of the Bonnor–Ebert theory to environments of molecular clouds by substituting in eq. (8.69) the relevant numbers. We obtain

$$\mathcal{M}_{BE} \simeq 21.9 \left( \frac{\mu}{2.3} \right)^{-2} \left( \frac{n_t}{10^2\,\mathrm{cm}^{-3}} \right)^{-1/2} \left( \frac{T}{15\,\mathrm{K}} \right)^{3/2} \mathcal{M}_\odot. \tag{8.70}$$

Note that $n_t = \rho_t/(\mu m_p) = 10^2\,\mathrm{cm}^{-3}$ is the truncation density of a cloud at $T = 15\,\mathrm{K}$ in pressure equilibrium with a medium at $P_{ISM}/k_B = 1.5 \times 10^3\,\mathrm{K\,cm}^{-3}$ (nearly the average pressure of the ISM in the Milky Way, see also §4.6.2). The mass in eq. (8.70) is orders of magnitude below the typical mass of a GMC. Thus the Bonnor–Ebert theory leads to a conclusion in line with that of the Jeans instability (§8.3.2): GMCs cannot be sustained by thermal pressure. On the other hand, as for the Jeans mass (eq. 8.49), the Bonnor–Ebert mass is of the order of the dense core masses (Tab. 4.4). However, an important added value is that Bonnor–Ebert analysis allows us to compare not only the mass but also the detailed density distribution of the gas. The comparison between the theoretical predictions and the observed density profiles in Galactic dense cores gives a remarkably good agreement, showing that these structures are very well described by Bonnor–Ebert spheres seen at the verge of the gravitational collapse.

### 8.3.5 Virial Theorem with Magnetic Field

Further insights on the dynamical state of GMCs can be gained by writing the virial theorem (§5.3.1) for a gas cloud including the magnetic field,[7]

$$\frac{1}{2}\frac{\mathrm{d}^2 I}{\mathrm{d}t^2} = 2K + 2U + W + E_{\mathrm{M}}, \tag{8.71}$$

where $I$ is the moment of inertia of the system, while $K$, $U$, $W$ and $E_{\mathrm{M}}$ are the total kinetic, internal, gravitational and magnetic energies, respectively. If none of the terms on the r.h.s. balances the gravitational energy ($W \approx -G\mathcal{M}^2/r$, with $\mathcal{M}$ and $r$ the mass and radius of the gas cloud, respectively), the second derivative of the moment of inertia ($I \approx \mathcal{M}r^2$) is negative. Thus the gas in the cloud acquires a negative radial acceleration and we find again the result that the collapse takes place in a free-fall time $t_{\mathrm{ff}}$ (§8.3.2). As mentioned in §4.6.2, Galactic GMCs appear to have lifetimes (ages) longer than $t_{\mathrm{ff}}$ by at least one order of magnitude, thus their collapse is likely balanced by one of the energy components on the r.h.s.

Given the low temperatures of GMCs, the internal energy fails to balance gravity. We can see this by evaluating the ratio

$$\frac{2U}{|W|} \approx \frac{3Nk_{\mathrm{B}}Tr}{G\mathcal{M}^2} = \frac{3k_{\mathrm{B}}Tr}{\mu m_{\mathrm{p}}G\mathcal{M}}, \tag{8.72}$$

where the total number of particles in the cloud is $N = \mathcal{M}/(\mu m_{\mathrm{p}})$. If we consider typical values for Galactic GMCs (Tab. 4.4) we find $2U/|W| \sim 10^{-2}$, which confirms the results obtained with the Bonnor–Ebert theory (§8.3.4) and the Jeans instability analysis (§8.3.2).

Let us now consider the kinetic energy. We should, in principle, make a distinction between ordered (rotation) and random (turbulent) kinetic energy. However, large molecular clouds do not appear to have significant rotation and we can restrict ourselves to turbulence. Postponing the discussion on turbulence to §8.3.7, here we simply assume that $K_{\mathrm{turb}} = \frac{1}{2}\mathcal{M}(\sqrt{3}\sigma_{\mathrm{gas}})^2$ with $\sigma_{\mathrm{gas}}$ the observed line broadening in molecular clouds along the

[7] This is the so-called *scalar* virial theorem, where we have implicitly neglected two surface integrals as they are typically subdominant with respect to the other terms (see Shu, 1992, for details).

line of sight (the factor $\sqrt{3}$ comes from assuming isotropic turbulence). The ratio between the turbulent and the gravitational terms is

$$\frac{2K_{\rm turb}}{|W|} \approx \frac{3r\sigma_{\rm gas}^2}{G\mathcal{M}} \simeq 1 \left(\frac{\sigma_{\rm gas}}{2\,{\rm km\,s^{-1}}}\right)^2 \left(\frac{\mathcal{M}}{10^5\,\mathcal{M}_\odot}\right)^{-1} \left(\frac{r}{50\,{\rm pc}}\right), \tag{8.73}$$

where in the second rough equality we have indicated typical values of Galactic GMCs. Eq. (8.73) shows that the turbulence present in GMCs appears to be of the right amount to stabilise them against gravitational collapse. This finding is referred to as the **second Larson law** (Larson, 1981).

Finally, we consider the magnetic energy

$$E_{\rm M} \approx \frac{4\pi r^3}{3}\frac{B^2}{8\pi} = \frac{B^2 r^3}{6}, \tag{8.74}$$

where $B^2/(8\pi)$ is the magnetic energy density (§D.2.6). It is convenient to make use of the magnetic flux $\phi_{\rm B}$, which is a conserved quantity (§8.3.6). Given a typical magnetic field strength inside the cloud of $B = |\boldsymbol{B}|$, we can expect that the magnetic flux threading the cloud is $\phi_{\rm B} \approx \pi B r^2$. We then rewrite the virial theorem with only magnetic and gravitational energies (i.e. we assume that all the other energies are negligible):

$$E_{\rm M} + W \approx \frac{\phi_{\rm B}^2}{6\pi^2 r} - \frac{G\mathcal{M}^2}{r} = \frac{G}{r}\left(\mathcal{M}_\phi^2 - \mathcal{M}^2\right), \tag{8.75}$$

where we have defined

$$\mathcal{M}_\phi \equiv \frac{\phi_{\rm B}}{\pi\sqrt{6G}} \approx 2\times 10^5 \left(\frac{B}{10\,\mu{\rm G}}\right)\left(\frac{r}{50\,{\rm pc}}\right)^2 \mathcal{M}_\odot \tag{8.76}$$

as the **magnetic critical mass**. Following eq. (8.75), in clouds with $\mathcal{M}$ smaller than $\mathcal{M}_\phi$, the magnetic energy can balance gravity, while for larger clouds gravity should prevail. Given that this mass is of the order of those of GMCs, this estimate shows that the magnetic field could have a role in opposing their self-gravity.

In the end, this exploration of the role of the different energetic contributions brings us to the conclusion that magnetic fields and turbulence can play important roles in stabilising molecular clouds against gravitational collapse. In the following we explore in a bit more detail these two contributors.

### 8.3.6 The Role of Magnetic Fields

The estimate in eq. (8.76) shows that the magnetic energy in GMCs can be of the same order as the gravitational energy (from eq. 8.75). Note that we have used a relatively high value of the magnetic field strength with respect to the average magnetic field strength in the ISM (§4.2.8). This is, in fact, a value typically determined through the Zeeman effect (§4.6.2) in GMCs (Tab. 4.4). This enhancement is expected given the so-called **flux freezing** of the magnetic field. This property can be written as the conservation of the magnetic flux across a generic surface $S$ that moves with the fluid,

$$\frac{{\rm d}\phi_{\rm B}}{{\rm d}t} = 0, \tag{8.77}$$

where $\phi_B$ is the **magnetic flux** ($\phi_B \equiv \int_S \boldsymbol{B} \cdot d\boldsymbol{S}$) and $d\boldsymbol{S}$ is the surface element vector locally normal to $S$; see also §D.2.6.

Eq. (8.77) attests to a tight relation between the magnetic field and the motion of the gas. Consider a cloud permeated by a magnetic field with an average strength $B = |\boldsymbol{B}|$. Given that the surface is proportional to $r^2$, eq. (8.77) implies that

$$B \propto \frac{1}{r^2} \tag{8.78}$$

and so, for instance, if the cloud undergoes compression, the strength of the magnetic field can increase greatly with respect to its initial value. As a consequence, we can expect denser regions to harbour stronger magnetic fields with respect to the mean value in the ISM.

An alternative way to depict the role of the magnetic field in molecular clouds is in terms of **magnetic waves**. In a medium with a negligible magnetic field, the typical speed that a gas perturbation achieves is the sound speed. In contrast, in a magnetised gas with density $\rho$ and magnetic field $\boldsymbol{B}$, perturbations also produce waves, called **Alfvén waves**, of very different kind that propagate at velocity

$$v_A \equiv \frac{\boldsymbol{B}}{\sqrt{4\pi\rho}}, \tag{8.79}$$

called the **Alfvén velocity**. The main property of Alfvén waves is that they are transverse waves (they propagate orthogonally to the direction of oscillation) unlike sound waves which are always longitudinal (compression waves). Moreover, the Alfvén speed can be very different from the sound speed.[8] In particular, in GMCs the typical Alfvén speed inferred from the measured magnetic field strength (Tab. 4.4) can be one order of magnitude higher than the sound speed, $v_{A,GMC} \sim 1\ \mathrm{km\,s^{-1}}$. This value is of the order of the measured dispersion $\sigma_{gas}$, thus again pointing to a potentially important role of the magnetic field in the energy budget.

The reader may find it surprising that the magnetic field can have any role in the molecular gas which is largely neutral. Indeed, the magnetic field only acts, through the Lorentz force, on ions and electrons (§D.2.6). However, due to collisions, these charged particles (ions in particular given their larger masses and cross sections) can efficiently transfer their momentum to the neutrals. As a consequence, the gas as a whole is influenced by the magnetic field through the **drag force** between ions and neutrals. The ratio between ions and neutrals in molecular clouds is extremely low ($n_i/n_n \sim 10^{-6}$–$10^{-7}$). It is, however, enough to keep the magnetic field linked to the bulk of the gas at least on the large scales (see also §8.3.8).

### 8.3.7 The Role of Gas Turbulence

Gas turbulence (see Elmegreen and Scalo, 2004, for a review) is a key ingredient of the ISM dynamics and of molecular clouds in particular. The onset of turbulent flow in a

[8] In general the maximum speed achievable in a magnetised medium is a combination of sound and Alfvén speeds: $\sqrt{c_s^2 + v_A^2}$. These are called **magnetosonic waves**, of which Alfvén waves are a subclass.

fluid is determined by the so-called **Reynolds number**, defined as the order-of-magnitude ratio between the inertial ($\boldsymbol{u} \cdot \boldsymbol{\nabla u}$) and the viscous ($\nu\nabla^2\boldsymbol{u}$) forces in the Navier–Stokes equation (§D.2.5)

$$\mathcal{R}e \equiv \frac{LV}{\nu}, \tag{8.80}$$

where $\nu$ is the kinematic viscosity coefficient, $L$ is the largest scale over which turbulence develops and $V$ is the typical (turbulent) speed within $L$. It has been established experimentally that whenever the Reynolds number is larger than a certain critical value ($\mathcal{R}e > \mathcal{R}e_{\rm crit} \sim 10^3$–$10^4$) the fluid is prone to the development of turbulence.

Let us estimate the Reynolds number in the ISM. The kinematic viscosity coefficient can be simply written as the thermal speed $v_{\rm T}$ times the mean free path $\ell$ ($\nu = v_{\rm T}\ell$), which leads to

$$\nu \approx \frac{c_{\rm s}}{An} = 10^{20}\left(\frac{c_{\rm s}}{1\,{\rm km\,s^{-1}}}\right)\left(\frac{A}{10^{-15}{\rm cm}^2}\right)^{-1}\left(\frac{n}{1\,{\rm cm}^{-3}}\right)^{-1}{\rm cm^2\,s^{-1}}, \tag{8.81}$$

where $A$ is the cross section and we have used $\ell = 1/(An)$ (with $n$ the number density). We have also substituted the thermal speed with the sound speed (eq. D.30), given that $c_{\rm s} \approx v_{\rm T}$. The value $A \sim 10^{-15}{\rm cm}^2$ is the typical cross section for encounters between H I atoms; molecules and ionised gas have larger cross sections leading to lower values of $\nu$. The expression for the Reynolds number in the ISM becomes

$$\mathcal{R}e_{\rm ISM} \approx 3\times10^5\left(\frac{L}{100\,{\rm pc}}\right)\left(\frac{n}{1\,{\rm cm}^{-3}}\right)\mathcal{M}_{\rm ISM}, \tag{8.82}$$

where $\mathcal{M}_{\rm ISM}$ is the Mach number (eq. D.32) of the ISM, typically of order unity or larger, and we have taken as typical size of the system the thickness of the H I layer (§4.6.2). Eq. (8.82) shows that it is reasonable to expect the development of turbulence in the ISM as $\mathcal{R}e_{\rm ISM} > \mathcal{R}e_{\rm crit}$. The same equation applied to molecular clouds leads to even larger values of $\mathcal{R}e$.

A general analytic theory of turbulence does not exist. However, under the assumption of *fully developed* and stationary turbulence, a simple analytic treatment has been developed by Kolmogorov (1941). The **Kolmogorov theory** describes a turbulent ($\mathcal{R}e \gg \mathcal{R}e_{\rm crit}$) and incompressible fluid[9] as a combination of chaotic motions at very different spatial scales. The turbulent entities are called **eddies** and they are characterised by physical sizes $\lambda$ and wavenumbers $k = 2\pi/\lambda$. Kinetic energy is transferred between these eddies to keep a stationary state, as we explain in §D.2.7. In the end, the system can be simply described by an **energy power spectrum** $\mathsf{E}(k)$ that has the form

$$\mathsf{E}(k) \propto k^{-\beta}, \tag{8.83}$$

where $\beta = 5/3$ is the slope of the Kolmogorov power spectrum and $\mathsf{E}(k){\rm d}k$ represents the specific (per unit mass) kinetic energy stored in eddies with wavenumbers between $k$ and $k + {\rm d}k$.

[9] An **incompressible fluid** is such that its density does not change during motion (${\rm D}\rho/{\rm D}t = 0$ using the Lagrangian derivative; §D.2.1). This approximation is very good for gases in subsonic conditions.

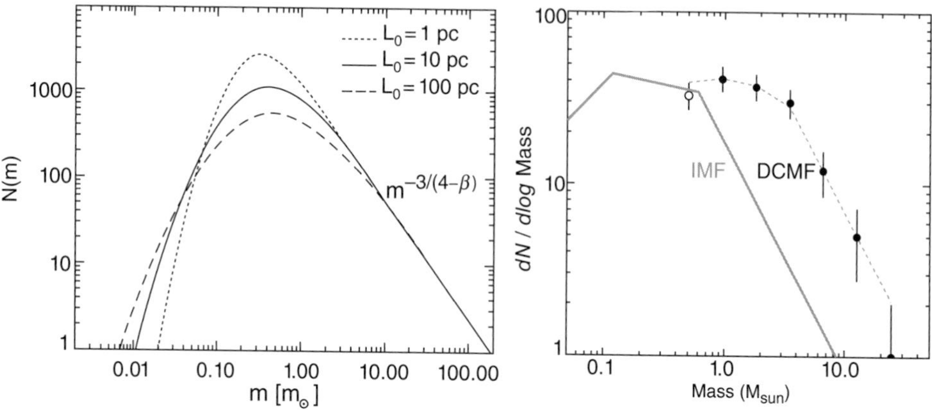


Fig. 8.10 *Left panel*. Distribution of masses of collapsing cores produced by a model of turbulent fragmentation. The turbulence is considered supersonic and super-Alfvénic and has a Kolmogorov-like power spectrum with $\beta = 1.74$ (eq. 8.83), which gives a slope at high masses compatible with the observed slopes. The turnover at low masses is determined by Jeans instability conditions on the dense cores and $L_0$ is the largest turbulent scale ($L$ in §D.2.7). From Padoan and Nordlund (2002). © AAS, reproduced with permission. *Right panel*. Observed mass function of dense cores (dots) derived from extinction maps (see eq. 4.16 in §4.2.7) in the nearby Pipe molecular cloud. The solid grey curve shows instead the stellar initial mass function (IMF) determined in the Trapezium cluster inside the Orion nebula (roughly a Kroupa IMF; §8.3.9). The dashed grey curve is a binned version of the same IMF shifted by a factor $\approx 4$ in mass. From Alves et al. (2007).

Revisions of the Kolmogorov turbulence obtained with more complex treatments that include the compressibility of the gas and with hydrodynamic simulations tend to find slightly steeper slopes of the power spectrum of $\beta \approx 2$ (especially for supersonic flows). However, apart from the exact shape of the power spectrum, the general prediction of turbulence theories is the production of a hierarchy of structures characterised by *power-law distributions*. As an example, in Fig. 8.10 (left) we show the prediction for the sizes of the gravitationally unstable dense cores obtained with a simple model of turbulence fragmentation with a nearly Kolmogorov-like power spectrum of $\beta = 1.74$. The right panel shows instead the observed mass distribution of dense cores in a nearby star-forming region in our Galaxy called the Pipe nebula (points). The similarity between these distributions points to a key role of turbulence in the formation of substructures inside GMCs. Fig. 8.10 (right) also shows the initial mass function (IMF; §8.3.9) observed in a young star cluster of the Milky Way. This appears as a shifted version of the dense core function, thus suggesting a role of turbulence in the shaping of the IMF itself.

If turbulence is so ubiquitous in GMCs, where does it come from and what are its energy sources? These sources are indeed required, as turbulence *dissipates* on timescales shorter than the typical lifetime of GMCs. To date, a consensus on the energy sources of turbulence in galaxy discs and in GMCs has not been reached. It is expected that an important role is played by stellar feedback (§8.7) and injection of energy is clearly taking place in GMCs around nearly formed protostars (Fig. 4.25, right). Other contributions may come from gas accretion from the external environment into the inner parts of GMCs, from magnetic energy, from galactic shear (§10.1.5) and from the conversion of gravitational energy during collapse.

Before proceeding, we give a brief summary on the equilibrium of molecular clouds. We have seen that, as the ages of GMCs ($\gtrsim 10^7$ yr) are longer than their free-fall times ($\approx 10^6$ yr), we should expect that their self-gravity is efficiently balanced by some form of energy and we have investigated various possibilities.

- Internal energy (thermal pressure) cannot oppose gravity, essentially because GMCs are too cold ($T \approx 10$–$15$ K).
- The equilibrium mass (Bonnor–Ebert mass) that one derives for a cloud in hydrostatic equilibrium is of the order (1–100 $\mathcal{M}_\odot$) of the smallest structures inside GMCs (dense cores). A dense core is therefore very close to equilibrium but it can also become unstable and collapse to form a star.
- Magnetic energy can play an important role in opposing gravity in GMCs for clouds with masses up to $\sim 10^5\ \mathcal{M}_\odot$.
- Most of all, every molecular cloud appears to have a large amount of supersonic ($\sigma_{\rm gas} \gg c_{\rm s}$) turbulence, which is thought not only to play a stabilising role but also to shape the hierarchy of structures in its interior.

In the next sections, we outline the phases that lead to the formation of stars inside dense cores (once they become unstable) and finally discuss possible shapes for the end product of all these processes: the stellar IMF.

### 8.3.8 Gravitational Collapse

When the 'internal' energies (thermal pressure, magnetic field and turbulence) in a dense core can no longer balance gravity, collapse begins. The theory of gravitational collapse is very complex and it is today mostly studied by employing hydrodynamic simulations. Here, we briefly discuss some aspects that involve the conservation of magnetic flux and angular momentum.

In §8.3.6 we saw how the magnetic field remains frozen in the fluid and its strength increases as the inverse of the radius squared. However, if we calculate the magnetic field that a star like the Sun should have starting from a dense core with typical size ($\sim 0.1$ pc) and magnetic field strength ($\simeq 30\ \mu$G; Tab. 4.4), we obtain a value of about $6 \times 10^8$ G, whereas the observed magnetic field of the Sun is only $B_\odot \approx 1$ G. This stunning difference shows that the field flux cannot be conserved during collapse from dense core to star, but it must diffuse outside the collapsing protostar. This phenomenon, called **ambipolar diffusion**, occurs because the coupling of ions and neutral particles described in §8.3.6 becomes ineffective on the scales of dense cores.

We can describe the process in simple terms as follows. During collapse, ions feel not only gravity but also the Lorentz force that keeps them linked to the large-scale magnetic field lines. The situation is different for neutral particles: they are subject to gravity and tend to collapse, but they also collide with the ions and get slowed down by these collisions (drag force). In the centres of dense cores, the density of ions is extremely low (typically one every $10^7$ neutral particles) and thus the mean free paths of neutrals (given as the length crossed before running into an ion) is quite large. It is therefore intuitive to imagine that there is a length scale below which neutral particles cannot be slowed down significantly by collisions with ions. This scale is the ambipolar diffusion length and it is about one-tenth of a parsec in the densest parts of molecular clouds. It follows that in structures of size comparable to or smaller than this length, ions and neutral particles can *drift apart*, and the neutrals can collapse without carrying with them the ions and, consequently, the magnetic field.

Let us now consider rotation and impose the conservation of angular momentum $J$ during collapse. The gravitational energy of a cloud collapsing at constant mass (no mass outflows or accretion from the environment) scales with $r^{-1}$ (§8.3.5). The rotational kinetic energy has instead the dependence

$$K_{\rm rot} \sim \mathcal{M}\Omega^2 r^2 = \frac{J^2}{\mathcal{M}r^2} \propto \frac{1}{r^2}. \tag{8.84}$$

Thus, if the initial dense core had even a slight rotation, the proceeding of the collapse may enhance its $K_{\rm rot}$ dramatically to the point that it overcomes the gravitational energy. This leads to the possibility that the collapse is halted by rotation. However, the observed rotation of stars (including the Sun) is much smaller than the values that we would obtain following this reasoning. This indicates that part of the angular momentum is transferred to the external medium, allowing the inner core to proceed with the collapse. This transfer likely takes place thanks to the magnetic field lines trapped into the collapsing core and connecting the core to the outer envelope. In practice, the magnetic field is slowing down the cloud rotation (**magnetic braking**) by transferring its angular momentum to the external envelope. The main consequences of magnetic braking are: (1) the collapse is slowed down with respect to predictions of non-rotating theories (nearly the free-fall time), in better agreement with the observed pre-main-sequence times (see below), and (2) the angular momentum of the collapsing cloud is efficiently transferred outwards, in agreement with the observed relatively small rotation velocities of most stars.

A general property of the gravitational collapse of a dense core is that it occurs *inside-out* in the sense that its inner parts collapse first and faster. The process is thought to proceed through different phases.

1. After $t \sim 10^4$–$10^5$ Myr, a protostar forms in the central collapsing region and part of the envelope settles in a circumstellar (protoplanetary) disc. This is often accompanied by the onset of **stellar jets**, whose formation is likely governed by the protostar magnetic field. These jets, observed in a number of dense cores, act to clear the way, opposing the infall of gas and ejecting part of the outer envelope.
2. At $t \sim 10^5$–$10^6$ Myr, we enter the so-called **T-Tauri phase**, where a protoplanetary disc is present and the jets have cleared up a large fraction of the envelope. The jets are still ejecting some material and gas accretion proceeds perpendicularly to them.

3. Finally, at $t \sim 10^6$–$10^7$ Myr, the protostar is cleared of the envelope and sets into its pre-main-sequence **Hayashi track** (Hayashi, 1961). The jets are now absent and the protoplanetary disc slowly cools, leading to the formation of a planetary system. This last phase may take times longer than 10 million years, although its timescale is an inverse function of the stellar mass (more massive stars collapse more quickly). The protostar eventually sets into the main sequence and we enter the realm of stellar evolution (§C.5).

In conclusion, after all the stages described in this section, the molecular gas contained in large molecular clouds is turned into stars. We currently lack an exact quantification, but it appears clear that during the lifetime of a molecular cloud only a very small fraction of its gas (few per cent) ends up in stars. This is referred to as a low star formation efficiency, to which a number of phenomena can contribute. As we have seen, given the high turbulent motions in GMCs, most of the gas is kept away from very dense regions where star formation can take place (Fig. 4.25). Moreover when stars start forming in abundance, the kinetic and radiative energy injected by these stars can disperse, heat and potentially destroy the molecular cloud. It is not clear whether this low efficiency in converting gas into stars is universal to all galaxies, as starburst galaxies may have higher efficiency. In a somewhat alternative scenario, GMCs are envisioned as transitory objects that keep forming and dissolve due to a complex interplay between turbulence, accretion and feedback. We do not go into the details of this theory and refer the interested reader to Mac Low and Klessen (2004).

### 8.3.9 The Initial Mass Function

The complex phenomena described in the previous sections are believed to lead to the birth of a stellar population with a characteristic number of stars as a function of their masses. The function that describes the distribution of masses of stars at birth is called the **initial mass function** (IMF). The shape of the IMF is observationally determined by observing single stars in young star clusters and inferring their masses (§4.6.1). Despite intrinsic uncertainties in the procedure, there is a general similarity in the clusters observed in our Galaxy. We cannot observe single stars in clusters in external galaxies, except nearby systems like the LMC, and thus it is presently not clear how 'universal' the IMF can be. It may, for instance, differ in starburst galaxies with respect to quieter and less dense environments like a present-day galaxy disc.

In general terms, the IMF $\phi(m)$, with $m$ the star mass in units of solar mass ($m \equiv \mathcal{M}/\mathcal{M}_\odot$), quantifies the number of stars (d$N$) with masses between $m$ and $m$ + d$m$. The first determination of the IMF was carried out by Salpeter (1955). The **Salpeter IMF** has the form of a single power law

$$\mathrm{d}N = \phi(m)\mathrm{d}m = \phi_0 m^{-2.35}\mathrm{d}m, \tag{8.85}$$

where $\phi_0$ is the normalisation. This formulation was originally determined for masses between $m$ = 0.4 and $m$ = 10. In the following we extend it to the range of stellar masses between $m=0.1$ and $m=100$.[10] For use in chemical evolution models (§8.5) the normalisation is often chosen such that

$$\int_{0.1}^{100} \phi(m) m \, \mathrm{d}m = 1, \tag{8.86}$$

which, for a Salpeter IMF, gives $\phi_0 \simeq 0.17$. However, the range in masses where the integration is performed can be different, as we see in §8.5.3. Note that $\phi(m) m \, \mathrm{d}m$ gives the fraction of mass contained between $m$ and $m + \mathrm{d}m$.

In the subsequent decades, it has been realised that the IMFs of stellar clusters in the Milky Way typically depart from the Salpeter slope at the low-mass end. Several formulations have then been proposed: here we report three among the most used. Scalo (1986) proposed an IMF with three different power laws that apply to different ranges of masses. We can write the **Scalo IMF**,[11] simply normalised at $m=1$, as

$$\phi(m) = \begin{cases} m^{-1.8} & (0.1 \le m < 1), \\ m^{-3.25} & (1 \le m < 10), \\ 0.16\, m^{-2.45} & (m \ge 10). \end{cases} \tag{8.87}$$

The so-called **Kroupa IMF** (Kroupa, 2002) is also described by three power laws:[12]

$$\phi(m) = \begin{cases} 2.0\, m^{-1.3} & (0.08 \le m < 0.5), \\ m^{-2.3} & (0.5 \le m < 1), \\ m^{-2.3(-2.7)} & (m \ge 1), \end{cases} \tag{8.88}$$

where at high masses two different possible slopes are given, although the most used in practice is the $-2.3$ slope. Note that the typical errors associated with these slopes are of $\approx 0.3$, something that should give an idea of the uncertainties involved.

Chabrier (2003) derived different IMFs for various Galactic environments. The formulation for disc stars (normalised at $m=1$) reads

$$\phi(m) = \begin{cases} 3.58 \, \dfrac{1}{m} \exp\left\{-1.050 \left[\log\left(\dfrac{m}{0.079}\right)\right]^2\right\} & (m < 1), \\ m^{-2.3} & (m \ge 1), \end{cases} \tag{8.89}$$

which is referred to as the **Chabrier IMF**.[13] Finally, the Larson IMF (eq. 9.33), that is often used for primordial Population III stars, is described in §9.5.4.

[10] The lower mass is very close to the **hydrogen burning limit** for the formation of a star of $0.08\, \mathcal{M}_\odot$. Below this mass, stars are classified as **brown dwarfs**. These latter are able to sustain the nuclear fusion of deuterium and, only the most massive, lithium. The demarcation between brown dwarfs and giant planets is set at $13\, \mathcal{M}_{\mathrm{Jupiter}} \simeq 0.0124\, \mathcal{M}_\odot$.

[11] The Scalo IMF is given for $m > 0.2$ but here we have extended it to $m = 0.1$ to match with the others.

[12] Note that Kroupa (2002) also gives a slope ($\propto m^{-0.3}$) for masses $0.01 < m < 0.08$, in the brown dwarf regime.

[13] As they are both used in practice, we also report a second version of the Chabrier IMF, given in Chabrier (2005) as

$$\phi(m) = \begin{cases} 2.24 \, \dfrac{1}{m} \exp\left\{-1.653 \left[\log\left(\dfrac{m}{0.2}\right)\right]^2\right\} & (m < 1), \\ m^{-2.35} & (m \ge 1). \end{cases} \tag{8.90}$$

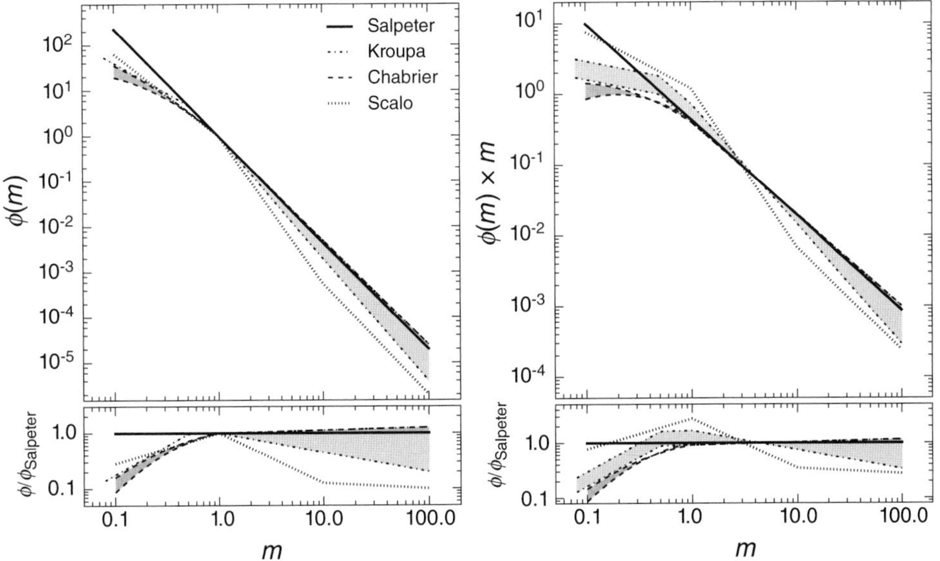


Fig. 8.11

Comparison between the four IMFs presented in the text. *Top left panel*. IMFs by number $\phi(m)$, showing the number of stars between $m$ and $m + dm$, with $m = \mathcal{M}/\mathcal{M}_\odot$, and normalised at $m = 1$. *Top right panel*. IMFs by mass showing the mass contained between $m$ and $m + dm$, normalised to the integral of the mass between $m = 1$ and $m = 100$. In both plots we show the two versions of Kroupa IMFs (light grey band) with different high-mass slopes (eq. 8.88) and of the Chabrier IMFs (dark grey band) (eq. 8.90 is the one that lies below at low masses). The *bottom panels* show the ratio between the various IMFs and the Salpeter IMF.

Compared to the other IMFs, a Salpeter IMF is called **bottom-heavy**, where 'bottom' refers to the region of low-mass stars (typically below 1 $\mathcal{M}_\odot$) and 'heavy' to the fact that the integrated mass of stars at low masses is maximised (with respect to the high masses). Conversely, an IMF that maximises the contribution of high-mass stars with respect to low-mass stars (like the Chabrier IMF compared to the Salpeter IMF for instance) is referred to as a **top-heavy** IMF.

Fig. 8.11 shows a comparison between the four IMFs described above. The left panels show the IMF *by number* as presented above, while the right panels show the IMF *by mass*. As anticipated, the general trend of all the revisions of the Salpeter IMF is to become shallower at low masses ($m < 1$). When we observe a galaxy either by resolving single stars or by measuring its total stellar luminosity, we are typically dominated by intermediate- and high-mass stars. This is particularly true for SFGs where there are young, massive ($\mathcal{M} >$ a few $\mathcal{M}_\odot$) and very bright stars. However, a large fraction of the mass lies in low-mass stars ($\mathcal{M} < 1\,\mathcal{M}_\odot$) and this fraction strongly depends on the assumed IMF. Given that the IMF of a system is not known *a priori*, assuming one or another formulation to estimate the total stellar mass introduces an unavoidable systematic uncertainty. For instance, at equal optical/IR luminosity, the stellar mass of a galaxy estimated with a Salpeter IMF is roughly 1.5–2 times higher than that obtained using a Chabrier IMF.

## 8.4 Gas Consumption and Evolution of the Interstellar Medium

With the passage of time, star formation turns the cold ISM of a galaxy progressively into stars and the material that goes into the formation of a star is removed from the gaseous phase. Part of this removal is however temporary as, in the course of their evolution, stars return a fraction of their mass back to the ISM. The amount, properties and timescales of this **gas return** depend very much on the type of star and its evolution, which, in turn, are governed by its initial mass and also by its metallicity. High-mass ($\mathcal{M} \gtrsim 8\,\mathcal{M}_\odot$) stars explode as Type II (core-collapse) SNe in timescales ($\sim 10\,\mathrm{Myr}$) which are very short compared to galaxy lifetimes. These explosions return to the ISM a significant fraction of the gas that had been removed when the star had formed. Moreover, this gas is now highly enriched in heavy chemical elements and it contributes to the pollution of the ISM (§8.5). Intermediate-mass ($2\,\mathcal{M}_\odot \lesssim \mathcal{M} < 8\,\mathcal{M}_\odot$) stars go through phases of intense mass loss. These phenomena release enriched gas into the ISM over longer timescales that go from tens of Myr to gigayears. Low-mass ($\mathcal{M} \lesssim 1\,\mathcal{M}_\odot$) stars have lifetimes comparable to or longer than the Hubble time and negligible contribution to the gas return. Finally, the explosion of Type Ia SNe returns gas and metals (iron in particular) into the ISM on timescales that can go from $\sim 100\,\mathrm{Myr}$ to several Gyr from the formation of the stellar population.

The complex interplay between stars and the surrounding ISM can be written in a relatively simple formalism. The baryonic mass ($\mathcal{M}_\mathrm{b}$) of a galaxy is the sum of the mass in stars ($\mathcal{M}_\star$) and the mass in gas ($\mathcal{M}_\mathrm{gas}$), where in $\mathcal{M}_\star$ we also include stellar remnants: white dwarfs, neutron stars and stellar black holes. We exclude the negligible mass contribution from dust (§4.2.7). The evolution in time of this baryonic mass will then simply be

$$\frac{\mathrm{d}\mathcal{M}_\mathrm{b}}{\mathrm{d}t} = \frac{\mathrm{d}(\mathcal{M}_\star + \mathcal{M}_\mathrm{gas})}{\mathrm{d}t} = \dot{\mathcal{M}}_\mathrm{acc} - \dot{\mathcal{M}}_\mathrm{out}, \tag{8.91}$$

where $\dot{\mathcal{M}}_\mathrm{acc}$ is the accretion (infall) rate of baryonic matter into the galaxy from the external environment and $\dot{\mathcal{M}}_\mathrm{out}$ is its outflow rate. If we assume that the baryonic accretion is entirely gaseous (negligible contribution from accretion of other stellar systems) and so is the outflow (§8.7.3), the evolution of the stellar mass is

$$\frac{\mathrm{d}\mathcal{M}_\star}{\mathrm{d}t} = \mathrm{SFR} - \dot{\mathcal{M}}_\mathrm{ret}, \tag{8.92}$$

where SFR is the star formation rate and $\dot{\mathcal{M}}_\mathrm{ret}$ is the rate at which gas is returned to the ISM by SN explosions and mass losses (§8.4.1). Then the equation that regulates the gas mass in the system is

$$\frac{\mathrm{d}\mathcal{M}_\mathrm{gas}}{\mathrm{d}t} = \dot{\mathcal{M}}_\mathrm{acc} - \dot{\mathcal{M}}_\mathrm{out} + \dot{\mathcal{M}}_\mathrm{ret} - \mathrm{SFR}. \tag{8.93}$$

Thus, apart from outflows and accretion, the gas mass varies because star formation removes gas and stellar evolution returns gas to the ISM.

### 8.4.1 Gas Return to the ISM from Stellar Evolution

The return term in eq. (8.93) is particularly important for chemical evolution because the returned gas is metal-enriched. To handle this term one needs to know what gas mass is returned by a star of mass $\mathcal{M}$ as a function of time. It is customary to use the approximation that a star loses mass in a short time after leaving the main sequence (see also §C.5). This is obviously a very good approximation for Type II SN explosions, but it is also acceptable for intermediate-mass stars.[14] With this prescription, we only need to know two functions of the mass of the star: the **main-sequence time** $\tau_{\rm MS}(m)$, which is roughly the lifetime of the star, and the mass of the **stellar remnant** $m_{\rm rem}(m)$, where, as above, $m = \mathcal{M}/\mathcal{M}_\odot$.

The main-sequence time is given by stellar evolution theory and there are a number of parameterisations involving broken power laws. A simple one is

$$\tau_{\rm MS}(m) = \begin{cases} A\, m^{-\alpha}\ {\rm Gyr} & m \lesssim 10, \\ B\, m^{-\beta}\ {\rm Gyr} & m > 10, \end{cases} \tag{8.94}$$

where the constants are in the ranges $A \approx 10$–$12$, $B \approx 0.11$–$0.12$, $\alpha \approx 2.5$–$2.8$ and $\beta \approx 0.75$–$0.86$. The stellar remnant mass can be assumed to have a linear dependence on the initial stellar mass going from $m_{\rm rem} \approx 0.7$–$0.8$ for a star of $3\,\mathcal{M}_\odot$ to $m_{\rm rem} \approx \mathcal{M}_{\rm Chan}/\mathcal{M}_\odot$ for a $8\,\mathcal{M}_\odot$ star, where the **Chandrasekhar mass** $\mathcal{M}_{\rm Chan} \simeq 1.39\,\mathcal{M}_\odot$ is the highest mass of a stable white dwarf star (stellar remnants with masses higher than $\mathcal{M}_{\rm Chan}$ become neutron stars or black holes). Note that we are making the simplification that the mass return to the ISM only depends on mass. In reality, there is also a dependence on metallicity, as more metal-rich stars tend to have more efficient winds and thus more mass losses (see also §8.5.1). For star masses between $8\,\mathcal{M}_\odot$ and $25\,\mathcal{M}_\odot$, the remnant is usually taken to be of the order of $\mathcal{M}_{\rm Chan}$. Stars with higher initial masses ($\mathcal{M} > 25\,\mathcal{M}_\odot$) likely generate black holes; in this case the remnant mass is assumed to have a linear dependence starting from $\mathcal{M}_{\rm Chan}$ at $\mathcal{M} \approx 25\,\mathcal{M}_\odot$ and reaching up some tens of $\mathcal{M}_\odot$ at $\mathcal{M} \approx 100\,\mathcal{M}_\odot$. This high-mass regime is rather unconstrained but, given the slope of the stellar IMF (§8.3.9), we expect a relatively minor contribution to the total returned mass from these stars.

Using the above prescriptions we can write the **mass return rate** to the gas phase by stellar evolution for a generic stellar population at a given time $t$ as

$$\dot{\mathcal{M}}_{\rm ret}(t) = \int_{m_{\rm min}(t)}^{m_{\rm max}} {\rm SFR}\,[t - \tau_{\rm MS}(m)]\,\phi(m)\,[m - m_{\rm rem}(m)]\,{\rm d}m, \tag{8.95}$$

where the integral upper bound is the most massive star that can form (typically $m_{\rm max} \sim 100$), whereas, at any given $t$, the lower bound $m_{\rm min}(t)$ is such that $\tau_{\rm MS}(m_{\rm min}) = t$. The term $\phi(m)$ in eq. (8.95) is the IMF (§8.3.9). As an example, let us suppose that we are considering a simple stellar population born with a single burst (see also §8.6.2). In this case, SFR($t$) is a Dirac function at a certain time in the past that we can take as $t = 0$. The

14 The main mass-loss phases of stellar evolution occur when a star crosses the horizontal branch and moves to the asymptotic giant branch (§C.5). Stars that most contribute to mass loss have masses between $3\,\mathcal{M}_\odot$ and $8\,\mathcal{M}_\odot$. The time that a star of $\mathcal{M} \gtrsim 3\,\mathcal{M}_\odot$ spends in going through the red giant, horizontal and asymptotic giant branches is $\lesssim 100$ Myr; thus it can be considered almost instantaneous if we are interested in the global galactic evolution.

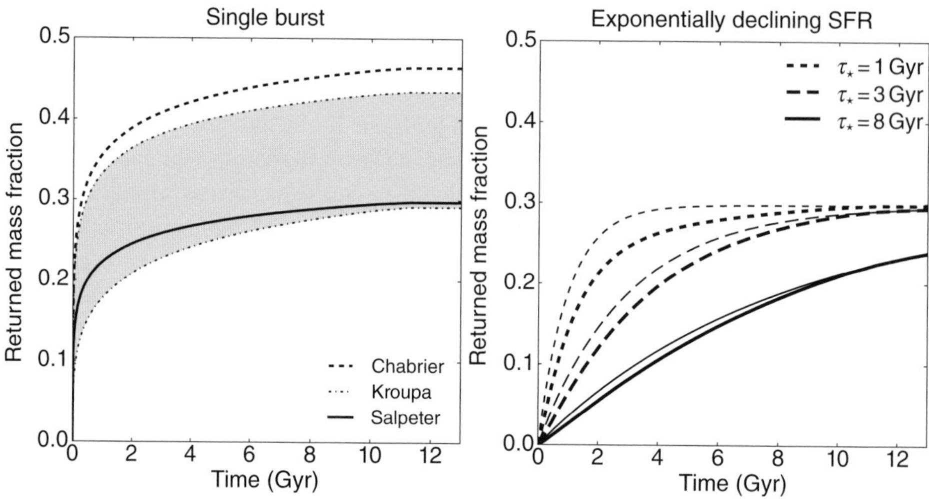


Fig. 8.12 Cumulative fraction of mass initially locked in stars that is returned to the ISM in the course of stellar evolution. *Left panel*. In this case the stellar population is born in a single burst at $t = 0$ for three IMFs (§8.3.9): Salpeter, Kroupa (grey band) and Chabrier (eq. 8.89). *Right panel*. This is the situation arising with an exponentially declining SFH (eq. 8.96) with three different star formation timescales, assuming a Salpeter IMF. The thin curves show models employing the instantaneous recycling approximation (§8.5.3).

stellar mass return at time $t$ will be contributed by all the stars that have had enough time to leave the main sequence, so all stars with masses such that $\tau_{\rm MS}(m) < t$. In particular if $t$ is larger than a few tens of million years, these will include all stars that exploded as Type II SNe.

Fig. 8.12 (left) shows the cumulative fraction of mass returned to the ISM by such a stellar population characterised by a single burst that occurred at $t = 0$. The curves displayed here are obtained by integrating $\dot{\mathcal{M}}_{\rm ret}(t)$ in eq. (8.95) between $t = 0$ and $t$. Independently of the IMF, the curves show a rapid increase in the first few tens of Myr, due to the explosion of Type II SNe. These contribute about one-third to one-half of the total mass return. The following phase is much shallower and reaches the asymptotic values (at $t = 10\,$Gyr) of $\approx 0.3$ for a Salpeter IMF and 0.46 for a Chabrier IMF. Note that using high-mass slopes of $-2.3$ or $-2.7$ for the Kroupa IMF (eq. 8.88) makes a big difference for the asymptotic return fraction (respectively 0.43 and 0.29). Note that all these curves do not include Type Ia SNe that return a large amount of metals but relatively little mass.

We can consider the case of a more continuous SFR that, for a galaxy, is much more realistic than a single burst. We use the commonly employed **exponentially declining SFH**, for which the SFR as a function of time is

$$\mathrm{SFR}(t) = \frac{\mathcal{M}_{\star,\infty}}{\tau_\star} \exp\left(-\frac{t}{\tau_\star}\right), \tag{8.96}$$

where $\tau_\star$ is the **star formation timescale** and $\mathcal{M}_{\star,\infty}$ is the stellar mass produced at $t = \infty$. The right panel of Fig. 8.12 shows the fraction of the total stellar mass (normalised at

$\mathcal{M}_{\star,\infty}$) returned to the ISM as a function of time for three different star formation timescales using a Salpeter IMF. The difference with respect to the single burst is apparent. The case for $\tau_\star = 1$ Gyr resembles the single burst because most of the stars do form in a relatively short time. However, the longer the timescale of the star formation, the lower the fraction of gas returned to the ISM. To see why this occurs, take, for instance, an intermediate time, say $t = 5$ Gyr. At this time about half of the stars of the system with $\tau_\star = 8$ Gyr have not yet been born, but also among those born there is a percentage of stars that are too young to have substantially contributed to the gas return. Exponential timescales of $\tau_\star \lesssim 1$ Gyr are considered typical of ETGs, while the case of $\tau_\star \approx 8$ Gyr well describes present-day SFGs (§4.1.5).

### 8.4.2 Structural Evolution of a Galaxy Disc

The equations presented above (in particular eqs. 8.92 and 8.93) can be used to follow the evolution of stars and gas in any type of galaxy, including elliptical and dwarf galaxies. In the case of disc galaxies, however, they can be conveniently rewritten in terms of surface densities $\Sigma$. For this purpose, it is customary to divide a galaxy disc into concentric and coplanar annuli of radius $R$. The analogue of eq. (8.93) is then

$$\frac{\partial \Sigma_{\rm gas}}{\partial t} = \dot{\Sigma}_{\rm eff} - \dot{\Sigma}_{\rm out} + \dot{\Sigma}_{\rm ret} - \Sigma_{\rm SFR}, \tag{8.97}$$

where each term depends on $R$ and $t$, and they are analogous to those in eq. (8.93) except for $\dot{\Sigma}_{\rm eff}$ that we describe below. The stellar component (see eq. 8.92) evolves as

$$\frac{\partial \Sigma_\star}{\partial t} = \Sigma_{\rm SFR} - \dot{\Sigma}_{\rm ret}, \tag{8.98}$$

assuming that the disc does not appreciably acquire/lose stars from/to the environment.

The first term on the r.h.s. of eq. (8.97) is the **effective accretion rate** of material into the annulus at radius $R$. This can be decomposed into two terms, one representing the *actual* accretion of gas from the environment ($\dot{\Sigma}_{\rm acc}$) and the second the flow of gas from the adjacent annuli within the disc,

$$\dot{\Sigma}_{\rm eff} \equiv \dot{\Sigma}_{\rm acc} - \frac{1}{2\pi R}\frac{\partial \mu}{\partial R}, \tag{8.99}$$

where

$$\mu \equiv 2\pi R \Sigma_{\rm gas} u_R \tag{8.100}$$

is the **gas mass flux** along $R$, with $u_R$ being the net radial velocity of the gas across the disc. Note that, at any given radius $R$, the second term on the r.h.s. of eq. (8.99) can be positive or negative depending on whether there is more material flowing into or out of the annulus. In the absence of radial motions ($u_R = 0$), it would be null (approximation of independent annuli). However, this is unlikely to be the case in most situations for the following reason. If some gas accretes onto the disc at a radius $R$, it will mix with the local disc material that has a certain specific angular momentum $j_0(R) \equiv R v_{\varphi,0}(R) \simeq R v_{\rm c}(R)$, with $v_{\varphi,0}$ its rotation velocity and $v_{\rm c}$ the circular speed (eq. 4.28); the last equality comes from the fact that the gas is kinematically cold (§4.3.4). The accreting gas is however unlikely to have *exactly* the same $v_\varphi$ as the disc (§10.7.2) and the resulting (after mixing) average gas velocity at $R$ will be $v_{\varphi,1}(R) \neq v_c(R)$ with specific angular momentum $j_1(R) \neq j_0(R)$. The gas must then move radially to find a new centrifugal equilibrium (conserving $j$). The most realistic case is that the accreting gas will rotate locally at $v_{\varphi,\mathrm{acc}}(R) < v_c(R)$. This is certainly the case if the accretion comes from the galactic corona (§10.7.2). In this situation, we will have $j_1(R) < j_0(R)$ and thus an inward flow.

In looking for a solution of eq. (8.97) we note that for galaxies like the present-day Milky Way, without evidence of powerful mass ejection, the term $\dot{\Sigma}_{\mathrm{out}}$ is often ignored.[15] Alternatively, one can assume a proportionality between the outflow rate and the SFR. The ratio between these two rates is called the mass loading factor and we further discuss it in §8.7.4. $\dot{\Sigma}_{\mathrm{ret}}$ is given by the analogue of eq. (8.95) with the SFR replaced by $\Sigma_{\mathrm{SFR}}$, which, in turn, is linked to the gas density given a star formation law (§4.2.9). In the absence of radial flows, eq. (8.97) can then be solved to obtain the accretion rate once a shape for the SFH has been assumed, for instance eq. (8.96). A solution of this kind is shown in §10.7 (Fig. 10.11). If radial flows are present, other constraints are needed.

A treatment like the one just described has a few caveats. First, it neglects the importance of mergers as a contribution to the stellar mass of a galaxy disc (but note that the gas acquired in mergers can be incorporated in the accretion term $\dot{\Sigma}_{\mathrm{acc}}$). This is however a reasonably good approximation for disc galaxies at least at $z \lesssim 1$ (§10.7.1). Second, only the radial flow of gas is taken into account, while stars can also move to different radii with respect to their formation radius (radial stellar migration; §10.7.5). In general, adding refinements to the model requires more constraints and these can be provided by the detailed chemical abundances of gas and stars using a **chemical evolution model**, as we see in the next section.

## 8.5 Chemical Evolution

The ISM of any galaxy is rich in elements that have not been formed by the primordial nucleosynthesis (§2.5) but are the result of stellar evolution and the return of gas to the ISM, discussed in §8.4.1. The release of these elements tends to produce a general **enrichment** of the ISM (i.e. to increase its metallicity; §4.1.4) and of the subsequent generations of stars. This effect is counteracted by the fact that galaxies also accrete gas from the environment (§4.2.10). Given that this gas has been less polluted by stellar evolution, it is *poor* in metals and tends to produce a **dilution** of the ISM metallicity. Solving the competition between these two processes is at the core of chemical evolution models, of which we give a brief outline in this section. For more details the reader is referred to Pagel (2009) and Matteucci (2012).

[15] The galactic fountain (§4.2.10) does not remove the gas permanently and the typical recycling timescale for the ejected gas to fall back to the disc is expected to be short ($\lesssim 10^8$ yr). Powerful galactic fountains may, however, have important long-term effects in the redistribution of angular momentum (§10.1.4).

### 8.5.1 Stellar Yields

In the course of their evolution, stars release into the ISM a significant amount of chemical elements. These elements either were part of the gas *clouds* from which the stars originally formed or have been synthesised in their interiors. The elements ejected from stars are generically called **stellar yields**. We indicate with $Y_i$ the fraction of the initial mass of the star that gets ejected as a consequence of stellar evolution in the form of the element (or the isotope of an element) $i$. This is a function of the stellar mass and is composed of two terms:

$$Y_i(m) = \frac{m - m_{\rm rem}(m)}{m} Z_{i,0} + y_i(m), \tag{8.101}$$

where $m = \mathcal{M}/\mathcal{M}_\odot$, $m_{\rm rem}$ is the mass of the stellar remnant in units of solar mass (§8.4.1), $Z_{i,0}$ is the mass fraction of element/isotope $i$ that was present at the formation of the star, and $y_i$ is the **net yield**.[16] For an element $i$ that does not get destroyed in stellar interiors, stellar evolution returns to the ISM at least all the mass that does not end up in the stellar remnant. The net yield is *positive* ($y_i > 0$) if the nuclear reactions in the star produce some amount of newly synthesised element $i$. Elements that get destroyed in stellar interiors (such as deuterium) have instead *negative* net yields.

The quantities $y_i(m)$ and $Y_i(m)$ are, in principle, known for any element $i$ and star of mass $m$ from the theory of stellar evolution (Tab. 8.1 and §C.5). However, there are a number of unsolved issues and these theoretical stellar yields should be considered uncertain by a factor of about 2 on average. The yields coming from the theory of stellar evolution are referred to as **true yields**.

In general, the yields will depend not only on the mass of the star but also on its metallicity, whose inclusion complicates the treatment. However, for a number of elements/isotopes of different masses up to iron (for instance $^{12}$C, $^{16}$O, $^{20}$Ne and $^{56}$Fe) the metallicity dependence is not too strong. These are called **primary elements** as they are characterised by chain productions that start essentially from hydrogen and helium. Elements/isotopes whose yields depend on non-primordial composition of the material are called **secondary elements**, (for instance $^{13}$C, $^{14}$N and $^{18}$O).

This situation is, in fact, more complicated for some of these elements when looked at in more detail. Important contributors to the release of metals in the ISM are intermediate-mass stars ($2\,\mathcal{M}_\odot < \mathcal{M} < 8\,\mathcal{M}_\odot$) that experience intense mass loss during the post-main-sequence evolution, in particular the asymptotic giant branch (AGB) phase (§C.5). AGB stars release a lot of carbon, but the precise amount and its isotopic composition depend on internal processes in the stars (mixing between layers) that, in turn, depend on metallicity. As a consequence, for instance, low-metallicity AGB stars release more $^{12}$C than those at solar metallicity and so the return of $^{12}$C, despite being a primary element, can depend on metallicity. Conversely, $^{14}$N and $^{13}$C, which are technically secondary elements, become essentially primary ones in low-metallicity stars. In Tab. 8.1 we give a summary of the most abundant metals released by stars in different mass ranges.

[16] Note that in some textbooks and articles the net yield is called simply yield.

**Table 8.1** Stellar yields

| Initial mass ($\mathcal{M}_{\rm ini}$) ($\mathcal{M}_\odot$) | Ejection phase | Species | $y_i \mathcal{M}_{\rm ini}$ ($10^{-3}\,\mathcal{M}_\odot$) |
|---|---|---|---|
| 11–40[a] | SN II | $^{16}$O | 53–5720 |
| | | $^{20}$Ne | 31–1240 |
| | | $^{28}$Si | 17–345 |
| | | $^{12}$C | 24–259 |
| | | $^{24}$Mg | $\leq 235$ |
| | | $^{32}$S | $\leq 159$ |
| | | $^{56}$Fe | 11–26 |
| | | $^{40}$Ca | $\leq 10$ |
| 7–8 | AGB | $^{14}$N | 68–88 |
| | | $^{13}$C | $\sim 1$ |
| 4–6[b] | AGB | $^{14}$N | 3–52 |
| | | $^{12}$C | $< 19$ |
| 2.5–4 | AGB | $^{12}$C | 4–20 |
| | | $^{14}$N | 0.5–7 |
| $\lesssim 1.4$[c] | SN Ia | $^{56}$Fe | 610 |
| | | $^{28}$Si | 160 |
| | | $^{54}$Fe | 140 |
| | | $^{24}$Mg | 90 |
| | | $^{32}$S | 80 |
| | | $^{58}$Ni | 60 |

In this table we give the mass of the main metals released by solar-metallicity stars.

[a] Values from Woosley and Weaver (1995). Pre-SN mass losses are not taken into account. Ranges indicate the minimum and maximum production from single stars in the range 11–40 $\mathcal{M}_\odot$.

[b] Values for intermediate-mass stars ($\mathcal{M} = 2.5$–$8\,\mathcal{M}_\odot$) are maximum variation ranges mostly taken from van den Hoek and Groenewegen (1997) and Karakas (2010).

[c] Values from Pagel (2009) estimated from a CO white dwarf of 1 $\mathcal{M}_\odot$ accreting material from a red giant companion. In this case, most of the metals are produced in the explosion of the star.

This table is highly simplified given the complexity mentioned above, but it is meant to give a general indication of the main yields of the different types of stars.

Finally, we briefly mention the production of elements heavier than iron synthesised with the so-called neutron capture mechanisms. These are divided into **s-processes** in which neutrons are added *slowly* to nuclei that thus have time to become stable with respect to $\beta$-decay and **r-processes** where instead neutrons are added *rapidly* to neutron-rich unstable nuclei. The former type takes place in AGB stars (§C.5) while the latter, most likely, in

SN explosions and in the merger of compact objects like neutron stars. Typical elements produced mainly with s-processes are Ba and Y, while a prototypical r-process element is Eu. Most heavy elements are formed by both s- and r-processes. All these heavy elements are quite rare with respect to the lighter ones but they are easily detectable in stellar spectra, which makes them unique probes of stellar and galactic chemical evolution.

In the end, to accurately treat all the above elements/isotopes in chemical evolution models one would need to include metallicity-dependent yields. However, in dealing with only primary elements we can neglect, to a first approximation, the $Z$ dependence. We adopt this approach in the following sections.

## 8.5.2 Chemical Evolution Models

Consider a generic stable (non-radioactive) element $i$ in the ISM that constitutes a fraction $Z_i$ of the total gas mass ($\mathcal{M}_{\rm gas}$). The equation that governs the evolution of its mass ($Z_i \mathcal{M}_{\rm gas}$) derives directly from eq. (8.93) and reads

$$\frac{\mathrm{d}(Z_i \mathcal{M}_{\rm gas})}{\mathrm{d}t} = Z_{i,\rm acc}\dot{\mathcal{M}}_{\rm acc} - Z_{i,\rm out}\dot{\mathcal{M}}_{\rm out} + \dot{\mathcal{M}}_{i,\rm ret} - Z_i \mathrm{SFR}, \tag{8.102}$$

where we have considered different abundances for the element in the accreting and outflowing material. It is indeed highly probable that gas accretion occurs from a medium that has not been polluted as much as the ISM. On the contrary, the outflow, driven by powerful winds (§8.7.3), may occur in star-forming regions where the gas tends to be more enriched. The third term on the r.h.s. of eq. (8.102) is the return rate of element $i$ by stellar evolution, which can be written, analogously to eq. (8.95), as

$$\dot{\mathcal{M}}_{i,\rm ret}(t) = \int_{m_{\rm min}(t)}^{m_{\rm max}} \{[m - m_{\rm rem}(m)]\, Z_i[t - \tau_{\rm MS}(m)] + m y_i(m)\} \times \mathrm{SFR}\,[t - \tau_{\rm MS}(m)]\, \phi(m)\mathrm{d}m, \tag{8.103}$$

where the first part is the yield, similar to eq. (8.101), but now with the time dependence of the abundance $Z_i$. Note that, unlike in §8.4.1, where we were interested in the bulk of the gas mass and we could neglect Type Ia SNe, it is now important to include them if we aim to follow the evolution of metals typically produced by them, Fe in particular (Tab. 8.1). Their return timescales however do not depend on $\tau_{\rm MS}$, but on the fraction of binaries, their orbital parameters and evolution. These parameters are not fully under control and are usually encapsulated into a function called the **delay-time distribution**.

In analogy with what was done in §8.4.2, one can write eq. (8.102) for a galaxy disc by expressing the quantities as surface density rates. These equations can then be solved in a way similar to that described in §8.4. The added complication and uncertainties that come from the stellar yields are compensated by the large number of constraints that are furnished by the observations of metallicities of stars and gas. Note that all the above equations can be written for a specific element $Z_i$, for example oxygen, or, by summing over all elements, we can obtain the total mass fraction in metals (metallicity), which we indicate with $Z$ (§C.6.1). The Sun has $Z_\odot \approx 0.013$–$0.014$, while the values of $Z_i$ for the specific elements are one or more orders of magnitude smaller (Tab. C.5).

### 8.5.3 Instantaneous Recycling Approximation

The **instantaneous recycling approximation** (IRA) is a simplification of the chemical evolution equations obtained by assuming that the return of metals into the ISM takes place *instantaneously* at the moment of star formation.[17] Under this approximation, the main-sequence time is $\tau_{\rm MS}=0$ for every star. This leads to a great simplification of the equations and to the possibility of building analytic models of chemical evolution. The justification of the IRA is that a large amount of gas return occurs typically on timescales that are much shorter than the characteristic times of galaxy evolution. We can see this by looking at Fig. 8.12. In particular, the return due to Type II SNe is almost instantaneous (sharp rise of the curves) and thus, as long as we deal with chemical elements typically released by them ($\alpha$-elements; Tab. 8.1), the approximation is very good. Elements like N, largely produced by AGB stars (timescales of a Gyr or more), and Fe, mostly produced by Type Ia SNe (timescales from hundreds of Myr to Gyr), are less accurately accounted for by models employing the IRA.

Under the IRA, the return term in eq. (8.95) is simply

$$\dot{\mathcal{M}}_{\rm ret}(t)=\mathcal{R}\,{\rm SFR}(t), \tag{8.104}$$

where

$$\mathcal{R}=\int_{m_{\rm min}}^{m_{\rm max}}\phi(m)\,[m-m_{\rm rem}(m)]\,{\rm d}m \tag{8.105}$$

is the **return fraction** of gas from the evolution of the entire stellar population, now happening instantaneously when the population forms. The return fraction can vary between 0.2 and 0.5 depending on the IMF and other assumptions. In Fig. 8.12 it is the asymptotic value to which the curves tend for $t\to\infty$. In the IRA, the evolution of the gas mass (eq. 8.93) simplifies to

$$\frac{{\rm d}\mathcal{M}_{\rm gas}}{{\rm d}t}=\dot{\mathcal{M}}_{\rm acc}-\dot{\mathcal{M}}_{\rm out}-(1-\mathcal{R}){\rm SFR}, \tag{8.106}$$

which can be rewritten, considering the evolution of the stellar mass,

$$\frac{{\rm d}\mathcal{M}_\star}{{\rm d}t}=(1-\mathcal{R}){\rm SFR}, \tag{8.107}$$

as

$$\frac{{\rm d}\mathcal{M}_{\rm gas}}{{\rm d}\mathcal{M}_\star}=\frac{\dot{\mathcal{M}}_{\rm acc}-\dot{\mathcal{M}}_{\rm out}}{(1-\mathcal{R}){\rm SFR}}-1. \tag{8.108}$$

Let us now add the chemistry by considering the modification introduced by the IRA to eq. (8.103). Given that $\tau_{\rm MS}=0$ for every star, SFR($t$) and $Z_i(t)$ can be taken out of the integral. Moreover, the integration in mass will start now from a minimum stellar mass that

[17] Note that, in the case of the IRA, as for delayed return (§8.4.1), one also makes the implicit assumption of **instantaneous mixing**. This implies that, once released into the ISM, the metals spread instantaneously and contribute to increase the local *average* metallicity. This assumption is justified by the rapid evolution of SNRs (§8.7.1) with respect to the typical timescales of chemical evolution.

can be taken as $m = 1$ given that stars with lower masses contribute very little to the mass return (§8.4.1). The return rate of element $i$ then becomes

$$\dot{\mathcal{M}}_{i,\mathrm{ret}}(t) = \mathrm{SFR}(t)\left[\mathcal{R}Z_i(t) + \int_1^{m_{\max}} y_i(m)\,\phi(m)\,m\,\mathrm{d}m\right]. \tag{8.109}$$

With this we can rewrite eq. (8.102) as

$$\frac{\mathrm{d}(Z_i\mathcal{M}_{\mathrm{gas}})}{\mathrm{d}t} = Z_{i,\mathrm{acc}}\dot{\mathcal{M}}_{\mathrm{acc}} - Z_{i,\mathrm{out}}\dot{\mathcal{M}}_{\mathrm{out}} + (1-\mathcal{R})(p_i - Z_i)\mathrm{SFR}, \tag{8.110}$$

where we have defined

$$p_i \equiv \frac{1}{1-\mathcal{R}}\int_1^{m_{\max}} y_i(m)\,\phi(m)\,m\,\mathrm{d}m \tag{8.111}$$

as the **net stellar yield of the entire population**. For elements that are not destroyed in stellar interiors, the stellar yield $p_i$ is the mass of the element $i$ that is produced and released into the ISM by a stellar population characterised by a certain IMF $\phi(m)$. This mass is given in units of the mass that *remains* in stellar objects (low-mass stars and stellar remnants) at the end of the stellar evolution: division by $(1-\mathcal{R})$. We remind the reader that, in chemical evolution models, the integral of $\phi(m)m\,\mathrm{d}m$ over the considered mass range is normalised to one (eq. 8.86). For example, $p_i$ calculated for oxygen and a Scalo IMF is about 0.006, so if we consider a star cluster that had a single burst of star formation a long time ago and has now a stellar mass of $10^5\,\mathcal{M}_\odot$, we can estimate that it has released $\approx 600\,\mathcal{M}_\odot$ of oxygen. In practical applications, given the uncertainties on the true stellar yields, the abundance of element $i$ is often normalised to the yield by dividing it by $p_i$.

We can further divide all terms of eq. (8.110) by the derivative in time of the stellar mass and use eq. (8.107) to obtain

$$\frac{\mathrm{d}(Z_i\mathcal{M}_{\mathrm{gas}})}{\mathrm{d}\mathcal{M}_\star} = \frac{Z_{i,\mathrm{acc}}\dot{\mathcal{M}}_{\mathrm{acc}}}{(1-\mathcal{R})\mathrm{SFR}} - \frac{Z_{i,\mathrm{out}}\dot{\mathcal{M}}_{\mathrm{out}}}{(1-\mathcal{R})\mathrm{SFR}} + p_i - Z_i, \tag{8.112}$$

where the first two terms on the r.h.s. describe accretion and outflow of element $i$ from and to the external medium, while the last two terms are the mass fractions that get released ($p_i$) and locked by the stars ($-Z_i$). Eq. (8.112) can be further simplified if we assume that the outflow is homogeneous with the ISM ($Z_{i,\mathrm{out}} = Z_i$) and we develop the derivative on the left-hand side (l.h.s.) using eq. (8.108) to finally obtain

$$\mathcal{M}_{\mathrm{gas}}\frac{\mathrm{d}Z_i}{\mathrm{d}\mathcal{M}_\star} = p_i - (Z_i - Z_{i,\mathrm{acc}})\frac{\dot{\mathcal{M}}_{\mathrm{acc}}}{(1-\mathcal{R})\mathrm{SFR}}, \tag{8.113}$$

independent of the mass outflow. Eq. (8.113) shows how the abundance of element $i$ (or the metallicity of the ISM if we sum over all elements) increases if $p_i > 0$ and decreases if $Z_i - Z_{i,\mathrm{acc}} > 0$. The latter is likely the case, as the abundance of the accreting gas is lower than the average ISM abundance $Z_i$. Thus we see, in one equation, the competition between enrichment ($p_i$) and dilution (second term on the r.h.s.) that we mentioned at the very beginning of this section.

### 8.5.4 Closed-Box Model

We conclude this section on chemical evolution models by considering their most extreme simplication: the **closed-box model**. In this case, the system is considered to be initially only gaseous, $\mathcal{M}_{\rm gas}(0)=\mathcal{M}_{\rm gas,0}$ and $\mathcal{M}_\star(0)=0$, with gas at initial zero metallicity, $Z_i(0)=0$, for every non-primordial element $i$. Most importantly, the system does not exchange material with the environment and thus its total baryonic mass ($\mathcal{M}_{\rm b}=\mathcal{M}_{\rm gas}+\mathcal{M}_\star$) does not change with time. We also assume the validity of the IRA (§8.5.3). Under these assumptions, eq. (8.113) greatly simplifies to

$$\mathcal{M}_{\rm gas}\frac{{\rm d}Z_i}{{\rm d}\mathcal{M}_\star}=-\mathcal{M}_{\rm gas}\frac{{\rm d}Z_i}{{\rm d}\mathcal{M}_{\rm gas}}=p_i, \tag{8.114}$$

where the first equality derives from the conservation of the total mass (${\rm d}\mathcal{M}_{\rm gas}=-{\rm d}\mathcal{M}_\star$). Eq. (8.114) is readily integrated with the boundary conditions stated above leading to

$$Z_i(t)=p_i\ln\left[\frac{\mathcal{M}_{\rm b}}{\mathcal{M}_{\rm gas}(t)}\right]=p_i\ln\left[\frac{1}{f_{\rm gas}(t)}\right], \tag{8.115}$$

where $\mathcal{M}_{\rm b}=\mathcal{M}_{\rm gas,0}$ and $f_{\rm gas}$ is the gas fraction (eq. 4.12).

Eq. (8.115) states that, in the closed-box model, the metallicity increases with time as the natural logarithm of the inverse of the gas fraction, which obviously decreases. The yield calculated using eq. (8.115) is called the **effective yield** as opposed to the true yield: the theoretical value coming from stellar evolution models (§8.5.1). Note that the abundance of the element $i$, $Z_i(t)$, should be regarded as that of the gas or, equivalently, of the population of stars that form at time $t$. In a galaxy we may be interested in the average abundance of the stars, which can be written as

$$\begin{aligned}\langle Z_i\rangle(t)&=\frac{1}{\mathcal{M}_\star(t)}\int_0^{\mathcal{M}_\star(t)}Z_i(\mathcal{M}'_\star){\rm d}\mathcal{M}'_\star\\&=p_i\left(1+\frac{f_{\rm gas}(t)\ln f_{\rm gas}(t)}{1-f_{\rm gas}(t)}\right)\xrightarrow[t\to\infty]{}p_i,\end{aligned} \tag{8.116}$$

where the second equality is obtained by substituting $f_{\rm gas}=1-\mathcal{M}_\star/\mathcal{M}_{\rm b}$ in eq. (8.115) and the limit is justified by the fact that the gas fraction decreases with time. Eq. (8.116) shows that, in a closed box, the average abundance of an element $i$ in stars approaches the net stellar yield for that element. In general, galaxies do not evolve as closed boxes, as they experience gas accretion and gas outflows. In §10.7.5 we see how the application of this simple model fails to reproduce the metallicity distribution of stars in our solar neighbourhood.

## 8.6 Theoretical Spectra of Evolving Galaxies

During their evolution, galaxies emit radiation that originates from different sources such as stars, gas and dust (§3.2). Hence, an essential ingredient of galaxy formation models is the calculation of spectra emerging from different galaxy types as a function of cosmic time. These theoretical spectra are crucial for two main reasons. First, they allow us to assign an SED/spectrum to each galaxy simulated within the numerical models of galaxy formation described in §10.11. Moreover, they provide us with the possibility to perform a systematic comparison of the SED/spectra with the observed properties of real galaxies as a function of redshift (e.g. luminosity, continuum shape, colours, line emission and absorption) in order to constrain the physical processes of the baryonic matter. Based on the ingredients contributing to galaxy SEDs (§3.3), the simulated spectra must include the stellar component resulting from the conversion of gas into stars (§8.3), and the ISM with its main phases: hot, warm and cold gas, and dust grains (§4.2; §8.4). An AGN component can also be included in the models in order to assess the effects of such a high-energy source on the emerging spectrum of the host galaxy. This section illustrates the theoretical basis of modelling galaxy spectra, focusing mostly on their stellar component and briefly describing the models which include also the evolution of the ISM.

### 8.6.1 Simulating Galaxy Stellar Spectra

The individual stars in galaxies can be resolved only within a distance of a few Mpc from our Galaxy (§6.3). This means that the light that we receive from more distant galaxies is the integral of the radiation emitted by all (unresolved) stars. In this case, since stars cannot be observed individually, it is not possible to place them in the Hertzsprung–Russell diagram (§C.5) to infer their properties and evolution. It is therefore necessary to develop a different approach for extracting meaningful information on the stellar populations from their integrated light (§3.3). This approach makes use of the so-called **stellar population synthesis** (SPS) models which predict theoretical (often called **synthetic**) spectra of stellar populations as a function of their properties and time.

#### Simple Stellar Populations

The basic unit of SPS models is the **simple stellar population** (SSP), which is an ideal case of an ensemble of coeval stars all born in an instantaneous burst of star formation and with the same metallicity. The monochromatic flux of an SSP with metallicity $Z$ at the time $t$ elapsed since $t=0$ when all stars are born is

$$F_{\lambda,\mathrm{SSP}}(t,Z)=\int_{m_1}^{m_2} F_{\lambda,\mathrm{star}}(m,t,Z)\phi(m)\mathrm{d}m, \tag{8.117}$$

where $m_1$ and $m_2$ are the masses of the lowest- and highest-mass star in the SSP (in units of $\mathcal{M}_\odot$), $F_{\lambda,\mathrm{star}}(m,t,Z)$ is the monochromatic flux of a single star with mass $m$, age $t$ and metallicity $Z$, and $\phi(m)$ is the IMF (§8.3.9). The adopted lowest and highest masses are usually the hydrogen burning limit ($m_1=0.08$ or $m_1=0.1$) and $m_2=100$, respectively. For a given IMF and metallicity, the aim of SPS models is to provide theoretical spectra of SSPs as a function of the age of the stellar population. The calculation of these spectra with eq. (8.117) requires the adoption of the following main ingredients (see Conroy, 2013, for a detailed review).

1. *Isochrones*. Isochrones are the curves describing, for a given metallicity, the luminosity ($L$) and effective temperature ($T_{\rm eff}$) of stars with a fixed age on the Hertzsprung–Russell diagram (§C.5). A set of isochrones allows us to derive the evolutionary tracks that are needed to describe the evolution of a stellar population. However, the computation of isochrones, even when done with state-of-the-art theoretical models of stellar evolution, is affected by several uncertainties, such as the effects of convection, rotation and mass loss. Several isochrones are available in the literature to fully sample the parameter space of stellar evolution (ages, metallicities, evolutionary phases).
2. *Library of stellar spectra.* The goal of SPS models is to produce synthetic spectra of stellar populations. Hence, it is necessary to have a library of stellar spectra that allows us to convert the quantities estimated with stellar evolution calculations (e.g. $T_{\rm eff}$ and surface gravity) into spectra as a function of metallicity. If the library is constructed using observed stellar spectra only,[18] a number of limitations are inevitably present. The main reason is that some spectral regions (e.g. UV at $\lambda < 3200$ Å and large fractions of the NIR from $\lambda \approx 1$ $\mu$m to $\lambda \approx 4$ $\mu$m) are not accessible with ground-based spectroscopy due to the opacity of the Earth's atmosphere. Moreover, stellar spectra obtained with different spectrographs can have heterogeneous calibrations, signal-to-noise ratios (§11.2) and spectral resolutions (§3.4). Finally, it is very challenging to obtain observed spectra that fully cover the entire parameter space of stellar physical properties. To circumvent these limitations, the gaps present in the libraries based on observed spectra are often filled with theoretical spectra calculated using models of stellar atmospheres. Some libraries are based entirely on theoretical stellar spectra.
3. *Initial mass function.* Since the initial distribution of newborn stars along the main sequence is not known beyond the Milky Way galaxy, it is always necessary to adopt an IMF (§8.3.9) and a range of stellar masses. It is usually assumed that the IMF is constant with time.

Once the above building blocks are defined, eq. (8.117) allows us to calculate synthetic spectra. Fig. 8.13 shows a few examples of SSP spectra for a wide range of ages. It is important to recall that each of the ingredients listed above is affected also by other uncertainties in addition to those already mentioned. For instance, despite the fact that binary stars are known to be common in galaxies, a full understanding of their contribution to galaxy spectra is rather incomplete, especially for the close binary systems. As a consequence, not all the SPS models include the effects of binary star evolution. Another source of uncertainty is the role of thermally pulsating AGB (TP-AGB) stars. This evolutionary phase concerns stars with typical masses of 1–8 $\mathcal{M}_\odot$ and ages between $\approx 0.3$ Gyr and 2 Gyr. Their NIR luminosity can be dominant with respect to other stars (Fig. 8.14 and Fig. 8.15). However, TP-AGB stars are difficult to model, and their detailed contribution as a function of age and metallicity is debated.

The evolution of SSP spectra $F_{\lambda,\rm SSP}$ (eq. 8.117) is driven by the evolution of stars in the Hertzsprung–Russell diagram. This means that the upper main sequence becomes gradually devoid of massive and luminous stars with increasing time (§C.5). As a

[18] An example is the STELIB spectral library based on observed spectra of Milky Way stars at 3200–9500 Å, with spectral resolution $R \approx 2000$ and metallicity $-2.0 < [\rm Fe/H] < 0.5$ (Le Borgne et al., 2003).

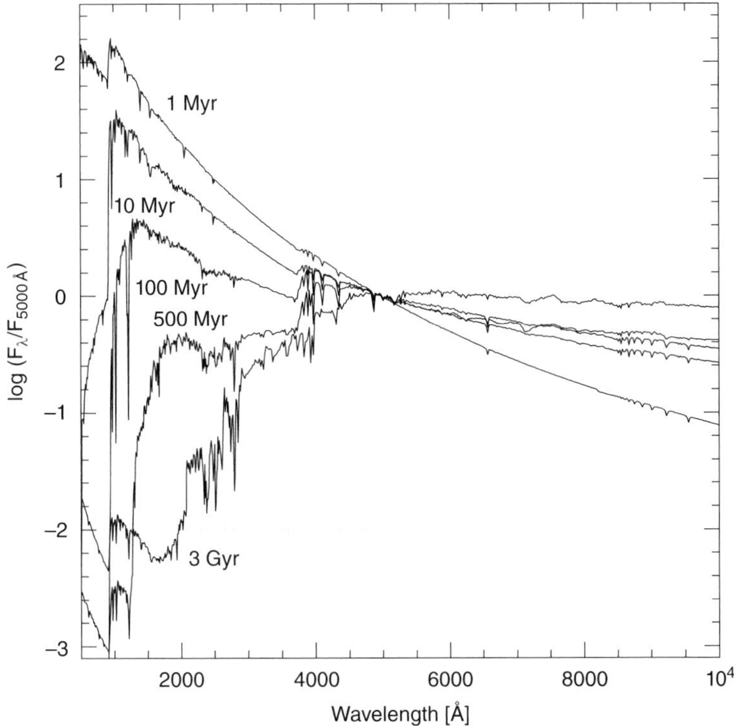


Fig. 8.13 SSP spectra obtained with SPS modelling based on the data of Bruzual and Charlot (2003). Solar metallicity and a Chabrier IMF (§8.3.9) are adopted for all SSPs. The examples display how the SSP spectra change as a function of the age of the stellar population from 1 Myr to 3 Gyr. The gradual decline of the UV—blue flux with increasing age is due to progressive disappearance of hot massive (i.e. UV-luminous) stars.

consequence, the luminosity of an SSP fades (Fig. 8.14, left panel) and its stellar mass-to-light ratio ($\mathcal{M}_\star/L$) increases (Fig. 8.15, right panel) as a function of time. In parallel, the colours of an SSP redden with the ageing of the stellar population because hot stars progressively abandon the main sequence (Fig. 8.15, left panel).

The observed evolution of an SSP depends strongly on the photometric filter of the observation. This occurs because the blue/UV filters are more sensitive to the hottest stars (e.g. main-sequence stars with ages $< 1$ Gyr), whereas red/NIR filters are more influenced by the light of cooler stars (low-mass main-sequence stars, red giants, supergiants and AGB stars). As a consequence, the fading of the luminosity is more rapid and pronounced with short-wavelength filters, because they are more sensitive to the disappearance of short-lived luminous main-sequence stars. For example, from $10^7$ to $10^9$ years, the $B$-band luminosity of an SSP decreases by nearly two orders of magnitude, whereas it fades by a factor of 10 in the $K$ band ($\lambda \approx 2.2\ \mu$m; Fig. 8.14, left panel).

Besides the evolutionary effects due to main-sequence stars, the development of the giant and supergiant phases plays an important role because of the high luminosity of these stars which can even dominate the global radiation output in some spectral ranges. This is illustrated in Fig. 8.14 (right panel) where the contributions of the main evolutionary phases is shown for the $B$ and $K$ photometric bands. For example, TP-AGB stars increase

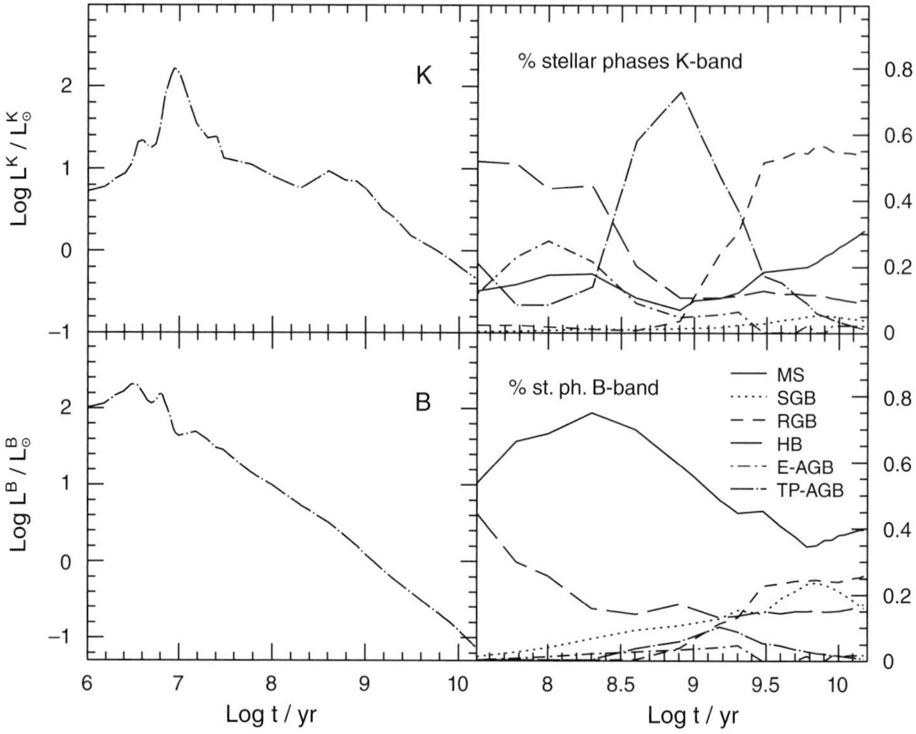


Fig. 8.14 *Left panels*. The time evolution of the luminosity of an SSP in the *B* band (*bottom*) and *K* band (*top*) based on the models of Maraston (2005). The peak in the *K* band at 10 Myr is due to the red supergiant phase of massive stars. *Right panels*. The evolution of fractional contributions (whose values are indicated in the right-hand *y*-axis) of the different phases of stellar evolution to the total *K*-band (*top*) and *B*-band (*bottom*) luminosity. MS, main sequence; SGB, subgiant branch; RGB, red giant branch; HB, horizontal branch; E-AGB, early asymptotic giant branch; TP-AGB, thermally pulsing asymptotic giant branch. A Kroupa IMF is assumed in all cases. Courtesy of C. Maraston.

the NIR luminosity temporarily for ages around $t \approx 0.5$–1 Gyr, and contribute up to 50–70% of the total luminosity in the $K$ band due to their location in the Hertzsprung-Russell diagram (high luminosity and low effective temperature, i.e. red colours). The evolutionary trend of an SSP depends significantly also on the metal abundance because of two main reasons. First, metals decrease the luminosity especially in the blue and UV, due to the increasing number and strength of metal absorptions at these wavelengths, and therefore make the colours redder (**blanketing effect**). Second, a higher metal abundance implies higher opacity and more absorption of the energy coming from the interior of the stars, hence causing a larger expansion of red giants, which become cooler and therefore redder. The two effects imply that the metallicity plays an important role in reddening the colours (Fig. 8.16). Thus, the colours of an SSP can be red because of the old age of the stars or their high metallicity, or both (§5.1.3). This ambiguity is the age–metallicity degeneracy discussed in §5.1.3 and §11.1.3. Last but not least, the characteristics of an SSP depend also on the IMF, which ultimately determines the abundance of massive stars with respect to low-mass ones, the $\mathcal{M}_\star/L$ of the stellar population and the relative importance of Type II SNe.

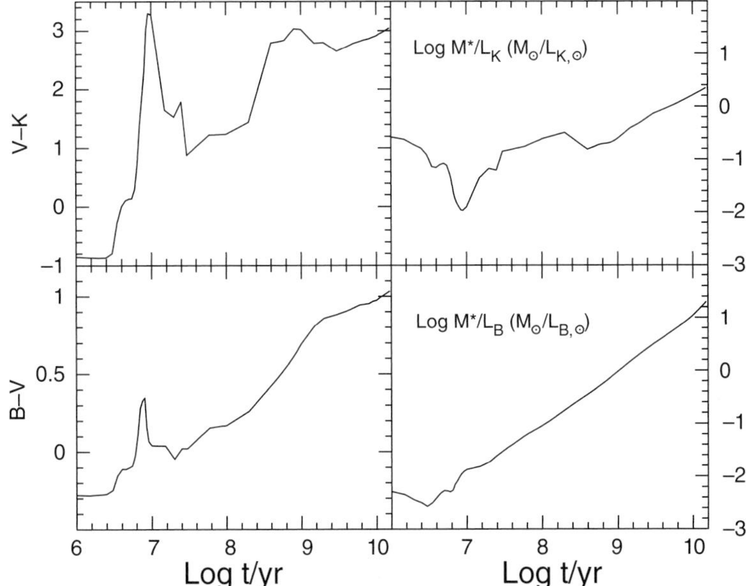


Fig. 8.15 *Left panels*. The time evolution of the $B-V$ (*bottom*) and $V-K$ (*top*) colour indices of the same SSP as in Fig. 8.14. The peak at 10 Myr is due to red supergiants, whereas the bump at 1 Gyr is caused by the luminous phase of TP-AGB stars. *Right panels*. The evolution of the stellar mass-to-light ratio in $B$ band and $K$ band. A Kroupa IMF is assumed in all cases. Courtesy of C. Maraston.

## Composite Stellar Populations

SSPs are unrealistic approximations because real stellar populations form during an extended time interval. An example is the disc of our Galaxy which includes stars with a wide range of ages. A collection of stars formed at different times and with different initial chemical compositions is called a **composite stellar population** (CSP). A mathematically convenient approach is to consider a CSP as the sum of individual SSPs, and to express its spectrum as a function of time as

$$F_{\lambda,\mathrm{CSP}}(t) = \int_{t'=0}^{t'=t} \int_{Z=0}^{Z=Z_{\max}} [\mathrm{SFR}(t-t')\mathcal{P}(Z,t-t')F_{\lambda,\mathrm{SSP}}(t',Z)]\mathrm{d}t'\,\mathrm{d}Z, \tag{8.118}$$

where $F_{\lambda,\mathrm{CSP}}$ is the integrated flux of the CSP at time $t$, $t'$ is the age of the SSP, $Z$ is the stellar metallicity, $\mathrm{SFR}(t-t')$ is the SFR at time $(t-t')$ (i.e. the SFH), $\mathcal{P}(Z,t-t')$ is the distribution of stellar metallicity at time $(t-t')$, and $F_{\lambda,\mathrm{SSP}}(t',Z)$ is the flux of an individual SSP with age $t'$ and metallicity $Z$. It is customary to apply eq. (8.118) assuming a single value of metallicity $Z$ for the entire CSP. This simplification means that $\mathcal{P}(Z,t-t')$ is assumed to be a Dirac delta function in $Z$, independent of time. The SFHs of galaxies are generally poorly constrained by the observations. Hence, in order to compute the integral in eq. (8.118), it is necessary to adopt an analytic function to describe $\mathrm{SFR}(t-t')$. The following parameterisations are typically adopted in the literature.

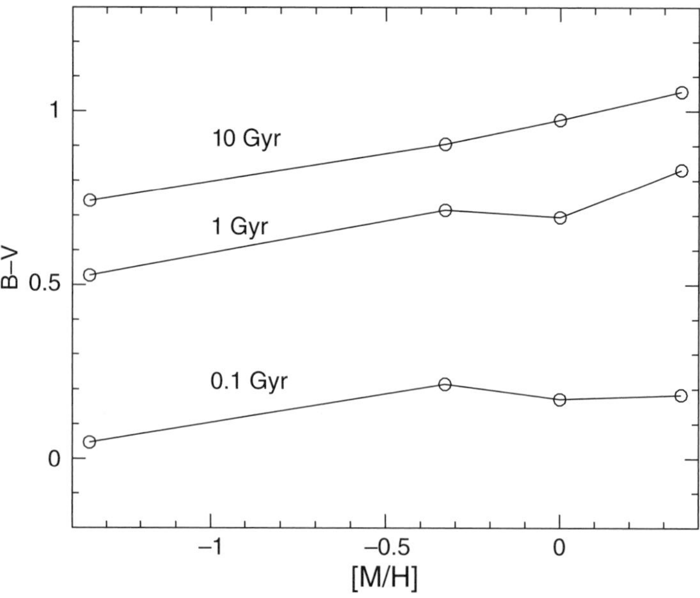


Fig. 8.16 The dependence of the $B-V$ colour on the abundance of all metals ([M/H]) for three SSPs with different ages. The age–metallicity degeneracy is evident: a given colour (e.g. $B-V=0.8$) is compatible with a range of ages and metal abundances. A Kroupa IMF is assumed. Courtesy of C. Maraston.

1. Constant star formation rate: $\mathrm{SFR}(t)=\mathrm{const}$.
2. Exponentially declining model with $\mathrm{SFR}(t)=A\exp[-(t-t_0)/\tau_\star]$, where $A$ is a constant, $t_0$ is the time at which the star formation started and $\tau_\star$ is the star formation timescale (eq. 8.96).
3. Exponentially delayed models with $\mathrm{SFR}(t)=At^d\exp[-(t-t_0)/\tau_\star]$, which attempt to take into account the rising part of the SFR that is generally expected in theoretical models of galaxy formation.

Once the above ingredients have been defined, the SPS models allow us to predict synthetic spectra of CSPs. It is clear that several assumptions and simplifications enter in this process. However, despite these intrinsic limitations, the SPS models are considered crucial tools in galaxy formation models to predict galaxy properties and interpret the observations when the individual stars of galaxies cannot be spatially resolved.

### 8.6.2 Simulating Galaxy Spectra Including the Interstellar Medium

Galaxies do not contain only stars, and their spectra are shaped also by other components (§3.2). Hence, theoretical SEDs must include also the other ingredients which play a role in the emission and absorption of radiation. The most advanced models of galaxy SEDs incorporate the physical treatment of both stellar and ISM ingredients, and produce synthetic spectra extended over a wide wavelength range. In these models, the stellar component is treated using the SPS approach described above, whereas the ISM can be added with different recipes. For instance, if the stars and dust are taken into account, the emerging spectrum can be written by modifying eq. (8.118) as

$$F_{\lambda,\mathrm{CSP}}(t) = \int_{t'=0}^{t'=t} \int_{Z=0}^{Z_{\max}} [\mathrm{SFR}(t-t')\mathcal{P}(Z,t-t')F_{\lambda,\mathrm{SSP}}(t',Z)e^{-\tau_{\mathrm{d}}(t')} + AF_{\lambda,\mathrm{d}}(t',Z)]\mathrm{d}t'\,\mathrm{d}Z, \tag{8.119}$$

where $\tau_{\mathrm{d}}(t')$ is the dust optical depth, $F_{\lambda,\mathrm{d}}$ is the dust emission spectrum and $A$ is a constant obtained by balancing the luminosity absorbed by dust grains with the reradiated one. **Nebular emission** (lines and continuum) can also be included through self-consistent photoionisation models which take into account the evolution of the radiation field due to O/B stars and include the relevant processes required for the calculation of the emitted spectrum (bremsstrahlung, free–bound, two-photon and line emission). The most detailed models allow one also to define the geometrical distribution of the stellar and ISM components and their radial profiles. The predicted theoretical SEDs must be validated through a comparison with the observations of different galaxy types. Fig. 8.17 shows an example of a model spectrum and its ingredients compared to the observed SED of an SFG.

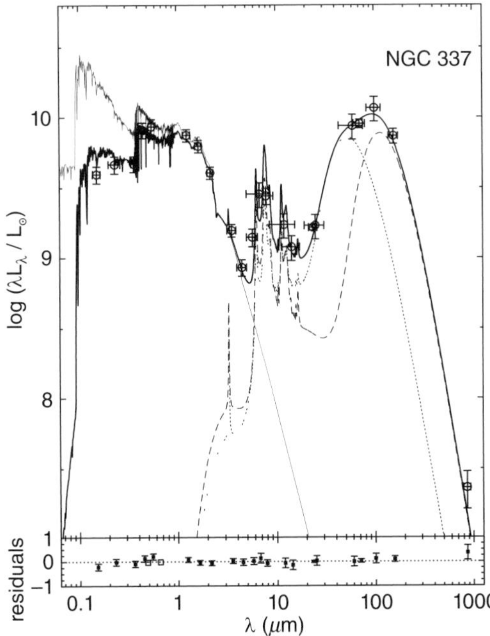


Fig. 8.17 *Top panel.* The observed SED of the galaxy NGC 337 (SB(s)d type) from the UV to the millimetre. The points with error bars indicate the measured photometric fluxes. Thick line: a model spectrum which reproduces the data as dust-attenuated stellar light plus dust emission. Thin solid line: the intrinsic stellar light unattenuated by dust extinction. Thin dashed line: the emission from dust in the diffuse ISM. Dotted line: the emission from dust in the regions where star formation takes place. *Bottom panel.* The residuals are the logarithmic values of the data-to-model ratio. Adapted from da Cunha et al. (2008).

## 8.7 Feedback from Stars

The term **feedback** applied to galactic astrophysics comprises *all* the effects that the evolution of an astrophysical object, in particular a star or a black hole, has on the surrounding medium. SN explosions and stellar winds eject a large amount of gas into the ISM. In §8.4.1 we have discussed how this affects the chemical evolution of a galaxy. In this section, we focus on the energy release as the ejection of mass from SNe and winds from young and massive stars occur at very high kinetic energies. The effect of this ejection on the surrounding ISM is called **stellar feedback** or, given the prominence of SNe in the energy release, **supernova feedback**. Note that another form of stellar feedback is photoionisation, which has been discussed in §8.1.2. Cosmic rays are also potentially important contributors. As we see for AGN feedback (§8.8), stellar feeback can remove gas from the ISM of a galaxy and reduce its star formation (negative feedback) but also locally compress the gas and enhance it (positive feedback). In the following sections, we mostly focus on the gas removal. For a description of these topics see also Dyson and Williams (1997).

### 8.7.1 Evolution of a Supernova Remnant

The explosion of an SN in the ISM produces a rapid (a few seconds) ejection of large quantities of gas at extremely high speeds $v_{\rm ej} \sim 10^3$–$10^4\,{\rm km\,s^{-1}}$. In the case of Type II SNe this gas is in the outer layers of massive stars and the mass ejected can be a significant fraction of the initial stellar mass (§8.4.1). The ejection of material at highly supersonic velocities produces the quick (a few hundred years) formation of a shock (§4.2.4 and §D.2.4). The expansion of this shock into a homogeneous ISM with number density $n_0$ generates a roughly spherical *shell* of shocked ISM gas: this shell and its interior are called a **supernova remnant** (SNR). The kinetic and internal energies nearly reach equipartition in the shell. We can write the internal energy of a particle in the shell as

$$U_{\rm p} = \frac{3}{2} k_{\rm B} T_{\rm sh} = \frac{9}{32} \mu m_{\rm p} v_{\rm sh}^2, \tag{8.120}$$

where $T_{\rm sh}$ is the temperature of the shocked medium (eq. D.37) and $v_{\rm sh}$ is the shock speed (roughly equal to the shell speed). The shell initially evolves adiabatically (no significant radiative losses) and the total energy released by the SN explosion ($E_{\rm SN}$) is conserved thus

$$E_{\rm SN} \approx 2U = 2\,\frac{9}{32} \mu m_{\rm p} v_{\rm sh}^2\, \frac{4\pi}{3} r_{\rm sh}^3 n_0, \tag{8.121}$$

where $U$ is the total internal energy and $r_{\rm sh}$ is the radius of the shell.[19] The factor of 2 in eq. (8.121) comes from energy equipartition, and to obtain the total internal energy we have multiplied the energy per particle (eq. 8.120) by the number of particles that have been enveloped by the shell ($n_0$ is the ISM density).

[19] It can be shown that the shell is thin enough not to make a distinction between the inner radius of the shell and its outer radius, where the shock is located.

We can now rewrite $v_{\rm sh}$ in eq. (8.121) as $\mathrm{d}r_{\rm sh}/\mathrm{d}t$ obtaining

$$r_{\rm sh}^{3/2}\frac{\mathrm{d}r_{\rm sh}}{\mathrm{d}t}=\left(\frac{4E_{\rm SN}}{3\pi\mu n_0 m_{\rm p}}\right)^{1/2}, \tag{8.122}$$

which can be integrated to obtain the evolution of the shell radius and speed,

$$r_{\rm sh}\simeq 14.1\left(\frac{E_{\rm SN}}{10^{51}\,\mathrm{erg}}\right)^{1/5}\left(\frac{\mu n_0}{1\,\mathrm{cm}^{-3}}\right)^{-1/5}\left(\frac{t}{10^4\,\mathrm{yr}}\right)^{2/5}\,\mathrm{pc} \tag{8.123}$$

and

$$v_{\rm sh}\simeq 552.9\left(\frac{E_{\rm SN}}{10^{51}\,\mathrm{erg}}\right)^{1/5}\left(\frac{\mu n_0}{1\,\mathrm{cm}^{-3}}\right)^{-1/5}\left(\frac{t}{10^4\,\mathrm{yr}}\right)^{-3/5}\,\mathrm{km\,s^{-1}}, \tag{8.124}$$

where we have indicated standard values for SN energy and ISM density, and normalised to $t=10^4$ yr, which, as we see below, is a typical time for this phase. Eq. (8.123) represents the **Sedov solution** (Sedov, 1959) for the evolution of a blast wave applied to the ISM.[20] Note the very weak dependence in eqs. (8.123) and (8.124) on both $E_{\rm SN}$ and $n_0$, i.e. the normalisations do not depend much on the initial conditions. Note also that this solution cannot be extrapolated to very early times as it is only valid for, say, $t>100$ yr.

The Sedov phase, also called the **adiabatic phase**, of the evolution of an SNR lasts until radiative losses can no longer be neglected. This occurs because the deceleration of the shock decreases the temperature of the shell and makes the cooling time (eq. 8.3) become of the order of the age of the system. For an ISM with solar metallicity and density $\mu n_0=1\,\mathrm{cm}^{-3}$, this time is $t_{\rm a}\approx 5\times 10^4$ yr. At $t_{\rm a}$, the SNR has reached a radius $r_{\rm sh}(t_{\rm a})\approx 27$ pc (eq. 8.123). In the subsequent evolution (**radiative phase**), a significant fraction of the energy is lost through radiation and the evolution of the shell radius can be obtained via momentum conservation:

$$\frac{\mathrm{d}}{\mathrm{d}t}(\mathcal{M}_{\rm sh}v_{\rm sh})=4\pi r_{\rm sh}^2(P_{\rm b}-P_0), \tag{8.125}$$

where $\mathcal{M}_{\rm sh}$ is the mass of the shell, and $P_{\rm b}$ and $P_0$ are the pressure of the shell interior (bubble) and of the external ISM, respectively. A reasonably accurate simplified solution can be obtained assuming negligible pressure contribution ($P_{\rm b}\approx P_0\approx 0$). In this situation, $\mathcal{M}_{\rm sh}v_{\rm sh}\approx$ const and so $r_{\rm sh}^3 v_{\rm sh}$ is also a constant that we can set at the time $t_{\rm a}$, the end of the adiabatic phase. We then integrate to obtain the behaviours of the shell radius and speed that, for $t\gg t_{\rm a}$, read

$$r_{\rm sh}\simeq 36.0\left(\frac{E_{\rm SN}}{10^{51}\,\mathrm{erg}}\right)^{1/5}\left(\frac{\mu n_0}{1\,\mathrm{cm}^{-3}}\right)^{-1/5}\left(\frac{t}{10^5\,\mathrm{yr}}\right)^{1/4}\,\mathrm{pc} \tag{8.126}$$

and

$$v_{\rm sh}\simeq 88.0\left(\frac{E_{\rm SN}}{10^{51}\,\mathrm{erg}}\right)^{1/5}\left(\frac{\mu n_0}{1\,\mathrm{cm}^{-3}}\right)^{-1/5}\left(\frac{t}{10^5\,\mathrm{yr}}\right)^{-3/4}\,\mathrm{km\,s^{-1}}. \tag{8.127}$$

[20] The full Sedov solution also describes the interior of the blast wave and its analytic derivation is rather involved (Shu, 1992). A numerical solution to the problem was derived by Taylor (1950), so the solution is sometimes referred to as Sedov–Taylor.

This is called the **snowplough solution** as the SNR is essentially slowed down by the accumulation of ISM mass in the shell. Note that nearly all the mass engulfed by the shock remains in the shell: the interior of the SNR (bubble) gains negligible mass.

The radiative phase ends when the velocity of the shell (eq. 8.127) becomes of the order of the typical random speeds of the ISM ($\sigma_{\rm gas} \sim 10\,{\rm km\,s^{-1}}$). This occurs at a time $t_{\rm r} \approx 2 \times 10^6$ yr when the SNR has a radius $r_{\rm sh}(t_{\rm r}) \approx 76$ pc from eq. (8.126). Around this time, the shell of the SNR loses coherence, disperses and its left-over kinetic energy is released into the surrounding ISM. This release of kinetic energy is a key contributor to the feeding of the ISM turbulence (§8.3.7). We can estimate the efficiency $\eta$ in transferring kinetic energy to the ISM by taking the ratio between the kinetic energy at the end of the radiative phase $K(t_{\rm r})$ and the initial energy $E_{\rm SN}$. Due to equipartition and energy conservation in the adiabatic phase, $E_{\rm SN} \approx 2K(t_{\rm a})$. In general, $K = \mathcal{M}_{\rm sh} v_{\rm sh}^2/2 \propto v_{\rm sh}^2 r_{\rm sh}^3$ and this leads to an efficiency

$$\eta = \frac{K(t_{\rm r})}{E_{\rm SN}} = \frac{1}{2}\left[\frac{r_{\rm sh}(t_{\rm r})}{r_{\rm sh}(t_{\rm a})}\right]^3 \left[\frac{v_{\rm sh}(t_{\rm r})}{v_{\rm sh}(t_{\rm a})}\right]^2 \simeq 0.02, \tag{8.128}$$

where $v_{\rm sh}(t_{\rm a})$ is calculated with eq. (8.124). This estimate can change slightly with different assumptions, but it is always of the order of a few per cent. We conclude that only a very small fraction of the initial SN energy is released as kinetic energy into the surrounding ISM. Most of the energy is radiated away at different wavelengths across the electromagnetic spectrum. This radiation includes thermal emission in X-rays and UV/optical bands (as the shell cools down), but also non-thermal (synchrotron; §D.1.7) emission at radio wavelengths. Spectacular examples of 'young' Galactic SNRs are Cassiopea A and the Crab nebula.

### 8.7.2 Stellar Wind Bubbles

The evolution of massive O/B stars is characterised by intense mass loss due to strong winds powered by radiation pressure pushing on ions in the outer layers of the stellar atmospheres. The wind kinetic energy per unit time is called the **mechanical luminosity** in analogy to the ordinary (radiation) luminosity and it can be written as

$$L_{\rm w} \equiv \frac{1}{2}\dot{\mathcal{M}}_{\rm w} v_{\rm w}^2 \simeq 1.3 \times 10^{36} \left(\frac{\dot{\mathcal{M}}_{\rm w}}{10^{-6}\,\mathcal{M}_\odot\,{\rm yr}^{-1}}\right)\left(\frac{v_{\rm w}}{2\times 10^3\,{\rm km\,s^{-1}}}\right)^2 \,{\rm erg\,s^{-1}}, \tag{8.129}$$

where $\dot{\mathcal{M}}_{\rm w}$ and $v_{\rm w}$ are the wind mass outflow rate and velocity, respectively, and $L_{\rm w}$ can be considered nearly constant during the wind phase of the star.

A stellar wind generates a bubble that expands into the ISM and whose structure is sketched in Fig. 8.18a. Since the wind is an ejection of gas particles at very high speed, it naturally interacts with the external ISM producing a shock (S2) that perturbs the ISM. However, in contrast with an SNR (§8.7.1), the particles of the wind are released as a continuous flow coming from the central star. This high-speed ($v_{\rm w}$) flow impacts on the shocked material behind S2 leading to the formation of a second shock, called the **reverse shock** (S1), that propagates *through the wind* and perturbs it. An equilibrium configuration is eventually reached formed by three concentric regions: (a) freely expanding wind, (b)

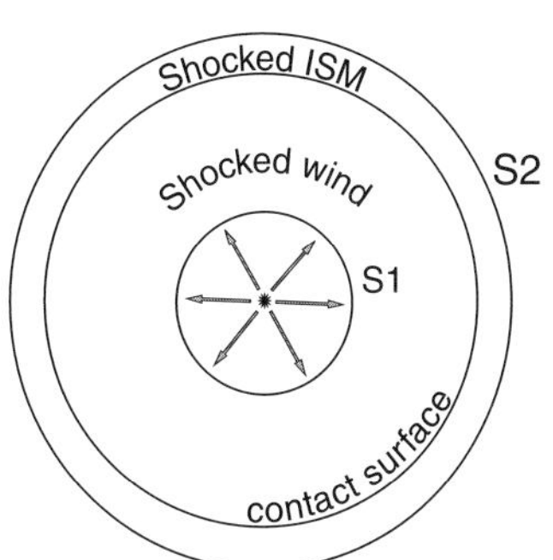


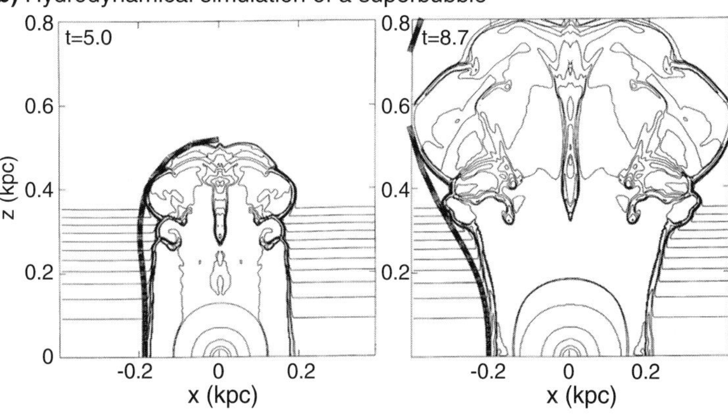


Fig. 8.18 *Panel a.* Interior of a stellar wind bubble. The star is blowing a wind (fast-moving particles) that gets shocked by the reverse shock S1. The outer shell is made of ISM shocked by the outward-going shock S2 (the various regions are not to scale). Figure inspired by Weaver et al. (1977). *Panel b.* Two snapshots (at $t = 5.0$ Myr and $t = 8.7$ Myr from the beginning) of a hydrodynamic simulation of a superbubble expanding in a stratified galactic ISM at the moment of the blow-out. The $x$-axis is along the plane of the galaxy disc, while the $z$-axis is perpendicular to it. Contours show the gas isodensity levels. Adapted from Mac Low et al. (1989). © AAS, reproduced with permission.

a bubble of shocked wind and (c) a shell of shocked ISM. Both S1 and S2 propagate at velocities that are much lower than $v_{\rm w}$ and this allows us to estimate the thermodynamic state of the wind bubble. If we imagine ourselves in the reference frame of the free wind escaping the star, we would see S1 coming towards us at a velocity nearly equal to $v_{\rm w}$. This produces a shock *in* the wind creating a bubble at a very high temperature ($T \sim 10^7$ K; eq. 4.10). Given that the cooling time is very long at these temperatures (eq. 8.3), we can expect that such a bubble evolves nearly adiabatically. On the contrary, the shell of shocked ISM, perturbed by S2, radiates away a large fraction of its energy. Thus, we end up with a system consisting of an adiabatic bubble and a radiative shell.

With these considerations in mind, we can study the evolution of the system by imposing the conservation of energy in the bubble:

$$\frac{dE_{\rm b}}{dt} = L_{\rm w} - P_{\rm b}\frac{dV}{dt}, \qquad (8.130)$$

where $E_{\rm b}$ is the total energy of the bubble, which, given its high temperature, we can assume to be essentially all internal energy ($E_{\rm b} \approx U_{\rm b}$). The mechanical luminosity $L_{\rm w}$ acts as an energy input and the last term of eq. (8.130) represents the work per unit time done by the bubble on the shell ($P_{\rm b}$ is the pressure of the bubble and $V$ is the volume). For the shell, we can write the conservation of momentum that is analogous to eq. (8.125), but this time $P_{\rm b}$ cannot be neglected and $P_{\rm b} \gg P_0$. Note that, given that the shell is thin, we can assume that the internal radius of the shell (contact surface between wind and ISM in Fig. 8.18a) and the S2 shock radius are the same. The relation between internal energy and pressure (eq. 8.160) allows us to write $E_{\rm b} = (3/2)P_{\rm b}V$ for $\gamma = 5/3$. Then using $V = (4\pi/3)r_{\rm sh}^3$ and $dr_{\rm sh}/dt = \dot{r}_{\rm sh} = v_{\rm sh}$, eq. (8.130) can be rewritten as a function of $P_{\rm b}$, $r_{\rm sh}$ and $v_{\rm sh}$. By assuming that all the mass of the ISM engulfed by the shock remains in the shell, eq. (8.125) gives an expression for the pressure of the bubble as

$$P_{\rm b} = \rho_0 \left( \frac{1}{3} r_{\rm sh} \ddot{r}_{\rm sh} + \dot{r}_{\rm sh}^2 \right), \tag{8.131}$$

where $\rho_0$ is the ISM density. Eq. 8.131 can be substituted in eq. (8.130), leading to a final expression with only $r_{\rm sh}$ and its time derivatives.

At this point, we can look for a self-similar solution of the type $r_{\rm sh} = At^{\alpha}$. After some calculations, we obtain

$$r_{\rm sh} = \left( \frac{125}{154\pi} \right)^{1/5} \left( \frac{L_{\rm w}}{\rho_0} \right)^{1/5} t^{3/5}. \tag{8.132}$$

Thus the radius and the expansion speed of the shell are

$$r_{\rm sh} \simeq 7.0 \left( \frac{L_{\rm w}}{10^{36}\ {\rm erg\ s^{-1}}} \right)^{1/5} \left( \frac{\mu n_0}{1\ {\rm cm^{-3}}} \right)^{-1/5} \left( \frac{t}{10^5\ {\rm yr}} \right)^{3/5} \ {\rm pc} \tag{8.133}$$

and

$$v_{\rm sh} \simeq 41.3 \left( \frac{L_{\rm w}}{10^{36}\ {\rm erg\ s^{-1}}} \right)^{1/5} \left( \frac{\mu n_0}{1\ {\rm cm^{-3}}} \right)^{-1/5} \left( \frac{t}{10^5\ {\rm yr}} \right)^{-2/5} \ {\rm km\ s^{-1}}, \tag{8.134}$$

with very weak dependence on mechanical luminosity and ISM density. We conclude that, in the presence of a constant source of kinetic energy (the stellar wind), the radius of the shell grows faster in time (power 3/5) than in the case of an instantaneous energy release (SN explosion, power 2/5). As a consequence, the shell decelerates at a slower rate in a wind bubble than in an SNR.

We now investigate the energy budget of the wind bubble by writing

$$E_{\rm b} \approx U_{\rm b} = \frac{3}{2} P_{\rm b} V = \frac{5}{11} L_{\rm w} t, \tag{8.135}$$

where we have substituted eq. (8.132) in $P_{\rm b}$ (given by eq. 8.131) and $V$. Eq. (8.135) shows that nearly half of the total energy of the wind at any time ($L_{\rm w}t$) goes into internal energy of the bubble. The other half ($E_{\rm s} = (6/11) L_{\rm w} t$) will then end up in the shell somewhat divided between kinetic and internal energies. We can calculate the kinetic energy of the shell as

$$K_{\rm sh} = \frac{1}{2} M_{\rm sh} v_{\rm sh}^2 = \frac{4\pi}{6} r_{\rm sh}^3 \rho_0 \left( \frac{3}{5} \frac{r_{\rm sh}}{t} \right)^2 = \frac{15}{77} L_{\rm w} t \approx 0.2 L_{\rm w} t. \tag{8.136}$$

Thus about 20% of the total energy released by the wind is transferred to the expansion of the shell. This fraction remains constant with time as long as the bubble does not lose too much internal energy and, at the end of the expansion (when $v_{\rm sh} \approx \sigma_{\rm gas}$), it will be transferred to the ISM. Thus, we have found that a stellar wind bubble can be more efficient than an SNR in transferring kinetic energy to the ISM (compared with eq. 8.128). An important consequence of this result is discussed in the following section.

### 8.7.3 Superbubbles and Galactic Winds

The vast majority of O and B stars (Tab. C.4) in the Milky Way reside in star clusters or loose groups inside GMCs called **O/B associations**. If the stars in these associations are nearly coeval we can expect the emission of powerful stellar winds by about 100 stars (typical number for a relatively large association) occurring *simultaneously* in a *small* region of space. Geometrical considerations and numerical calculations show that the combined action of these 100 stellar winds quickly produce a large bubble surrounding the whole stellar association, called a **superbubble**. The structure and evolution of a superbubble are just rescaled versions of those of the wind bubble of a single star (§8.7.2). We can therefore use eqs. (8.133) and (8.134) to calculate the evolution of the radius and the velocity of its **supershell**. For $L_{\rm w} = 10^{38}$ erg s$^{-1}$ (100 O/B stars) and $t = 10^6$ yr we obtain $r_{\rm sh} \approx 70$ pc and $v_{\rm sh} \approx 41$ km s$^{-1}$.

After a timescale of ~ 1 Myr, the most massive stars in the O/B association stop producing winds and explode as SNe. This constitutes a new energy input that contributes to the heat of the bubble and the expansion of the shell. With the passage of time, progressively more stars explode and we enter a **supernova phase** of the evolution (as opposed to the previous **stellar wind phase**), with the expansion of the supershell essentially driven by the SN explosions. We can obtain a rough estimate of the energy input by taking as a timescale of this phase the main-sequence lifetime of a star of 8 $\mathcal{M}_\odot$ (last star to explode as SN), $\tau_{\rm MS}(8\,\mathcal{M}_\odot) \approx 3 \times 10^7$ yr. The average mechanical luminosity provided by the SNe is then

$$L_{\rm SN} \approx \frac{N_{\rm SN} E_{\rm SN}}{\tau_{\rm MS}(8\,\mathcal{M}_\odot)} \approx 10^{38} \left(\frac{N_{\rm SN}}{100}\right) \left(\frac{E_{\rm SN}}{10^{51}\,{\rm erg}}\right) {\rm erg\,s^{-1}}, \tag{8.137}$$

where $N_{\rm SN}$ is the number of SNe. Thus SNe produce an average energy input per unit time similar to that of stellar winds and the expressions for the evolution of the radius and the speed of the supershell are analogous. We can write them, respectively, as

$$r_{\rm sh} \approx 70 \left(\frac{N_{\rm SN}}{100}\right)^{1/5} \left(\frac{E_{\rm SN}}{10^{51}\,{\rm erg}}\right)^{1/5} \left(\frac{\mu n_0}{1\,{\rm cm^{-3}}}\right)^{-1/5} \left(\frac{t}{10^6\,{\rm yr}}\right)^{3/5} {\rm pc} \tag{8.138}$$

and

$$v_{\rm sh} \approx 41 \left(\frac{N_{\rm SN}}{100}\right)^{1/5} \left(\frac{E_{\rm SN}}{10^{51}\,{\rm erg}}\right)^{1/5} \left(\frac{\mu n_0}{1\,{\rm cm^{-3}}}\right)^{-1/5} \left(\frac{t}{10^6\,{\rm yr}}\right)^{-2/5} {\rm km\,s^{-1}}. \tag{8.139}$$

Two important points are worth noting. First, the SN phase lasts about 10 times longer than the stellar wind phase. Thus, given that the mechanical luminosities in the phases are similar, SNe contribute much more than winds to the overall energetics of a superbubble. Second, superbubbles are more efficient than single SNRs in transferring kinetic energy to the ISM (eq. 8.136). This is essentially due to the fact that SNe that explode *inside* a superbubble encounter less resistance from the surrounding medium (because it is more rarefied). Given that an important fraction of SNe in a galaxy occur in O/B associations, the global efficiency can then be higher than the few per cent given by eq. (8.128).

From the above calculations we also conclude that superbubbles can, in principle, keep expanding for tens of Myr and reach dimensions of hundreds of parsecs. However, if we now consider the evolution of one of these structures in a realistic galaxy gaseous disc, whose scaleheight is 100–200 pc (§4.6.2), we realise that the expansion of a superbubble must depart from spherical symmetry and accelerate in the vertical direction where the resistance of the ISM is lower (decreasing density). This eventually leads to a phenomenon called **blow-out**, as a consequence of which a fraction of the dense and cold gas in the supershell and the hot gas inside the bubble are ejected out of the galactic disc into the halo region: see the hydrodynamic simulation in Fig. 8.18b. In real galaxies, superbubbles are observed in both neutral atomic and ionised gas. In Fig. 8.19 we give two examples: one in the Milky Way and one in a nearby spiral galaxy.

In general, we can distinguish two *regimes* of gas ejection from stellar feedback: the galactic fountain and the galactic wind regimes (see also §4.2.10). With the term 'galactic fountain' one usually indicates ejection of gas at velocities large enough to escape the ISM but much lower than the escape speed from the galaxy (eq. 4.21). For instance, in a present-day disc galaxy, $v \lesssim 100\,\mathrm{km\,s^{-1}}$. The material ejected at these velocities travels through the halo and falls back onto the disc at a different location. This causes a redistribution of gas, metals and angular momentum that can affect the evolution of the disc. Galactic winds are instead characterised by speeds of the order of few/several hundreds of $\mathrm{km\,s^{-1}}$. These can

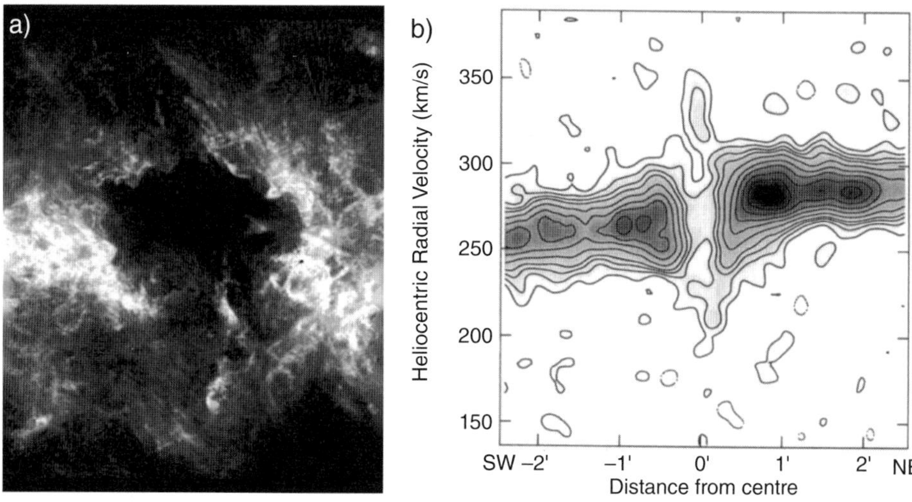


Fig. 8.19 *Panel a.* A superbubble in the Milky Way seen as a large ($\sim$ 600 pc in diameter) hole in the H I emission. This image displays the neutral hydrogen at line-of-sight velocity $v \approx 40\,\mathrm{km\,s^{-1}}$ (the Galactic plane runs horizontally). Filaments of gas are stretching towards the halo region. The gas ejected by the superbubble becomes extraplanar (§4.2.10) and eventually falls back to the disc (galactic fountain). Adapted from McClure-Griffiths et al. (2003). © AAS, reproduced with permission. *Panel b.* A superbubble in the nearby spiral galaxy M 101 shown as a position–velocity diagram (§4.3.3) of the H I emission, taken along a line intersecting a large H I hole in the disc; 1 arcmin $\simeq$ 2 kpc. The horizontal darker emission shows the normal disc gas. Corresponding to the hole (centre of the diagram), two blobs of H I emission appear at line-of-sight velocities of $\pm$(50–80) km s$^{-1}$ with respect to the disc gas velocity at that location. These are thought to be the approaching and receding sides of the shell of the expanding superbubble. From Kamphuis et al. (1991).

be achieved in the central regions of some galaxies (starbursts), where the star formation rate density is so high that the combined action of stellar winds and SNe is able to drive these fast outflows. An example of galactic wind is shown in Fig. 4.2. These winds are considered crucial for several aspects of galaxy formation and evolution, in particular for the removal of gas from low-mass galaxies (§10.7.3). We describe their efficiency in the next section.

### 8.7.4 Global Models of Galactic Winds

There have been several attempts to link the power of a galactic wind to the global properties of a starburst galaxy, in particular to its SFR. The amount of energy per unit time released by a starburst galaxy can be estimated by multiplying the energy of a single SN by the SN rate. The **supernova rate** $R_{\rm SN}$ can be related to the SFR by estimating the number of stars with $\mathcal{M} \geq 8\,\mathcal{M}_\odot$ per unit SFR. For a Chabrier IMF (§8.3.9) we obtain

$$R_{\rm SN} \approx 10^{-2} \left( \frac{\rm SFR}{\mathcal{M}_\odot\,{\rm yr}^{-1}} \right) {\rm yr}^{-1}. \tag{8.140}$$

A fraction of the SN energy can be transferred to the ISM of the galaxy in the form of kinetic energy and produce a galactic wind. We can then write the kinetic energy per unit time ($\dot{K}$) available for the galactic wind as

$$\dot{K} \approx 3 \times 10^{40} \left( \frac{\eta}{0.1} \right) \left( \frac{E_{\rm SN}}{10^{51}\,{\rm erg}} \right) \left( \frac{\rm SFR}{\mathcal{M}_\odot\,{\rm yr}^{-1}} \right) {\rm erg\,s}^{-1}, \tag{8.141}$$

where $\eta$ is the efficiency in transferring energy to the wind.[21]

If the wind leaves the galaxy with a velocity $v_{\rm out}$, the kinetic power is simply $\dot{K} = \dot{\mathcal{M}}_{\rm out}\, v_{\rm out}^2/2$, where $\dot{\mathcal{M}}_{\rm out}$ is the mass outflow rate. A galactic wind driven in this fashion is referred to as an **energy-driven wind**. The mass outflow rate is

$$\dot{\mathcal{M}}_{\rm out,ED} \approx 1 \left( \frac{\eta}{0.1} \right) \left( \frac{\rm SFR}{\mathcal{M}_\odot\,{\rm yr}^{-1}} \right) \left( \frac{v_{\rm out}}{300\,{\rm km\,s}^{-1}} \right)^{-2} \mathcal{M}_\odot\,{\rm yr}^{-1}, \tag{8.142}$$

where we have used $E_{\rm SN} = 10^{51}$ erg. Eq. (8.142) shows that star formation has the power to eject gas out of a galaxy at a rate comparable with its SFR for a wind speed $v_{\rm out} \lesssim 300\,{\rm km\,s}^{-1}$. In general, the ratio between the mass outflow rate and the SFR of a galaxy,

$$\beta \equiv \frac{\dot{\mathcal{M}}_{\rm out}}{\rm SFR}, \tag{8.143}$$

is called the **mass loading factor** of the wind. In §10.7.3, we see that powerful outflows ($\beta \gtrsim 1$) of gas at velocities close to the escape speed (eq. 4.21) seem to be needed to explain a number of properties of galaxies. For instance, the escape speed from the centre of the Milky Way is $v_{\rm esc} \approx 800\,{\rm km\,s}^{-1}$. These velocities are very difficult to reach with SN feedback unless $\beta \ll 1$, which, in turn, significantly reduces the outflow rate.

[21] Note that, instead of taking the pure SN efficiency from eq. (8.128), we have indicated an intermediate choice (10%) between that and the superbubble efficiency (eq. 8.136).

The difficulties encountered in driving fast outflows with large $\beta$ using energy-driven models have brought researchers to postulate the possibility of the so-called **momentum-driven winds** (see Murray et al., 2005, for details). In this scenario, the gas is supposed to acquire a momentum from star formation at a rate $\dot{p} \approx \dot{\mathcal{M}}\, v_{\rm out}$, which is then conserved despite the presence of radiative losses (see discussion in §8.7.1 for the evolution of a single SNR). The momentum could be given to the gas by radiation pressure from massive stars onto dust grains, which then drag along the rest of the gas. Cosmic rays can also contribute with a comparable amount of momentum. For radiation pressure to be efficient, the medium around the starburst must be optically thick with $\tau \gtrsim 1$ (§D.1.1). The mass outflow rate of a momentum-driven galactic wind is expected to be

$$\dot{\mathcal{M}}_{\rm out,MD} \approx 1 \left( \frac{\rm SFR}{\mathcal{M}_\odot\, {\rm yr}^{-1}} \right) \left( \frac{v_{\rm out}}{300\, {\rm km\, s}^{-1}} \right)^{-1} \mathcal{M}_\odot\, {\rm yr}^{-1}, \qquad (8.144)$$

similar to the one in eq. (8.142), but with a milder dependence on the wind speed. Given this difference, at velocities approaching the escape speed, momentum-driven winds could be more efficient than energy-driven winds in ejecting large masses of gas.

## 8.8 Feedback from Active Galactic Nuclei

The observed properties of AGNs (§3.6) suggest that, when gas is present in the central regions of a galaxy hosting a central SMBH, the black hole and this ambient gas interact significantly. On the one hand, the gas in the very central regions of a galaxy is believed to be accreted by the SMBH and thus to fuel its activity. On the other hand, the emission and outflows from the AGN interact with the surrounding gas. This interaction is usually called **feedback from AGNs** or **AGN feedback**. Given that the mass of the SMBH is three orders of magnitude lower than the host bulge mass (§5.4.4), the gas infalling onto the SMBH is negligible in terms of the overall energy, mass and momentum budget of the host galaxy. Nevertheless, AGN feedback processes are powerful and affect dramatically the evolution of the host galaxy. An important question is what is the indirect effect of an AGN on star formation. In this respect, AGN feedback is classified as **negative feedback** when its effect is to suppress star formation (for instance, by heating or expelling cold gas from the galaxy core), and as **positive feedback** when its effect is to enhance star formation (for instance, by compressing gas clouds of the ISM and triggering their collapse; see §8.3.2, §8.3.3 and §8.3.8).

Feedback from AGN is a phenomenon characterised by an extremely large dynamic range, because it involves coupling between processes occurring on scales of the order of the **Schwarzschild radius** (Schwarzschild, 1916)

$$r_\bullet \equiv \frac{2G\mathcal{M}_\bullet}{c^2} \simeq 9.57 \times 10^{-6} \left( \frac{\mathcal{M}_\bullet}{10^8 \mathcal{M}_\odot} \right) {\rm pc}, \qquad (8.145)$$

where $\mathcal{M}_\bullet$ is the black hole mass, and other processes on the characteristic scales of the host galaxies, groups and clusters of galaxies (from ~ kpc to hundreds of kpc). AGN feedback is a widely studied, but still poorly understood, phenomenon. Here we try to report some fundamental properties and general principles that should be borne in mind when considering the possible effects of AGN feedback.

Feedback from AGNs is believed to occur in two principal forms: **radiative feedback**, mediated by photons, and **mechanical** (or **kinetic**) **feedback**, mediated by outflows of particles such as electrons and protons. In §3.6 we have seen that, based on their observational properties, AGNs are classified in several different categories, which can be partly explained with projection effects (depending on the line of sight, the same object can be classified differently, because of the non-spherical geometry; §3.6.2). The interpretation of the observational classification of AGNs is further complicated by the fact that black hole accretion and AGN emission are typically highly time-variable phenomena: over cosmological timescales an SMBH can alternate between periods of quiescence and activity, and it can be characterised by different modes of accretion.

From the point of view of feedback processes, the most important distinction is between phases of high and low accretion rate, where 'high' and 'low' must not be intended in absolute terms, but relative to the maximum accretion rate for a given black hole mass $\mathcal{M}_\bullet$. Under the assumption of spherical symmetry, the maximum luminosity produced by accretion of a pure hydrogen plasma onto a black hole of mass $\mathcal{M}_\bullet$ is the **Eddington luminosity** (Eddington, 1921)

$$L_{\rm Edd} \equiv \frac{4\pi c G m_{\rm p} \mathcal{M}_\bullet}{\sigma_{\rm T}} \simeq 1.26 \times 10^{46} \left( \frac{\mathcal{M}_\bullet}{10^8 \mathcal{M}_\odot} \right) \,{\rm erg\,s^{-1}}, \tag{8.146}$$

which is such that the magnitude of the force due to radiation pressure

$$F_{\rm rad} = \frac{L \sigma_{\rm T}}{4\pi c r^2} \tag{8.147}$$

equals the magnitude of the gravitational force

$$F_{\rm grav} = \frac{G \mathcal{M}_\bullet m_{\rm p}}{r^2}, \tag{8.148}$$

where $L$ is the luminosity, $r$ is the distance from the black hole and $\sigma_{\rm T}$ is the Thomson cross section[22] (eq. 6.9). Due to the Coulomb force, electrons and protons drag each other, so the forces in eqs. (8.147) and (8.148) can be thought of as acting on an electron–proton pair with cross section $\sigma_{\rm T}$ (because the proton cross section is negligible) and mass $m_{\rm p}$ (because the electron mass is negligible). It is worth noting that $F_{\rm rad}$ and $F_{\rm grav}$ have the same dependence on $r$, so $F_{\rm rad} = F_{\rm grav}$ at all radii when $L = L_{\rm Edd}$. Given the **radiative efficiency** of the AGN $\epsilon_{\rm rad}$ (§3.6.1), the **Eddington mass accretion rate** is

$$\dot{\mathcal{M}}_{\rm Edd} \equiv \frac{L_{\rm Edd}}{\epsilon_{\rm rad} c^2}. \tag{8.149}$$

Under the above hypotheses, $\dot{\mathcal{M}}_{\rm Edd}$ is the maximum accretion rate for a black hole of mass $\mathcal{M}_\bullet$, because higher accretion rates correspond to $F_{\rm rad} > F_{\rm grav}$, which is inconsistent

[22] In eq. (8.147) we have assumed that the gas opacity is due only to Thomson scattering. In general the cross section, depending on temperature, density and composition, can be larger than $\sigma_{\rm T}$, but it is close to $\sigma_{\rm T}$ for a fully ionised plasma.

with accretion. In most cases, the observed luminosity of AGNs is found to be sub-Eddington ($L < L_{\rm Edd}$), but super-Eddington luminosities ($L > L_{\rm Edd}$) and mass accretion rates ($\dot{\mathcal{M}}_{\rm acc} > \dot{\mathcal{M}}_{\rm Edd}$) are possible when accretion is not spherically symmetric, because the assumption of spherical symmetry maximises the effect of radiation pressure. For instance, super-Eddington accretion can occur when the accreting gas is distributed in a thin accretion disc.

When the black hole is hosted in a spheroidal gas distribution, the accretion rate is usually more limited by the thermal pressure of the host galaxy gas than by radiation pressure. The accretion rate onto a black hole of mass $\mathcal{M}_\bullet$ at rest at the centre of a spherically symmetric gas distribution is the **Bondi mass accretion rate** (Bondi, 1952)

$$\dot{\mathcal{M}}_{\rm Bondi} = \frac{4\pi\lambda_{\rm c}(G\mathcal{M}_\bullet)^2\rho_\infty}{c_\infty^3}, \tag{8.150}$$

where $c_\infty$ and $\rho_\infty$ are, respectively, the sound speed and density of the gas far from the black hole, where the distribution is assumed uniform and isothermal, and $\lambda_{\rm c}$ is a dimensionless factor of the order of unity. In present-day elliptical galaxies, groups and clusters of galaxies, the main gas reservoir is the hot, roughly spherical, gaseous halo, so the central SMBH is expected to accrete at a rate of the order of $\dot{\mathcal{M}}_{\rm Bondi}$. When $\dot{\mathcal{M}}_{\rm Bondi}$ and $\dot{\mathcal{M}}_{\rm Edd}$ are estimated for these systems, one finds $\dot{\mathcal{M}}_{\rm Bondi} \ll \dot{\mathcal{M}}_{\rm Edd}$, which is an indication of the reason why the corresponding AGNs have significantly sub-Eddington luminosities.

The potential importance of AGN feedback in galaxy evolution can be quantified by the following argument. During its growth, an SMBH of mass $\mathcal{M}_\bullet$ has released an energy $E_\bullet = \langle\epsilon_{\rm rad}\rangle\mathcal{M}_\bullet c^2$, where $\langle\epsilon_{\rm rad}\rangle$ is the average radiative efficiency (§3.6.1). If the SMBH is hosted by a bulge with stellar mass $\mathcal{M}_\star$ and velocity dispersion $\sigma$, the binding energy of the bulge $E_\star$ is of the order of its kinetic energy (eq. 5.13): $E_\star \approx \mathcal{M}_\star\sigma^2$. The ratio between these two energies is

$$\frac{E_\bullet}{E_\star} \approx \frac{\langle\epsilon_{\rm rad}\rangle\mathcal{M}_\bullet c^2}{\mathcal{M}_\star\sigma^2} \simeq 100\left(\frac{\langle\epsilon_{\rm rad}\rangle}{0.1}\right)\left(\frac{\sigma}{300\,{\rm km\,s^{-1}}}\right)^{-2}\left(\frac{\mathcal{M}_\bullet/\mathcal{M}_\star}{0.001}\right), \tag{8.151}$$

so the energy released by the SMBH can be much larger than the binding energy of the host bulge (see §5.4.4 for observational values of $\sigma$ and $\mathcal{M}_\bullet/\mathcal{M}_\star$).

For a black hole with luminosity $L$ and accretion rate $\dot{\mathcal{M}}_{\rm acc}$, it is useful to define the **Eddington ratio** $f_{\rm Edd} \equiv L/L_{\rm Edd} = \dot{\mathcal{M}}_{\rm acc}/\dot{\mathcal{M}}_{\rm Edd}$. The prototypes of AGNs with, respectively, high and low Eddington ratios are quasars and radio galaxies (§3.6). Therefore, the high and low accretion rate modes are often referred to as **quasar mode** (or QSO mode) and **radio mode**, respectively. We now discuss in some more detail the properties of AGN feedback in these two different regimes.

### 8.8.1 Quasar Mode Feedback

An SMBH is said to be in a quasar mode phase when it is accreting gas at a high rate and it is emitting a substantial amount of energy per unit time in the form of photons. In other words, in the quasar mode the radiative luminosity of the AGN is comparable to $L_{\rm Edd}$ (the typical values of the Eddington ratio are $10^{-2} \lesssim f_{\rm Edd} \lesssim 1$). In this regime both radiative and mechanical feedback processes are expected to be important, but, given the high luminosity, radiative feedback is dominant. The electromagnetic emission is thought to be due to a small ($\approx 10$–$100r_\bullet$; see eq. 8.145) accretion disc surrounding the black hole (§3.6.2). When a mass $\mathcal{M}_{\rm acc}$ is accreted, the energy emitted as radiation by the black hole is (eq. 3.13)

$$E = \epsilon_{\rm rad}\mathcal{M}_{\rm acc}c^2, \tag{8.152}$$

where $\epsilon_{\rm rad} \approx 0.1$ (as estimated on the basis of the Soltan argument; §3.6.1) is the radiative efficiency. Photons can deposit energy into the surrounding gas via photoionisation (§8.1.2), that is, by ionising atoms or partially ionised ions, with consequent heating of the gas. Photoionisation heating is dominated by moderately high-energy X-ray photons that heat the gas at temperatures $\sim 10^7$ K.

The feedback mechanism described above is not the only feedback process produced by the photons. For accreted mass $\mathcal{M}_{\rm acc}$ the emitted photons carry momentum $p = E/c = \epsilon_{\rm rad}\mathcal{M}_{\rm acc}c$, where we have used eq. (8.152). This momentum can be at least partly transferred to the surrounding gaseous medium, thus driving an **AGN wind**, that is, an AGN-driven outflow of gas from the galaxy centre (analogous to a wind from stellar feedback; §8.7.2–§8.7.4).

Here we present an analytic description of some properties of these AGN winds, based simply on the conservation of mass, momentum and energy (see Ostriker et al., 2010, for more details). The net accretion rate $\dot{\mathcal{M}}_{\rm acc}$ onto the SMBH is given by the difference between the inflowing ($\dot{\mathcal{M}}_{\rm inflow}$) and the outflowing (wind, $\dot{\mathcal{M}}_{\rm w}$) mass flows:

$$\dot{\mathcal{M}}_{\rm acc} = \dot{\mathcal{M}}_{\rm inflow} - \dot{\mathcal{M}}_{\rm w} = \frac{\dot{\mathcal{M}}_{\rm inflow}}{1+\eta}, \tag{8.153}$$

where

$$\eta \equiv \frac{\dot{\mathcal{M}}_{\rm w}}{\dot{\mathcal{M}}_{\rm acc}} \tag{8.154}$$

is the ratio between wind and accretion rates. The mechanical luminosity of the wind (§8.7.2) is

$$L_{\rm w} = \frac{1}{2}\dot{\mathcal{M}}_{\rm w}v_{\rm w}^2 = \epsilon_{\rm w}\dot{\mathcal{M}}_{\rm acc}c^2, \tag{8.155}$$

where $v_{\rm w}$ is the wind speed and $\epsilon_{\rm w} \equiv L_{\rm w}/(\dot{\mathcal{M}}_{\rm acc}c^2)$ is the **mechanical efficiency** of the wind. Using eqs. (8.153)–(8.155), the wind mechanical luminosity can be written as

$$L_{\rm w} = \frac{\epsilon_{\rm w}}{1+\eta}\dot{\mathcal{M}}_{\rm inflow}c^2, \tag{8.156}$$

and the momentum flow of the wind $\dot{p}_{\rm w} = \dot{\mathcal{M}}_{\rm w}v_{\rm w}$ as

$$\dot{p}_{\rm w} = \frac{\eta}{1+\eta}\dot{\mathcal{M}}_{\rm inflow}v_{\rm w}. \tag{8.157}$$

From eqs. (8.154) and (8.155) it follows that

$$\eta = 2\epsilon_{\rm w}\frac{c^2}{v_{\rm w}^2}, \tag{8.158}$$

so, for a given gas inflow rate $\dot{M}_{\rm inflow}$, $L_{\rm w}$ and $\dot{p}_{\rm w}$ are fully determined by $\epsilon_{\rm w}$ and $v_{\rm w}$. Theoretical and observational estimates suggest typical values $\epsilon_{\rm w} \sim 10^{-3}$ for the mechanical efficiency and $v_{\rm w} \sim 10^4\,{\rm km\,s^{-1}}$ for the wind speed. For these values, $\eta$ is of the order of unity, which means that, even if $\epsilon_{\rm w} \ll \epsilon_{\rm rad}$, the effect of the wind is important (for $\eta = 1$, half of the inflowing mass is accreted onto the central black hole and the other half is ejected by the wind). Thus, even if the quasar mode is characterised by substantial radiative feedback, mechanical feedback is by no means negligible.

In summary, in the quasar mode phase of AGN activity, the central regions of the galaxy are characterised by the presence of relatively dense and cold gas (responsible for the accretion), but both radiative and mechanical feedback are effective in heating and displacing this gas from the galaxy centre, thus interrupting the fuelling of the AGN itself. In this phase the AGN is believed to self-regulate, alternating between powerful outbursts and more quiescent periods, as apparent from Fig. 8.20, which shows $f_{\rm Edd}$ as a function of time for a simulated quasar mode AGN. The fraction of time spent in outburst, known as the **duty cycle** of the AGN, is believed to be relatively small (of the order of $10^{-3}$–$10^{-2}$).

### 8.8.2 Radio Mode Feedback

Radio mode feedback is at work when the black hole is accreting at rates much lower than the Eddington rate ( $f_{\rm Edd} \lesssim 10^{-2}$) and when the ISM of the host system is mainly hot gas at the virial temperature (§8.2.1). The extreme manifestations of the radio mode feedback are the radio sources in the central galaxies of groups and clusters of galaxies, often referred to as central radio galaxies (§3.6 and §6.4.1), an example of which is shown in Fig. 8.21. In this case, radiative feedback is negligible, because the radiative luminosity of the AGN is very low, in terms of $L_{\rm Edd}$. Radio mode feedback is thus dominated by mechanical feedback. Radio galaxies are characterised by powerful **jets** and **radio lobes** produced by the activity of the central black hole. Jets and lobes, which are composed of relativistic plasma, are often found in pairs, roughly symmetric with respect to the central source. Jets are highly collimated outflows of material from the central regions of the galaxy. Though the mechanism producing the jets is not well understood (several possible models have been envisaged, in which the magnetic field plays a fundamental role), it appears clear that the jets interact with the surrounding medium mechanically. As a consequence of the interaction, the jets inflate lobes, which are extended cocoons with diffuse radio emission produced by synchrotron radiation (§D.1.7) from relativistic electrons.

In the massive elliptical galaxies that host radio sources, the ambient medium in which the AGN is immersed is mostly ionised hot gas associated with the host galaxy, group or cluster (§5.2.1, §6.2 and §6.4.2), which is detected as a diffuse X-ray emission. Spatially resolved observations in the X-rays have revealed that depressions in the X-ray surface brightness of these systems are often found to overlap with the radio lobes. The interpretation of these observational findings is that the relativistic plasma of the radio jets makes it way through the ambient medium by inflating **X-ray cavities** (or **bubbles**) in the ISM (or ICM; Fig. 8.21). The gas outside the cavities is at the virial temperature (§8.2.1)

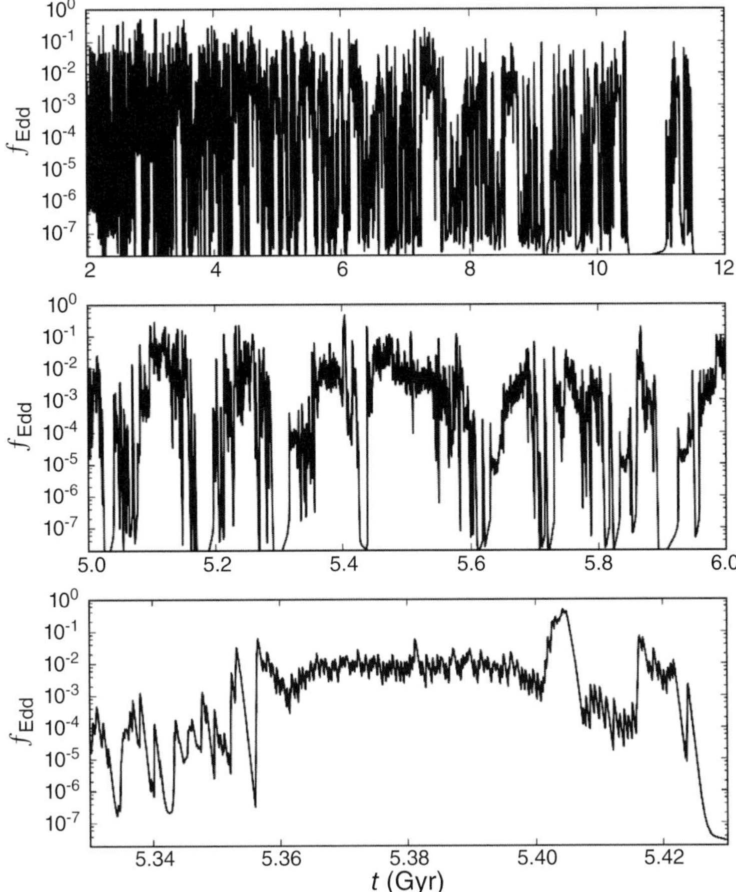


Fig. 8.20 *Top panel.* Eddington ratio $f_{Edd} = L/L_{Edd}$ as a function of time for an AGN in the centre of an elliptical galaxy, according to a 2D (axisymmetric) hydrodynamic simulation. *Middle and bottom panels.* Zoom-ins of the top panel. In this case the luminosity is always sub-Eddington, with peaks at $f_{Edd} \approx 0.6$. We note that the AGN spends only about 1% of the time at $f_{Edd} > 0.1$. From Novak et al. (2011). © AAS, reproduced with permission.

of the host system ($T_{vir} \sim 10^7$–$10^8$ K). Despite their name, the cavities are not empty, but contain relativistic plasma (emitting in radio; Fig. 8.21), which is confined by the pressure of the ambient medium.

In the framework of the study of AGN feedback, cavities are extremely useful because they allow us to estimate the mechanical energy transferred from the jet to the ambient medium and therefore to quantify the mechanical feedback of the radio mode. Here we report a simple estimate of the transfer of energy between the bubble and the ambient gas, following the treatment of Churazov et al. (2002), where the reader can find more details. The energy spent by the AGN to produce a cavity with volume $V$ and pressure $P$ is

$$E = U + W = \frac{\gamma}{\gamma - 1} PV, \tag{8.159}$$

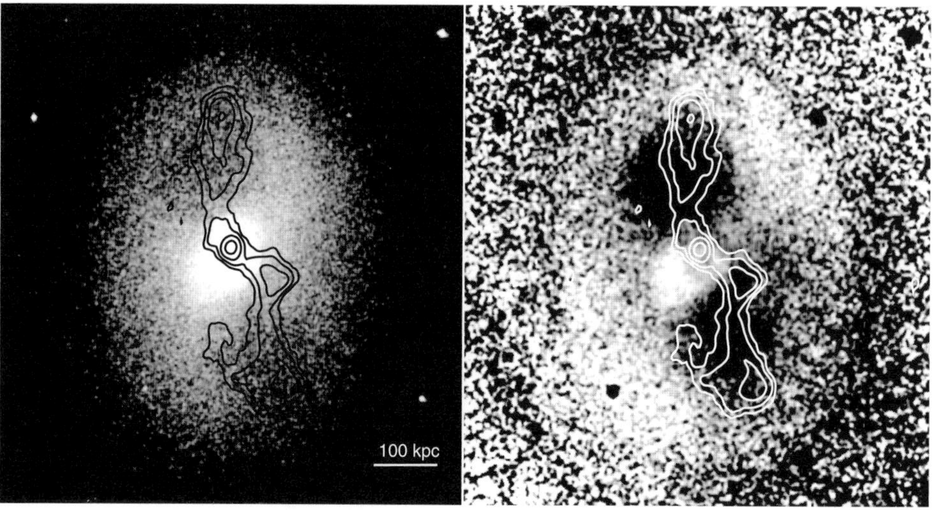


Fig. 8.21 *Left panel.* X-ray image of the cluster of galaxies MS 0735.6+742 at redshift $z \simeq 0.216$ (lighter regions have higher X-ray surface brightness). The contours indicate the 327 MHz emission produced by the central radio galaxy. *Right panel.* Same as the left panel, but with X-ray image obtained after subtraction of the $\beta$-model (§6.4.2) that best fits the smooth surface brightness distribution. The so-called X-ray cavities are the dark regions overlapping with the radio lobes. See McNamara and Nulsen (2007) and Vantyghem et al. (2014) for details. Data courtesy of B. McNamara and A. Vantyghem. Figure courtesy of M. Gitti.

where

$$U = \frac{1}{\gamma - 1} PV \tag{8.160}$$

is the internal energy of the bubble and $W = PV$ is the work done to expand the ambient medium ($\gamma$ is the adiabatic index of the relativistic gas filling the cavity). Assuming that the ambient medium is spherically symmetric, the bubble is lifted by a radial buoyancy force

$$\boldsymbol{f}_{\rm buoy} = \rho V \frac{{\rm d}\Phi}{{\rm d}r} \frac{\boldsymbol{x}}{r}, \tag{8.161}$$

where $\boldsymbol{x}$ is the position vector, $r = |\boldsymbol{x}|$, $\rho(r)$ is the density of the ambient gas and $\Phi(r)$ is the gravitational potential of the host system. While rising, the bubble experiences from the ambient medium a drag force $\boldsymbol{f}_{\rm drag}$ opposite to the direction of motion. If the drag is efficient, the motion of the bubble is soon determined by the condition $\boldsymbol{f}_{\rm drag} = -\boldsymbol{f}_{\rm buoy}$. When the bubble is at $\boldsymbol{x}$, the mechanical energy transferred to the ambient gas is

$$\Delta E = \int_{\boldsymbol{x}_0}^{\boldsymbol{x}} \boldsymbol{f}_{\rm drag} \cdot {\rm d}\boldsymbol{x} = -\int_{\boldsymbol{x}_0}^{\boldsymbol{x}} \boldsymbol{f}_{\rm buoy} \cdot {\rm d}\boldsymbol{x} = -\int_{r_0}^{r} \rho V \frac{{\rm d}\Phi}{{\rm d}r} {\rm d}r, \tag{8.162}$$

where $\boldsymbol{x}_0$ is the initial (i.e. when the bubble is close to the centre of the host system) position and $r_0 = |\boldsymbol{x}_0|$. Let $P_0$ and $V_0$ be, respectively, the initial pressure and volume of the bubble. Assuming that the bubble expands adiabatically ($PV^\gamma$ = const) and that the ambient medium is in hydrostatic equilibrium ($\rho\, \mathrm{d}\Phi/\mathrm{d}r = -\mathrm{d}P/\mathrm{d}r$), eq. (8.162) gives

$$\Delta E = -V_0 P_0 \int_{P_0}^{P} \left(\frac{P'}{P_0}\right)^{-1/\gamma} \frac{\mathrm{d}P'}{P_0} = E_0 \left[1 - \left(\frac{P}{P_0}\right)^{(\gamma-1)/\gamma}\right], \qquad (8.163)$$

where $E_0 = \gamma P_0 V_0/(\gamma - 1)$. As soon as the bubble is far enough from the centre that $P/P_0 \ll 1$, $\Delta E \approx E_0$, which means that the entire mechanical energy responsible for inflating the bubble (eq. 8.159) is transferred to the ISM or ICM.

The analysis of the X-ray cavities thus suggests that the mechanical energy provided in the radio mode can be efficiently transferred to the ambient medium. The efficiency is expected to be maximal when the AGN lies at the bottom of the deep potential well of a cluster or group of galaxies, because the ambient medium is relatively dense (§6.2 and §6.4.2), while the efficiency might be somewhat reduced in isolated galaxies, in which the ambient medium is less dense and powerful relativistic jets can propagate easily out of the core. As happens for the quasar mode, also the radio mode feedback process is not continuous: the production of relativistic jets is intermittent (with duty cycle $10^{-3}$–$10^{-2}$), thus subsequent generations of radio lobes and cavities are produced. At least in the case of central galaxies in galaxy clusters, the average mechanical luminosity of the central AGN is of the order of the X-ray luminosity of the ICM; thus mechanical heating from AGN feedback is believed to be the main mechanism responsible for halting cooling flows in cool-core clusters (§6.4.3 and §10.6.4).

## 8.9 Merging of Galaxies

The process of **galaxy merging** is an **encounter** of two or more galaxies that leads to the formation of a single galaxy. We have seen that there is observational evidence that this process occurs (§6.1), and it is also predicted theoretically to be a fundamental mechanism for structure formation in a CDM Universe (§7.4.4).

First of all, it must be stressed that not all encounters of galaxies end up with the formation of a new single galaxy. The fate of an encounter depends not only on its orbital parameters (§8.9.5), but also on the internal properties of the interacting galaxies. Broadly speaking, when the relative speed is high, the encounter is classified as a **fly-by** (or **high-speed encounter**): the effect of the interaction is a temporary perturbation and the interacting galaxies do not merge. When the relative speed is sufficiently low, the interacting galaxies merge to form a single galaxy in a relatively short time (§8.9.5), and the encounter is classified as a **merger**. Here we focus on mergers, which are more important than fly-bys for galaxy evolution. We note however that the fly-bys can be important for the structural evolution of galaxies in galaxy clusters: the cumulative effect of several high-speed encounters on a cluster galaxy is usually referred to as **galaxy harassment**.

It is useful to introduce a classification of galaxy merging, which is widely used in the literature. Based on the number of galaxies involved, we make a distinction between **binary merging**, when only two galaxies are involved, and **multiple merging**, when more than two galaxies are involved. In the case of binary merging, based on the mass ratio of the merging systems, we can have either minor or major merging (§7.4.4). Mergers can also be classified according to the matter content of the merging systems. If all the involved galaxies are gas-poor (i.e. they are mainly composed of stars and dark matter), the process is called **dissipationless** (or **dry**) **merging**. Otherwise, the process is called **dissipative** (or **wet**) **merging**. In the presence of gas the process is dissipative because gas can dissipate energy by cooling radiatively (§8.1.1). An interaction among galaxies composed only of stars and dark matter is dissipationless, because there are variations of potential and kinetic energy, but the energy is not dissipated.

## 8.9.1 The Physics of Galaxy Merging

Galaxy merging involves the interaction of stars, dark matter and gas. As far as the evolution of the stellar component is concerned, we must keep in mind that the stellar density is so low that star–star physical collisions in practice do not occur. Moreover, the stellar and dark matter components are collisionless, so two-body relaxation (§8.9.2) is not at work. Instead, the evolution of merging stellar systems made only of stars and dark matter is driven by collisionless relaxation (§8.9.2). The behaviour of gas during galaxy interactions is more complex. Gas is dissipative and collisional, so during the merging process it can be compressed, shock heated and dissipate energy radiatively. As a consequence of gas inflows, compression and cooling, it is also possible that the physical conditions of the gas become such that it can fragment and form stars (§8.3). Numerical $N$-body and hydrodynamic simulations (§10.11.1) are thus fundamental tools to describe quantitatively the effect of merging and to predict the properties of the merger remnant, for given properties of the progenitor galaxies and of the encounter. In the following sections we give an overview of the main physical processes involved in galaxy merging.

## 8.9.2 Collisionless Relaxation

Galaxy merging is an example of a dynamical process that starts from non-equilibrium initial conditions (two or more colliding galaxies) and ends up in equilibrium (a single virialised galaxy). The mechanisms by which a system can reach equilibrium are usually called **relaxation processes**. Let us, for simplicity, focus on a non-equilibrium dynamical system made only of stars and dark matter particles, which interact only gravitationally. One possibility is that this system reaches equilibrium via **two-body relaxation**, which is relaxation following exchange of energy in two-body interactions between particles. The **two-body relaxation time** $t_{2b}$, which is the time necessary for an $N$-body system to reach equilibrium thanks to two-body interactions, increases with increasing number of particles $N$. Based on estimates of $t_{2b}$, $N$-body systems are classified as **collisional** (when $t_{2b} \lesssim t_H$, where $t_H$ is the Hubble time) and **collisionless** (when $t_{2b} \gtrsim t_H$).

Two-body relaxation is effective for collisional systems (star clusters) but is negligible for collisionless systems (galaxies and dark matter halos), so it cannot be responsible for virialisation in galaxy merging. Analytic calculations and numerical $N$-body simulations show that collisionless systems, for which collisional (two-body) relaxation is *not* at work, can nevertheless virialise in sufficiently short timescales thanks to **collisionless relaxation**. The fundamental collisionless relaxation processes are **violent relaxation** (Lynden-Bell, 1967), that is the redistribution of orbital energies in a strongly time-varying gravitational potential, and **phase mixing**, that is the fact that the orbits of stars (and dark matter particles) tend to be spread in phase space (§C.8) and thus mix (see Binney and Tremaine, 2008). In practice, during merging, similar to what happens in collisionless gravitational collapse (§7.3.2 and §10.3.1), phase mixing and violent relaxation co-operate and lead, in a timescale comparable to the dynamical times (eq. 8.24) of the merging galaxies, to the virialisation and formation of the merger remnant.

### 8.9.3 Dynamical Friction

**Dynamical friction** is the deceleration of a massive body orbiting through a distribution of much less massive particles due to the gravitational interactions between the massive body and the low-mass particles. In the case of a very minor merger (a massive galaxy of mass $\mathcal{M}_{\rm host}$ accreting a much less massive satellite galaxy of mass $\mathcal{M}_{\rm sat} \ll \mathcal{M}_{\rm host}$), the process of merging can be described in terms of dynamical friction, where the population of low-mass particles consists of the stars and dark matter particles of the more massive galaxy, while the satellite of mass $\mathcal{M}_{\rm sat}$ is the massive body. While decelerating, the satellite transfers energy and angular momentum to the host galaxy: this is an example of **dynamical friction heating**, a process that has important implications for the structure and kinematics of dark matter halos (§7.5.5). Also the interaction of a galactic bar with the host dark halo is described by the dynamical friction formalism. The bar, which plays the role of the massive body, is decelerated and the halo is heated (§10.2.1).

Dynamical friction is important in several aspects of the galaxy merging phenomenon: minor mergers, evolution of galaxies in common dark matter halos, galactic cannibalism in galaxy clusters (§6.4.1) and evolution of SMBHs in merging galaxies (§8.9.7). However, merging is not always driven by dynamical friction. It must be stressed that the dynamical friction formalism does *not* apply in the case of major mergers. When the merging galaxies have similar masses, none of them can be considered either host or satellite, and the process can be conveniently described in terms of violent relaxation and phase mixing.

Though in general it is not possible to compute analytically the dynamical friction force, it is useful to report the expression that is obtained in the idealised case in which the background particles (stars and dark matter particles of the host) have uniform spatial distribution with density $\rho$ and Maxwellian velocity distribution (eq. D.12) with 1D velocity dispersion $\sigma$ (see Binney and Tremaine, 2008, for details). In this case the deceleration of the massive body of mass $\mathcal{M}_{\rm sat}$, travelling with velocity $\boldsymbol{v}$, is

$$\frac{\mathrm{d}\boldsymbol{v}}{\mathrm{d}t} = -4\pi \ln \Lambda \frac{G^2 \rho \mathcal{M}_{\rm sat}}{v^2} \mathcal{F}\left(\frac{v}{\sqrt{2}\sigma}\right)\left(\frac{\boldsymbol{v}}{v}\right), \tag{8.164}$$

where

$$\mathcal{F}(x) = \mathrm{erf}(x) - \frac{2x}{\sqrt{\pi}} e^{-x^2}, \tag{8.165}$$

$\mathrm{erf}(x) = (2/\sqrt{\pi}) \int_0^x \exp(-s^2)\mathrm{d}s$ is the error function, and the dimensionless quantity $\ln\Lambda$ is the **Coulomb logarithm**, which is the logarithm of the ratio between the maximum and minimum impact parameters of the encounters between the massive body and the background particles (the impact parameter is, broadly speaking, a measure of the distance at closest approach). Eq. (8.164) is known as the **Chandrasekhar dynamical friction formula** (Chandrasekhar, 1943). We note that when the individual masses of the background particles are much lower than $\mathcal{M}_{\rm sat}$, the deceleration $\mathrm{d}\boldsymbol{v}/\mathrm{d}t$ depends only on the background particle density ($\rho$ in eq. 8.164) and not on their individual masses. The deceleration $\mathrm{d}\boldsymbol{v}/\mathrm{d}t$ is proportional to $\mathcal{M}_{\rm sat}$, so the effect is more important for more massive satellites.

Though eq. (8.164) has been derived for a uniform background distribution, we expect it to be approximately applicable, with a suitable value of $\ln\Lambda$, to a body of mass $\mathcal{M}_{\rm sat}$ orbiting in a spherical system with density $\rho(r)$ and total mass $\mathcal{M}_{\rm host} \gg \mathcal{M}_{\rm sat}$. Let us assume, for instance, that the mass density distribution of the host system is that of a singular isothermal sphere (eq. 5.37) with 1D velocity dispersion $\sigma = v_{\rm c}/\sqrt{2}$, where $v_{\rm c}$ is the (radius-independent) circular speed. In this case, for a satellite of mass $\mathcal{M}_{\rm sat}$ orbiting on a circular orbit ($v = v_{\rm c}$) at radius $r$, eq. (8.164) gives

$$\frac{\mathrm{d}v}{\mathrm{d}t} = -A \ln\Lambda \frac{G\mathcal{M}_{\rm sat}}{r^2} = -2\pi A \ln\Lambda \frac{G^2 \rho \mathcal{M}_{\rm sat}}{\sigma^2}, \tag{8.166}$$

where $A = \mathcal{F}(1) \simeq 0.4276$ (eq. 8.165) and $\rho \propto r^{-2}$. The dynamical friction timescale

$$t_{\rm fric} \equiv \frac{v}{|\mathrm{d}v/\mathrm{d}t|} = \frac{1}{\sqrt{2}\pi A \ln\Lambda} \frac{\sigma^3}{G^2 \rho \mathcal{M}_{\rm sat}} \tag{8.167}$$

is inversely proportional to $\mathcal{M}_{\rm sat}$ and $\rho$, for given velocity dispersion $\sigma$ of the host. Alternatively (using eq. 5.37), eq. (8.167) can be written as

$$t_{\rm fric} = \frac{\sqrt{2}}{A \ln\Lambda} \frac{\sigma r^2}{G\mathcal{M}_{\rm sat}} \simeq 3.0 \left(\frac{\ln\Lambda}{5}\right)^{-1} \left(\frac{\sigma}{200\,\mathrm{km\,s^{-1}}}\right) \left(\frac{r}{10\,\mathrm{kpc}}\right)^2 \left(\frac{\mathcal{M}_{\rm sat}}{10^9 \mathcal{M}_\odot}\right)^{-1} \mathrm{Gyr}, \tag{8.168}$$

which means that, for instance, in the Milky Way, over a time span of a few Gyr, the orbit of a massive satellite, such as the Sagittarius dwarf galaxy (§4.6.1), has substantially shrunk, while the orbits of lower-mass satellites, such as globular clusters (§4.6.1), have not been significantly modified by dynamical friction.

### 8.9.4 Tidal and Ram-Pressure Stripping

A galaxy (satellite) orbiting close to or within another galaxy (host) is subject to **tidal forces**, because the host system exerts gravitational forces with different strengths in different parts of the satellite. As a consequence of these tidal forces, the satellite can lose matter, a phenomenon known as **tidal stripping**. The process of tidal stripping depends on the structure of the host and of the satellite, as well as on the satellite's orbit. Specific predictions for tidal stripping can be obtained with $N$-body simulations. However, it is possible to obtain a useful analytic approximation to the effect of tidal stripping. The size of the tidally stripped satellite can be approximately described by a **tidal radius** $r_t$ such that particles within a sphere of radius $r_t$ (centred at the centre of the satellite) are gravitationally bound to the satellite, while particles outside this sphere are not bound to the satellite. The latter form the so-called **tidal tails**, that are leading and trailing narrow structures formed from material stripped from the satellite (Fig. 8.22).

The simplest estimate of the tidal radius can be obtained by means of the **circular restricted three-body problem**, in which we study the motion of a test particle of mass $m$ (representing, for instance, a star of the satellite) in the joint gravitational potential of two point masses (the host, with mass $\mathcal{M}_{\rm host} \gg m$, and the satellite, with mass $\mathcal{M}_{\rm sat} \gg m$) orbiting each other in circular orbits. The circular restricted three-body problem is characterised by five **Lagrangian points**, which are stationary points in the rotating reference frame in which the host and the satellite are at rest. The two Lagrangian points

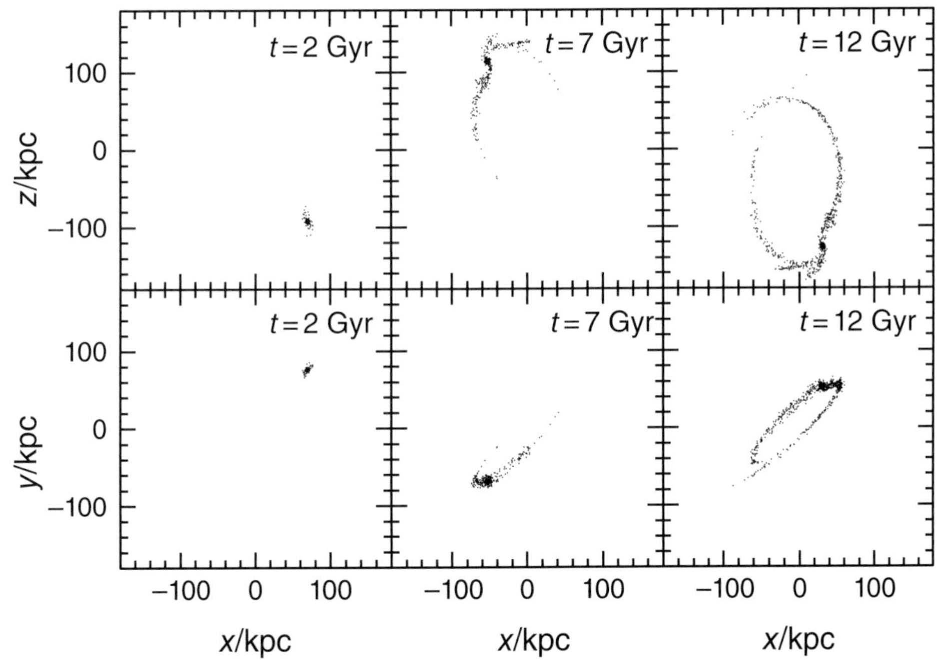


Fig. 8.22 Projected distributions in the Cartesian $x$–$z$ (*top panels*) and $x$–$y$ (*bottom panels*) planes of the stars of a simulated dwarf satellite galaxy of mass $1.5 \times 10^8 \mathcal{M}_\odot$ orbiting a Milky Way-like galaxy (with centre of mass at $x=y=z=0$) after 2 Gyr (*left panels*), 7 Gyr (*middle panels*) and 12 Gyr (*right panels*) of evolution. In this $N$-body simulation the satellite is modelled as an $N$-body system with $N = 51\,200$ stellar particles. The host galaxy is modelled as a fixed analytic axisymmetric gravitational potential, where $z$ is the axis of symmetry. The satellite, which is spherically symmetric at the beginning of the simulation ($t=0$), develops around its core prominent tidal tails that become more extended with time. Adapted from Battaglia et al. (2015).

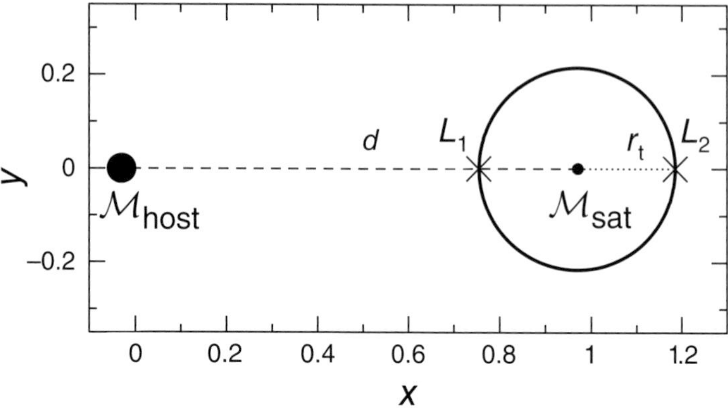


Fig. 8.23 Representation of the tidal interaction between a host galaxy of mass $\mathcal{M}_{host}$ and a satellite galaxy of mass $\mathcal{M}_{sat} \simeq 0.03\mathcal{M}_{host}$ in circular orbit. In the framework of the Hill approximation of the circular restricted three-body problem, the two mass distributions are represented by point masses. Here $(x, y)$ is a non-inertial Cartesian reference system in the plane of the orbit, corotating with the galaxies, with origin at the centre of mass (in this system the positions of the two galaxies do not change in time). As a consequence of tidal stripping, a satellite in circular orbit at a distance $d$ (dashed line) from the centre of the host is tidally truncated at the tidal radius $r_t$ (dotted line; eq. 8.169). $L_1$ and $L_2$ are the two Lagrangian points closest to the satellite.

closest to the satellite ($L_1$ and $L_2$; Fig. 8.23) mark the size of the volume around the satellite within which particles are bound to the satellite. In the limit $\mathcal{M}_{sat} \ll \mathcal{M}_{host}$ (known as the **Hill approximation**) this volume is almost spherical and these two Lagrangian points lie approximately at the same distance

$$r_t = \left(\frac{\mathcal{M}_{sat}}{3\mathcal{M}_{host}}\right)^{1/3} d \qquad (8.169)$$

from the satellite's centre, where $d$ is the separation between the centres of the satellite and of the host system (Fig. 8.23). The quantity $r_t$, known as the **Hill radius** or **Roche radius**, can therefore be considered as the tidal radius of a satellite of mass $\mathcal{M}_{sat}$ orbiting a host system of mass $\mathcal{M}_{host}$ at a distance $d$ from the centre of the host. As the host and satellite galaxies are in fact extended systems, we can identify $\mathcal{M}_{sat}$ with the mass of the satellite within $r_t$ and $\mathcal{M}_{host}$ with the mass of the host within $d$, and rewrite eq. (8.169) as

$$\langle\rho\rangle_{sat} = 3\langle\rho\rangle_{host}, \qquad (8.170)$$

where $\langle\rho\rangle_{sat}$ is the average density of the satellite within $r_t$ and $\langle\rho\rangle_{host}$ is the average density of the host within $d$. Eq. (8.170) shows that for a satellite to survive against tidal stripping its characteristic density must be at least of the order of the average density of the host within its orbit. When a satellite is completely disrupted by tidal stripping, the remnant of its tidal tails is called a **tidal stream**, which is a very elongated stellar structure with no evident nucleus (Fig. 4.4, right, and Fig. 10.9).

If both the satellite galaxy and the host system have a gaseous component, in addition to tidal stripping, also **ram-pressure stripping** can contribute to deprive the satellite of

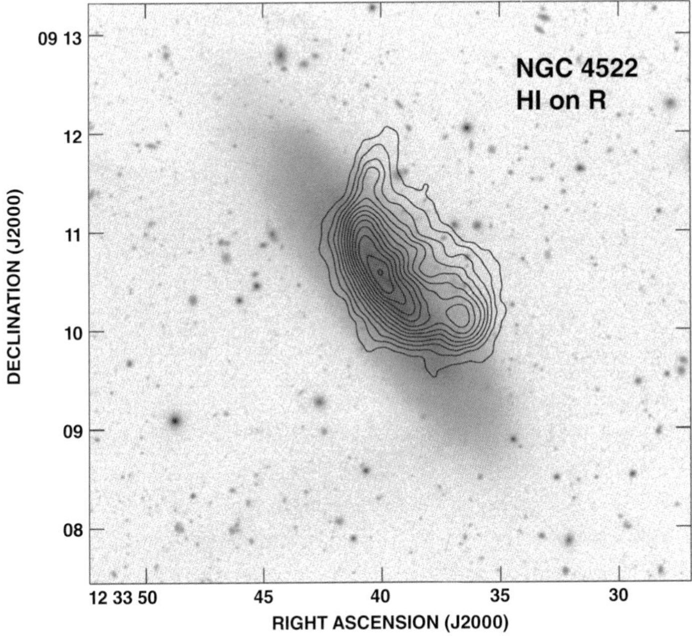


Fig. 8.24 A 21 cm map (contours; tracing the neutral atomic hydrogen) and *R*-band image (greyscale; tracing the stars) of the spiral galaxy NGC 4522 in the Virgo cluster of galaxies. It is apparent that the gas distribution is distorted by ram pressure from the ICM, while the stellar disc is undisturbed. The morphology suggests that the galaxy is moving in the plane of the sky towards the bottom-left of the image. The frame is about 25 kpc × 25 kpc. From Kenney et al. (2004). ©AAS, reproduced with permission.

its gas. For instance, let us consider a disc galaxy orbiting in a cluster of galaxies. If the galaxy moves at speed $v$, the ICM exerts on the gaseous disc of the galaxy a ram pressure

$$P \approx \rho_{\rm ICM} v^2, \tag{8.171}$$

where $\rho_{\rm ICM}$ is the ICM density. If this ram pressure exceeds the gravitational force per unit area exerted by the galaxy onto its gaseous disc, the gas is stripped by the galaxy and becomes part of the ICM. For instance, for a bulge-dominated galaxy with stellar velocity dispersion $\sigma$ and ISM gas density $\rho_{\rm ISM}$, the condition to have ram-pressure stripping is $\rho_{\rm ICM} v^2 \gtrsim \rho_{\rm ISM} \sigma^2$. A manifest example of a galaxy subjected to ram-pressure stripping from the ICM is the spiral NGC 4522 in the Virgo cluster of galaxies (Fig. 8.24).

### 8.9.5 Orbital Parameters and Merging Timescale

Fundamental quantities to characterise galaxy encounters are the orbital parameters. Let us consider an encounter between two galaxies of masses $\mathcal{M}_1$ and $\mathcal{M}_2$. Assuming that the galaxy pair is sufficiently isolated from other galaxies, when the separation between the two galaxies is large, their orbit can be described as a **two-body orbit**, approximating each of the two systems as a point mass. In this two-body point-mass approximation the orbit of the galaxy encounter is fully described by the orbital energy

$$E_{\rm orb} = \frac{1}{2}\mathcal{M}_{\rm red}v^2 - \frac{G\mathcal{M}_{\rm red}\mathcal{M}_{1+2}}{d}, \tag{8.172}$$

and by the orbital angular momentum

$$\boldsymbol{J}_{\rm orb} = \mathcal{M}_{\rm red}\boldsymbol{d}\times\boldsymbol{v}, \tag{8.173}$$

where $\boldsymbol{d}$ is the separation vector, $\boldsymbol{v}$ is the relative velocity, $\mathcal{M}_{1+2} \equiv \mathcal{M}_1 + \mathcal{M}_2$ is the total mass, and $\mathcal{M}_{\rm red} \equiv \mathcal{M}_1\mathcal{M}_2/\mathcal{M}_{1+2}$ is the **reduced mass**. Based on the orbital energy, an encounter between two galaxies is classified as **elliptic** ($E_{\rm orb} < 0$), **parabolic** ($E_{\rm orb} = 0$) or **hyperbolic** ($E_{\rm orb} > 0$). Based on the orbital angular momentum, we make a distinction between **head-on encounters** ($J_{\rm orb} \approx 0$), when the closest approach distance is much smaller than the characteristic sizes of the galaxies, and **off-axis encounters** ($J_{\rm orb} \neq 0$), when the closest approach distance is larger.

We define the **merging timescale** $\tau_{\rm merg}$ as the time elapsing from the first close passage between the two galaxies and the virialisation of the merger remnant. Thus, we can say that an encounter leads to a merger only if $\tau_{\rm merg}$ is shorter than the Hubble time. Clearly, $\tau_{\rm merg}$ depends not only on the orbital parameters, but also on the properties of the involved galaxies. For given properties of the two galaxies, $\tau_{\rm merg}$ increases for increasing $E_{\rm orb}$ and for increasing $J_{\rm orb}$. Merging is thus favoured for elliptic or parabolic orbits ($E_{\rm orb} \leq 0$) and for head-on collisions ($J_{\rm orb} \approx 0$). Rapid mergers occur when the orbits are bound, $E_{\rm orb} < 0$, but also for unbound orbits ($E_{\rm orb} \geq 0$), provided the orbital angular momentum magnitude $J_{\rm orb}$ is sufficiently low. Given the variety of internal structures, kinematics and components (stars, gas and dark matter) of interacting galaxies, it is difficult to quantify the effect of the internal galaxy properties on $\tau_{\rm merg}$. However, a useful rule of thumb is that galaxy merging is possible only if the relative speed at the time of closest approach is of the order of (or lower than) the characteristic internal velocities (rotation speeds or velocity dispersions) of the colliding galaxies, because for higher collision speeds the gravitational perturbation is too rapid to have strong effects on the internal dynamics of the interacting galaxies.

To give an idea of the order of magnitude of the merging timescale, it is useful to note that for parabolic equal-mass mergers $\tau_{\rm merg}$ is typically of the order of 10 internal dynamical times $t_{\rm dyn}$ (eq. 8.24). For instance, for Milky Way-size galaxies (with virial mass $\sim 10^{12}\mathcal{M}_\odot$), $t_{\rm dyn} \sim 10^8$ yr, so $\tau_{\rm merg} \sim 10^9$ yr (see Fig. 10.7). For minor mergers $\tau_{\rm merg}$ is of the order of the dynamical friction timescale $t_{\rm fric}$ (§8.9.3), which, for given mass of the main galaxy, scales approximately as $1/\zeta$, where $\zeta$ is the mass ratio between the satellite and the main galaxy (§7.4.4). The results of $N$-body simulations suggest that, for typical orbital parameters and galaxy properties, $\tau_{\rm merg}$ is shorter than the Hubble time when $\zeta \gtrsim 0.01$.

In the case of encounters between galaxies that have significant spin (i.e. internal angular momentum), typically disc galaxies, besides $E_{\rm orb}$ and $J_{\rm orb}$, additional orbital parameters that can influence the merging timescale are the orientations of the spins of the galaxies. When the spins and the orbital angular momentum point in the *same* direction the encounter is called **prograde**, otherwise it is called **retrograde**. Merging tends to be *favoured* by prograde orbits and *disfavoured* by retrograde orbits.

### 8.9.6 Properties of the Merger Remnants

An important piece of information is how galaxies are transformed by mergers. For instance, in a binary merger we would like to be able to predict, for given properties and orbital parameters of the colliding galaxies, the properties of the merger remnant. Of course this question is extremely complex. In the relatively simple case of merging between collisionless systems, a robust answer is obtained with $N$-body simulations. In the presence of gas, it is not easy to make quantitative predictions, even with hydrodynamic simulations, because the properties of the merger remnant depend on not fully understood processes such as star formation (§8.3) and feedback from stars (§8.7) and AGNs (§8.8). Nevertheless, some important global properties can be predicted without resorting to numerical simulations.

#### Dissipationless (Dry) Mergers

Let us first consider the simplest possible case, that is the merging of galaxies with no gas. These dry mergers are encounters between collisionless stellar systems, made of stars and dark matter (for instance, two elliptical galaxies[23]). Given the dissipationless nature of the systems, the total energy is conserved in the merging. Ordered kinetic energy (orbital energy) is converted into random kinetic energy: in this sense a collision between galaxies can be thought of as an inelastic collision. Neglecting mass loss, the conservation of total energy in a merger with orbital energy $E_{\rm orb}$ (eq. 8.172) can be written as

$$E_{1+2} = E_1 + E_2 + E_{\rm orb}, \tag{8.174}$$

where $E_1$ and $E_2$ are the total energies of the two interacting galaxies (before the collision), and $E_{1+2}$ is the total energy of the remnant. The galaxy labelled with $i$ (where $i$ = '1', '2' or '1+2') has virial mass $\mathcal{M}_{{\rm vir},i}$, virial velocity dispersion $\sigma_{{\rm vir},i}$ and gravitational radius $r_{{\rm g},i}$ (§5.3.1). For the sake of clarity, in this section we indicate these quantities simply as $\mathcal{M}_i$, $\sigma_i$ and $r_i$, respectively. Using the virial theorem (eq. 5.12), the total energy $E_i = K_i + W_i$, where $K_i$ is the kinetic energy (eq. 5.13) and $W_i$ is the gravitational potential energy (eq. 5.14) of a galaxy of total mass $\mathcal{M}_i$, can be written as

$$E_i = -K_i = -\frac{1}{2}\mathcal{M}_i\sigma_i^2. \tag{8.175}$$

Combining eq. (8.174) with eq. (8.175), the virial velocity dispersion of the merger remnant $\sigma_{1+2}$ is given by

$$\sigma_{1+2}^2 = \frac{\mathcal{M}_1\sigma_1^2 + \mathcal{M}_2\sigma_2^2 - 2E_{\rm orb}}{\mathcal{M}_{1+2}}. \tag{8.176}$$

[23] Elliptical galaxies do contain gas (§5.2), but most of their gas is hot and located in the outskirts. Thus, the main features of a merger between two ellipticals are expected to be captured by a collisionless model.

In the case of a merger between two identical systems ($\mathcal{M}_1 = \mathcal{M}_2$ and $\sigma_1 = \sigma_2$), the final system has mass $2\mathcal{M}_1$ and virial velocity dispersion given by

$$\frac{\sigma_{1+2}^2}{\sigma_1^2} = 1 - \frac{E_{\rm orb}}{\mathcal{M}_1\sigma_1^2}, \tag{8.177}$$

and the evolution of the gravitational radius (eq. 5.15)

$$r_i = \frac{G\mathcal{M}_i}{\sigma_i^2} \tag{8.178}$$

is given by

$$\frac{r_{1+2}}{r_1} = 2\left(1 - \frac{E_{\rm orb}}{\mathcal{M}_1\sigma_1^2}\right)^{-1}. \tag{8.179}$$

Thus, for equal-mass parabolic encounters ($E_{\rm orb} = 0$), $M_{1+2} = 2\mathcal{M}_1$, $r_{1+2} = 2r_1$ and $\sigma_{1+2} = \sigma_1$ (the final system has twice the mass and the size, but the same velocity dispersion as the progenitors). The final velocity dispersion is higher for elliptic orbits ($E_{\rm orb} < 0$) and lower for hyperbolic orbits ($E_{\rm orb} > 0$).

Let us now consider the parabolic ($E_{\rm orb} = 0$) merging of two systems with mass ratio $\mathcal{M}_2/\mathcal{M}_1 = \zeta \leq 1$ and squared velocity dispersion ratio $\sigma_2^2/\sigma_1^2 = \zeta^\alpha$. The final mass is $\mathcal{M}_{1+2} = (1+\zeta)\mathcal{M}_1$ and eq. (8.176) implies that the final virial velocity dispersion is given by

$$\frac{\sigma_{1+2}^2}{\sigma_1^2} = \frac{1+\zeta^{1+\alpha}}{1+\zeta}. \tag{8.180}$$

Combining eqs. (8.178) and (8.180), we find that the final gravitational radius $r_{1+2}$ is given by

$$\frac{r_{1+2}}{r_1} = \frac{(1+\zeta)^2}{1+\zeta^{1+\alpha}}. \tag{8.181}$$

When $\zeta < 1$, for $\alpha > 0$ (as expected; §5.4.2 and §5.4.3), the velocity dispersion of the remnant is lower[24] than that of the main progenitor and the gravitational radius of the remnant is such that $r_{1+2}/r_1 \geq \mathcal{M}_{1+2}/\mathcal{M}_1$ (the increase of the gravitational radius with mass is superlinear).

The above relationships allow us to predict the evolution of the virial velocity dispersion and of the gravitational radius. How these quantities are related to the observable properties of the galaxies (such as, for instance, the effective radius $R_{\rm e}$ and the central velocity dispersion $\sigma_0$; §3.1.2 and §5.1.2) depends on the structure and kinematics of the merger remnant. The general trend is that dry mergers tend to make systems less dense, so the remnant will have, in proportion to mass, larger size and lower velocity dispersion than the progenitors.

The morphology of the merger remnant can be very different from the progenitors. For instance, in the case of major mergers, stellar discs are likely to be disrupted and the remnant is typically characterised mainly by spheroidal components (see also §10.3.2).

[24] We recall that, in general, higher mass $\mathcal{M}$ does not imply higher virial velocity dispersion $\sigma$, because $\sigma$ depends not only on $\mathcal{M}$, but also on the gravitational radius $r$ ($\sigma^2 \propto \mathcal{M}/r$; eq. 5.15).

### Dissipative (Wet) Mergers

In the case of wet mergers (i.e. in which at least one of the merging galaxies is gas-rich), the process is dissipative, so the total energy is not conserved and the relations derived above do not apply. However, it is possible, at least qualitatively, to understand the effect of dissipative processes on the properties of the remnant. As a consequence of dissipation the gas tends to accumulate at the centre of the remnant, with possible associated star formation (§8.3), thus deepening the gravitational potential well. As a result the remnant tends to be more concentrated (smaller size and higher velocity dispersion) than in an equivalent dissipationless case. However, additional complications derive from the fact that, in the presence of gas, also stellar and AGN feedback (§8.7 and §8.8) processes are at work, which can be crucial in determining the properties of the remnant. When the merging systems are very gas-rich, the compression of the gas can lead to very intense episodes of star formation, which are believed to be associated with the observed phenomenon of starburst (§4.5 and §11.2.9).

## 8.9.7 Supermassive Black Holes in Merging Galaxies

It is common for ETGs and early spirals to host at their centre an SMBH with a mass of the order of 1/1000th of the stellar mass of the host ETG or bulge (§3.6.1, §4.6.3 and §5.4.4). Therefore, when two galaxies merge it is possible that a binary SMBH forms at the centre of the merger remnant. Let us consider, for instance, the case of a *minor* merger. The satellite galaxy (with its SMBH) spirals in down to the centre of the more massive galaxy as a consequence of dynamical friction (§8.9.3). In the mean time the satellite gradually loses its dark matter and baryons because of tidal stripping (§8.9.4), so that the SMBH of the satellite ends up orbiting 'naked' around the SMBH of the host. Then the orbit of the binary black hole shrinks, first because of dynamical friction, then via three-body interactions with the stars in the core of the host. If the orbit becomes sufficiently bound, the two black holes eventually coalesce due to emission of gravitational waves and form a single remnant SMBH at the centre of the remnant galaxy. In a *major* merger the stellar nuclei (i.e. the stars in the innermost regions) of the progenitors coalesce to form the stellar nucleus of the remnant, so also in this case the two SMBHs are expected to form in the centre of the remnant a binary SMBH, which then evolves in a way similar to the case of a minor merger.

# 9 From Recombination to Reionisation

This chapter describes the main topics related to the early evolution of baryonic matter during the first billion years after the Big Bang, the formation of the first luminous objects and their influence on the IGM. At these cosmic epochs (say from $z \sim 1000$ to $z \sim 10$) the Universe goes through crucial transition phases. The first is the cosmological recombination, when baryonic matter changes from a fully ionised plasma (§2.4) to an almost completely neutral gas composed of H, D, He and Li atoms and a few simple molecules. This gas is called pregalactic because neither luminous stars nor galaxies were present at that time, and therefore this era is named the **dark ages**. At these early times, primordial molecules are fundamental to allow the efficient radiative cooling of the gas despite the low gas temperatures and the lack of metals. This cooling is essential to promote the gravitational collapse and the formation of the first stars in the history of the Universe at $z \approx 20$–$30$ within dark matter halos with masses $\sim 10^6\ \mathcal{M}_\odot$. The formation of the first galaxies occurs later ($z \approx 10$) in halos of $\sim 10^8\ \mathcal{M}_\odot$ where atomic hydrogen cooling becomes possible. It is thought that also the seeds of SMBHs start to form at these epochs at the centres of primordial galaxies. The formation of the first sources of UV radiation marked the end of the dark ages. The energetic photons emitted by the first luminous objects gradually ionise the surrounding IGM and lead to the second major transition called cosmological reionisation. As a consequence, an essential ingredient of this chapter is also the description of the IGM, its physical properties and its evolution due to cosmological reionisation. More details on the topics of this chapter can be found in Stiavelli (2009) and in the reviews suggested in the next sections.

## 9.1 The Cosmological Recombination

A few thousand years after the Big Bang, the Universe consists of a fully ionised plasma composed of protons, electrons and the stable atomic nuclei formed during the primordial nucleosynthesis (i.e. $^4$He, $^3$He, D and $^7$Li; §2.5). This plasma is sometimes called the **primeval fireball**. Its temperature is too hot ($k_B T \gg 13.6$ eV) to allow the stable recombination of the electrons with protons and nuclei to form neutral atoms. Photons and matter are strongly coupled due to very frequent scattering interactions, and the fireball radiates as a perfect black body. Due to the incessant scattering by free electrons, the Universe is opaque because photons are not free to propagate and the scattering rate is much higher than the expansion rate of the Universe (§2.4). Neutrinos and dark matter particles coexist in the primeval fireball. Due to the expansion of the Universe, this hot plasma progressively cools and decreases its density (§2.1.2). In this section we focus on the epoch marking the transition from the fully ionised plasma to the neutral pregalactic gas. This starts when, thanks to the gradual cooling of the photon–baryon fluid, the free electrons begin to recombine with protons and nuclei to form neutral atoms. In the case of hydrogen, this occurs through the interaction

$$p^{+} + e^{-} \rightarrow H + \gamma, \tag{9.1}$$

where $p^{+}$ and $e^{-}$ are, respectively, the proton and the electron involved in the formation of a neutral hydrogen atom (H) followed by the emission of a photon $\gamma$.

The transition from ionised to neutral gas is called **cosmological recombination**. During this phase, the progressively smaller number of free electrons and the lower matter density make Thomson scattering less important. Around $z \approx 1000$, when the recombination is nearly completed, the Universe becomes transparent because photons can start to propagate freely through the neutral gas without being scattered. These are the photons of the CMB already described in §2.4.

## 9.1.1 Hydrogen Recombination with Equilibrium Theory

We now examine the physics of the cosmological recombination by treating the whole Universe as a cloud of photons and baryons. The key question addressed in this section is how to derive the redshift range during which this process takes place. We first focus on the case of hydrogen (the most abundant element in the Universe) under the assumption of thermodynamic equilibrium and baryon number conservation. We anticipate that the equilibrium assumption is not correct because a recombining plasma is not in equilibrium by definition. However, this simplified approach is instructive because it shows an interesting cosmological application of the **Saha equation** (§D.1.4) and provides results not too different from those of more realistic methods. In the case of pure hydrogen, we define the number densities of neutral and ionised H ($n_{\rm H}$ and $n_{\rm p}$), the total density $n_{\rm t} = n_{\rm H} + n_{\rm p}$, and the ionisation fraction $x = n_{\rm p}/n_{\rm t} = n_{\rm e}/n_{\rm t}$ (a pure hydrogen gas implies $n_{\rm p} = n_{\rm e}$).

The Saha equation in this case is

$$\frac{n_{\rm H}}{n_{\rm e}n_{\rm p}} = \left(\frac{2\pi m_{\rm e} k_{\rm B} T}{h^2}\right)^{-3/2} \exp\left(\frac{\chi}{k_{\rm B}T}\right), \tag{9.2}$$

where $m_{\rm e}$ is the mass of the electron and $\chi$ is the hydrogen ionisation potential ($\chi \simeq 13.6$ eV). Based on the definition of $x$,

$$\frac{n_{\rm H}}{n_{\rm p}} = \frac{1-x}{x}, \tag{9.3}$$

and hence the Saha equation can be rewritten as

$$\frac{1-x}{x^2} = n_{\rm t}\left(\frac{2\pi m_{\rm e} k_{\rm B} T}{h^2}\right)^{-3/2} \exp\left(\frac{\chi}{k_{\rm B}T}\right). \tag{9.4}$$

This equation can also be expressed as a function of the cosmological ratio of baryon-to-photon total densities ($\eta = n_t/n_\gamma$). In this regard, we recall that the number density of photons with wavelengths between $\lambda$ and $\lambda + d\lambda$ is $n_{\gamma,\lambda} d\lambda = u_\lambda d\lambda/(hc/\lambda)$, with $u_\lambda$ the black-body energy density

$$u_\lambda d\lambda = \frac{4\pi}{c} B_\lambda d\lambda, \tag{9.5}$$

where $B_\lambda$ is the Planck function (eq. D.8), and $hc/\lambda$ is the energy of a photon with wavelength $\lambda$. Hence, the total photon number density $n_\gamma$ is

$$n_\gamma = \int_0^\infty n_{\gamma,\lambda} d\lambda = \int_0^\infty \left(\frac{u_\lambda}{hc/\lambda}\right) d\lambda = 16\pi\zeta(3)\left(\frac{k_B T}{hc}\right)^3, \tag{9.6}$$

where $\zeta$ is the Riemann zeta function ($\zeta(3) \simeq 1.202$). As a reference, at $z = 0$, the total number density of the black-body radiation ($T \simeq 2.73$ K) that is the relic of the Big Bang is $n_\gamma \approx 400$ cm$^{-3}$. Combining the two previous formulae, the Saha equation can be further rewritten as

$$\frac{1-x}{x^2} \simeq 3.84\eta \left(\frac{k_B T}{m_e c^2}\right)^{3/2} \exp\left(\frac{\chi}{k_B T}\right), \tag{9.7}$$

where $\eta \simeq 2.7\times10^{-8}(\Omega_{b,0}h^2)$ and $\Omega_{b,0}$ is the baryon density parameter. The very small value of $\eta$ shows that photons outnumber baryons by a huge amount. Based on this modified Saha equation, it is possible to derive the temperature at which a given fraction $x$ of hydrogen is ionised. At redshifts greater than a few hundred, gas and radiation temperatures are strongly coupled by Compton scattering ($T_{rad} \sim T_{gas}$; §9.2). The radiation temperature $T_{rad}$ varies with redshift as $T_{rad} = T_{rad,0}(1+z) \simeq 2.73(1+z)$ (§2.4), and thus it is possible to trace the evolution of $x$ as a function of $z$. It can be found that $x \approx 0.5$, 0.1 and 0.01 for $T_{rad} \approx 3740$, 3420 and 3100 K or $z \approx 1370$, 1250 and 1140, respectively. This implies that about 50% of the hydrogen is still ionised at $z \approx 1370$ ($\approx 0.25$ Myr after the Big Bang) and that the recombination is basically completed by $z \approx 1100$, when the Universe is $\approx 0.36$ Myr old. However, it is important to remind ourselves that these results are not rigorous, being based on the inappropriate assumption of thermodynamic equilibrium.

### 9.1.2 Advanced Modelling of Hydrogen Recombination

A more realistic treatment shows that the hydrogen recombination did not occur so easily because several processes made it less efficient than in the simplistic scenario of equilibrium theory. A full description of these topics can be found in Peebles (1968), Zeldovich et al. (1968) and Sunyaev and Chluba (2009). The main processes acting against hydrogen recombination can be summarised as follows.

1. The cosmological recombination refers to the whole Universe treated as a closed system. This implies that the ionising photons cannot escape anywhere, and can influence the evolution of the primordial gas as long as they are present.
2. The huge number density of photons compared to that of baryons (i.e. the high value of $\eta^{-1}$) implies that there is a significant number of ionising photons in the Wien tail of the black-body radiation field even if the temperature is below the temperature

($T > 10\,000$ K) at which hydrogen starts to become fully ionised. For instance, $T_{rad}$ is only $\approx 2700-5500$ K in the redshift range when most of the cosmological recombination takes place ($z \approx 1000-2000$).

3. The recombination of an electron to a given energy level with quantum number $n$ produces photons with energy $h\nu = E_{kin}(e^-) + 13.6/n^2$ eV, where $E_{kin}(e^-)$ is the kinetic energy of the free electron before recombining with the H nucleus. It can be shown that, given the plasma temperatures (and thus the electron kinetic energies) at these cosmic epochs, if the recombination occurs directly to the ground state ($n = 1$), this process can produce photons with $h\nu > 13.6$ eV capable of further photoionising hydrogen and therefore hampering its recombination.
4. The cascade recombination can produce Ly$\alpha$ photons that are efficiently absorbed by nearby neutral hydrogen due to the large cross section of Ly$\alpha$ absorption. This implies that the absorbed Ly$\alpha$ photons can excite the neutral H atoms to energy levels with $n > 1$ and make them easier to photoionise by the abundant photons with $h\nu < 13.6$ eV.

Altogether, these processes can inhibit the cosmological hydrogen recombination, making it too slow. However, the problem can be mitigated thanks to a radiative process called hydrogen **two-photon emission**. This occurs when an electron initially at the level 2s decays spontaneously to 1s emitting two photons within a broad spectral range from the UV to the optical (Fig. 9.1, left):

$$\mathrm{H(2s)} \rightarrow \mathrm{H(1s)} + \gamma + \gamma, \tag{9.8}$$

where the sum of the energies of the two photons is equal to the energy difference between the 2s and 1s levels. This decay has a rate of about 8.2 $s^{-1}$ and applies to a quarter of the H atoms excited to the energy level with $n = 2$. The two emitted photons do not have energies high enough to ionise the neutral H atoms. Fig. 9.1 (left) shows the continuum spectrum produced by this process. Thus, these photons are harmless and allow a higher efficiency of the hydrogen recombination to the ground level. In addition to the two-photon emission, the recombination is facilitated also by the expansion of the Universe because each H atom receives redshifted photons that are not energetic enough to ionise or excite H atoms.

If all these processes are taken into account, the hydrogen ionisation fractions $x \approx 0.5$, 0.1 and 0.01 are achieved at $z \approx 1210$, 980 and 810, respectively. This means that the recombination takes more time and is completed later than in the case of Saha equilibrium. Another important result is that, unlike in the equilibrium scenario, the fraction of free electrons never goes to zero. For $z < 800$, the ionisation fraction remains constant at the low level of $x \approx 2 \times 10^{-4}$ even when the recombination is completed. This occurs because, for small $x$, the rate at which the electrons recombine with protons becomes lower than the expansion rate of the Universe (this is called the **freeze-out** effect). The availability of residual electrons and protons turns out to be essential for the formation of the first luminous objects (§9.4). Fig. 9.1 (right) shows the abundance of ionised and neutral H and He as a function of redshift. The steep decline of ionised hydrogen around $z \approx 1000$ is evident. It is during this cosmic epoch that the Universe becomes optically thin by reaching a low optical depth ($\tau \ll 1$) for Thomson scattering. As a consequence, the rate of Thomson scattering is so low that the photons propagate freely. For this reason, $z \approx 1000$ is called the

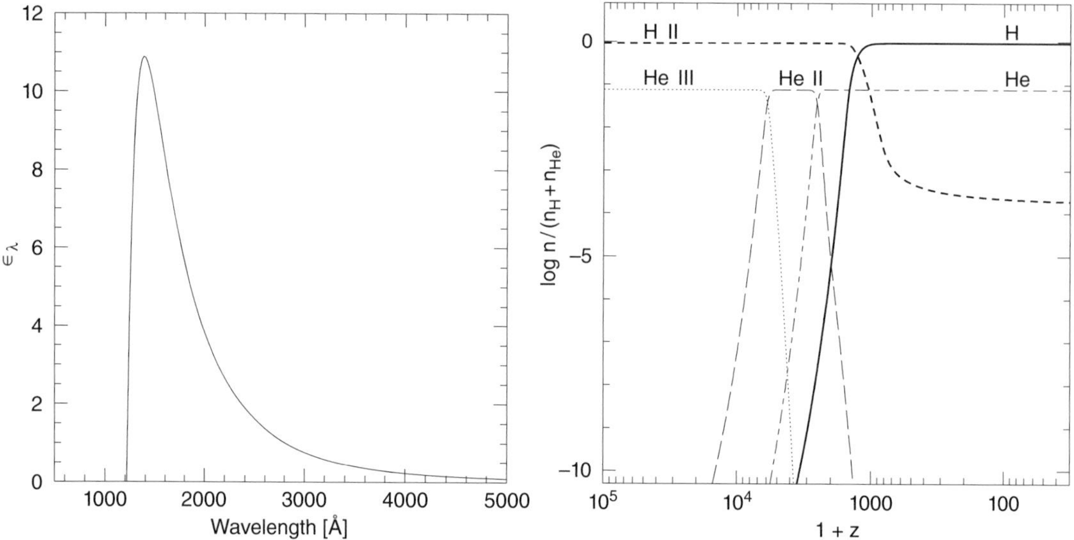


**Fig. 9.1** *Left panel*. The emissivity of the hydrogen two-photon emission calculated according to equation (4) of Nussbaumer and Schmutz (1984). The emitted photons have $\lambda > 912$ Å and therefore cannot photoionise hydrogen. The units of $\epsilon_\lambda$ are $10^{-14}$ erg s$^{-1}$ Å$^{-1}$ per particle. *Right panel*. The abundances of ionised and neutral hydrogen and helium as functions of redshift. The abundance is defined as the ratio of the number density $n$ of a given species to the total number density of H and He. Data from Galli and Palla (2013).

surface of last scattering, and the map of the CMB represents the true image of the diffuse black-body thermal emission of the Universe just after the recombination when radiation finally decoupled from matter (§2.4).

### 9.1.3 Helium and Lithium Recombination

Helium and lithium were also present in the primordial plasma, and they also experienced the cosmological recombination. As in the case of hydrogen, their recombination starts when $k_B T < \chi$, where $\chi$ is their ionisation potential. The complete recombination of helium requires two steps: He III $\rightarrow$ He II $\rightarrow$ He I. However, since the ionisation potentials of He I (24.6 eV) and He II (54 eV) are larger than that of hydrogen (13.6 eV), helium recombined before hydrogen when the Universe is hotter (Fig. 9.1). A detailed modelling of these processes, similar to that illustrated in the previous section, shows that the He III $\rightarrow$ He II transition occurs at $5500 < z < 7000$, when the age of the Universe was between 13.5 kyr and 21.1 kyr. The full recombination of helium (He II $\rightarrow$ He I) is completed at $1500 < z < 3500$, in a period from 50 kyr to 0.2 Myr since the Big Bang. In the case of lithium (ionisation potentials of 122.4 eV, 75.6 eV and 5.4 eV), the Li IV $\rightarrow$ Li III recombination occurs around $z \approx 14\,000$ (i.e. the first recombination ever in the Universe), whereas the Li III $\rightarrow$ Li II transition takes place at $z \approx 8600$. The last recombination (Li II $\rightarrow$ Li I) never occurs because the Ly$\alpha$ photons emitted in the hydrogen recombination are energetic enough (10.2 eV) to maintain lithium singly ionised.

## 9.2 The Pregalactic Gas in the Dark Ages

The period from the completion of the hydrogen recombination ($z \approx 800$, $\approx 0.6$ Myr after the Big Bang) and the formation of the first luminous objects is called the dark ages. The end of the dark ages is thought to occur when the age of the Universe is around 100–200 Myr ($z \approx 20$–30). During the dark ages, baryonic matter consists of a diffuse neutral gas, which is called **pregalactic gas** to indicate that it is relative to a cosmic time when luminous objects and galaxies are not present. The evolution of the pregalactic gas is driven by two main processes. On the one hand, the expansion of the Universe causes adiabatic cooling, with the gas temperature declining as $T_{gas} \propto a^{-2}$, where $a$ is the scale factor (§2.1.1). On the other hand, Compton scattering between baryons and CMB photons can either heat or cool the gas depending on whether the electrons increase their energy through the recoil, or give part of their energy back to the CMB (inverse Compton scattering). The other cooling processes (recombination, forbidden line emission, free–free continuum) and heating mechanisms (photoionisation, cosmic rays, shocks) become important only at later cosmic epochs when the first luminous objects form and influence their surrounding environment with a variety of feedback effects (§8.1).

In the case of a homogeneous Universe, the adiabatic expansion and Compton scattering can be included in a single equation describing the time evolution of the gas temperature:

$$\frac{\mathrm{d}T_{gas}}{\mathrm{d}t} \approx -2T_{gas}\left(\frac{\dot{a}}{a}\right) + \frac{2}{3k_B n_b}\dot{\mathcal{E}}, \tag{9.9}$$

where $\dot{\mathcal{E}}$ is the net Compton heating (i.e. heating minus cooling) per unit volume (§8.1.1) and

$$n_b = \frac{\Omega_{b,0}\rho_{crit,0}}{\mu m_p}(1+z)^3 \approx 1.9 \times 10^{-7}(1+z)^3 \text{ cm}^{-3} \tag{9.10}$$

is the baryon number density, where $\Omega_{b,0}$ is the baryon density parameter and $\rho_{crit,0}$ is the critical density of the Universe (eq. 2.23). The first term on the r.h.s. of eq. (9.9) describes the adiabatic cooling, whereas the second term is the net transfer of energy from the CMB photons to the electrons. When the second term dominates ($z > 300$), the radiation and gas temperatures are equal ($T_{rad} \approx T_{gas}$) because the ionisation fraction is high enough to allow an efficient Compton scattering. When the Compton scattering term becomes negligible, the gas temperature decouples from the CMB photons and declines as $T_{gas} \propto a^{-2}$. We recall that the radiation temperature always decreases as $T_{rad} \propto a^{-1}$, or $T_{rad} \simeq 2.73(1+z)$ K. A practical formula to estimate the pregalactic gas temperature at $z < 100$ is

$$T_{gas} \approx 0.02(1+z)^2. \tag{9.11}$$

For example, at $z \approx 50$, the gas temperature is $T_{gas} \approx 50$ K. In comparison, the radiation temperature at the same redshift is $T_{rad} \approx 139$ K. The above relation is valid on average for the diffuse gas. In this context, $T_{gas}$ is also called the **background cosmological temperature** of the gas.

## 9.3 The Collapse of the Pregalactic Gas

In the theory of density perturbation evolution, dark matter halos form through the gravitational collapse of overdensities (§7.3). Baryonic matter is supposed to collapse within the potential wells of dark matter halos and form the first gravitationally bound gaseous objects (§8.2.2). This occurs only when the internal pressure of the gas does not prevent the gravitational collapse. However, the classical Jeans instability criterion (eq. 8.44) must be reformulated in the cosmological framework taking into account the expansion of the Universe, which changes the gas temperature and density as a function of time (see the review of Barkana and Loeb, 2001, for details). The Jeans criterion is usually derived starting from a sinusoidal density perturbation superposed on a uniformly expanding background and considering a mixture of dark matter and baryons. It can be shown that at $z < 100$, when Compton scattering between CMB photons and the partially ionised gas becomes negligible, the **cosmological Jeans mass** can be written as

$$\mathcal{M}_{\rm J} \simeq 5.73 \times 10^3 \left(\frac{\Omega_{\rm m,0}h^2}{0.15}\right)^{-1/2} \left(\frac{\Omega_{\rm b,0}h^2}{0.022}\right)^{-3/5} \left(\frac{1+z}{10}\right)^{3/2} \mathcal{M}_\odot. \tag{9.12}$$

This equation is valid in the linear phases of the evolution of density perturbations when $\delta \ll 1$ (§7.2). A more realistic treatment of these processes shows that it is more appropriate to use a time-averaged mass threshold because the Jeans mass of eq. (9.12) can change significantly during the collapse. This mass, called the **filtering mass**, allows one to better estimate the expected mass threshold above which the gas collapse can occur (Gnedin, 2000). The filtering mass is

$$\mathcal{M}_{\rm F} = \frac{4\pi}{3}\overline{\rho}_0 \left(\frac{\lambda_{\rm F}}{2}\right)^3, \tag{9.13}$$

where $\overline{\rho}_0$ is the average matter density at $z = 0$, and $\lambda_{\rm F}$ is the filtering wavelength, which depends on redshift as

$$\lambda_{\rm F}^2(z) = \frac{3}{1+z}\int_z^\infty \lambda_{\rm J}^2 \left[1 - \left(\frac{1+z}{1+z'}\right)^{1/2}\right] dz', \tag{9.14}$$

where $\lambda_{\rm J}$ is the classical Jeans length (§8.3.2). A comparison between the Jeans and the filtering masses shows that the former is higher at $z > 50$, whereas the two masses are comparable around $z \approx 50$. At lower redshifts, the filtering mass becomes lower than the Jeans one.

The Jeans and filtering masses set only minimum masses above which baryons can fall and collapse into a dark matter halo at a given redshift. However, it is complicated to estimate the actual mass of collapsed bound objects with respect to $\mathcal{M}_{\rm J}$ or $\mathcal{M}_{\rm F}$. This depends on the evolution of the accretion of baryonic matter onto the final gravitational potential well of a dark matter halo. The gas is cold when it starts to infall, but it can be rapidly heated to a temperature at which the pressure support can prevent further collapse. If shock heating is effective, once the collapse is completed, the gas in the inner regions of the dark matter halo settles in a configuration close to hydrostatic equilibrium at the halo virial temperature (§8.2.1). Rewriting eq. (8.19), we get

$$T_{\rm vir} \simeq 1.98 \times 10^4 \left(\frac{\mu}{0.6}\right)\left(\frac{\mathcal{M}_{\rm vir}}{10^8 h^{-1}\,\mathcal{M}_\odot}\right)^{2/3}\left[\frac{\Omega_{\rm m,0}}{\Omega_{\rm m}(z)}\frac{\Delta_{\rm c}(z)}{18\pi^2}\right]^{1/3}\left(\frac{1+z}{10}\right)\,\mathrm{K}, \qquad (9.15)$$

where $\mathcal{M}_{\rm vir}$ is the halo virial mass, and $\Delta_{\rm c}(z)$ is the overdensity given by eq. (7.21). In $\Lambda$CDM cosmology, $\Omega_{\rm m}(z) \approx 1$ and $\Delta_{\rm c}(z) \approx 18\pi^2$ for $z > 6$ (§7.3.2). This allows us to simplify eq. (9.15) because the term in square brackets reduces to $\Omega_{\rm m,0}^{1/3}$, and eq. (9.15) can be rewritten as

$$T_{\rm vir} \approx 10^4 \left(\frac{\mathcal{M}_{\rm vir}}{10^8 h^{-1}\mathcal{M}_\odot}\right)^{2/3}\left(\frac{1+z}{10}\right)\,\mathrm{K}, \qquad (9.16)$$

which shows that the gas collapsed and virialised within dark matter halos is much hotter than the diffuse one. For instance, for $z \approx 50$ and $\mathcal{M}_{\rm vir} \approx 10^5\ \mathcal{M}_\odot$, $T_{\rm vir} \approx 500$ K, a factor of 10 higher than the temperature of the diffuse gas derived with eq. (9.11). For $z < 100$, the virial temperature of a halo and the background cosmological average temperature of the gas ($\overline{T}_{\rm gas}$) are related to the overdensity of the baryons ($\delta_{\rm b}$) settled in a configuration of hydrostatic equilibrium within a halo through the following relation:

$$\delta_{\rm b} = \frac{\rho_{\rm b}}{\overline{\rho}_{\rm b}} - 1 \approx \left(1 + \frac{6}{5}\frac{T_{\rm vir}}{\overline{T}_{\rm gas}}\right)^{3/2} - 1, \qquad (9.17)$$

where $\rho_{\rm b}$ and $\overline{\rho}_{\rm b}$ are the baryon mass density of the collapsed gas and the cosmological background density. This equation holds for $z < 100$ and is approximated because it neglects other effects such as the role of the infalling gas and the Hubble expansion at the interface between the collapsed system and the background IGM. As a reference, for $\delta_{\rm b} \approx 100$, a typical value expected as the threshold for the baryon collapse, eq. (9.17) implies $T_{\rm vir} \approx 17\overline{T}_{\rm gas}$.

## 9.4 The Cooling of Primordial Gas

The collapse of pregalactic gas within dark matter halos discussed in the previous section is only one of the two necessary conditions to form a luminous object. The second is that the collapsed gas cools and condenses into clouds rapidly enough to allow the formation of the first stars. Based on analytic calculations and numerical simulations, the general consensus is that the first luminous objects form at $z \approx 20$–$30$ within dark matter halos with masses around $10^6\ \mathcal{M}_\odot$ called **minihalos**. These minihalos are originated around $z \approx 20$–$30$ from the collapse of overdensities deviating at 2–3$\sigma$ level from the average density background, and therefore are quite abundant at these cosmic epochs. According to eq. (9.16), the virial temperature of the gas hosted by these minihalos is $T_{\rm vir} \lesssim 1000$ K. It is worth recalling that this gas has a primordial composition (mostly H, D, $^4$He and $^3$He) without metals. The key question is therefore to understand how this gas can cool efficiently. In §8.2.3, it was shown that the key requirement for a fast collapse is the efficient radiative cooling of the gas. This efficiency is tightly linked to the cooling function and its competition with the heating processes (§8.1.1). However, for $T_{vir} \lesssim 1000$ K there are no atomic processes able to cool the primordial gas because hydrogen becomes an efficient coolant only for $T \gtrsim 10^4$ K, when it is fully ionised (§8.1.1). The same argument holds for helium at even higher temperatures. Thus, how can the primordial gas cool radiatively fast enough (i.e. with $t_{cool} \ll t_{dyn}$; §8.2.3) to allow a rapid collapse? The answer is that a few simple molecules make the cooling possible through rotational and vibrational transitions. The origin of these primordial molecules is described in the next section.

### 9.4.1 The Formation of Primordial Molecules

The formation of primordial molecules in the pregalactic era is possible thanks to the residual fraction ($x \sim 10^{-4}$) of free electrons and protons left over (frozen out; §9.1.2) by the recombination. According to the theory of primordial chemistry, the first molecule to form in the history of the Universe is $HeH^+$ (helium hydride ion). Its formation starts at very high redshifts ($800 < z < 2000$) during the end of the hydrogen recombination, when the ionisation fraction rapidly decreases from $\approx 10\%$ to $\approx 1\%$. In an environment still populated by energetic CMB photons, $HeH^+$ forms through the following **radiative association reaction**:

$$H^+ + He \rightarrow HeH^+ + \gamma. \tag{9.18}$$

Radiative association reactions occur also in the ISM of present-day galaxies, and are a channel to form larger molecules from the collision of smaller species in the gas phase. The kinetic energy is temporarily stored in an excited state, and the resulting molecule stabilises through the subsequent emission of photons through electronic and/or rovibrational transitions. The frequency of the emitted radiation (typically in the UV–IR range) depends on the energy of the collision.

At $z \approx 10$, the $HeH^+$ molecule has a very low abundance relative to atomic hydrogen, $n(HeH^+)/n(H) \approx 2 \times 10^{-14}$. However, it plays a significant role in allowing the formation of $H_2^+$ which is relevant in the subsequent formation of $H_2$:

$$HeH^+ + H \rightarrow He + H_2^+. \tag{9.19}$$

Molecular hydrogen ($H_2$) is the most abundant molecule in the Universe. However, its formation during the dark ages is completely different from what occurs today. At low redshifts, the formation of $H_2$ through gas-phase reactions is slow and inefficient. In fact, the majority of $H_2$ forms with adsorption reactions taking place on the surface of interstellar dust grains (§8.3.1). However, during the dark ages, the metallicity of the gas is zero and no dust grains are present, and therefore $H_2$ can form only through gas-phase reactions. The first step for $H_2$ formation is the radiative association reaction

$$H + H^+ \rightarrow H_2^+ + \gamma, \tag{9.20}$$

whereas the second step is the charge transfer reaction

$$H_2^+ + H \rightarrow H_2 + H^+. \tag{9.21}$$

Reactions (9.20) and (9.21) are important at $300 < z < 800$ when the age of the Universe is between 0.6 Myr and 3 Myr. An additional source of $H_2^+$ (important to form $H_2$) comes from the conversion of $HeH^+$ in reaction (9.19), although the most efficient production of $H_2^+$ occurs through reaction (9.20).

At redshifts below $z \approx 100$ ($\approx 16$ Myr after the Big Bang) two new reactions play a major role for the formation of $H_2$:

$$H + e^- \rightarrow H^- + \gamma \tag{9.22}$$

and

$$H^- + H \rightarrow H_2 + e^-. \tag{9.23}$$

These reactions are delayed with respect to those involving $H_2^+$ because $H^-$ (binding energy of 0.754 eV) is easily ionised by the photons of the CMB radiation at $z > 300$. The previous reactions produce a gradual increase of $H_2$ abundance which reaches a stable value at $z \approx 60$ when the Universe is about 36 Myr old. As a reference, the abundance of $H_2$ relative to atomic hydrogen at $z \approx 10$ is $n(H_2)/n(H) \approx 6 \times 10^{-7}$.

$H_2$ is not the only molecule formed at these times. The theory of primordial chemistry predicts that about 250 reactions occur in this cosmic epoch, and nearly 30 molecular species are formed during the dark ages. Some of them are particularly relevant for gas cooling. An important case is represented by hydrogen deuteride (HD) formed by an atom of hydrogen and an atom of deuterium. The formation of HD is driven by two main reactions involving a deuterium nucleus exchange with $H_2$:

$$D + H_2 \rightarrow HD + H \tag{9.24}$$

and

$$D^+ + H_2 \rightarrow HD + H^+. \tag{9.25}$$

The first reaction is more important in the conditions of high temperature and density typical of gravitational collapse or shocks. The second is more relevant in the low-density gas before the formation of the first objects. The abundance of HD relative to atomic hydrogen at $z \approx 10$ is $\approx 4 \times 10^{-10}$. Other important molecules form through reactions involving lithium. After the recombination of $Li^{3+}$ and $Li^{2+}$, the remaining species (Li, $Li^+$) interact with neutral H, D and He and ions to form a variety of molecules (e.g. LiH, $LiH^-$, LiD, $LiD^-$, $LiHe^+$) and $Li^-$. These molecules are very fragile to photodissociation, and their abundance is therefore very low. The first molecule including lithium (lithium hydride) forms at $z < 200$ through the radiative association reaction:

$$Li + H \rightarrow LiH + \gamma. \tag{9.26}$$

At later times ($20 < z < 80$), the other route of LiH formation is

$$Li^- + H \rightarrow LiH + e^-. \tag{9.27}$$

At redshifts $z < 40$, when the energy of the CMB photons is no longer sufficient for photodissociation, the $LiH^+$ molecule (binding energy of only 0.14 eV) is also formed. At $z \approx 10$, the number density abundances of LiH and $LiH^+$ relative to atomic hydrogen

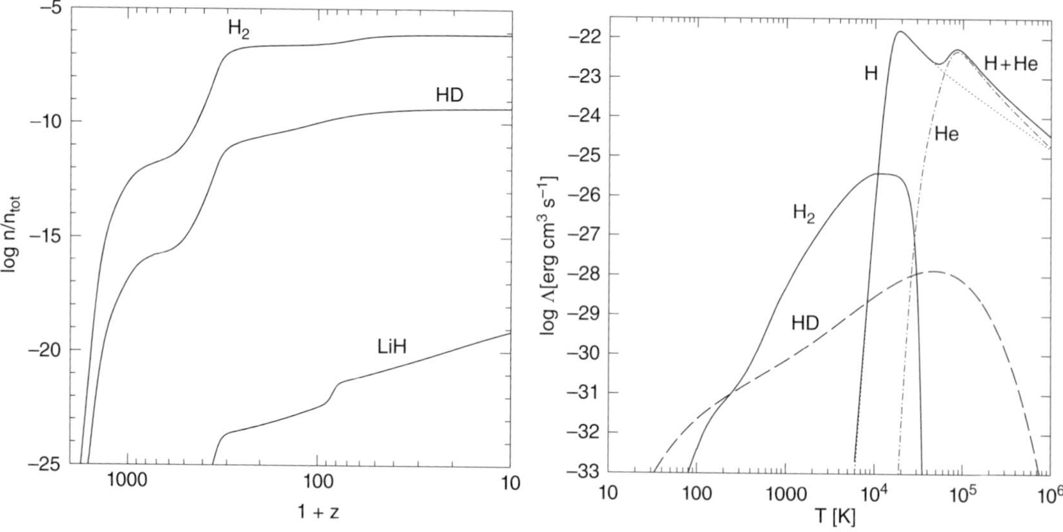


**Fig. 9.2** *Left panel*. The redshift evolution of the number density $n$ of a given species relative to the total number of baryons. Data from Galli and Palla (2013). *Right panel*. The cooling functions of $H_2$ and HD for $n_{H_2}/n_H = 10^{-4}$ and $n_{HD}/n_H = 10^{-7}$, respectively, in the case of low-density gas ($n_H = 1$ cm$^{-3}$). A comparison with $\Lambda(T)$ of H and He is also shown to illustrate the relative efficiency of molecular and atomic cooling as a function of the temperature. Data from Galli and Palla (2013).

were $\approx 9 \times 10^{-20}$ and $\approx 4.6 \times 10^{-20}$, respectively. Fig. 9.2 shows the redshift evolution of the molecular species which play a key role in gas cooling during the dark ages.

### 9.4.2 The Cooling Function of Primordial Molecules

As described in §8.1.1, gas can cool efficiently by emitting radiation only if three conditions are satisfied.

1. The gas must be optically thin to allow a large fraction of the emitted photons to escape and carry out energy.
2. The temperature of the gas must be high enough to allow collisional excitations of electrons to higher energy levels and promote subsequent spontaneous emission. In other words, $k_B T_{gas}$ must be larger than the typical energy level spacing of a given atom or molecule.
3. For a given coolant line, the density of the gas must be lower than the critical density (§4.2.3) of that line in order to make spontaneous emission dominant over collisional de-excitation.

At temperatures $\lesssim 1000$ K, primordial molecules are the only possible coolants in the dark ages. Molecular hydrogen is the most abundant molecule. However, being a homonuclear molecule, it is difficult to excite because it does not have a permanent dipole moment (§4.2.5). For instance, the energy difference between $J = 2$ and $J = 0$ corresponds to a temperature $\Delta E/k_B \simeq 510$ K. This makes it difficult to collisionally excite $H_2$ during the dark ages when the temperatures are generally much lower. In addition, the excited rotational and vibrational transitions have small radiative probabilities (radiative lifetimes $\gtrsim 10^6$ s) and collisional de-excitations become competitive with spontaneous transitions at densities $n \gtrsim 10^4$ cm$^{-3}$. However, other molecules, such as HD and LiH, help to compensate the inefficiency of $H_2$ cooling at low temperatures. Despite their very low abundance, HD and LiH have the advantage of being heteropolar molecules with a permanent electric dipole ($8.3 \times 10^{-4}$ debye for HD). This implies that rotational transitions with $\Delta J = \pm 1$ are allowed. In the case of HD, the energy of the $J = 1 \rightarrow 0$ transition corresponds to a temperature $\Delta E/k_B \simeq 128$ K. Moreover, the radiative lifetimes of HD transitions are 100 times shorter than those of $H_2$. The LiH electric dipole is much larger (5.89 debye) and this makes this molecule a relevant coolant at low temperatures despite its low abundance. Fig. 9.2 (right) shows the cooling functions of the two most important coolants (HD and $H_2$) in a low-density case corresponding to the typical densities in the phase preceding gravitational collapse. It is evident that HD plays a significant role at low temperatures ($T < 100$ K), whereas $H_2$ becomes important at higher temperatures.

## 9.5 Population III Stars

### 9.5.1 The Collapse and Formation of the First Luminous Objects

The first luminous objects of the Universe are called **Population III** (or **Pop III**) stars as an extension of the historical classification of Population I and Population II stars of the Milky Way (§4.6.1). Present-day stars form in cold, dense, inhomogeneous clouds of molecular gas enriched with metals and dust grains. These clouds are supported against gravity mostly by turbulence, and magnetic fields also play a significant role through complex magnetohydrodynamic processes (§8.3.6). Instead, the formation of Pop III stars occurs in a gas with primordial composition (H, D, $^4$He, $^3$He and $^7$Li), without dust grains and probably with very weak magnetic fields. Depending on the ionisation state of the gas cloud, two main evolutionary paths are expected for the collapse and formation of Pop III stars.

#### Nearly Neutral Gas

If the initial ionisation fraction is similar to that left by the cosmological recombination ($x \sim 10^{-4}$, i.e. nearly neutral), the collapse proceeds with the following phases as a function of increasing density. These phases are mainly driven by the availability of $H_2$ and its efficiency as radiative coolant, whereas LiH and HD are thought to also play a role at low temperatures (§9.4.2).

1. At very high redshifts, the $H_2$ abundance relative to H is low ($< 10^{-6}$), the gas collapses nearly adiabatically and $H_2$ is formed mainly through the $H^-$ channel (eq. 9.23).

2. A critical change occurs at $z \approx 20$ when the $H_2$ abundance reaches about $10^{-3}$ and when the gas density in the collapsing system is $n_H \sim 10$ cm$^{-3}$. The $H_2$ rovibrational emission produces a rapid enhancement of the cooling efficiency. This counteracts the gravitational heating of the gas. During this phase, the gas temperature stabilises around 200 K.
3. When the gas density reaches the $H_2$ critical density ($n_{crit} \sim 10^4$ cm$^{-3}$), the vibrational and rotational spontaneous decays are reduced by the frequent collisional de-excitations.
4. At $n > n_{crit}$, cooling is further reduced.
5. At $n > 10^8$ cm$^{-3}$, the abundance of $H_2$ increases thanks to the three-body reactions

$$H + H + H \rightarrow H_2 + H \tag{9.28}$$

and

$$H + H + H_2 \rightarrow 2H_2. \tag{9.29}$$

In each reaction, the $H_2$ binding energy of 4.48 eV is converted into heat. However, despite the increase of the gas thermal energy, the increase of the $H_2$ fraction by about one order of magnitude produces an increment of the cooling rate that is enough to compensate for the compressional heating. As a consequence, the temperature of the collapsing cloud stabilises around 1500 K.
6. At $n > 10^{10}$ cm$^{-3}$, the $H_2$ rotational and vibrational lines become optically thick, i.e. they are self-absorbed by the gas itself due to the large number of $H_2$ molecules. In these conditions, the $H_2$ cooling has a low efficiency.
7. At $n > 10^{12}$ cm$^{-3}$, cooling is dominated by $H_2$ through a process called **collision-induced emission**. This occurs in high-density conditions when the frequent collisions between $H_2$ molecules make them temporary non-zero dipoles for the duration of the collision and allow them to emit through dipole transitions with much higher probabilities than for an $H_2$ quadrupole. However, at the same time, $H_2$ also becomes easily dissociated and radiative cooling decreases its efficiency.
8. At this point, the gas temperature continues to increase adiabatically until the pressure force succeeds in halting the gravitational collapse. This marks the formation of a stable **protostellar core** which continues to accrete additional gas from the surrounding cloud.

## Preionised Gas

A different evolution occurs if the primordial cloud is preionised in environments exposed to external UV radiation and/or cosmic-ray fields produced by nearby objects previously formed. In these cases, the ionisation fraction can easily become $x \gg 10^{-4}$, and the gas temperature at the start of the collapse is higher. Thanks to the higher ionisation fraction, the formation of $H_2$ is more rapid and efficient, and the critical threshold of $n_{H_2}/n_H \sim 10^{-3}$ (Fig. 9.3, left) is achieved at much lower densities ($n_H < 10$ cm$^{-3}$). Due to the more efficient cooling, the temperature decreases to values ($T < 150$ K) critical for the formation of the HD molecule and leading to a further drop of the temperature to a minimum value around $\approx 30$ K. After this point, the evolution is similar to the case of the nearly neutral cloud described previously.

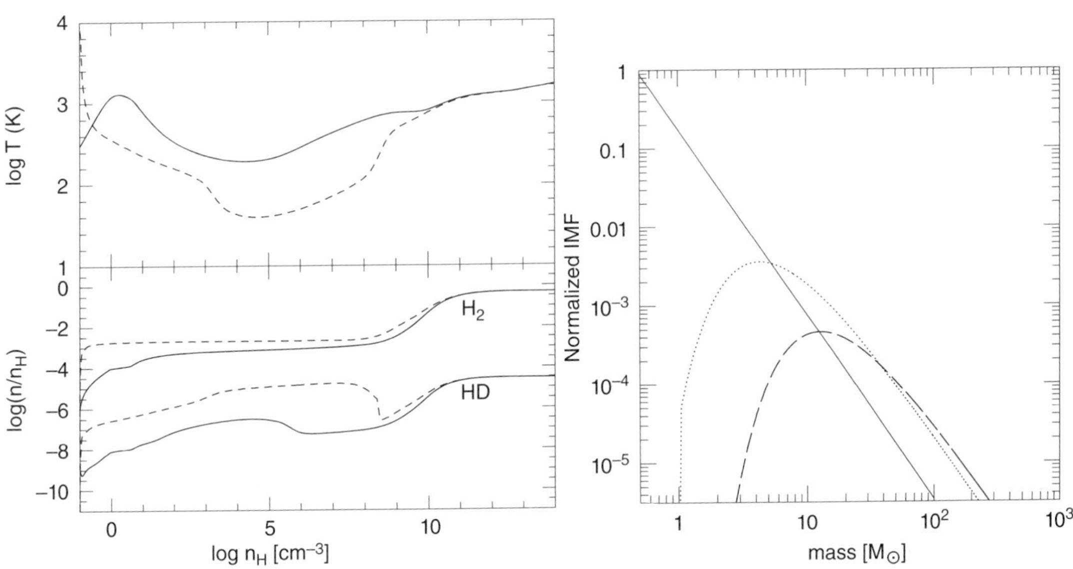


Fig. 9.3 *Left panel.* The evolution of gas temperature (*top*) and molecule abundances ($H_2$ and HD) (*bottom*) as a function of hydrogen density during the collapse of a Pop III protostar. The solid and dashed curves indicate the cases of nearly neutral ($x = 10^{-4}$) and preionised ($x \gg 10^{-4}$) gas, respectively. Data from Omukai (2012). *Right panel.* Two examples of Larson IMFs (eq. 9.33) predicted for Pop III stars with a mass range of $1 \leq m \leq 1000$. The characteristic masses are $m_{ch} = 10$ (dotted line) and $m_{ch} = 30$ (dashed line). A Salpeter IMF ($0.1 \leq m \leq 100$) is shown (solid line) to allow a comparison with a typical present-day IMF. All IMFs are normalised so that $\int_{m_{min}}^{m_{max}} \phi(m) m \, dm = 1$. Data courtesy of S. Salvadori.

Fig. 9.3 summarises the thermochemical evolution of a neutral and a preionised cloud. The formation of Pop III objects is a complex ensemble of physical processes. Some of them can be treated analytically with reasonable accuracy. However, the analytic approach is not sufficient to model the whole interplay of gas cooling and heating, thermal balance, chemical evolution, feedback effects and the potential role of magnetic fields. In the early 2000s, a major step forward was made possible with the advent of high-resolution hydrodynamic simulations. This approach allows us to explore the formation and evolution of Pop III objects in great detail and to predict their observable properties.

## 9.5.2 The Properties and Fate of Population III Stars

The protostars grown from the gas collapse described in §9.5.1 are the seeds for the formation of Pop III stars. These protostars are surrounded by the gas that was not involved in the gravitational collapse. This gas is then accreted onto the core object, increasing the mass of the protostar until the gas reservoir is exhausted. The gas accretion can be halted by mechanical and/or radiative feedback processes due to the protostar itself. In the mechanical case, the protostar transfers energy and momentum to gas outflows. In the radiative case, the protostar radiation transfers energy and momentum to the infalling gas.

The simplest approach to estimate the final mass of the star is to consider the idealised case of a smooth spherical (i.e. radial) accretion. If fragmentation (§8.3.2) of the accreting gas and feedback processes are neglected, the accretion rate can be approximated by

$$\dot{\mathcal{M}}_{\rm acc} \approx \frac{\mathcal{M}_{\rm J}}{t_{\rm dyn}} \propto \frac{T^{3/2}\rho^{-1/2}}{\rho^{-1/2}} \sim T^{3/2}, \tag{9.30}$$

where $\mathcal{M}_{\rm J}$ is the Jeans mass, $t_{\rm dyn} \propto \rho^{-1/2}$ is the dynamical timescale of the collapse (like the free-fall time of eq. 8.45), $T$ is the gas temperature and $\rho$ is the mass density of the gas. The relation $\dot{\mathcal{M}}_{\rm acc} \propto T^{3/2}$ shows that the expected accretion rates are very high. This can be understood by taking star formation at $z \approx 0$ as a reference (§8.3). In the case of present-day protostars, the temperatures are $T \approx 10$ K and the accretion rates $\dot{\mathcal{M}}_{\rm acc} \sim 10^{-5}$ $\mathcal{M}_\odot {\rm yr}^{-1}$. Instead, for Pop III objects, the typical temperatures are $T \approx 300-1000$ K, and this implies much higher accretion rates with $\dot{\mathcal{M}}_{\rm acc} \sim 10^{-3}-10^{-2}$ $\mathcal{M}_\odot$ ${\rm yr}^{-1}$. These estimates of $\dot{\mathcal{M}}_{\rm acc}$ for Pop III stars should be considered strict upper limits because feedback and fragmentation have been neglected. However, even including these processes, the accretion rates remain much higher than in present-day star formation. As long as the accretion rate does not decline, the mass can increase substantially. The bottom line is that Pop III stars are expected to have masses higher than those of typical stars in the present-day Universe. Once the accretion is completed and a critical mass is achieved to trigger thermonuclear reactions in the core, the protostar moves to the main sequence and starts to shine as a Pop III star.

The main properties of Pop III stars can be estimated using the equations of stellar structure (see Bromm, 2013, for a review). As a reference, the typical radius of a Pop III star with 100 $\mathcal{M}_\odot$ is $\sim 5$ $R_\odot$, and the central temperature is of the order of $10^8$ K. Pop III stars are hotter than the hottest O stars because their photospheric temperature is around $10^5$ K. For a Pop III star of 100 $\mathcal{M}_\odot$, the typical luminosity is of the order of $10^6$ $L_\odot$. The luminosity of massive Pop III stars is expected to be close to the Eddington luminosity (§8.8), i.e. the limiting luminosity below which the radiation pressure does not exceed the gravitational force (eq. 8.139). Similar to the massive stars in the present-day Universe, Pop III stars are supported against gravity primarily by radiation pressure. The photons produced in the hot interior are diffused radiatively towards the stellar surface. However, the main source of opacity is Thomson scattering because no metals are present in Pop III stars. The lifetime of Pop III stars is given by $t_\star \approx 0.007\mathcal{M}c^2/L$, where 0.7% is the efficiency of the thermonuclear reactions which convert hydrogen into helium. This time is very short, of the order of $10^6$ yr. Pop III stars terminate their life cycle rapidly, exploding as SNe of different types depending on the initial mass of the star. Let us consider the cases of Pop III stars with masses above $\approx 8$ $\mathcal{M}_\odot$, i.e. the lower limit for a core collapse SN (§8.7.1).

1. If the mass is between 8 and 100 $\mathcal{M}_\odot$, Pop III stars explode as normal Type II SNe through the usual process of core collapse. This occurs at the end of the whole sequence of thermonuclear reactions in the stellar core started with the fusion of hydrogen into helium, continued with helium into carbon and oxygen, and completed with the burning of carbon, oxygen and silicon. The final product of this chain of reactions is the fusion

of silicon into iron. This is the beginning of the end of the star. No further thermonuclear reactions are possible with iron because this element absorbs energy in order to fuse into heavier elements. This produces an inert core of iron which cannot counterbalance the gravity for the lack of energy sources, and the star collapses. When the core exceeds the Chandrasekhar limit ($\approx 1.4\mathcal{M}_\odot$), a rapid implosion of the core takes place. This causes a large increase of the temperature to $10^{10}$ K with a subsequent emission of gamma-ray photons. These photons destroy the iron nuclei in a process called **photodisintegration** and produce helium, $\alpha$-particles, protons (p) and neutrons (n) through the reaction

$$\gamma + {}^{56}\text{Fe} \rightarrow 13\,{}^{4}\text{He} + 4\text{n} \tag{9.31}$$

and

$$\gamma + {}^{4}\text{He} \rightarrow 2\text{p} + 2\text{n}. \tag{9.32}$$

These are endothermic (energy-absorbing) reactions which reduce the internal pressure of the star and trigger its immediate collapse. During this phase, protons and electrons combine to form neutrons, releasing a large amount of neutrinos. These neutrinos escape from the core, carrying away energy, and generate a further collapse of the stellar structure followed by a final explosion. The remnant is a neutron star or a black hole depending on the mass of the progenitor star.

2. Pop III stars with masses between 100 and 260 $\mathcal{M}_\odot$ end their evolution as **pair instability supernovae**. This process occurs if the core temperature is high enough to efficiently convert gamma-ray photons into electron–positron pairs. Gamma-ray photons and their associated radiation pressure are essential to maintain the stellar interior in equilibrium between the competing gravitational and pressure forces. The rapid decrease of gamma-ray photons due to pair creation originates a reduction of the internal pressure and the star collapses. In pair instability SNe, the original star is thought to be completely disrupted.
3. For masses higher than 260 $\mathcal{M}_\odot$, the stars explode due to the photodisintegration of the $\alpha$-particles that were previously produced by the photodisintegrated iron nuclei in the core. The lack of inner support causes the rapid collapse of the outer regions of the star. In this scenario, it is thought that the final remnant is a black hole.

Irrespective of the progenitor mass, all these SN explosions have a profound influence on the surrounding medium because they produce the first metal enrichment in the history of the Universe and change the chemical composition of the pristine pregalactic gas. As a consequence, the subsequent generation of objects form from metal-enriched gas (i.e. $Z > 0$) and cannot be qualified any more as pure Pop III stars. Thus, the transition from Pop III to Population II occurred very rapidly because of the very short lifetimes of Pop III stars.

### 9.5.3 Feedback Processes Caused by Population III Stars

The birth and evolution of Pop III stars is a major perturbation for the primordial IGM due to processes like mass accretion, gas heating and cooling, energy injection, emission of UV radiation and SN explosions. These processes are collectively called feedback (§8.7).

In the case of Pop III stars, feedback processes can be either positive or negative towards star formation (§8.7). An example of negative feedback is when SN explosions expel a large fraction of the gas from a galaxy and make it unavailable for star formation. Instead, a case of positive feedback is the compression and fragmentation of the gas caused by the propagation of the SNRs in the ISM which can trigger new star formation (self-regulated star formation; §4.2.10). In the case of Pop III objects, three main feedback processes are expected to influence the evolution of the star formation and the IGM. Each of them can be subdivided into negative and positive towards star formation.

## Radiative Feedback

Due to their high photospheric temperatures, Pop III stars are powerful emitters of UV photons. As a reference, the rate of hydrogen-ionising photons averaged over the stellar lifetime is $Q(\mathrm{H}) \sim 10^{50}\ \mathrm{s}^{-1}$ for a mass $\mathcal{M} \approx 120\ \mathcal{M}_{\odot}$. In comparison, for the same mass, the rates of photons capable of largely ionising also helium are $Q(\mathrm{He\ I}) \approx 8 \times 10^{49}\ \mathrm{s}^{-1}$ and $Q(\mathrm{He\ II}) \approx 5 \times 10^{48}\ \mathrm{s}^{-1}$. Each Pop III star produces a large Strömgren sphere (§4.2.3) where the gas is fully ionised and heated by the collisions with the emitted electrons which carry off the residual kinetic energy $(h\nu - E_i)$, where $E_i$ is the ionisation energy of a given atom. This process is called photoheating (§8.1.2). If the thermal speed of the photoionised gas exceeds the escape velocity, the gas is dispersed away from the ionising source, a process called **photoevaporation**. Photoionisation is the major negative feedback process and is expected to radically change the IGM physical conditions. The other radiative feedback process with a negative character is **$\mathbf{H_2}$ photodissociation**. Ultraviolet photons with $h\nu < 13.6$ eV emitted by Pop III stars can travel and penetrate through the gas without ionising the neutral atomic hydrogen. However, photons with $11.2 \lesssim h\nu \lesssim 13.6$ eV can interact with $H_2$ by exciting the ground-state electrons to the Lyman and Werner electronic states. For this reason, these photons are also called **Lyman–Werner photons**. Once the Lyman and Werner levels are populated, the excited electrons can spontaneously decay radiatively with rapid timescales. However, about 10–15% of the decays occur into the repulsive vibrational state (vibrational quantum number $\upsilon > 14$) of the ground state, and $H_2$ photodissociates. This is called the **Solomon process**. This feedback process is negative because the $H_2$ fraction is reduced and therefore the $H_2$ cooling becomes less efficient. Moreover, the mean energy released per photodissociation (0.4 eV) is non-negligible and is converted into kinetic energy of the free H atoms which then heat the gas through collisions. Although the radiative feedback is generally negative, simulations show that $H_2$ can form again when the ionised gas recombines in regions (called H II relics) where star formation was previously quenched by negative feedback processes. In this case, the new formation of $H_2$ produces an increase of cooling. Thus, this feedback process is positive and can partly compensate the negative one by promoting again star formation.

## Mechanical Feedback

This class of feedback includes all cases concerning the motion of the gas and its kinetic energy (§8.7). A first example is given by the stellar winds (§8.7.2). Stellar winds are very important in the evolution of massive stars in the present-day Universe, where the gas is accelerated mostly by the absorption of the stellar UV photons by the metal line transitions. However, being metal-free Pop III stars, these winds are much less relevant, and they become important only in the subsequent generations of stars born from a metal-enriched gas. The most important effects of mechanical feedback come from the explosion of Pop III SNe. The overall energy budget of SNe ranges from the extreme cases of pair instability SNe ($E \sim 10^{53}$ erg) to normal Type II SNe ($E \sim 10^{51}$ erg). Two phenomena can provide important negative feedback. The first is called **blow-away** and consists in the partial removal of the gas from the dark matter halo due to SN explosions producing high-energy shock waves which propagate through the surrounding region sweeping up the diffuse gas. The second is called **blow-out**[1] and is an extreme version of blow-away because it causes the complete removal of the gas from the halo. Simulations suggest that blow-out can occur only in halos with $\mathcal{M}_{\rm vir} \lesssim 5 \times 10^6\ \mathcal{M}_\odot$. The feedback of SN explosions can also be positive when the expansion of the SN remnants compresses the surrounding gas, increases its density and promotes fragmentation and collapse.

#### Chemical Feedback

After the very first generation of Pop III stars with $Z \approx 0$, the primordial IGM gas becomes gradually enriched by metals. As a consequence, cooling becomes more efficient due to the emission produced by metal lines (e.g. C, N and O; §8.1.1). This implies that the formation of new stars can proceed more easily than in the case of Pop III objects. The timescale over which the gas is polluted by metals depends on the IMF, on the rate of SN explosions, on their energies, and on how efficiently metals are transported and mixed with the pristine primordial gas. At first sight, this is clearly a positive feedback because it promotes star formation. However, it also changes substantially the chemical composition of the primordial gas and determines the end of the era of pure metal-free Pop III stars. From this point of view, the chemical feedback is negative towards the formation of pure Pop III stars by marking their rapid disappearance and defines a point of no return.

### 9.5.4 The Initial Mass Function of Population III Stars

Theoretical models suggests that Pop III stars are, on average, more massive than present-day young stars. However, due to the uncertainties on the feedback processes occurring during the accretion phase, it is unclear what are the highest achievable masses. The **Larson IMF** is often used to describe the IMF of Pop III stars. This IMF is defined as (Larson, 1998)

$$\phi(m) = Am^{\alpha} \exp\left(-\frac{m_{\rm ch}}{m}\right), \tag{9.33}$$

where $m = \mathcal{M}/\mathcal{M}_\odot$, $A$ is a constant, $\alpha = -2.35$ and $m_{\rm ch}$ is a characteristic mass. Fig. 9.3 shows two examples of Pop III IMFs. Advanced numerical simulations show that Pop III

[1] Not to be confused with the same word used when a superbubble escapes from the disc of a spiral galaxy (§8.7.2).

stars tend to form in groups due to fragmentation of the gaseous disc that is thought to form around the protostellar object during the accretion of gas with angular momentum. However, the final distribution of stellar masses varies significantly from halo to halo. The presence of massive binaries and multiple systems within the Pop III stellar population is also possible depending on the fragmentation properties. In all cases, the resulting IMF is very broad and can extend also down to $\approx 1\ \mathcal{M}_\odot$. Hence, some of these low-mass Pop III stars may have survived until the present-day Universe. At the high-mass end, Pop III stars can easily reach tens to hundreds of solar masses, but the simulations show that it is unlikely that they exceed $\sim 1000\ \mathcal{M}_\odot$ probably due to negative feedback processes. The top-heavy IMF of Pop III stars implies a large number of SNe and, consequently, a fast chemical enrichment of the IGM.

## 9.6 From First Stars to First Galaxies

Pop III stars cannot be considered the first galaxies. The reason is that Pop III stars are located in minihalos with masses $\sim 10^6\ \mathcal{M}_\odot$. These halos have shallow gravitational potential wells, and therefore are very sensitive to the effects of negative feedback from star formation (photoionisation, SNe). Numerical simulations show that the combination of strong UV radiation and SN explosions can cause a strong heating of the gas and its expulsion from minihalos. This implies that a second generation of new stars is not allowed to form within minihalos with masses $\sim 10^6\ \mathcal{M}_\odot$ where the formation of Pop III objects took place.

A galaxy can be considered as a longer-lived system hosted by a dark matter halo capable of gravitationally retaining a substantial fraction of the gas. It is clear that minihalos cannot host the first galaxies. In order to estimate the requirements for a dark matter halo to host a first galaxy, it is useful to consider the virial temperature and the binding energy of the halo. At redshifts around $z \approx 10$, the virial temperature reaches $10^4$ K for a halo mass around $\sim 10^8\ \mathcal{M}_\odot$ (eq. 9.16). This temperature is crucial because the primordial (low-metallicity) gas can start to cool radiatively with high efficiency through the recombination lines of ionised hydrogen. The binding energy of a dark matter halo is of the order of

$$E_{\rm bind} \sim \frac{G\mathcal{M}_{\rm vir}^2}{r_{\rm vir}}. \tag{9.34}$$

This binding energy is important to assess if a halo is capable of retaining its gas when SNe explode. For $\mathcal{M}_{\rm vir} \sim 10^8\ \mathcal{M}_\odot$, $E_{\rm bind}$ is of the order of $10^{53}$ erg, much higher than the typical energy released by a Type II SN ($E_{\rm SN} \sim 10^{51}$ erg). Thus, this halo mass should be sufficient to retain a substantial fraction of the gas, whereas the effects of SNe can be devastating for lower-mass halos. Numerical simulations show that halos with $\mathcal{M}_{\rm vir} \sim 10^8\ \mathcal{M}_\odot$ and $T_{\rm vir} \sim 10^4$ K at $z \approx 10$ can indeed be considered the sites where the **first galaxies** formed and evolved without losing their gas. Moreover, these halos were massive enough to accrete the gas that was affected by the feedback of previous star formation in minihalos. Thus, the generally accepted definition of first galaxies is based on the presence of dark matter

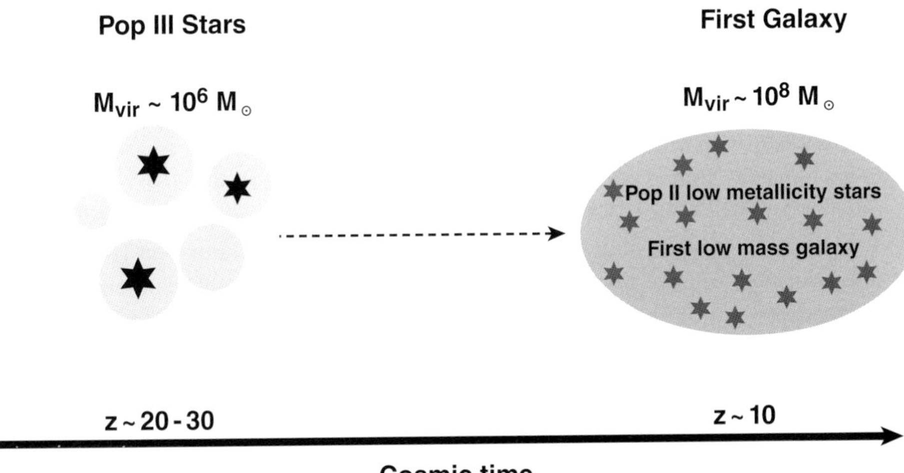


Fig. 9.4 A sketch of the transition from Pop III stars to first galaxies. At $z \approx 20$–30, Pop III stars are predicted to form in dark matter halos with typical masses of $\sim 10^6\ \mathcal{M}_\odot$ and when the radiative cooling of the gas from primordial molecules is efficient enough. The first galaxies are expected to form at lower redshifts ($z \approx 10$) in halos with masses (and therefore virial temperatures) high enough to make radiative cooling from atomic hydrogen efficient.

halos where atomic cooling dominates as opposed to molecular cooling. These halos are usually called **atomic cooling halos** (see the review of Bromm and Yoshida, 2011). It is worth noting that the cosmic time between $z \approx 20$–30 (the supposed epoch of Pop III star formation) and $z \approx 10$ is only $\approx 0.3$–0.4 Gyr. This implies that the transition from the first stars to the first galaxies must have been quite rapid (Fig. 9.4).

## 9.7 The Formation of the First Massive Black Holes

In the present-day Universe, galaxies show a remarkable correlation between the mass (or velocity dispersion) of the spheroidal/bulge component and the mass of the black holes hosted in their centres (§5.4.4). The mere existence of such a correlation suggests that galaxies and SMBHs shared an intimate coevolution across cosmic time. SMBHs can manifest themselves through the emission of powerful non-thermal radiation during their accretion phase in galaxies which host AGN activity (§3.6). Fossil SMBHs (i.e. where the accretion has terminated) can be detected through their gravitational effects in the nuclear regions of nearby galaxies. Luminous QSOs have been discovered up to $z > 7$. Their SMBH masses are surprisingly very large ($\mathcal{M}_\bullet \sim 10^{8-9}\ \mathcal{M}_\odot$) and comparable with those of SMBHs at $z \approx 0$. If SMBHs acquired their mass through accretion, this process must have been very efficient and fast to achieve $\mathcal{M}_\bullet \sim 10^{8-9}\ \mathcal{M}_\odot$ at $z \approx 6$–7, when the age of the Universe was less than one billion years. For example, if an SMBH seed had an initial mass of 200 $\mathcal{M}_\odot$, it had to accrete continuously at the Eddington rate (§8.8) for roughly 500 Myr to achieve $\mathcal{M}_\bullet \sim 10^9\ \mathcal{M}_\odot$ at $z \approx 7$. Our understanding of these processes and their evolution is still in its infancy, and no firm observational constraints on SMBH seeds are available yet. We refer the reader to reviews such as Volonteri (2012). The main models proposed so far can be collected into two main categories.

1. *SMBH seeds from Pop III remnants*. In the first scenario, the SMBH seeds are expected to have masses around 100 $\mathcal{M}_\odot$ and to be the fossil remnants of massive Pop III stars of a few hundred solar masses. The main problem of these models is the inefficient gas accretion of the black hole due to the feedback of the progenitor Pop III star which is expected to remove a substantial fraction of the gas through photoionisation and photoevaporation. For this reason, the SMBH seeds would be located in low-density regions where the gas infall cannot be as efficient and fast as needed to rapidly build up an SMBH. However, a possibile solution to these problems is to consider that galaxies hosting luminous QSOs are a very small fraction of the whole population of high-$z$ galaxies. This suggests that an efficient SMBH accretion may have occurred only in the rare cases where the requirements of Eddington (or even super-Eddington) rates were possible thanks to special local conditions. Another limitation of these models is that the Pop III progenitors are required to be very massive (at least a few hundred $\mathcal{M}_\odot$, i.e. very rare) in order to produce SMBH seeds with masses around 100 $\mathcal{M}_\odot$.
2. *SMBH seeds from direct gas collapse*. In these models, the black hole seeds were originated from the direct collapse of a huge cloud of gas without any intermediate star formation (**direct collapse black holes**). Simulations show that this option is viable only in the special conditions of primordial gas clouds composed only of hydrogen and helium and without metal line cooling. If this primordial gas cools rapidly through $H_2$ line emission, the cloud collapses, fragments and forms stars. However, if the system is surrounded by strong Lyman–Werner UV radiation ($h\nu < 13.6$ eV; §9.5.3) emitted by neighbouring galaxies, these photons photodissociate $H_2$, therefore preventing the gas from fragmenting and forming stars. In this case, only atomic cooling (mainly through Ly$\alpha$ emission) can keep the gas in the collapsing core below $10^4$ K. Thanks to this effect, the gas in the galaxy is kept just hot enough to avoid star formation, but cool enough to allow the collapse of a substantial fraction into a black hole (provided that the Jeans instability criterion is locally satisfied). The resulting free-fall collapse of the atomic cooling gas could then produce an SMBH directly.

In other scenarios, a rapid collapse may occur when the gas in the inner region of a galaxy becomes globally unstable because of an insufficient rotational support, leading to a rapid accumulation of the gaseous matter in the centre. If the gas accumulation timescale is shorter than the thermonuclear timescale, this dynamics-driven gas collapse could form a single star with a very high mass, up to $10^6$ $\mathcal{M}_\odot$ (**supermassive star**). Simulations show that in the centre of such supermassive stars the contraction of the core occurring after the exhaustion of the hydrogen can lead to the formation of a black hole with a few tens $\mathcal{M}_\odot$. This small object could then accrete the stellar envelope from the inside (such a system is called a **quasi-star**) and grow up to a massive SMBH seed of $10^{5-6}$ $\mathcal{M}_\odot$. Another possibility is that the dense and locally unstable gas in the centres of galaxies triggers star formation and leads to the creation of a dense star cluster. Stars and/or their black hole remnants may have subsequently merged into a very massive star ($> 10^3$ $\mathcal{M}_\odot$), which then collapses into an SMBH seed with mass $10^3\ M_\odot$. The main scenarios outlined above do not exclude each other, and they could have been at work at different cosmic times and/or in different physical and structural conditions of the primordial galaxies where they occurred.

## 9.8 The Intergalactic Medium

The first stars and galaxies form out of only a small fraction of the gas present in the early Universe. The rest of the gas remains located in the space between collapsed structures: this material is generically referred to as the intergalactic medium (IGM; §6.7). The boundaries between galaxies and the IGM are somewhat undefined. In this book, we use the term circumgalactic medium (CGM) to identify the material close to galaxies (§6.7), typically within their virial radii (e.g. $r_{\rm vir} \approx 280$ kpc for the Milky Way today; §7.3.2). This is also referred to as **halo gas**. In this section we mostly focus on gas that is found beyond the virial radii of galaxies and makes up the low-density **baryonic cosmic web**. The average baryon density of the Universe is given by eq. (9.10). This is a reasonable approximation of the average density that we can expect in the IGM as a function of redshift. Note that it is valid independently of the ionisation state of the medium.

### 9.8.1 The Lyman-$\alpha$ Forest

The best probes to study the IGM out to high redshift are absorption lines against bright distant QSOs (§3.6). A QSO typically shines light across the whole electromagnetic spectrum peaking at UV wavelengths (in its rest frame). A clear spectral feature of QSOs is their broad Ly$\alpha$ emission line at $\lambda_{\rm Ly\alpha} \simeq 1215.67$ Å ($\nu_{\rm Ly\alpha} \simeq 2.466 \times 10^{15}$ Hz) that falls in the optical band for $z > 2$. The intrinsic spectrum of a QSO around the Ly$\alpha$ emission is characterised by a very strong continuum, well described by a power law $F(\nu) \propto \nu^{-\alpha}$ with $\alpha \approx 1$. These photons can be absorbed by intervening gas located between the QSO and us. In general, we can expect that an atom will absorb the QSO light at the frequency $\nu_{ij} \equiv |\Delta E_{ij}|/h$, where $\Delta E_{ij}$ is the difference in binding energies between the lower (initial) and upper (final) level of the transition. Let us consider a 'cloud' of gas, located at redshift $z_{\rm abs}$ along the line of sight to the QSO and containing atoms of neutral hydrogen. If these atoms are at their ground level, they will absorb incident photons at, or nearly at (see below), their rest-frame Ly$\alpha$ wavelength ($\lambda_{\rm Ly\alpha}$), as this transition has a very large cross section. The subsequent re-emission of the photons has an extremely high probability but occurs in a random direction: *not* towards the observer.[2] This process produces a decrement (reduction of the flux) in the observed QSO spectrum at

$$\lambda_{\rm obs} = \lambda_{\rm Ly\alpha}(1 + z_{\rm abs}). \tag{9.35}$$

[2] In reality, if the density of the absorbing cloud is high enough, the photon is reabsorbed by other hydrogen atoms and promptly re-emitted many times, thus it ends up following a 'random walk' until it eventually escapes the absorbing region. For this reason this process is called **Lyman-$\alpha$ resonant scattering**.

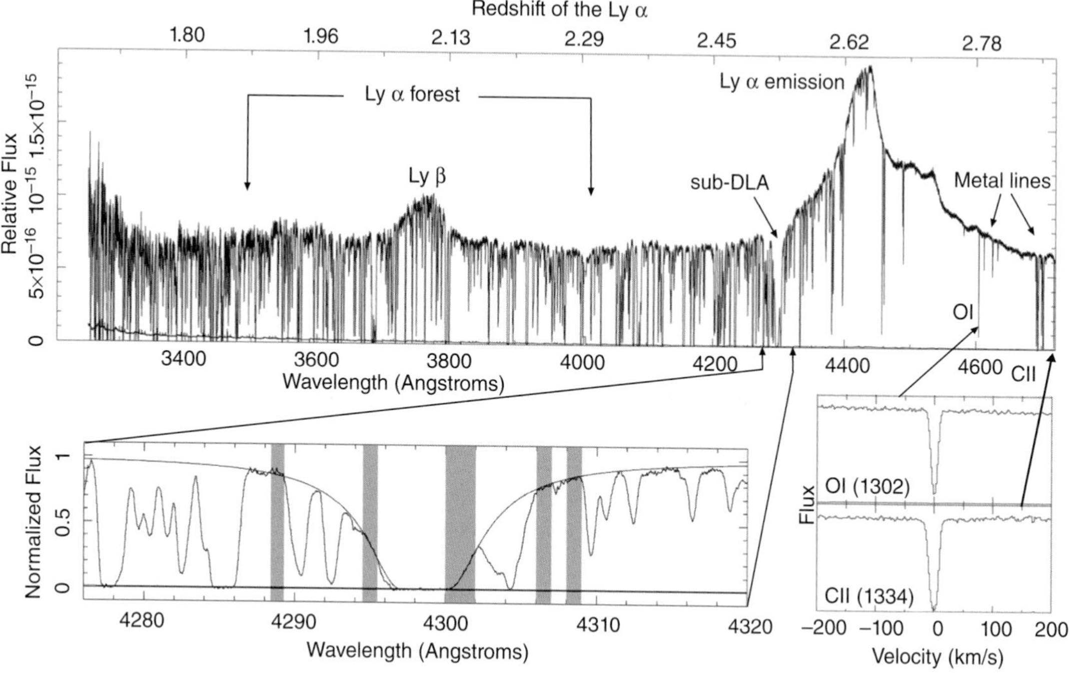


Fig. 9.5 *Top panel*. High-resolution spectrum of the QSO HS 0105+1619 at $z \simeq 2.64$. *Bottom left panel*. Blow-up of the spectral region around the sub-DLA with a fit of the wings (the shaded regions indicate the intervals used in the fit). *Bottom right panels*. Blow-up of two metal lines (also indicated in the top panel) with the *x*-axis showing line-of-sight velocity centred at the sub-DLA. The number in parentheses indicates the rest-frame wavelength of the lines. Adapted from O'Meara et al. (2001). © AAS, reproduced with permission.

Fig. 9.5 shows the spectrum of a QSO at $z_q \simeq 2.64$, where a strong and broad Ly$\alpha$ line is seen redshifted to $\lambda_{\mathrm{Ly}\alpha,\mathrm{q}} \simeq 4425$ Å. Lyman-$\beta$ (Ly$\beta$) is also visible as a bump around $\lambda_{\mathrm{Ly}\beta,\mathrm{q}} \simeq 1026(1+z_q) \simeq 3734$ Å. While at wavelengths longer than $\lambda_{\mathrm{Ly}\alpha,\mathrm{q}}$ the spectrum shows only a few absorption features, at shorter wavelengths it is 'pierced' by a multitude of absorption lines. The vast majority of these are caused by Ly$\alpha$ absorption from intervening material between us and the QSO. Because necessarily $z_{\mathrm{abs}} < z_q$, all these features must be at wavelengths $\lambda_{\mathrm{obs}} < \lambda_{\mathrm{Ly}\alpha,\mathrm{q}}$. In the presence of internal motions within the absorbing features (see §9.8.2), with velocities of order $\pm\Delta v$, we observe a flux reduction in the spectral range $\langle\lambda_{\mathrm{obs}}\rangle \pm \Delta\lambda$, where $\Delta\lambda = \langle\lambda_{\mathrm{obs}}\rangle \Delta v/c$ and $\langle\lambda_{\mathrm{obs}}\rangle$ is the average wavelength of the absorption (§C.7.1). Collectively, these absorption lines are called the **Lyman-$\alpha$ forest**.

The bottom left panel of Fig. 9.5 shows a saturated Ly$\alpha$ absorption line at $z \simeq 2.54$. This is called a **damped Lyman-$\alpha$ absorber** (DLA) and it is characterised by a much higher column density than the other lines. Note that the broadening of this line is *not* due to internal motions, as we see below. In the specific example of Fig. 9.5 the column density of this absorber is $\log(N_{\mathrm{HI}}/\mathrm{cm}^{-2}) \simeq 19.4$, so the feature is classified as sub-DLA. These types of absorptions are much rarer than other features in the Ly$\alpha$ forest, which have much lower column densities (Tab. 9.1). The lines on the right side of the QSO Ly$\alpha$ emission are metal lines. Some of them are from the medium surrounding the QSO itself, others from intervening material. The chemical species that are absorbing the QSO light are easily identifiable if they have a counterpart in the Ly$\alpha$ forest. The O I and C II absorptions shown in the blow-ups in Fig. 9.5 (bottom right) are at the same redshift as the sub-DLA ($z \simeq 2.54$). This tells us that at the location between us and the QSO where there is intervening material with high hydrogen column densities there are also metals, testifying to an important chemical enrichment of this material (§9.8.3).

### 9.8.2 Physics of the Lyman-$\alpha$ Forest

When QSO spectra are observed at high spectral resolution, the absorption features can be used to derive key properties of the intervening gas. Typically we can expect an absorption (or an emission) line to be broadened by three phenomena: (1) thermal broadening, (2) turbulence and (3) the natural broadening due to the Heisenberg uncertainty principle. For a Maxwellian distribution of particle speeds (§D.1.4), the 1D thermal profile has a probability of absorbing photons between speeds $v$ and $v + \mathrm{d}v$ of

$$\mathcal{P}(v)\mathrm{d}v = \frac{1}{\sqrt{\pi} b} \exp\left(-\frac{v^2}{b^2}\right)\mathrm{d}v, \tag{9.36}$$

where $v = 0$ corresponds to the redshift $z_{\mathrm{abs}}$ of the absorber (§C.7). In eq. (9.36), $b = \sqrt{2}\sigma$, with $\sigma$ the 1D (e.g. line-of-sight) velocity dispersion of the gas, is the **Doppler parameter**. If the broadening is purely thermal we have $b^2 = 2k_{\mathrm{B}}T/m$ (with $m$ the mass of the absorbing atom). More generally, $b^2 = 2k_{\mathrm{B}}T/m + b_{\mathrm{turb}}^2$, with $b_{\mathrm{turb}}$ the turbulent broadening.

The natural broadening of a line gives a **Lorentzian profile** characterised by pronounced wings. The convolution of a Lorentzian profile with a Gaussian leads to the **Voigt profile**, which has an inner Gaussian shape and broad wings. The relative importance of natural and thermal broadening of a line depends on the type of absorption/emission (in particular its Einstein coefficients) and on its physical condition (in particular its temperature).

A drastic modification of the line profiles occurs at high column densities where one enters the saturated regime. In this regime, all the radiation is absorbed at the central wavelengths of the line, while prominent Lorentzian wings (**damping wings**) emerge on the sides. The unsaturated and saturated regimes of absorption lines correspond to the two steep parts of the so-called **curve of growth** (Fig. 9.6 for the specific case of the Ly$\alpha$), which links the column density of the absorbing material to the measured equivalent width of the lines

$$W \equiv \int_0^\infty \left[1 - e^{-\tau(\lambda)}\right] \mathrm{d}\lambda, \tag{9.37}$$

where $\tau(\lambda)$ is the optical depth of the absorbing medium (§D.1.1). Note that eq. (9.37) is a convenient way to rewrite eq. (3.6) where we have used eq. (D.4) with $S_\nu = 0$.

The above considerations are valid in general, while we now look at the case of Ly$\alpha$ absorption. Fig. 9.6 shows the curve of growth for Ly$\alpha$ lines with three different Doppler parameters. Low-column-density lines in the Ly$\alpha$ forest belong to the linear part of the

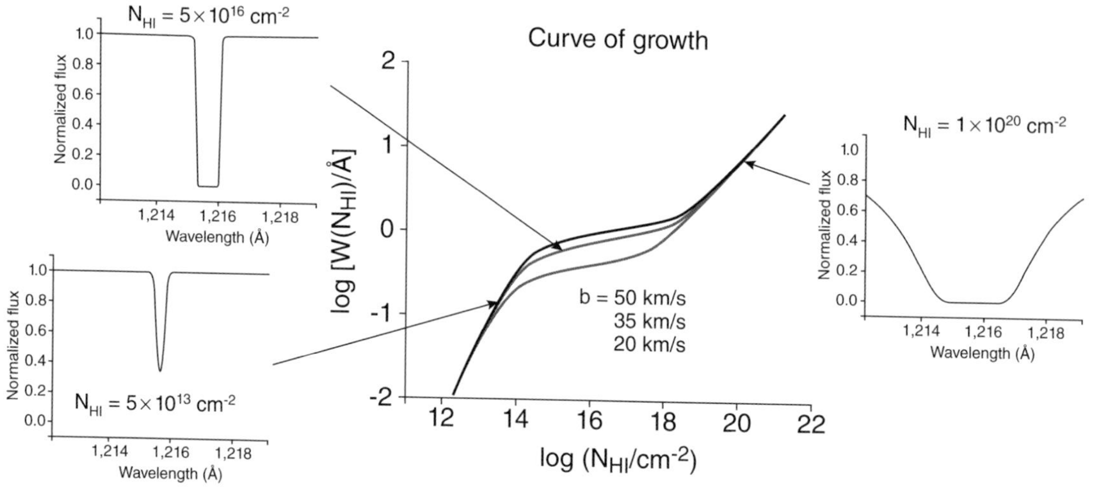


Fig. 9.6 Curve of growth for Ly$\alpha$ absorption lines with different thermal broadenings; the values of the Doppler parameter $b$ refer to the curves from top (50 km s$^{-1}$) to bottom (20 km s$^{-1}$). The line profiles in the insets are calculated at the Ly$\alpha$ rest frame for three different column densities of neutral hydrogen and for $b = 35$ km s$^{-1}$. Courtesy of Y.-S. Ting.

curve at $N_{\rm H\,I} \lesssim 10^{14}\,{\rm cm}^{-2}$. There, the optical depth is $\tau < 1$ and the H I column density is simply[3]

$$N_{\rm H\,I} \simeq 1.84 \times 10^{13} \left(\frac{W_{{\rm Ly}\alpha}}{0.1\text{Å}}\right)\,{\rm cm}^{-2}, \tag{9.38}$$

where $W_{{\rm Ly}\alpha}$ is the Ly$\alpha$ equivalent width. In the fully saturated ($\tau \gg 1$) regime at $N_{\rm H\,I} \gtrsim 10^{19}\,{\rm cm}^{-2}$, the line profile is dominated by the damping wings (see rightmost inset in Fig. 9.6) and the equivalent width becomes proportional to the square root of the H I column density,

$$N_{\rm H\,I} \simeq 1.87 \times 10^{20} \left(\frac{W_{{\rm Ly}\alpha}}{10\text{Å}}\right)^2\,{\rm cm}^{-2}. \tag{9.39}$$

The intermediate regime ($10^{14} < N_{\rm H\,I} < 10^{19}\,{\rm cm}^{-2}$) is hard to study and requires a more detailed modelling. In this regime the density of neutral atoms in the intervening material is such that the optical depth to Ly$\alpha$ absorption at the line centre is $\tau \gtrsim 1$. Most photons with $\lambda < 912$ Å (at the frequency of the absorber) are absorbed to ionise hydrogen atoms (§4.2.3) and, for this reason, absorbers with $N_{\rm H\,I} > 10^{16}\,{\rm cm}^{-2}$ are called **Lyman limit systems** (LLSs).[4] Note that in this regime, the equivalent width is rather insensitive to changes of the column density, but it is quite sensitive to changes in the Doppler parameter

[3] For a generic transition between level $i$ and $j$ at wavelength $\lambda_{ij}$ the optically thin column density is $N \simeq 1.13 \times 10^{12} (W/\lambda)/(f_{ij}\lambda_{ij})\,{\rm cm}^{-2}$ where $f_{ij}$ is the oscillator strength of the transition. We refer the reader to Draine (2011) for details on the derivation of the equations in this section.

[4] The complete absorption of ionising photons occurs for $\log(N_{\rm H\,I}/\,{\rm cm}^{-2}) > 17.2$; thus the systems between this value and $\log(N_{\rm H\,I}/\,{\rm cm}^{-2}) = 16$ are sometime referred to as partial LLSs.

**Table 9.1** Absorption lines in QSO spectra

| Name | $N_{\rm HI}$ ($\rm cm^{-2}$) | Neutral fraction | [M/H][a] | Location |
|---|---|---|---|---|
| Ly$\alpha$ forest | $10^{12}-10^{16}$ | $10^{-5}-10^{-4}$ | $-2.82\pm0.75$ | Filaments/voids |
| LLS | $10^{16}-10^{19}$ | $10^{-3}-10^{-2}$ | $-2.10\pm0.84$ | CGM/filaments |
| sub-DLA[b] | $10^{19}-10^{20.3}$ | $10^{-1}-1$ | $-1.92^{+0.76}_{-1.04}$ | CGM/galaxies |
| DLA | $>10^{20.3}$ | $\approx 1$ | $-1.39\pm0.52$ | Galaxies |

[a] Metallicities (see C.6.1) at $2.3<z<3.3$. Values are median and standard deviations from Lehner et al. (2016).

[b] These are also called super Lyman limit systems.

(three curves in Fig. 9.6). For a summary of the nomenclature of Ly$\alpha$ absorbing features, see Tab. 9.1.

The column density distribution of Ly$\alpha$ forest absorbers follows the power law

$$\Phi(N_{\rm HI})\propto N_{\rm HI}^{-\beta}, \tag{9.40}$$

with $\beta\approx1.5$. This slope is well determined below $N_{\rm HI}\sim10^{15}\,\rm cm^{-2}$, but it appears to extend to the regime of the LLSs. The observed Doppler parameters of Ly$\alpha$ forest absorbers are typically $b=15$–$50\,\rm km\,s^{-1}$ corresponding to temperatures of 1–few$\times10^4$ K (if we consider only thermal broadening). It is important to note that the neutral hydrogen is *just a tracer* of the gas contained in the Ly$\alpha$ forest: the neutral fraction is $x_{\rm HI}\lesssim10^{-4}$ (§9.9.1) but this low density of neutral atoms is enough to produce the absorption. The vast majority of the material is ionised and, in particular, photoionised by the local (at $z_{\rm abs}$) extragalactic UVB (§8.1.2).

We believe that the Ly$\alpha$ forest absorptions take place in the IGM far away from galaxies. The formation and evolution of these intervening absorbers have been extensively studied using cosmological hydrodynamic simulations (§10.11.1). Based on these investigations, there is a general agreement that the Ly$\alpha$ forest is essentially the baryonic counterpart of the dark matter cosmic filaments and sheets that are naturally produced in a $\Lambda$CDM cosmology (Fig. 7.7). The typical temperatures of the gas in the filaments are $T\approx10^4$ K and the column densities are in agreement with the one observed in the Ly$\alpha$ forest. Simulations with and without feedback (§8.7 and §8.8) produce slightly different properties of the absorbers especially around $z\approx2$ when stellar/AGN feedback is most effective (§11.3.14). Finally, the statistics of the Ly$\alpha$ forest is quite sensitive to variations in the cosmological parameters and it has been used to constrain WDM models (§7.2).

The observed number density $\mathcal{N}$ of the absorbers (Ly$\alpha$ forest lines) strongly evolves with redshift as

$$\frac{d\mathcal{N}}{dz}\propto(1+z)^{\alpha} \tag{9.41}$$

with $\alpha\approx1$–3 (depending on the range of column densities considered) at $2<z<4$, becoming much shallower at $z\lesssim1$. If we use these measurements to estimate the *total* gas density we find that, at $z \approx 3$–4, most of the baryons (roughly 80% of the total) were in the Ly$\alpha$ forest, while at $z \approx 0$, the Ly$\alpha$ forest appears to contain only $\approx 30\%$ of the baryons (§6.7). This difference suggests that a fraction of the gas in the Ly$\alpha$ forest at high redshift has made its way from the IGM into collapsed structures like galaxies and galaxy clusters. This accretion of gas has accompanied the hierarchical build-up of the dark matter halos (§7.4.4) that go from containing $\approx 20\%$ of the total dark matter at $z \approx 3$–4 to nearly half at $z \approx 0$. However, part of the IGM gas may have also been heated to temperatures higher than $10^4$ K, as discussed in §6.7. For details on the physics of the IGM we refer the reader to Meiksin (2009).

### 9.8.3 Lyman Limit and Damped Lyman-$\alpha$ Systems

When the radiation from a distant QSO is intercepted by material with column densities $N_{\rm HI} > 10^{16}\,{\rm cm}^{-2}$ we enter the regime of LLSs. These are systems that absorb *every* photon with $\lambda < 912$ Å. In this regime, the equivalent width of the Ly$\alpha$ line is essentially insensitive to $N_{\rm HI}$ (Fig. 9.6). However, the column densities of this material are high enough that a number of metal lines can also be detected in the QSO spectra. Some of these fall at wavelengths longer than the Ly$\alpha$ of the QSO and are therefore easily identified (Fig. 9.5).

The presence of these metal lines allows a more detailed characterisation of the ionisation state of the intervening material, which is usually obtained with **photoionisation models**. The calculations are carried out numerically with specific codes that, given as input an extragalactic UVB flux $F(\lambda)$ as a function of wavelength (§8.1.2) and the column densities of neutral hydrogen ($N_{\rm HI}$), calculate the column densities of metal lines as functions of the metallicity ($Z$) and of the **ionisation parameter** $U \equiv F_{\rm ion}/(n_{\rm H}c)$, where $F_{\rm ion}$ is the total (integrated over all the relevant wavelengths) flux of ionising photons and $n_{\rm H}$ is the total hydrogen volume density. Media with similar ionisation parameters are expected to have similar properties. The column densities predicted by the photoionisation code can then be compared with those measured from the available absorption lines to determine the best-fitting values of $Z$ and $U$ or, equivalently, $n_{\rm H}$. Typically, in these calculations, one assumes (1) ionisation balance (equilibrium between photoionisation and recombination; §4.2.3), (2) that all the elements (hydrogen and metal species) are cospatial and (3) that the medium has a simple geometry (for instance a slab). In general, the final estimate of the total gas density should become more reliable the larger the number of available metal features and when non-saturated hydrogen lines are present (the Ly$\alpha$ line is typically saturated, but Ly$\beta$ or other lines of the series may not be). Uncertainties on the exact shape of the extragalactic UVB and its evolution with redshift (§8.1.2) can play a significant role.

The general results of photoionisation models is that the ionisation fraction (§4.2.3) of the IGM becomes lower for increasing column densities and essentially vanishes in the regime of DLAs (Tab. 9.1). Moreover, as the column density increases, the sizes of the absorbing structures tend to decrease and their clustering (§6.6) to increase. This indicates that LLSs occur in confined regions of the IGM, likely associated with the CGM of galaxies

(§6.7), and they may be tracing flows of material from (outflows) or towards (inflows) galaxies (§4.2.10). Results on the metallicity of the intervening material are discussed in §9.8.4.

We conclude by briefly describing the fully saturated intervening features in QSO spectra: the DLAs, typically observed at $z > 2$. These have column densities $N_{\rm H\,I}$ that are similar to those of local SFGs (§4.2.1), supporting the idea that the absorption occurs in the ISM of gas-rich high-$z$ galaxies. Current radio telescopes are not sensitive enough to reveal 21 cm H I emission at these redshifts and thus DLAs provide a unique way to study the atomic ISM of high-$z$ galaxies.

The distribution of column densities of DLA and sub-DLA follows a Schechter function (§3.5.1) similarly to the H I column densities in local galaxies,

$$\Phi(N_{\rm H\,I}, z)\mathrm{d}N_{\rm H\,I} = \Phi^* \left(\frac{N_{\rm H\,I}}{N^*}\right)^{\gamma} \exp\left(-\frac{N_{\rm H\,I}}{N^*}\right)\frac{\mathrm{d}N_{\rm H\,I}}{N^*}, \tag{9.42}$$

where $-2 < \gamma \lesssim -1$, and $\Phi^*$ and $N^*$ are parameters that are almost constant with $z$ ($\Phi^*$ is a dimensionless normalisation and $N^* \sim 10^{21}\ \mathrm{cm}^{-2}$). The integration of eq. (9.42) over the DLA and sub-DLA regimes returns the total mass of *neutral* gas (less than 5% of neutral gas is contained at $N_{\rm H\,I} < 10^{19}\ \mathrm{cm}^{-2}$) and allows us to calculate the H I contribution ($\Omega_{\rm H\,I}$) to the cosmic density (see also Tab. 6.4). Interestingly, $\Omega_{\rm H\,I}$ calculated at $z \approx 3$–$4$ in this way is within a factor of $\approx 2$ the same as that at $z = 0$, calculated using H I directly observed in 21 cm emission: $\Omega_{\rm H\,I} \approx 0.001$. This indicates that the reservoir of neutral atomic gas in galaxies did not change significantly with time. Note that this contrasts with the relatively strong evolution of the molecular gas (§11.3.11).

Apart from these similarities, there are also some differences between present-day galaxies and DLAs. One is the fact that the Doppler parameters (and therefore the velocity dispersions; see §9.8.2) of high-redshift DLAs tend to be higher than those derived from H I observations of present-day SFGs. Moreover, the estimated sizes of DLAs appear slightly larger than local H I discs. However, the comparison between saturated Ly$\alpha$ absorption lines and observations of H I in emission are not at all straightforward. A further difference between DLAs and galaxies is that the former tend to have lower metallicities even at low redshifts. An effect that one should take into account is, however, that H I discs potentially extend to large radii also at $z > 0$ (as is the case in present-day galaxies; §4.2.1) and that, on average, the outer discs have a larger cross section than the inner discs. Given that galaxy discs have negative metallicity gradients (§4.1.4), this may bias DLAs towards more metal-poor material, partially explaining the discrepancy.

### 9.8.4 Metals in the IGM

A number of metal species are detected in the spectra of QSOs associated with absorbing systems with relatively high H I column density (typically $N_{\rm H\,I} > 10^{15}\ \mathrm{cm}^{-2}$). This testifies to the ubiquitous enrichment of the IGM/CGM and poses questions as to when and how this enrichment has occurred (§10.7.3). The most common elements are C, Si and O, detected in a variety of ionisation states. Examples of so-called low-ionisation species, typically present in gas at temperatures $T \sim 10^4$ K, are C II, Si II, C III and Si III. High-ionisation species are also often detected, in particular C IV and O VI, this latter requiring a hotter medium at temperatures around a few $\times 10^5$ K. Other useful species for the determination of the metallicity are Fe II, Ni II and Zn II. A potential uncertainty is the depletion of these elements into dust grains (§4.2.7) and one must correct for this effect. However, some elements such as zinc are thought to be weakly affected by this problem.

The metallicity of absorbing systems with $N_{\rm HI} > 10^{15}$ cm$^{-2}$ is quite well determined. At $z \approx 3$, the average metallicity is around $0.01\, Z_\odot$, but with a large spread over three orders of magnitude (Tab. 9.1). There is a general trend for increasing metallicity at increasing column densities, but DLAs with $Z \lesssim 0.01 Z_\odot$ do exist at $z \approx 2$–3. Moving to lower redshifts, the general trend is an increase in metallicity for all systems by about 1 dex and a decrease in the spread as only a few systems are found at supersolar metallicity.

As mentioned, high-ionisation species like O VI are also detected. Unlike the low-ionisation species, these are likely produced by collisional ionisation and therefore by shocks (§4.2.4 and §8.1.1). There are a number of reasons to expect collisional ionisation to be at play in the IGM. The formation of filaments and sheets in the cosmic web is expected to produce shocks that can ionise the medium. Moreover, when the gas flows into dark matter potential wells, shock heating may take place (§8.2.2). Finally, strong feedback from galactic winds and AGNs can also heat up a large volume around some starburst and active galaxies. The collisionally ionised IGM (and CGM) could be at temperatures even higher than those traced by O VI (see §6.7).

## 9.9 The Cosmological Reionisation

In the previous section, we showed that the IGM gas at $z \approx 3$ is largely ionised. However, from §2.4 and §9.1 we know that just after recombination the vast majority of the gas in the Universe was neutral; thus the IGM gas must have been ionised again at some point between $z_{\rm rec} \approx 1000$ and $z \approx 3$. The phenomenon that stripped away again the electrons from the hydrogen atoms is called **cosmological reionisation** or simply **reionisation** (§2.4). The cosmic time when this reionisation took place is called the **epoch of reionisation** (EoR).

### 9.9.1 Evidence for Reionisation

The amount of radiation from a QSO that is absorbed by the intervening material can be parameterised as follows. Given the expected continuum flux from the QSO $F_{\nu,\rm cont}(\nu)$ and the observed flux $F_{\nu,\rm obs}(\nu)$, we can define the **decrement**

$$D_{\rm A} \equiv \left\langle 1 - \frac{F_{\nu,\rm obs}(\nu)}{F_{\nu,\rm cont}(\nu)} \right\rangle = \langle 1 - e^{-\tau_\nu} \rangle \equiv 1 - e^{-\tau_{\rm eff}}, \tag{9.43}$$

where the average is taken across narrow ranges usually around the absorption frequencies of Ly$\alpha$ and/or Ly$\beta$. The last definition in eq. (9.43) of **effective optical depth** $\tau_{\rm eff}$ is an efficient way to parameterise the absorption.

Let us consider the spectrum of a QSO at redshift $z_q$ where we observe a decrement due to Ly$\alpha$ (or another Lyman series line) absorption. This must have occurred along the line of sight at some redshift $z < z_q$. The cross section for Ly$\alpha$ absorption is

$$\sigma_{\mathrm{Ly}\alpha}(\nu) = \frac{\pi e^2}{m_e c} f_{\mathrm{Ly}\alpha} \phi(\nu), \tag{9.44}$$

where $f_{\mathrm{Ly}\alpha} \simeq 0.416$ is the oscillator strength of this transition and $\phi(\nu)$ is the line profile of the intervening absorber determined by its internal motions.

In general, the probability of a photon of frequency $\nu$ to be absorbed in the infinitesimal distance $\mathrm{d}x_{\mathrm{los}}$ is described by the optical depth $\mathrm{d}\tau_\nu = \kappa_\nu \mathrm{d}x_{\mathrm{los}}$ (eq. D.2), where $\kappa_\nu$ is the absorption coefficient (§D.1.1). In the case of Ly$\alpha$ absorption at cosmological distances, we have $\mathrm{d}\tau_\nu = n_{\mathrm{H\,I}}(z)\sigma_{\mathrm{Ly}\alpha}(\nu)\mathrm{d}r$, with $n_{\mathrm{H\,I}}(z)$ the proper (non-comoving) number density and $\mathrm{d}r$ the infinitesimal proper distance (eq. 2.9). Note that we are neglecting stimulated Ly$\alpha$ emission (eq. D.7) as it is important only at temperatures much higher than those of the Ly$\alpha$ forest. Then the total Ly$\alpha$ optical depth at a given observed frequency $\nu_0$, such that $\nu = \nu_0(1 + z)$, is

$$\tau(\nu_0) = \int_0^{z_q} n_{\mathrm{H\,I}}(z)\sigma_{\mathrm{Ly}\alpha}\left[\nu_0(1+z)\right] \frac{\mathrm{d}r}{\mathrm{d}z}\mathrm{d}z \tag{9.45}$$

$$= \int_0^{z_q} n_{\mathrm{H\,I}}(z)\sigma_{\mathrm{Ly}\alpha}\left[\nu_0(1+z)\right] \frac{c\,\mathrm{d}z}{(1+z)H(z)}, \tag{9.46}$$

where, for the second equality, we have used eqs. (2.9) and (2.38). Using eq. (9.44) and assuming that the cross section is highly peaked at $\nu_{\mathrm{Ly}\alpha}$, i.e. $\phi(\nu) = \delta(\nu - \nu_{\mathrm{Ly}\alpha})$, with $\delta$ the Dirac delta function, we can then rewrite eq. (9.46) as

$$\tau(\nu_0) = \frac{\pi e^2 f_{\mathrm{Ly}\alpha}}{m_e} \int_0^{z_q} n_{\mathrm{H\,I}}(z)\delta\left[\nu_0(1+z) - \nu_{\mathrm{Ly}\alpha}\right] \frac{\mathrm{d}z}{(1+z)H(z)}$$

$$= \frac{\pi e^2 f_{\mathrm{Ly}\alpha}}{m_e} \int_{\nu_0}^{\nu_q} \frac{n_{\mathrm{H\,I}}(z)}{H(z)} \frac{\delta(\nu - \nu_{\mathrm{Ly}\alpha})}{\nu}\mathrm{d}\nu, \tag{9.47}$$

where, in the second equality, we have made the substitution $\nu = \nu_0(1 + z)$. Note that, given that the cross section is a delta function, the absorption occurs exactly at $z = z_{\mathrm{abs}}$ such that $\nu_0 = \nu_{\mathrm{Ly}\alpha}(1 + z_{\mathrm{abs}})^{-1}$ and thus we can integrate eq. (9.47) to obtain

$$\tau(\nu_0) \approx \frac{\pi e^2 f_{\mathrm{Ly}\alpha}}{m_e \nu_{\mathrm{Ly}\alpha}} \frac{n_{\mathrm{H\,I}}(z_{\mathrm{abs}})}{H(z_{\mathrm{abs}})} \simeq \frac{5.9 \times 10^{10}}{\left[\Omega_{\mathrm{m},0}(1+z_{\mathrm{abs}})^3 + \Omega_{\Lambda,0}\right]^{1/2}} \left(\frac{h}{0.7}\right)^{-1} \left[\frac{n_{\mathrm{H\,I}}(z_{\mathrm{abs}})}{\mathrm{cm}^{-3}}\right], \tag{9.48}$$

where we have used eq. (2.37) for $H(z)$.

Eq. (9.48) can be further simplified if we take absorbers at $z \gtrsim 2$ for which the $\Omega_{\Lambda,0}$ part in the denominator can be neglected. We obtain

$$\tau_{\mathrm{GP}} \simeq 1.1 \times 10^{11} \left(\frac{h}{0.7}\right)^{-1} \left(\frac{\Omega_{\mathrm{m},0}}{0.3}\right)^{-1/2} \left[\frac{n_{\mathrm{H\,I}}(z)}{\mathrm{cm}^{-3}}\right] (1+z)^{-3/2}, \tag{9.49}$$

where, for simplicity of notation, we substituted a generic $z$ for $z_{\mathrm{abs}}$. Eq. (9.49) is a formulation of the so-called **Gunn–Peterson optical depth** (Gunn and Peterson, 1965) for Ly$\alpha$ absorption at the observed frequency $\nu_0 = \nu_{\mathrm{Ly}\alpha}(1 + z)^{-1}$.

We now write the average density of neutral hydrogen ($n_{\rm H\,I}$) as a function of $z$ (see eq. 9.10) and use it to express the optical depth as a function of the average neutral fraction of hydrogen $x_{\rm H\,I} \equiv n_{\rm H\,I}/n_{\rm H}$ in the IGM. We obtain

$$\tau_{\rm GP} \approx 2 \times 10^5 x_{\rm H\,I}(z) \left(\frac{1+z}{4}\right)^{3/2}. \tag{9.50}$$

Eq. (9.50) shows that, if the neutral fraction of the IGM were of order one, the optical depth would be very high and thus all radiation from QSOs (at wavelengths shorter than the redshifted Ly$\alpha$ emission) should be absorbed: we call this prediction the **Gunn–Peterson effect**. As we saw in §9.8.1, however, absorption from a QSO at say $z \approx 3$ typically occurs only in individual features that carve the QSO continuum without suppressing it completely (Ly$\alpha$ forest). Thus the average optical depth of the medium must be $\tau \lesssim 1$. If we substitute this value in eq. (9.50) we immediately see that the average neutral fraction of the IGM is extremely low: $x_{\rm H\,I} < 10^{-5}$. This is the most compelling evidence that the IGM is predominantly ionised.

Fig. 9.7 shows the optical spectra of two high-$z$ QSOs. If we compare these spectra with that in Fig. 9.5 it is apparent that the decrement of the QSO flux is much more pronounced here. This is a common feature of spectra of QSOs beyond $z \approx 5.5$ and, in fact, in the spectrum of the $z \simeq 6.28$ QSO (bottom panel) the absorption is almost complete. This is called the **Gunn–Peterson trough** and it indicates a high optical depth of the IGM at those redshifts. An increase of $\tau_{\rm GP}$ with $z$ is expected from eq. (9.50) because of the increasing density of the IGM; however this is not enough to explain the troughs observed in spectra like those in Fig. 9.7. Instead, they indicate that the neutral fraction of the IGM is increasing much faster with $z$.

The Gunn–Peterson optical depth can be measured directly by estimating the flux decrements due to Ly$\alpha$ absorption (eq. 9.43). This measure is shown in Fig. 9.8 where we see a strong increase beyond $z = 3$. The dashed line is a fit for $z < 5.5$ returning $\tau_{\rm GP}^{\rm eff} \simeq 0.85\,[(1+z)/5]^{4.3}$. Remarkably, the data points at $z \gtrsim 6$ are all above this line, testifying to an ever faster increase of the optical depth. From these optical depths one can estimate the fraction of neutral gas, obtaining values of $x_{\rm H\,I} \sim 10^{-3}$ at $z \gtrsim 6$. The neutral fraction increases by more than an order of magnitude between $z \simeq 5.7$ and $z \simeq 6.4$ and this indicates that the Universe is rapidly becoming more neutral for increasing $z$. In other words, if we imagine that we start from a condition in which the Universe *was* neutral (dark ages; §9.2), we are observing the completion of its reionisation. For these reasons, $z \approx 6$ is taken to mark the end of the EoR.

The above technique does not allow one to probe much earlier in time to trace the beginning of the reionisation because there are relatively few known QSOs at $z > 6$ and at these redshifts it is difficult to estimate $x_{\rm H\,I}$ from the Gunn–Peterson trough. Further constraints on the EoR come from the study of Ly$\alpha$-emitting galaxies as a function of $z$ (§11.2.14). There appears to be a decrease of Ly$\alpha$-emitting galaxies at $z \approx 7$–$8$ with respect to what is expected by extrapolating the counts at $z < 6$. This gives upper limits on the optical depth (or, equivalently, the neutral fraction) of the IGM at these redshifts.

An independent constraint on reionisation comes from the study of the CMB power spectrum (§2.4).The photons of the CMB, emitted at the recombination era, go through

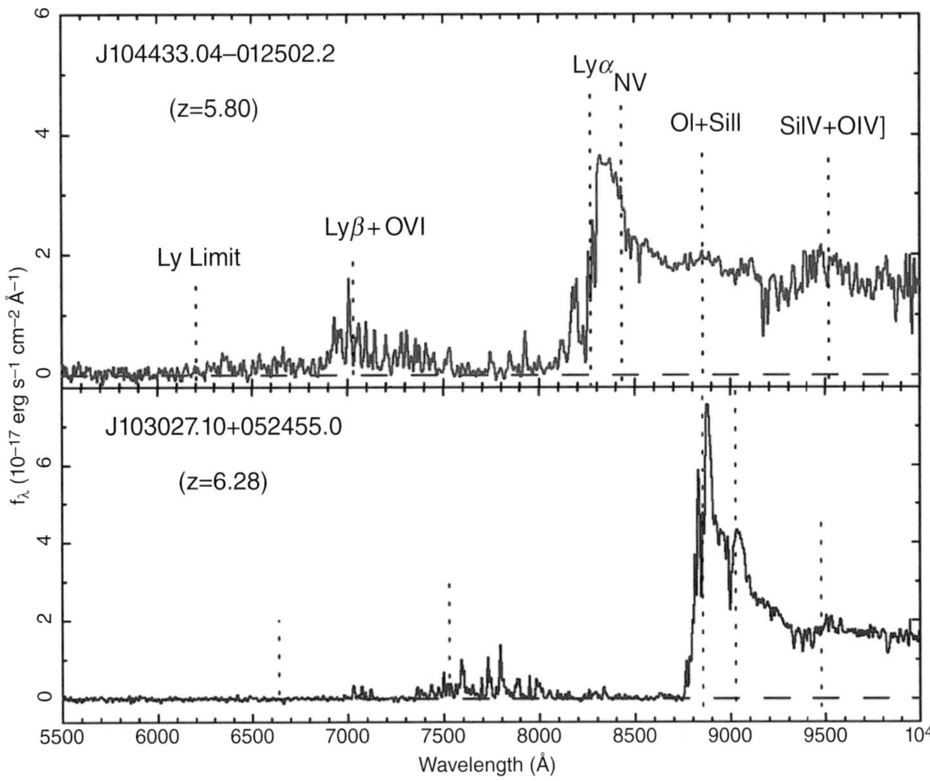


Fig. 9.7 SDSS spectra of two QSOs at redshifts $z \simeq 5.8$ (*top panel*) and $z \simeq 6.28$ (*bottom panel*). Expected locations of the most important lines and of the Lyman limit are indicated by vertical dashed lines. Adapted from Becker et al. (2001). 

a neutral IGM whose density is decreasing with time following eq. (2.30). When this IGM starts to be reionised, the photons are subject to Thomson scattering (§8.1.3) and this has two consequences: (1) it linearly polarises the CMB radiation, and (2) it damps the CMB power spectrum at high $\ell$. By fitting the power spectrum of the CMB combined with the polarised emission, one can determine the optical depth for Thomson scattering $\tau_T \approx 0.05$–$0.06$. In order to turn this into a constraint on the EoR, one needs to assume a redshift evolution for the reionisation. By assuming a very fast reionisation we obtain that the reionisation occurred at $z \approx 7$–$9$. Most likely, as we see in §9.9.3, the reionisation was a gradual process that started at $z \approx 10$ and ended at $z \approx 6$, thus taking a time of about 0.5 Gyr.

## 9.9.2 The Reionisation of Helium

The second constituent of the IGM is helium, which can be neutral (He I), singly (He II) or fully ionised (He III). The ionisation potential of He II is 54.4 eV, four times the one of hydrogen. This energy corresponds to a wavelength $\lambda \simeq 228$ Å, which falls in the extreme UV (§C.1). Stars do not have sufficient emission in this range of the spectrum

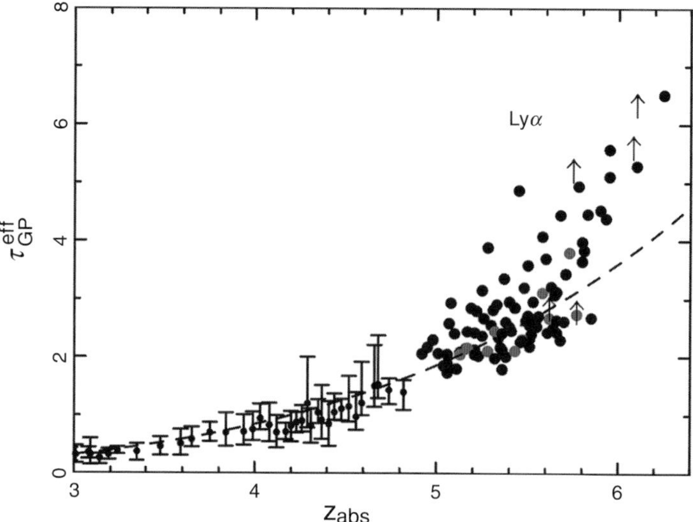


Fig. 9.8 Gunn–Peterson effective optical depth measured from Ly$\alpha$ decrements (eq. 9.43) as a function of the redshift of the absorbers. The dashed line is a fit with the equation described in the text. From Fan et al. (2006). © AAS, reproduced with permission.

and thus to fully ionise helium one requires powerful QSOs. These have their maximum activity at $z \approx 2$–3 and this is indeed the time at which the helium reionisation appears to take place.

The reionisation of the helium component of the IGM can be studied in a way analogous to that of hydrogen. The main absorption line is the He II Ly$\alpha$ at wavelength $\lambda_{\text{He II,Ly}\alpha} \simeq 304\,\text{Å}$. These lines are observed in QSO spectra producing a helium Ly$\alpha$ forest and, for high enough redshifts, a corresponding Gunn–Peterson effect (§9.9.1). In practice, to estimate the reionisation redshift, one measures the decrement of the QSO emission (eq. 9.43) corresponding to the He II Ly$\alpha$ and constructs a plot analogous to the one shown in Fig. 9.8. These investigations show that the reionisation of helium starts at $z \gtrsim 4$ and is completed around $z \approx 2.7$.

### 9.9.3 Model Predictions for Reionisation

There is a general consensus that the reionisation of the gas in the Universe has been produced by photons, i.e. there was no significant contribution from collisional ionisation (§4.2.4 and §8.1.1). This implies the existence of powerful sources of ionising photons at $z > 6$. These sources were probably *not* QSOs or AGNs in general, as their activity sharply declines with redshift for $z > 3$. The currently favoured sources of ionising photons are massive stars. These could have been Pop III stars (§9.5) or second-generation stars that formed in the first galaxies at $z > 6$. Alternative sources for the reionisation have also been considered like mini-QSOs powered by intermediate-mass black holes or annihilation of dark matter particles.

The difficulty to determine the sources of ionising photons at $z > 6$ is due to several factors, including the poorly constrained SFR density of the Universe at these redshifts and the amount of UV flux that early stars can actually emit. For instance, we do not know whether they have the same IMF as observed in local galaxies (§9.5.4). Moreover, a largely unknown quantity is the **escape fraction** $f_{\rm esc}$, which is the amount of ionising ($\lambda < 912$ Å) radiation that can escape from a galaxy into the IGM (if $f_{\rm esc} = 1$ *all* photons escape). We can expect that, given the high gas densities of high-$z$ galaxies, a considerable fraction of the UV radiation is absorbed within their ISM, reducing the number of photons available for the reionisation of the IGM.

The reionisation is studied theoretically with models of expanding ionised bubbles. The analytic treatment describes these bubbles as large-scale Strömgren spheres (§4.2.3) powered by UV-emitting massive stars, whose ionisation front increases with time. The non-linear regime of the evolution is best followed with hydrodynamic simulations that include radiative transfer (§D.1.1). The general findings of these simulations are that the expansion of these bubbles causes a progressively larger volume of the Universe to become ionised until all bubbles merge with each other at $z \approx 6$. Thus during the EoR, the IGM is expected to be quite 'patchy' and inhomogeneous, with almost completely neutral regions surviving in between the ionised bubbles. The volume ionisation fraction is usually called $Q_{\rm HII}$ and evolves from a value of $Q_{\rm HII} = 0$ in the dark ages to unity at the end of the EoR. The rate of IGM ionising photons $n_{\rm ion}$ (in photons s$^{-1}$ Mpc$^{-3}$) can be written as

$$\frac{dn_{\rm ion}}{dt} = f_{\rm esc}\, \zeta_{\rm Q}\, \rho_{\rm SFR}, \tag{9.51}$$

where $\zeta_{\rm Q}$ is the number of hydrogen-ionising photons produced per second per unit SFR (in units of s$^{-1}$ $\mathcal{M}_\odot^{-1}$ yr) and $\rho_{\rm SFR}$ is the comoving SFR density in units of $\mathcal{M}_\odot$ yr$^{-1}$ Mpc$^{-3}$ (§11.3.4). To fully reionise the IGM, the ionising sources have to provide at least one photon per hydrogen atom and ideally more to overcome recombination.

## 9.10 Observing the Primeval Universe

Sections §9.1 to §9.7 illustrated how baryonic matter evolved from the hot primordial plasma to the very first luminous objects which illuminated the dark ages. Each step of this evolution was characterised by specific radiative processes. The aim of this section is to describe how telescopes can observe these phases as a function of cosmic time, from the cosmological recombination to the reionisation.

### 9.10.1 Cosmological Recombination

The radiation emitted during the cosmological recombination can in principle be observed by detecting the emission lines and continuum produced by hydrogen (H II $\rightarrow$ H I) at

$500 \lesssim z \lesssim 2000$ and helium (He II $\rightarrow$ He I at $1600 \lesssim z \lesssim 3500$ and He III $\rightarrow$ He II at $5000 \lesssim z \lesssim 8000$). These radiative transitions should be observable as redshifted features superimposed on the black-body spectrum of the CMB radiation. The predicted recombination spectrum for hydrogen is derived by taking into account the free–bound and bound–bound transitions among thousands of atomic levels, as well as continuum emission processes such as the Balmer continuum and the 2s–1s two-photon continuum. Similar calculations can be done also for helium (§9.1.2). The predicted spectrum is expected to show up as ripples on the CMB spectrum (Fig. 9.9, left). These ripples, called **spectral distortions**, are broad because the cosmological recombination is not instantaneous, so the lines are emitted during an interval of cosmic time, i.e. within a redshift range. The expected signal of the recombination on the CMB spectrum is at the $\sim \mu$K level with typical amplitude of $\approx 1030$ nK. From the width of the features ($\Delta\nu/\nu \approx 0.1$) it would be possible to derive empirically the duration of the recombination. Highly sensitive observations of the CMB are required to detect this weak signal.

### 9.10.2 The 21 cm Signal from Atomic Hydrogen

The observation of the redshifted H I emission line ($\lambda_{\rm rest} \simeq 21$ cm, $\nu_{\rm rest} \simeq 1420$ MHz; §4.2.1) offers a unique opportunity to study the Universe during the transition from recombination to reionisation (see Furlanetto et al., 2009, and Pritchard and Loeb, 2012, for details). The expected evolution of the 21 cm brightness temperature ($T_{\rm b}$; §C.2) depends on four main variables: (1) the kinetic temperature of the gas ($T_{\rm K}$), (2) the density of neutral hydrogen, (3) the ionisation fraction of hydrogen, and (4) the Ly$\alpha$ photon background radiation intensity. Another relevant ingredient is the **spin** (or **excitation**) **temperature** $T_{\rm spin}$, that is, the excitation temperature of the 21 cm line defined as the ratio of the number densities $n_1$ and $n_0$ of H I atoms in the two hyperfine levels ($F = 1$ and $F = 0$):

$$\frac{n_1}{n_0} = \frac{g_1}{g_0} \exp\left(-\frac{T_*}{T_{\rm spin}}\right), \tag{9.52}$$

where $g_1/g_0 = 3$ is the ratio of the statistical degeneracy factors of the two levels and $T_* = \Delta E_{10}/k_{\rm B} = 0.068$ K. Three processes are important to determine $T_{\rm spin}$. The first is the absorption and spontaneous emission of the CMB photons (or another background source). The second is the collisions of H I atoms with other H I atoms, but also with electrons and protons if the gas is partly ionised (collisions are important in the early Universe when the gas density was higher). The third is the resonant scattering of Ly$\alpha$ photons emitted by the first luminous sources (**Wouthuysen–Field effect**, after Wouthuysen, 1952, and Field, 1958). The expected evolution of the 21 cm signal depends on the relative importance of these processes as a function of cosmic time. The detailed treatment of this topic is beyond the scope of this textbook, and we refer the reader to specialised reviews such as Pritchard and Loeb (2012). The main results of these studies can be summarised as follows. First of all, the 21 cm signal is expected to become observable only when $T_{\rm spin}$ deviates from the photon background temperature $T_{\rm rad}$. A practical formula to predict the observability of the 21 cm signal is

$$\delta T_{\rm b} \approx 25 x_{\rm HI}(1+\delta)\left(\frac{1+z}{10}\right)^{1/2}\left[1-\frac{T_{\rm rad}(z)}{T_{\rm spin}}\right]\left[\frac{H(z)/(1+z)}{{\rm d}v_{\rm los}/{\rm d}x_{\rm los}}\right]\ {\rm mK}, \qquad (9.53)$$

where $\delta T_{\rm b}$ is the brightness temperature of an H I cloud relative to the CMB temperature, $x_{\rm HI}$ is the neutral fraction, $\delta = (\rho - \overline{\rho})/\overline{\rho}$, where $\overline{\rho}$ is the average IGM density, is the fractional IGM overdensity in units of the mean density, $H(z)$ is the Hubble parameter and ${\rm d}v_{\rm los}/{\rm d}x_{\rm los}$ is the line-of-sight velocity gradient.

After the recombination ($200 \lesssim z \lesssim 1000$), the ionisation fraction is high enough to make Compton scattering effective to maintain thermal coupling of the gas to the CMB ($T_{\rm rad} = T_{\rm K}$), and the frequent collisions of the gas particles imply $T_{\rm rad} = T_{\rm spin}$. As a consequence, no deviations of $T_{\rm spin}$ from $T_{\rm rad}$ occur, and no 21 cm signal is expected. At $z \lesssim 200$, the following phases are expected depending on the relative evolution of $T_{\rm spin}$, $T_{\rm rad}$ and $T_{\rm K}$ (Fig. 9.9, right).

I. At $40 < z < 200$, the IGM has high density. As a consequence, there is a uniform coupling of the spin temperature with the kinetic gas temperature. We recall that, after the decoupling from the CMB (§9.2), the gas cools adiabatically as $T_{\rm K}(z) \propto (1+z)^2$, more rapidly than the CMB ($T_{\rm rad}(z) \propto (1+z)$). Thus, in this redshift range $T_{\rm K} = T_{\rm spin} < T_{\rm rad}$, and the signal is expected to be in absorption (negative).

II. This is the transition phase between $z \approx 40$ and the redshift $z_*$ at which the first luminous sources switch on. Due to the expansion of the Universe, the gas becomes less dense and the collisions become inefficient. As a result, the spin temperature gradually decouples from the kinetic temperature and approaches the CMB temperature ($T_{\rm K} < T_{\rm spin} \leq T_{\rm rad}$). Note that this decoupling from $T_{\rm K}$ depends on the local gas density (underdense regions decouple first). During this phase, $\delta T_{\rm b}(z)$ rises towards zero until there is little or no signal at $z_* \approx 30$, when the IGM is entirely decoupled.

III. The Ly$\alpha$ background produced by the first luminous sources causes the re-coupling of $T_{\rm spin}$ and $T_{\rm K}$ through the Wouthuysen–Field effect. Thus, $T_{\rm spin} \approx T_{\rm K} < T_{\rm rad}$, and this corresponds to an absorption signal.

IV. As in present-day SFGs, primeval luminous objects (which are forming stars) are expected to emit X-rays. This high-energy radiation field has the effect to heat the IGM. As a consequence, the spin temperature becomes increasingly coupled to the gas temperature ($T_{\rm spin} \sim T_{\rm K}$) and rises. The signal gradually increases, becoming positive when the gas temperature surpasses $T_{\rm rad}$.

V. The last phase occurs during the reionisation epoch (§9.9). The IGM becomes gradually ionised and the cosmic 21 cm positive signal decreases, approaching zero. The details of this phase depend also on the nature and spatial distribution of the UV ionising sources. The consequence of reionisation is to gradually remove neutral hydrogen. However, H I can survive in regions where the density is high enough not to be affected by the radiation field. The DLAs observed in the spectra of background QSOs (§9.8.1) could be H I gaseous systems with $N_{\rm H} > 10^{20}$ cm$^{-2}$ that survived to the reionisation. Thus, after the reionisation, the only residual signal from 21 cm is expected to be associated with the neutral hydrogen present in individual galaxies.

Fig. 9.9 shows the expected evolution of $\delta T_{\rm b}$ as a function of cosmic time from the dark ages to the reionisation. The Low-Frequency Array (LOFAR) and the new generation of

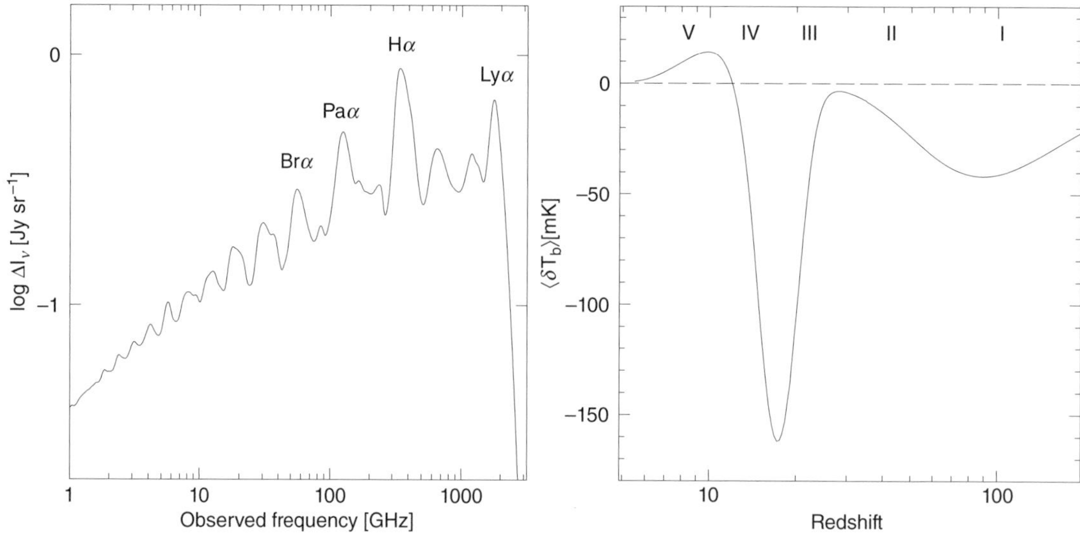


Fig. 9.9 *Left panel*. The simulated spectrum originating from the cosmological recombination of hydrogen and helium. $\Delta I_\nu$ is the departure of the CMB intensity from that of a perfect black body (also called spectral distortions). The labels indicate the Ly$\alpha$, H$\alpha$, Pa$\alpha$ (Paschen-$\alpha$) and Br$\alpha$ (Brackett-$\alpha$) emission lines expected from the hydrogen recombination. Data courtesy of J. Chluba. *Right panel*. The expected evolution of the average $\delta T_{\mathrm{b}}$ (brightness temperature deviation from the CMB) as a function of redshift. The Roman numerals refer to the main evolutionary phases described in the text. From Mesinger et al. (2016).

interferometers are designed to detect $\delta T_{\mathrm{b}}$ fluctuations using the CMB as the background source. Larger facilities such as the Square Kilometre Array (SKA) are designed to provide detailed maps of H I in the transition from the dark ages to the reionisation. The combination of data taken in redshift slices can allow us also to reconstruct the 3D map of the hydrogen distribution (**21 cm tomography**). A potentially important possibility is to exploit high-redshift radio-loud QSOs as background sources to observe the 21 cm forest in absorption in a way similar to what is done in the visible bands. However, this approach is feasible only if radio-loud QSOs are identified at very high redshifts (say $z > 8$).

### 9.10.3 Intensity Mapping

A promising observational approach to observe the distant Universe is **intensity mapping**. A full description of this method can be found in the review by Kovetz et al. (2017). This technique aims at mapping the intensity fluctuations of a given emission line on wide angular scales on the sky in order to reconstruct the large-scale spatial distribution of galaxies (and hence matter) emitting that line. The key difference with respect to traditional surveys is that this method exploits the integrated (low angular resolution) emission coming from many unresolved galaxies rather than resolving individual galaxies

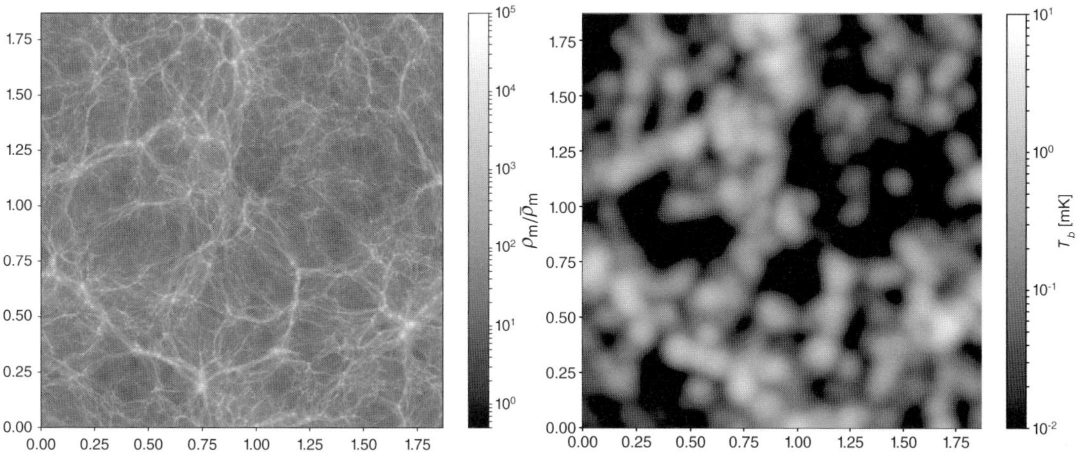


Fig. 9.10 *Left panel.* A simulation of the large scale distribution of matter (dark+baryonic) at $z = 1$ projected over a sky area of $\sim 4$ deg$^2$. The units in $x$ and $y$ axes are degrees. The greyscale is in units of the mass density contrast displayed in the y-axis on the right. *Right panel.* A simulation showing how the same structure would appear if observed with the intensity mapping of the 21-cm line emitted by neutral hydrogen redshifted at 710 MHz and observed with a bandwidth of 1 MHz and an angular resolution of 3 arcsec. It is evident that the main features of the large scale distribution are reconstructed exploiting the fluctuations of 21-cm brightness temperature as tracers of the underlying matter field. The maps in both panels have been generated from the IllustrisTNG hydrodynamic cosmological simulation (Pillepich et al., 2018). From Villaescusa-Navarro et al. (2018). © AAS, reproduced with permission.

identified one by one. Fig. 9.10 shows a simulation of intensity mapping. Thus, intensity mapping does not require large telescopes and can be efficiently performed with modest-aperture facilities (e.g. a single-dish radio telescope for the 21 cm emission line from H I). Intensity mapping surveys can target a variety of emission lines depending on the redshift range of interest. Notable examples for the distant Universe are Ly$\alpha$ redshifted in the optical/NIR at $z > 6$, [C II]$\lambda 158$ $\mu$m and CO lines at $z > 5$–6 in the submm/mm, and the H I 21 cm emission at $z > 6$ in the radio. Once a given emission line is targeted, its redshift is known from the observed wavelength (or frequency) at which the intensity mapping survey is performed. For instance, to map the distribution of H I on large scales at $z = 10$, one should scan a wide region of the sky in the radio at 1420 MHz/$(1 + z) = 129.091$ MHz, where 1420 MHz is the rest-frame frequency of the 21 cm line. The observation of the same sky region at other frequencies (i.e. redshift slices) allows us to obtain a datacube that can be exploited to reconstruct a 3D map of neutral hydrogen as a function of cosmic time. In intensity mapping, the problem of contamination from lines different from the targeted one can be mitigated by cross-correlating the emission in two different lines in order to isolate the emission line of interest. Intensity mapping can have several applications in a variety of science cases such as BAOs (§2.4), the epoch of reionisation (§9.9.1) and the evolution of the cosmic SFR density (§11.3.4).

### 9.10.4 Primordial Molecules in the Pregalactic Era

Primordial molecules could be observed through a variety of processes due to their interaction with the CMB. A relevant case is the resonant scattering occurring when a CMB photon is absorbed by a molecule and then re-emitted at the same frequency, but in a direction different from the initial one. In the observer frame, this causes anisotropies in the CMB. The amplitude of this process on the CMB temperature fluctuations ($\Delta T_{\rm rad}/T_{\rm rad}$) depends on the density (or optical depth) of the molecular gas cloud and its velocity. However, the predicted amplitudes are very small ($\mu$K level).

Another possibility is to observe directly the emission lines due to molecule formation occurring through radiative association reactions. In this case, the emitted radiation could be detected in excess over the black-body spectrum of the CMB. For instance, in the case of $H_2$ formation through associative detachment at $z < 100$, the vibrational emission line at 2.12 $\mu$m (rest frame) is expected to be redshifted in the FIR at $\lambda < 200\ \mu$m. Other opportunities are provided by the new generation of infrared space telescopes for the detection of $H_2$ rovibrational lines (e.g. 9.66 $\mu$m) at $z > 10$. Finally, an additional possibility to detect primordial molecules is to focus on the protostellar phase of Pop III stars. Some $H_2$ lines such as $J = 5 \rightarrow 3$ emitted by collapsing primordial clouds are redshifted in the FIR–mm region. Other examples are the $H_2$ $J = 2 \rightarrow 0$ ($v = 0$, $\nu_{\rm rest} = 5.33179 \times 10^{11}$ Hz) and HD $J = 4 \rightarrow 3$ ($v = 0$, $\nu_{\rm rest} = 5.33388 \times 10^{11}$ Hz) lines that are in principle detectable at $z \approx 10$–$40$ in the submm/mm spectral range.

### 9.10.5 First Stars and First Galaxies

Once Pop III stars and first galaxies start to shine, it becomes possible to observe their strong UV continuum and emission lines redshifted in the NIR. These objects are metal-free and not affected by dust reddening, and their properties open two main possibilities to unambiguously detect them. The first is to search for objects with hard (rest-frame) UV spectra. A Pop III star of $\approx 300\ \mathcal{M}_\odot$ should have a continuum spectrum very similar to a black body with $T \approx 10^5$ K. Compared to the hottest and most massive stars in the present-day Universe, the ionising photon emission would be larger by a factor of $\approx 10$ for H and He I, and $\approx 100$ for He II. Even in the case of lower masses ($< 100\ \mathcal{M}_\odot$), the ionising photon rate of a pure Pop III object would still be higher than in any Pop II stars also because of the absence of metal blanketing. Such powerful UV sources are expected to photoionise the surrounding medium and therefore to produce strong recombination lines with different properties compared to present-day H II regions. Fig. 9.11 shows the comparison of predicted spectra of zero-age SSPs in the case of $Z = 0$ (Salpeter-slope IMF with 50–500 $\mathcal{M}_\odot$) and $Z = 0.02$ (Kroupa IMF with $0.1$–$100\ \mathcal{M}_\odot$). It is evident that the He II $\lambda$1640 and He II $\lambda$4686 are expected to be much stronger due to the harder UV spectrum of the ionising photons. However, large uncertainties are present in these predictions due to the unknown shape of the Pop III IMF (§9.5.4). An additional possibility is to observe

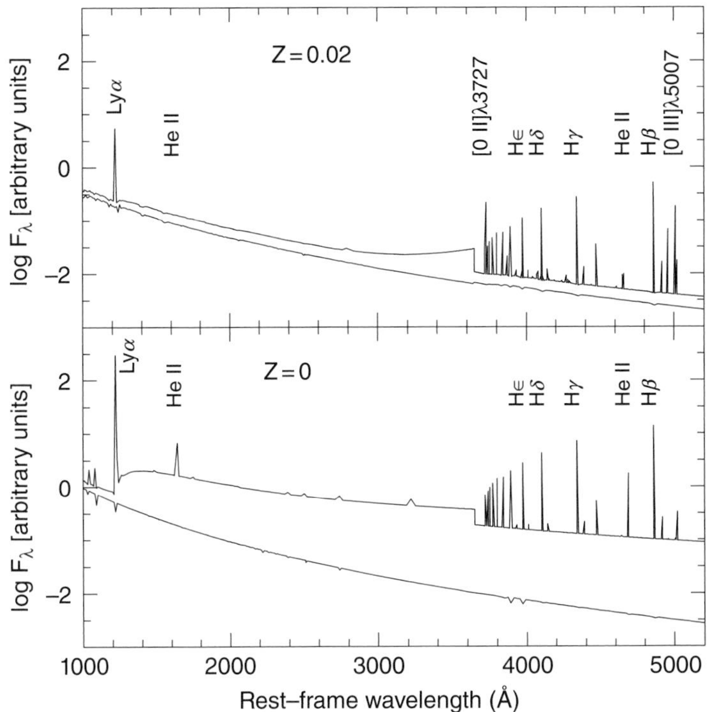


Fig. 9.11 *Top panel.* Lower spectrum: the photospheric continuum of a zero-age SSP with nearly solar metallicity ($Z = 0.02$), Kroupa IMF and a stellar mass range of $0.1-100\ \mathcal{M}_{\odot}$. The upper spectrum (rescaled by +0.1 dex for clarity) includes the nebular emission from the surrounding gas assuming no escape of Lyman continuum photons. *Bottom panel.* Same as in the top panel, but relative to a zero-age SSP with no metals ($Z = 0$) and with a top-heavy IMF (Salpeter slope, $50-500\ \mathcal{M}_{\odot}$). Compared to the top panel spectrum, the stronger He II emission lines are due to the harder UV photons emitted by more massive (hotter) stars. The lack of metal lines is also evident. Data from the Yggdrasil model of Zackrisson et al. (2011).

the final steps of the Pop III stars evolution by detecting the $H_2$ rovibrational emission line at 2.12 $\mu$m (rest frame) redshifted in the mid-infrared. This line is expected to be emitted by the gas ejected by SN explosions, where $H_2$ formation could occur behind the shocks and where the temperatures are high enough to produce vibrational radiative transitions.

# 10 Theory of Galaxy Formation

Understanding the formation and the evolution of galaxies is a very ambitious endeavour that has occupied astrophysicists for decades. Analytic calculations (such as those presented in Chapter 8), though very useful to describe the basic principles of the theory of galaxy formation, cannot capture the details of the involved processes and must therefore be complemented by the results of sophisticated numerical models.

The starting point of galaxy formation is the collapse of matter overdensities in the early Universe (§7.2). These overdensities are dominated by dark matter, but a fraction $f_b \simeq 0.16$ (§2.5) of their matter is baryonic, namely primordial gas (§9.2). While the dark matter halos assemble hierarchically (§7.4.4), the gas hosted in these halos cools (§8.1.1) and forms stars (§8.3), leading eventually to the galaxies that we observe today. On the one hand, the extraordinary variety of galaxies in the present-day Universe indicates that this conversion of infalling gas into galaxies follows different paths. On the other hand, the existence of specific classes of galaxies and tight scaling relations suggests that galaxy formation must be regulated by a few dominant physical mechanisms.

Throughout this book, we have followed the general classification of present-day galaxies that divides them into SFGs and ETGs, with SFGs being essentially discs and ETGs spheroids (§3.1). However, when looked at in detail, the properties of a single object are more complex and a large fraction of galaxies comprise more than one baryonic component. For instance, several disc galaxies host bulges with properties similar to ellipticals, as well as bars or pseudobulges (§4.1.2); some ETGs (in particular S0s) have disc components and significant rotation (§5.1.2). Moreover both SFGs and ETGs can be embedded in stellar halos (§4.1.3) that may have their own formation mechanism.

In the light of the above, in this chapter we first investigate the formation mechanisms of the different *galaxy components* (§10.1–§10.4), which we summarise in Tab. 10.1. Then, after introducing some characteristic scales of galaxy formation (§10.5) and the concept of quenching of star formation (§10.6), we describe possible *assembly histories* of different classes of present-day galaxies (§10.7–§10.9). Finally, we discuss the origin of the *demographics* of galaxies (§10.10) and give an overview of the numerical models of galaxy formation (§10.11).

## 10.1 Formation of Galaxy Components: Discs

A distinctive property of present-day galaxy discs with respect to spheroids (which we describe in §10.3) is that they are rotation-supported in all their baryonic matter

**Table 10.1** Main mechanisms relevant to the formation of galaxy components

| Galaxy component | Formation mechanisms | Sections |
|---|---|---|
| Disc | Dissipative collapse | §10.1.2 |
| | Angular momentum from tidal torques | §10.1 |
| | Angular momentum redistribution | §10.1.3–§10.1.5 |
| | Interactions (spiral arms) | §10.1.6 |
| | Instability (spiral arms) | §10.1.6 |
| Bar | Global instability | §10.2.1 |
| | Interactions | §10.2.1 |
| Pseudobulge | Buckling instability | §10.2.2 |
| Spheroid/bulge | Violent relaxation | §10.3.1–§10.3.3 |
| | Merging | §10.3.2–§10.3.3 |
| | Violent disc instability | §10.3.3 |
| Stellar halo | Dissipative collapse | §10.4 |
| | Accretion of stars from satellites | §10.4 |

components. As discussed in §7.5.4, dark matter halos acquire spin via interactions with the tidal field of the surrounding density distribution. Consider a protohalo that is forming and is surrounded by structures that are also coming together at the same time. In general, the external density distribution, as seen from the protohalo, will not be homogeneous. The gravitational field of these neighbouring structures exerts tidal torques on the protohalo making it acquire rotation. Obviously, the global angular momentum must be conserved, so halos will effectively acquire spins in different directions preserving the (null) angular momentum of the Universe.[1] The details of the so-called **tidal-torque theory** (Peebles, 1969) are rather complex and we do not repeat them here. The interested reader is referred to other texts like Mo et al. (2010).

The starting point to understand disc formation is the assumption that the baryonic matter that takes part in the formation of a galaxy experiences the *same* gravitational field and hence torques as the dark matter. As a consequence, the initial *specific* (per unit mass) angular momenta of dark matter and baryons (only gas at this point) are roughly the same. We can write this as

$$j_{\rm b} = \frac{J_{\rm b}}{\mathcal{M}_{\rm b}} \approx j_{\rm DM} = \frac{J_{\rm DM}}{\mathcal{M}_{\rm vir}}, \tag{10.1}$$

where $J_{\rm b}$ and $J_{\rm DM}$ are the angular momentum magnitudes of baryons and dark matter, $j_{\rm b}$ and $j_{\rm DM}$ are the respective specific angular momenta, $\mathcal{M}_{\rm b}$ is the baryonic mass and we have assumed that the dark matter mass ($\mathcal{M}_{\rm DM}$) dominates the total mass so $\mathcal{M}_{\rm DM} \approx \mathcal{M}_{\rm vir}$. Theoretical calculations and hydrodynamic simulations show that the condition in eq. (10.1) is already fulfilled at the turnaround time (before virialisation; §7.3.2). Later acquisition of angular momentum, due to hierarchical build-up and cosmological

[1] In a homogeneous and isotropic Universe the angular momentum on large scales (much larger than single halos) is expected to be null.

accretion of matter (§7.4.4), also takes place roughly preserving eq. (10.1). However, as the evolution progresses, neighbouring structures draw apart and the effect of torques is lessened. The reader may have noticed that we explicitly refer to *specific* angular momenta. This is because the masses of dark matter and baryons, in both primordial and present-day galaxies, are very different. As a consequence, the angular momentum of the dark matter will always be larger than that of the baryons only because its mass is larger; however this is not very meaningful for the calculations that we carry out in this section.

Let us now define two ratios that have great importance in investigating the processes of galaxy formation and evolution. The first is the **stellar-to-halo mass ratio**

$$f_{\mathrm{m},\star} \equiv \frac{\mathcal{M}_\star}{\mathcal{M}_{\mathrm{vir}}} < f_{\mathrm{b}}, \tag{10.2}$$

where $\mathcal{M}_\star$ is the stellar mass and $f_{\mathrm{b}}$ is the cosmic baryon fraction (§2.5); the inequality is valid for every galaxy and typically $f_{\mathrm{m},\star} \ll f_{\mathrm{b}}$ (§6.7 and §10.10.1). The second quantity is the **stellar-to-halo (specific) angular momentum ratio**

$$f_{\mathrm{j},\star} \equiv \frac{j_\star}{j_{\mathrm{DM}}} \approx \frac{1}{f_{\mathrm{m},\star}} \frac{J_\star}{J_{\mathrm{DM}}}, \tag{10.3}$$

where $J_\star$ is the angular momentum magnitude of the stars and $j_\star = J_\star/\mathcal{M}_\star$ is the specific angular momentum. This ratio is sometimes referred to as the **retained fraction of specific angular momentum**, although note that it can also exceed unity. In the following section, we focus on $f_{\mathrm{j},\star}$ and we see how its value is linked to the sizes of present-day galaxy discs.

### 10.1.1 The Size of Galaxy Discs

Stars in a galaxy disc have orbits close to circular, thus the region they span (the size of the disc) is set by their angular momentum. To first order, a larger size means a higher specific angular momentum, at least for discs with comparable rotation velocities (and thus with comparable gravitational potentials, eq. 7.26). In this section, we get the first insights about disc formation by estimating the angular momentum in present-day discs using the observed disc size as our main constraint. We start from the definition of the spin parameter (eq. 7.64) that, using eq. (10.1), we can rewrite as

$$\lambda = \frac{j_{\mathrm{DM}}}{\sqrt{2} r_{\mathrm{vir}} v_{\mathrm{vir}}}, \tag{10.4}$$

which is essentially the fraction of specific angular momentum of the dark matter halo with respect to the maximum specific angular momentum achievable by a fully rotation-supported system ($v_{\mathrm{vir}}$ is the circular speed at the virial radius $r_{\mathrm{vir}}$, eq. 7.26). *N*-body simulations tell us that the typical spin parameter of a virialised dark matter halo is $\lambda \approx 0.035$ with relatively little spread and no significant dependence on halo mass nor redshift (§7.5.4). Thus the specific angular momentum of a typical halo is largely below its centrifugal support capability, that is, rotation is dynamically unimportant for the dark matter. This is radically different for discs.

The specific angular momentum of an exponential disc is given by eq. (4.40). If we combine this with eq. (10.4) using the ratio in eq. (10.3), and solve for the disc scale radius, we obtain

$$R_{\rm d} = \frac{1}{\sqrt{2}\alpha}\lambda f_{\rm j,\star}\left(\frac{v_{\rm vir}}{v_{\rm flat}}\right) r_{\rm vir} \simeq 0.018\left(\frac{\lambda}{0.035}\right)\left(\frac{f_{\rm v}}{1.4}\right)^{-1} f_{\rm j,\star} r_{\rm vir}, \tag{10.5}$$

where, in the last passage, we have indicated a realistic ratio ($f_{\rm v} \equiv v_{\rm flat}/v_{\rm vir}$) between $v_{\rm flat}$ and the circular speed[2] of the dark matter halo at $r_{\rm vir}$ and used $\alpha = 1$. Eq. (10.5) shows that the size of the stellar disc of a galaxy (few$\times R_{\rm d}$) is much smaller than $r_{\rm vir}$ (for $f_{\rm j,\star} \approx 1$ as we see below). We can now rewrite eq. (10.5) in terms of the halo's circular speed. By combining eq. (7.25) and eq. (7.26) we obtain an expression for the virial radius,

$$r_{\rm vir} = \frac{1}{H(z)}\sqrt{\frac{2}{\Delta_{\rm c}(z)}}\, v_{\rm vir}, \tag{10.6}$$

where $\Delta_{\rm c}$ is the critical overdensity for virialisation (eq. 7.21) and substituting it into eq. (10.5) using eq. (2.37) we find

$$\begin{aligned} R_{\rm d} &= \frac{\lambda f_{\rm j,\star}}{\alpha H_0\sqrt{\Omega_{\rm m,0}(1+z)^3 + \Omega_\Lambda}\sqrt{\Delta_{\rm c}(z)}}\left(\frac{v_{\rm vir}}{v_{\rm flat}}\right) v_{\rm vir} \\ &\simeq 5.3\left(\frac{h}{0.7}\right)^{-1}\left(\frac{\lambda}{0.035}\right)\left(\frac{f_{\rm v}}{1.4}\right)^{-1} f_{\rm j,\star}\left(\frac{v_{\rm vir}}{150\,{\rm km\,s^{-1}}}\right)\,{\rm kpc}, \end{aligned} \tag{10.7}$$

where the second equality is given at $z = 0$ for the standard cosmological model (§2.3) with $\Delta_{\rm c} \simeq 101$ (eq. 7.21) and $\alpha = 1$. Present-day galaxies with $v_{\rm flat} \approx 150 f_{\rm v}\,{\rm km\,s^{-1}}$ have exponential scalelengths in the range 2–7 kpc. Thus eq. (10.7) shows that to achieve disc sizes comparable to those of real galaxies one needs to have $f_{\rm j,\star} \sim 1$. Let us now see the implications of this result.

Typically, the stellar-to-halo mass ratio ($f_{\rm m,\star}$) in a disc galaxy is much lower than the universal baryon fraction (§10.10.1). This means that stars form out of a rather small portion ($< 25\%$ for the Milky Way; §6.7) of the primordial gas associated with a dark matter halo. We have now found that the stellar-to-halo specific angular momentum ratio ($f_{\rm j,\star}$) is instead close to unity. Thus the specific angular momentum of the stellar disc, the final product of the disc formation, is similar to that of the dark halo and, as mentioned, to that of the primordial gas. This result was not necessarily expected because the small fraction of the initial gas that ends up in the stellar disc could have had a specific angular momentum very different from that of the whole gas. What eq. (10.7) suggests is that the gas that forms stars is *representative* of the pregalactic gas in terms of global angular momentum content and that it *conserves* angular momentum (does not transfer it to the dark matter) during collapse. The latter request turns out to be very important for disc formation and we discuss it further in the next section.

[2] We remind the reader that observations that give us $v_{\rm flat}$ (§4.3.3) typically probe radii $r \ll r_{\rm vir}$. One can therefore expect that $v_{\rm flat}$ would be larger than $v_{\rm vir}$ because: (1) the circular speed of an NFW profile declines at large radii (Fig. 4.15, bottom right) and (2) the contribution of the baryons may increase the inner $v_{\rm c}$ (§4.3). Alternatively, one could also take the peak (maximum) circular speed of NFW halos $v_{\rm max}$ instead of $v_{\rm vir}$, as $v_{\rm flat}/v_{\rm max} \approx 1$. Note however that both this ratio and $f_{\rm v}$ may depend on galaxy mass.

### 10.1.2 Dissipative Collapse and Global Angular Momentum Problem

One key process for the formation of a galaxy disc is **dissipative collapse**. With this term one refers to the fact that the gas, unlike the dark matter that collapses conserving its total energy, tends to radiate away a fraction of its energy (§8.1.1). Thus, the gas *loses* energy but, at the same time, *conserves* angular momentum, and the consequence of this is to settle into a rotating disc, whose plane is perpendicular to the original direction of the angular momentum. The more efficient the gas is at radiating away its energy, the *colder* it becomes and thus the closer it gets to a fully centrifugally supported disc, which rotates at a velocity $v_{\rm rot} = v_{\rm c}$, where $v_{\rm c}$ is the circular speed (eq. 4.23). In fact, it is the original specific angular momentum of the gas ($j_0$) that determines the radius at which centrifugal equilibrium is reached ($R = j_0/v_{\rm c}$) and thus the size of the final disc (eq. 10.7). The gas in this disc will form stars (§8.3) and these will inherit its angular momentum, eventually leading to present-day stellar discs.

The simple comparison between halo and disc angular momentum contents, carried out in §10.1.1, points to the necessity for the baryonic material to retain most of its angular momentum during collapse ($f_{\rm j,\star}$ close to 1; eq. 10.7). As mentioned, at the turnaround time, the specific angular momenta of dark matter and gas are nearly the same and they have the same spin parameter ($\lambda \approx 0.035$) much lower than unity. The rotation velocity of the gas initially must then be much lower than its circular speed, $v_{\rm rot,i} \sim \lambda v_{\rm c}$. However, we said that by collapsing and conserving its angular momentum, this gas leads to a rotation-supported disc with a final velocity $v_{\rm rot,f} \approx v_{\rm c}$. Thus the initial and final radii must have a ratio $r_{\rm f}/r_{\rm i} \sim v_{\rm rot,i}/v_{\rm rot,f} \sim \lambda$, meaning that the pregalactic medium must collapse to the centre of the halo with compression factors of 10 or more as found in their seminal paper by Fall and Efstathiou (1980). Present-day galaxies have indeed stellar discs with sizes much smaller than their virial radii (eq. 10.5).

Until about 2010, cosmological hydrodynamic simulations (§10.11.1) had problems reproducing the sizes of observed galaxy discs. In these simulations, the gas collapsing into the centre of the halo was transferring a large fraction of its angular momentum to the dark matter and the discs were far smaller than those of real spiral galaxies. This problem has been referred to as the **angular momentum catastrophe** or **global angular momentum problem**. Later on, it was realised that these shortcomings were partially caused by the formation of large clumps of gas (inside subhalos) that were transferring angular momentum to the dark matter because of dynamical friction (§8.9.3). These clumps were, however, mostly artifacts produced by the poor resolution of the simulations. The higher mass/spatial resolution of modern simulations has largely solved the issue. However strong stellar feedback (§8.7) seems also to be needed to make the collapsing gas a more diffuse medium and further reduce the effect of dynamical friction.

In summary, the observed sizes of present-day galaxy discs give us important information about their formation and assembly history. Stellar discs have sizes that are much smaller than those of dark matter halos (eq. 10.5). This is fully expected if they have similar specific angular momentum ($j \sim r v_{\rm rot}$) because the discs are rotation-supported ($v_{\rm rot} \approx v_{\rm c}$) while the dark matter is not ($v_{\rm rot} \ll v_{\rm c}$). In fact, the sizes of the observed present-day discs imply that their specific angular momentum is nearly exactly the same as that of the dark matter ($f_{j,\star} \approx 1$; eq. 10.7). This result suggests that stellar discs formed from gas that (1) acquired angular momentum from tidal torques (as the dark matter) and (2) *conserved it* during galaxy and star formation processes. These are the two pillars of our general understanding of disc formation. Note, however, that the above considerations refer to the *global* values of the specific angular momentum of dark matter and baryons. In the following, we investigate the distribution of angular momentum *within* halos and galaxies. By doing this, we encounter a more subtle angular momentum problem in disc formation and discuss its possible solutions.

### 10.1.3 The Angular Momentum Distribution

We now consider a more detailed comparison between the angular momenta of dark matter halos and of present-day discs that will give us further insights on disc formation mechanisms. Let us first introduce the **angular momentum distribution** (AMD) defined as the amount of mass in a system per unit specific angular momentum:

$$\psi(j) \equiv \frac{\mathrm{d}\mathcal{M}}{\mathrm{d}j}. \tag{10.8}$$

The first two moments of the AMD are

$$\mathcal{M} = \int_0^\infty \psi(j)\mathrm{d}j \qquad \text{and} \qquad j_{\text{tot}} = \frac{1}{\mathcal{M}} \int_0^\infty j\psi(j)\mathrm{d}j, \tag{10.9}$$

where $\mathcal{M}$ is the total mass of the system and $j_{\text{tot}}$ is its total specific angular momentum.

The AMD of a system can be obtained by knowing its mass profile and the distribution of its specific angular momentum with radius (see, for instance, Fig. 4.15 for an exponential disc) and it is a powerful tool to investigate the evolution of a system. The main reason for this is that if all the particles of the system *individually* conserve their angular momentum then the AMD of the system does not change with time. We have mentioned that the general starting point for galaxy formation is that, at turnaround, both dark matter and baryons have nearly the same specific angular momentum: in fact, they also have the same AMD. This occurs because *each element* of dark matter receives the same tidal torque as the associated element of baryonic matter. A cosmologically motivated AMD for dark matter halos, found using $N$-body simulations, can be parameterised as

$$\psi_{\text{DM}}(j) = \begin{cases} \mathcal{M}_{\text{vir}} \dfrac{\mu j_0}{(j_0 + j)^2} & 0 \le j \le j_{\max}, \\ 0 & j > j_{\max}, \end{cases} \tag{10.10}$$

where $j_{\max}$ is the $j$ at which the AMD is truncated to avoid divergence, and $\mu$ and $j_0$ are parameters linked to $j_{\max}$ by $j_0 = (\mu - 1) j_{\max}$. The parameter $\mu$, called the shape parameter, is by definition larger than unity with a typical value $\mu = 1.25$ and relatively large spread (see Bullock et al., 2001, for details). Eq. (10.10) also describes the AMD of the primordial (pregalactic) gas associated with a generic dark matter halo once one substitutes $\mathcal{M}_{\text{vir}}$ with $f_{\text{b}}\mathcal{M}_{\text{vir}}$.

We now derive the AMD of the stellar disc of a present-day galaxy. This can be easily calculated if we assume that the disc is exponential and the galaxy has a flat rotation curve.

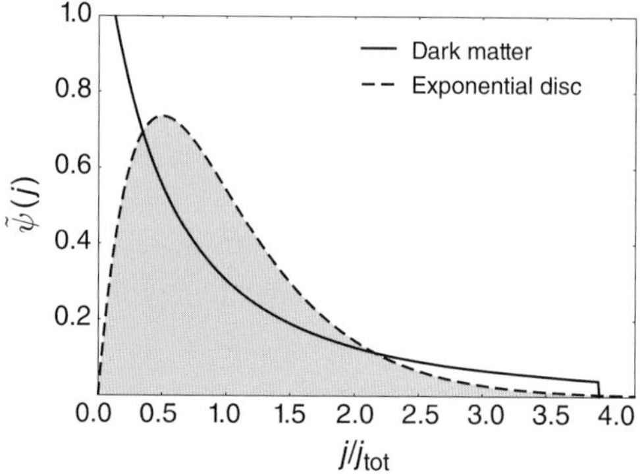


Fig. 10.1 Angular momentum distributions of dark matter halos at virialisation (eq. 10.10; solid curve) and of a present-day exponential disc with a flat rotation curve (eq. 10.11; dashed curve). Both curves are normalised ($\tilde{\psi} = \psi/\mathcal{M}$) such that the integrals in eq. (10.9) return unity. The dark matter curve is calculated for $\mu = 1.25$ and $j_{max} = 3.9$.

If we use the equation for the mass of an exponential disc within the radius $R = j/v_{flat}$ (analogous to eq. 4.4 with mass and surface density in place of $L$ and $I$, respectively) and take the derivative in $j$, we obtain

$$\psi_{d}(j) = \frac{4\mathcal{M}_{d}}{j_{flat}} \left(\frac{j}{j_{flat}}\right) \exp\left(-\frac{2j}{j_{flat}}\right), \tag{10.11}$$

where $j_{flat} = 2R_{d}v_{flat}$ (eq. 4.40) and $\mathcal{M}_{d}$ is the total mass of the disc. The reader may verify that the zeroth and first moments of eq. (10.11) return total mass and specific angular momentum, respectively.

Fig. 10.1 compares normalised versions of the two AMDs that we just described and shows a striking difference. The dark matter distribution has a considerable amount of material at very low angular momentum while the exponential disc's AMD totally lacks this low angular momentum material, peaking instead at intermediate values. Moreover, the dark matter appears to have a slight overabundance of high angular momentum material with respect to a stellar disc. Given that the AMD of the dark matter should be considered the same as that of the pregalactic baryons, the fact that the distributions are so different tells us that stellar discs like those of present-day galaxies cannot form from a randomly selected fraction of these baryons. In this sense, Fig. 10.1 captures one of the most serious challenges that models of disc formation face. We refer to this as the **detailed** or **local angular momentum problem**, of which we provide possible solutions in §10.1.4.

### The Orientation of the Angular Momentum Vector

Before proceeding, we briefly discuss a couple of assumptions on the orientation of the angular momentum vector that we have implicitly employed. First, we have assumed that the orientations of the angular momenta of dark matter and baryons are exactly the same. This is not necessarily true as the build-up of the two components is not fully simultaneous and the dissipative nature of the gas may play a role. However, cosmological hydrodynamic simulations show that the misalignment should not exceed $\approx 20$ degrees in most cases. In the rare cases of strongly misaligned halos, this can affect the evolution of the gas and stellar discs. For instance, it may drain part of the angular momentum from the disc to the dark matter due to tidal interactions. In this respect, we recall that the halos are generally triaxial (§7.5.2), which makes gravitational tides effective.

A second simplification is that the orientation of the angular momentum vector does not change with time. This is also not fully justified as the variation of the external torques causes this orientation to change. However, in the inner regions of halos, where dynamical processes are faster (the dynamical time is shorter; eq. 8.24), there is a tendency over time to reach a coherent spin direction whereas the outer regions may be misaligned. The misalignment of outer halos has been proposed as an explanation for the formation of the warps observed in the outer discs of many SFGs (§4.3.3 and §4.6.2). These are often seen in the neutral atomic gas component, which extends out to larger radii with respect to the stellar disc. Following this interpretation, in the inner parts of the gaseous disc the potential is dominated by the stellar disc (and inner halo) and the gas follows the spin of the stars, whereas beyond a certain radius it tends to align with the spin direction of the outer dark matter halo and/or the later infalling material. Although these effects are probably important in some cases, for the purpose of understanding the generalities of galaxy disc formation, we will consider tilts in the orientation of the angular momentum second-order effects.

### 10.1.4 Evolution of the Angular Momentum Distribution

In §10.1.3 we have outlined an important problem for galaxy disc formation, namely that the AMD of the gas from which the disc forms is very different from that of the final stellar disc[3] (Fig. 10.1). In the following we describe the mechanisms that can produce this dramatic transformation between the initial and the final AMDs.

1. *Selective accretion*. In this case, the disc forms out of a small fraction of the available gas having an AMD markedly different from the average. This is possible given that only a small fraction of the initial baryons typically end up in the stellar disc ($f_{m,\star} \ll f_b$). This scenario requires a mechanism capable of performing the 'selection' of the accreting material. The most obvious mechanism is radiative cooling, which is indeed selective in the sense that not all the gas cools at the same rate, but the denser gas cools faster, because the cooling time is inversely proportional to the density (eq. 8.4). However, radiative cooling alone does not appear to solve the problem of the AMD. This can be understood by considering that the densest gas will always be in the centre of the halos and thus at low specific angular momentum (small radii). So if we form the disc out of this densest, low-$j$ gas, $f_{j,\star}$ would be much lower than unity, differently from what is needed to reproduce the observed disc sizes (eq. 10.7). Galaxy discs must gather material *also* from the outer parts of the halos (and somehow prevent the accretion from the very inner parts). We return to this fundamental topic in §10.7.2.

[3] Note that if we also include the gaseous disc, the picture does not change appreciably.

2. *Strong selective feedback.* If the gas that accumulates in the central parts of the disc is removed very efficiently by stellar (or AGN) feedback, this can have a great impact on the AMD. As this gas tends to have specific angular momentum lower than the average (because it is at small radii), its disappearance can produce an AMD similar to that of present-day discs (Fig. 10.1). Ideally, one would like this low angular momentum material to be ejected by feedback and *never* come back to the disc, thus producing a permanent modification of the AMD. However, the feedback power needed to move material from the disc out of the virial radius of the halo is very high (§8.7.4) and this phenomenon is likely to work efficiently only in low-mass galaxies (§10.10.1).
3. *Galactic fountain.* This takes place when the ejection from stellar feedback is not powerful enough to expel the gas from the halo, but expels it (temporarily) from the disc, making it circulate in the halo region until it falls back (§8.7.3). As mentioned in §6.7 and §9.8, the space around galaxies contains large amounts of ambient gas, in particular the galactic corona (§10.7.2), with which the gas ejected from the centre of a galaxy interacts and mixes. If the galactic fountain can reach large radii and if the ambient gas has *some* rotation,[4] the mixing and interaction of the low-$j$ gas, ejected by the fountain, with the high-$j$ (because of the large $r$) ambient material will then produce gas with intermediate specific angular momentum that, once reaccreted, can make the overall AMD of the baryons similar to that of the observed discs (Fig. 10.1). Hydrodynamic simulations of galaxies with prominent galactic fountains confirm these expectations.

In the end, the solution to the local angular momentum problem of §10.1.3 may be a combination of the above. The situation is however complicated by the fact that there are other mechanisms that can produce a modification of the AMD of the baryons. Leaving aside the effect of mergers (§8.9), most of these mechanisms represent *slow modifications* that usually occur due to internal processes in the disc and are referred to as **secular processes**. One such mechanism is dynamical friction (§8.9.3) between the baryonic component and the dark matter. This is effective in the presence of non-axisymmetric structures, in particular an inner stellar bar (§10.2). Spiral arms (§4.1.1) and galactic shear (§10.1.5) can also redistribute and transfer angular momentum. Unfortunately, these processes typically transfer angular momentum away from the inner disc, thus going in the direction of aggravating the local angular momentum problem.

We conclude with a brief summary of what we have learned so far.

- We have seen that galaxy discs must form out of gas with relatively high specific angular momentum.
- This angular momentum does ultimately come from the gravitational torques of the external field (tidal-torque theory) and it is acquired *nearly equally* by the baryons and by the dark matter.
- To produce discs of the observed sizes, the baryons must collapse conserving (i.e. not transferring it to the dark matter) most of their angular momentum to avoid the global angular momentum problem.

4 There are theoretical and observational indications that a galactic corona should have significant rotation.

- The AMD of the dark matter, which in turn we can assume is similar to that of the baryons before disc formation, is very different from that of the present-day discs. This is the local angular momentum problem, of which we discussed possible solutions.

In the following sections we look at three key properties of galaxy discs. First, we briefly discuss the formation of the observed exponential surface brightness profiles. Second, we analyse the development of spiral arms that are very important for the dynamical evolution of the discs and, finally, we discuss their vertical structure and the mechanisms that can lead to the formation of thick discs.

### 10.1.5 Formation of Exponential Discs

The radial distribution of the stellar surface brightness of the discs of SFGs tends to follow an exponential profile (§4.1.1). To some extent, also the gas radial profiles, if one includes atomic and molecular gas, are roughly exponential although with larger scalelengths (Fig. 4.6, left). Most remarkably, galaxies of very different kinds (Hubble types) and masses (from dwarfs to massive spirals) follow the same exponential profiles (Fig. 4.1), albeit with different scalelengths and central surface brightnesses. The reasons behind this very general pattern are not clearly understood. We briefly outline here some of the theories that have been proposed.

One of the first attempts uses the concept of shear viscosity that we now briefly explain. In general, **shear** takes place when parts of a fluid, *in contact* with each other, move at different velocities. In galactic discs, this occurs naturally as a consequence of differential rotation (§4.3.3) as two adjacent annuli do not share the same angular speed. For instance, if the disc has a flat rotation curve, the angular speed $\Omega = v_{\rm rot}/R$ is a hyperbolically decreasing function of radius (Fig. 8.8, top panel). This means that gas at a certain radius will complete a rotation in a longer time with respect to the gas in the adjacent inner radius (Fig. 10.2b). On the contrary, if $\Omega = \text{const}$, i.e. solid-body rotation, there would be no shear. Now suppose that, in the presence of shear, two adjacent annuli can exchange momentum with each other, as if there were a *friction* between them: we call this phenomenon **shear viscosity**. In hydrodynamics, **viscosity** operates at microscopic levels (gas particles; Fig. 10.2a) in so-called **viscous fluids**, described by the Navier–Stokes equation (§D.2.5). However, in galactic dynamics, one typically invokes *other forms* of 'viscosity' that can take place for a variety of reasons. One such reason may be that the ISM is turbulent (§4.2.2), something that makes the gas in different annuli exchange momentum. Another reason is the presence of magnetic fields, in which case we speak of **magnetically induced shear**.

Leaving aside the exact nature and the effectiveness of viscosity in galaxy discs, for which we refer the reader to Shu (1992), let us see what effect it could have in the context of exponential disc formation. Lin and Pringle (1987) proposed that viscous redistribution of the gas in the initial gaseous disc is responsible for the final stellar exponential disc. Assuming that the viscous radial redistribution has a characteristic timescale $t_\alpha$ and the star formation in discs proceeds on a timescale $t_{\rm SF}$, they showed that if $t_{\rm SF} \sim t_\alpha$, the final radial distribution of gas and stars should follow exponential laws. This model predicts

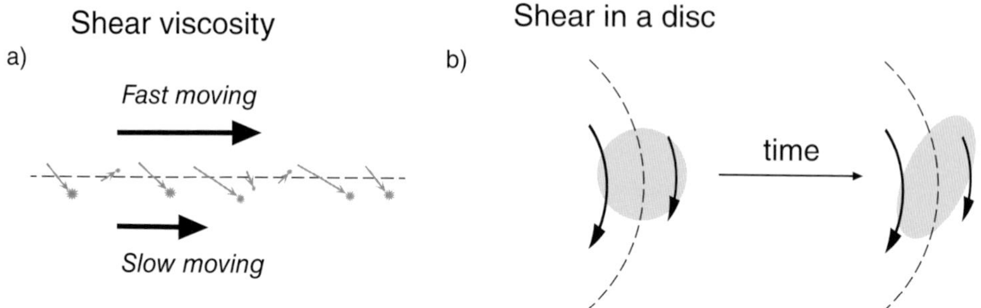


Fig. 10.2 *Panel a.* Sketch of hydrodynamic shear viscosity. Two portions of a fluid in contact and moving at different velocities exchange momentum due to the collisions of particles that cross the contact surface. *Panel b.* In a galactic disc with differential rotation (varying angular speed) two adjacent rings experience shear. A material structure like a gas cloud (grey circle) can be distorted by shear.

that exponential discs are produced starting from arbitrary distributions of the gaseous disc, as long as the latter is more extended than the final stellar disc.[5] Unfortunately, this theory has the problem that the original gaseous disc should have an AMD skewed to even higher $j$ than the final disc, which aggravates the local angular momentum problem described in §10.1.3.

Starting from a more realistic AMD, other models have tried to explore the effect of feedback and star formation in the production of exponential profiles. These effects must be investigated using zoom-in cosmological hydrodynamic simulations (§10.11.1). It appears that, to finally produce nearly exponential profiles, one needs a combination of several factors. The most crucial one is efficient feedback to remove low angular momentum material in the inner parts (§10.1.4). However, inefficient star formation in the outer parts of the disc is also important. In general, cosmological hydrodynamic simulations that employ these effects produce discs that are close to exponentials, at least in a portion of their profiles.

Instead of starting from the gaseous disc, other theories propose that the stellar discs *become* exponential due to internal evolution. A possibility is provided by the formation of spiral arms (§10.1.6) and their role in radially redistributing the angular momentum. More specifically, one can invoke **radial stellar migration**. This is the radial diffusion of stars through a galactic disc produced by a variety of phenomena like transient spiral arms and the interaction between the bar and the spiral structure (§10.1.6). In general, radial migration produces a change in the angular momenta of single stars, but is thought to have little effect on the overall AMD of the disc (§10.1.3). However, it was found that if the radial migration is so strong that it can shuffle the angular momenta of individual stars in a timescale much shorter than the age of the galaxy, the angular momenta can reach a 'maximum-entropy' state and produce radial density profiles that resemble exponentials. Finally, exponential profiles could be generated by the scattering of stars as they interact

[5] Note that, as the gas in a disc is typically in centrifugal equilibrium, radial motions towards the centre are possible if the disc transfers part of its angular momentum to some other component. In the case of an isolated disc instead, a significant mass inflow can only be achieved if accompanied by a flow of a (small) fraction of gas to (very) large distances to assure angular momentum conservation.

with high-mass clumps, in particular GMCs. This last model may work not only for large spiral galaxies, but also for dwarf irregular galaxies where other theories could fail, in particular because some dwarf galaxies have very little shear ($\Omega \approx$ const) and no spiral arms. There is a rather vast literature on the above models for which we refer the reader to Herpich et al. (2017) and references therein.

### 10.1.6 The Development of Spiral Arms

One of the most striking features of present-day galaxy discs is the presence of extended spiral arms that often start from the inner regions developing to the outer edges of the discs (grand design, §4.1.1). In any attempt to understand the formation mechanisms of the spiral structure, the following established facts must be kept in mind.

1. Spiral arms are observed in every baryonic component both in the cold (atomic and molecular) gas and in the stars, including the old stellar population (Fig. 4.2).
2. Lenticular (S0) galaxies without discs of cold gas have much less prominent or absent spiral arms, even though their stars rotate and reside in a disc.
3. Due to differential rotation, if a spiral arm is made of gas and stars that are locked to the arm, it should wind up (be stretched along the direction of rotation; see Fig. 10.2b) rapidly and eventually disappear (winding problem; §4.1.1).

Observations 1 and 3 suggest that spiral arms are not 'material' structures but instead **density waves**, which means that they are locations in the disc where stars and gas accumulate. A pictorial way to describe these waves is with the analogy with car traffic jams. A traffic jam can remain localised at a certain position (say at a road junction) despite the fact that cars flow in and out of it: as new cars enter on one side, others exit from the other. In the same way, stars and gas can continuously go into and out of spiral arms, slowing down and reaching the highest densities when they are inside. If this is the case, the winding problem is avoided provided that the spiral pattern (the wave) rotates with angular speed (pattern speed $\Omega_{sp}$; §8.3.3) roughly independent of radius. Observation 2 tells us that the cold gaseous disc has a key role in the formation and/or the maintenance of the spiral arms. In §10.8 we see that lenticular galaxies are likely former spiral galaxies that have lost their cold ISM due to some process such as ram-pressure stripping. This indicates that the loss of the cold gas leads to a rapid disappearance of the spiral arms also from the stellar component.

Despite decades of investigation, we do not yet have a definitive theory to account for spiral arms in disc galaxies. A detailed description of the various proposals would be very complex and beyond the scope of this book.[6] In this section, we give a brief account of some general ideas that have been put forward and we end by discussing the role of spiral arms in the evolution of the disc. The hypothesis that spiral arms are not material entities was first formulated in the context of **density wave theory** (Lin and Shu, 1964). This is a linear perturbation analysis of the stability of a gaseous disc (later also found for a stellar disc) in the tight-winding approximation (§8.3.3). In its original formulation, density waves

[6] We refer the interested reader to more specialised textbooks such as Binney and Tremaine (2008) and Bertin (2014).

were **stationary waves** and persisted for times comparable with the Hubble time $t_{\rm H}$. These density waves propagate through the galaxy disc similarly to waves in the sea but they are absorbed and damped beyond two particular radii called **Lindblad resonances**. The outer and inner Lindblad resonances are defined as the radii ($R_{\rm LR}$) at which

$$m\left[\Omega(R_{\rm LR})-\Omega_{\rm sp}\right]=\pm\kappa, \tag{10.12}$$

where $m$ is the mode of the perturbation (the number of spiral arms) and $\kappa$ is the epicycle frequency (eq. 8.53). These radii occur in the inner (+ sign) and outer (− sign) disc and, because of their damping effect, a density wave spiral is expected to develop only between them. The ideal conditions for the development of a quasi-stationary wave is to have a disc where the $Q$ parameter (eq. 8.58) is slightly larger than 1 nearly everywhere except in the inner regions where it rises steeply.

In contrast with the idea of quasi-stationary waves, other authors proposed that spiral arms are, instead, short-lived. A mechanism that can efficiently generate them is the so-called **swing amplifier** (Toomre, 1981). This requires the initial formation of a leading spiral pattern (Fig. 8.7) that then becomes unstable and quickly 'swings' into a trailing spiral. Leading waves are acceptable solutions of the density wave theory, but they are not observed in real galaxies (§8.3.3). The swing amplifier not only turns a leading pattern into a trailing wave, but also highly amplifies the intensity (density contrast between arms and intra-arm regions) of the wave. This mechanism is also expected to be effective when $Q$ is greater than but rather close to 1, so the disc is stable but self-gravity is important, which makes it prone to respond to gravitational perturbations (Fig. 8.8). Thanks to the swing amplification, *any* generic perturbation can potentially lead to a prominent trailing spiral. This is because a generic perturbation can be seen as a combination of trailing plus leading modes (of the type shown in eq. 8.51) and the leading modes will always be unstable. Structures formed in this fashion are called **transient spirals** and have timescales much shorter than $t_{\rm H}$. If spirals in real galaxies are transient then there should be mechanisms that produce them easily (recurrent transients) and/or that regenerate the instability as the wave propagates (recurrent instability). The reader can find more details on these possibilities in Sellwood (2013).

The formation of spiral arms has been extensively investigated using numerical ($N$-body) simulations. These showed that spiral patterns in differentially rotating discs can be produced by both external and internal disturbances. External perturbations are caused by encounters (fly-bys, §8.9) with satellite galaxies or dark matter substructures, and the development of a grand design spiral is normal in such circumstances. There are some galaxies, e.g. the nearby spiral M 51, in which there is little doubt that the spiral pattern is produced by the interaction with a companion galaxy. However, it is unlikely that this phenomenon is the main driving mechanism of spiral arm formation. Internal perturbations can instead have a variety of causes. *Some* grand design spirals (e.g. in NGC 1300) are likely excited by stellar bars (§10.2). However, in $N$-body simulations, even a plain Poisson noise can easily generate spiral structures. Given that galaxy discs are typically not smooth, as they host star clusters and massive clouds like GMCs (§4.6.2), such a simple mechanism could produce spiral arms in most cases.

Fig. 10.3 shows a simulation in which a spiral pattern develops due to the presence of GMCs that are continuously perturbing the disc. These spirals persist for a long time (several Gyr), but they are different from those predicted by the classical density wave theory. Their persistence is due not to a constant pattern speed but to the fact that the arms are continuously breaking up and reconnecting: $\Omega_{\rm sp}$ is, in this case, decreasing with radius similarly to the angular speed of gas and stars (Fig. 8.8). These types of spirals, which are the easiest to form in simulations, are referred to as **dynamical spirals**. Despite the rather different pattern speeds and arm morphology, it is presently difficult to establish whether spiral patterns in real galaxies behave in this fashion or are more akin to density waves.

Independently of their exact formation mechanism and lifetimes, spiral arms are expected to play a key role in the dynamical evolution of a galaxy disc. For instance, a trailing spiral, especially if grand design, can efficiently transfer angular momentum from the inner to the outer disc through gravitational torques. Most importantly, spiral arms damp part of their energy and transfer it to random motions (increasing the radial velocity dispersion) of the stars, effectively *heating* the stellar disc. This may help explain the lack

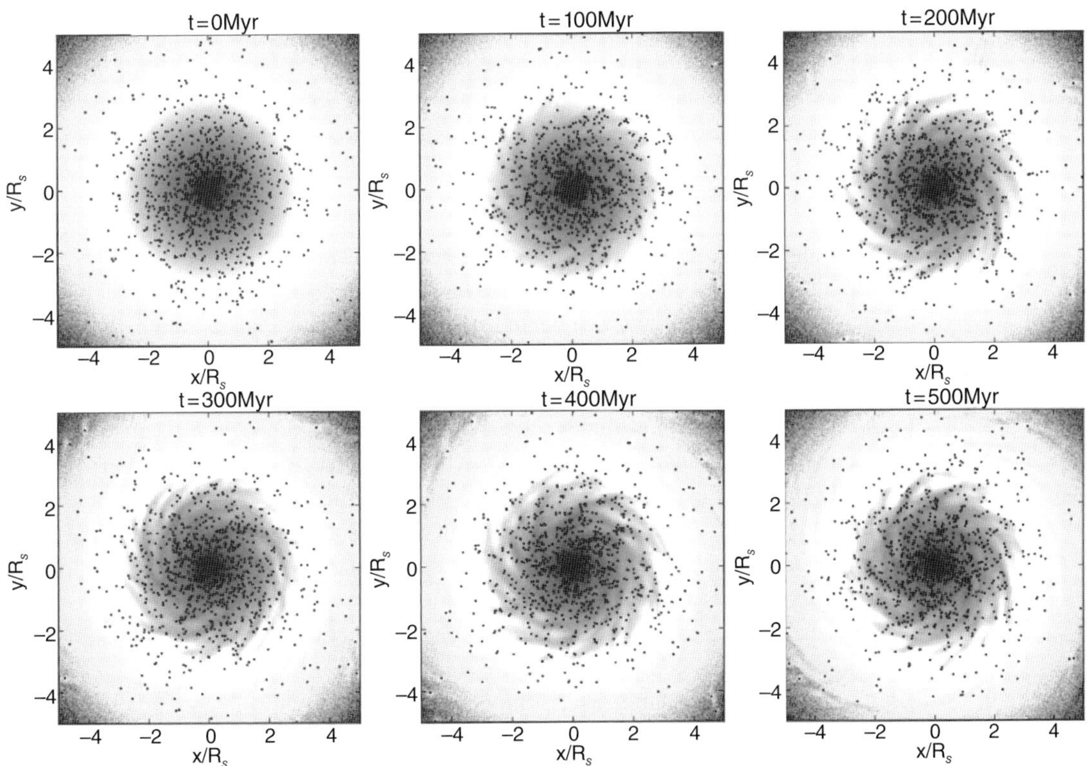


Fig. 10.3 Snapshots of an $N$-body simulation of a stellar disc embedded in a dark matter halo of $\mathcal{M}_{\rm vir} \simeq 10^{12}\,\mathcal{M}_\odot$. One thousand giant molecular clouds (black dots) are added to the simulation and act as perturbers for the development of spiral arms. Each panel shows a region 30 kpc $\times$ 30 kpc; $R_s$ is the initial scalelength of the disc. Time is indicated on the top of each panel. From D'Onghia et al. (2013). © AAS, reproduced with permission.

of spiral arms in lenticular galaxies. When a galaxy is stripped of its cold ISM and stops forming stars, the heating of the stellar disc is no longer counteracted by the formation of new stars with low velocity dispersion (§4.6.1). This can cause the $Q$ parameter to increase and the spiral pattern can be washed away. Finally, spiral arms can pave the way to the formation of stellar bars, as we discuss in §10.2.

### 10.1.7 The Vertical Structure: Thin and Thick Discs

The last feature of galaxy discs that we discuss is the vertical distribution of the stars. Both in the Milky Way and in most external galaxies, this distribution cannot be reproduced by a single exponential, but two components are required, namely a thin and a thick disc (§4.6.1). We briefly consider here the possible formation mechanisms of this double-exponential structure, but, before that, let us summarise the observational findings that give us clues in this direction.

1. Thick discs are common in spiral galaxies.
2. Thick disc stars have higher velocity dispersion and lower rotation velocity with respect to thin disc stars.
3. The stars in the thick disc of the Milky Way are, in general, older than those of the thin disc; a possible separation in ages has been suggested at $\approx 8$ Gyr.
4. In the Milky Way, thick disc stars have, on average, lower metallicities and, at least within the solar circle, higher $[\alpha/\mathrm{Fe}]$ with respect to the thin disc stars (Fig. 4.21).
5. Thick discs are radially extended structures; their scalelengths, as measured in external edge-on galaxies, are comparable to those of the thin discs.

We can divide the scenarios of thick disc formation into two main categories: those in which the thick disc formed thick in the first place and those in which it thickened in the later evolution. The main idea behind the first scenario is that stars in early galaxies formed from more turbulent and unsettled gas, with a higher velocity dispersion (§11.2.5) and thus in a thicker layer. This high turbulence could have been produced by a high rate of gas accretion, high rate of minor mergers, strong stellar feedback or a combination of these. In the second scenario, the stars now in the thick disc originally formed in a thinner component that later experienced an increase of its (vertical) velocity dispersion and got 'puffed up'. Note that both these scenarios can be consistent with the older age of thick disc stars as long as the turbulence drops with time (in the first case) or the heating occurs at early times (in the second case).

Dynamical heating can occur through external or internal mechanisms. In the former case, the original disc could have been heated by mergers with satellite galaxies[7] and the infall of dark matter subhalos at a time when these events were common. Later (say after $z \approx 1$), these phenomena may have become less frequent (§11.3.2) and the subsequent evolution proceeded as a slow growth of the thin disc (§10.7.2). The main internal heating mechanisms are three: GMCs (§4.6.2), spiral arms and the bar (bar formation or buckling;

[7] Note that, in this case, a fraction of the stellar mass of the thick disc could come from the satellites, i.e. have formed *ex situ* (§10.3.3).

§10.2). Among these, there is evidence that dynamical friction heating by a GMC is effective to puff up a stellar disc but it is not clear whether it is sufficient to create a thick disc. Note that the efficiency of this mechanism was higher at early times ($z \gtrsim 1$) as the self-gravity of the stellar disc was much lower. Spiral arms should not work effectively as they tend to increase the radial velocity dispersion instead of the vertical dispersion, while on the effect of bars there is no general consensus.

## 10.2 Formation of Galaxy Components: Bars and Pseudobulges

Roughly 60% of present-day disc galaxies have non-axisymmetric stellar components called stellar bars (§4.1.2). These bars are mostly seen in the old stellar population (Fig. 4.2) and they are hosted by disc galaxies of all Hubble types. Bars appear less common at higher redshifts (say $z \approx 1$) with respect to the local galaxies and this gives us clues on their formation timescale. In this section, we describe the disc gravitational instabilities that are thought to lead to the formation of the stellar bars. We then see how a bar can undergo a further transformation and generate a pseudobulge (§10.2.2).

### 10.2.1 Bar Instability

Stellar bars in the central regions of spiral galaxies are thought to form mainly via some gravitational disc instabilities. These are referred to as **global instabilities** as opposed to the types discussed in §8.3.3, which are instead *local* and lead to the formation of smaller structures. Global instabilities involve a large portion of the disc and, because of this, are rather difficult to treat analytically as they depend on both the detailed distribution of matter and the disc dynamics. They can be due to internal processes, in particular the evolution of a spiral perturbation[8] (§10.1.6), or to external processes like galaxy encounters or fly-bys. Here, we mainly focus on internal formation mechanisms.

A large number of numerical simulations have been performed to study the development of bar instabilities in galaxy discs. These are mostly **simulations of individual galaxies** performed at very high resolution (difficult to achieve in cosmological simulations; §10.11.1). Increasing levels of complexity have been included to explore the various mechanisms that can affect the development of the bar instability. Most numerical experiments have been carried out with only stars and dark matter ($N$-body), but in some cases the effects of a gas disc and GMCs have also been investigated (Fig. 10.4). We can summarise the general results as follows.

1. Bar instability appears to be easily excited in galaxy discs especially if the disc is thin, the dark matter does not dominate the inner potential and the gas fraction is low.

[8] The phenomenon that leads from the propagation of spiral density waves to a bar instability is complex and we refer the interested reader to Binney and Tremaine (2008).

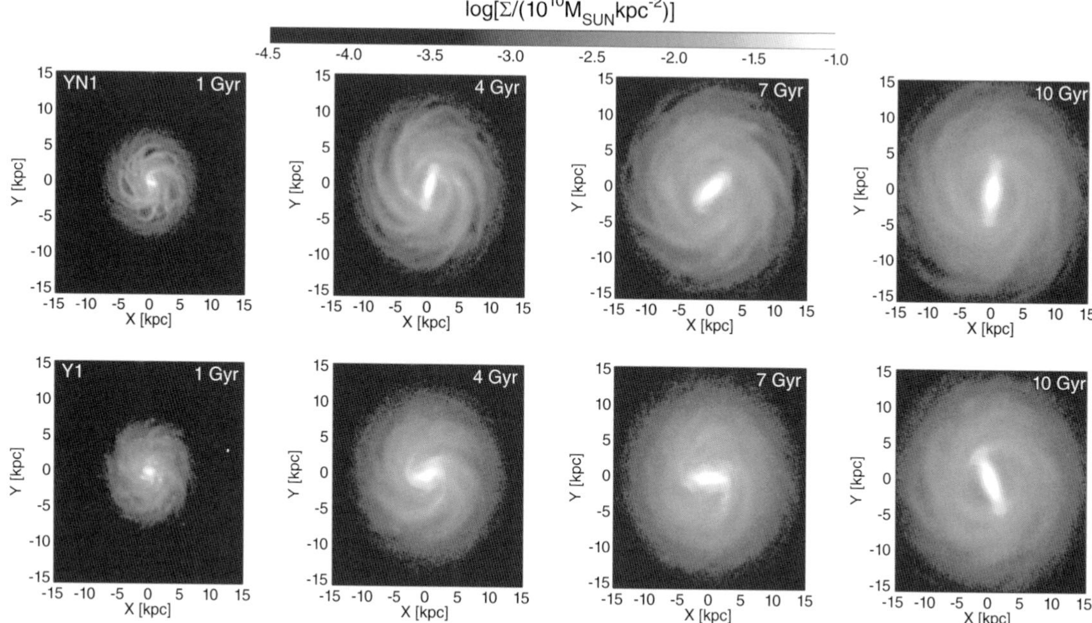


Fig. 10.4 Snapshots of two numerical simulations for the evolution of a stellar disc. The greyscale shows the stellar surface density and the time of the snapshot is given in the top right corners. The *bottom* and *top panels* are with and without the presence of GMCs, respectively; the differences are only subtle. Note the development and growth of an inner stellar bar, not present in the first snapshot. From Aumer et al. (2016).

2. High gas fraction in the disc, a massive dark matter halo and also a hotter (thicker) stellar disc tend to delay or suppress the bar formation.
3. Despite this attenuation, bars tend to develop in discs with typical timescales of a few Gyr. Thus, perhaps paradoxically, what appears more difficult to explain is *not* the presence of stellar bars but rather their *absence* in a relatively large fraction of present-day spiral galaxies.
4. The *figures* of bars rotate as rigid bodies with the same spin orientation as the disc and slowly transfer part of their angular momentum to the dark matter halo.
5. A bar can become unstable to the buckling instability (§10.2.2) that leads to the formation of a pseudobulge.
6. A bar could be destroyed by mergers or by other processes (e.g. by the accumulation of matter in the nuclear region).

The dynamics of individual stars within a bar is rather complex and characterised by the presence of two families of closed (in the rest frame rotating with the bar) orbits called $x_1$ and $x_2$. The $x_1$ orbits have larger sizes and are elongated as the bar long axis (they roughly give the bar its shape); the $x_2$ orbits are more internal and elongated orthogonally to the bar long axis. Apart from this internal kinematics, the bar rotation described in item 4 occurs with a pattern speed ($\Omega_{\rm bar}$) that can be measured in real galaxies. The radius at which the pattern speed equals the rotation of material in centrifugal equilibrium is called the **bar**

**corotation radius** ($R_{\rm bar,cr}$). Observations show that the corotation radius in real bars is always very similar to the semi-major axis ($a_{\rm bar}$)[9] of the bar ($0.9 \lesssim R_{\rm bar,\,cr}/a_{\rm bar} \lesssim 1.3$). This demonstrates that all real bars should be classified as 'fast' in terms of rotation (a 'slow' bar would have $R_{\rm bar,cr}/a_{\rm bar} \gg 1$).

Cold ISM (§4.2) is present in the region of the bar. Strong non-circular (radial) motions develop in the gas and this is thought to lead to an overall inflow of gas towards the nuclear regions of the galaxy. The mass inflow rates, estimated from simulations, are such that, in the course of the whole lifetime of a galaxy, a substantial fraction of the inner ISM of a galaxy could be transported to the centre. The non-axisymmetric motion also produces, at some locations, shocks that make the gas denser. This is manifested by the long (often straight) dust lanes observed parallel to the long axis of the bar (Fig. 4.3).

Numerical simulations show that, once a bar is formed in a galactic disc, it persists for a long time, several gigayears (Fig. 10.4). The slow (secular) evolution of a bar is characterised by an exchange of angular momentum (due to dynamical friction, §8.9.3) with the dark matter halo and also with the stellar disc. As a consequence, the inner halo acquires some spin and the disc can slightly increase its radial scalelength. However, the observed ratio $R_{\rm bar,cr}/a_{\rm bar} \sim 1$ (fast bars) for all galaxies puts strong constraints on the amount of momentum that real bars can give away. This creates some tension with the results of cosmological simulations (§10.11.1) that tend to produce slow bars. Simulations of individual galaxies perform better and allow us to study the long-term dynamical evolution of a bar and in particular an important instability that eventually leads to the formation of galactic bulges as we see in the next section.

### 10.2.2 Buckling Instability and Pseudobulge Formation

Most of the disc instabilities that we have encountered until now (e.g. the gravitational instability in §8.3.3 and the bar instability) develop in the planes of galactic discs. There are however other types of instabilities that can cause the material (stars or gas) to be displaced out of the plane. Some of these are associated with the so-called **bending modes** of oscillations. The study of these bending modes can be performed analytically in simple cases like a razor-thin isothermal slab. The calculation is however involved and we refer the interested reader to e.g. Fridman et al. (1984). The general result is that the system becomes prone to the development of an instability if the ratio between the vertical velocity dispersion and the total velocity dispersion ($\sigma_z/\sigma$) is small. Quantitative determinations of this ratio for an axisymmetric rotating (self-gravitating) disc lead to the following condition for instability: $\sigma_z/\sigma \lesssim 0.6$. The presence of a dark matter halo makes the disc more stable, but the main concept remains: if the disc material is relatively 'hot' (high total velocity dispersion) but 'thin' (low vertical velocity dispersion), an instability can develop that causes the thin disc to 'buckle' and develop into a much thicker structure. We call this phenomenon **buckling instability**.

[9] The semi-major axis is the distance from the galaxy centre to the edge of the bar along its long axis. The bar's edge can sometimes be difficult to measure precisely.

Consider now a galactic disc in which a bar instability has already developed. The motions of stars in the bar can be seen as a large 'in-plane' velocity dispersion ($\sigma_R$ and $\sigma_\phi$ and thus $\sigma$) while, given that the bar has originated from the stellar disc, one can expect it to have relatively low $\sigma_z$. Thus, this is the typical situation in which the buckling instability can develop. Numerical simulations confirm this expectation that it is a rather natural evolution for a bar to eventually buckle and form an out-of-plane thick structure. These structures resemble very much the bulges of many disc galaxies, belonging in particular to the class of bulges that we have called pseudobulges (§4.1.2); thus bar buckling is considered one of the main mechanisms of bulge formation (for the formation of the so-called classical bulges, see §10.3). Note that only the inner half of the bar buckles, so the pseudobulge becomes part of a larger thin bar. Fig. 10.5 shows 12 snapshots from an $N$-body simulation, where we see the edge-on view of a barred disc. In this simulation, the bar buckles twice in the course of the evolution; the first episode occurs at $t \sim 2.3$ Gyr, after which the overall structure thickens significantly and also extends radially. Real galaxies seen edge-on do show boxy or peanut-shaped bulges (Fig. 4.3, right), which are considered evidence of bar/bulges that have evolved in the aforementioned fashion.

For decades, the Milky Way has been thought to have a classical bulge (§4.1.2) potentially formed by mergers (§10.3). Near-infrared surveys of stars in the inner Galaxy have, however, shown that the Galactic bulge has a boxy (Fig. 4.22) or even an X shape. This points to the presence of a pseudobulge and to formation via disc instabilities. Another observation that points to this formation mechanism is the fact that the kinematics of the stars in the Galactic bulge region tends to be dominated by coherent rotation and not by random motions. However, it is still not clear if *all* the stars in the bulge are part of this pseudobulge structure. In particular, a small fraction of metal-poor stars ($[\mathrm{Fe/H}] < -1$) show no rotation and a high velocity dispersion, more akin to the expectation for a classical bulge (see §4.6.1). It is presently not clear what role, if any, violent disc instability (§10.3.3) could have played.

Before investigating the formation of classical bulges and spheroidal components in the next sections, we briefly summarise the main conclusions that we have reached about discs, bars and pseudobulges (see also Tab. 10.1).

- We have seen that disc galaxies form from the collapse of gas that has efficiently acquired angular momentum from tidal torques. This gas settles in an extended disc where massive molecular clouds can form from local instabilities and star formation can take place.
- These discs may initially be turbulent, prone to instabilities and/or dynamical heating by mergers, phenomena that can produce the formation of a thick stellar disc.
- Spiral arms quickly emerge in discs and contribute significantly to their subsequent dynamical evolution.
- In a large fraction of galaxies, global disc instabilities create non-axisymmetric stellar bars.
- The buckling instability of, in particular, the inner regions of bars produces puffed-up structures that we call pseudobulges.

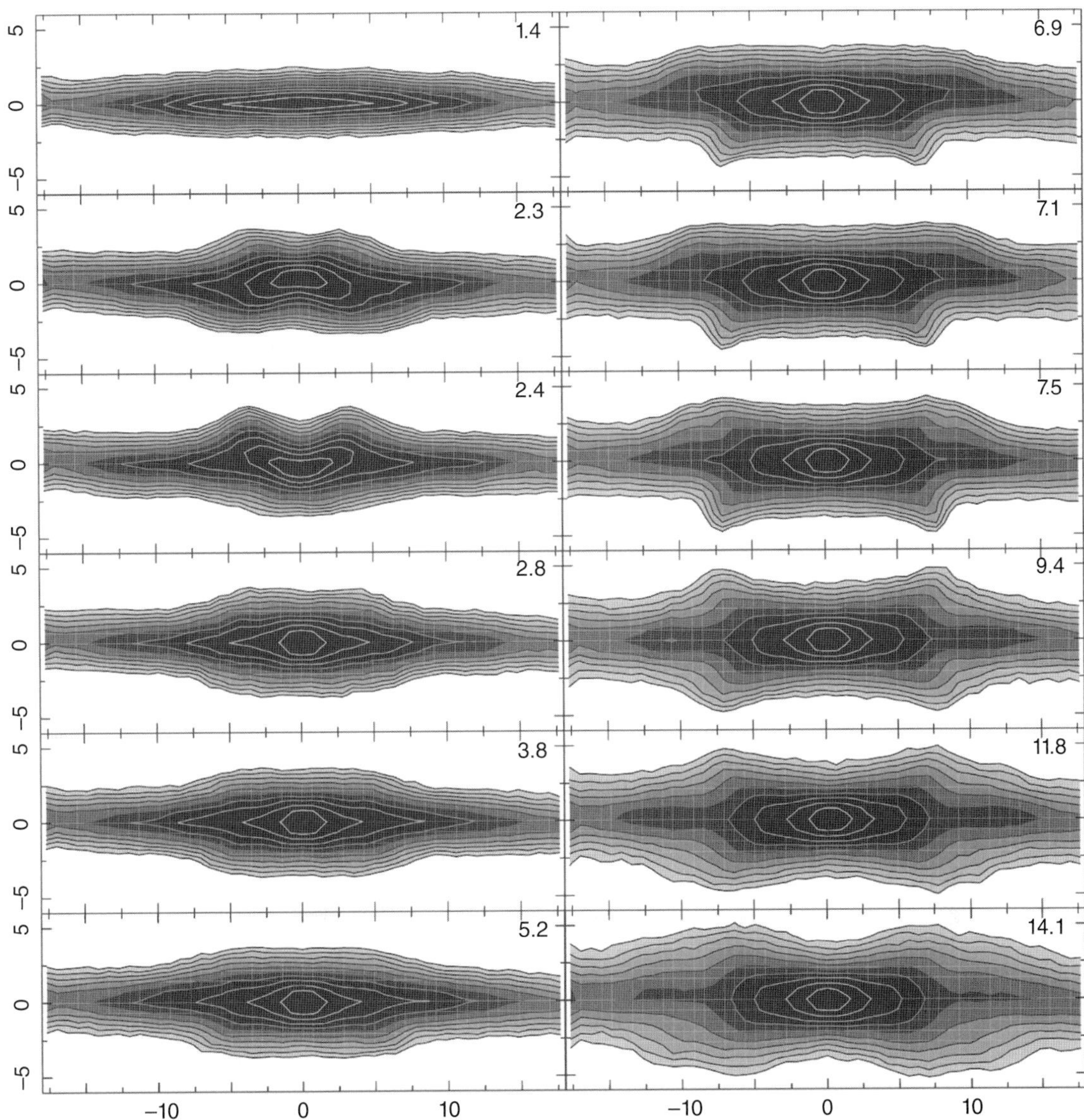


Fig. 10.5 $N$-body simulation of a barred disc seen edge-on. The different panels show the projected stellar density (greyscale and contours) at different times (top right corners of each panel, in Gyr). The labels on the $x$- and $y$-axes are in kpc. The bar evolution encounters two episodes of buckling instability at $t \simeq 2.3$ Gyr and at $t \approx 7$ Gyr. From Martinez-Valpuesta et al. (2006). © AAS, reproduced with permission.

## 10.3 Formation of Galaxy Components: Spheroids

The classical bulges of disc galaxies (§4.1.2) share their structural and kinematic properties with the main stellar components of ellipticals (§5.1). Both classical bulges and ellipticals are pressure-supported systems characterised by ellipsoidal stellar density distributions, and with projected surface density profiles well represented by the $R^{1/n}$ law (eq. 3.1) with index $n \gtrsim 2$ (often by the classical $R^{1/4}$ law; eq. 5.2). Therefore, in the context of the study of the formation mechanisms, we can use the term 'spheroids' or bulge-like systems to refer to both classical bulges of spirals and S0s, and to the main stellar components of ellipticals. Note that, in contrast, pseudobulges (§4.1.2) have different structural properties (for instance, different values of the Sérsic index), and are believed to be produced by other formation processes (§10.2).

The models proposed to describe the formation of a spheroid have been traditionally classified in two families: 'monolithic', based on the collapse of a single protogalaxy, i.e. a progenitor system made of dark matter and gas, and 'hierarchical', based on merging of pre-existing galaxies. While historically these two scenarios were considered as opposite paradigms, the actual formation mechanism of spheroids may have ingredients of both. Before describing the currently favoured scenario of spheroid formation (§10.3.3) it is useful to illustrate simple models based on either collapse (§10.3.1) or merging (§10.3.2).

### 10.3.1 Collapse

Though hierarchical merging of structures is a fundamental ingredient in the standard cosmological model, it is very instructive to describe simple models, based on ideas dating back to Eggen et al. (1962), in which galaxy spheroids form as a consequence of the gravitational collapse of a single protogalaxy.

The simplest of these models is the **dissipationless collapse** (van Albada, 1982), which is the purely gravitational evolution of an initial roughly spherical distribution of stars with very low kinetic energy. This is an idealised model in which, for simplicity, one assumes that the stars have already formed before virialisation, so gravity is the only force at work. Numerical $N$-body simulations have shown that the virialised end products of dissipationless collapses reproduce very well a few observed properties of bulges and ellipticals. In particular, these end products have ellipsoidal stellar density distributions with projected density profiles well represented by the Sérsic law with $n \approx 4$ (Fig. 10.6). In detail, the properties of the end product of a dissipationless collapse (for instance, axis ratios and Sérsic index) are found to depend on the properties of the initial conditions, such as angular momentum, virial ratio (i.e. ratio of kinetic to gravitational potential energy), clumpiness and radial slope of the density distribution. Surface brightness profiles resembling those of bulges and ellipticals can be produced by dissipationless collapses of star-only $N$-body systems (with no dark matter). However, similar results are obtained with more realistic dissipationless collapses in the presence of massive dark halos. The dissipationless collapse is driven by collisionless relaxation processes (§8.9.2), and in particular by violent relaxation, because the collisionless stars and dark matter particles orbit in a rapidly varying gravitational potential. Thus, the finding that simple dissipationless collapses reproduce the surface brightness profiles of spheroids suggests that violent relaxation might be important for spheroid formation.

Clearly, a strong limitation of the dissipationless collapse model is that it neglects the all-important dissipative effects: when these effects are accounted for, we speak of dissipative

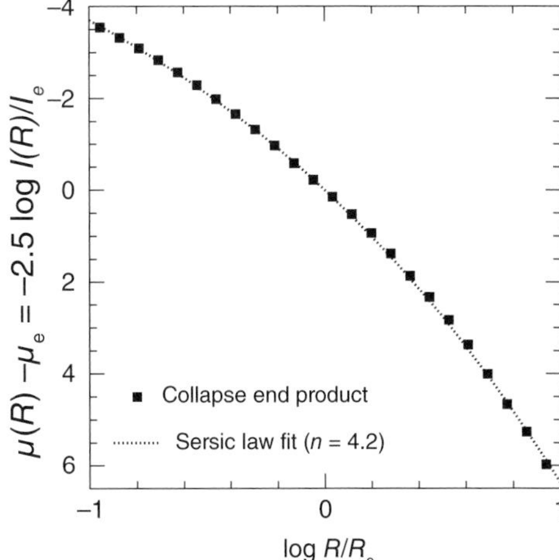


Fig. 10.6 Circularised surface brightness profile of the end product of an $N$-body simulation of cold dissipationless collapse (squares) and best-fitting Sérsic law (with Sérsic index $n = 4.2$; dotted curve). All the particles in the simulation are assumed to have the same mass-to-light ratio. The profile is normalised to the effective radius $R_e$: $\mu_e \equiv \mu(R_e)$ and $I_e \equiv I(R_e)$. Adapted from Nipoti et al. (2006).

collapse, which is a substantially more complex process, whose description must include explicitly hydrodynamics and star formation. Given that they do not have the appealing simplicity of dissipationless collapses, without being necessarily fully realistic, dissipative collapses have not been explored in the context of the theory of spheroid formation as much as in the context of disc formation (§10.1.2). However, the results of the available simulations suggest that systems with properties compatible with bulges and ellipticals can be formed also in collapses allowing for the presence of dissipation, but only with very low values of the initial angular momentum, as might be the case for almost radial filamentary cold gas accretion (§8.2.4).

### 10.3.2 Merging

One of the earliest models proposed for the formation of ellipticals is the **disc-galaxy merging** model (Toomre, 1977). In this model, two roughly equal-mass spiral galaxies merge and the merger remnant is a spheroid-dominated stellar system. In the simplest implementation of the disc-galaxy merging model, the process is dissipationless, in the sense that the merging galaxies are gas-poor, with negligible associated star formation. During mergers the ordered kinetic energy of the discs is transformed into random kinetic energy, so mergers can turn rotation-supported stellar systems (discs) into pressure-supported stellar systems (bulges and ellipticals; §8.9.6). Numerical $N$-body simulations of equal-mass mergers of disc galaxies have shown that the merger remnant is similar to bulges and ellipticals in terms of shape and density profile (§8.9.6). However, it must be stressed that present-day ellipticals cannot be produced by collisionless mergers of Milky Way-like disc-dominated galaxies, because the phase-space density (§C.8) cannot increase in collisionless processes (eq. C.37) and ellipticals have phase-space densities much higher than discs[10] (§4.1.1 and §5.1). Simulations have shown that remnants with high phase-space density can be produced in dissipationless mergers of disc galaxies, provided the merging systems have substantial bulge components.

As an alternative to the collisionless disc-galaxy merging model, we can consider the dissipative (wet) merging of disc galaxies (§8.9.6). Hydrodynamic and $N$-body simulations of disc-galaxy merging (Fig. 10.7) suggest that, also in this case, it is possible to produce spheroidal remnants with $R^{1/4}$-like surface density profiles. As a consequence of gas dissipation and star formation, the stellar phase-space density can increase in wet mergers: realistic ellipticals are obtained in mergers of progenitor disc galaxies with gas fractions of the order of 30%. For this mechanism to be consistent with the old stellar populations of most ellipticals and bulges (§4.1.2 and §5.1.3), it must operate at sufficiently high redshift, because new stars form during dissipative mergers.

Another process believed to have a role in the late evolution of massive spheroids is the **spheroid merging**, that is the dissipationless (dry) merging of pre-existing spheroids (§8.9.6). In this case the merging systems are by definition poor in cold gas, so dissipation and star formation can be safely neglected. The detailed characteristics of the remnant (for instance, shape, size, velocity dispersion and Sérsic index) depend on the properties of the merger (for instance, mass ratio and orbital parameters of the encounter), but, overall, the numerical results show that the remnant of the dry merging of spheroids maintains the morphology of a spheroid (§8.9.6).

### 10.3.3 Build-up of Spheroids in a Cosmological Framework

A key aspect of the structure formation in a CDM Universe is that many scales collapse simultaneously. When a big structure is turning around (§7.3.2), smaller structures are collapsing within it. It follows that neither the pure collapse of a single protogalaxy (§10.3.1) nor the merging of pre-existing virialised systems (§10.3.2) is a realistic description of spheroid formation. Models combining features of collapse and merging can be termed hybrid formation models. It is now widely accepted that mergers have an important role in the formation history of present-day bulges and ellipticals. It is also well established that mergers easily destroy discs, transforming them into spheroids, and maintain the main properties of bulge-like systems. Dry mergers between spheroids or bulge-dominated systems are likely more important at lower redshift, while wet mergers of disc-dominated galaxies are at work at higher redshift, when the bulk of the old stars of the spheroid formed (§10.8).

Theoretical models predict that, especially at high redshift, accretion of gas occurs mainly through narrow cold filamentary streams (§8.2.4). The consequence of this cold accretion is the formation of central gaseous discs. A possibility is that spheroids are

[10] Though the phase-space density is not directly observable, it can be estimated using observable quantities such as stellar mass surface density, velocity dispersion and disc scaleheight.

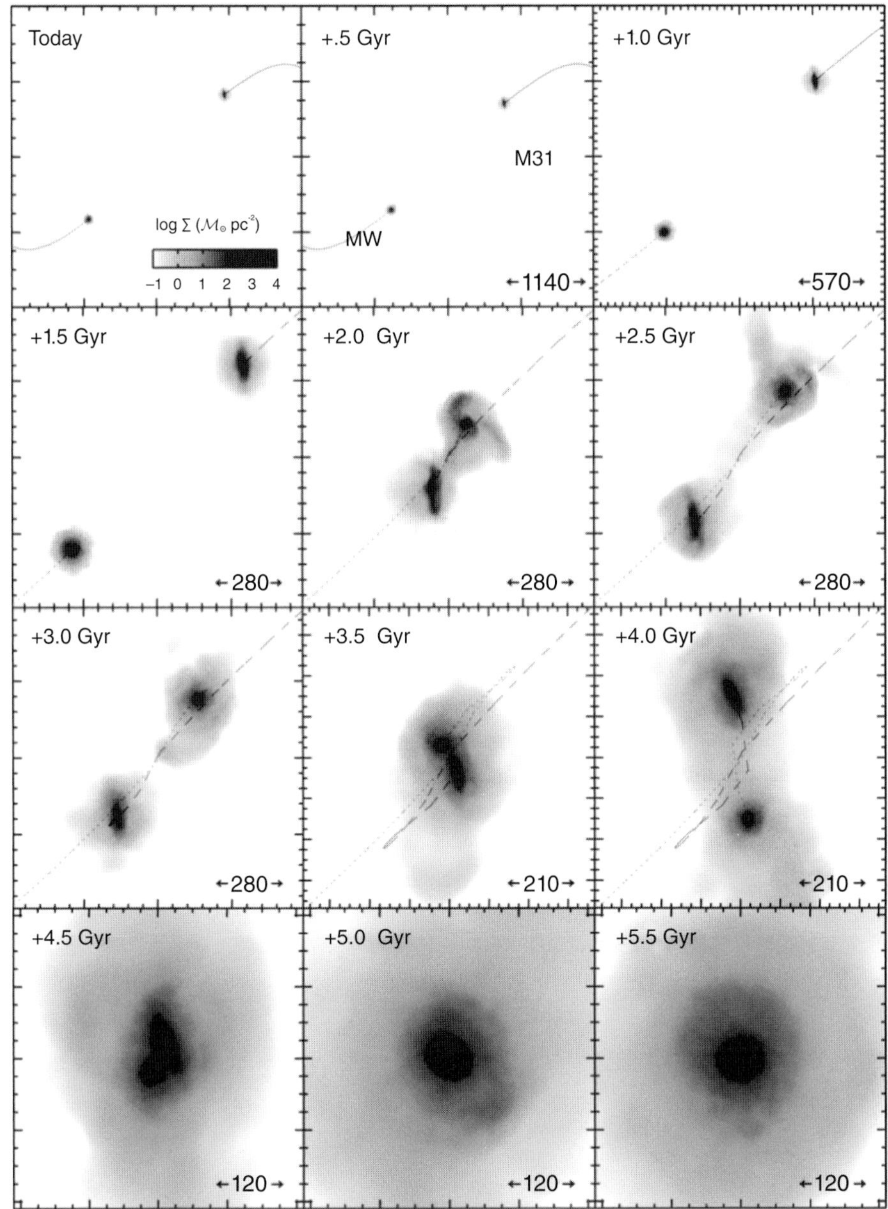


Fig. 10.7

*N*-body and hydrodynamic simulation of a disc-galaxy merging, with initial conditions that today reproduce the pair Milky Way–M 31, evolving until 5.5 Gyr in the future. The merger remnant qualitatively resembles an elliptical galaxy. Each panel shows the stellar surface density distribution (with scale given in the top left panel) and (until 4 Gyr in the future) the trajectories of the two galaxies (dotted and dashed curves). The numbers in the top left and bottom right corners of the panels indicate, respectively, the time elapsed from today in Gyr and the physical scale of the frame in kpc. From Cox and Loeb (2008).

produced by the instability of these stream-fed discs. In particular, in the so-called **violent disc instability** model, the bulk of the stars of high-redshift bulges and ellipticals form as a consequence of gravitational instabilities in gaseous discs. In this framework, protogalaxies at redshift $z \gtrsim 2$ are characterised by the presence of relatively thick gaseous discs, which, due to the local gravitational instability (§8.3.3), fragment into massive gaseous clumps (Fig. 10.8). Because of dynamical friction (§8.9.3) against the dark matter, these big clumps transfer angular momentum to the host halo and rapidly spiral in towards the centre of the system, where they merge and form stars (§8.3), thus building up a compact spheroid. The main uncertainty in the violent disc instability model is the survival time of the giant clumps. Stellar feedback (§8.7) can in principle blow apart the clumps themselves before they reach the centre of the host halo. Moreover, for an instability to occur, the system must first reach equilibrium: it is not clear whether such equilibrium is actually attained by protogalaxies, for which mergers are believed to be frequent. In any case, protospheroids, once formed, are not believed to evolve undisturbed: given the high merger rate at high redshift (Fig. 7.5, right), we expect dissipative mergers to contribute significantly to this early formation phase.

In summary, it is now generally believed that bulges and ellipticals are built in two phases: (1) a first phase at high redshift, characterised by ***in situ* star formation**, in which the stars formed in the main progenitor halo (§7.4.4) via wet mergers, or violent disc instability; and (2) a second phase at lower redshift, characterised by ***ex situ* star formation**, in which stars that formed in other galaxies are accreted via mostly dry mergers. The relative contributions of the two phases are in general different in different systems, the second phase being more important in more massive galaxies. We discuss this scenario in more detail in §10.8.

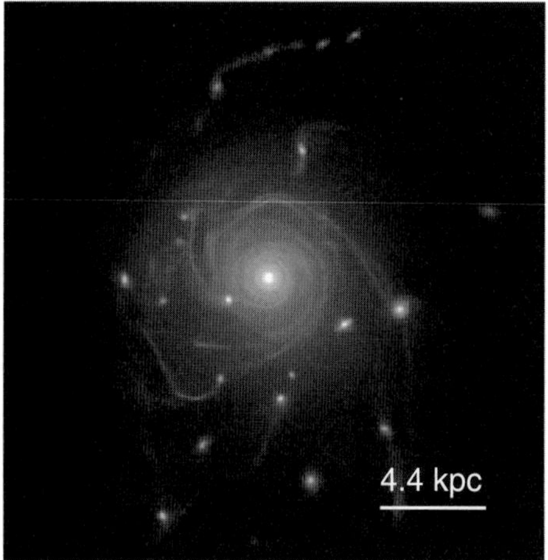


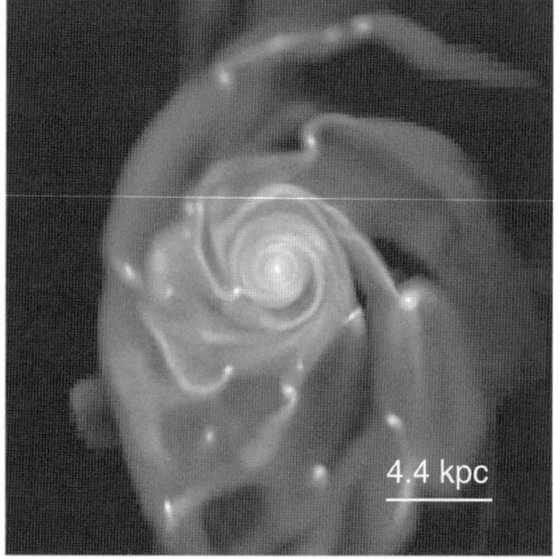


Fig. 10.8 Face-on projections of the stellar (*left panel*) and gaseous (*right panel*) density distributions of a simulated disc galaxy at $z \simeq 2.7$, in which large clumps of mass $10^7$–$10^9$ $\mathcal{M}_\odot$ have formed as a consequence of local gravitational instabilities. From Agertz et al. (2009).

## 10.4 Formation of Galaxy Components: Stellar Halos

Stellar halos are diffuse, low-density spheroidal stellar components that are detected in SFGs (§4.1.3), including the Milky Way (§4.6.1), as well as in the outskirts of normal ETGs (§5.1.1). Also the extended luminous halos of cD galaxies in clusters (sometimes referred to as intracluster light) might be considered extreme examples of stellar halos (§6.4.1).

The constituents of stellar halos (i.e. diffuse stars, globular clusters and stellar streams) tend to be characterised by old stellar populations; thus the formation of halo stars is believed to occur early with respect to the lifetime of the host galaxy. Similar to the case of elliptical galaxies and bulges (§10.3), there has been a long-standing debate on whether the formation of stellar halos is 'monolithic', with stars formed *in situ*, or 'hierarchical', with stars formed *ex situ*. In the *in situ* scenario, the stellar halo is the product of a fast dissipative collapse in the very early phases of the formation of the galaxy (with formation timescales of the order of $10^8$ yr for Milky Way-like galaxies). In the *ex situ* scenario, the stellar halo is built by accretion of stars tidally stripped (§8.9.4) from satellite galaxies (with formation timescales of the order of Gyr for Milky Way-like galaxies). In both cases the halo stars are old: in one case stars are formed early *in situ*; in the other case the accreted stars are formed early, but in other galaxies.

Recent deep observations of stellar halos in the Milky Way (§4.6.1) and in external galaxies (Figs. 4.4 and 5.4) have revealed that these systems are characterised by the presence of numerous stellar streams, strikingly resembling the tidal streams (§8.9.4) produced in simulations of interacting galaxies (Fig. 10.9). These observations leave no doubt that hierarchical accretion of satellites contributes to the formation of stellar halos, at least in the galaxy outskirts, but it is still not clear how important this contribution is. It is also possible that the inner and outer parts of stellar halos have different formation histories, with a predominance of dissipative processes in the formation of the inner halo.

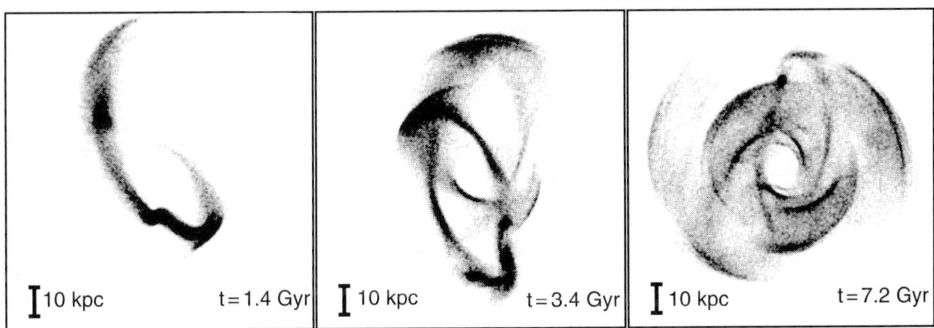


**Fig. 10.9** Formation of shell-like structures of debris in an *N*-body simulation as a consequence of tidal stripping. In the simulation an initially spherical *N*-body stellar system orbits in a static spherical potential. The three panels show, at different times from the beginning of the simulation, the projected distribution of the particles assuming line of sight orthogonal to the orbital plane. From Hendel and Johnston (2015).

Essentially due to the dependence of tidal stripping on the density of the accreted satellite (eq. 8.170), in the *ex situ* scenario the building blocks of stellar halos are expected to be low-density dwarf galaxies such as the dSphs observed in the Local Group (§6.3.1). A still open question is whether such a model is able to reproduce quantitatively the observed features of the stellar halos, such as their shape, density profile, kinematics, and age and metallicity of their stellar populations. This question can be addressed, on the one hand, by studying the properties of stellar halos in hydrodynamic and $N$-body simulations, and, on the other hand, by comparing the properties of the halo stars with those of the stellar populations of the candidate building blocks (Fig. 6.7). However, present-day dSphs are not necessarily representative of the dwarfs that contributed to the build-up of the stellar halo, which might have been more massive satellites, with a different abundance pattern. In this picture some of the globular clusters observed today in stellar halos once belonged to now disrupted satellite dwarf galaxies. But globular clusters do not necessarily form *ex situ*: several models have been proposed, including formation during major mergers and formation in very high-mass GMCs in gaseous discs of galaxies.

## 10.5 Characteristic Scales in Galaxy Formation

Having illustrated the mechanisms responsible for the formation of galaxy components, and before addressing the question of how present-day galaxies have assembled, we give here an overview of the characteristic mass, velocity and temperature scales of galaxy formation.

*Maximum mass* $\mathcal{M}_{\rm max}$. As we have seen in §8.2.3 and §8.2.4, the possibility to form a galaxy within a dark matter halo is strongly influenced by the halo virial temperature (or mass). In §8.2.3 we have used estimates of the cooling and dynamical times to explain why we expect to have an upper limit $\mathcal{M}_{\rm max} \sim 10^{13}\mathcal{M}_\odot$ to the mass of halos that can form galaxies efficiently.

*Minimum mass* $\mathcal{M}_{\rm min}$. Moving to lower-mass halos, we consider the minimum mass of halos in which star formation is possible. In §9.6 we have seen that the first galaxies likely formed, before cosmological reionisation ($z \gtrsim 10$; §9.9), in halos with $T_{\rm vir} \sim 10^4$ K, which are the lowest-mass halos able to retain some gas after hosting the formation of the first stars. Even after reionisation ($z \lesssim 6$), $T_{\rm vir} \sim 10^4$ K remains a characteristic virial temperature for galaxy formation, though for somewhat different reasons. The ionising background radiation (§9.5.3 and §9.8.2) responsible for the cosmological reionisation heats the gas above $10^4$ K, thus preventing most gas from being bound to the shallow potential wells of halos with $T_{\rm vir} \lesssim 10^4$ K (corresponding to circular speeds at the virial radius $v_{\rm vir} \lesssim 17\,{\rm km\,s^{-1}}$; eq. 8.18). These values of $v_{\rm vir}$ and $T_{\rm vir}$ can be translated into virial masses $\mathcal{M}_{\rm vir}$, but we recall that $\mathcal{M}_{\rm vir}$ depends on $z$ for given $v_{\rm vir}$ (eq. 7.26) or $T_{\rm vir}$ (eq. 8.19). For instance, at $z = 6$ and $z = 2$, halos with $v_{\rm vir} = 17\,{\rm km\,s^{-1}}$ have $\mathcal{M}_{\rm vir} \simeq 1.7 \times 10^8\mathcal{M}_\odot$ and $\mathcal{M}_{\rm vir} \simeq 6 \times 10^8\mathcal{M}_\odot$, respectively. More detailed theoretical calculations show that, just after reionisation, the photoheated gas in halos with circular speeds in the range $17\,{\rm km\,s^{-1}} \lesssim v_{\rm vir} \lesssim 35\,{\rm km\,s^{-1}}$, though confined, is not able to settle at the bottom of the potential well, and therefore cannot form stars. From $z \approx 6$ to $z \approx 2$, only halos with $v_{\rm vir} \gtrsim 35\,{\rm km\,s^{-1}}$ (i.e. $T_{\rm vir} \gtrsim 5 \times 10^4$ K, corresponding to $\mathcal{M}_{\rm vir} \gtrsim 1.5 \times 10^9 \mathcal{M}_\odot$ and $\mathcal{M}_{\rm vir} \gtrsim 5 \times 10^9 \mathcal{M}_\odot$ at $z = 6$ and $z = 2$, respectively) have potential wells deep enough to make the gas fragment and form stars. At redshifts lower than $z \approx 2$ the intensity of the UVB decreases rapidly with cosmic time (§8.1.2), so at $z \lesssim 2$ star formation becomes possible also in halos with $v_{\rm vir} \lesssim 35\,{\rm km\,s^{-1}}$. From all the above, we conclude that an order-of-magnitude estimate of the minimum halo mass for galaxy formation is $\mathcal{M}_{\rm min} \sim 10^8 - 10^9 \mathcal{M}_\odot$.

*Supernova-feedback mass* $\mathcal{M}_{\rm SN}$. Protogalaxies hosted in halos with $\mathcal{M}_{\rm vir} > \mathcal{M}_{\rm min}$ can be vulnerable to stellar feedback: following a burst of star formation, feedback from SNe (§8.7.1) can effectively remove the gas from the halo, thus inhibiting, at least temporarily, further star formation. The inability of small halos to retain SN-heated gas can have implications for galaxy formation in general: in a hierarchical scenario massive galaxies are the product of the coalescence of small substructures, so outflows from dwarf building blocks delay star formation also in higher-mass galaxies. More quantitatively, there is a characteristic velocity scale $v_{\rm SN} \approx 100\,{\rm km\,s^{-1}}$ associated with SNe. SN explosions naturally produce collisionally ionised gas (§4.2.4) with typical temperatures $\gtrsim 10^6$ K (§8.7.1), which heats any surrounding gas and produces winds with speeds of the order of $100\,{\rm km\,s^{-1}}$. Similar speeds are also reached by the cold gas in the shells of superbubbles that achieve blow-outs (§8.7.3). It follows that SN feedback in halos with $v_{\rm vir} \lesssim v_{\rm SN}$ could efficiently remove part of the ISM. The mass of halos with circular speed $v_{\rm SN}$ is $\mathcal{M}_{\rm SN} \approx 5 \times 10^{11} \mathcal{M}_\odot$ at $z = 0$ and $\mathcal{M}_{\rm SN} \approx 1 \times 10^{11} \mathcal{M}_\odot$ at $z = 2$ (eq. 7.26).

*Shock-heating mass* $\mathcal{M}_{\rm sh}$. In §8.2.4 we have introduced another characteristic halo mass $\mathcal{M}_{\rm sh} \approx 5 - 7 \times 10^{11} \mathcal{M}_\odot$, essentially independent of $z$, below which shock heating never occurs, and thus cold accretion is at work.

*Critical mass* $\mathcal{M}_{\rm crit}$. As $\mathcal{M}_{\rm SN}$ and $\mathcal{M}_{\rm sh}$ are of the same order, we can conclude that there is a critical mass $\mathcal{M}_{\rm crit} \sim \mathcal{M}_{\rm SN} \sim \mathcal{M}_{\rm sh}$, which is a reference scale of galaxy formation. Halos with $\mathcal{M}_{\rm vir} > \mathcal{M}_{\rm crit}$ are able to retain SN-heated gas and accumulate hot gas, while halos with $\mathcal{M}_{\rm vir} < \mathcal{M}_{\rm crit}$ lack this hot atmosphere. This has crucial consequences for the build-up and evolution of galaxies.

In summary, in this section we have seen that galaxy formation can be characterised in essence by three fundamental scales, which, expressed in terms of halo virial mass, are: $\mathcal{M}_{\rm min} \sim 10^8 - 10^9 \mathcal{M}_\odot$, $\mathcal{M}_{\rm crit} \sim 10^{11} - 10^{12} \mathcal{M}_\odot$ and $\mathcal{M}_{\rm max} \sim 10^{13} \mathcal{M}_\odot$. In the following sections we see how these characteristic scales contribute to determine the formation and evolution of galaxies over cosmic time.

## 10.6 Quenching of Star Formation

Galaxies that are now quiescent were able to form stars in the past, but at some point in their lifetime something happened that prevented further star formation. This phenomenon is usually referred to as **quenching of star formation**. Any process that prevents gas from cooling and forming stars, or removes star-forming gas, is thus called a **quenching**

**mechanism**. In §10.6.1–§10.6.3, we define the main modes of quenching that are believed to be responsible for shutting off star formation in galaxies. In §10.6.4 we address the question of how a galaxy can maintain its status as a quenched system.

### 10.6.1 Mass Quenching

When the mass of the host dark halo becomes higher than the critical mass $\mathcal{M}_{\rm crit}$ (§10.5), a hot atmosphere builds up, because a fraction of the accreted gas can be heated to the virial temperature, and the potential well is deep enough to retain SN-heated gas. In the presence of such a virial-temperature atmosphere, the cold mode accretion becomes inefficient, because infalling filaments of cold gas can be ablated by the corona and thus assimilated by the hot phase (§10.7.2). Moreover, when the virial temperature is high, the hot gaseous halo can shield the galaxy from accretion of cold gas in mergers (§10.7.2). If a gas-rich satellite galaxy is accreted, its cold gas can join the hot phase as a consequence of tidal and ram-pressure stripping (§8.9.4), turbulent boundary layer ablation (the cold and hot media mix; §10.7.2, Fig. 10.12) and **thermal evaporation** (heat is transferred from the hot to the cold phase via thermal conduction).

The inefficiency of cold mode accretion does not necessarily imply the shut-off of star formation, because the virial-temperature gas can still cool and form stars. However, as we see in §10.6.4, this is unlikely to occur in very massive systems, ultimately due to AGN feedback (§8.8). The fact that galaxies hosted in halos with $\mathcal{M} \gg \mathcal{M}_{\rm crit}$ stop forming stars is a quenching mechanism usually called **mass quenching** (or **halo quenching** or **internal quenching**), because it is at work in all sufficiently massive galaxies, independent of external phenomena, such as interactions with the environment.

### 10.6.2 Environmental Quenching

**Environmental quenching**, sometimes called **satellite quenching**, is a phenomenon that can inhibit star formation in galaxies that are insufficiently massive for mass quenching to be effective. In this context it is important to distinguish between **central galaxies** and **satellite galaxies**. A galaxy is defined as central if it resides at the centre of a dark halo that is not a subhalo of a more massive halo (§7.5.3); otherwise it is classified as satellite. Let us focus for simplicity on a cluster of galaxies with a BCG (§6.4.1) at its centre: in this case all galaxies are satellites, with the exception of the central BCG. The internal properties of each satellite galaxy are affected by the interaction with the environment (the cluster). In particular, the gravitational field of the host cluster and the ICM can remove gas from the satellite galaxies via processes such as tidal and ram-pressure stripping (§8.9.4) and ablation (§10.7.2). The effect of this interaction is twofold. On the one hand, it can lead to a rapid quenching of star formation by directly depriving the galaxy of its cold gas. On the other hand, even without removing the inner cold gas (ISM), it can strip the extended lower-density outer gas (CGM), which is the main gas reservoir of the galaxy: in this case star formation will cease on longer timescales (the latter process is usually referred to as **starvation** or **strangulation**). The term 'environmental quenching' refers to the ensemble of processes by which the host system induces quenching of star formation in a satellite galaxy. Environmental quenching is most effective in clusters of galaxies, but it is at work also for satellites of smaller (group-size or massive galaxy-size) halos. For instance, dSphs might have experienced a similar effect (§10.9.1).

### 10.6.3 Quenching by Negative Feedback

Examples of quenching mechanisms that we have already encountered are negative stellar (§8.7) and AGN (§8.8) feedback, which can heat the gas and/or expel it from the central regions of galaxies, thus inhibiting star formation and directly contributing to make galaxies quiescent. If this feedback-driven quenching is effective, star formation can be stopped in a galaxy also following a starburst and/or a merger. Let us consider, for instance, a major merger between two disc galaxies rich in cold gas. During the merger, compression of the gas and tidal torques (§7.5.4 and §10.1) can lead to enhanced central star formation and accretion of gas onto the central SMBH. However, the consequent stellar feedback (§8.7) and possibly associated quasar mode AGN feedback (§8.8.1) can produce shocks and winds that can effectively heat and expel the star-forming cold gas, thus preventing, at least temporarily, further star formation. This phenomenon is sometimes called merger quenching. Given the complexity of the involved physical processes, the details of this mechanism are highly uncertain. In fact, AGN feedback might also be positive, so not all merger remnants necessarily stop forming stars, and the disc could form anew.

### 10.6.4 Maintaining the Gas Hot

In §10.6.1 we have seen that in sufficiently massive galaxies essentially all the accreting cold gas is assimilated by the hot phase. However, *per se*, this is not sufficient to quench star formation forever. The hot gas is subject to radiative cooling (§8.1.1), so in principle it can cool and create the conditions for star formation. The question is thus what prevents the hot gas in massive halos from cooling and forming stars. This is essentially the classical problem of cooling flows (§5.2.1 and §6.4.3). In the centre of the virial-temperature atmosphere of a massive galaxy (as well as of a group or cluster of galaxies), there is a **cooling region**, in which the gas is sufficiently dense to have cooling time shorter than the Hubble time. If cooling is not balanced by some form of heating, the inevitable consequence is catastrophic cooling in the very centre of the halo followed by intense star formation, which in fact are not observed.

Important sources of heating are stellar (§8.7) and AGN (§8.8) feedback. While stellar feedback, mainly in the form of explosions of Type Ia SNe (§4.6), could be sufficient to heat the gas in relatively low-mass galaxies, in the most massive systems AGN feedback is believed to be responsible for halting cooling flows and preventing cooling catastrophes. When the cooling gas flows towards the system's centre, it triggers AGN feedback, which counteracts further gas accretion. In the most massive systems (groups and clusters of galaxies) there is evidence that radio mode feedback (§8.8.2) is at work: powerful radio jets blow in the X-ray gas cavities (Fig. 8.21) that can transfer, on average, sufficient energy to stop the cooling flow. The observational evidence is weaker in galaxy-size systems, but there are reasons to believe that a similar process of radio mode feedback quenching is at work in all galaxies with hot atmospheres. We recall that AGN jets have small duty cycles (they turn on and off on short timescales), so the absence of jets does not imply that there has not been heating in the recent past, over timescales shorter than the cooling time of the central gas. Similarly, when the accretion rate is higher, subsequent outbursts of quasar mode AGN (§8.8.1; Fig. 8.20) can, on average, counterbalance the cooling flow.

If the hot gaseous halo were thermally unstable (§8.1.4), gas would not cool monolithically in the system's centre, but gas condensation, that is the formation of small cold gas clouds out of a hot ambient medium, and star formation could occur throughout the cooling region. Whether the gas cools only in the centre or throughout the system is an important question, because the cooling can be effectively counteracted by AGN feedback only in the central regions. The thermal stability properties of gaseous halos are very complex, depending, for instance, on the stratification in a gravitational field, thermal conduction and magnetic field. In general, distributed gas condensation is not theoretically expected far from the centre of hot atmospheres. Whether these cold gas clouds condense in the very central regions of galactic halos is a matter of debate. Cold gas is sometimes observed in the central regions of massive ETGs (§5.2.2), but the origin of this gas is unclear.

## 10.7 Assembly History of Present-Day Star-Forming Galaxies

A typical present-day SFG is made up of several components (gas and stellar thin discs, stellar thick disc, bulge, bar and halo) and their individual formation mechanisms have been discussed in §10.1–§10.4. The purpose of the present section is, instead, to describe the assembly history that produces an SFG as the end product at $z = 0$. In particular, we concentrate on a few key features that characterise SFGs and differentiate them from ETGs (discussed in §10.8). Present-day SFGs are, by definition, forming stars today as a consequence of the fact that they have relatively large amounts of cold gas. However, they also have old stellar populations of different ages, which tells us that they must have had cold gas in the past. Finally, they have a disc morphology and thus processes that transform discs into spheroids (§10.3) must have not occurred or must have been inefficient.

The evolution of any present-day galaxy, and an SFG in particular, can be investigated following two main approaches. First, one can observe the detailed properties of galaxies today and use them to reconstruct their past evolution: we can call this a **backward approach** and it is akin to the archaeological approach described in §11.3.10. For instance, the presence in SFGs of stars of very different ages can be used to show that they have had a prolonged SFH (§4.1.5). If we model this SFH with an exponential function (eq. 8.96), the typical star formation timescale that we obtain is of several Gyr. More thoroughly, we can measure the amount and distribution of gas, stars, SFR and metallicity in a present-day

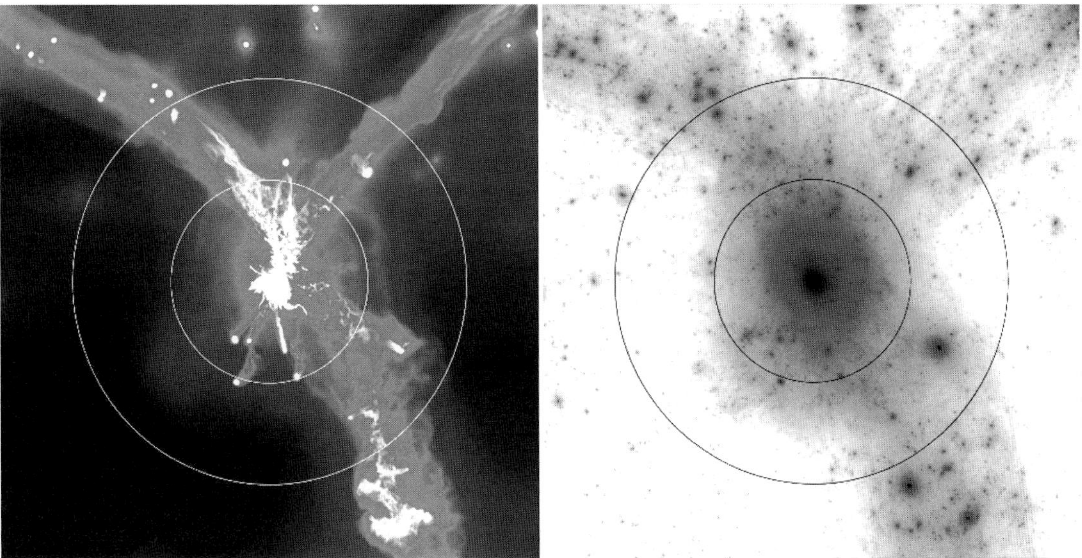

Fig. 10.10 Snapshot of a cosmological hydrodynamic simulation at $z = 2$. The *left panel* shows the column density of the gas (lighter is denser), while the *right panel* shows the dark matter (darker is denser). The virial mass is $\mathcal{M}_{\rm vir} \simeq 8 \times 10^{11}\, \mathcal{M}_\odot$ and the virial radius $r_{\rm vir} \simeq 92$ kpc. The two circles indicate $r_{\rm vir}$ and $2r_{\rm vir}$ of the halo on which the snapshot is centred. Both gas and dark matter are flowing into the inner potential well along filaments. Note that the gas filaments tend to be thinner than the dark matter ones. Adapted from Nelson et al. (2016).

SFG and use them as boundary conditions for structural and chemical evolution models that allow us to reconstruct the evolution of the galaxy back in time (§10.7.5). Such models, for instance, give us a detailed reconstruction of the gas accretion rate that must have occurred from the environment onto the SFG (§10.7.2).

The second approach, called a **forward approach**, is usually adopted using cosmological hydrodynamic simulations and semi-analytic models that we describe in §10.11. Cosmological simulations are particularly important to investigate the evolution of the dark halos hosting SFGs in the so-called cosmological context, i.e. following their evolution in a realistic large-scale environment. They give us crucial information on the rates of matter accretion onto the halos and the large-scale properties in general, while on the small scales the situation may be more complex as we see below. According to these simulations, the early progenitors of disc galaxies mostly grow through cold mode accretion from filaments (§8.2.4 and Fig. 10.10). Feedback acts to prevent very fast accretion and star formation by heating the infalling gas and ejecting part of the ISM into the surrounding environment (§10.7.3). The accretion history of a dark halo should change when the halo reaches the critical mass $\mathcal{M}_{\rm crit}$ (§8.2.4 and §10.5), for instance around $z \approx 1$–$2$ for a Milky Way-type galaxy ($\mathcal{M}_{\rm vir} \approx 1$–$2 \times 10^{12}\, \mathcal{M}_\odot$ at $z = 0$). After this time, the SFG becomes surrounded by an extended hot gaseous halo (or corona) at temperatures close to the virial temperature ($T \gtrsim 10^6$ K; §6.7). The gas filaments infalling from the IGM will then struggle to reach the disc directly, but accretion can still proceed from the cooling of the corona (§10.7.2).

In the following, we report results obtained with both the backward and forward approaches focusing on specific physical processes that, on the one hand, make SFGs grow and evolve and, on the other hand, allow them to *remain* star-forming and retain their gaseous and stellar discs down to $z=0$. The three most important requirements to produce a $z=0$ SFG appear to be that: (1) the galaxy does not experience major mergers at late times, (2) gas accretion from the IGM keeps occurring onto the star-forming disc and (3) efficient stellar feedback is at play. After discussing these three points in the following sections, we describe the inside-out growth of galaxy discs and two applications of (backward approach) chemical evolution models.

### 10.7.1 Lack of Late Major Mergers

If a major merger takes place between two disc galaxies, the discs are likely destroyed or profoundly disrupted (§10.3). If this happens at early stages (say $z>1$) when gas accretion is fast and the stellar mass is low, a disc may re-form and the evolution can proceed. Instead, if a major merger occurs at lower redshifts, it will likely lead to the production of a spheroid and, potentially, the quenching of the star formation (§10.6.3). In some simulations, strong stellar feedback or gravitational tides can momentarily eject a fraction of the gas that is then later reaccreted to re-form a disc. However, also in this case the initial stellar disc is partially or totally destroyed due to violent relaxation (§8.9.2); thus the final product of the merger will be a spiral galaxy with a large classical bulge and a small disc, of Hubble type Sa or a lenticular if the gas is later consumed (§4.1.2).

At the present time, most disc galaxies have relatively low $B/T$ ratios and most of the bulges seem to be pseudobulges testified by the low Sérsic index (§4.1.2). As we have seen, pseudobulges are formed not by mergers but by global disc instabilities (§10.2.2); thus the lack of major mergers at late times for these galaxies is quite compelling. Moreover, most stellar discs are 'cold' in dynamical terms (§4.1.1), meaning that the $V/\sigma$ ratio is much higher than unity, reaching for instance values of 10 in the thin disc of the Milky Way. In conclusion, major mergers must be scant in the history of present-day discs, at least at $z \lesssim 1$, and their evolution should be mostly dominated by slow, non-violent processes (for observations of merger rates, see §11.3.2).

### 10.7.2 Gas Accretion

Typically, a present-day spiral galaxy has more stars than cold gas in its disc, the gas-to-stellar-mass ratio being around 1/10 for massive spirals (§4.2.6). The depletion time (§4.2.10) of this gas in the inner parts of the discs is short (of the order of a Gyr) with a weak dependence on redshift (§11.3.11). This tells us that, in general, SFGs do not have enough gas to keep forming stars for much longer than a Gyr and yet they have continued to do so for a Hubble time. This simple observation shows the need for substantial gas accretion to take place onto galaxy discs at any epoch. Unfortunately, gas accretion is very difficult to observe directly as it likely occurs at very low column densities (§4.2.10). However, given that star formation is the main consequence of this accretion, there are ways to estimate

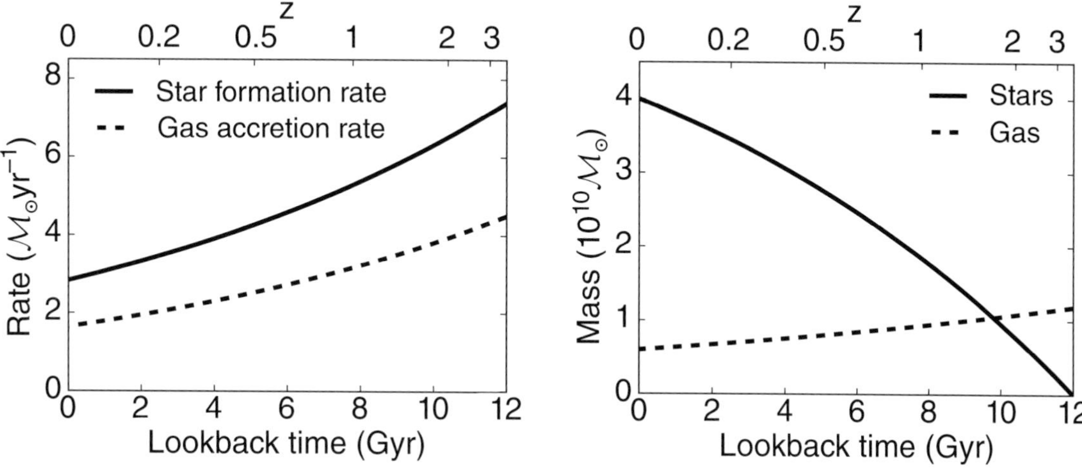


Fig. 10.11

*Left panel*. SFR of a spiral galaxy with properties similar to the Milky Way where the gas accretion rate onto its disc is reconstructed using a backward approach (§8.4.2). This accretion has been derived assuming an exponential SFH (eq. 8.96) starting from look-back time 12 Gyr, the standard star formation law (eq. 4.18) and the IRA with $\mathcal{R} = 0.3$ (§8.5.3). Delayed feedback (§8.4.1) would increase the accretion and decrease the SFR at early times, but it has only a minor effect at later times. Outflows of gas are neglected. If one allowed for significant outflow, the accretion rate would increase accordingly. *Right panel*. Resultant stellar and gas masses as functions of time in this disc model.

it indirectly. We have seen in §10.7.1 that galaxy discs are likely to evolve, for a large fraction of their life, without a contribution from major mergers; thus we can use models of discs evolving in isolation to estimate the gas accretion rate. Note that by saying 'in isolation' we do not mean that the disc does not exchange matter with the environment (gas accretion is included), but only that it is not evolved in a fully cosmological context (§10.11.1). Fig. 10.11 shows the result of one such model (described in §8.4.2) for a galaxy with properties similar to the Milky Way. The rate of accretion closely follows the SFR, but it is lower mostly because stars also form out of the gas returned to the ISM by stellar evolution (§8.4.1).

The SFH (and the accretion history) in the above galaxy has a timescale of 8 Gyr and thus a decrease of about a factor of 2–3 from $z \simeq 2$ to the present day. In §11.3.3 we see that cosmic SFR density declines by a factor of 3–4 faster than this in the same redshift range (Fig. 11.33). The difference between the SFH of a typical present-day SFG and that of the cosmic average is due to the quenching of the star formation (§10.6) in massive galaxies. Indeed, a significant fraction of galaxies that were contributing to the cosmic SFH at $z = 1$–$2$ have, since then, experienced quenching and do not contribute any more at $z = 0$. Their reconstructed SFHs are, not surprisingly, steeper than the cosmic average (§11.3.10).

Gas accretion onto SFGs must ultimately come from the CGM/IGM or from satellite galaxies. In §7.4.4, we have seen that dark matter halos keep growing through mergers and diffuse accretion of dark matter. As a consequence, gas is also accreted at roughly the universal baryon fraction. However, we have also seen (§6.7) that only a small fraction of the gas that we should expect to have made its way into galaxy gravitational potentials (crossed the virial radius) is observed today in the discs and bulges of SFGs. The remaining gas (missing baryons) is probably largely inside the virial radius but *not* in the galaxy. We remind the reader that the virial radius is typically, at least, one order of magnitude larger than the disc size (§10.1.1). To be more quantitative, a halo hosting a Milky Way-like galaxy is expected to accrete baryons at rates going from tens of $\mathcal{M}_\odot\,\mathrm{yr}^{-1}$ at $z=2$–$4$ to $\approx 10\,\mathcal{M}_\odot\,\mathrm{yr}^{-1}$ at $z=0$. However, the accretion of gas that feeds that star formation in the disc (Fig. 10.11), in the same period, is an order of magnitude lower. This simple consideration tells us that accretion onto a dark matter potential well may be happening at the universal baryon fraction, but accretion *onto the galaxy discs* does not follow straightforwardly. This latter is, in fact, regulated by a complex interplay of competing mechanisms such as gravity, gas heating, gas cooling, stellar and AGN feedback. In the following, we discuss the three main channels of gas accretion onto the discs of SFGs.

## Minor Mergers

Minor mergers (§7.4.4) are not disruptive for galaxy discs (the disc survives almost intact), although their cumulative effect can be significant. For instance, minor mergers can kinematically *heat* a galaxy stellar disc and possibly form thick discs (§10.1.7). If the merging galaxy is a gas-rich satellite, it can provide a significant amount of gas to be used to feed the star formation in the main galaxy. Cosmological simulations show that gas accretion from minor mergers should be quite important at high redshifts, but it becomes less and less important after $z \sim 1$. Estimates of the amount of gas available in satellites and of the merger timescales confirm this by showing that the present-day satellites are not enough to feed the star formation in large spirals at the observed rate. In the end, it is unlikely that minor mergers constitute a dominant gas accretion channel for the whole population of spiral galaxies, at least at low redshifts.

Gas accretion can also occur from satellites that do not end up merging. Tidal and especially ram-pressure stripping can, in fact, remove the gas from a flying-by satellite, which, in time, can fall onto the main galaxy. One well known interaction of this kind is taking place between the Magellanic Clouds and our own Milky Way (§6.3.1). The Magellanic Clouds are moving too fast to merge with the Milky Way anytime soon, but their gas is being stripped and slowed down by the interaction with the Galactic corona. This gas currently forms the Magellanic stream that, if it survives its journey through the corona (see below), could be incorporated into the disc of the Milky Way in a timescale of a few Gyr.

## Cold Accretion

In galaxies with halos that are less massive than $\mathcal{M}_{\rm crit}$ (§8.2.4 and §10.5), typical of low-mass spirals and dwarf galaxies, gas accretion from the IGM is expected to take place in the cold mode (§8.2.4). The accretion from cold filaments (Fig. 10.10) is predicted to occur at the cosmological rate, which is the dark matter accretion rate (see Fig. 7.6) multiplied by the baryon fraction ($f_{\rm b}$). How the gas from cold filaments joins with the disc of the galaxy at the centre of the potential well is a complex problem given the multi-phase nature of the CGM (§6.7) and the different scales involved. The process is often studied using zoom-in cosmological hydrodynamic simulations (§10.11.1). In these simulations, gas accretion onto the discs tends to occur in the outer parts, as in the inner regions it is blocked or counteracted by powerful outflows from stellar feedback (§10.7.3).

Galaxy halos with masses higher than $\mathcal{M}_{\rm crit}$ are expected to have thermalised their surrounding gas (§8.2.4). This means that a hot corona has formed around these galaxies and any incoming filaments will impact against this hot gas. A recurring question is whether these cold filaments can survive and reach the star-forming discs. The problem of mixing and interaction between different gas phases is difficult to study in cosmological simulations due to their relatively poor resolution (§10.11.1) and one must resort to idealised simulations of gas clouds and single gas filaments performed at very high (parsec or sub-parsec scale) resolution. One of the key effects that these simulations need to capture is the **Kelvin–Helmholtz instability**. This takes place when two portions of a fluid in contact are in motion with respect to each other, in this case the cloud/filament and the corona. If the conditions for the onset of the instability are met, ripples develop at the contact surface, first into small vortices and then into a turbulent boundary layer (§8.3.7) that mixes the gases and 'peels away' the outer parts of the cloud. This phenomenon is called **turbulent boundary layer ablation**. An example of such hydrodynamic simulations, including also radiative cooling (§8.1.1) and photoionisation (§8.1.2), is shown in Fig. 10.12. Finally, another process that can play a key role in a multi-phase medium (gas at very different temperatures in contact) is thermal conduction (§8.1.4). The comparison between the right and left panels of Fig. 10.12 shows how the shape of the so-called **turbulent wake** that develops behind the cloud changes dramatically when thermal conduction is included. The effectiveness of thermal conduction also depends on the geometry of the magnetic fields which are poorly constrained in the CGM.

Simulations like the above suggest that the survival times of infalling clouds depend on the size of the cloud and the properties of the hot corona. Small structures are very quickly destroyed and only the large ones (kpc scale or more) can survive for longer times ($\gtrsim 100$ Myr) and potentially make it to the galaxy. In general, with the development of a galactic corona, the supply of cold gas from the IGM decreases greatly and it is eventually shut off. However, accretion onto the disc can proceed via the cooling of the corona as we discuss in the next section.

We conclude this section by reminding the reader that low-mass present-day SFGs tend to have much higher gas fractions than massive spirals (Fig. 4.8). Typically the gas fraction increases with decreasing stellar mass and reaches values close to unity in dwarf irregular galaxies (§4.2). Much of this gas is, however, stored at large radii in extended discs where it is in centrifugal equilibrium, while the gas in the inner disc is slowly consumed by star formation. The depletion times (eq. 4.20), if one considers the whole H I disc, can be of the order of the Hubble time or more. This means that these galaxies have enough gas to feed their star formation and do not require further accretion from the IGM. However, they keep their reservoir in the outer discs and radial inflows are required to bring it to the star-forming (central) part of the galaxy. Whether or not these inflows take place and as a consequence of what mechanism is yet to be established.

## Cooling of the Corona

The formation of a virial-temperature ($T \sim 10^6$ K) corona (§6.7) during the evolutionary history of a present-day disc galaxy can produce a change in the way gas accretion onto the disc takes place. When the temperature and density of the corona become high enough, cold clouds and filaments falling from the IGM cannot easily penetrate the inner halo as they tend to be destroyed by hydrodynamic instabilities and thermal conduction (Fig. 10.12). They can then disperse and thermalise, thus joining the hot phase. At this point, gas accretion onto the disc can continue to take place from the cooling of the hot gas in the corona. As mentioned, a galaxy like the Milky Way is expected to start forming its hot corona at $z \approx 1$–$2$, when its stellar mass was between one-fifth and one-half of the current mass (Fig. 10.11); thus cooling of the corona could have been an important channel of gas accretion, in particular for the evolution of the thin disc (§4.6.1). Unfortunately, the properties of galactic coronae are not known in great detail and these uncertainties make a big difference when we try to estimate cooling and accretion rates. This is illustrated in Fig. 10.13, where we show the density profiles (left) and cooling times (right) of two models of isothermal coronae in hydrostatic equilibrium (eq. 6.15) in the same potential, but at two different temperatures. We see that a mere factor of 2 in

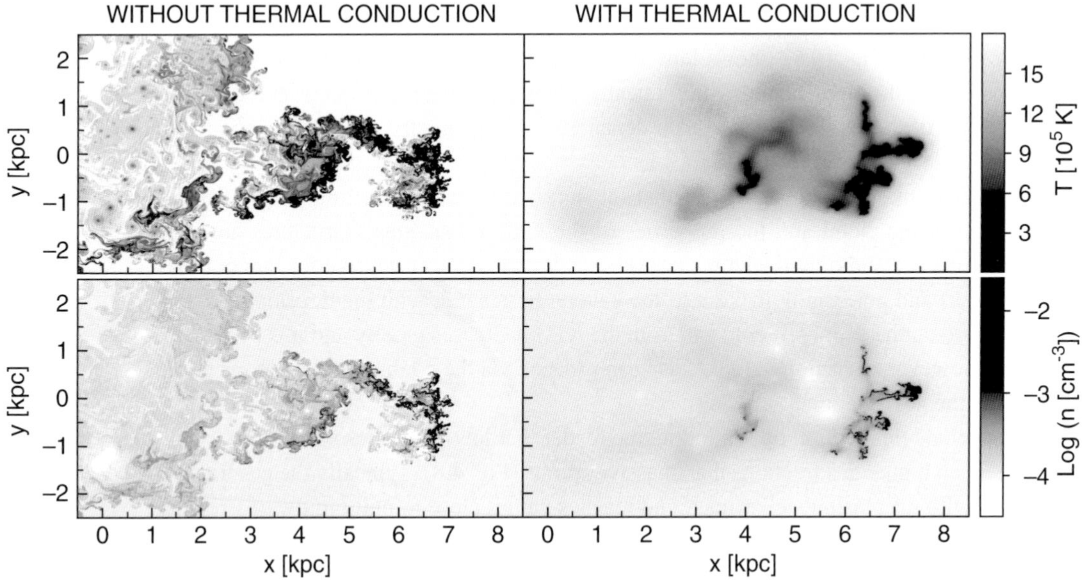


Fig. 10.12 Snapshots of two high-resolution 2D hydrodynamic simulations of initially round clouds of gas (with diameter of 500 pc) at $T = 10^4$ K and $n = 2 \times 10^{-2}$ cm$^{-3}$ moving at $v = 100$ km s$^{-1}$ (towards the right) with respect to a hot medium with $n = 10^{-4}$ cm$^{-3}$ and $T = 2 \times 10^6$ K. The *top* and *bottom panels* show, respectively, temperatures and densities after 200 Myr. The *left* and *right panels* show simulations without and with thermal conduction, respectively. Thermal conduction smooths the temperature gradients largely removing the small-scale eddies in the turbulent wake. Adapted from Armillotta et al. (2017).

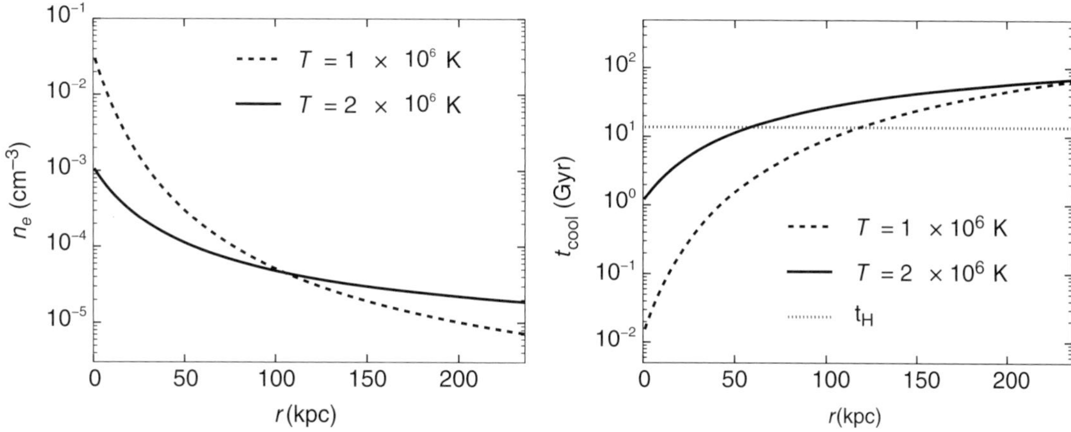


Fig. 10.13 *Left panel.* Electron number density profiles of an isothermal hot corona in hydrostatic equilibrium in an NFW gravitational potential (§7.5.2) with virial mass $\mathcal{M}_{200} = 1.5 \times 10^{12} \mathcal{M}_\odot$ and concentration $c_{200} = 7.5$. The dashed curve shows the profile for a gas at $T = 1 \times 10^6$ K (the halo virial temperature; eq. 8.20), while the solid curve is at $T = 2 \times 10^6$ K, which is the measured temperature for the corona of the Milky Way. In both curves the total mass of the corona is $0.2 f_b \mathcal{M}_{vir}$, where $f_b \simeq 0.16$ is the cosmic baryon fraction (§2.5). *Right panel.* Cooling time (eq. 8.3) as a function of radius for the same two coronae as in the left panel assuming a metallicity of 10% solar. The horizontal dotted line shows the Hubble time ($t_H$). Courtesy of A. Afruni.

temperature makes the inner densities and cooling times change by more than an order of magnitude.[11]

The process of gas accretion from the corona is rather complex as competing mechanisms are at work to heat and cool the hot gas. Heating is provided by photoionisation (§8.1.2) and feedback from stars (§8.7) and AGN (§8.8). Cooling is due to the emission of radiation (§8.1.1), which strongly depends on the temperature of the gas, in particular at $T \sim 10^6$ K (Fig. 10.13). In the regions close to the gaseous disc, the temperature of the corona is also affected by the presence of the disc and the fact that the two media are *in contact*, which produces an interface layer at intermediate temperatures.[12] Moreover, the disc–corona contact surface is greatly increased as the interface is continuously 'stirred up' by the galactic fountain (§4.2.10). The consequence is a *mixing* of disc and coronal gas that can reduce the temperature and increase the metallicity of the corona (the disc gas is typically at a higher metallicity) enhancing its cooling rate (eq. 8.3). In the end, the system is likely to reach an equilibrium state between competing cooling and heating mechanisms that allows for continuous gas accretion and the feeding of the star formation for a long time (Fig. 10.11).

We conclude with a short remark on the angular momentum. Observations show that galaxy discs form inside-out, meaning that the sizes of the discs grow as a function of time

[11] Changing the metallicity to a value different from the 10% solar adopted here would produce further variations (§8.1.1).

[12] The same process can take place at the interface between cosmological filaments or accreting clouds that succeed in penetrating down the inner corona.

(§10.7.4). In §10.1 we have seen how the size is related to the specific angular momentum $j$ and how a larger $j$ tends to imply a larger size. The obvious consequence of this is that, in general, for a disc to grow radially, it is important that the accreting gas has a relatively high specific angular momentum. This, in turn, implies that the corona must have significant rotation in the regions where it is in contact with the disc. For more information on observations and theory of gas accretion in galaxies, we refer the reader to Fox and Davé (2017).

### 10.7.3 Stellar Feedback

Stellar feedback (§8.7) is one of the most important ingredients of galaxy formation and evolution. Although the details of feedback remain somewhat unclear, the general conclusion reached by both cosmological hydrodynamic simulations and semi-analytic models (§10.11) is that stellar feedback should be very efficient in SFGs especially at high redshift. The main reason for this is that without strong feedback there is too much cold gas accretion in galaxies at early times that produces a number of shortcomings with respect to the observed galaxies. The most relevant are: (1) stellar masses become too large, (2) stars form in a concentrated structure and not in an extended disc and (3) star formation declines too fast with time. In most cosmological simulations (§10.11.1) these problems are solved by a very efficient stellar feedback, which ejects gas from the galaxy, thus making it unavailable for star formation. This mechanism is referred to as **ejective feedback** in which baryons are first accreted at a rate comparable to the universal baryon fraction ($f_b \simeq 0.16$) and then largely ejected. When such a strong feedback is implemented in simulations, the comparison with observations improves for several galaxy properties. As an example, Fig. 10.14 shows a comparison between SFHs obtained in cosmological hydrodynamic simulations using two different values for the efficiency of SN feedback. The simulation with strong feedback significantly delays the star formation, producing an SFH much more similar to that of galaxies like the Milky Way (Fig. 10.11).

Despite these successes, the inclusion of strong stellar feedback in cosmological simulations and theoretical models has two potential problems that are worth a brief discussion. The first is that there are not self-consistent ways to include feedback effects in large-scale cosmological models and one has to resort to different 'recipes' and 'calibrations'. These implementations have free parameters that are tuned to reproduce some observables; among these, one often used is the galaxy SMF at $z = 0$ (Fig. 10.24).We discuss this further in §10.11.1. The second issue is the energy requirement. Typical implementations of stellar feedback in cosmological simulations have efficiencies of conversion between the energy released by SN explosions and the energy acquired by the outflowing gas close to unity. This efficiency is referred to as a high **energy coupling** between SN feedback and the gas. In §8.7 we saw that typical values achieved by models of SNR and/or superbubble expansions are between a few per cent and $\approx 20\%$, as most of the initial energy is radiated away and lost. Thus, large-scale cosmological simulations seem to require significantly higher efficiencies than those typically predicted by small-scale ISM models.

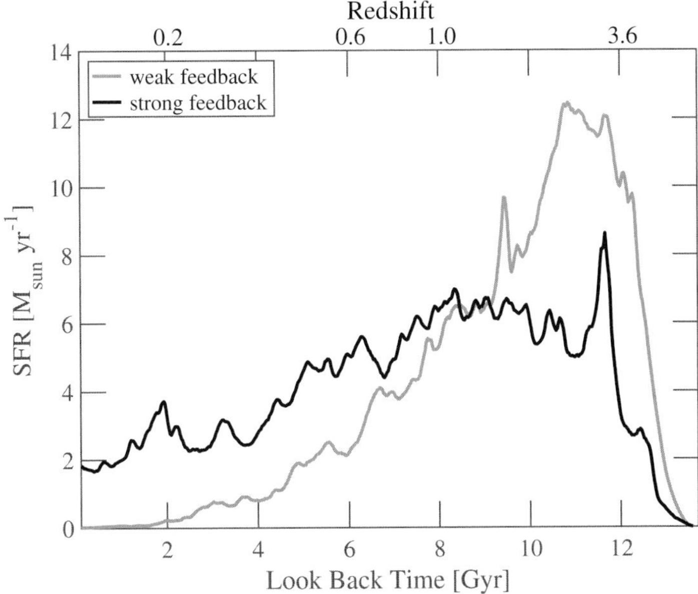


Fig. 10.14 SFHs of simulated galaxies extracted from the cosmological hydrodynamic simulations of the IllustrisTNG project with two different implementations of stellar feedback: one a factor of 4 more energetic than the other (Springel et al., 2018; Pillepich et al., 2018). Each of the two curves has been obtained by averaging the SFHs of 10 galaxies with $z = 0$ stellar masses in the range $3 \times 10^{10}\ \mathcal{M}_\odot < \mathcal{M}_\star < 7 \times 10^{10}\ \mathcal{M}_\odot$. The strong feedback SFH is more prolonged and typical of a present-day disc galaxy (§4.1.5 and Fig. 10.11), while the weak feedback SFH has a prominent peak at high redshift with very little residual SFR at $z = 0$. Courtesy of A. Pillepich.

Independently of the exact efficiency, there is little doubt that stellar feedback must play a key role in dwarf galaxies, where it can permanently remove part of their ISM (§10.5 and §10.9). Strong feedback from dwarf galaxies becomes even more relevant given that, at high enough redshift, all galaxies tend to have shallower potential wells (Fig. 7.5, right). Thus, it is natural to expect that early stellar feedback from galaxies had a strong impact on their subsequent evolution. For instance, massive outflows from low-mass galaxies are thought to explain the slope of the stellar mass function at low masses (§10.10.1). Early feedback is also required to explain the presence of metals in the IGM at high redshift (§9.8.4). A potentially efficient mechanism for the gas to escape galactic potential wells is buoyancy (see also §8.8.2). If a considerable amount of gas is heated to $T \sim 10^5 - 10^6 K$, this becomes buoyant in the circumgalactic media of low-mass galaxies and it can *slowly* make its way out of their potential wells.

We conclude by also mentioning the scenario of **preventive feedback** or **preheating** that can go in the direction of solving the aforementioned problem of energy requirement. In this picture, the intergalactic gas is heated at early times ($z > 4$) to a non-negligible entropy level before it can be accreted into dark matter halos. Proposed sources for this heating are SN and AGN winds, intergalactic turbulence and blazars (§3.6). The density of the virial-temperature gas in a halo is related to the entropy index $K$ (eq. 6.20) by $\rho \propto (T_{\rm vir}/K)^{3/2}$

so, if the IGM has high $K$, the accretion of baryons can be vastly reduced (they fall at a rate lower than $f_b \dot{M}_{DM}$, where $\dot{M}_{DM}$ is the dark matter growth rate; Fig. 7.6) or, in general, delayed. As a consequence, the need to eject baryons from the ISM of galaxies is suppressed. Note that some form of preheating is present in any model of galaxy formation as the cosmological reionisation (§9.9) effectively brings the IGM to $T \sim 10^4$ K, drastically reducing its ability to collapse into halos with $M_{vir} \lesssim 10^8 M_\odot$ (§9.6 and §10.5).

### 10.7.4 Inside-Out Growth

The stellar populations of disc galaxies are generally younger in the outer than in the inner parts (§4.1.4). This is interpreted in terms of the disc assembly being faster in the inner regions in the past and having progressively moved to the outer regions. This phenomenon is called **inside-out growth** of the discs. One of its consequences is that the stellar discs were smaller than now in the past (see also §11.3.1). Note also that the most evolved present-day disc galaxies, for instance those of type Sa, appear to have come to an end of their disc assembly in the inner regions as they have there little star formation. This may be evidence for *also* quenching (§10.6) happening inside-out, potentially connected to the formation of stellar bulges (§4.1.2 and §10.3).

From a theoretical point of view, inside-out growth of discs is an expected phenomenon. As we have seen in §10.7.2, a galaxy disc evolves by accreting gas initially located in the inner halo and then progressively at larger distances. The specific angular momentum ($j$) of the halo gas, as of any stable system, must increase with radius;[13] thus the disc should be able to acquire progressively material with higher $j$. More precisely, to assure inside-out growth, the average specific angular momentum of the infalling material has to be higher than the average specific angular momentum of the disc *at any time*. The details of the process are rather complex and depend on the AMD of the accreting gas (§10.1.3), on the effect of feedback (§10.7.3) and on the way accretion takes place (§10.7.2).

Ultimately, the angular momentum growth rate of galaxies is related to the angular momentum growth rate of dark matter halos (§10.1.1). However, in general, the growth of galaxies is not simply the consequences of the assembly of the dark matter halos, as more complex baryonic processes are active to regulate gas accretion and star formation, as well as their inside-out growth.

### 10.7.5 Chemical Evolution of Galaxy Discs

We have seen that present-day disc galaxies have had an evolution that is largely dominated by internal processes rather than mergers. Thus, we can model a galactic disc assuming that stars form and evolve *in situ* (no significant contribution from stars formed

[13] This is the so-called **Rayleigh criterion** for rotational stability, which can be written

$$\frac{\mathrm{d}j^2}{\mathrm{d}R} = \frac{\mathrm{d}[(R^2\Omega)^2]}{\mathrm{d}R} > 0, \qquad (10.13)$$

where $\Omega(R)$ is the angular speed. The reader can verify that this criterion is satisfied in galaxies rotating with both solid-body and flat rotation curves, as well as in a Keplerian disc ($\Omega \propto R^{-3/2}$).

elsewhere). These models of disc evolution are extremely useful as one can include a number of physical processes, in particular the chemical evolution, and compare them in detail with observations of nearby disc galaxies. They can be carried out with $N$-body/hydrodynamic simulations (see for instance models of bar/bulge formation in §10.2) or with analytic/numerical treatments. In the following sections we see two examples in which the predictions of chemical evolution models are compared with the abundance gradients in galaxy discs and the metallicity distribution of stars in the Milky Way.

## The Development of Metallicity Gradients

In §8.4.2 we have outlined the main equations of models of structural disc evolution. In these types of models, one follows the evolution of gas and stars in a disc subject to an external inflow of gas (in massive spiral galaxies, gas outflows are usually neglected; §8.7.3). At each time step in the evolution, the star formation law (eq. 4.18) determines how the gas is turned into stars. If we further assume that the SFH is exponentially declining[14] (eq. 8.96), then eq. (8.97) is solvable for the net accretion rate density as a function of time. The accretion rate obtained with this model for a galaxy similar to the Milky Way is shown in Fig. 10.11. Using the nomenclature introduced at the beginning of §10.7, this is an example of the backward approach as the boundary conditions of the differential equations are given using observations at $z = 0$.

The introduction of the chemistry in such a model is done by rewriting eq. (8.112) or eq. (8.113) using surface densities instead of masses. An observable that can be used both in the Milky Way and in external galaxies is the ISM metallicity[15] as a function of galactocentric radius (§4.1.4). Fig. 10.15 shows the prediction of two chemical evolution models for the abundance profiles (as a function of time) in the disc of a spiral galaxy with properties similar to the Milky Way. Assuming that the accreting material has zero metallicity[16] and falls at each radius $R$ with exactly the same specific angular momentum (thus the same azimuthal velocity) as the disc material, one obtains a model without radial flows and the radial distribution of abundances shown in the left panel of Fig. 10.15, i.e. at the present time the metallicity is essentially independent of $R$ (there is no clear metallicity gradient). Observations instead tell us that, in present-day disc galaxies, the metallicity declines with $R$ (there is a negative gradient, §4.6.1). A way to produce this decline is to consider that the material falls with a deficit of specific angular momentum. This produces radial flows that can be taken into account in models using the prescription outlined in §8.4.2. The right panel of Fig. 10.15 shows the metallicity as a function of radius that one obtains by considering that the accreting gas is rotating at 80% of the disc circular speed at each radius (corresponding to a 20% deficit of specific angular momentum, which

14 Using different functional forms for the SFH would make little difference.

15 Here we are considering the gas and not the stellar metallicity. In Fig. 4.21 we show that, in the Milky Way, the metallicity gradients of the ISM and *young* stars are similar. This, however, is not necessarily true for older stars (or for an averaged stellar population), as they could be affected by other evolutionary processes such as radial stellar migration (§10.1.5).

16 Very similar final results are obtained for non-zero, but low, metallicity, e.g. $Z \sim 0.1 Z_{\odot}$.

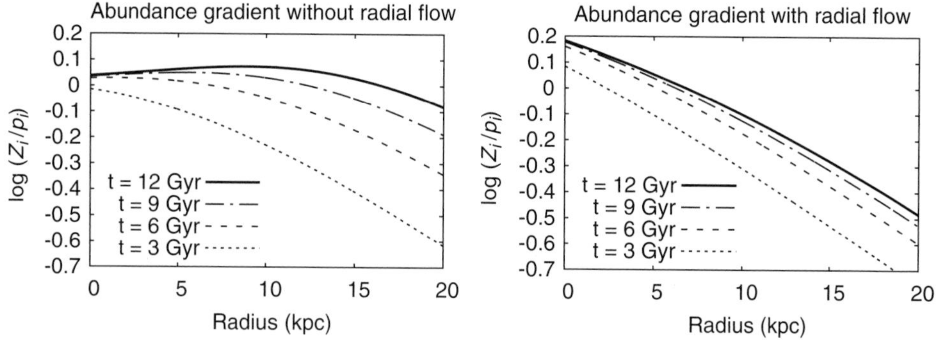


Fig. 10.15 Evolution of the abundance profiles predicted by a chemical evolution model of a disc galaxy akin to the Milky Way that forms stars according to the Schmidt–Kennicutt law (§4.2.9). The abundances of a generic element $Z_i$ are normalised to the stellar yields $p_i$ (eq. 8.111). The evolution of the abundances is followed in the case of gas accretion having the same (*left panel*) specific angular momentum as the disc at every radius or a deficit of 20% (*right panel*). The curves at $t = 12$ Gyr can be considered representative of the present time. From Pezzulli and Fraternali (2016).

is a plausible value; §10.7.2). The metallicity gradient that one obtains in this way is compatible with those observed in the Milky Way today (Fig. 4.21).

The lower angular momentum of the accreting gas is not the only phenomenon that can generate or modify metallicity gradients. A gradient is also the natural outcome of the inside-out growth of the discs (§10.7.4). However, it has been shown that, if the Schmidt–Kennicutt law in its classical form (§4.2.9) is valid, then pure inside-out growth without radial flows does not reproduce simultaneously the abundance gradient and the radial profile of star formation rate in the disc, at least for the Milky Way. Conversely, inside-out growth and radial flows could both be present. We can indeed build a model like the one in Fig. 10.15 that includes inside-out growth and this will require only a slightly smaller deficit of angular momentum to reproduce the observed gradients. In general, in all these models, metallicity gradients develop because the peak of the gas accretion profile is often not at the centre of the galaxy. In models with radial flows, the material accreted at large radii is then transported internally towards the central regions of the disc. In models including inside-out growth, the effect is furthermore accentuated at late times, as the position of the accretion peak moves farther and farther away from the centre with increasing time. In the end, at $z = 0$, in the model in Fig. 10.15, gas accretion peaks at $R \approx 5-10$ kpc. We remind the reader that we expect the accreting gas to be more metal-poor than the gas present in the disc. Therefore, in the outer parts, where most of the accretion takes place at later times, the new generations of stars can form from a diluted medium, while in the centre the metallicity of stars and ISM can grow more efficiently. This is the main mechanism for the development of a negative metallicity gradient and it is further accentuated, in the presence of radial flows, by the continuous motion of metals towards the central regions of the disc.

The redistribution of gas in the disc due to the effect of bars, spiral arms and stellar feedback can also have an impact on the radial metallicity gradient. However, we can expect that these phenomena go in the direction of flattening the abundance gradients.

Finally, radial stellar migration (§10.1.5) can play a role, although it should not have any effect in the abundance of $\alpha$-elements in the ISM. This is because these elements are released by Type II SNe (§8.5.1) that explode on short timescales so the stars do not have enough time to leave the location where they were born. Radial migration instead affects stellar abundances and also the ISM abundances of specific elements, like Fe, due to the redistribution of the progenitors of Type Ia SNe. As a consequence of all this, when comparing chemical models with real galaxies, one should take into account the combination of all these effects. Nevertheless, the simple example shown in Fig. 10.15, restricted to $\alpha$-elements in the ISM, gives an illustration of the power of chemical and structure evolution models to study the evolution of galaxies and *infer* the detailed properties of the accreting gas.

### A Test Case of the Backward Approach: the Milky Way

The Milky Way is the only large spiral for which we have detailed metallicity measurements of large samples of individual stars (§4.6.1). This, since the early works on chemical evolution (e.g. Pagel and Patchett, 1975; Tinsley, 1980), has given us the unique opportunity to try to reproduce with our models the **metallicity distributions**, which contain a greater wealth of information than the average metallicities of gas or stars. The modern approach is to have **chemo-dynamical models** that reproduce the chemistry and the dynamic patterns of the stars simultaneously. In this section, we compare the performance of one of these models with the simple closed-box model described in §8.5.4. We recall that, in this latter, the evolution starts from a gas with zero metallicity, no outflow or inflow of gas is allowed and the system evolves progressively, forming stars, decreasing its gas content and increasing its metallicity.

Fig. 10.16 shows the distribution of the metallicity of nearby G-dwarf stars that we take as representative of stars in the thin disc of the Milky Way at the solar circle.[17] These are long-lived stars (Tab. C.4) that formed during the entire evolution of the Galaxy. The peak of the distribution is around solar metallicity, with a relatively narrow distribution slightly skewed towards low metallicity. The shape of this distribution is in clear contradiction with the prediction of a closed-box model that instead produces much wider tails. To understand the long tail at low metallicity, consider that, in a closed-box model, a very large mass of low-metallicity gas is present at the beginning of the evolution. Therefore, a large number of stars form from this unpolluted gas before an appreciably high gas metallicity is reached. This manifests in the predicted metallicity distribution of G-dwarfs as a large number of stars with very low metallicities, which are not observed. This is called the **G-dwarf problem** and it is one of the classical indications that the Galactic disc did not evolve as a closed box. An obvious solution to this problem is that the amount of gas in place at the beginning was much less than the stellar mass today, and most of the gas was accreted at later times and mixed with enriched material before contributing to the next generation

[17] Remember that, as explained in §4.6.1, a clear *distinction* between the various components of the Milky Way is difficult to achieve. As a consequence, in any sample of disc stars, bulge stars, halo stars, etc., there is always some *contamination* by other components and this should be taken into account in the models.

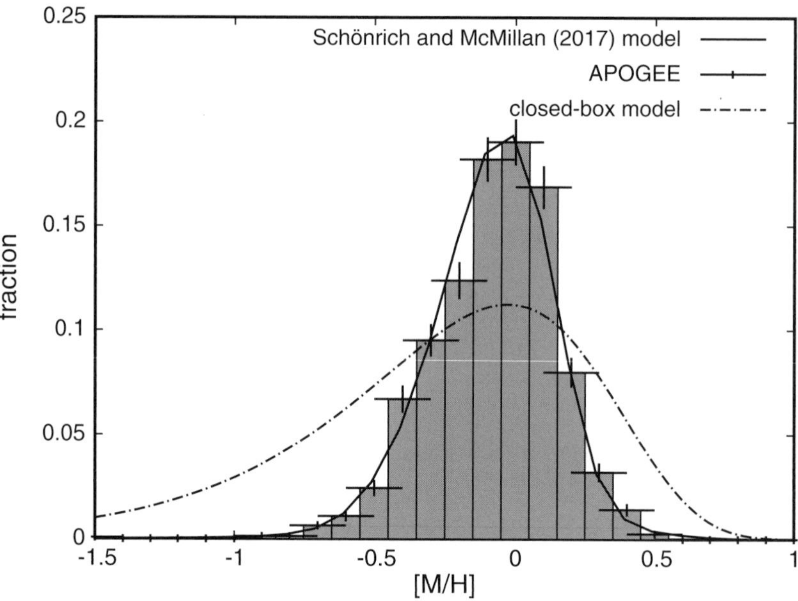


Fig. 10.16 Distribution of the metallicities of G-type stars in the thin disc of the Milky Way at the solar circle. A closed-box model (dot-dashed line) predicts a much wider distribution and, in particular, many more low-metallicity stars than observed. A chemo-dynamical model including gas accretion, gas flows and radial migration describes very well the whole metallicity distribution. Data from the APOGEE survey (Majewski et al., 2017); chemo-dynamical model from Schönrich and McMillan (2017). Figure courtesy of R. Schönrich.

of stars. Such a model produces metallicity distributions much more in agreement with observations, although the distribution becomes narrower than observed. A better match with the data is found by also considering radial flows of gas and stars (stellar migration). Fig. 10.16 shows a chemo-dynamical model that includes all these effects. In this model, stellar migration from the inner Galaxy is needed to explain the stars at metallicities higher than the Sun.

In conclusion, in the application to galaxies in general, chemical evolution models require gas accretion and radial flows. Stellar migration should be taken into account to reproduce the detailed stellar abundances, while gas outflows could be important in low-mass galaxies. As for gas accretion, chemical evolution models allow one to infer some properties of the accreting gas indirectly from the chemistry of gas and stars (see for instance the constraints on the angular momentum of the accreting gas described in the previous section). In general, all chemical evolution models of the Milky Way require the accreted gas to be metal-poor ($Z$ lower than solar by a factor of at least a few). This accretion must take place throughout the evolution of the Galaxy at a slowly decreasing rate (Fig. 10.11). Different episodes of gas accretion may have contributed to the growth of the different components of the Milky Way like the thick and the thin disc.

## 10.8 Assembly History of Present-Day Early-Type Galaxies

Present-day ETGs are often called 'red and dead', where 'red' refers to their optical colour and 'dead' to the fact that they are currently not forming stars. Any successful theory of the formation of ETGs must explain that the bulk of their stars were formed at relatively high redshift and that these systems are now essentially quiescent (§5.1.3). Each of these old stars formed either *in situ* in the main progenitor of the galaxy where it is observed today, or in another galaxy (*ex situ*) and was later accreted by its present-day host galaxy. It is then important to distinguish between the SFH of a galaxy and its **stellar mass assembly history**. A galaxy can have an extremely peaked SFH (say, for instance, that all stars formed at $z > 2$), but a very prolonged stellar mass assembly history (it has kept accreting stars from dry merging, and thus increasing its stellar mass, until $z \approx 0$, without forming new stars).

Though the theory of the formation of ETGs is not fully understood, a widely accepted framework is a hybrid two-phase formation model, in which a first rapid phase of *in situ* star formation is followed by a prolonged 'hierarchical' assembly of stellar (and dark matter) mass (§10.3.3). We have seen in §5.1.3 and §5.4 that several properties of present-day ETGs vary gradually and systematically with the galaxy stellar mass (or velocity dispersion). For instance, the SFH of the most massive ellipticals (slow rotators; §5.1.2) is strongly peaked: the bulk of their stars were formed about 10 Gyr ago, over a short time span. Lower-mass ETGs (fast rotators; §5.1.2) tend to have slightly less old stellar populations and more extended SFHs (Figs. 5.9 and 11.39). In §10.8.1 and §10.8.2 we describe separately the formation history of prototypical members of these two families of ETGs: a high-mass slow rotator with very old stellar populations and a lower-mass fast rotator with not so old stellar populations. Galaxies with intermediate properties can have formation and assembly histories which combine those of these two extreme examples.

### 10.8.1 Slow Rotators

The prototypes of slow rotators (§5.1.2) are the most massive ellipticals, which are now hosted by massive ($\mathcal{M}_{\rm vir} \gtrsim 10^{13}\mathcal{M}_\odot$) dark matter halos, being often central galaxies in groups and clusters of galaxies. Tracing back the accretion history of their host halos in the standard cosmological model (§7.4.4), we find that the $z \gtrsim 2$ progenitors of these halos had masses $\mathcal{M}_{\rm vir} \gtrsim 10^{12}\mathcal{M}_\odot$ (Fig. 7.6, left). Though these masses are above the shock heating mass $\mathcal{M}_{\rm sh}$ (§8.2.4), at these redshifts cosmic cold filamentary streams of gas should be able to penetrate through the hot medium and reach the halo centre (Fig. 8.6). These cold filaments are believed to give rise to the first phase of formation, which is then followed by a second phase of growth via accretion at lower redshift (§10.3.3).

#### First Phase of Formation

The high-redshift ($z \gtrsim 2$) phase of formation of a massive ETG is characterised by very intense star formation and QSO activity, leading to the build-up of a spheroidal, compact bulge-like component, which is the progenitor of the central stellar distribution of the present-day massive elliptical. The details of the formation of this stellar spheroid at high redshift are unclear (§10.3.3). An infalling cold stream forms a gaseous disc, which can turn itself into a bulge via violent disc instability. Alternatively, this disc can merge with another disc and form a spheroid via highly dissipative mergers of disc galaxies (§10.3.2), with substantial associated star formation: provided these wet mergers occur at sufficiently high redshift, they should succeed in forming a compact spheroid with roughly coeval stellar populations. Moreover, in order to produce systems as metal-rich as observed, the process responsible for the formation of spheroids at high redshift must be able to pollute the star-forming medium with heavy elements and to lose little gas.

Independent of the specific mechanism involved (violent disc instability or wet merging), the *in situ* star formation responsible for the early spheroid formation, being characterised by a starburst phase, is not believed to proceed undisturbed. Observations of starburst galaxies and theoretical models provide evidence that starbursts are associated with strong feedback from the newly formed stars (radiative and mechanical feedback from SNe and massive stars; §8.7.1 and §8.7.2). As the halo mass is above the critical mass $\mathcal{M}_{\rm crit}$ (§10.5), the gas heated by SNe is retained and a hot atmosphere builds up.

Moreover, the detection of AGNs in some starburst galaxies suggests that quasar mode feedback from the central SMBH (fed by the available cold gas; §8.8.1) is also important in this early phase of massive ETG formation, and can co-operate with stellar feedback in regulating star formation. Feedback-driven quenching mechanisms (§10.6.3) can halt star formation in this phase, but it is also possible that positive AGN feedback (§8.8) triggers further star formation. It is worth noting that, as long as cold gas flows are at work, a protospheroid can also be transformed into a disc galaxy, as a consequence of cold accretion of high angular momentum gas (see also §10.7.1). Such a disc could again be destroyed via violent disc instability or merging.

## Second Phase of Formation

The second phase of formation of a massive ETG is believed to start around $z \approx 2$, when mass quenching, due to the accumulation of hot gas in the deep potential well of the halo, makes cold accretion inefficient (§10.6.1). As most cold gas of the satellites is eliminated by mass quenching processes, in this second phase of elliptical formation most mergers are effectively dry (with negligible associated star formation). Therefore, the stellar component of a massive elliptical grows at $z \lesssim 2$ mainly via assembly, through merging, of stars formed *ex situ* (Fig. 10.17, left). The overall effect of this growth driven by dry merging is to increase the stellar mass of the galaxy, without the formation of new stars. If, as expected on the basis of cosmological $N$-body simulations (Fig. 7.5, right), a massive elliptical grows substantially via several mergers, the galaxy will eventually have relatively low angular momentum (§5.1.2), because, on average, the orbital and intrinsic angular momenta of satellites accreted from different directions cancel each other out. While dry minor mergers tend to deposit mass and angular momentum in the outskirts (§8.9.6), a major merger, possibly with some gas, could leave an imprint in the central regions,

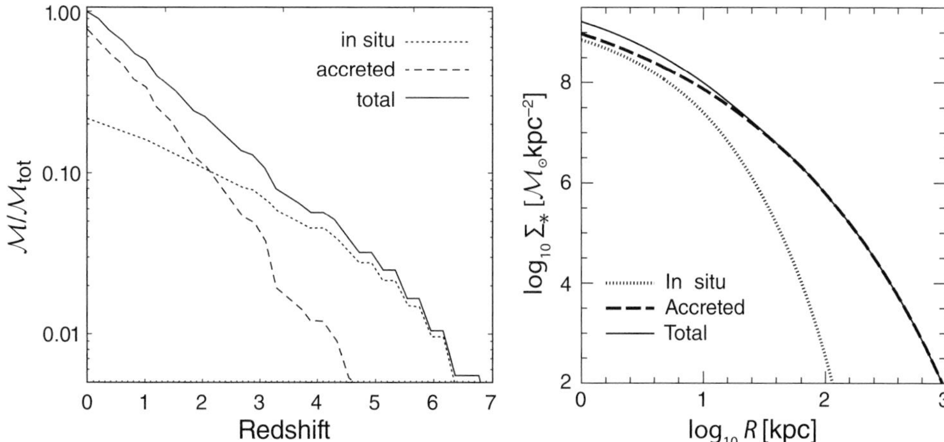


Fig. 10.17

*Left panel.* Average fractions of the present-day stellar mass formed *in situ* (dotted curve) and *ex situ* (accreted; dashed curve) as functions of redshift for massive ETGs, with virial masses $12.7 \lesssim \log(\mathcal{M}_{200}/\mathcal{M}_\odot) \lesssim 13.4$ at $z = 0$, in a hydrodynamic cosmological simulation (§10.11.1). The normalised average total stellar mass growth is represented by the solid curve. Adapted from Oser et al. (2010). © AAS, reproduced with permission. *Right panel.* Average contributions of *in situ* (dotted curve) and accreted (dashed curve) stars to the average stellar mass surface density profile (solid curve) of massive ETGs at $z = 0$, with virial masses $13.3 \lesssim \log(\mathcal{M}_{200}/\mathcal{M}_\odot) \lesssim 14.2$, according to a semi-analytic model (§10.11.2) applied to a dark matter-only cosmological simulation (§7.5.1). Data from Cooper et al. (2013).

producing a kinematically distinct core. This is in agreement with observations (§5.1.2): the absence of substantial rotation support is, by definition, a fundamental feature of slow rotators, some of which also have kinematically distinct cores. Provided the stars accreted in the dry mergers are relatively old, the resulting galaxy will have old stellar populations, consistent with observations (§5.1.3).

We have seen in §8.9.6 that the effect of dry merging is to make galaxies less compact. In dry mergers, the increase of galaxy size, as measured for instance by the effective radius $R_{\rm e}$, with stellar mass $\mathcal{M}_\star$ can be approximately described with a power law $R_{\rm e} \propto \mathcal{M}_\star^a$, with $a \geq 1$, while the stellar velocity dispersion $\sigma_0$ remains almost constant, or even slightly decreases, for increasing $\mathcal{M}_\star$. This can be inferred from eqs. (8.180) and (8.181), under the assumption that $\mathcal{M}_\star$, $R_{\rm e}$ and $\sigma_0$ scale with the virial quantities $\mathcal{M}$, $r$ and $\sigma$, respectively. Thus, the dry merging-driven growth of massive ETGs can, at least partly,[18] explain the fact that high-redshift ($z \gtrsim 2$) massive quiescent galaxies are much more compact than present-day massive ETGs (§10.10.3 and §11.3.1). Minor dry merging is particularly effective in making galaxies less compact (§8.9.6), because most of the stars in a small accreted satellite are less bound than the stars of the main galaxy, so they are easily stripped and contribute mainly to build the outer, extended stellar envelope of the remnant. It follows that the stars formed *in situ* and *ex situ* will be spatially

[18] Part of the observed size evolution of the population of ETGs can also be due to the so-called progenitor bias (§11.2.2).

segregated in the present-day descendant galaxy: this two-phase model of ETG formation predicts that the stars formed *in situ* occupy mainly the central parts of the galaxies (Fig. 10.17, right).

Extreme objects in which the growth via dry mergers is especially important are the BCGs (§6.4.1), the central galaxies in galaxy clusters, which, thanks to galactic cannibalism (§6.4.1) driven by dynamical friction (§8.9.3), manage to reach the highest stellar masses, up to $\mathcal{M}_\star \sim 10^{12}\mathcal{M}_\odot$. Merging is also invoked as the most popular explanation for the origin of the cored surface brightness profiles of slow rotators (§5.1.1; Fig. 5.2; Tab. 5.2). Following a binary merger, the SMBHs are expected to form a binary system in the galactic nucleus of the remnant. This SMBH binary shrinks because of the interaction with the stars of the galactic nucleus (§8.9.7). The energy lost by the binary is transferred to the stars of the nucleus, which then become less bound, leave the very central parts of the galaxy, eventually transforming an originally coreless profile into a cored profile (this process is called **core scouring**). The resulting core is a permanent feature of massive slow rotators, because, as the star formation is quenched by the mechanisms described above, it cannot be refilled by central starbursts.

In this second ($z \lesssim 2$) phase of massive ETG formation, the effects of stellar feedback are still important, though less dramatic than at higher redshift, because in these later stages there is little ongoing star formation. Longer-term phenomena such as explosion of Type Ia SNe (§4.6 and §8.7.1) and stellar mass loss (§8.4.1) contribute significantly to the continuous build-up and heating of the hot gaseous halo. However, ETGs produced in cosmological simulations including stellar feedback, but not AGN feedback, have by far too high stellar masses. This indicates that the quenching mechanisms included in these simulations are not sufficient and that AGN feedback is necessary to keep present-day massive ETGs red and dead, and limit their stellar mass. While quasar mode AGN feedback (§8.8.1) is dominant in the first phase of formation, at lower redshift also the mechanical input from radio mode AGN feedback (§8.8.2) is believed to be an important quenching mechanism, especially in the most massive ellipticals.

### 10.8.2 Fast Rotators

Fast rotators (§5.1.2) are relatively low-mass (say $\mathcal{M}_\star \lesssim 10^{11}\mathcal{M}_\odot$) ETGs (both ellipticals and S0s), characterised by non-negligible rotation support and stellar populations younger than those of slow rotators. Let us focus here on the formation and assembly history of an ETG that lives today in a halo with $\mathcal{M}_{\rm vir} \lesssim 10^{12}\mathcal{M}_\odot$. The $z \gtrsim 2$ progenitor of such a halo has mass lower than $\mathcal{M}_{\rm sh}$ (Fig. 7.6, left, and Fig. 8.6), so all the infalling gas is immediately available for star formation. This *first phase* of formation of the bulk of the stellar component of a fast rotator is in many respects similar to that of a more massive slow rotator (§10.8.1) and, especially in the case of S0s, shares also properties with the high-$z$ phase of the formation of early spirals (§10.7). Cold accretion (§8.2.4) is expected to form a protobulge via violent disc instability or disc-galaxy merging (§10.3). The main difference with respect to the assembly history of slow rotators is that, as long as $\mathcal{M}_{\rm vir} \lesssim \mathcal{M}_{\rm crit}$ (§10.5), the SN-heated gas is not effectively retained. This has important implications for the *second phase* of formation ($z \lesssim 2$): gas accretion can keep occurring in the cold mode, mass quenching is not at work and some accreted satellite galaxies can manage to bring cold gas down to the central regions. In other words, in these lower-mass ETGs *in situ* star formation can be non-negligible also in the second formation phase, consistent with the fact that the present-day stellar populations are not as old as those of more massive ETGs (§5.1.3). It is possible that, even at relatively low redshift, a new disc is formed as a consequence of fresh gas accretion (§10.7.2), but, in order for the galaxy to be an ETG at $z = 0$, any such disc must neither be dominant nor keep forming stars for long.

The results of cosmological and binary merging simulations (§10.11.1) suggest that different pathways can lead to the formation of fast rotators. End products with properties comparable to present-day fast rotators are obtained with late wet major mergers that result in a spin-up of the system, but, especially for S0s, also in assembly histories with no recent major mergers and relatively few minor mergers (similar to disc galaxies; §10.7.1). The core scouring mechanism described in §10.8.1 can be at work also during the assembly of fast rotators, but, given that in these lower-mass galaxies star formation is not completely inhibited at low redshift, small central starbursts can easily refill any central core in the stellar distribution, consistent with the fact that present-day fast rotators tend to have coreless surface brightness profiles (§5.1.1 and §5.1.2).

While the most massive galaxies (stellar mass $\mathcal{M}_\star > 10^{11}\mathcal{M}_\odot$) are invariably ETGs (slow rotators), at lower stellar mass (and correspondingly lower halo mass) star-forming and early-type (fast rotators) galaxies coexist. A natural question to ask is why some of these lower-mass halos host spiral galaxies and others host ETGs (see also §10.10.2). An important factor is the detailed merging history: for instance, a necessary condition for a galaxy to have an extended disc today is not to have experienced a late major merger (§10.7.1). Another fundamental ingredient is environment: at fixed stellar mass, due to environmental quenching (§10.6.2), a galaxy in a cluster of galaxies is more likely to be passive than a galaxy in the field. Environmental quenching is the key factor in the formation of S0s (§5.1.1), which are fast rotators that have many properties in common with disc galaxies, but have little cold gas and do not form stars significantly. If a star-forming disc galaxy happens to become a satellite of a galaxy cluster, it stops forming stars, because of environmental quenching, and becomes a quiescent S0. Environmental quenching is thus at the origin of the morphology–density relation (§6.5.1).

Above, for simplicity, we have described the formation and assembly histories of extreme examples of a very massive slow rotator and a lower-mass fast rotator. In fact, these two families of objects are not neatly separated, and there are intermediate ETGs which, for some of their properties, cannot be completely identified with either of these two cases. The detailed formation and assembly history of each ETG could be the combination of different aspects of the above two scenarios. However, overall we expect that the two models bracket the possible formation histories of ETGs. This variety of formation modes is believed to be at the basis of the observed trends of ETG properties with stellar mass or velocity dispersion $\sigma_0$ (§5.1.3 and §5.4). In particular, the very existence of the metallicity–

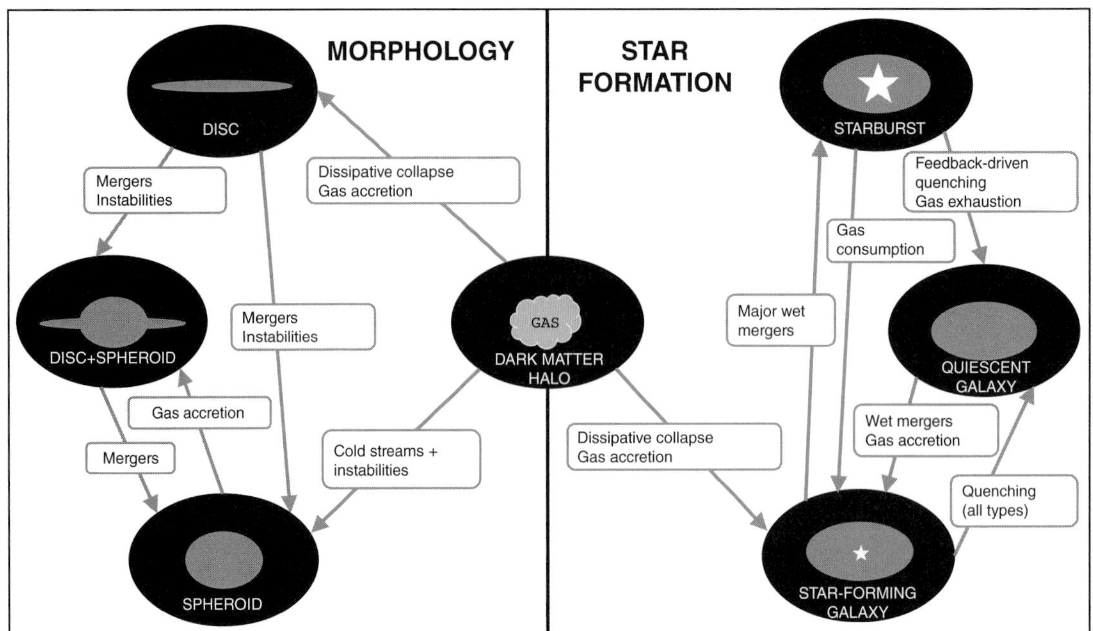


Fig. 10.18 A schematic representation of the main processes and transformations occurring during the formation and assembly histories of galaxies. *Left panel.* Morphological transformations. Gas infalling into a dark matter halo can form either a stellar disc, via dissipative collapse (§10.1.2) or gas accretion (§10.7.2), or a stellar spheroid, via cold streams and violent disc instability (§10.3.3). Galaxy mergers can transform discs into spheroids (§10.3.2 and §10.3.3), thus producing spheroid-dominated galaxies. Also instabilities, such as the buckling instability (§10.2.2), can form spheroidal galaxy components. It is also possible that gas accretion onto a spheroid leads to the formation of a stellar disc (§10.7.2 and §10.8.2). *Right panel.* Transformations in terms of star formation activity. Star-forming galaxies (§10.7) can be fed by dissipative collapse (§10.1.2) or gas accretion (§10.7.2). Starbursts are produced by major mergers of gas-rich galaxies (§10.3.2, §10.3.3 and §10.8.1). Galaxies can become quiescent (§10.8) via mass (§10.6.1), environmental (§10.6.2) or feedback-driven (§10.6.3) quenching. In some cases, probably infrequent, accretion of fresh gas (or of a gas-rich satellite) by a quiescent galaxy can rejuvenate it by forming new stars (§10.10.2). A starburst can become a normal SFG when it starts consuming its gas reservoir, but also directly a quiescent galaxy if the gas is rapidly exhausted as a consequence of stellar and AGN feedback (§10.6.3).

$\sigma_0$ relation of ETGs (Fig. 5.9, middle panel) indicates that most of the (metal-rich) stars of massive (high-$\sigma_0$) ETGs formed in relatively massive progenitors, while the trend between $\alpha$-enhancement and $\sigma_0$ (Fig. 5.9, top panel) implies that star formation was more prolonged in less massive ETGs (see also Fig. 11.39).

Before moving to the question of the formation of dwarf galaxies, the reader can find in Fig. 10.18, as an extremely synthetic overview of §10.1–§10.8, a scheme that summarises the main processes and transformations, in terms of morphology (left panel) and star formation (right panel), occurring during the formation and assembly histories of luminous (non-dwarf) galaxies.

## 10.9 Formation of Dwarf Galaxies

Here we address the question of the formation of dwarf galaxies, broadly defined as galaxies at least one order of magnitude less luminous than the Milky Way (§3.1).

### 10.9.1 Origin of the Different Types of Dwarfs

The family of dwarf galaxies includes very different classes of galaxies: both SFGs, such as the dIrrs (Tab. 4.1) and the BCDs (§4.5), and passive galaxies such as the dEs (Tab. 5.1), the dSphs (§6.3.1) and the UFDs (§6.3.2). On the one hand, BCDs and dIrrs are believed to be different manifestations of the same family of galaxies. If a dIrr, during its lifetime, undergoes an episode of starburst, it will temporarily appear as a BCD with high luminosity and SFR. On the other hand, dEs (compact and relatively luminous) and dSphs (diffuse and faint) are intrinsically different stellar systems. The origin of dEs is not well understood and still a matter of debate. The fact that they are preferentially found in clusters of galaxies (though with remarkable exceptions, such as M 32 in the Local Group; §6.3.1) suggests that environmental processes, such as tidal and ram-pressure stripping (§8.9.4), and galaxy harassment (§8.9), are important in their evolution. However, the nature of their severely stripped progenitor galaxies is unclear: both ellipticals and spirals have been proposed as candidates. Moreover, the very existence of M 32 implies that even less extreme environmental conditions, such as being a satellite of a massive spiral, are effective in shaping dEs.

The formation history of dIrrs, dSphs and UFDs is believed to be distinct from that of more luminous (non-dwarf) galaxies, essentially due to the fact that their dark halos have very low mass. The virial masses of the observed present-day dIrrs, which can be inferred directly from the analysis of their rotation curves (§4.3.3), are typically lower than $10^{11}\mathcal{M}_{\odot}$ (Tab. 4.1). The known dSphs and UFDs are satellite galaxies (most of them belong to the Local Group; §6.3): as a consequence, their dark halos must have been severely truncated by tidal interactions with their host galaxies. However, the observationally inferred properties of their dark halos are such that, if not truncated, they would have virial masses $\lesssim 10^{10}\mathcal{M}_{\odot}$, similar to those of dIrrs. The main progenitors (§7.4.4) of such halos just before the epoch of reionisation ($z \gtrsim 10$; §2.4 and §9.9) are the so-called atomic cooling halos (with $\mathcal{M}_{\rm vir} \sim 10^{8}\mathcal{M}_{\odot}$ and $T_{\rm vir} \sim 10^{4}$ K; §9.6), candidate hosts of the first galaxies. As discussed in §10.5, due to the presence of the UVB (§8.1.2), star formation is expected to be inhibited in dwarf-size ($\mathcal{M}_{\rm vir} < \mathcal{M}_{\rm min}$) halos from reionisation to $z \approx 2$. Some of the present-day passive dwarf galaxies could then be **fossils**[19] of these first galaxies, formed before reionisation. While this is possibly the case for the UFDs, which have very old stellar populations, the more luminous dSphs are not fossils because they have more extended SFHs. In fact, though theoretically there is general consensus that reionisation has an important role in the formation of dwarf

[19] Fossil galaxies are systems that completed their formation in the early Universe, but are observed in the present-day Universe.

galaxies, the observational evidence for a signature of reionisation in the SFHs of dwarfs is lacking (see bottom left panel of Fig. 6.6, showing that the SFH of the Cetus dSph peaks at $2 \lesssim z \lesssim 5$).

dIrr, dSphs and UFDs are thus believed to have in common a significant part of their formation and evolution history. The main distinction is that, while present-day dIrrs are still forming stars, at some point in the life of dSphs and UFDs some process must have quenched star formation and removed their gas. The responsible mechanism is believed to be a form of environmental quenching (§10.6.2). If the dwarf becomes a satellite of a massive galaxy (such as the Milky Way or M 31, in the Local Group), the combination of tidal (due to the host's gravitational field) and ram-pressure (due to the host's gaseous corona) stripping (§8.9.4) is expected to remove essentially all the gas from the dwarf, thus shutting off star formation.

In conclusion, the most popular explanation for the origin of dSphs is that they once were dIrrs later transformed by the interaction with the host galaxy, consistent with the observational finding that, in the Local Group, gas-rich satellites are predominantly found at larger distances from the massive galaxies (§6.3.2; Fig. 6.4). One potential problem with this scenario is that, in some cases, the stellar components of dIrrs are more rotation-supported than typical dSphs, which are mainly pressure-supported systems. This should not be an issue for the lowest-mass passive dwarfs, because the lowest-mass dIrrs, which are their candidate progenitors, have relatively low $V/\sigma$ ratios (§4.3.4 and §6.3.2). For higher-mass dSphs, it has been proposed that in the interaction of the dwarf with the host galaxy part of the ordered kinetic energy of the stars is transformed by tidal forces into random kinetic energy: this process is named **tidal stirring**. Alternatively, it is also possible that dwarf–dwarf galaxy merging (§8.9) has a role in the late formation stage of dSphs, effectively destroying coherent internal rotation of the stars.

### 10.9.2 Dark Matter on the Scale of Dwarfs

As dIrrs and dSphs are dark matter-dominated even in their central regions, they are ideal systems to study the density distribution of dark halos. In at least some of these systems, the observational data (§7.1) are not easily reconciled with the presence of a central cusp in the dark matter distribution. We have seen in §7.5.2 that cuspy halos are found in dark matter-only ΛCDM cosmological $N$-body simulations. Though the absence of dark matter cusps in observed dwarf galaxies is sometimes interpreted as a failure of the ΛCDM model, it must be stressed that the ΛCDM cosmology does not necessarily imply that dwarfs have cuspy dark matter profiles. In fact, it is widely believed that baryon physics can turn dark matter cusps into cores. Even if some dwarf galaxies are now locally dark matter-dominated down to their centre, this could not be the case in the past, because a necessary condition for star formation is that the self-gravity of the gas is important (§8.3.2). A self-gravitating gaseous disc is expected to form gas clumps via local gravitational instability (§8.3.3 and §10.3.3; Fig. 10.8). Before forming stars, these clumps can transfer kinetic energy to the central parts of the dark halo via dynamical friction (§8.9.3), thus flattening any original cusp. Moreover, strong feedback from episodes of intense star formation can expel gas from the dwarf protogalaxy, thus producing a sudden modification of the gravitational potential of the system and further flattening the central dark matter profile.

Another potential challenge posed to the standard ΛCDM model on the scales of dwarfs is the so-called **missing satellite problem**: the number of dwarfs observed in the Local Group is much smaller than the predicted number of dwarf-size dark subhalos. A possible solution, within the ΛCDM paradigm, is to assume that on the scales of dwarfs the efficiency of star formation drops dramatically for decreasing halo mass. In this picture, many low-mass ($\lesssim 10^9 \mathcal{M}_\odot$) dark halos did not form stars at all and could have survived as completely dark substructures of their host galaxy. It is still debated whether such a solution is quantitatively consistent with the properties of the observed dwarfs. In particular, further observational constraints on the low-mass end ($\mathcal{M}_\star \lesssim 10^6 \mathcal{M}_\odot$) of the SMF appear necessary to assess the actual importance of the missing satellite problem. We refer the reader to Bullock and Boylan-Kolchin (2017) for a general review of the potential problems of the ΛCDM paradigm on small scales.

## 10.10 Origin of the Demographics of Present-Day Galaxies

In the above sections we have focused on the theoretical models for the formation and evolution of individual galaxies, differing in mass, morphological properties and environment. Here we show how, combining these models, it is possible to explain theoretically the main features of the observed demographics of present-day galaxies. In other words, we consider globally the origin of the properties of present-day galaxies as a population of objects.

### 10.10.1 Stellar Mass Function and Stellar-to-Halo Mass Relation

As we have seen in §3.5, fundamental quantities that characterise the present-day galaxy population are the distribution functions of galaxy properties, such as the LF, the SMF and the baryonic mass function. Here we focus on the theoretical interpretation of the SMF, but analogous considerations hold for the LF and the baryonic mass function.

In Fig. 10.19 (left) we compare the observationally inferred SMF of present-day galaxies (for all morphological types; Fig. 3.11) with the HMF predicted at $z = 0$ in the standard cosmological model (Fig. 7.4). Let us consider for simplicity a model in which each halo hosts one and only one galaxy. In such a model the typical halo mass $\mathcal{M}_{\rm vir}$ of a galaxy of given stellar mass $\mathcal{M}_\star$ can be inferred by *matching* the galaxy SMF and the HMF, that is, finding the halo mass $\mathcal{M}_{\rm vir}$ such that the number density of the HMF is the same as the number density of the SMF at mass $\mathcal{M}_\star$. For example, according to the galaxy SMF and HMF shown in the left panel of Fig. 10.19, galaxies with stellar mass $\log(\mathcal{M}_\star/\mathcal{M}_\odot) = 11.5$ would be associated with dark halos with $\log(\mathcal{M}_{\rm vir}/\mathcal{M}_\odot) \simeq 13.3$, because they have the same number density $\log({\rm d}n/{\rm d}\mathcal{M}) \simeq -16.5$ (in units of ${\rm Mpc}^{-3}\mathcal{M}_\odot^{-1}$). In fact, satellite galaxies are hosted by subhalos, thus more realistic models take into account not only the HMF, but also the subhalo mass function (§7.5.3). This technique is called **halo abundance matching** or **subhalo abundance matching**.[20]

As expected, at any given mass $\mathcal{M}$ there are, per unit volume, many more halos with $\mathcal{M}_{\rm vir} = \mathcal{M}$ than galaxies with $\mathcal{M}_\star = \mathcal{M}$, or, vice versa, at given number density the halo mass is much higher than the stellar mass. This is primarily due to the fact that in the Universe there is much more dark matter than baryonic matter (the cosmic baryon fraction is $f_{\rm b} \simeq 0.16$; §2.5). It is then useful to rescale the HMF by multiplying the halo masses by $f_{\rm b}$. If each halo had converted its cosmic share of baryons into stars, the SMF would coincide with this rescaled HMF (plotted in the left panel of Fig. 10.19). In fact, there are a few apparent differences between the SMF and the rescaled HMF. (1) At all masses the SMF lies below the rescaled HMF, which means that star formation is never 100% efficient. (2) The knee of the rescaled HMF occurs at masses $\gtrsim 10^{13}\mathcal{M}_\odot$, at least two orders of magnitude higher than the knee of the SMF: in other words there is a dearth of galaxies with high stellar mass. (3) At the low-mass end the slope of the SMF is shallower than that of the HMF, reflecting the fact that there is a dearth of galaxies with low stellar mass.

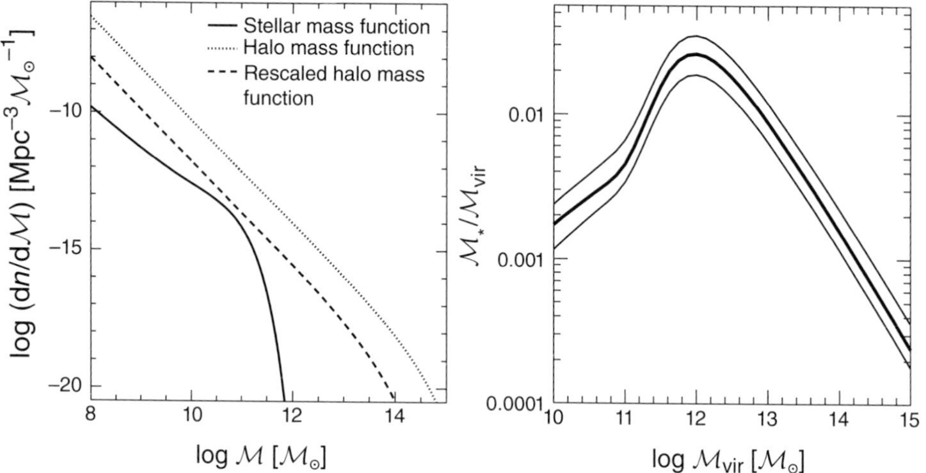


**Fig. 10.19** *Left panel.* Comparison between the observed SMF of galaxies and the predicted mass function of dark matter halos. The solid curve (corresponding to the solid curve in Fig. 3.11) represents the best fit to the observed SMF of present-day galaxies (all morphological types), as found by Kelvin et al. (2014a). The dotted curve (corresponding to the solid curve in Fig. 7.4; here we adopt $h = 0.7$) is the $z = 0$ HMF as computed by Angrick and Bartelmann (2010), extrapolated to lower masses as a power law. The dashed curve is the rescaled $z = 0$ HMF, obtained by multiplying the halo masses by the cosmic baryon fraction $f_{\rm b} = 0.16$ (§2.5). *Right panel.* Stellar-to-halo mass relation at redshift $z = 0.1$ assuming a Chabrier IMF (§8.3.9). The thick curve is the best fit and the thin curves indicate the $1\sigma$ scatter. Data from Behroozi et al. (2013).

[20] Subhalo abundance matching is not the only method used to link the distribution functions of galaxies and dark matter halos: widely used alternative tools are the so-called halo occupation distribution and conditional luminosity function (see Mo et al., 2010, and Wechsler and Tinker, 2018, for descriptions).

The **stellar-to-halo mass relation** (SHMR) of galaxies, that is the ratio $f_{m,\star} = \mathcal{M}_\star/\mathcal{M}_{vir}$ (eq. 10.2) as a function of $\mathcal{M}_{vir}$, can be quantitatively obtained with different methods, ranging from direct estimates of the halo mass of galaxies (for instance using weak gravitational lensing data; §5.3.4) to statistical methods (such as the abundance matching technique described above), which exploit the knowledge of the HMF for a given cosmological model. An estimate of the SHMR at $z = 0.1$ is shown in the right panel of Fig. 10.19. The peak of the stellar-to-dark mass ratio occurs in halos with virial mass $\approx 10^{12}\mathcal{M}_\odot$, of the order of the virial mass of the Milky Way. Note however that, even at the peak of the SHMR, $f_{m,\star} \ll f_b$: only a small fraction of the available gas is converted into stars (the ratio $f_{m,\star}/f_b$ can be seen as a measure of the star formation efficiency). More generally, as we have seen in §6.7, the total baryon budget of galaxies falls short of the cosmic fraction: a significant part of the cosmic baryons are in the intergalactic and circumgalactic media. As done for the SMF (§3.5.2), also the SHMR can be estimated separately for different galaxy types. The total SHMR, shown in Fig. 10.19 (right), is dominated by SFGs at low masses and ETGs at high masses.

Overall, it is evident from Fig. 10.19 that star formation is most efficient (or, better, least inefficient) in galaxies with stellar mass $\mathcal{M}_\star \approx \mathcal{M}^*$, where the horizontal offset between the SMF and the HMF is minimum ($\mathcal{M}^* \approx 5 \times 10^{10}\mathcal{M}_\odot$, about the stellar mass of the Milky Way, corresponds to the knee of the galaxy SMF; Fig. 3.11 and left panel of Fig. 10.19). Instead, star formation is extremely inefficient at both the high-mass and low-mass ends of the galaxy SMF.

In the most massive (group- and cluster-size) halos ($\mathcal{M}_{vir} > \mathcal{M}_{max}$; §10.5), the cooling time $t_{cool}$ of the virial-temperature gas can exceed the Hubble time $t_H$ (Fig. 8.5): this is the essential reason why there are not galaxies with stellar masses $10^{13}$–$10^{14}\mathcal{M}_\odot$. However, simple timescale arguments such as those illustrated in Fig. 8.5 are not sufficient to explain in detail the dearth of luminous galaxies apparent from Fig. 10.19. In fact, the condition $t_{dyn} < t_{cool} < t_H$ is a necessary, but not sufficient, condition for gas in a halo to form stars efficiently. When dark matter halos are massive enough ($\mathcal{M}_{vir} > \mathcal{M}_{crit}$; §10.5) that a hot atmosphere builds up, mass quenching (§10.6.1) and AGN feedback (§8.8) inhibit the formation of stars from cooling virial-temperature gas. Though the details of AGN feedback are not completely understood, the results of different hydrodynamic cosmological simulations (§10.11.1) and semi-analytic models (§10.11.2) suggest that AGN feedback is a key factor to reproduce, at least qualitatively, the shape of the high-mass end of the present-day galaxy SMF.

The reasons for the dearth of galaxies at the low-mass end of the SMF are quite different. In this case, at least for $T_{vir} \gtrsim 10^4$ K (corresponding to $\mathcal{M}_{vir} \gtrsim 2 \times 10^9\mathcal{M}_\odot$ at $z = 0$; eq. 8.20), the cooling time would be short enough to realise physical conditions favourable for star formation (Fig. 8.5). However, the formation of dwarf galaxies is strongly limited by the fact that, from cosmological reionisation to $z \approx 2$, halos with $\mathcal{M}_{vir} < \mathcal{M}_{min}$ (i.e. circular speeds lower than $\approx 35\,\mathrm{km\,s^{-1}}$) cannot effectively trap the cosmic gas, which is at temperatures comparable to or higher than their virial temperature (§10.5 and §10.9.1). This is a first reason why star formation is inefficient in dwarfs. Moreover, in all halos with $\mathcal{M}_{vir} \lesssim \mathcal{M}_{SN} \sim \mathcal{M}_{crit}$ (circular speed $v_{vir} \lesssim 100\,\mathrm{km\,s^{-1}}$), SNe are effective in removing gas (§10.5), so also in halos with $35 \lesssim v_{\rm vir} \lesssim 100\,{\rm km\,s^{-1}}$ the gas available for fuelling star formation falls short of the cosmic budget (though some of these galaxies have significant reservoirs of gas in their outer regions; §10.7.2). We recall that the Milky Way has a circular speed at the virial radius $v_{\rm vir} \approx 130–150\,{\rm km\,s^{-1}}$ (see Tab. 4.1 and eq. 7.28).

In summary, the theoretical explanation of the fact that the star formation efficiency peaks at $\mathcal{M}_{\rm vir} \sim 10^{12}\mathcal{M}_\odot$ (about the virial mass of the Milky Way) can be schematically synthesised as follows.

1. The formation of dwarf galaxies is hampered first by cosmological reionisation, which prevents early gas confinement and fragmentation, and then, when stars form, by stellar feedback.
2. The formation of galaxies less massive than the Milky Way is largely limited by feedback from SNe.
3. The formation of galaxies more massive than the Milky Way is regulated by their hot atmospheres, built up mainly by SN ejecta and kept hot by AGN feedback, which limit star formation from cold gas accretion.
4. In cluster-size halos, the cooling time of the gas exceeds the Hubble time throughout the system, with the exception of the central regions, where, however, AGN feedback manages to prevent substantial cooling.

### 10.10.2 Bimodality

The population of present-day galaxies is characterised by a bimodality (§3.4.1): on the one hand there is the family of red, passive, spheroid-dominated ETGs; on the other hand the family of blue, disc-dominated SFGs. The bimodality is most evident in the colour–stellar mass diagrams (Fig. 3.9), in which the two families are nicely segregated into the red sequence and the blue cloud. In this section we briefly discuss the theoretical origin of this bimodality.

In Fig. 10.20 the present-day red sequence and blue cloud are schematically represented in a colour–stellar mass diagram. Though this is a snapshot of the present-day population, surveys of galaxies at higher redshifts suggest that also in the past (at least back to $z \approx 2$) the galaxy population was characterised by a similar bimodality (§11.2.6). Therefore, provided rest-frame colours are considered, it makes sense to study evolutionary tracks of individual galaxies in the stellar mass–colour plane and compare them with the present-day galaxy distribution.

Though some red sequence galaxies are reddened by dust absorption (§3.4.1), most of them are red because they are passive. Galaxies in the blue cloud are instead star-forming: their blue colour is due to the presence of young stars. Clearly, the progenitors of present-day red sequence galaxies once were blue cloud galaxies (when they were forming stars). The question posed by the observed bimodality is thus how galaxies move from the blue cloud to the red sequence. There is little doubt that the key process is quenching of star formation (§10.6). In the diagram of Fig. 10.20 the evolution induced by quenching can be represented by almost vertical arrows. Let us consider two extreme cases. (1) If star formation stops abruptly, as a consequence, for instance, of environmental

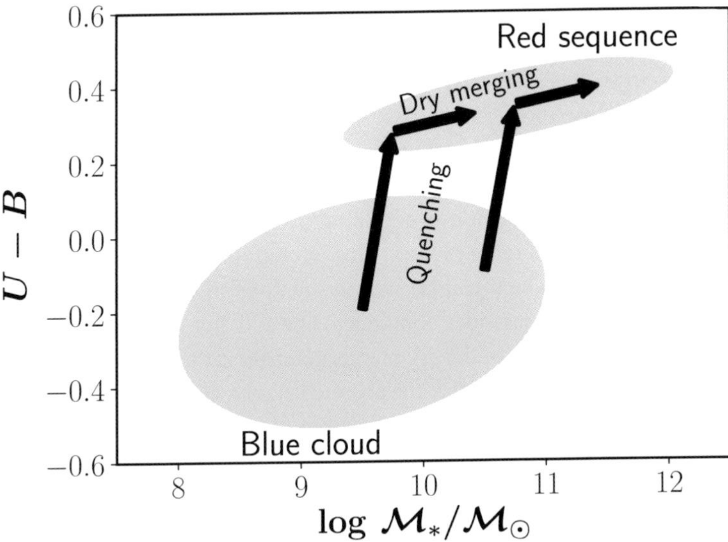


Fig. 10.20 Schematic illustration of the evolution of representative individual galaxies in the galaxy rest-frame colour–stellar mass diagram. The ellipses indicate the loci of the present-day observed red sequence (higher colour index $U - B$) and blue cloud (lower $U - B$). The arrows indicate possible evolutionary tracks composed of almost vertical quenching tracks and almost horizontal dry merging tracks. The truncation of the blue cloud at $\mathcal{M}_\star \sim 10^{11}\mathcal{M}_\odot$ is interpreted as a consequence of mass quenching (§10.6.1). Figure inspired by Faber et al. (2007).

quenching (§10.6.2), the arrow is vertical: the stellar mass remains essentially constant[21] and the galaxy colour gradually becomes redder because of passive evolution (the stars become older and then redder; §8.6.2; Fig. 8.15). As before quenching the galaxy is likely disc-dominated, the quenched galaxy will appear morphologically as an S0 when on the red sequence. (2) If a blue cloud galaxy is involved in a major merger, we expect star formation to occur during the merger. However, if feedback-driven quenching (§10.6.3) is effective, the final consequence of the merger can be to stop star formation. The overall effect of this process is represented, in the diagram in Fig. 10.20, by a diagonal, but almost vertical, arrow: the stellar mass increases, while the galaxy becomes red because of quenching. Discs are likely destroyed in major mergers, so in case 2 there is also a morphological transformation, consistent with the fact that red sequence galaxies tend to be bulge-dominated. In general, the transition from the blue cloud to the red sequence can have intermediate properties between cases 1 and 2.

In the observed stellar mass–colour plane, the area between the red sequence and the blue cloud, called the green valley (§3.4.1), is not empty. At least some of the green valley galaxies are believed to be objects transitioning from the blue cloud to the red sequence: the fact that the green valley is scarcely populated implies that the transition is a relatively quick process.

[21] Rigorously speaking, when a galaxy evolves passively, the stellar mass slowly diminishes because of stellar mass loss due to stellar evolution (§4.2.10 and §8.4.1).

When a galaxy has reached the red sequence, it can still evolve in the stellar mass–colour plane in two different ways. In principle it is possible that the galaxy is rejuvenated by accretion of fresh gas (either diffuse or belonging to a merging gas-rich galaxy) and subsequent star formation: in this case the galaxy could go back to the blue cloud. However, this is not so often the case, because some quenching mechanisms, such as mass or environmental quenching, are 'irreversible' and prevent further accretion of cold gas (§10.6). A more likely phenomenon is evolution of the galaxies within the red sequence as a consequence of dry mergers (§8.9.6). This evolution is represented in Fig. 10.20 by almost horizontal arrows, which can be slightly inclined towards the direction of redder colour because the stellar populations continue to age and then become redder. However, it must be stressed that the colour of the dry merger remnant is the combination of the colours of the original stellar population and of the accreted stars, which might influence the actual direction of the evolution in the stellar mass–colour plane.

An important feature of the distribution of present-day galaxies in the stellar mass–colour plane (Fig. 3.9) is that the blue cloud is truncated: essentially all the most massive ($\mathcal{M}_\star > 10^{11}\mathcal{M}_\odot$) galaxies are in the red sequence. This behaviour is nicely explained by mass quenching (§10.6.1): for halo masses $\mathcal{M}_{\rm vir} \gg \mathcal{M}_{\rm crit}$ (§10.5), independent of the galaxy environment, mass quenching inhibits star formation, eventually making these giant galaxies red and dead. At $\mathcal{M}_\star \sim 10^{10}\mathcal{M}_\odot$ a galaxy can belong to either the blue cloud or the red sequence, depending on its specific growth history and environment (§10.8.2).

### 10.10.3 Scaling Relations

An important test for any model aimed at describing the formation and evolution of galaxies is the comparison with the observed empirical scaling relations (§4.4, §5.1.3 and §5.4). These seemingly simple correlations (some of them are well approximated by power laws) are believed to be the outcome of the complex interplay of baryon physics and cosmological evolution of dark matter halos. For this reason, in most cases they cannot be explained by simple analytic models, and we can gain insight into the question of their origin with the aid of hydrodynamic cosmological simulations (§10.11.1) or sophisticated semi-analytic models (§10.11.2). At least some of the observed scaling relations of galaxies are expected to be linked to the properties of the host dark halos. Before considering separately SFGs and ETGs, it is then useful to recall some fundamental structural and kinematic scaling laws theoretically expected for dark matter halos.

#### Dark Matter Halos

The characteristic size of a dark matter halo of mass $\mathcal{M}_{\rm vir}$ is its virial radius $r_{\rm vir}$ (§7.3.2). By definition (eq. 7.25), $r_{\rm vir}$ and $\mathcal{M}_{\rm vir}$ are related by

$$r_{\rm vir} = A\mathcal{M}_{\rm vir}^{1/3}, \tag{10.14}$$

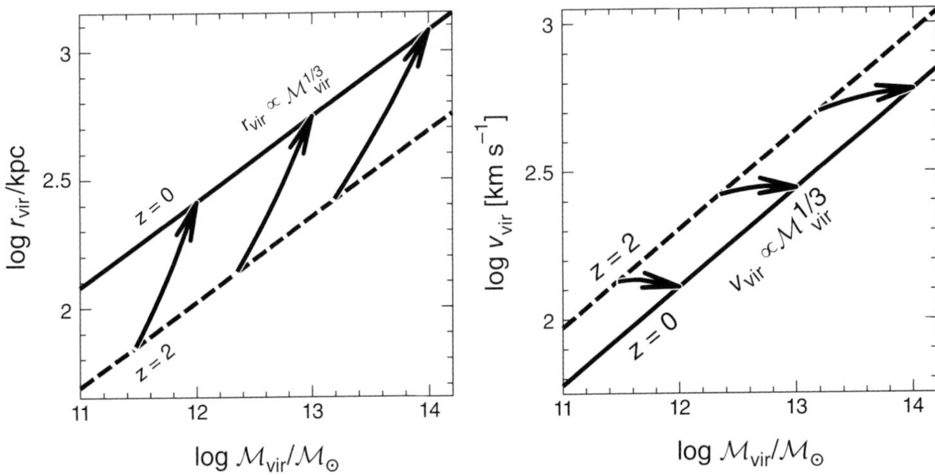


Fig. 10.21

*Left panel*. Virial radius $r_{vir}$ as a function of virial mass $\mathcal{M}_{vir}$ of dark matter halos at $z = 0$ (solid line) and $z = 2$ (dashed line). The arrows indicate the evolution of representative individual halos in the $\mathcal{M}_{vir}$–$r_{vir}$ plane from $z = 2$ to $z = 0$, calculated using the average mass growth histories estimated from cosmological *N*-body simulations by Fakhouri et al. (2010; see left panel of Fig. 7.6). *Right panel*. Same as left panel, but in the plane $\mathcal{M}_{vir}$–$v_{vir}$, where $v_{vir}$ is the halo circular speed at the virial radius (eq. 7.24).

where

$$A(z) = \left[\frac{4\pi}{3}\Delta_c(z)\rho_{crit}(z)\right]^{-1/3}. \tag{10.15}$$

Here $\Delta_c$ is the critical overdensity for virialisation (§7.3.2) and $\rho_{crit}$ is the critical density of the Universe (eq. 2.23). It follows that, at given $z$ (i.e. at fixed $A$), all halos lie on the correlation $r_{vir} \propto \mathcal{M}_{vir}^{1/3}$. At given $\mathcal{M}_{vir}$, $r_{vir}$ is smaller at higher redshift, because $A$ is a monotonically decreasing function of $z$ (both $\rho_{crit}$ and $\Delta_c$ increase with $z$; §7.3.2). This is visualised in the left panel of Fig. 10.21, showing $r_{vir}$ as a function of $\mathcal{M}_{vir}$ at $z = 0$ and $z = 2$. In the same diagram, the evolution of individual dark matter halos is indicated with arrows: individual halos move in the $\mathcal{M}_{vir}$–$r_{vir}$ plane approximately along power laws $r_{vir} \propto \mathcal{M}_{vir}^a$ with $a \approx 0.8$–$1.1$, much steeper than the $r_{vir} \propto \mathcal{M}_{vir}^{1/3}$ scaling relation at given $z$: the net effect is that, at fixed mass, halos have larger sizes at lower $z$.

Similar considerations can be done for $v_{vir}$, the circular speed at the virial radius. By definition (eq. 7.26),

$$v_{vir} = B\mathcal{M}_{vir}^{1/3}, \tag{10.16}$$

where $B(z) = \sqrt{G/A(z)}$ (with $A$ given by eq. 10.15) is a monotonically increasing function of redshift (we recall that, at given $z$, $v_{vir} \propto r_{vir}$; eq. 10.6). For instance, at both $z = 0$ and $z = 2$ the halos lie on the correlation $v_{vir} \propto \mathcal{M}_{vir}^{1/3}$, but the normalisation of this correlation is lower at $z = 0$ than at $z = 2$ (Fig. 10.21, right). Individual halos evolve in the $\mathcal{M}_{vir}$–$v_{vir}$ plane almost horizontally (arrows in the right panel of Fig. 10.21). Halos grow in mass at almost constant or slightly increasing circular speed, following a power law $v_{\rm vir} \propto \mathcal{M}_{\rm vir}^{b}$, with $0 \lesssim b \lesssim 0.1$, shallower than the $v_{\rm vir} \propto \mathcal{M}_{\rm vir}^{1/3}$ correlation at given redshift. The overall effect is that, at fixed mass, halos have lower circular speeds at lower $z$.

It is also useful to derive a correlation theoretically expected for the halo angular momentum. If we take eqs. (7.26) and (10.6), substitute them into the definition of the spin parameter $\lambda$ (eq. 10.4) and solve for the specific angular momentum, we obtain

$$j_{\rm DM} = \frac{\lambda}{[\Delta_{\rm c}(z) H^2(z)]^{1/6}} (2G\mathcal{M}_{\rm vir})^{2/3}, \tag{10.17}$$

which, given that $\lambda$ is essentially independent of $\mathcal{M}_{\rm vir}$ and $z$ (§7.5.3), is a power-law relation between the specific angular momentum and the mass of the halo, with redshift-dependent normalisation. At $z=0$

$$j_{\rm DM} \simeq 1.65 \times 10^3 \left(\frac{\lambda}{0.035}\right)\left(\frac{\mathcal{M}_{\rm vir}}{10^{12}\,\mathcal{M}_\odot}\right)^{2/3}\left(\frac{h}{0.7}\right)^{-1/3} \text{kpc km s}^{-1}. \tag{10.18}$$

## Star-Forming Galaxies

We now make a few intuitive considerations regarding the origin of some of the scaling relations of SFGs described in §4.4.

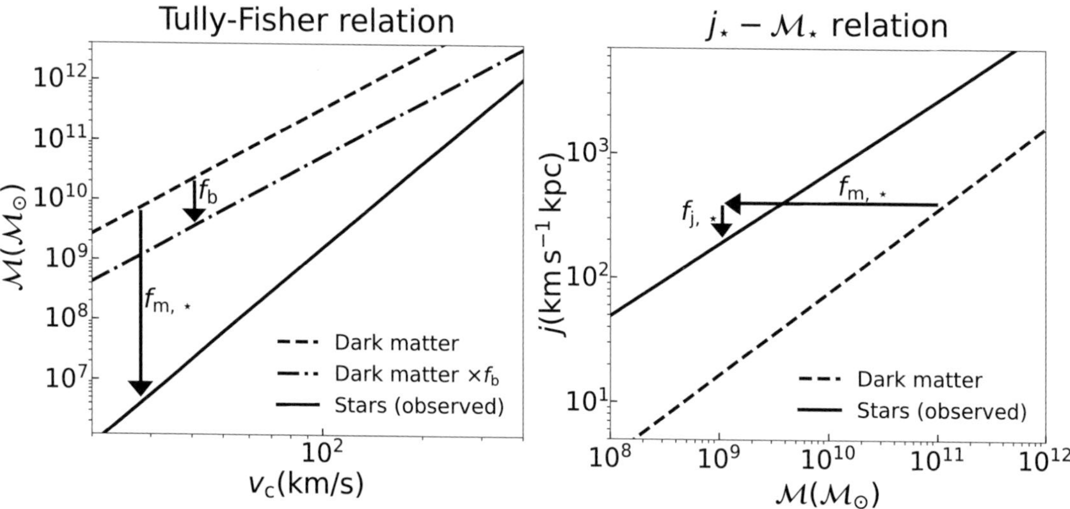


**Fig. 10.22** *Left panel.* Observed Tully–Fisher relation (solid line; same as in Fig. 11.41) between stellar mass and rotation velocity ($v_{\rm flat}$) for local disc galaxies compared with the relation between virial mass ($\mathcal{M}_{200}$) and circular speed ($v_{200}$) of dark matter halos (dashed line) and that expected if stars were at the cosmological fraction $f_{\rm b}$ and rotating at $v_{200}$ (dotted line). *Right panel.* Observed $j_\star - \mathcal{M}_\star$ relation (solid line) between the specific angular momentum of the stellar disc and the stellar mass in local disc galaxies compared to the $j_{\rm DM} - \mathcal{M}_{\rm vir}$ relation for dark matter halos (dashed line; eq. 10.18). In both plots, the arrows indicate the direction along which the various ratios described in the text operate.

*Tully–Fisher relation* (TFR; §4.4.1). Based on the above properties of the halos, we can ask ourselves what slope for a relation between mass and rotation velocity we would expect if the observed $v_{\rm rot}$ scaled with $v_{\rm vir}$. Assuming, for simplicity, that in the flat part of the rotation curve the baryonic matter rotates at $v_{\rm vir}$ ($f_{\rm v} = 1$; §10.1.1), we can use, as a proxy for the TFR, the relation between the circular speed at the virial radius and the virial mass $\mathcal{M}_{\rm vir} \propto v_{\rm vir}^3$ (eq. 10.16; dashed line in Fig. 10.22, left). If we further assume that each galaxy contains the cosmological baryon fraction ($f_{\rm b}$) we obtain the dotted line shown in Fig. 10.22 (left). Comparing this relation with the observed stellar mass TFR (solid line) we see that the two relations have different normalisations and slopes. The lower normalisation of the observed relation is not surprising, because the stellar mass contained in dark matter halos is lower than that expected from the baryon fraction (stellar-to-halo mass ratio typically $f_{\rm m,\star} \ll f_{\rm b}$; eq. 10.2). The slope of the observed relation is more difficult to understand. As we have seen in §10.10.1, $f_{\rm m,\star}$ is a *non-monotonic* function of halo mass (Fig. 10.19, right). As a consequence, a power law in Fig. 10.22 (left) for the mass–velocity relation of halos naturally turns into a curved function for the stellar mass–velocity relation. The fact that the observed TFR is instead well approximated by a single power law requires other mechanisms to take place. Zoom-in cosmological hydrodynamic simulations (§10.11.1) with very efficient stellar feedback reproduce reasonably well the global shape of the observed TFR although the theoretical relation tends to be less straight and to have a larger scatter than the observed one.

*$j_\star$–$M_\star$ relation* (§4.4.2). This is another relation of disc galaxies that has a cosmological relevance. We can use the definitions in eqs. (10.2) and (10.3) to write eq. (10.17) in terms of observable quantities and obtain

$$j_\star \simeq 76.7 f_{\rm j,\star} f_{\rm m,\star}^{-2/3} \left(\frac{\lambda}{0.035}\right) \left(\frac{\mathcal{M}_\star}{10^{10}\,\mathcal{M}_\odot}\right)^{2/3} \ \mathrm{kpc\,km\,s^{-1}}, \tag{10.19}$$

which is very reminiscent of the observed relation (eq. 4.41). The above seems to indicate that, as long as the quantity $f_{\rm j,\star} f_{\rm m,\star}^{-2/3}$ is nearly constant with stellar mass, the relation is a power law with roughly the right slope (we recall that $\lambda$ is essentially independent of halo mass; §7.5.4). The amount of angular momentum of the stellar discs is then determined by the spin produced by the tidal torques and by the value of this constant. Fig. 10.22 (right) shows the comparison between eq. (10.19) and the observed relation. In this case, $f_{\rm m,\star}$ brings galaxies towards the left, while, if for instance $f_{\rm j,\star} < 1$, this reduces their specific angular momentum and thus brings them down. The arrows in the plot indicate their effects. Because $f_{\rm m,\star}$ is a non-monotonic function of the halo mass the above simple calculation would imply that also $f_{\rm j,\star}$ must depend in a non-trivial way on the halo mass.

*Star formation main sequence* (SFMS; §4.4.4). The SFMS is a relation between the SFR and the stellar mass in SFGs. As mentioned in §4.4.3, the stellar mass of a galaxy is proportional to the integral of the past SFR, neglecting the role of mergers. If we assume that the delay of gas return is small (IRA; §8.5.3) we can write the stellar mass of a galaxy at redshift $z$ as

$$\mathcal{M}_\star(z) = (1-\mathcal{R}) \int_0^{t(z)} \mathrm{SFR}(t)\mathrm{d}t \approx (1-\mathcal{R})\langle\mathrm{SFR}\rangle t(z), \tag{10.20}$$

where $t(z)$ is the age of the Universe at $z$, $\mathcal{R}$ is the return fraction, $\langle\text{SFR}\rangle$ is the average past SFR (see eq. 4.45) and we are assuming that the star formation starts at the Big Bang ($t=0$), which is a reasonably good approximation for $t(z) \gtrsim$ a few Gyr. We use eq. (10.20) to rewrite the sSFR (eq. 4.44) as

$$\text{sSFR}(z) \equiv \frac{1}{t_*} \approx \frac{\text{SFR}(z)}{(1-\mathcal{R})\langle\text{SFR}\rangle t(z)}, \tag{10.21}$$

with $t_*$ a characteristic time. We can then also write the sSFR using the SFMS relation (eq. 4.43) and obtain

$$\text{sSFR} \propto \mathcal{M}_\star^{\kappa-1}, \tag{10.22}$$

which, for $\kappa \approx 0.8$ (§4.4.4), becomes a very mild function of the stellar mass and we can consider it constant for simplicity. At $z=0$, $t(z) \approx t_\text{H}$, we have $\text{sSFR} \approx 10^{-10}\,\text{yr}^{-1}$ (Fig. 4.18) and thus $t_* \approx t_\text{H}$. Eq. (10.21) then tells us that $\langle\text{SFR}\rangle \approx \text{SFR}(0)/(1-\mathcal{R})$, i.e. galaxies that are still in the main sequence at $z=0$ are forming stars at a rate which is comparable with their average past SFR. In other words, they have a slowly declining (or nearly flat) SFH (see also Fig. 10.11). Starburst galaxies have $\text{sSFR} \gtrsim 10^{-9}\,\text{yr}^{-1}$ and thus are forming stars today at much higher rate than in the past. The same reasoning applies at higher redshifts. At $z=1$ for instance, $t_* \approx 1.5\,\text{Gyr}$ for galaxies with $\mathcal{M}_\star \sim 10^{10}\,\mathcal{M}_\odot$ (Fig. 11.35), about a quarter of the age of the Universe at that redshift. Thus galaxies at $z=1$ are experiencing a peak of their stellar mass build-up.

The normalisation of the SFMS increases from $z=0$ to $z \approx 2$ (Fig. 11.35). This increase should not come as a surprise; in fact it would be expected even if the SFR did not increase with $z$. Consider the extreme case of a completely flat SFH for all galaxies. In this case, the total stellar mass would increase linearly with time starting from zero at $t \approx 0$ and reaching its maximum today. At $z \approx 1$–$2$ the stellar mass should then be $\approx 25$–$40\%$ of the present value and thus the normalisation of an SFMS with $\kappa=1$ (eq. 10.22) should be a factor of 2.5–4 larger (simply because $\mathcal{M}_\star$ is smaller). In reality, the observed normalisation increases by a factor of $\sim 10$ and this is due to the fact that, on average, galaxies have had their peak of star formation at $z=1$–$2$ (Fig. 11.33).

*Mass–metallicity relation*. In §4.4.3 we have presented two versions of this relation, one for stars and one for the ISM. The gas measurement gives us an 'instantaneous' (at the time of the observations) abundance of the galaxy, while the stellar metallicity gives a time-averaged value. Note that the timescale of this average depends on the lifetime of the stars that are used to determine the metallicity. Using bright young stars is nearly the same as using the gas. Both relations have similar shapes with an increase at low masses and a plateau (or change in slope) above $\mathcal{M}_\star \sim 10^{10}\,\mathcal{M}_\odot$ (Figs. 4.17 and 11.40).

The metal content in a galaxy is built up by generations of stars that enrich the ISM, raising the overall metallicity provided that the enrichment is not balanced by dilution of low-metallicity gas (§8.5). Thus, it is qualitatively expected that galaxies with higher masses, which also have deeper potential wells, can retain their metals more easily and hence enhance their metal content. The actual shape of the relation is however more challenging to explain. We can get some clues by considering the prediction of a closed-box model, with no exchange of material with the external medium (§8.5.4). We use eq. (8.115), but to do so we need a relation between gas mass and stellar mass. From the top left panel of Fig. 4.8 we can approximate this relation as

$$\mathcal{M}_{\rm gas} = A\mathcal{M}_\star^\alpha, \tag{10.23}$$

where $A \approx 10^6$, $\alpha \approx 0.35$ and $\mathcal{M}_{\rm gas}$ and $\mathcal{M}_\star$ are in solar masses. We then obtain

$$Z_i = p_i \ln\left[1 + \frac{1}{A}\left(\frac{\mathcal{M}_\star}{\mathcal{M}_\odot}\right)^{1-\alpha}\right], \tag{10.24}$$

where $Z_i$ and $p_i$ are, respectively, the abundance and stellar yield of a generic metal. Eq. (10.24) gives a somewhat steeper relation than the observed mass–metallicity relation and does not significantly flatten at the high masses. Thus, assuming a simple closed-box model we cannot reproduce the shape of the mass–metallicity relation at any mass range. This failure should not come as a surprise and we can see it as further evidence that a fundamental part of the evolution of SFGs is the exchange of gas with the ambient in terms of both outflows and inflows (§10.7).

## Early-Type Galaxies

Here we attempt to address the question of the origin of the empirical scaling relations of ETGs. For simplicity we limit ourselves to the correlations among stellar mass, velocity dispersion and size (size–stellar mass, velocity dispersion–stellar mass and stellar mass fundamental plane relations; §5.4.3), whose theoretical interpretation is more straightforward than correlations involving properties of the stellar populations, such as luminosity, colour and metallicity. We present the results of relatively simple models and compare them with observations of present-day ETGs.

The correlations shown in Fig. 10.21 for cosmological dark matter halos are *qualitatively* similar to the empirical size–stellar mass ($R_{\rm e}$–$\mathcal{M}_\star$) and velocity dispersion–stellar mass ($\sigma_0$–$\mathcal{M}_\star$) relations of observed $z = 0$ ETGs (§5.4.3). For both cosmological dark matter halos and the stellar components of ETGs, the size and the velocity dispersion (or the circular speed) increase with mass. However, *quantitatively*, the slopes of the correlations are different: the $R_{\rm e}$–$\mathcal{M}_\star$ relation is steeper than the $r_{\rm vir}$–$\mathcal{M}_{\rm vir}$ relation, while the $\sigma_0$–$\mathcal{M}_\star$ relation is shallower than the $v_{\rm vir}$–$\mathcal{M}_{\rm vir}$ relation. This comes as no surprise, because we have seen in §10.10.1 that $\mathcal{M}_\star$ does not scale linearly with the host halo mass. Moreover, while both theoretical (analogous to those presented in §10.1.1) and empirical (based on the abundance matching technique; §10.10.1) arguments suggest that $R_{\rm e} \propto r_{\rm vir}$, it is not necessarily the case that $\sigma_0$ scales linearly with $v_{\rm vir}$.

More quantitative theoretical explanations of the $R_{\rm e}$–$\mathcal{M}_\star$ and $\sigma_0$–$\mathcal{M}_\star$ relations can be obtained as follows. Take the $z = 0$ $r_{\rm vir}$–$\mathcal{M}_{\rm vir}$ relation shown in Fig. 10.21 and convert $\mathcal{M}_{\rm vir}$ into $\mathcal{M}_\star$ using the $z = 0$ SHMR (§10.10.1). Taking $R_{\rm e} \propto r_{\rm vir}$ with $R_{\rm e}/r_{\rm vir} \approx 0.01-0.02$, we get an $R_{\rm e}$–$\mathcal{M}_\star$ relation consistent with the one observed for present-day massive ($\mathcal{M}_\star \gtrsim 3 \times 10^{10}\mathcal{M}_\odot$) ETGs (Fig. 10.23, top). In a similar spirit, *assuming* that the central stellar velocity dispersion $\sigma_0$ scales as $\sigma_0^2 \propto \mathcal{M}_\star/R_{\rm e}$, we are able to reproduce reasonably well the slope of the $\sigma_0$–$\mathcal{M}_\star$ relation of present-day massive ETGs (Fig. 10.23, bottom). Quantitatively, taking for instance the results shown in Fig. 10.23, this simple model

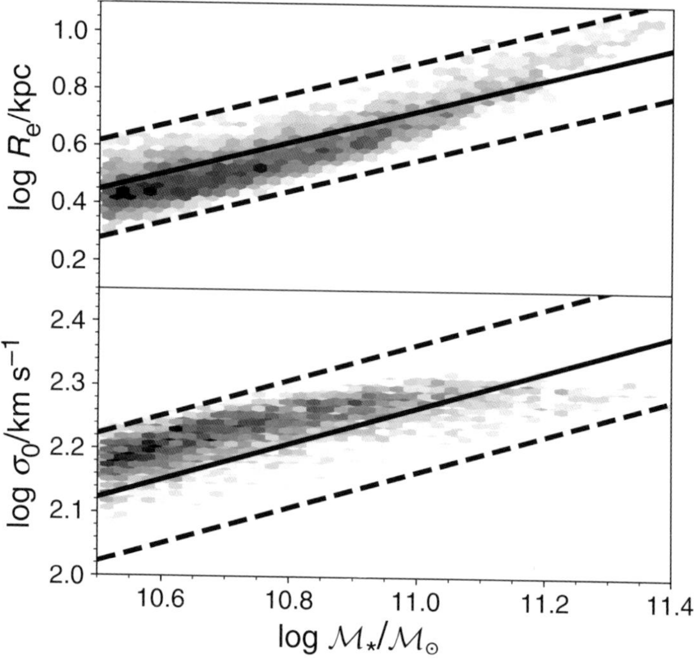


Fig. 10.23 *Top panel*. Effective radius $R_e$ as a function of the stellar mass $\mathcal{M}_\star$ of simulated ETGs at $z = 0$. The greyscale (from light to dark) is proportional to the logarithm of the number of ETGs. The simulated ETGs are the results of a model that combines a dark matter-only cosmological *N*-body simulation with an SHMR (§10.10.1) computed by extrapolating to $z = 0$ the SHMR of Leauthaud et al. (2012). The solid line is the best fit of the correlation observed at $z = 0$ for SDSS ETGs (Shen et al., 2003), with the corresponding $1\sigma$ scatter (dashed lines). *Bottom panel*. Same as the top panel, but for the central stellar velocity dispersion $\sigma_0$ as a function of $\mathcal{M}_\star$. Here the best fit and $1\sigma$ scatter of the observed $z = 0$ correlation for SDSS ETGs are from Hyde and Bernardi (2009). Adapted from Posti et al. (2014). Courtesy of L. Posti.

gives $R_e \propto \mathcal{M}_\star^{0.6}$ and $\sigma_0 \propto \mathcal{M}_\star^{0.2}$, close to the observed slopes (§5.4.3), though with some differences at the high-mass end ($\mathcal{M}_\star \gtrsim 1.5 \times 10^{11}\mathcal{M}_\odot$). This suggests that the observed $z = 0$ structural and kinematic scaling relations can be partly explained by the physical mechanisms that shape the galaxy SMF (§10.10.1), combined with the structural and kinematic evolution of the host dark halos.

The redshift evolution of the correlations shown in Fig. 10.21 is similar to the evolution of the ETG size–stellar mass and velocity dispersion–stellar mass trends observed with redshift (§11.3.1). Under the assumption that the halo gravitational radius $r_g$ and virial velocity dispersion $\sigma_{vir}$ (§5.3.1) scale linearly with $r_{vir}$ and $v_{vir}$, respectively, the evolution of individual halos in the $\mathcal{M}_{vir}$–$r_{vir}$ and $\mathcal{M}_{vir}$–$v_{vir}$ planes (Fig. 10.21) can be described by simple dissipationless merging models (§8.9.6). Numerical studies confirm that the evolution indicated by the arrows is consistent with these simple models, provided the merger mass ratio $\zeta$ and orbital energy $E_{orb}$ are taken into account. For instance, using eqs. (8.180) and (8.181), it can be shown that, for parabolic dry mergers ($E_{\rm orb} = 0$), $\sigma_{\rm vir} \propto \mathcal{M}_{\rm vir}^{a}$ with $a \leq 0$ and $r_{\rm g} \propto \mathcal{M}_{\rm vir}^{b}$ with $b \geq 1$, where $a$ and $b$ depend on $\zeta$. The time evolution of individual ETGs in the $\mathcal{M}_\star$–$R_{\rm e}$ and $\mathcal{M}_\star$–$\sigma_0$ planes is believed to be analogous to the evolution of individual dark matter halos in the $\mathcal{M}_{\rm vir}$–$r_{\rm vir}$ and $\mathcal{M}_{\rm vir}$–$v_{\rm vir}$ planes (arrows in Fig. 10.21), especially at relatively late times ($z \lesssim 1$–$2$), when dry mergers are believed to drive the evolution of massive ETGs (§10.8). Whether and how the observed properties of high-$z$ quiescent galaxies (§11.2.6 and §11.3.1) can be quantitatively explained in the framework of the evolution of dark matter halos in a ΛCDM cosmology is still a matter of debate.

In conclusion, we stress again that the origin of the scaling relations of ETGs is a complex question, which eludes any simple theoretical explanation. For instance, by construction, in the model shown in Fig. 10.21 the dynamical mass $\mathcal{M}_{\rm dyn}$ (eq. 5.17) scales linearly with the stellar mass $\mathcal{M}_\star$, so this model is unable to reproduce the observed tilt of the stellar mass fundamental plane (eq. 5.62), whose origin is not easy to explain with simple models such as the one described above. The question of the origin of the scaling relations of ETGs is mainly addressed with sophisticated numerical models of galaxy formation (§10.11). The observed empirical correlations, their scatter and their redshift evolution represent fundamental benchmarks for these numerical experiments.

## 10.11 Numerical Models of Galaxy Formation

Our current understanding of galaxy formation theory largely relies on numerical models. In this section, we give a brief overview of hydrodynamic simulations and semi-analytic models, which are the major techniques of computational galaxy formation. For a more detailed treatment we refer the reader to specialised reviews (Somerville and Davé, 2015; Naab and Ostriker, 2017).

### 10.11.1 Hydrodynamic Simulations

Though complex, the process of galaxy formation is a well defined physical problem, which, at least in principle, could be modelled self-consistently *ab initio*, i.e. from cosmological initial conditions. This is the approach adopted in **hydrodynamic cosmological simulations** in which a cubic portion of the Universe (usually referred to as a cosmological box) is simulated, following the evolution of both the dark matter and the baryons. Clearly, these simulations are substantially more sophisticated than the dark matter-only cosmological simulations (§7.5.1). In practice, the hydrodynamic cosmological simulations cannot follow self-consistently all the physical processes involved in galaxy formation for essentially two reasons. (1) The simulations are limited by their finite resolution: state-of-the-art large-scale simulations (with box size of the order of 100 Mpc) have spatial resolution of the order of a few hundred parsecs, so, for instance, they cannot resolve the scales of GMCs (§4.6.2), in which star formation occurs, not to mention single stars. (2) Our understanding of some fundamental physical processes (e.g. star formation, stellar and AGN feedback) is incomplete, so, even with *ideal* spatial resolution, these processes could not be modelled self-consistently at the moment. For these reasons, in the simulations one is forced to resort to the so-called **subgrid prescriptions** (or **subresolution recipes** or **effective models**), which are physically motivated analytic prescriptions aimed at accounting for the processes that occur on scales smaller than the simulation spatial resolution or that we are unable to model from first principles.

Hydrodynamic cosmological simulations of galaxy formation start at high redshift (say $z \gtrsim 100$) with initial conditions similar to those of dark matter-only cosmological simulations (§7.5.1), but including a gaseous component with mass determined by the cosmic baryon fraction (§2.5). The evolution of the gaseous component is followed by solving the equations of hydrodynamics (§D.2) with different numerical methods. In the **Eulerian hydrodynamic codes** the hydrodynamic equations are solved on a **grid** (or **mesh**), often with **adaptive mesh refinement** (AMR), which is a technique that allows one to vary the grid resolution in time and space, depending on the properties of the gas distribution. In the **Lagrangian hydrodynamic codes**, usually based on the **smoothed particle hydrodynamics** (SPH) method, the hydrodynamic equations are solved by modelling the fluid with particles. Note that these particles (with typical masses $\gtrsim 10^6 \mathcal{M}_\odot$) by no means represent microscopic gas particles, but are instead tracers that carry information on the macroscopic properties of the fluid (for instance, density, pressure and velocity) at their position. There are also codes that combine the Eulerian and Lagrangian approaches, being based on an **unstructured moving mesh**, in which not only the resolution but also the shape of the mesh vary with time, depending on the properties of the gas distribution.

A fundamental ingredient of the hydrodynamic simulations of galaxy formation is the gravitational field of both the dark matter and the baryons, which is computed self-consistently with the same numerical methods used in dark matter-only simulations (§7.5.1). Similarly, all the resolved (i.e. on scales sufficiently larger than the spatial resolution) hydrodynamic processes can be computed self-consistently. Convergence tests (i.e. comparison of simulations with the same initial conditions, but different resolution) are necessary to assess the reliability of such self-consistent calculations. We now give an overview of the main subresolution processes: star formation, stellar feedback, SMBH growth and AGN feedback.

1. *Star formation.* Stars form as a consequence of radiative cooling (§8.1.1) and self-gravity of the gas, so the code must allow for the conversion of gas into stars (§8.3). In hydrodynamic cosmological simulations the process of star formation is implemented with subresolution prescriptions: gas elements satisfying some specified star formation criteria (e.g. when the gas density exceeds some critical value) generate stellar particles, which are then followed in the simulation in a way similar to the dark matter particles. We stress that in these simulations the stellar particles (with typical masses $\gtrsim 10^6 \mathcal{M}_\odot$) are tracers of the collisionless stellar distribution and do not represent individual stars. In most implementations the SFR is taken to be proportional to the gas density divided by

a characteristic timescale (either the dynamical or the cooling time), with normalisation tuned to match the observed Schmidt–Kennicutt law (§4.2.9).

2. *Stellar feedback.* A fundamental consequence of star formation is the associated feedback on the ISM (§8.7). Simulations cannot resolve simultaneously the small scales of single SN explosions and the large scales of the CGM of galaxies or, even less, of a significant volume of the Universe, so one resorts to subgrid stellar feedback prescriptions. In general terms, galactic-scale stellar feedback is implemented as either **thermal feedback**, in which the gas surrounding the region where star formation has occurred is heated to high temperatures, or kinetic feedback (§8.8), in which the gas is given high outflowing speed. In addition to injecting energy and momentum into the ISM, stellar feedback has the fundamental role of modifying the chemical composition of the ISM by polluting it with metals. In hydrodynamic simulations, metal enrichment of the ISM by mass return from stars (§8.4.1) is accounted for with subgrid prescriptions based on sophisticated stellar evolution models.
3. *Growth of the SMBHs.* In the simulations, seed black holes are placed, following some empirical prescription, at the centre of dark matter halos. The black holes then grow via accretion of gas and mergers with other black holes (§8.9.7). The accretion of gas onto the SMBHs is usually implemented following the Bondi mass accretion rate (eq. 8.150) with an upper limit given by the Eddington mass accretion rate (eq. 8.149).
4. *AGN feedback.* We have seen in §8.8 that, depending on the accretion rate, feedback from central black holes comes in the form of either quasar mode or radio mode feedback. Estimates, based on theory and observations, of the mass, energy and momentum output from black holes accreting in these two modes can be used to build subgrid prescriptions that link such output to the black hole mass and accretion rate. Such prescriptions are implemented as source terms in the hydrodynamic equations (§D.2). In practice, given the complexity and our incomplete understanding of the problem, the effect of AGN feedback is often accounted for with relatively simple prescriptions that suffer from limitations and uncertainties similar to those of subgrid modelling of stellar feedback.

Additional physical processes that are sometimes included at some level in hydrodynamic cosmological simulations are the magnetic field (the magnetohydrodynamic equations are used in place of the hydrodynamic equations; §D.2.6) and the ionisation by the UVB radiation field (§8.1.2, §9.5.3 and §9.8.2), possibly including a treatment of radiative transfer (§D.1.1). The main physical processes implemented, either self-consistently or with subgrid prescriptions, are listed in Tab. 10.2. To give an idea of the impact of the subgrid prescriptions on the outcome of hydrodynamic cosmological simulations, in Fig. 10.24 we show some results of three simulations differing only in the implementation of stellar feedback. The subgrid parameters of a simulation are usually *calibrated* to reproduce some reference observed quantity, for instance the $z=0$ SMF.

A problem inherent in hydrodynamic cosmological simulations is that, for given computational resources, one is forced to find a compromise between the spatial resolution and the size of the simulated cosmological box. Thus the highest-resolution simulations have poor statistics, because they explore a relatively small cosmological volume, while the simulations with the largest cosmological boxes have poor spatial resolution. To mitigate this problem it is possible to use the so-called **zoom-in simulation** technique, in which a portion of a large-scale cosmological simulation is rerun at much higher resolution, while the surrounding volume is followed at lower resolution to keep track of the large-scale tidal field and accretion of matter. Having information on the lower-resolution simulation, one can select the volume of the zoom-in simulation in order to include at $z = 0$ the desired target (for instance, a Milky Way-like galaxy).

In addition to zoom-in simulations, also hydrodynamic and $N$-body simulations of individual galaxies (§10.2) and **simulations of binary galaxy mergers** (§10.3.2) are very useful to improve our understanding of the physical processes on small scales. Though even in these simulations subresolution models are required, the better spatial resolution allows us to carefully test these models, also taking for comparison detailed properties of observed galaxies. In particular, due to the presence of the subgrid prescriptions, hydrodynamic simulations have a number of free parameters (entering, for instance, the star formation and feedback effective models) that can be tuned in order to produce simulated galaxies with properties as similar as possible to real galaxies. The results of this calibration of the subgrid prescriptions can be used as reference for the calibration of hydrodynamic cosmological simulations, which we have mentioned above.

**Table 10.2** Main physical processes and phenomena treated in hydrodynamic cosmological simulations and semi-analytic models

| | Hydrodynamic simulations | Semi-analytic models |
|---|---|---|
| Black hole growth | prescription | prescription |
| Disc angular momentum | self-consistent | prescription |
| Disc instability | self-consistent | prescription |
| Dynamical friction | self-consistent | prescription |
| Feedback from AGNs | prescription | prescription |
| Feedback from stars | prescription | prescription |
| Galaxy colour | prescription | prescription |
| Galaxy luminosity | prescription | prescription |
| Galaxy morphology | self-consistent | prescription |
| Gas cooling | self-consistent | prescription |
| Gas distribution | self-consistent | prescription |
| Halo contraction | self-consistent | prescription |
| Halo density distribution | self-consistent | self-consistent |
| Halo merger history | self-consistent | self-consistent |
| Mass return from stars | prescription | prescription |
| Metal enrichment | prescription | prescription |
| Photoionising radiation | prescription | prescription |
| Ram-pressure stripping | self-consistent | prescription |
| Star formation law | prescription | prescription |
| Star formation threshold | prescription | prescription |
| Tidal stripping | self-consistent | prescription |

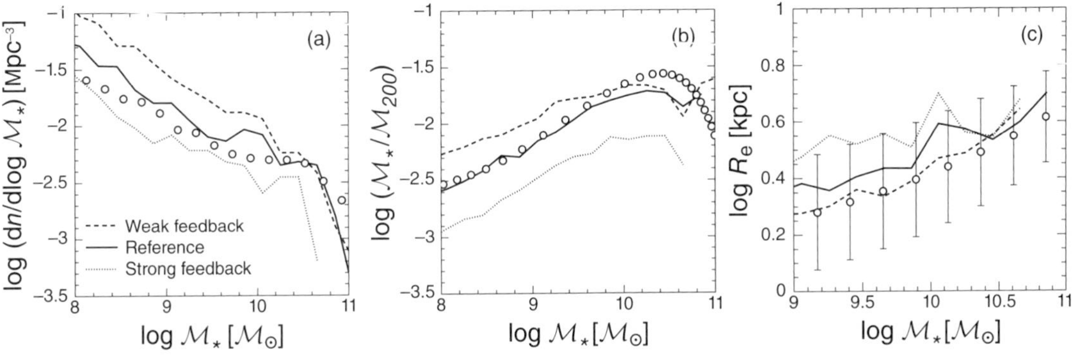


Fig. 10.24 Comparison with observational data (empty circles) of the $z = 0$ galaxy SMF (*panel a*), SHMR (*panel b*) and size–stellar mass relation (*panel c*) produced by three hydrodynamic cosmological simulations of the EAGLE project (Schaye et al., 2015) differing in the implementation of stellar feedback. Adapted from Crain et al. (2015). Data courtesy of R. Crain.

## 10.11.2 Semi-analytic Models

A technique in a sense complementary to the hydrodynamic cosmological simulations takes the name of semi-analytic modelling of galaxy formation. **Semi-analytic models** (SAMs) are useful to improve, from a statistical point of view, the comparison of galaxy formation models with observational data. The starting point of a SAM is the halo merger tree either built with the extended Press–Schechter formalism (§7.4.3) or taken from dark matter-only cosmological $N$-body simulations (§7.5.1). The evolution of the baryons within these evolving halos is modelled on the scale of an entire galaxy, without solving the hydrodynamic equations, but resorting to analytic or simple numerical prescriptions, typically built and calibrated using the results of *ad hoc* simulations.

At a given initial high redshift, each halo is assigned a dark matter density profile, based on the results of dark matter-only cosmological $N$-body simulations, and is populated by a gas component at the virial temperature[22] (§8.2.1). The gas is allowed to cool depending on an estimate of its cooling time (§8.1.1): the cooling gas conserves the angular momentum (originally acquired by tidal torques; §7.5.4 and §10.1) and thus forms a gaseous disc. Given some empirical star formation prescription, similar to those implemented in hydrodynamic simulations (§10.11.1), a fraction of this gas can be converted into stars, leading to the formation of a stellar disc. When a merger occurs in the merger tree, an estimate of the merging timescale (usually the dynamical friction timescale for minor mergers; §8.9.3) determines when the stellar components of the two galaxies merge. In SAMs, stellar spheroids are produced in mergers (§8.9.6), but also as a consequence of internal processes, such as disc instability (§10.2 and §10.3), which occurs when some specified criteria are satisfied.

[22] In some SAMs only a fraction of the gas, depending on the halo mass, is assumed to be shock heated to the virial temperature, the remainder being cold (§8.2.4).

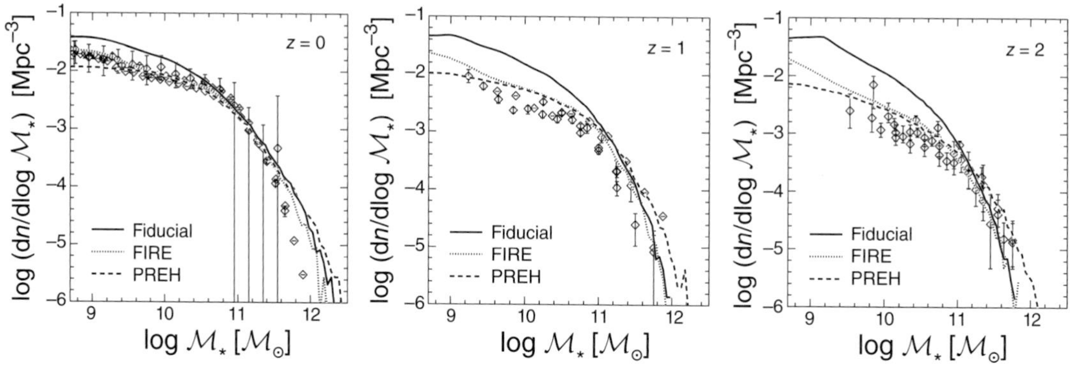


Fig. 10.25 Effect of different stellar feedback prescriptions on the galaxy SMFs at $z = 0$ (*left panel*), $z = 1$ (*middle panel*) and $z = 2$ (*right panel*) in a semi-analytic model of galaxy formation. The solid curve refers to a reference (fiducial) prescription, the dashed curve to a preventive feedback (PREH) prescription (§10.7.3) and the dotted curve to a prescription parameterised from cosmological zoom-in simulations (FIRE; Hopkins et al., 2014). The PREH and FIRE prescriptions are improvements of the reference (fiducial) prescription. The diamonds indicate observational estimates of the SMF. Adapted from Hirschmann et al. (2016). Data courtesy of M. Hirschmann.

In a similar way, other global properties of the galaxies (such as luminosity, colour, metallicity and dust content) are calculated by applying physically motivated, but empirical, prescriptions. For instance, once the SFH of a model galaxy is known, its luminosity and colour are computed based on SPS models (§8.6.2). It must be noted that some physical processes, such as tidal and ram-pressure stripping (§8.9.4), harassment (§8.9) and halo contraction (§7.5.5), which are implicitly accounted for in hydrodynamic simulations, must be implemented with specific prescriptions in SAMs. The effects of stellar feedback, black hole growth and AGN feedback are included in SAMs using relatively simple recipes, some of which are similar to those used in the subresolution models of hydrodynamic simulations (§10.11.1). Most SAMs also account for the process of reionisation due to the extragalactic UVB (§9.5.3 and §9.8.2). A list of the main physical processes and phenomena typically treated in SAMs is given in Tab. 10.2.

In summary, the output of a SAM is, at each sampled redshift in the explored redshift range, a catalogue of galaxies, each characterised by several global quantities, including dark matter mass, circular speed, total stellar mass, cold gas mass, hot gas mass, SFR, bulge-to-total stellar mass ratio, disc size, bulge size, metallicity, luminosity as a function of wavelength, dust absorption and black hole mass. The main advantage of SAMs is that they are computationally much cheaper than hydrodynamic simulations, so several different models, for instance for star formation and feedback, can be considered and the space of the model parameters can be explored extensively. The results of SAMs can be efficiently compared with observational data, also exploiting the fact that large samples of model galaxies can be studied over wide redshift ranges. However, SAMs are clearly limited by the fact that even physical processes that could be followed self-consistently (for instance hydrodynamics) are accounted for with relatively simple approximations.

Moreover, the inclusion of several empirical prescriptions implies that SAMs have a large number of free parameters (typically more than hydrodynamic simulations) that must be tuned in order to reproduce some well constrained observed properties of the galaxy population, such as, for instance, the present-day galaxy SMF. As an example of the dependence of the results of SAMs on specific prescriptions, Fig. 10.25 compares with observational data the galaxy SMFs predicted at $z = 0$, $z = 1$ and $z = 2$ by a SAM with different implementations of the stellar feedback.

# 11 Observing Galaxy Evolution

Previous chapters presented a global view of galaxy properties in the present-day Universe and a general description of the physical processes of galaxy formation. This chapter illustrates what the observations can tell us about the evolution of galaxies across cosmic time. This is a young research field characterised by a very fast development. Before the mid-1990s, the spectroscopic identification of galaxies at cosmological distances (say $z > 1$) was limited to a few cases, or to AGNs thanks to their high luminosities. Since the mid-1990s, this limitation was overcome when ground-based telescopes with 8–10 metre diameters became available (Keck and VLT) and allowed the spectroscopic identification of normal galaxies out to $z > 3$. This milestone, together with *HST* deep and high-resolution imaging, and the synergy with multi-wavelength data from the gamma-rays to the radio, opened a brand new research field allowing the direct observation of evolving galaxies at cosmological distances. However, despite this major progress, our understanding of galaxy formation and evolution based on observations is still incomplete. For this reason, this chapter is focused only on the most robust observational results. The topics are divided into three main sections. The first deals with the galaxy physical properties that can be derived from observational data. The second describes how to identify high-redshift galaxies and illustrates their main characteristics. The third discusses the key results inferred from the observation of galaxy samples as a function of redshift. Specialised reviews are cited for each topic in order to guide readers who may want to have more details.

## 11.1 The Main Observables of Galaxy Evolution

The aim of this section is to review some of the main galaxy properties that can be derived from the observations (the so-called **observables**). Here the focus is on distant galaxies ($z > 1$), but the same methods can be applied to galaxies at lower redshift, as they were implicitly used in Chapters 3–6. The spectral regions mentioned in this chapter (e.g. UV, optical, IR) are defined in §C.1. Unless otherwise specified, magnitudes are given in the AB photometric system (§C.3; §C.4). Tabs. 11.1 and 11.2 give a summary of the galaxy properties that can be derived from the observational data.

### 11.1.1 Redshifts

When a galaxy spectrum is available and spectral lines are detected, the **spectroscopic redshift** (spec-$z$) is derived as $z = (\lambda_{\rm obs}/\lambda_{\rm rest}) - 1$, where $\lambda_{\rm obs}$ and $\lambda_{\rm rest}$ are the observed

**Table 11.1** Main observables of galaxy evolution – I

| Observable | Data/method | Sections |
|---|---|---|
| General | | |
| Spectroscopic redshift | Spectra with detected lines | §11.1.1 |
| Photometric redshift | Multi-band photometry + SED fitting | §11.1.1 |
| Rest-frame luminosity | Redshift; K correction | §11.2.3 |
| Morphology | | |
| Morphological class | Imaging | §3.1 |
| Surface brightness profile | Isophote fitting with Sérsic function | §3.1.2 |
| Non-parametric indices (e.g. $C, A, S$) | Imaging | §11.1.6 |
| Stellar populations | | |
| Stellar age | Spectra and/or SED fitting | §11.1.3 |
| Stellar metallicity | Spectra and/or SED fitting | §11.1.3 |
| Stellar mass | Spectra and/or SED fitting | §11.1.3 |
| Star formation rate (SFR) | $L_{\mathrm{H}\alpha}$, $L_{\mathrm{UV}}$, $L_{\mathrm{IR}}$, $L_{\mathrm{X}}$, $L_{\mathrm{radio}}$ | §11.1.2 |
| Star formation duration | Spectra (abundance of $\alpha$-elements) | §11.1.3 |
| Star formation history (SFH) | Spectra and/or SED fitting | §11.1.3 |

and rest-frame wavelengths of the identified lines, respectively (§2.1.2). It is useful to express the redshift uncertainty as $\sigma_z/(1+z)$ (where $\sigma_z$ is the redshift error) because, for a given spectral resolution, the ratio $\sigma_z/(1+z)$ is constant with redshift. Spectroscopic redshifts are accurate, with typical uncertainties $0.0001 < \sigma_z/(1+z) < 0.001$. When galaxies are too faint for spectroscopy, it is possible to estimate the so-called **photometric redshift** (photo-$z$). A photo-$z$ is obtained by comparing the observed photometric SED (§3.3) of a given galaxy with a library of template spectra of different galaxy types or SPS models (§8.6.2). These templates are left free to vary in redshift and normalisation until the best match is found (e.g. through $\chi^2$ fitting; Fig. 11.1). Photo-$z$ accuracy improves with the number of photometric filters available across the SED and with the accuracy of the photometry. Photo-$z$ must be tested and calibrated with samples of galaxies with known spec-$z$. Despite this calibration, a fraction of photo-$z$ can remain discordant with respect to the spectroscopic ones. The error of a photo-$z$ is called catastrophic when $|\text{photo-}z - \text{spec-}z|/(1+\text{spec-}z) > 0.15\text{–}0.20$. However, catastrophic failures can also occur for spec-$z$ when, for instance, a single emission line is detected in the spectrum, but it is misidentified with another line. In the very best cases, photo-$z$ accuracy can be as good as $\sigma_z/(1+z) \approx 0.01$, i.e. still 1–2 orders of magnitude lower than for spec-$z$ values. Spectroscopic redshifts require a large amount of telescope observing time because the signal-to-noise ($S/N$) ratio (eq. 11.16) must be high enough to detect emission and/or absorption lines. Thus, despite the lower accuracy, photo-$z$ values are very important to derive approximate redshift estimates when galaxy samples are too large and/or too faint for spectroscopy. However, the photo-$z$ accuracy is usually insufficient to reconstruct the large-scale distribution of galaxies (the cosmic web; §6.6) as illustrated by the simulation in Fig. 11.2.

**Table 11.2** Main observables of galaxy evolution – II

| Observable | Data/method | Sections |
|---|---|---|
| Interstellar medium | | |
| Dust extinction | Spectra (H$\alpha$/H$\beta$), UV continuum slope | §11.1.5 |
| Gas ionisation | Spectra (emission line ratios) | §11.1.4 |
| Gas temperature | Spectra and emission line ratios | §4.2.1; §4.2.3 |
| Electron density | Spectra (emission line ratios) | §11.1.4; §4.2.3 |
| Gas metallicity | Spectra (emission line ratios, absorption lines) | §11.1.4 |
| Gas mass | Spectra (e.g. H I, CO) | §11.1.4; §4.2.1 |
| Dust temperature and mass | Infrared SED fitting with grey body | §11.1.5 |
| Molecular gas excitation | Spectral line energy distribution (SLED) | §11.1.4 |
| Kinematics | | |
| Rotation curves | Spatially resolved spectra | §4.3.2 |
| Velocity dispersion | Spectral lines | §4.3.2; §5.1.2 |
| Mass | Rotation curve (discs) | §4.3.2 |
| | Velocity dispersion and size (spheroids) | §5.1.2 |
| | Gravitational lensing | §5.3.4 |
| Outflows/inflows | Spectra (line shifts and profiles) | §4.3.2 |
| Turbulence | Spectra (line velocity dispersion) | §4.3.2 |
| Thermal broadening | Spectra (line velocity dispersion) | §4.3.2 |
| Distribution functions | | |
| Luminosity function | Galaxies with redshifts and multi-band SEDs | §3.5.1 |
| Stellar mass function | Galaxies with redshifts and multi-band SEDs | §3.5.2 |
| Environment | | |
| Galaxy density field | Galaxies with sky coordinates and redshifts | §6.6; §11.1.9 |
| Clustering | | |
| Angular correlation function | Galaxies with sky coordinates | §6.6 |
| Spatial correlation function | Galaxies with sky coordinates and redshifts | §6.6; §11.1.9 |

## 11.1.2 Star Formation Rate

One of the most important properties of SFGs is the current (also called instantaneous, i.e. at the time of the observation) SFR (§4.1.5). This is not a quantity directly derived from the data, as it requires several assumptions, such as the IMF (§8.3.9). The SFR is determined using a variety of indicators based on the luminosity ($L$) of a component of the galaxy that is sensitive to the presence of young massive stars:

$$\mathrm{SFR} = C L_{\mathrm{indicator}}, \tag{11.1}$$

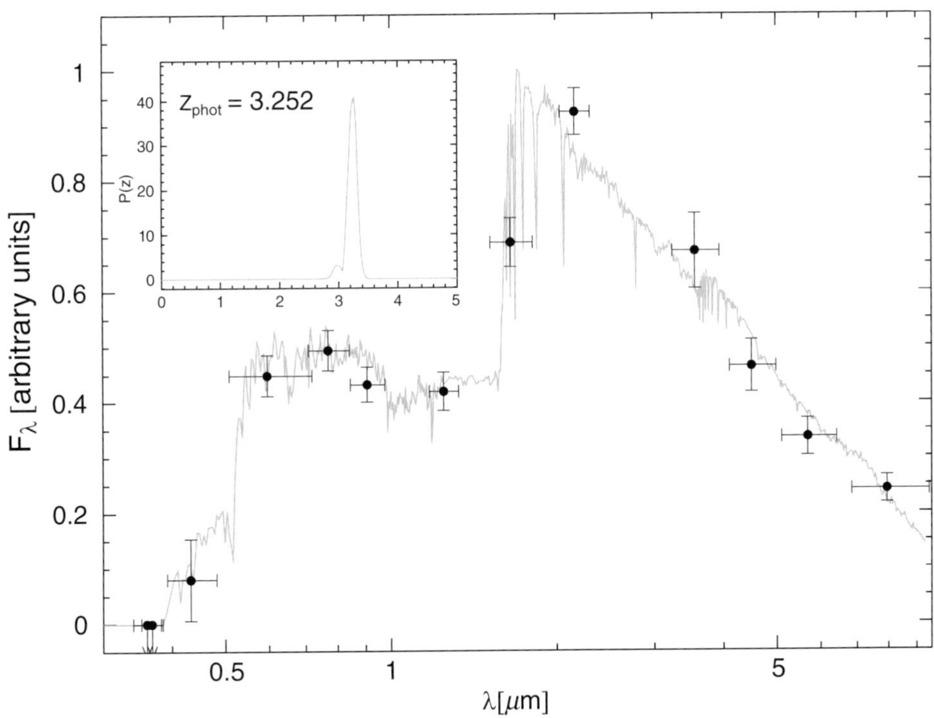


Fig. 11.1 Example of SED fitting and photometric redshift estimate. The black dots are the observed fluxes obtained with 12 filters (§C.4) which provide a photometric SED from the optical to the NIR. The vertical and horizontal error bars indicate the uncertainties on the flux and the widths of the filters, respectively. The fluxes, shown as a function of the observed wavelength $\lambda$, are fitted with an SPS model (§8.6.2) spectrum (thin grey line). The inset shows the probability distribution of the photo-$z$ values and indicates that the most likely photometric redshift is $z = 3.252$. Courtesy of M. Bolzonella.

where $C$ is a conversion factor to be determined under a number of assumptions. For more details, we refer the readers to the reviews of Kennicutt and Evans (2012), Calzetti (2013) and Madau and Dickinson (2014).

One of the primary indicators is the **H$\alpha$ luminosity**. In the case of star formation, due to the very short lifetimes of O and B stars (e.g. approximately $\lesssim 30$ Myr for masses higher than 8 $\mathcal{M}_\odot$), H II regions are present as long as these stars are present. If such stars were suddenly switched off, the timescale during which ionised hydrogen would recombine entirely is $t_{\rm rec} \approx (n_{\rm e}\alpha_{\rm H})^{-1}$, where $n_{\rm e}$ is the electron density and $\alpha_{\rm H}$ the recombination rate coefficient in units of cm$^3$ s$^{-1}$ (§4.2.3). For a typical density $n_{\rm e} \approx 100$ cm$^{-3}$, this timescale is extremely short ($\sim 10^3$ yr) and implies that the existence of an H II region is possible only if star formation is ongoing (i.e. if hot massive stars are present). To derive the SFR from H$\alpha$ luminosity, one first assumes that all the H$\alpha$ emission is produced in H II regions around O/B stars. Then one has to convert the H$\alpha$ luminosity into the number of photoionising photons with $\lambda < 912$ Å per unit time using prescriptions from photoionisation theory

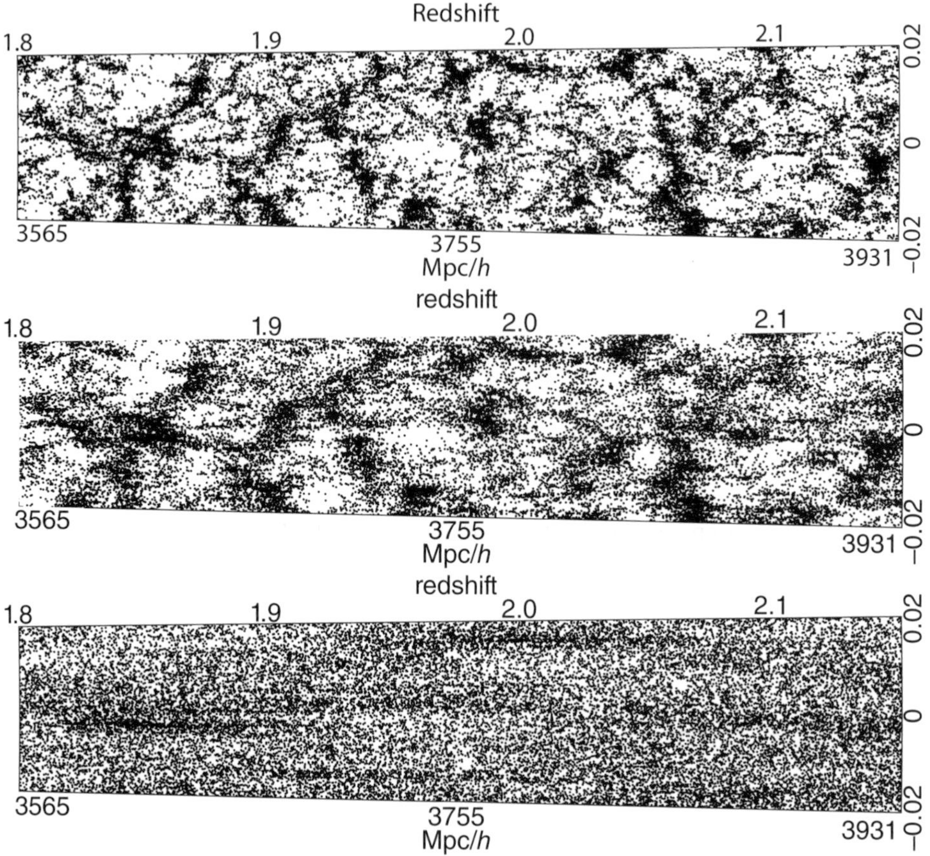


Fig. 11.2 Simulation of how the galaxy large-scale distribution can be reconstructed depending on redshift uncertainty parameterised as $\sigma_z = A(1 + z)$, where $A = 0.0001$ (*top*), $A = 0.001$ (*middle*) and $A = 0.01$ (*bottom*). Each dot is a galaxy, and the units on the $y$-axes are radians. The details of the LSS (e.g. voids, overdensities, filaments; see also Fig. 6.14) are clearly visible only with the accuracy of spectroscopic redshifts (top and middle panels). Courtesy of A. Orsi.

(§4.2.3) and assuming that all the photons produced by the stars end up ionising a hydrogen atom around the star. From the total photoionising flux it is possible to derive the number and total mass of the massive stars emitting Lyman continuum photons, and then to estimate the SFR. Based on this method, the final formula that relates the H$\alpha$ luminosity to the SFR is

$$\mathrm{SFR} = C_{\mathrm{H}\alpha}\left(\frac{L_{\mathrm{H}\alpha}}{\mathrm{erg\ s^{-1}}}\right) \mathcal{M}_{\odot}\mathrm{yr}^{-1}, \tag{11.2}$$

where $L_{\mathrm{H}\alpha}$ refers to the luminosity of the line corrected for dust extinction (§4.2.7; §11.1.5) and $C_{\mathrm{H}\alpha} \simeq 7.9 \times 10^{-42}$ is valid for a Salpeter IMF. At longer wavelengths, the hydrogen Paschen lines are much less affected by dust attenuation, but their observation is difficult because for $z > 1$ they are redshifted in the MIR where spectroscopy is challenging or impossible from the ground (§11.2). The SFR estimated from the luminosity of recombination lines has a few drawbacks. First, the assumption that all the UV photons from the stars are absorbed in the surrounding photoionised region is questionable as we know that a fraction of them must escape (§4.2.3). In this sense, the SFR derived through eq. (11.2) should be considered a lower limit. On the other hand, if some of the emission is due to collisionally ionised gas not included in this calculation, or if non-stellar photoionising photons are also present (e.g. from an AGN), the SFR is overestimated.

When hydrogen recombination lines are not available in the observed spectral range, the luminosity of the [O II]$\lambda$3727 forbidden line ($L_{[\mathrm{O\,II}]}$) can be used as a secondary (i.e. less reliable) indicator of the SFR. This line is a less accurate SFR estimator because the luminosity of forbidden lines depends also on the gas density. This problem is partly circumvented using the average ratio $R_{[\mathrm{O\,II}]/\mathrm{H}\alpha} = L_{[\mathrm{O\,II}]}/L_{\mathrm{H}\alpha}$ calibrated with observations of galaxies where both lines are detected (corrected for dust extinction) to derive an empirical relation

$$\mathrm{SFR} = C_{\mathrm{H}\alpha}\left(\frac{L_{[\mathrm{O\,II}]}}{R_{[\mathrm{O\,II}]/\mathrm{H}\alpha}}\right) = C_{[\mathrm{O\,II}]}\left(\frac{L_{[\mathrm{O\,II}]}}{\mathrm{erg\ s^{-1}}}\right) \mathcal{M}_{\odot}\mathrm{yr}^{-1}, \tag{11.3}$$

where $C_{[\mathrm{O\,II}]} \simeq 1.4 \times 10^{-41}$ is valid for a Salpeter IMF. The main limitation of this method is that $R_{[\mathrm{O\,II}]/\mathrm{H}\alpha}$ has a large scatter because its value varies significantly from galaxy to galaxy.

Other than emission lines, an alternative SFR indicator is the monochromatic **UV continuum luminosity** at a wavelength chosen within the range 1250–2500 Å (typically $\lambda_{\mathrm{rest}} = 1500$ Å, $\nu_{\mathrm{rest}} \approx 2 \times 10^{15}$ Hz). The radiation emitted by SFGs at these wavelengths is dominated by the youngest and short-lived stars because the main-sequence lifetimes of O and B stars are less than a few tens of Myr and of the order of 100 Myr, respectively (§C.5). Thus, the UV luminosity is related to the total number of O/B stars and therefore to the instantaneous (or very recent) SFR. The relation between the SFR and the UV luminosity can be written as

$$\mathrm{SFR} = C_{\mathrm{UV}}\left(\frac{L_{\nu,\mathrm{UV}}}{\mathrm{erg\ s^{-1}\ Hz^{-1}}}\right) \mathcal{M}_{\odot}\mathrm{yr}^{-1}, \tag{11.4}$$

where the monochromatic luminosity $L_{\nu,\mathrm{UV}}$ is corrected for dust extinction. The conversion factor $C_{\mathrm{UV}}$ is derived through SPS models (§8.6). As a reference, $C_{\mathrm{UV}} \simeq 1.3 \times 10^{-28}$ for $Z = Z_{\odot}$, Salpeter IMF, stellar mass range 0.1–100 $\mathcal{M}_{\odot}$, SFR($t$) = constant and a stellar population age $\gtrsim$ 300 Myr. For younger ages, the conversion factor changes significantly, and thus it is important to adopt the value of $C_{\mathrm{UV}}$ that is appropriate for the galaxies of interest (e.g. very young starbursts with respect to more mature SFGs). As the UV radiation is heavily attenuated by interstellar dust, the main uncertainty in the application of eq. (11.4) is the correction of $L_{\nu,\mathrm{UV}}$ for dust extinction.

A further possibility is to estimate the SFR from the **infrared luminosity** emitted by the dust grains that are heated at temperatures $T_{\mathrm{d}} \approx 20$–60 K by the absorption of the UV photons coming from massive stars:

$$\mathrm{SFR} = C_{\mathrm{IR}}\left(\frac{L_{\mathrm{IR}}}{\mathrm{erg\ s^{-1}}}\right) \mathcal{M}_{\odot}\mathrm{yr}^{-1}, \tag{11.5}$$

where $L_{IR}$ is the IR luminosity integrated from $\lambda_{rest} = 8$ $\mu$m to $\lambda_{rest} = 1000$ $\mu$m, and $C_{IR} \simeq 1.73 \times 10^{-10}$. The IR luminosity has the advantage of not being affected by dust extinction, but the IR-derived SFR can be overestimated if an AGN is present, or when the circumstellar IR emission from evolved stars is not negligible.

The **radio luminosity** (e.g. at $\nu_{rest} = 1.4$ GHz) is another SFR indicator. In SFGs, the radio continuum originates primarily from (1) the synchrotron radiation emitted by relativistic electrons produced in SNRs (indicators of recent star formation; §8.7.1) and (2) the bremsstrahlung emission from the H II plasma at $T \sim 10^4$ K (an indicator of ongoing star formation; §4.2.8). Fig. 11.3 shows the average shape of the SED of SFGs from the radio to the IR. The intimate link between the radio continuum emission and SFR is empirically supported by a tight correlation that is present between the radio and IR luminosities of SFGs (the so-called **radio–infrared correlation**). The radio luminosity is not affected by dust extinction, but the radio-derived SFRs can be severely overestimated if AGN emission is present at radio wavelengths.

Also the **X-ray luminosity** at $\approx 2-10$ keV (rest frame) can be related to star formation activity. At these energies, the X-ray emission of SFGs is dominated by massive X-ray binaries, massive stars and supernova remnants, i.e. by components all associated with young stars and star formation activity. In particular, high-mass binaries contain a massive star ($> 8$ $\mathcal{M}_\odot$) and a compact object such as a neutron star or a black hole. The X-ray photons are emitted from the hot matter accreting onto the compact star. In our Galaxy, the prototypical high-mass X-ray binary is Cygnus X-1, where the compact object is a black

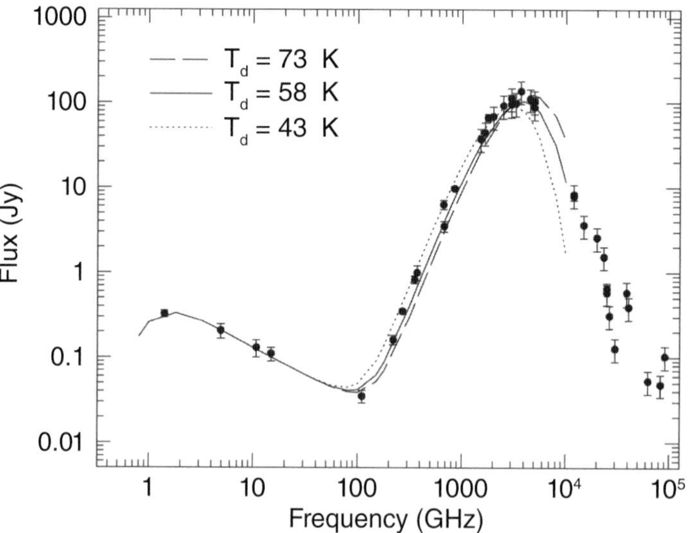


Fig. 11.3 The average SED of SFGs from the radio to the IR. The black points are the observed fluxes. The curves show the sum of grey-body emission (§D.1.3) with a range of compatible dust temperatures ($T_d$) and radio continuum. The radio emission is due to the sum of bremsstrahlung (§D.1.6) and synchrotron (§D.1.7) radiation, both dominant at frequencies below $\approx$ 100 GHz with respect to dust thermal emission. From Yun and Carilli (2002). © AAS, reproduced with permission.

hole. Due to the high mass of one of the two stars, these systems are short-lived and thus are present only in the case of ongoing or very recent star formation. The strong correlation between the X-ray and the IR luminosities of SFGs is an empirical proof of the connection between the SFR (traced by the IR luminosity) and the X-ray emission from massive (i.e. short-lived) binaries. Instead, in the case of more evolved galaxies, the X-ray luminosity is dominated by the hot gas emission (§5.2.1) and the integrated radiation of low-mass ($< 1$ $\mathcal{M}_{\odot}$) X-ray binaries. The latter is more correlated with the galaxy stellar mass because these binaries are tracers of the long-lived stellar populations. If AGN activity is present, the use of X-ray luminosity as an SFR indicator requires a careful subtraction of the strong emission from the active nucleus in this spectral region.

In practice, the choice among the various SFR indicators depends largely on the redshift range of the galaxy sample. For $z > 0.5$, H$\alpha$ is redshifted into the NIR where spectroscopy of large samples of distant galaxies can be challenging (§11.2). Thus, the most practical SFR indicators for high-$z$ galaxies are often based on the luminosity of the UV (redshifted in the optical for $z > 1$) and/or IR continuum (redshifted in the FIR/submm for $z > 1$). The former can be measured from rest-frame UV spectra or photometric SEDs. The latter requires space-based data in the FIR, whereas submm/mm data can also be obtained with ground-based telescopes in a few spectral windows where the atmosphere is reasonably transparent (e.g. at $\lambda \approx 450$ $\mu$m and $\lambda \approx 850$ $\mu$m).

When UV and IR data are both available, it is customary to express the total SFR of a galaxy as

$$\mathrm{SFR_{tot}} = C_{\mathrm{IR}} L_{\mathrm{IR}} + C_{\mathrm{UV}} L_{\mathrm{UV_{obs}}}, \tag{11.6}$$

where $L_{\mathrm{IR}}$ and $C_{\mathrm{IR}}$ are defined in eq. (11.6), $L_{\mathrm{UV_{obs}}}$ is the observed (i.e. not corrected for dust extinction) UV luminosity (e.g. at 1500 Å rest frame), and $C_{\mathrm{UV}}$ is defined in eq. (11.4). The first term of eq. (11.6) takes into account the fraction of star formation obscured by dust that is observed indirectly through the emission by dust grains heated by hot massive stars. The second term adds the component (if present) of unobscured star formation for which the UV radiation emitted by O/B stars is directly observable.

Finally, the SFR can also be derived as one of the best-fitting parameters through the analysis of photometric SEDs with SPS models, provided that an SFH has been assumed (§8.6; §11.1.3; Fig. 11.4). As a final remark, we recall that, irrespectively of the method used to estimate the SFR, the final value depends on the arbitrary choice of the IMF (§8.3.9). For instance, if a Chabrier IMF is adopted, the SFR will be lower by a factor of $\approx 1.7$ than with a Salpeter IMF.

When the SFR and the stellar mass $\mathcal{M}_{\star}$ of a galaxy are known (§11.1.3; eq. 11.7), it is possible to estimate the sSFR defined in eq. (4.44). This quantity allows us to assess the importance of the ongoing SFR with respect to the stellar mass previously formed in a given galaxy. The inverse of this quantity ($\mathrm{sSFR}^{-1}$) is a timescale which indicates how fast the stellar mass can increase through the observed SFR. As a reference, our Galaxy has a low sSFR ($\approx 2$–$6 \times 10^{-11}$ yr$^{-1}$, depending on the adopted SFR; Tab. 4.1), whereas galaxies caught in the act of forming a major fraction of their stellar mass have much higher sSFRs, up to $\sim 10^{-8}$ yr$^{-1}$.

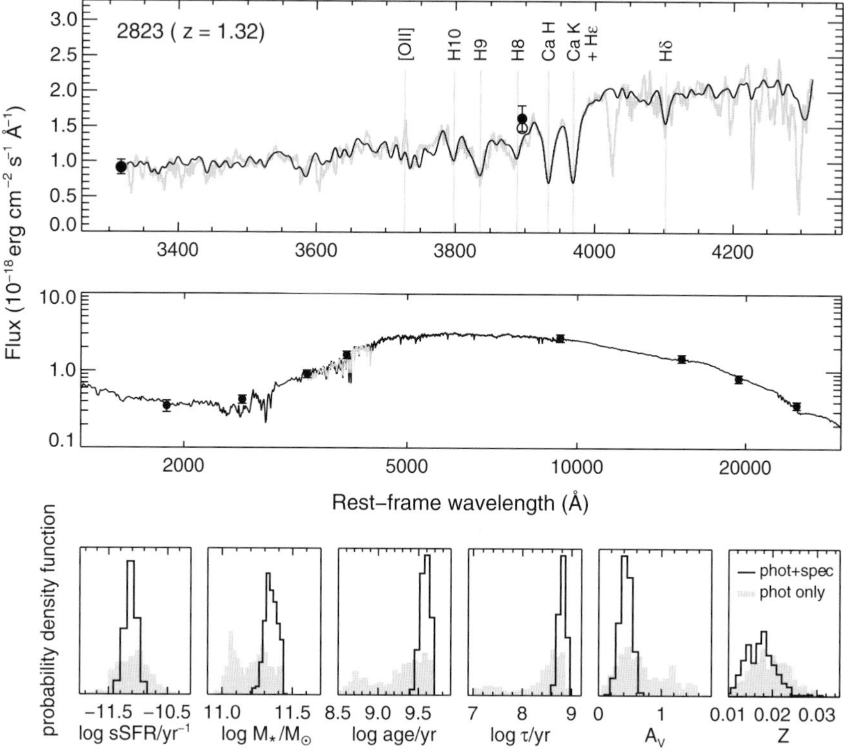


Fig. 11.4 Example of spectral and SED fitting for a quiescent galaxy at $z \simeq 1.32$. *Top panel.* The observed (grey) and SPS best-fitting (black) spectra. The three strong absorption lines in the observed spectrum at $\lambda > 4000$ Å are not real, but they are noise features originated by the subtraction of strong emission lines of the terrestrial atmosphere (Fig. 11.10). The two filled circles indicate the fluxes obtained through broad-band filters, and are the black points of the middle panel that fall into the wavelength range of the spectrum. The fluxes expected in these two filters from the best-fitting model are shown as empty circles. At $\lambda \approx 3900$ Å the expected flux is slightly underestimated, whereas at $\lambda \approx 3320$ Å the agreement is perfect as it coincides with the observed one. *Middle panel.* The photometric SED (black points), the observed spectrum (light grey) and the best-fitting SPS spectrum (black). Note the much wider wavelength coverage allowed by the photometric SED compared to the spectrum. *Bottom panel.* The probability distributions of the best-fitting parameters. From left to right: specific star formation rate (sSFR), stellar mass, stellar age, $\tau_\star$ (SFR($t$) $\propto \exp(-t/\tau_\star)$; eq. 8.96), dust extinction ($A_V$) and metallicity ($Z$). The grey and black histograms are relative to the SED-only and the (SED + spectrum) fitting, respectively. The improved accuracy of the estimated parameters is evident in the latter case because the combination of SED + spectrum provides more stringent constraints on SPS models. Adapted from Belli et al. (2015). © AAS, reproduced with permission.

### 11.1.3 Stellar Populations, Stellar Mass and Star Formation History

The stellar content of galaxies can be inferred from the analysis of the observed spectra and/or photometric SEDs through a comparison with SPS model spectra (§8.6.2). Fig. 11.4

shows an example of this method. The SPS model fitting provides information on the stellar age, metallicity and mass, as well as on the SFH. In particular, the **stellar mass** is derived as

$$\mathcal{M}_\star = L\left(\frac{\mathcal{M}_\star}{L}\right)_{\rm SPS}, \tag{11.7}$$

where $L$ is the observed luminosity of the galaxy in a given filter and $(\mathcal{M}_\star/L)_{\rm SPS}$ is the stellar mass-to-light ratio (§4.1.2) of the SPS best-fitting model spectrum in the same filter. Rest-frame NIR data ($\lambda_{\rm rest} \approx 1\text{–}4\ \mu$m) are particularly important to estimate stellar masses because the luminosity in this spectral range is dominated by low-mass evolved stars which make up the bulk of the total $\mathcal{M}_\star$ of a galaxy. In this regard, the *Spitzer* space telescope played a key role thanks to the photometry at $3\ \mu{\rm m} < \lambda_{\rm obs} < 8\ \mu$m which allowed the rest-frame NIR light of high-redshift galaxies to be observed.

The SPS model fitting method is affected by unavoidable systematic uncertainties such as the assumption of stellar isochrones, the choice of the observed and theoretical stellar spectra, the adopted IMF and the assumed SFH parameterisation (§8.6.2). Moreover, any SPS fitting result is biased towards the light emitted by the youngest stellar population. Young stars are much more luminous and dominate the integrated light of a galaxy (especially in the blue and UV) even when they contribute very little to its total stellar mass. For example, a 1 Myr old SSP contributing just 5% to the total stellar mass of a galaxy is sufficient to dominate the total spectrum, and to hide a 10 Gyr old population at $\lambda < 4000$ Å (Fig. 11.5). This **overshining** effect, also called **frosting**, is strongly dependent on wavelength. Thus, the quantities derived from fitting galaxy spectra or SEDs with SPS models are **luminosity-weighted**. This means that they are biased towards the youngest

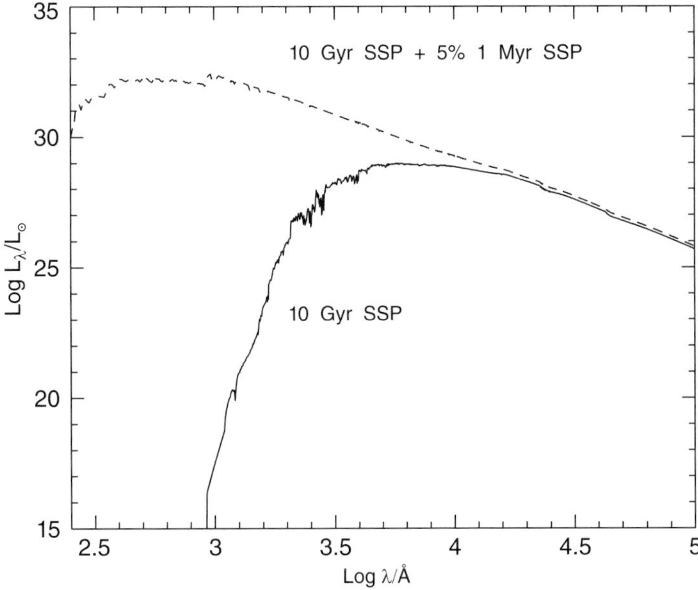


Fig. 11.5 An example of the overshining (or frosting) effect for a 10 Gyr old SSP to which a small fraction (5% of the total mass) of young (1 Myr old) stars is added. Despite the negligible mass contribution, the blue–UV light is dominated by young stars, whereas the two SEDs are basically indistinguishable in the NIR. Courtesy of C. Maraston.

stellar population, unless an accurate decomposition of the stellar populations is done (e.g. using high-resolution spectra).

In the case of low spectral resolution (§3.4), or when only photometric SEDs are available, another source of uncertainty is the age–metallicity degeneracy (§5.1.3). This makes it difficult to unambiguously separate the effects of age from those of metallicity on the observed galaxy colours or SEDs. For instance, a galaxy may look red because of old stars with low metallicity, or because of young stars with high metallicity (Fig. 8.16; Fig. 11.6, top). It is important to note that the estimate of the stellar mass (eq. 11.7) is less influenced by this degeneracy because both age and metallicity affect $\mathcal{M}_\star/L$ in similar ways. A possibility to mitigate this degeneracy is to exploit suitable stellar absorption lines such as the Lick indices (§5.1.3). These are stellar absorption lines in the range of $\lambda_{\rm rest} = 4200–6400$ Å whose equivalent widths have been calibrated against age and metallicity (Fig. 11.21). Balmer absorptions (e.g. H$\beta$, H$\gamma$, H$\delta$) are good tracers of the

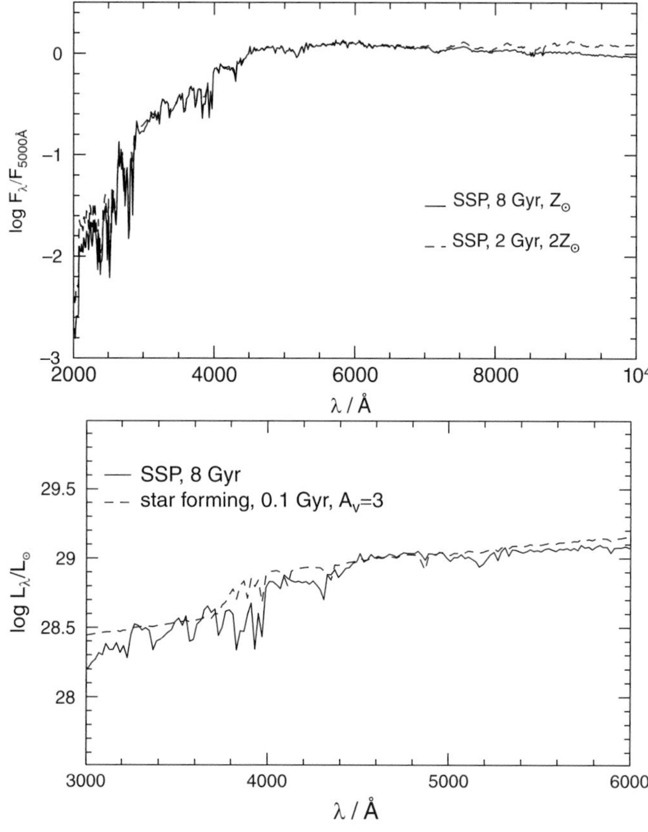


**Fig. 11.6** *Top panel.* Example of the age–metallicity degeneracy. Despite the completely different ages and metallicities (see legend), the two SSPs look globally very similar at 3000 Å $< \lambda <$ 7000 Å. *Bottom panel.* Example of the age–extinction degeneracy. Two SPS model spectra are shown. An old (8 Gyr) SSP has a spectral shape (and therefore colours) similar to that of a young (0.1 Gyr) galaxy reddened by dust extinction with $A_V = 3$. Figures courtesy of C. Maraston.

age, whereas Fe and Mg lines are good estimators of the global metallicity and of the abundance of $\alpha$-elements, respectively. However, the application of this method requires spectra with sufficiently high resolution and $S/N$ ratio. Additional constraints on the stellar content are provided by **continuum discontinuities** (also called breaks) at 2640 Å, 2900 Å, 3200 Å and 4000 Å (the D4000 break; §3.4) whose amplitudes depend on age and metallicity.

Reconstructing the SFHs (§4.1.5) is very challening. One possibility is to fit the spectra and/or SEDs with linear combinations of SSP model spectra and to derive how the total stellar mass of a galaxy at a given redshift was gradually formed as a function of time. As explained in §4.6.1 (Fig. 4.24), the stellar abundance of $\alpha$-elements relative to iron ([$\alpha$/Fe]) can be exploited to estimate the duration of star formation. In the case of ETGs (§5.1.3), this can be done with the relation

$$[\alpha/\mathrm{Fe}] \approx 0.2 - 0.17 \log \Delta t, \tag{11.8}$$

where $\Delta t$ (in units of Gyr) is the FWHM of the Gaussian curve assumed as the shape of the SFH. This relation has been obtained through a calibration with SPS models (§8.6.2). The higher the value of [$\alpha$/Fe], the shorter the duration of star formation ($\Delta t$). When these approaches are not possible, an analytic SFH must be adopted (§8.6.2; Fig. 11.4).

### 11.1.4 Interstellar Gas

The physical and chemical properties of the ISM are derived from spectroscopic observations of the gas phases and dust components. The case of neutral atomic hydrogen is described in §4.2.1. Regarding the ionised gas, its temperature and density can be inferred from the flux ratios of selected pairs of emission lines (§4.2.3). If the incident spectrum of the ionising source is known (e.g. O/B stars or AGN), the metallicity of the ionised gas can be estimated from the flux ratios of optical emission lines whose values have been calibrated against metal abundance. Two examples of metallicity estimators are the ratios $R_{23} \equiv$ ([O II]$\lambda$3727 + [O III]$\lambda$4959 + [O III]$\lambda$5007)/H$\beta$ and [N II]$\lambda$6583/H$\alpha$. In the absence of emission lines, constraints on metal abundances can be derived from the numerous ISM absorption lines (e.g. Si, C, O, Al, Fe) present in the rest-frame UV spectra of SFGs (§3.4; Fig. 3.8). Emission line ratios allow us also to constrain the source of gas ionisation through **diagnostic diagrams**. At optical wavelengths, an important example is that of Baldwin–Phillips–Terlevich (**BPT diagram**; Baldwin et al., 1981) based on the [N II]/H$\alpha$ and [O III]/H$\beta$ flux ratios (Fig. 11.7). These ratios measure the relative importance of high-ionisation and lower-ionisation lines. For instance, the [O III]/H$\beta$ flux ratio is higher for AGNs than for normal SFGs. As shown in Fig. 11.7, galaxies can be segregated into three regions of the BPT plane: star-forming systems dominated by stellar photoionisation (i.e. H II regions); AGNs; and galaxies where gas is ionised by shocks (§4.2.4).

In addition to the ionised gas, the colder phases of the ISM in distant galaxies allow us to place stringent constraints on galaxy evolution. As a matter of fact, the amount of molecular gas is a key property because it represents the reservoir of cold gas that can be converted into stars through star formation (§8.3). The molecular gas phase is accessible

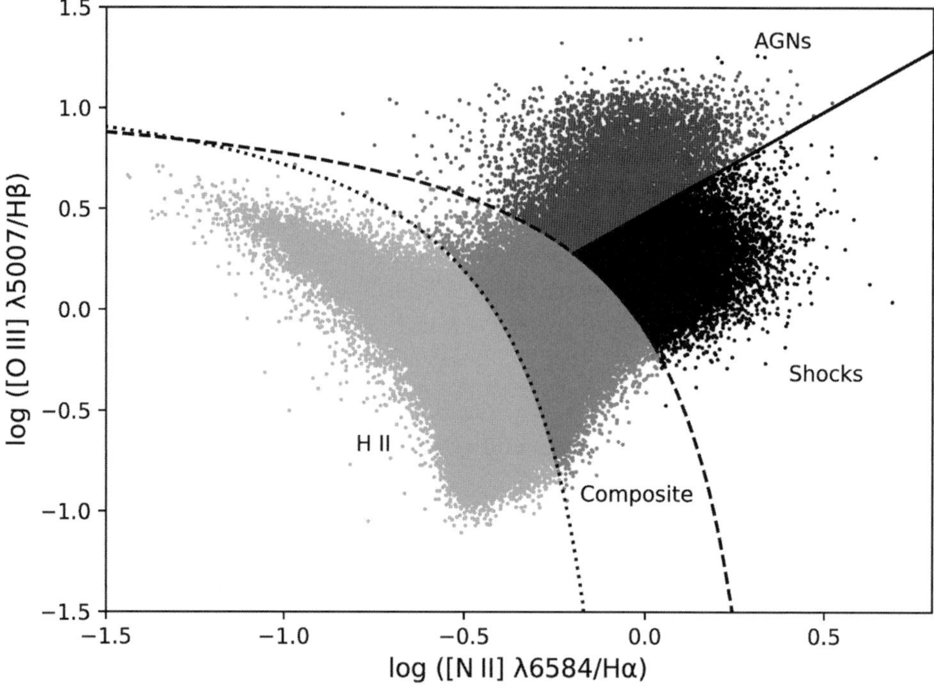


Fig. 11.7 The BPT diagram of a sample of galaxies at $0.04 < z < 0.21$ with $\mathcal{M}_\star > 10^9\ \mathcal{M}_\odot$ extracted from the SDSS. The dotted and dashed lines separate SFGs (i.e. H II regions) from AGNs according to the definitions of Kauffmann et al. (2003) and Kewley et al. (2001), respectively. The solid line segregates AGNs and shock-ionised systems (e.g. Kewley et al., 2006). Courtesy of A. Citro.

with IR to mm observations of vibrational and rotational spectral lines. This research field has been revolutionised in the IR by the *Herschel* space telescope (see Lutz, 2014, for a review), and in the submm/mm by ALMA and NOEMA. These facilities opened the unprecedented possibility to study the cold ISM of galaxies out to high redshifts. Ideally, the molecular gas should be studied through the observation of $H_2$ because this is the most abundant molecule in the Universe. However, $H_2$ has no permanent electric dipole, making it very hard or impossible to detect in emission when the gas is cold ($T_{gas} < 100$ K). Thus, CO plays a crucial role because it is the second most abundant molecule and can be collisionally excited also at low temperatures. As explained in §4.2.5, the total molecular gas mass (i.e. the $H_2$ mass) can be estimated from the CO ($J = 1 \rightarrow 0$) line luminosity through an assumed conversion factor (eq. 4.11). The mass of molecular gas can be used to estimate the gas fraction,

$$f_{H_2} \equiv \frac{\mathcal{M}_{H_2}}{(\mathcal{M}_{H_2} + \mathcal{M}_\star)}, \tag{11.9}$$

of a given galaxy (see also eq. 4.12) in order to assess the relative importance of cold gas and baryonic masses (§4.2.6; eq. 4.12). When CO ($J = 1 \rightarrow 0$) is not available in the observed spectral range, one has to resort to the use of other transitions and yet different calibrations. If more than one transition is available, one can build the so-called ***J*-ladder**,

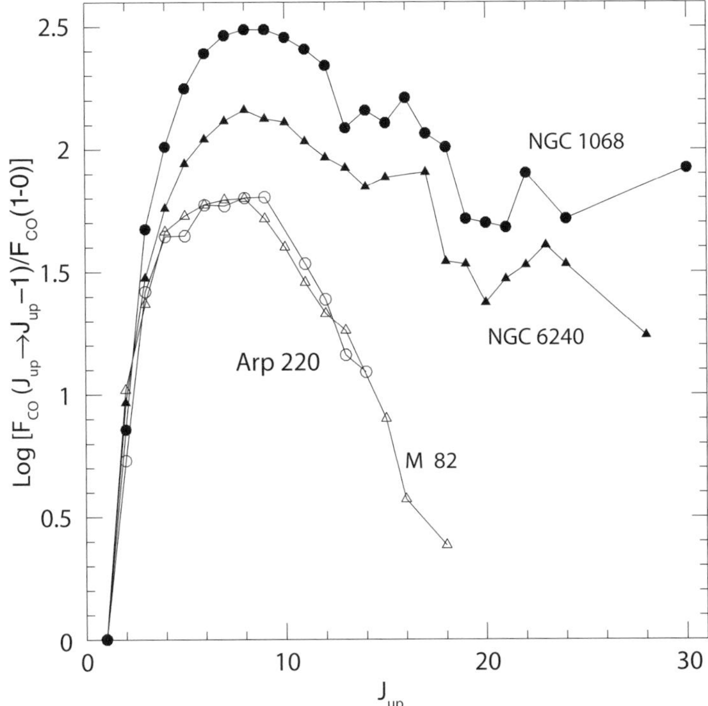


Fig. 11.8

The observed CO SLEDs of two starburst galaxies (Arp 220 and M 82) and two AGNs (NGC 1068 and NGC 6240). The CO emission line fluxes ($F_{\rm CO}$) from excited upper rotational levels ($J_{\rm up}$) are normalised to the CO ($J = 1 \rightarrow 0$) line flux. In the case of AGNs, the SLEDs are clearly extended to higher excitations. Data from Mashian et al. (2015). Courtesy of E. Sturm.

also known as the **spectral line energy distribution** (SLED). The SLED is the flux of high-$J$ lines (normalised to the $J = 1 \rightarrow 0$ flux), as a function of the upper level $J_{\rm up}$ (Fig. 11.8). Different galaxies have different SLEDs. In particular, starburst galaxies and AGNs have SLEDs markedly different from the one of Milky Way-like SFGs. Analysing the SLED helps us to choose the optimal $X_{\rm CO}$ to use in the different situations and to constrain the excitation process of the molecular gas.

### 11.1.5 Interstellar Dust

The cold ISM can also be investigated by studying the interstellar dust component in emission and through its extinction effects.

*Dust emission.* The observed SED produced by the dust continuum emission (Fig. 11.3) allows us to estimate the dust temperature ($T_{\rm d}$) and the slope $\beta$ of the efficiency factor of dust grain emission in the FIR, $Q_\lambda \propto \lambda^{-\beta}$, where $\beta$ depends on the grain composition and structure (§D.1.3). The chemical composition and structure of the grains can also be derived from IR spectroscopy. For example, silicates produce a strong absorption band at $\lambda \approx 10\ \mu$m, and the PAHs are characterised by strong emission features at $6\ \mu$m $< \lambda < 15\ \mu$m (§3.3). The total dust mass of a galaxy can be estimated from the grey-body emission of dust grains (§D.1.3). Let us consider an idealised cloud containing $N$ spherical grains, each with the same radius $a$, temperature $T_d$ and composition. If the cloud is at a distance $d$, and $\pi a^2/d^2$ is the solid angle subtended by one grain, the total received flux is

$$F_\lambda = N\left(\frac{\pi a^2}{d^2}\right) Q_\lambda B_\lambda(T_d), \tag{11.10}$$

where $Q_\lambda$ is the efficiency factor for emission in the FIR and $B_\lambda(T_d)$ is the Planck function. Eq. (11.10) is valid only if dust is optically thin to its own radiation, a condition that, for interstellar dust, is met at $\lambda_{rest} > 100$–$150\ \mu$m. Each grain has a volume $v = (4/3)\pi a^3$ and is composed of material with mass density $\rho_d$. Thus, the total volume of dust in the cloud is $V = Nv$ and the total mass is

$$\mathcal{M}_{dust} = V\rho_d = \frac{4\rho_d F_\lambda d^2}{3B_\lambda(T_d)} \frac{a}{Q_\lambda}. \tag{11.11}$$

If $T_d$ is derived from SED fitting (e.g. Fig. 11.3) the total mass of interstellar dust ($\mathcal{M}_{dust}$) can be determined assuming $\rho_d$ and $\beta$ (see Whittet, 1992, for more details). The dust mass can be exploited to estimate the mass of cold gas when this cannot be derived from CO observations (§4.2.5; eq. 4.11). In this case, the gas mass is determined from $\mathcal{M}_{dust}$ assuming an appropriate gas-to-dust mass ratio. If the ratio of our Galaxy is adopted (§4.2.7), then the total gas mass is simply $\mathcal{M}_{H_2} \approx 100\mathcal{M}_{dust}$.

*Dust extinction.* Interstellar dust extinction attenuates the intrinsic luminosity and causes the reddening of galaxy spectra (see Calzetti, 2001, for a review). Thus, the age–metallicity degeneracy can be exacerbated if dust extinction is also present: a galaxy can appear equally red because of old stars, high metallicity, dust extinction, or a combination of these causes (Fig. 11.6). It is therefore essential to estimate the amount of dust extinction (§4.2.7) to reliably recover the intrinsic galaxy properties. Two main methods can be applied to estimate dust extinction. The first exploits the flux ratio of recombination lines for which the theoretical (unextincted) value is known. For instance, photoionisation models show that the ratio $F_{H\alpha}/F_{H\beta} \approx 2.86$ is expected in the case of H II regions where 100% of the stellar photons with $\lambda < 912$ Å are used to ionise the hydrogen atoms. This method allows the derivation of the colour excess of the ionised gas (also called nebular gas) $E(B-V)_{neb}$. However, there are indications that stars are extincted differently from the nebular gas depending on the relative spatial distributions of stars, ionised gas and dust. Several studies suggest that $0.44 \lesssim E(B-V)_{star}/E(B-V)_{neb} \lesssim 1$.

The second method to estimate dust extinction is possible for SFGs, and it is based on the slope $\beta_{UV}$ of the UV spectra. The UV continuum of an SFG can be parameterised as $F_\lambda \propto \lambda^{\beta_{UV}}$, where $F_\lambda$ is the continuum flux. The slope $\beta_{UV}$ depends first of all on stellar age and metallicity, and has a typical value around $-2.2$ in the case of a galaxy with constant SFR, solar metallicity and no dust extinction. However, the UV spectra become redder ($\beta_{UV} > -2.2$) if dust extinction is present. The link between $\beta_{UV}$ and dust extinction is supported by the positive correlation between $\beta_{UV}$ and the ratio of the galaxy IR to UV flux. Thus, assuming an intrinsic (unextincted) spectrum of a newborn stellar population and a metallicity, the slope $\beta_{UV}$ can be calibrated against $E(B-V)$.

Both the above methods are affected by major limitations. The first is that the dust extinction curve in individual galaxies is poorly known (§4.2.7). This implies that the correction for dust extinction requires the assumption of an extinction curve, and this introduces systematic uncertainties. Regarding the UV-slope method, a value of $\beta_{\rm UV}$ must also be assumed for the dust-free slope. Moreover, the information derived from the UV continuum and emission lines provides a partial view because these components represent only the fraction of galaxy radiation least affected by dust extinction, whereas a significant fraction of the emission may remain largely obscured and detectable only in the IR as radiation reprocessed by dust grains.

### 11.1.6 Morphology and Structure

A major limitation in the study of distant galaxies is caused by their small angular sizes. For example, the typical half-light radii (§3.1.4) at $z > 1$ are smaller than $1''$ (1 arcsec). This means that observations with high angular resolution are required to properly study galaxy morphologies at cosmological distances. In space observations, this is feasible because the angular resolution is **diffraction-limited**, i.e. it depends only on the telescope size and the wavelength of the observation. In particular, the light emitted by an infinitely small point source produces a disc on the image plane (the detector) with an angular diameter $\theta \approx 1.22\lambda/D$, where $\theta$ is expressed in radians, $\lambda$ is the observed wavelength and $D$ is the telescope diameter. The size $\theta$ quantifies the width (e.g. the FWHM) of the **point spread function** (PSF), which describes how the flux of a point source is spread by an imaging system as a function of position on the image plane. Thus, the smaller $\theta$, the sharper the PSF, and the higher the angular resolution. As a reference, a point source observed with *HST* (2.4 metre diameter) at $\lambda = 5000$ Å produces a circular image with a size $\theta \approx 0.05''$. This is clearly sufficient to discern the structure and morphology of galaxies out to large distances. For instance, $0.05''$ corresponds to $\approx 0.4$ kpc at $z \approx 2$. However, the angular resolution reachable from the ground is much worse because the turbulent motions of the terrestrial atmosphere prevent the formation of sharp images. In particular, the thermal turbulence of the atmosphere causes several effects on scales of a few arcseconds such as scintillation due to the high atmosphere ($\approx 12$ km altitude) and other distortions originated by the lower layers. As a net result, the image of a point source appears fuzzy on the ground and has an angular size as large as the diameter of the so-called **seeing disc** (or simply **seeing**). Thus, the seeing is basically the PSF of a point source produced by the atmospheric turbulence, and it measures the optical quality of the atmosphere in a given site. The smaller the seeing, the better the atmospheric conditions. A point source observed at $\lambda = 5000$ Å with an 8 metre diameter telescope on the ground during typical seeing conditions of $1''$ produces a circular image whose size is also $1''$. If the same 8 metre telescope were in space, the PSF would be $0.015''$, and the angular resolution would be $(1''/0.015'') \approx 70$ times better than on the ground! Even in the best atmospheric conditions, the highest angular resolution achievable from the ground in the optical and NIR is rarely better than $\approx 0.3''$–$0.4''$, corresponding to $\approx 2.5$–$3.5$ kpc at $z \approx 2$. This implies that high-$z$ galaxies are barely resolved or unresolved in ground-based imaging observations. In comparison, low-redshift galaxies are much larger. For instance, the starburst galaxy M 82 (distance $\approx 3$ Mpc) has an angular size of $\approx 300'' \times 100''$. This is why the morphological study of distant galaxies requires telescopes in space such as *HST*.

A possibility to increase the resolving power of ground-based observations is provided by **adaptive optics**. These optical systems allow us to correct the image distortions caused by the atmospheric turbulence thanks to deformable mirrors controlled by fast computers. Adaptive optics works better in the NIR and relies on the availability of one or more bright point sources close to the scientific target. These sources allow us to monitor the temporal variation of the local turbulence and compensate for its effects. These bright reference sources can be real stars (**natural guide stars**) if they are available close to the target. If natural stars are not available, it is possible to create **laser guide stars** by shining powerful lasers into the sodium layer present in the terrestrial atmosphere at an altitude of $\approx 80-100$ km. The excited sodium atoms emit monochromatic light at 5892 Å and produce an **artificial star**.

The galaxy structure and the morphological components are investigated by measuring the shapes and the profiles of their surface brightness (§3.1.2 and §3.1.3). However, when this technique cannot be applied due to the small angular size and/or faintness of distant galaxies, it is still possible to derive information on galaxy structure through the **non-parametric morphological indices** (also called **structural indices**). These indices measure the spatial distribution of light within a galaxy. Examples of widely used indices are the **concentration** ($C$), **asymmetry** ($A$) and **clumpiness** ($S$) (Fig. 11.9).

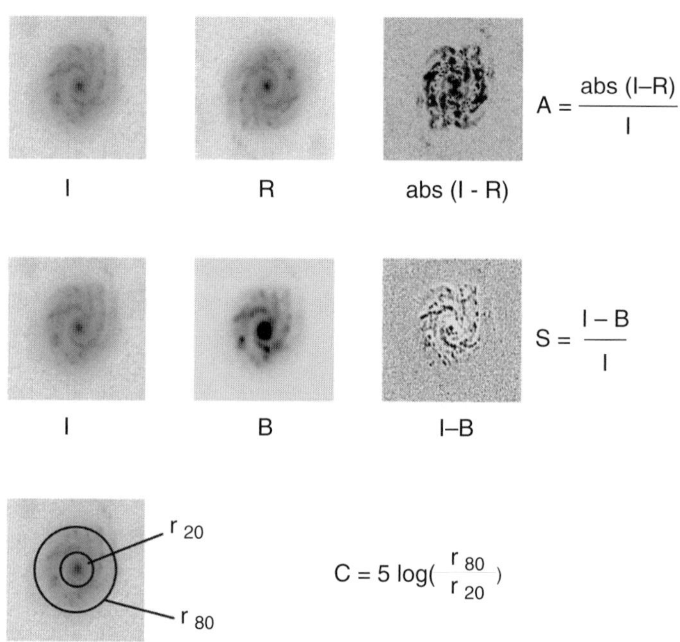


Fig. 11.9 Graphical representation of the $C$, $A$ and $S$ parameters for a face-on spiral galaxy. From Conselice (2003). ©AAS, reproduced with permission.

The concentration parameter measures the amount of light in the inner region compared to the outer region:

$$C \equiv 5 \log \left( \frac{r_{80}}{r_{20}} \right), \tag{11.12}$$

where the radii $r_{20}$ and $r_{80}$ contain 20% and 80% of the total light, respectively.

The asymmetry parameter quantifies the importance of non-symmetric components within the distribution of a galaxy light. If we call $I$ the original image of the galaxy and $R$ the same image rotated by 180° from its centre, the asymmetry parameter is defined as

$$A \equiv \frac{\mathrm{abs}(I - R)}{I}, \tag{11.13}$$

where $I - R$ is the image resulting from the subtraction of $R$ from $I$, and abs$(I - R)$ and $I$ have to be intended as the sums (performed on the images) of the absolute values of the pixel intensities over the area covered by the galaxy.

The clumpiness parameter measures how much the light is distributed in clumps (e.g. the H II regions in the arms of spiral galaxies). If $I$ is the original image and $B$ is the same image convolved with a Gaussian filter in order to produce a smoother (blurred) image, the clumpiness parameter is defined as

$$S \equiv \frac{I - B}{I}, \tag{11.14}$$

where $I - B$ and $I$ have to be intended as sums over the pixels of the images.

In addition to $C$, $A$ and $S$ (often referred to as $CAS$ parameters), two other parameters are also frequently used. The first is the **Gini coefficient** ($G$), which measures how the light is distributed from pixel to pixel, and ranges from 0 (homogeneous distribution) to 1 (all light concentrated in one pixel). The second is **M20**, which is the second-order moment of the brightest 20% of the galaxy flux. $M20$ is similar to the concentration index, and it has been found to be particularly sensitive to the morphological features typical of galaxy mergers. The $CAS$, $G$ and $M20$ parameters are complementary to each other and are broadly correlated with the Hubble morphological types. For instance, ellipticals are characterised by the highest $C$ and lowest $A$ and $S$, whereas discs and irregular SFGs have lower $C$ and higher values of $A$ and $S$ than ETGs.

### 11.1.7 Kinematics and Dynamical Masses of Distant Galaxies

Information on stellar and gas kinematics is of paramount importance to understand the dynamical properties of galaxies. Because of the small angular size of distant galaxies, the best information with ground-based observations is derived with integral field spectroscopy assisted by adaptive optics (§4.3.2) in the NIR (rest-frame optical). With 8–10 metre diameter telescopes, this technique allows one to derive spatially resolved kinematic data with ~ kpc resolution for galaxies at $z \approx 2$ using bright emission lines emitted from ionised gas and redshifted in the NIR (e.g. H$\alpha$ and [O III]$\lambda$5007). The kinematics of the cold gas component requires observations of molecular (e.g. CO) or atomic (e.g. [C II]$\lambda$158 $\mu$m) lines redshifted in the submm/mm and at radio wavelengths. These observations are performed with interferometers such as ALMA (in the southern hemisphere) and NOEMA (in the northern hemisphere), and with the VLA in the radio.

The datacubes (§4.3.2) obtained with these facilities allow us to perform a kinematic analysis of distant galaxies and to understand whether a given system is supported by rotation or velocity dispersion, or if it is undergoing a merger event. This kinematic information is also of primary importance to estimate the dynamical masses based on the rotation curves in the case of disc galaxies. When adaptive optics observations are not feasible, or galaxies remain unresolved, basic information on kinematics can still be extracted from spatially integrated spectra. For instance, the rotation speed can be estimated when a double-peaked emission line is present (§4.3.3). Moreover, the shape and profiles of the spatially integrated spectral lines, as well as their velocity offsets with respect to the galaxy systemic velocity, are essential to place constraints on the bulk motions of stars or the inflows and outflows of gas. In the case of ETGs, the velocity dispersion of stellar absorption lines, combined with the effective radius, allows us to estimate the dynamical mass (eq. 5.17) of these systems and study their scaling relations.

### 11.1.8 Statistical Distributions

The statistical distributions of galaxy properties (e.g. luminosity, mass, SFR, size and structural components) as a function of redshift are fundamental evolutionary tracers (§3.5.1). If we take the LF (eq. 3.7) as a working example, three cases can be envisaged for the evolution of the number density and luminosity of a galaxy population.

1. The characteristic number density remains constant ($\Phi^*(z) = \text{const}$), and only the characteristic luminosity changes ($L^*(z) \neq \text{const}$). This is called **pure luminosity evolution**.
2. $L^*(z)$ does not evolve, and only $\Phi^*(z)$ changes with redshift. This is the **pure density evolution**.
3. Galaxies evolve by changing both their number density and luminosity. This is the most realistic case because, on the one hand, the number density of a given galaxy type is expected to evolve due to the morphological transformations and/or mergers (§11.3.1) and, on the other hand, the luminosities of galaxies inevitably change with redshift due to the ageing of the stellar populations, the evolution of the SFR and ISM properties, and, if present, the AGN activity.

### 11.1.9 Galaxy Environment and Clustering

The environment plays a key role in galaxy evolution (§6.5). Several methods have been developed to estimate the local galaxy density in a given region of the Universe (**density field**). One approach to measure the density around a galaxy is to count the surrounding galaxies within a fixed (arbitrary) distance scale, e.g. within 1 Mpc, and derive the projected density (number of galaxies Mpc$^{-2}$). For instance, $\Sigma_N$ defines the surface density of galaxies around a galaxy based on the projected distance to the $N$th nearest neighbouring galaxy in the 2D or 3D space around a central galaxy considered as $N = 0$ or

$N=1$ (see the top $x$-axis of Fig. 6.13). Different values of $N$ are adopted in the literature, typically $N = 10$ or smaller. The 2D method can be applied within a given interval of redshift ($\Delta z$) or radial velocity (e.g. $\pm 1000$ km s$^{-1}$) centred on the considered galaxy. The accuracy and reliability of this method depend on the depth of the data, the choice of the number of neighbours and the search radius around a given galaxy. Another approach is to estimate the 3D density within a given volume, where two dimensions are given by the coordinates on the sky plane, and the third is provided by the redshift, which allows us to measure the galaxy distances. Other methods can be used (e.g. counts-in-cells and Voronoi tessellation), but their description is beyond the scope of this book. Clearly, all methods require that the galaxy redshifts are known, and the accuracy of the results is highest if the redshifts are spectroscopic (§11.1.1; Fig. 11.2).

As illustrated in §6.6, the spatial distribution of galaxies on large scales (galaxy clustering; Fig. 6.14) is a crucial probe of galaxy evolution and has several links with cosmology (see Coil, 2013, for a general review). One of the main tools to study galaxy clustering is the two-point correlation function $\xi_{\rm gal}(r)$ (§6.6; eq. 6.24). This function can be derived from the observation of a sample of galaxies within a given volume that must be large enough to minimise the influence of cosmic variance (§11.2.2). The first step is to create a catalogue of artificial objects with positions randomly distributed within the identical boundaries of the observed sample (sky coverage, volume, number of galaxies and redshift distribution). Once this random catalogue has been generated, and the number densities of observed galaxies ($n_{\rm D}$) and random objects ($n_{\rm R}$) are derived, $\xi_{\rm gal}(r)$ can be estimated as

$$\xi_{\rm gal}(r) = \frac{1}{RR(r)}\left[DD(r)\left(\frac{n_{\rm R}}{n_{\rm D}}\right)^2 - 2DR(r)\left(\frac{n_{\rm R}}{n_{\rm D}}\right) + RR(r)\right], \quad (11.15)$$

where $RR(r)$ is the number of pairs of random objects with separation within $r - {\rm d}r/2$ and $r + {\rm d}r/2$ ($r$ is calculated in comoving space in units of $h^{-1}$ Mpc), $DD(r)$ is the number of pairs of observed galaxies and $DR(r)$ is the number of pairs of observed galaxies and random objects within the same interval. It is important to notice that, if the galaxies have a random distribution, $\xi_{\rm gal}(r) = 0$. As dark matter halos cannot be observed directly, their correlation function can be reliably derived (for a given cosmological model such as $\Lambda$CDM) from their spatial distribution within an $N$-body simulation of the dark matter component only (§7.3.2). The study of clustering is essential to investigate the interplay between dark and baryonic matter across cosmic time (§10.10.1). For instance, the observed clustering and bias (eq. 6.26) of a given galaxy population, in combination with the results of cosmological simulations of dark matter halos, allow us to estimate the mean mass of the halos which host such galaxies. For more details, we refer interested readers to the review of Wechsler and Tinker (2018).

## 11.2 The Difficult Observation of Distant Galaxies

The observation of distant galaxies is challenging because they become rapidly very faint with increasing redshift due to the dependence of the observed flux on the luminosity distance (flux $\propto d_{\rm L}^{-2}$ at fixed luminosity; §2.1.4). At $z > 1$, most galaxies have optical apparent magnitudes fainter than $r \approx 23$–$24$. The major difficulty in the observation of such faint objects is due to the **sky background**. The night sky is not dark, and its surface brightness can be much higher than the fluxes of high-redshift galaxies. For instance, in nights without moonlight, the sky surface brightness is $\approx 21.6$, $\approx 20.3$ and $\approx 14.8$ mag arcsec$^{-2}$ in the $V$, $I$ and $K$ bands, respectively, i.e. much brighter than the typical magnitudes of distant galaxies. The sources of the sky background can be extraterrestrial and terrestrial.

*Sources of extraterrestrial background.* An important source of sky background in the optical/NIR is the **zodiacal light** originated by the solar radiation scattered by dust particles in the inner solar system. In addition, the Milky Way itself produces a diffuse Galactic light made of starlight scattered by interstellar dust, ISM emission lines, IR thermal continuum and emission bands emitted by ISM dust grains. Moreover, the diffuse extragalactic light due to faint undetected galaxies can also play a role (§11.3.15).

*Sources of terrestrial background.* In the optical and NIR, the sky surface brightness is actually the sum of several contributions (e.g. Fig. 11.10). The **airglow** (also called **nightglow**) plays a dominant role in ground-based observations and is mainly due to line emission from excited atoms (e.g. O and Na in the green–yellow spectral region)

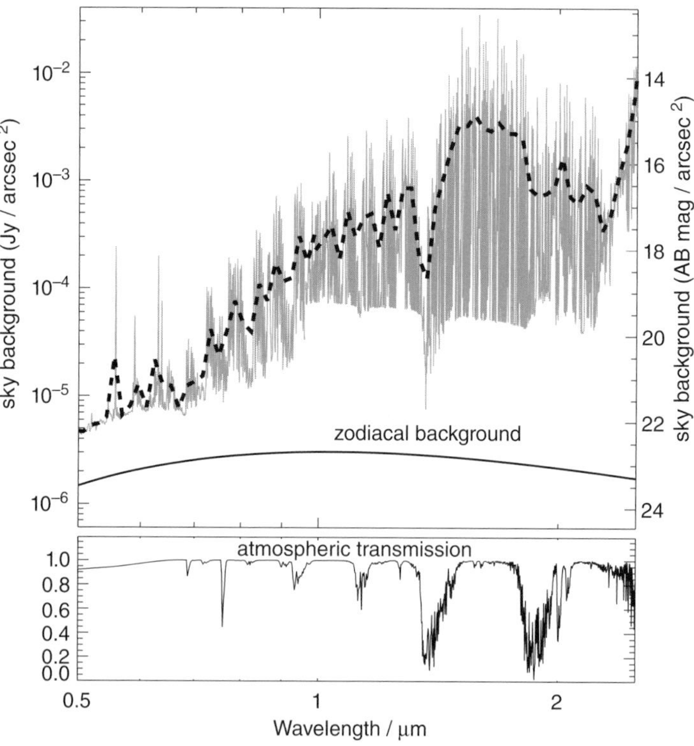


Fig. 11.10 *Top panel.* The surface brightness of the optical/NIR airglow as a function of wavelength at spectral resolutions (eq. 3.4) $R = 2000$ (grey) and $R = 43$ (dashed black curve). The zodiacal light is also displayed as a reference to show the strong reduction of the background allowed by space-based observations. *Bottom panel.* The terrestrial atmospheric transmission curve showing the broad telluric absorption bands ($\lambda > 1\ \mu$m) where the atmosphere is too opaque to allow high-quality observations. Courtesy of I. Baldry.

and molecules (e.g. OH in the optical/red and NIR) located in the upper atmosphere at 80–100 km altitude. The airglow intensity is highly variable in space and time on scales of $\sim 10$ km and $\sim 10$ minutes, respectively. A diffuse background is also present due to the starlight scattered by the atmosphere. The terrestrial atmosphere produces also several absorption lines and bands (called **telluric lines**) in the optical/red and NIR (Fig. 11.10). Being absorption features, they do not contribute to the luminous sky background, but they make observations nearly impossible in these spectral regions due to the high opacity of the atmosphere. In the IR, the Earth itself contributes to the sky background with its thermal emission. Last but not least, local sources of background such as thermal emission by the telescope and **straylight** (unintended scattered light in optical systems) can also be relevant. Obviously, a strong reduction of the terrestrial background is achieved with space-based observations, where the zodiacal light remains the dominant source of natural background at optical and NIR wavelengths (Fig. 11.10). As an example to illustrate the major gain of the observations from space, the sky surface brightness at the north ecliptic pole due to the zodiacal light is $\approx 20.6$ mag arcsec$^{-2}$ at $\lambda \approx 2.2$ $\mu$m ($K$ band), i.e. almost $\approx 6$ magnitudes fainter than in ground-based observations.

Due to the faintness of high-$z$ galaxies, long integration times are needed to obtain high-quality data. For instance, some of the deepest *HST* observations imaged the same region of the sky ($\approx 5$ arcmin$^2$) for about three weeks to detect galaxies as faint as $\approx 29$–$30$ mag. Spectroscopy is more challenging than imaging and cannot reach the same depth for a given telescope because photons are dispersed as a function of wavelength on the detector. The performances of astronomical observations can be quantified using the **signal-to-noise ratio**

$$S/N \equiv \frac{F}{\sigma_F}, \tag{11.16}$$

where $F$ is the flux of a given astronomical source and $\sigma_F$ is the total uncertainty associated with the flux measurement. This uncertainty is due to the contributions of noise from the sky background, the detector and the Poissonian fluctuations of the source flux itself. For instance, $S/N = 3$ means that the source is detected with a statistical significance of $3\sigma$ ($F/\sigma_F = 3$). As a reference example, to reach $S/N \approx 5$ (i.e. $\approx 5\sigma$) for a point source with magnitude $I \approx 24.5$, the FORS2 instrument at the VLT requires $\approx 125$ seconds in imaging and $\approx 21$ hours in spectroscopy at spectral resolution $R \approx 660$ (§3.4). When the sky background is higher than the galaxy flux, the signal-to-noise ratio increases slowly with the integration time $t$ as $S/N \propto t^{1/2}$. This is why the observation of distant (i.e. faint) galaxies requires large amounts of telescope time.

### 11.2.1 Surveys for Distant Galaxies

In the astronomical language, a **survey** is an observational campaign producing a set of imaging and/or spectroscopic data aimed at selecting a sample of objects from a given sky area. In **imaging surveys**, images of the same sky field are obtained with one or more photometric filters. In **spectroscopic surveys**, the observations are done with spectrographs able to obtain the spectra of several objects simultaneously. Searching for distant galaxies requires surveys designed for this purpose. In order to have a complete census of all galaxies, from the faintest to the rarest and most luminous, an ideal survey should cover a wide sky area and detect very faint galaxies. However, telescope time is always limited, and a given amount of observing time is inevitably used to perform either a very deep survey on a small sky field, or a shallower one on a wide area. Thus, a practical approach is to combine together surveys with different depths and sky coverages.

At a given observed wavelength $\lambda$, any survey can only *select* galaxies with fluxes higher than a given **limiting flux** ($F_\lambda > F_{\lambda,\mathrm{lim}}$) or with apparent magnitudes brighter than a **limiting magnitude** ($m_\lambda < m_{\lambda,\mathrm{lim}}$). The limiting flux (or magnitude) defines the so-called **depth** of a survey. An important characteristic of a survey is its **completeness**. This indicates the percentage of galaxies that a survey can select with respect to the total population. The completeness can refer to observable quantities (e.g. fluxes or apparent magnitudes) or physical properties derived from the data (e.g. luminosity, stellar mass and SFR). For instance, an imaging survey that is 90% complete to $K = 21$ means that 90% of the total galaxy population with apparent magnitude brighter than $K = 21$ has been detected. The accurate evaluation of the completeness is extremely important to avoid systematic errors and for a meaningful interpretation of the results. In order of increasing sky coverage and decreasing depth (i.e. limiting magnitude), notable examples of imaging surveys for high-$z$ galaxies are the Hubble Ultra Deep Field ($\approx 10$ arcmin$^2$; $m_{\mathrm{lim}}(\mathrm{NIR}) \approx 29$–$30$), GOODS/CANDELS ($\approx 800$ arcmin$^2$; $m_{\mathrm{lim}}(\mathrm{NIR}) \approx 26.5$–$27$) and COSMOS (2 deg$^2$; $m_{\mathrm{lim}}(\mathrm{NIR}) \approx 24$–$26$).

### 11.2.2 Main Biases in Galaxy Surveys

A **bias**[1] is a systematic effect which leads to an incorrect interpretation of the observational results. The evaluation of the biases is vital to any study based on galaxy surveys. Every galaxy survey is affected by two main biases caused by the unavoidable limiting flux. The first is the **Malmquist bias** (Malmquist, 1922). Galaxies with luminosities $L < L_{\mathrm{lim}} = 4\pi d_{\mathrm{L}}^2 F_{\mathrm{lim}}$ are not detected because they are fainter than the limiting flux ($F < F_{\mathrm{lim}}$). Clearly, the larger the distance, the higher the limiting luminosity $L_{\mathrm{lim}}$. This implies that at high redshift only the most luminous galaxies are selected in a given survey. Thus, the fraction of 'missed' galaxies (i.e. those with $L < L_{\mathrm{lim}}$) increases with redshift simply because their observed flux is too faint to be detected. Another important bias is the **Eddington bias** (Eddington, 1913). The origin of this bias lies in the inevitable measurement errors which introduce statistical fluctuations of the derived quantities around their true values. For instance, if a galaxy sample is divided into luminosity bins, the photometric errors move a fraction of galaxies from one bin to the adjacent ones. Since faint galaxies are much more numerous than bright ones, relatively few bright sources enter the fainter bin, whereas a larger fraction of faint sources enters the brighter bin. A correction for these effects is therefore required in order to derive reliable luminosity or stellar mass functions (§3.5.1).

An additional source of systematic uncertainties is the so-called **cosmic variance**. Galaxies are clustered in 3D space (§6.6) and, for any survey covering a finite sky area

[1] Not to be confused with the galaxy bias described in §6.6.

or volume of the Universe, this causes variations in the galaxy number density larger than those expected for pure Poissonian noise. This underlying LSS can make the sky surface density of galaxies highly inhomogeneous (§6.6). In particular, if the surveyed sky area is smaller than the projected clustering scale length ($r_0$; §6.6) of the targeted galaxy population at a given redshift, then the number of selected galaxies varies significantly depending on whether the observed sky field includes low- or high-density regions. This can have a strong impact on the galaxy number counts, and consequently on the statistical distributions such as the luminosity and mass functions. The cosmic variance depends on the observed sky area, the redshift range and the clustering of the observed galaxies. As a reference, for galaxies with $\mathcal{M}_\star > 10^{11}\ \mathcal{M}_\odot$ at $1.5 < z < 2.5$, the fractional error in the galaxy number counts due to the cosmic variance is $\approx 38\%$ for a sky field of 160 arcmin$^2$, and it decreases to $\approx 12\%$ for 2 deg$^2$. For less massive galaxies ($\mathcal{M}_\star \sim 10^{10}\ \mathcal{M}_\odot$), the cosmic variance is smaller by about a factor of 2, due to the weaker clustering of these galaxies (§6.6). The cosmic variance can be mitigated by observing sky areas significantly wider than the clustering scalelength of the targeted galaxies. Fig. 11.11 shows the LSS at $z = 3$ as projected on the sky and compared to the areas covered by two surveys. It is evident that wide-area surveys are needed to mitigate the cosmic variance.

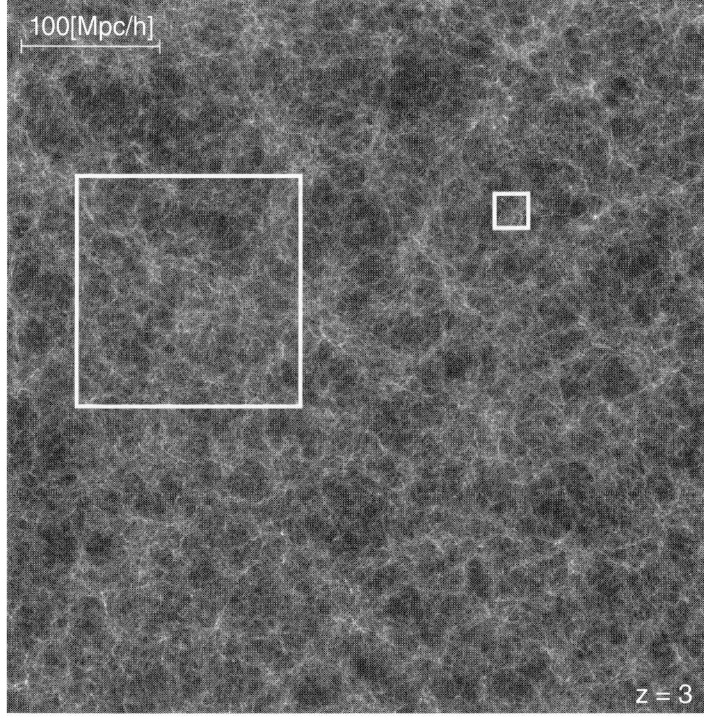


**Fig. 11.11** The spatial distribution of dark matter halos at $z = 3$ based on the cosmological $N$-body simulation of Lacey et al. (2016). The boxes show how a survey of 2 square degrees (large box) or 160 square arcminutes (small box) would sample the LSS projected on the sky plane. Courtesy of A. Orsi.

Other systematic effects in the interpretation of survey results can be introduced by the so-called **progenitor bias**. This is relevant when the evolution of a given galaxy population is investigated as a function of redshift. A typical case is the evolution of ETGs derived through the comparison of low- and high-redshift samples of these galaxies. The progenitor bias arises from the assumption that the properties of distant ETGs (say at redshift $z_2$) can be compared directly with those at lower redshift ($z_1 < z_2$). However, during the cosmic time interval between $z_2$ and $z_1$, some galaxies that were not ETGs at $z_2$ (for instance they were discs or merging systems) may have evolved into ETGs by $z_1$ and entered in the low-$z$ sample. In other words, if only ETGs are considered at $z_2$, this would introduce a bias towards galaxies that were already ETGs at $z_2$, and cause the exclusion of those galaxies that will become ETGs by $z_1$. The general effect of this bias is to provide an apparent evolution of ETGs which can be significantly different from the true one. Thus, this bias should be taken into account in the evolutionary studies where the properties of distant galaxies (e.g. morphology, size, luminosity and colours) are compared with those at $z \approx 0$.

### 11.2.3 The K Correction

If galaxies at different redshifts are observed through an observed passband (typically a photometric filter; §C.4), different wavelength ranges are sampled in the rest-frame galaxy spectra depending on redshift. For instance, the $V$-band filter (centred at $\lambda_{\rm obs} \approx 5500$ Å for $z = 0$) samples spectral regions centred at $\lambda_{\rm rest} \approx 2700$ Å and $\lambda_{\rm rest} \approx 1800$ Å for $z \approx 1$ and $z \approx 2$, respectively. It is therefore necessary to correct for this effect to perform a meaningful comparison between high- and low-redshift galaxies. This can be achieved through the so-called **K correction** which allows one to convert the observed fluxes (or magnitudes) at different redshifts to the equivalent measurements in a common spectral band in the rest frame of the galaxies. The estimate of the K correction requires the knowledge of the intrinsic spectral shape of the observed galaxies. Let us suppose that we observe a galaxy at a given redshift $z$ and measure its apparent magnitude $m_X$ through a photometric filter $X$. The absolute magnitude $M_Y$ (§C.3.1) of this galaxy in the $Y$ filter is given by

$$m_X = M_Y + 5\log\left[\frac{d_{\rm L}(z)}{10\ {\rm pc}}\right] + k_{YX}(z), \tag{11.17}$$

where $d_{\rm L}$ is the luminosity distance (eq. 2.11) and $k_{YX}$ is the term called K correction (see Hogg et al., 2002, for more details). Eq. (11.17) is relative to the general case when the $X$ and $Y$ filters are different. For instance, we may measure the apparent magnitude of a given galaxy with filter $X = R$ and derive the absolute magnitude relative to filter $Y = K$. When $X$ and $Y$ are the same filter (i.e. $X = Y$), the K correction term becomes $k_{XX}(z)$. Examples of K corrections are given in Fig. 11.12 for the case of $k_{XX}(z)$.

### 11.2.4 Searching for Distant Galaxies

The identification of galaxies at cosmological distances requires efficient methods capable of selecting them from all galaxies present in a given sky field. Two examples show that searching for high-$z$ galaxies is equivalent to finding a needle in a haystack. For

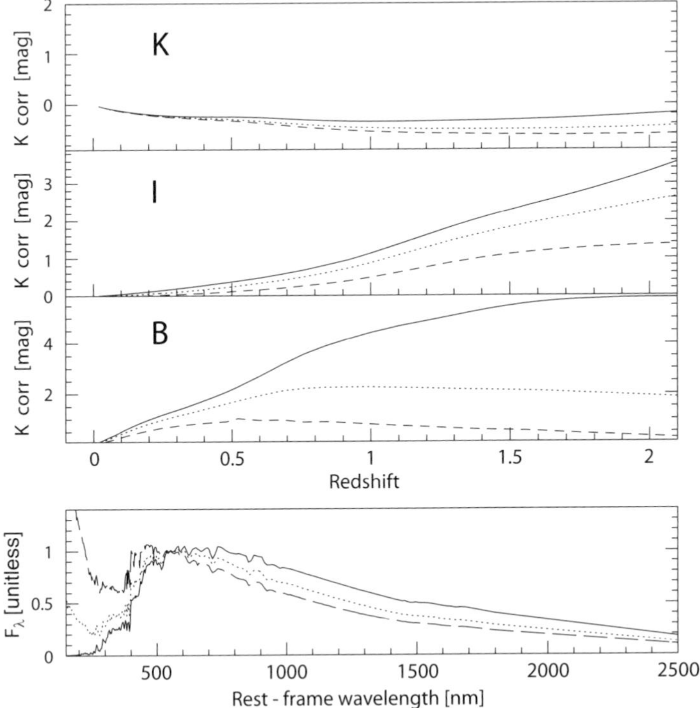


Fig. 11.12 *Top panel*. K corrections ($k_{XX}(z)$; eq. 11.17) as a function of redshift for the filters $K$, $I$ and $B$, and for galaxies of different Hubble types (ellipticals, solid; Sa, dotted; Sc, dashed). *Bottom panel*. SEDs of the galaxies used to compute the K corrections in the top panel. Data from Poggianti (1997).

instance, Lyman-break galaxies at $z \approx 3$ (§11.2.5) have a sky surface number density of $\approx 1.2$ arcmin$^{-2}$ for a limiting magnitude of $R < 25.5$, corresponding to about 5% of the total galaxy counts at the same depth. At $z \approx 9$–10, the same type of galaxies has a surface number density of $\approx 0.01$–0.03 arcmin$^{-2}$ for an NIR limiting magnitude $H < 26$–26.5, about 0.01% of the surface number density of all galaxies at the same depth. Despite the rarity and faintness of high-redshift galaxies, several methods have been successfully developed to select them based on their characteristic colours. The definition of the optimal **colour selection** is based on two main criteria: (1) the shape of the SED of the target galaxies, and (2) the redshift range at which the SED features are expected to influence the observed colours. This approach is clarified in the next sections, where the selection criteria for distant galaxies, as well as their main physical and structural properties, are illustrated as a function of the observed wavelength.

### 11.2.5 Selection in the Optical (Rest-Frame Ultraviolet)

At $z > 1$, the optical filters (§C.4) sample the rest-frame UV. This implies that the selection of distant galaxies at optical wavelengths favours young SFGs (i.e. luminous in the rest-frame UV) with low dust extinction. The presence of old stars or strong dust extinction would rapidly redden the UV continuum and make these galaxies fainter and difficult to detect in the observed optical range.

An efficient technique to select distant SFGs is called the **Lyman-break** or **dropout technique**. At rest-frame wavelengths shortward of Ly$\alpha$ ($\lambda_{\rm rest} < 1216$ Å), the UV continuum of SFGs is strongly depressed by the absorption of UV photons due to the neutral hydrogen present within the galaxy (Lyman break at $\lambda_{\rm rest} < 912$ Å) and in the IGM (§9.8.1) along the line of sight to the observer. At $z \approx 3$, the Lyman break is redshifted in the $U$ band, making these galaxies very faint and basically undetectable with this filter, but detectable with redder filters (e.g. $B$ and $V$) (Fig. 11.13 and Fig. 11.14). Thus, at $z \approx 3$, galaxies of this type (called **Lyman-break galaxies**, LBGs) can be easily segregated in a $U - B$ versus $B - V$ colour–colour diagram due to their very red $U - B$ colours. The Lyman-break criterion can be extended to $z > 3$ simply using combinations of redder filters. For instance, LBGs at $z > 4$ and $z > 5$ can be selected as $B$-band and $V$-band dropouts, respectively (Fig. 11.15). However, at $z < 3$ the Lyman break is redshifted blueward of

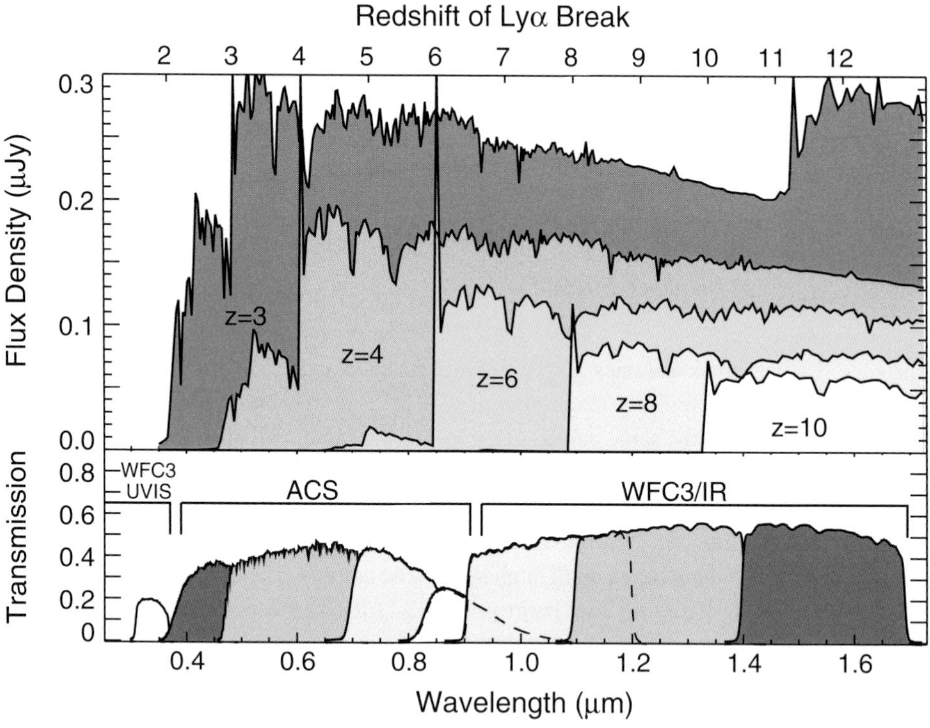


Fig. 11.13 The Lyman-break selection of SFGs. *Top panel*. The theoretical SSP spectrum for a galaxy with $\mathcal{M}_\star = 10^9\ \mathcal{M}_\odot$, age of $10^8$ yr and $E(B-V) = 0.03$ redshifted to $z = 3, 4, 6, 8$ and 10. The Lyman break is progressively shifted to longer observed wavelengths ($x$ axis) with increasing redshift. The flux density gradually decreases due to the increasing luminosity distance. At $z < 6$ the Lyman break is redshifted to optical wavelengths (ACS filters), whereas it moves to the NIR at $z > 7$ (WFC3 filters), and the galaxy becomes completely invisible at optical wavelengths (see also Fig. 11.14). *Bottom panel*. The transmission curves of *HST* filters in the observed UV/optical (ACS instrument) and NIR (WFC3 instrument). Adapted from Finkelstein (2016).

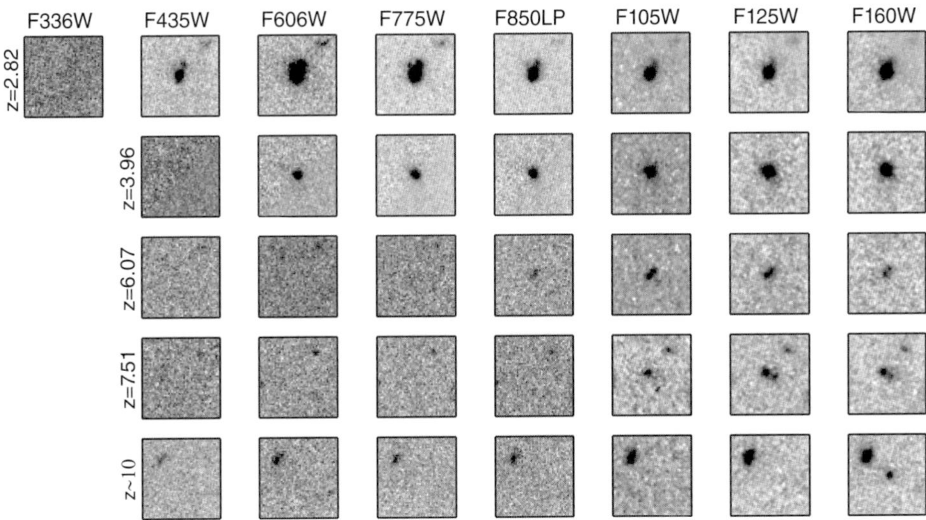


Fig. 11.14 Examples of five LBGs with increasing redshift from the *top row* to the *bottom row*. Each cutout *HST* image has a size of $3'' \times 3''$. This corresponds to $\approx 24 \times 24$ kpc and $\approx 12.8 \times 12.8$ kpc for $z = 2.82$ and $z = 10$, respectively. The labels on the top indicate the name of the *HST* filters shown in Fig. 11.13 (see Fig. C.3). For instance, the galaxy at $z = 6.07$ is invisible with filters F435W, F606W and F775W, but it becomes visibile at longer wavelengths (filter F850LP and redward). Adapted from Finkelstein (2016).

the $U$-band filter (e.g. $\approx 2700$ Å for $z \approx 2$), where the terrestrial atmosphere is opaque. Thus, space-based UV imaging is needed to apply the LBG selection at $z < 3$. In order to circumvent the need for space telescope data, other criteria have been developed to select UV-luminous SFGs at $1 < z < 3$ (see Shapley, 2011, for a review).

## Main Properties of Optically Selected Galaxies

At the limiting magnitudes accessible with ground-based spectroscopy with 8–10 metre diameter telescopes (typically $R < 26$), the vast majority of optically selected galaxies at $z > 1$–$2$ are star-forming. These systems show UV spectra with several ISM absorption lines produced by elements with a wide range of ionisations, as well as photospheric lines of young massive stars (§3.4 and §3.8). By construction, the LBG selection is based on the assumption of low dust reddening, and therefore dust extinction is necessarily low for these galaxies, typically with $E(B-V) < 0.2$. Some LBGs show Ly$\alpha$ emission, and the equivalent width of this line is anticorrelated with $E(B-V)$. This is ascribed not only to the absorption of Ly$\alpha$ photons by dust grains, but also to the resonant scattering with the neutral hydrogen present in the ISM (§9.8.2), and whose column density is proportional to $E(B-V)$. Due to the ongoing star formation, the rest-frame optical spectra show also the recombination and forbidden emission lines typical of H II regions (Balmer lines, [O II]$\lambda$3727 and [O III]$\lambda$4959,5007). The emission line ratios confirm that star formation is the dominant source of ionisation, although AGN activity is present in a few per cent of these galaxies. In some cases, the relative strength of the emission lines can be reproduced

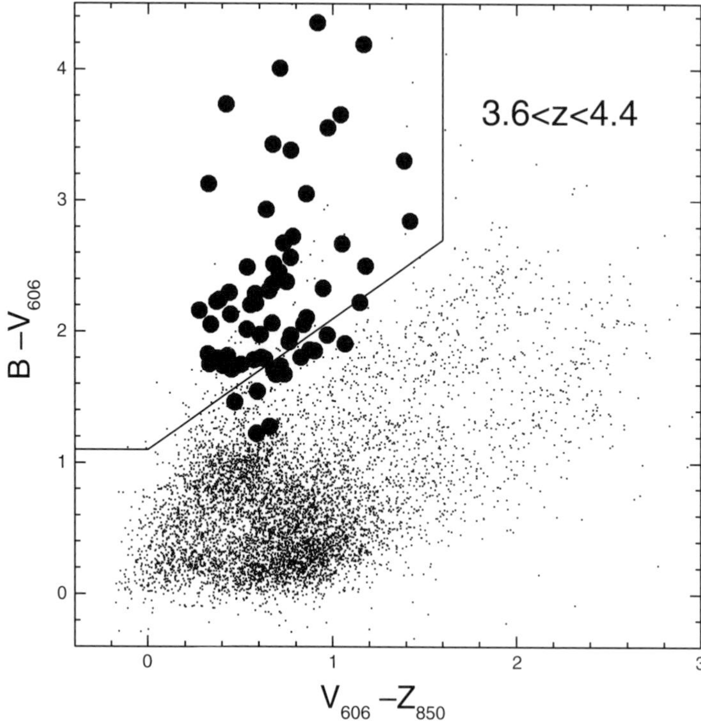


Fig. 11.15 Example of a colour–colour diagram to select LBGs at $z \approx 4$ as dropout galaxies in the $B$ band. LBGs at $3.6 < z < 4.4$ (large filled symbols) are preferentially located in the upper left selection region. The smaller dots show the colours of the other galaxies in the same field of 170 arcmin$^2$ down to a limiting magnitude $Z_{850} = 25.0$. The subscripts indicate the central wavelengths (in nm) of the $V$ and $Z$ filters. Courtesy of A. Grazian.

only if the photoionisation is caused by stellar populations emitting UV photons harder than in typical low-redshift SFGs, therefore suggesting the presence of young massive stars with higher masses and/or lower metallicities. The metallicity of the ionised gas of LBGs is typically subsolar in the range of $1/10 < Z/Z_\odot < 1$ and correlates with the stellar mass. The estimate of the stellar metallicity is challenging due to the weakness of the UV photospheric absorption lines, but the most accurate results suggest that it is comparable to the metal abundance of the ionised gas. The rest-frame UV spectra show blueshifted ISM absorption lines indicating large-scale motions due to gas outflows with typical velocities of 100–200 km s$^{-1}$. These outflows are thought to be generated by SN explosions and powerful stellar winds from hot massive stars (§8.7). These speeds increase up to $\approx 1000$ km s$^{-1}$ when AGN activity is present. At limiting magnitudes $R < 25$–$26$, optically selected SFGs at $z \approx 2$–$4$ display a wide range of SFR $\approx 10$–$100$ $\mathcal{M}_\odot$ yr$^{-1}$, although the bulk is around 10–30 $\mathcal{M}_\odot$ yr$^{-1}$. The typical stellar masses are $\mathcal{M}_\star \approx 1$–$5 \times 10^{10}$ $\mathcal{M}_\odot$, and the rather high sSFRs of $\approx 1$–$6 \times 10^{-9}$ yr$^{-1}$ suggest that these galaxies are observed during an important phase of stellar mass assembly. The rest-frame IR continuum luminosities $L_{\rm IR}$ are mostly below $10^{12}$ $L_\odot$ and indicate SFRs $\lesssim 100\mathcal{M}_\odot$ yr$^{-1}$, consistent with those estimated with the UV or H$\alpha$ luminosities. The infrared SEDs show grey-body emission from dust grains with typical temperatures 30–50 K. The observation of CO rotational transitions indicates that these galaxies have molecular gas with masses up to $\mathcal{M}_{H_2} \sim 10^{11}\ \mathcal{M}_\odot$, implying gas fractions $f_{H_2} \approx 0.4$–$0.6$. The molecular gas is spatially extended on scales of a few kpc with sizes comparable with the UV emission. Based on *HST* imaging, the morphologies are mostly irregular, multi-component and with several clumps contributing individually up to ≈20–50% of the total UV luminosity. These clumpy structures are interpreted as the result of minor mergers and/or instabilities in the gaseous component. The morphologies do not strongly depend on wavelength and are similar in the rest-frame UV and optical. The effective radii are typically in the range $R_e \approx 0.7$–$3$ kpc and correlate with the stellar mass, but, for a given stellar mass, these sizes are smaller than those of SFGs at $z \approx 0$ (Fig. 11.31). Important clues on the nature of these galaxies come from the kinematics of emission lines (e.g. H$\alpha$ or [O III]$\lambda$5007) exploited as tracers of the velocity fields. At $1 < z < 3$, the majority (≈ 70–80%) of galaxies with $\mathcal{M}_\star > 10^{10}\ \mathcal{M}_\odot$ have kinematics dominated by rotation as indicated by $V/\sigma \gg 1$ (§4.1.1). The remaining fraction is made of mergers or systems with complex kinematics (Fig. 11.16). The $V/\sigma$ ratio tends to be smaller than in present-day discs, and the velocity dispersions higher than at $z \approx 0$. A possible interpretation is that high-redshift discs are more turbulent than those in present-day galaxies.

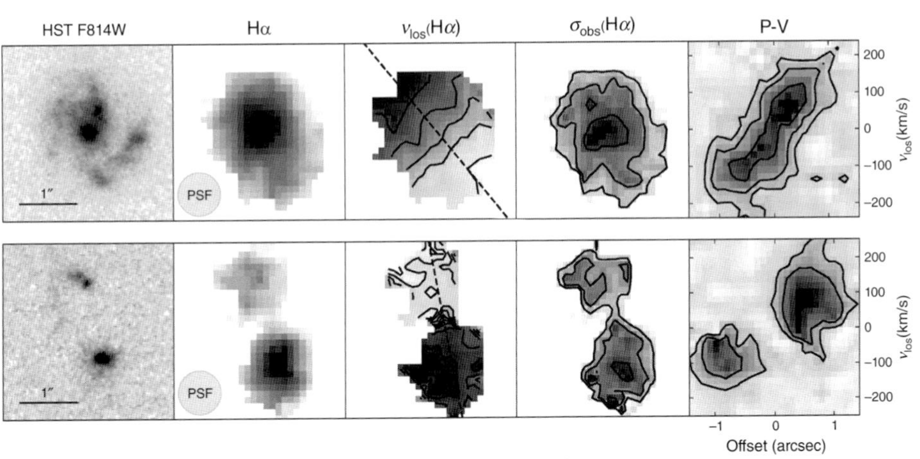


Fig. 11.16 Examples of kinematics of high-$z$ galaxies. A rotating disc (*top row*) at $z \approx 1$ and a merger (*bottom row*) at $z \approx 1.6$ are shown. Kinematic maps are derived from integral field observations (VLT+KMOS) of the H$\alpha$ emission line. From left to right: *HST* broad-band image with F814W filter, H$\alpha$ intensity map, velocity field, velocity dispersion map and position–velocity diagram (§4.3.3). Contours on the velocity fields vary between −100 and 100 km s$^{-1}$ with 50 km s$^{-1}$ steps for the disc and between −75 and 125 km s$^{-1}$ with 25 km s$^{-1}$ steps for the merger ($v = 0$ indicates the systemic velocity). Contours on the velocity dispersion maps are at 25, 50 and 75 km s$^{-1}$. The position–velocity cut is taken along the dashed line highlighted in the velocity field. Both the velocity field and the position–velocity diagram of the disc show the patterns of a gas kinematics dominated by circular motions, while no obvious rotation can be seen in the merger. Data of the disc from Di Teodoro et al. (2016). Data of the merging system courtesy of E. Wisnioski. Courtesy of E. Di Teodoro.

The two-point correlation function (§6.6) of LBGs at $z \approx 3$–5 has a scalelength $r_0 \approx 3$–6 $h^{-1}$ Mpc. This is comparable to the clustering of present-day SFGs. The inferred masses of their dark matter halos are $\mathcal{M}_{\rm vir} \sim 10^{11.5-12}$ $\mathcal{M}_{\odot}$, i.e. similar to or slightly lower than the halos hosting Milky Way-like galaxies at $z \approx 0$.

### 11.2.6 Selection in the Near-Infrared (Rest-Frame Optical)

The advent of NIR detectors for astronomy in the early 1990s opened a new window to study distant galaxies with ground-based observations in the $J$, $H$ and $K$ bands. Selecting galaxies in the NIR is important for three key reasons.

1. The K correction is smaller than in the optical bands because of the similarity of the SEDs of different galaxy types in the rest-frame red/NIR (Fig. 3.7; Fig. 11.12). This implies that the NIR selection helps to include galaxies with a broader variety of properties, and therefore it mitigates the bias which affects the optical selection towards UV-luminous SFGs. Typical examples are dust-reddened SFGs or passive galaxies with no star formation. These galaxies have red SEDs with very faint UV continuum, and they are largely missed by optical surveys, except those with very deep limiting magnitudes (e.g. $R \gg 26$). Instead, these galaxies are more easily selected in the NIR because of their SED shape and the brighter fluxes at these wavelengths. As an example, Fig. 11.17 shows how the $R$ and $K$ filters sample the SEDs of a young and an old stellar population both redshifted at $z = 2$.
2. The rest-frame red/NIR luminosity is a good tracer of the galaxy stellar mass because it correlates with the integrated light of evolved stars rather than with the UV radiation emitted by young short-lived massive stars. This means that the NIR selection allows us to select galaxy samples according to the stellar mass rather than to the star formation activity. As a reference, in the case of a passively evolving SSP (§8.6.2) with an age equal to the age of the Universe at all redshifts, an observed magnitude of $K = 24.5$ corresponds to a stellar mass of $\mathcal{M}_{\star} \approx 10^{9.5}$, $10^{10}$ and $10^{11}$ $\mathcal{M}_{\odot}$ at $z \approx 1$, $z \approx 2$ and $z \approx 4$, respectively.
3. NIR samples are less affected by dust attenuation thanks to the lower extinction at longer wavelengths. This clearly helps to select dust-reddened galaxies that would be otherwise easily missed in optical samples.

NIR surveys opened also the possibility to exploit colour selection criteria extended to galaxies with stellar populations older than those typically selected in the optical. The most prominent spectral features expected for stellar populations older than $\approx 0.5$ Gyr are the Balmer break at 3646 Å and the D4000 discontinuity (§3.4). These continuum jumps are redshifted in the NIR for $z > 1$, and they cause significant changes in the observed colours. These breaks can be exploited to devise specific selection criteria. Fig. 11.18 shows how the $R - K$, $I - K$ and $J - K$ colours evolve for an old stellar population. For instance, a colour cut of $R - K > 3.5$ allows one to select passively evolving galaxies at $z > 1$. This is the case of the so-called **extremely red objects** (EROs). Their extreme colours are due to the intrinsically red SED of old stellar populations and to the strong K correction which

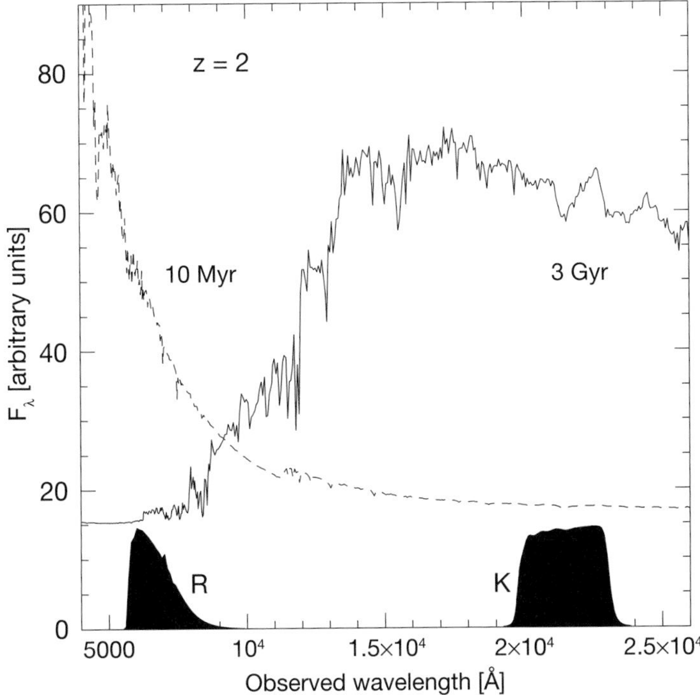


Fig. 11.17

The spectra of young (10 Myr old) and evolved (3 Gyr old) SSPs redshifted at $z = 2$. The SSPs are based on the Bruzual & Charlot (2003) SPS model with solar metallicity and Chabrier IMF. The filled curves at the bottom of the figure show the transmission profiles (in arbitrary units) of the *R* and *K* filters as a function of the observed wavelength. It is evident that optical surveys (*R* band in this example) are more sensitive to galaxies with young stars (i.e. SFGs) because their prominent rest-frame UV emission is redshifted at optical wavelengths for $z > 1$. Instead, NIR surveys (*K* band in this example) allow us to select more efficiently older galaxies because the bulk of their emission is redshifted in the NIR for $z > 1$, thus making them very faint in the observed optical.

shifts the bulk of the emitted light in the $K$ band for $z > 1$ and makes them very faint in $R$ (rest-frame UV). Another example is the $J - K > 1.6$ colour cut that has been successfully used to select mature galaxies at $z > 2$ (also called **distant red galaxies**). While EROs and distant red galaxies can be selected based on a single colour, the combination of more colours allows us to identify a wider range of galaxy types. A successful example is the so-called ***BzK*** **criterion** which exploits the $B - z$ and $z - K$ observed colours to select galaxies at $1.4 < z < 2.5$ and, at the same time, classify them as star-forming or quiescent (Fig. 11.19).

## Main Properties of Near-Infrared Selected Galaxies

In the early 2000s, the first NIR surveys with limiting magnitudes around $K \approx 22$ showed that distant galaxies were more diverse than previously thought. In particular, it was found that at $z \approx 2$ (about 10 Gyr ago), the population of SFGs included also systems with higher

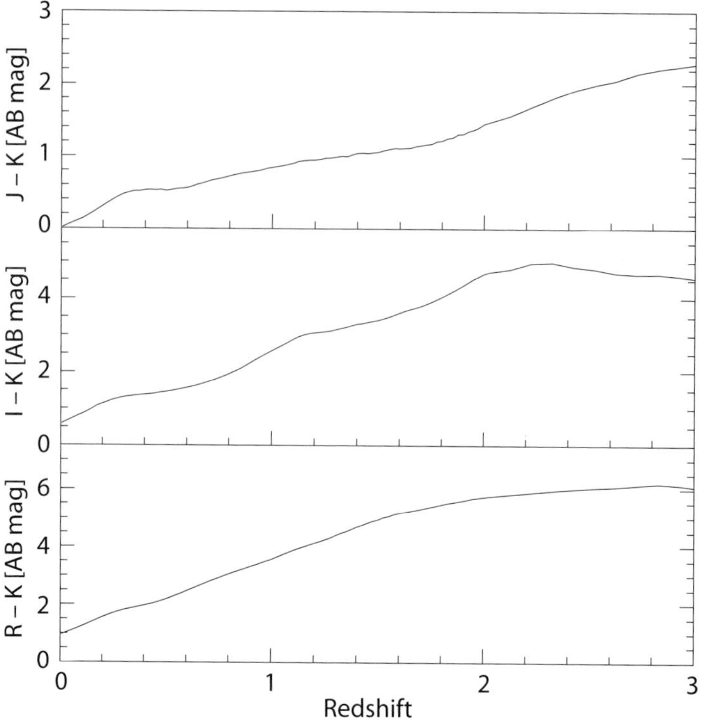


**Fig. 11.18** The evolution of $R - K$, $I - K$ and $J - K$ colours as a function of redshift based on a Bruzual & Charlot (2003) SPS model with an exponentially declining SFH ($\mathrm{SFR}(t) \propto \exp(-t/\tau_\star)$, with $\tau_\star = 0.1$ Gyr), formation redshift $z_{\rm form} = 6$, solar metallicity, Chabrier IMF and an age of 12.5 Gyr at $z = 0$. Data courtesy of L. Pozzetti.

masses ($\mathcal{M}_\star \gtrsim 10^{11}\ \mathcal{M}_\odot$) than those typically found in samples selected in the optical (i.e. rest-frame UV) (§11.2.5). Moreover, it was discovered that a substantial number of massive quiescent galaxies were also present at $z \gtrsim 1$–$2$ and therefore coexisting with the population of SFGs. At that time, these results represented a major challenge because models of galaxy formation did not predict the existence of such massive galaxies in the young Universe. The main properties of the NIR-selected galaxies can be summarised as follows.

*Star-forming galaxies*. At limiting magnitudes $K \lesssim 22$–$23$, the majority of NIR-selected SFGs is at $1 < z < 3$. Compared to optically selected SFGs (§11.2.5), they have a wider range of properties. Their UV, H$\alpha$ and IR luminosities imply higher SFRs up to $\approx 100$–$200$ $\mathcal{M}_\odot\ \mathrm{yr}^{-1}$ on average. The stellar masses can exceed $\mathcal{M}_\star \sim 10^{11}\ \mathcal{M}_\odot$. The metallicity of ionised gas and stars can reach solar values, and also dust extinction is on average larger than in optically selected SFGs. In general, the rest-frame UV morphologies are rather irregular and clumpy, but bulge-like components become in some cases apparent in the rest-frame optical, suggesting the presence of an older stellar population distributed in a spheroidal structure. Kinematic observations assisted by adaptive optics (§11.1.6) show that the majority of these SFGs are discs, and that some of them have rotation velocities as high as in the most massive discs at $z \approx 0$.

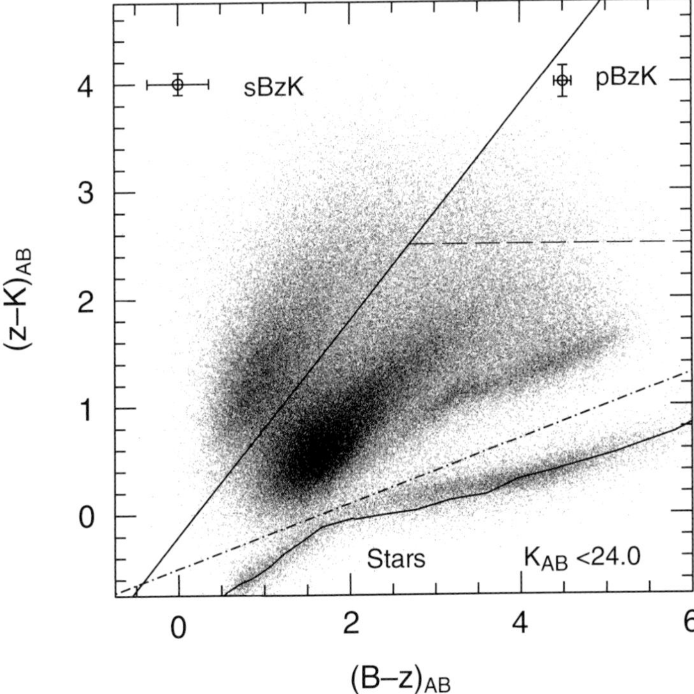


Fig. 11.19 The *BzK* diagram (see Daddi et al., 2004, for details) allows one to simultaneously select in the same redshift range ($1.4 < z < 2.5$) SFGs (sBzK; located on the left of the solid diagonal line) and quiescent (passive) galaxies (pBzK; located on the right of the solid diagonal line and above the long-dashed line). The figure is relative to an NIR-selected sample with limiting magnitude $K < 24$. In this diagram, foreground Galactic stars are segregated below the dash-dotted line. The region between sBzK/pBzK galaxies and Galactic stars is populated mostly by galaxies at $z < 1.4$. Adapted from McCracken et al. (2010).

Regarding the large-scale distribution at $z \approx 2$, the clustering scalelength increases with mass from $r_0 \approx 5\ h^{-1}$ Mpc for $\mathcal{M}_\star \approx 10^{10.3}\ \mathcal{M}_\odot$ to $r_0 \approx 11\ h^{-1}$ Mpc for $\mathcal{M}_\star \approx 10^{11.4}\ \mathcal{M}_\odot$. The inferred dark matter halo masses are therefore higher than for optically selected galaxies, and can be as high as $\mathcal{M}_{\rm vir} \approx 10^{13.5}\ \mathcal{M}_\odot$ for the galaxies with the largest stellar masses ($\mathcal{M}_\star > 10^{11}\ \mathcal{M}_\odot$).

*Quiescent galaxies*. NIR surveys unveiled also a previously unknown population of quiescent galaxies at $z \approx 1-2$ with negligible or absent star formation (see McCarthy, 2004, for a review). The morphologies are spheroidal and most surface brightness profiles have Sérsic indices $n > 2$. This implies that these systems can be broadly classified as the high-redshift equivalent of present-day ETGs. However, a fraction shows disc-like structures and, based on their stellar kinematics, significant rotation. At $z > 1$, massive quiescent galaxies are a substantial population compared to SFGs with the same mass. For instance, at $1.5 \lesssim z \lesssim 2$, the number density $n$ of quiescent galaxies with $10^{11}\,\mathcal{M}_\odot < \mathcal{M}_\star < 10^{11.6}\ \mathcal{M}_\odot$ ($n \approx 1 \times 10^{-4}$ Mpc$^{-3}$) is similar to that of SFGs with the same mass. However, at higher redshifts, quiescent galaxies become rapidly much rarer than SFGs. The rest-frame colours

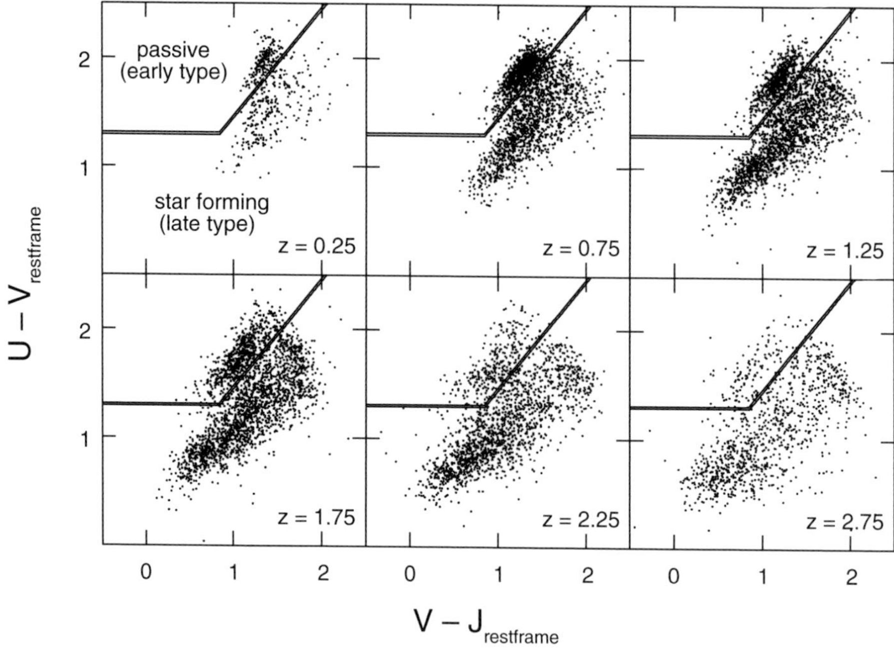


Fig. 11.20 An example of a rest-frame colour–colour diagram where star-forming and passive galaxies are efficiently segregated because the colour bimodality persists out to $z \sim 2$. Adapted from van der Wel et al. (2014). © AAS, reproduced with permission.

of quiescent galaxies are different from those of SFGs at the same redshift, and a colour bimodality (§3.4.1) persists out to at least $z \approx 2$ (Fig. 11.20). In turn, this offers another method to select distant SFGs and quiescent galaxies with an approach complementary to the other techniques based on the observed colours described in §11.2.5 and in this section. However, its application requires the availability of high-quality multi-band photometric SEDs and the knowledge of redshifts in order to derive the reliable colours in the galaxy rest frame. The spectra and SEDs indicate that the dominant stellar populations are old, with ages up to a few Gyr (Fig. 11.21 and Fig. 11.22). These old ages, together with the high stellar masses (up to $\mathcal{M}_\star > 10^{11}\ \mathcal{M}_\odot$) and the very low sSFR $< 10^{-11}$ yr$^{-1}$, indicate that the bulk of the stellar mass was formed at $z > 2$–4. The total metallicities are solar/supersolar. Most importantly, the supersolar abundance of $\alpha$-elements implies that the star formation occurred through short-lived ($\approx 0.1$–0.3 Gyr) starbursts (§11.1.3; eq. 11.8). The size of these galaxies correlates with stellar mass. However, at a fixed mass, they are more compact than at $z \approx 0$, their effective radii can be as small as $R_e < 1$ kpc and they have higher stellar velocity dispersions. This implies that they have stellar mass densities much higher than present-day ETGs, as clearly shown in Fig. 11.23. At a fixed stellar mass, the clustering of these galaxies at $z \approx 2$ is strong and $r_0$ is similar to that of NIR-selected massive SFGs at the same redshift. The similarity of the clustering amplitude of distant quiescent galaxies with that of ETGs at $z \approx 0$ is also suggestive of an evolutionary link between these systems.

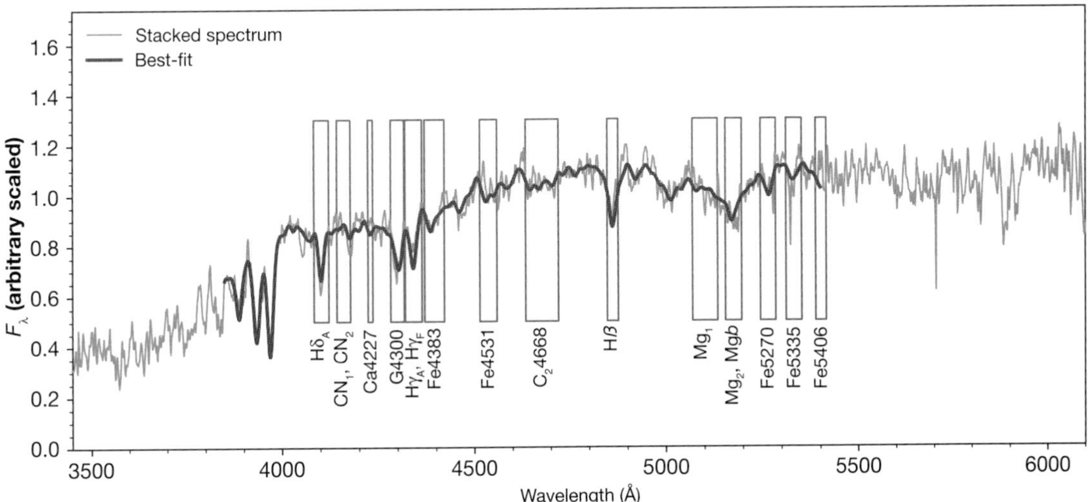


Fig. 11.21 The average rest-frame optical spectrum of galaxies with no evidence of star formation (passive galaxies) at $1.25 < z < 2.09$ ($\bar{z} \simeq 1.6$) based on NIR spectroscopy. The thin grey line shows the observed spectrum. The thick black line is a best-fitting SPS model spectrum of a 1.1 Gyr old stellar population with supersolar abundances of metals ([M/H] $\approx 0.24$) and $\alpha$-elements ([$\alpha$/Fe] $\approx 0.31$). The rectangles indicate the wavelength regions which define some of the Lick indices (§11.1.3) used in the analysis of this spectrum to constrain the stellar population properties. From Onodera et al. (2015). © AAS, reproduced with permission.

### 11.2.7 Selection with Narrow-Band Imaging

The colour selection criteria described in previous sections are based on the traditional broad-band filters (e.g. $UBVRIJHK$) with a passband of several hundred Ångströms. However, it is also possible to search for galaxies with emission lines (i.e. SFGs and AGNs) using **narrow-band filters** with passbands of $\approx 50$–$100$ Å. If an emission line is redshifted within the narrow-band filter, the galaxy would appear much brighter than in the broad-band one because of the dominant contribution of the emission line flux in the narrow-band filter (Fig. 11.24). This technique is very efficient in finding objects with strong emission lines such as SFGs and AGNs, and has been successfully exploited in the optical and NIR to search for Ly$\alpha$, [O II]$\lambda$3727, [O III]$\lambda$5007 and H$\alpha$ emitters. The narrow-band filters must be carefully designed to avoid the strong emission lines from the terrestrial atmosphere (Fig. 11.10). For instance, a filter centred at $\lambda_{\rm obs} \approx 9152$ Å with a width of $\Delta\lambda \approx 100$ Å falls in a spectral region where the atmospheric OH lines are very weak, and can be used to search for Ly$\alpha$ emitters at $z \approx 6.5$, as shown in Fig. 11.24. The main limitation of this technique is the narrow redshift range that can be covered with one filter (typically $\Delta z \approx 0.1$). However, narrow-band imaging is an essential complement to other methods to discover galaxies with emission lines when the continuum is too faint to be detected with broad-band imaging.

### 11.2.8 Selection with Integral Field or Slitless Spectroscopy

Another efficient method to search for distant galaxies is to observe a sky field with an IFU spectrograph (§4.3.2). These instruments produce datacubes where each spatial pixel

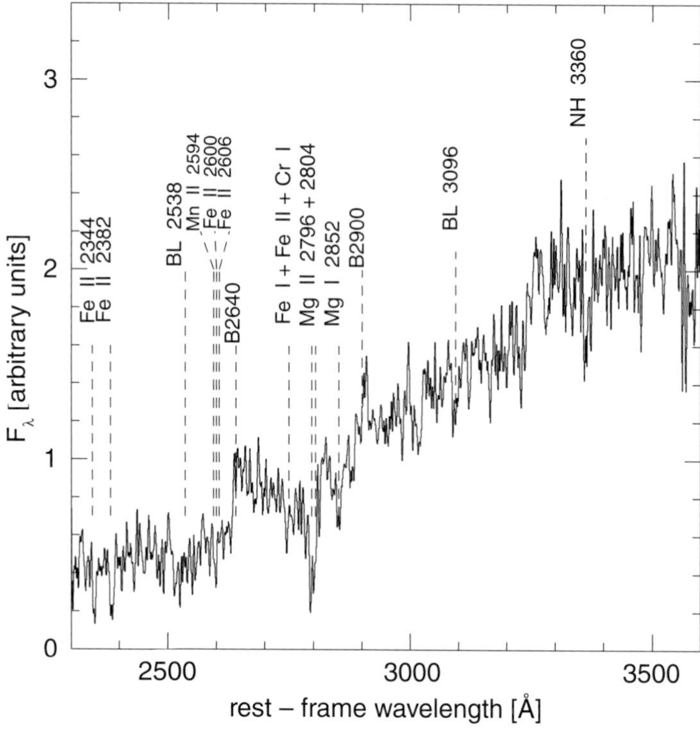


Fig. 11.22 The average rest-frame UV spectrum of passive galaxies at $1.39 < z < 1.99$ ($\bar{z} \simeq 1.6$) based on optical spectroscopy (GMASS survey; Kurk et al., 2013). The spectrum is characterised by the very red continuum, absorption lines and discontinuities typical of old and passive stellar populations with ages around 1 Gyr (§11.1.3). BL indicates blends of unresolved absorption lines (e.g. BL 3096 is a blend centred at 3096 Å). B indicates a discontinuity (break) of the continuum spectrum (e.g. B2640). From Cimatti et al. (2008).

(called spaxel) can be explored as a function of wavelength across the entire spectral range (and hence redshift range) covered by the spectrograph. The main limitation is the small field of view of these instruments, typically of the order of 1 arcmin$^2$ in the optical and NIR. However, the key advantage is the full spectroscopic sampling of the entire field without any preselection of galaxies. Such datacubes can be obtained also with observations in the submm/mm (e.g. with ALMA and NOEMA) and in the radio (e.g. with the VLA) to perform blind searches of ISM molecular and atomic emission lines of distant galaxies.

Another method to search for distant galaxies is offered by **slitless spectroscopy**. The name 'slitless' refers to the lack of the small apertures (slits) that are traditionally used in astronomical spectroscopy to select and isolate the light of the target objects (§4.3.2). Slitless spectroscopy provides a dispersed image of the entire field of view, and therefore a large number of spectra simultaneously (Fig. 11.25). The main disadvantages of this technique are that some spectra overlap with each other and that the sky background is much higher than for slit spectroscopy. Despite this limitation, slitless spectroscopy has been successfully exploited to identify distant galaxies, especially with space-based

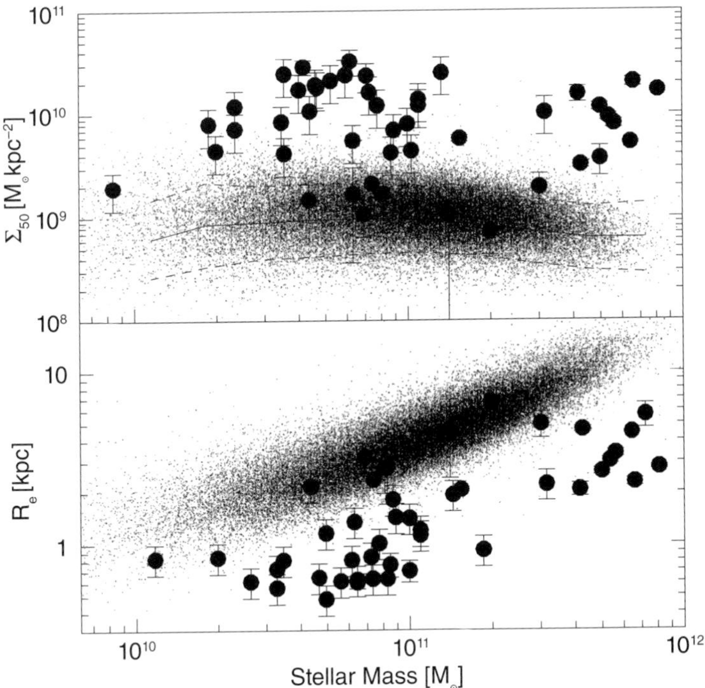


Fig. 11.23

*Top panel.* The stellar mass surface density ($\Sigma_{50} = 0.5\mathcal{M}_\star/\pi R_e^2$, where $R_e$ is the effective radius) as a function of the stellar mass $\mathcal{M}_\star$ of passive galaxies at $1.25 < z < 2.09$ (large filled circles; data from Cimatti et al., 2008) compared with a large sample of ETGs at $z \approx 0$ (smaller dots; data from Hyde and Bernardi, 2009), *Bottom panel.* The relation between the effective radius and stellar mass for the same galaxies as shown in the top panel.

observations (e.g. *HST*), where the sky background is largely reduced by the lack of the atmospheric airglow. The *Euclid* and *WFIRST* space missions have been designed to perform massive surveys of several thousand square degrees using slitless spectroscopy in the NIR and to obtain spectra of several tens of million galaxies.

### 11.2.9 Selection in the Infrared–Millimetre

The SEDs of SFGs show a peak at $\lambda_{\rm rest} \approx 100$ $\mu$m due to the grey-body emission from the interstellar dust heated by the UV radiation of hot massive stars (Fig. 3.6; Fig. 11.3). At $z > 1$, this peak is redshifted in the submm/mm (e.g. at $\lambda_{\rm obs} \approx 400$ $\mu$m for $z \approx 3$), causing a substantial increase of the flux density at these observed wavelengths. The consequence of this K correction effect is truly spectacular: in contrast to the usual trend of galaxies getting fainter with increasing redshift, SFGs become brighter in the submm/mm for $z > 1$ and, for a given IR luminosity, their observed flux density remains nearly constant up to $z \approx 10$ (Fig. 11.26). Since the mid-1990s, the advent of SCUBA at the JCMT and MAMBO at the IRAM telescope opened the possibility to do the first surveys at $\lambda_{\rm obs} \approx 450–1000$ $\mu$m down to typical fluxes of a few mJy. These observations unveiled a population of high-$z$ SFGs

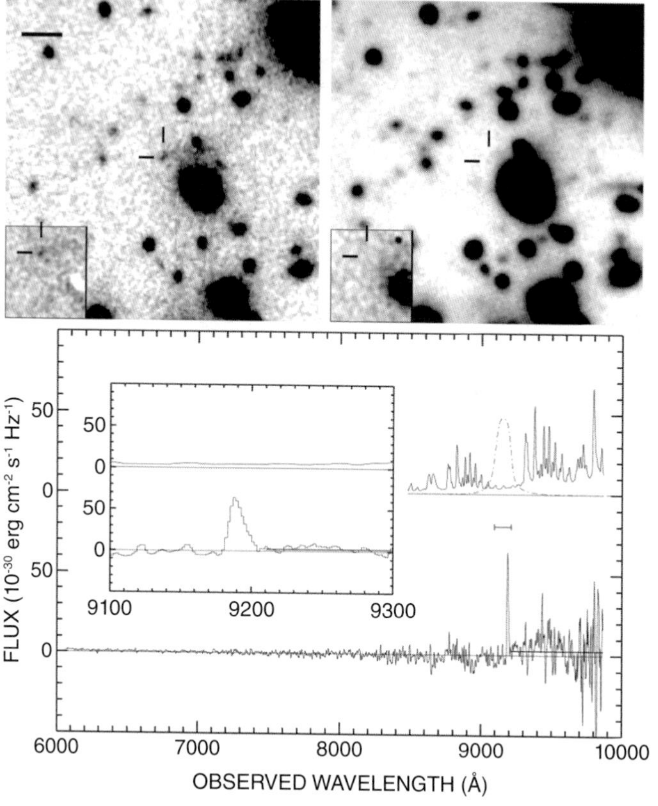


Fig. 11.24 *Top panel*. An SFG with Ly$\alpha$ emission line galaxy at $z \simeq 6.56$ selected with a narrow-band filter with a central wavelength of 9152 Å and a width of 118 Å. The *left* and *right* images show the narrow-band and $R$-band imaging data, respectively. The galaxy is invisible in the $R$ band. The bar in the upper left corner of the narrow-band image shows a 5″ scale, corresponding to $\simeq$27.7 kpc at $z = 6.56$. From Hu et al. (2002). © AAS, reproduced with permission. *Bottom panel*. The optical spectrum of the same galaxy as in the top panel, displaying Ly$\alpha$ emission redshifted at 9187 Å. The insets show the enlarged emission line and the night-sky background spectrum (*left*), and the transmission curve of the filter compared with the spectrum of the night-sky airglow (Fig. 11.10) spectrum (*right*). From Hu et al. (2002). © AAS, reproduced with permission.

with high IR luminosities ($L_{IR} > 10^{12}\ L_{\odot}$). These galaxies are often termed **submillimetre galaxies** (SMGs).

The search for SFGs based on dust thermal emission was then expanded with the *Spitzer* and *Herschel* IR space telescopes which operated in the spectral regions (24–160 $\mu$m and 55–500 $\mu$m, respectively) where the terrestrial atmosphere is opaque and/or excessively bright. The sensitivity of these IR surveys was deep enough to detect also SFGs with lower IR luminosities than SMGs. This allowed the unprecedented possibility to derive the IR photometric SEDs of SFGs previously selected in the optical/NIR and characterised by SFRs (and hence $L_{IR}$) too low to be detected as SMGs with ground-based submm/mm telescopes. Thus, the contributions of *Spitzer* and *Herschel* were essential to constrain the IR-based SFRs and dust masses of large galaxy samples for which only UV estimates were previously available.

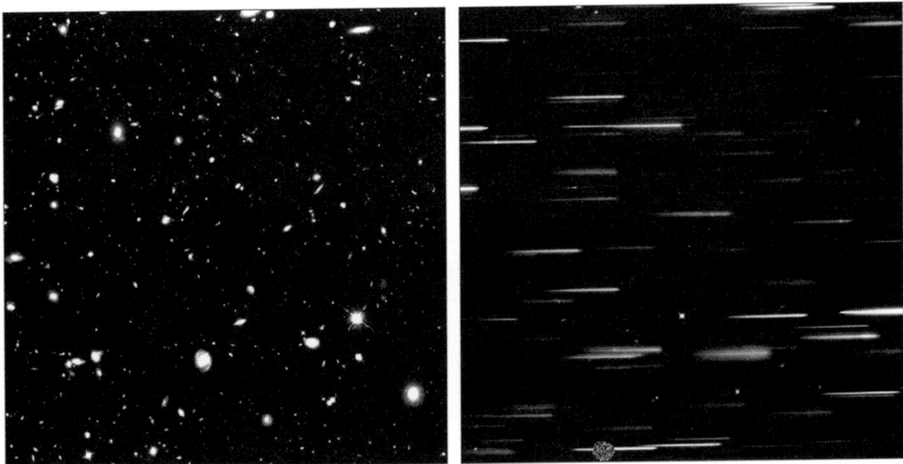

Fig. 11.25 A sky field (2.2 × 2.2 arcmin$^2$) observed with *HST* and the WFC3 instrument in imaging (*left panel*) and slitless spectroscopy (*right panel*). Each horizontal segment in the right panel is a spectrum of an object visible in the left panel. Each spectrum in the right panel covers a wavelength range of 8000–11 500 Å, and the wavelength increases from left to right. Slitless spectroscopy allows us to efficiently obtain spectra of several objects simultaneously, but the price to pay is that some of them inevitably overlap with each other (as apparent in the right panel). Courtesy of N. Pirzkal.

## Dusty Starbursts Selected in the Infrared–Millimetre

One of the major outcomes of IR–mm surveys is the identification of powerful SFGs ($L_{\rm IR} > 10^{12}\ L_\odot$) up to $z \approx 6$ (§11.2.14). We refer the interested readers to the review by Casey et al. (2014). The SFRs derived from these IR luminosities can be as high as $> 1000\ \mathcal{M}_\odot\ \rm yr^{-1}$. Large amounts of dense molecular gas have also been detected in these galaxies, with masses $\mathcal{M}_{\rm H_2}$ up to $\sim 10^{11}\ \mathcal{M}_\odot$ concentrated in small regions down to $< 1$ kpc. The interstellar dust temperatures are typically warmer than in SFGs with lower $L_{\rm IR}$, and they are in the range $\approx 40$–$80$ K. The total dust mass is typically $\sim 100$ times lower than $\mathcal{M}_{\rm H_2}$. The stellar masses can reach values of the order of $\mathcal{M}_\star \sim 10^{11}\ \mathcal{M}_\odot$. The sSFRs ($\sim 10^{-8}\ \rm yr^{-1}$) are typically higher than in optically and NIR-selected SFGs, indicating that these systems are starbursts caught in the act of forming a large amount of stars. Their extreme SFRs imply short depletion times (eq. 4.20) of the order of $\sim 100$ Myr, suggesting that the dense gas reservoir hosted in these galaxies is being rapidly exhausted. At $z \approx 2$, the number density of SMGs selected at $\lambda_{\rm obs} \approx 850\ \mu$m with fluxes higher than $\approx 3$–$5$ mJy is $\sim 10^{-5}\ \rm Mpc^{-3}$. These systems are therefore rarer by an order of magnitude than the optically and NIR-selected SFGs at the same redshift, consistently with a scenario where these extreme starbursts are sporadic and short-lived episodes of vigorous star formation triggered by major mergers. This picture is supported by the disturbed and multi-component morphologies of SMGs. At $z \approx 2$–$4$, the correlation scalelength is in the range $r_0 = 8$–$14\ h^{-1}$ Mpc. This suggests that these starbursts are located in massive halos ($\mathcal{M}_{\rm vir} > 10^{13}\ \mathcal{M}_\odot$), and that some of them could be the progenitors of the massive

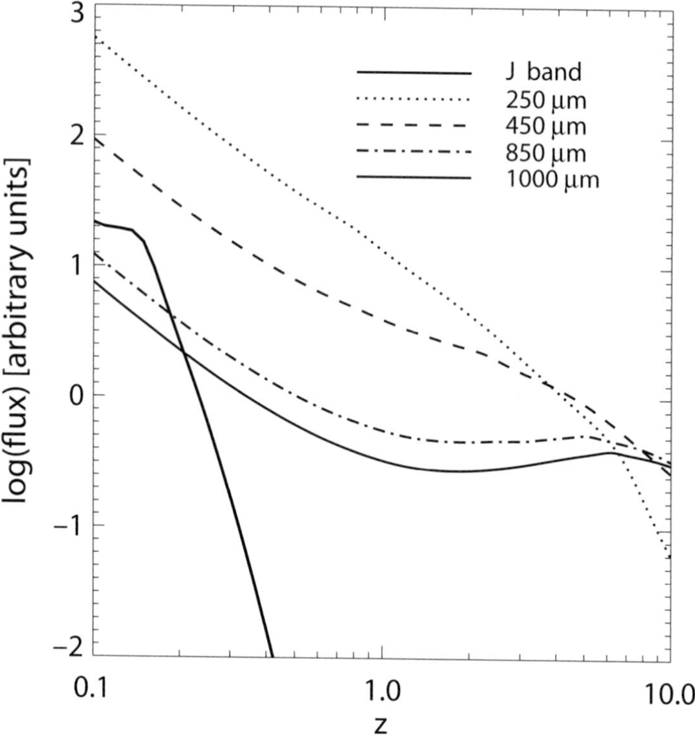


Fig. 11.26 The dimming of flux (in arbitrary units) expected from moving a main-sequence SFG (template SED of Magdis et al., 2012) to increasing redshifts. If the galaxy is observed in the *J* band ($\lambda_{\rm obs} \approx 1.25\ \mu$m, i.e. a wavelength range dominated by stellar light), its flux rapidly decreases (thick solid line). However, at 250 $\mu$m $\lesssim \lambda_{\rm obs} \lesssim$ 1000 $\mu$m, the flux decreases much slower (or not at all for $z > 1$ and 850 $\mu$m $\lesssim \lambda_{\rm obs} \lesssim$ 1000 $\mu$m) because the peak of dust thermal emission (placed at $\lambda_{\rm rest} \approx 100\ \mu$m; Figs. 3.6 and 11.3) enters in the submm/mm for increasing redshifts. This K correction effect compensates the flux dimming and makes dusty SFGs bright enough to be detected out to high redshifts. Courtesy of G. Rodighiero.

quiescent galaxies located in dense regions at lower redshifts and hosted by halos with a similar clustering (§11.2.6). About 10–20% of these galaxies host dust-enshrouded AGNs, therefore suggesting a relationship between the starburst event and the accretion of matter onto the central SMBH.

### 11.2.10 Selection in the Radio

Radio observations provide a complementary possibility to find distant galaxies. One approach is to search for the optical/NIR counterparts of radio-loud AGNs preselected in the radio. With this approach, a key role is played by **high-redshift radio galaxies**. In these systems, the direct emission of the AGN is not visible at optical/NIR wavelengths because of the orientation with respect to the line of sight (§3.6.2). This allows us to perform detailed studies of the host galaxies that would be otherwise impossible in the case of QSOs, where the outshining due to the active nucleus is dominant. It has been found that an efficient criterion to find distant radio galaxies is to preselect radio sources with steep synchrotron spectra ($\alpha < -1$, where $F_\nu \propto \nu^\alpha$; see McCarthy, 1993, for a review). With this approach, radio galaxies have been identified out to $z \approx 6$. The origin of the steepening of radio spectra at high redshift is usually ascribed to the energy losses of relativistic electrons due to inverse Compton scattering with CMB photons (a process which increases rapidly with redshift), and to K correction effects.

When AGN activity is absent, the radio continuum emission is originated by the star formation activity (Fig. 11.3), and the detection of normal SFGs requires very deep surveys. As a reference, observations at 3 GHz with the VLA down to limiting fluxes of the order of $\approx 10\ \mu$Jy allow us to detect SFGs out to $z > 3$. Radio surveys have the notable advantage not to be affected by dust extinction and represent an important complement to other approaches to investigate the cosmic evolution of star formation and AGN activity. SKA is designed to reach sub-$\mu$Jy sensitivities and to detect also the H I 21 cm emission line from distant galaxies (§9.10).

### 11.2.11 Selection of Galaxies and Active Galactic Nuclei in the X-Rays

X-ray surveys provide another essential complement for the selection and characterisation of distant galaxies. After the pioneering *ROSAT* mission, the deepest X-ray surveys have been performed with *XMM-Newton* and *Chandra* operating at energies of 0.1–15 keV and 0.1–10 keV, respectively. Their instruments provide images with average PSFs of $\approx 15$ arcsec and $\approx 2$ arcsec for *XMM-Newton* and *Chandra*, respectively. With integration times of a few million seconds (note that 1 million seconds is about 11.6 days!), the achievable fluxes are deep enough to detect the X-ray emission from distant galaxies. As a reference, an integration time of 7 million seconds with *Chandra* allows us to reach a limiting flux of $\approx 4.5 \times 10^{-18}$ erg s$^{-1}$ cm$^{-2}$ at 0.5–2 keV. At fluxes $F(0.5\text{–}2\ \text{keV}) > 10^{-17}$ erg s$^{-1}$ cm$^{-2}$, the X-ray samples become dominated by AGNs over normal SFGs. For this reason, X-ray surveys are considered a very efficient tool to select high-$z$ AGNs thanks to the high luminosity of these objects at these energies. At $F(0.5\text{–}2\ \text{keV}) < 10^{-17}$ erg s$^{-1}$ cm$^{-2}$, the sky surface number density of normal galaxies becomes comparable to that of AGNs. In the deepest surveys, it has been possible to identify normal galaxies (i.e. without AGNs) up to $z \gtrsim 3$. High-$z$ galaxies are generally too faint for X-ray spectroscopy. However, useful constraints on the X-ray spectra can be derived from the **hardness ratio** defined as $\text{HR} \equiv (H - S)/(H + S)$, where $H$ and $S$ are the photon counts in a high (hard; typically 2–8 keV) and a low (soft; typically 0.5–2 keV) rest-frame energy bands, respectively. The HR can be seen as an X-ray 'colour' which helps to discriminate between AGNs and normal galaxies. SFGs have typically $-1 \lesssim \text{HR} \lesssim -0.5$. Unobscured AGNs (i.e. Type 1; §3.6.2) have $-0.6 \lesssim \text{HR} \lesssim -0.2$. Finally, obscured AGNs (§3.6.2; §3.6.3) have $\text{HR} \gtrsim 0.2$. In ambiguous cases, the X-ray luminosity (§3.6) helps to better understand if the dominant source is an AGN ($L_X > 10^{42}$ erg s$^{-1}$) or star formation ($L_X < 10^{42}$ erg s$^{-1}$).

New-generation X-ray missions such as the *Athena X-ray Observatory* are designed to improve our understanding of the X-ray-emitting objects at high redshifts.

### 11.2.12 Selection of Active Galactic Nuclei at Other Wavelengths

In addition to the radio and X-ray observations, AGNs can be selected also at other wavelengths (see Padovani et al., 2017, for a review). In the optical and NIR, several colour criteria have been successfully developed to discover Type 1 AGNs up to $z > 7$. These methods exploit the blue power-law shape of the continuum which allows us to segregate these AGNs with respect to the colours of foreground stars and normal galaxies. For instance, QSOs up to $z \approx 2.5$ can be selected as star-like objects with UV excess (**UVX method**) in colour–colour diagrams where the $U$ filter is used (e.g. $U - B$ versus $B - R$). At higher redshifts, QSO surveys are based on the synergy of optical and NIR photometry. Fig. 11.27 shows an example of a colour–colour diagram designed to segregate QSOs at $z > 5.8$ with respect to the colours of foreground stars and brown dwarfs of our Galaxy. It is important to recall that the Lyman-break criterion described in §11.2.5 applies also to high-$z$ AGNs. This means that, for instance, the requirement for a QSO to be at $z > 7$ is to be undetected in the filters blueward of the $z$ band ($\lambda_{\rm obs} \approx 0.9$ $\mu$m). However, some level of contamination from Galactic white dwarfs or very cool stars is often present in all colour criteria to select AGNs. For instance, brown dwarfs have colours which can mimic the ones expected in the case of AGNs at $z > 5$ (Fig. 11.27). The problem of stellar contamination is exacerbated in the case of QSOs because they appear as point-like as stars. If dust obscuration is present in the host galaxy or around the active nucleus, the colours are strongly affected and it can be difficult to select AGNs based only on optical and NIR photometry. In these cases, MIR colours are very helpful for two reasons. First, they are not affected by dust extinction, and, second, the AGN SEDs are very different from those of normal galaxies because of the dusty torus (§3.6.2) which emits mostly in the rest-frame MIR. When spectroscopic data are available, a complementary method to select AGNs is to identify galaxies with the emission line flux ratios expected in the case of an AGN radiation field. The BPT criterion is an example of this approach (§11.1.4).

AGNs can be selected also based on the flux **variability**. In fact, one of the main observed properties of AGNs is their aperiodic and irregular flux variability at all wavelengths with a wide range of amplitudes and timescales, from minutes to years (§3.6). AGN variability is believed to be caused by the intrinsic variations during the accretion of matter onto the SMBH and by the variable emission from relativistic jets. Being a common feature of AGNs, variability has been successfully exploited to select these objects in imaging surveys extended over a suitable time baseline.

### 11.2.13 Selection of Distant Galaxy Clusters

Galaxy clusters (§6.4) are key probes of cosmic structure formation, and the evolution of their number density and total mass allows us to constrain some of the cosmological

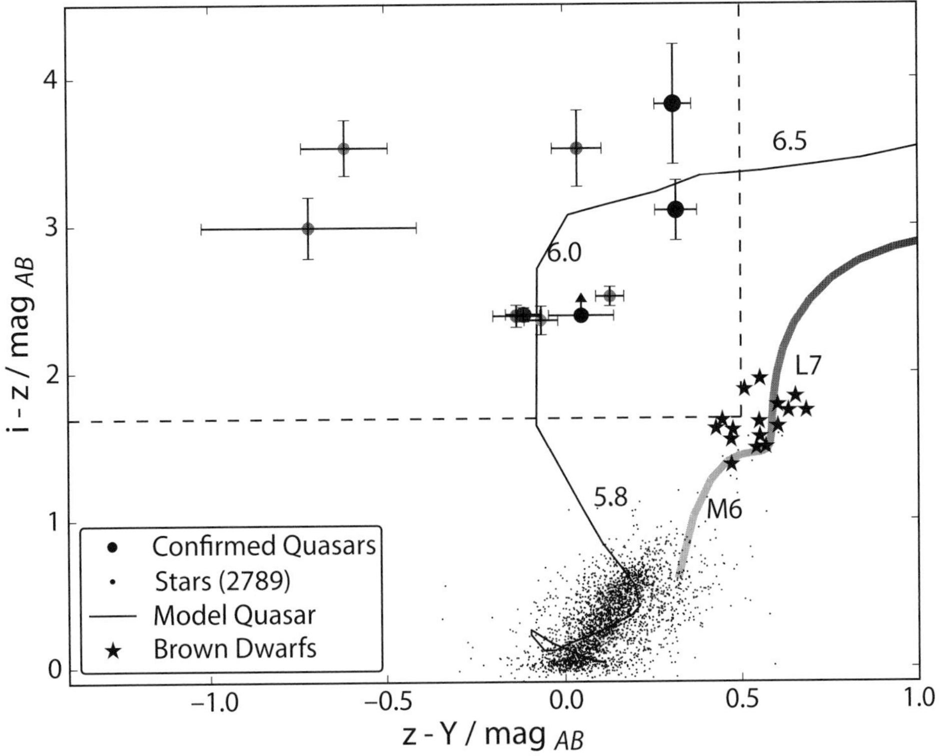


Fig. 11.27 A colour–colour diagram to select high-$z$ QSOs at $z > 5.8$. The thin solid line indicates the expected QSO colours as a function of redshift. The numbers 5.8, 6.0 and 6.5 indicate three reference redshifts. The thick curve shows the expected contamination by cool stars and brown dwarfs of spectral types M6 to L7. The dashed rectangle indicates the region where QSOs are selected. From Reed et al. (2017).

parameters (§7.4.3). There are several methods to identify galaxy clusters at high redshifts. The most obvious one is to search for galaxy overdensities in imaging data (§6.4.1). For $z > 1$, imaging surveys in the NIR are essential because ETGs (the dominant galaxy population within massive clusters) emerge more clearly thanks to their red colours with respect to the other galaxies (Fig. 11.18). Clusters can be found also by searching for overdensities of galaxies around known objects that are exploited as redshift markers. In spectroscopic surveys, clusters are also identified as concentrations of galaxies at the same redshift. Another possibility is offered by the strong X-ray luminosity emitted by the hot ICM (§6.4.2). This allows us to identify clusters as sources of extended flux in X-ray imaging surveys. An additional method is based on the SZE (§6.4.2). In this case, mm imaging surveys can detect the location of a cluster in the sky by searching for 'spots' where the observed CMB flux deviates from that of the cosmological CMB black-body spectrum. The advantage of the SZE is that its amplitude is independent of redshift. Thanks to the methods described above, it has been possible to unveil virialised clusters with $\mathcal{M}_{\rm vir} > 10^{14}\ \mathcal{M}_\odot$ out to $z \approx 2$ (Fig. 11.28). These structures often host mature galaxy populations with high fractions of massive ETGs similar to the clusters at lower

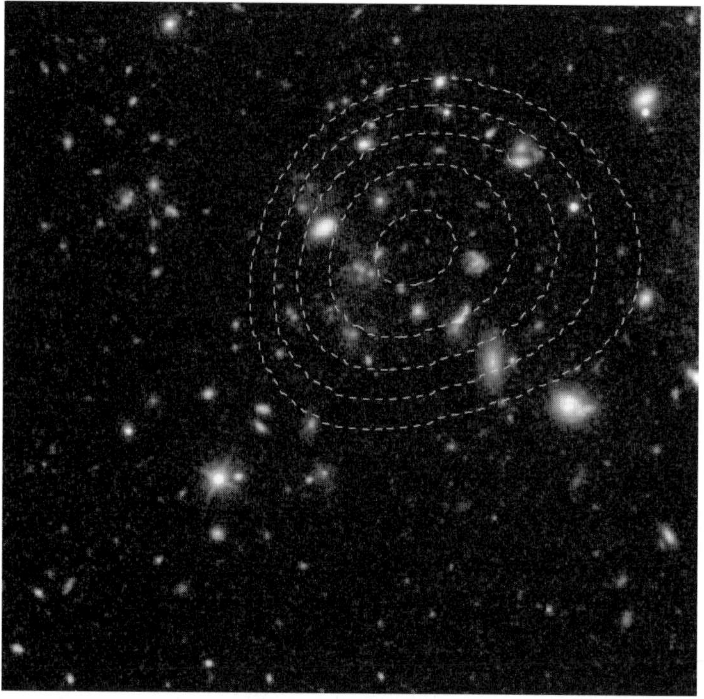

Fig. 11.28 A very distant and massive galaxy cluster ($z \simeq 1.995$) with a total mass ($\sim 10^{14} \mathcal{M}_{\odot}$) comparable to that of the nearby Virgo cluster. Background image: combination of *HST* WFC3 images with F140W, F105W and F606W filters covering $48'' \times 48''$ ($\approx 400 \times 400$ kpc). Dashed curves: contours of constant X-ray surface brightness at 0.5–2 keV as derived from *Chandra* observations. From Gobat et al. (2011).

redshifts. At higher redshifts, **protoclusters** have been identified as serendipitous galaxy concentrations or around known galaxies and AGNs at high $z$. These protoclusters are mostly populated by SFGs, and are thought to be the precursors of the mature clusters at $z < 2$ (see Overzier, 2016, for a review). A novel approach to search for galaxy overdensities is based on gravitational lensing surveys (e.g. with *Euclid*). In this case, wide sky areas are imaged to measure weak and strong lensing (§5.3.4) in order to reconstruct the maps of the total mass distribution. In these maps, clusters can be identified as the highest-density peaks.

### 11.2.14 The Last Frontier: Galaxies at Very High Redshifts

Our knowledge of distant galaxies decreases rapidly at $z > 6$, i.e. during the first billion years of cosmic time. The faintness of galaxies at these redshifts makes their spectroscopic identification very challenging (see Finkelstein, 2016, and Stark, 2016, for comprehensive reviews). Some of the most distant galaxies have in fact been found with the help of the flux amplification provided by the strong gravitational lensing (§5.3.4) due to intervening

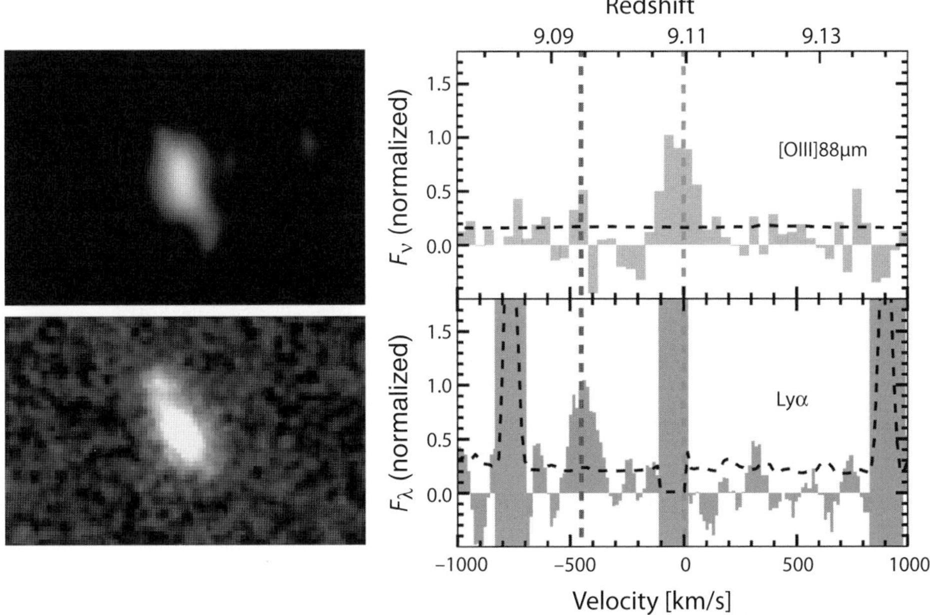


Fig. 11.29 A galaxy at $z \simeq 9.1$ preselected as a photometric dropout galaxy (§11.2.5) through a lensing galaxy cluster at $z \simeq 0.54$ which provides a flux amplification of a factor of $\sim 10$. *Top row*. ALMA map (*left*) and spectrum (*right*) of the [O III]$\lambda 88$ $\mu$m emission line. *Bottom row*. *HST* NIR image (filter F160W) (*left*) and VLT X-Shooter spectrum of the Ly$\alpha$ emission line (*right*). The vertical bands indicate the regions contaminated by terrestrial OH emission lines. The blueshift of Ly$\alpha$ with respect to [O III] is interpreted as due to inflowing gas with a high column density of hydrogen. Adapted from Hashimoto et al. (2018). Courtesy of N. Laporte.

galaxy clusters. Fig. 11.29 shows an example of a galaxy at $z \simeq 9.1$ identified with the aid of gravitational amplification. Galaxies at $z > 6$ have been found as SFGs, QSOs and hosts of gamma-ray bursts.

*Star-forming galaxies*. All galaxies spectroscopically identified at $z > 6$ are star-forming. Most of them were preselected as LBGs with broad-band imaging (§11.2.5) or as Ly$\alpha$ emitters with narrow-band imaging (§11.2.7) or slitless spectroscopy (§11.2.8). When present, Ly$\alpha$ is the strongest emission line, but also other lines have sometimes been detected (e.g. C III]$\lambda 1909$ and C IV $\lambda 1548$). He II $\lambda 1640$ emission, one of the signatures of Pop III stars (§9.5), is usually weak or absent, meaning that the galaxies identified so far are not truly primordial. In some galaxies, the fine-structure [C II]$\lambda 158$ $\mu$m and/or the [O III]$\lambda 88$ $\mu$m emission lines redshifted in the submm have been detected with ALMA (Fig. 11.29). The fraction of galaxies with Ly$\alpha$ emission declines at $z > 6.5$, suggesting that the increasing neutral fraction of the IGM (§9.9.1) and/or the physical conditions within the galaxies cause a deficit of Ly$\alpha$ photons. The galaxies identified as LBGs are usually very blue, and their UV continuum slopes ($-2.5 \lesssim \beta_{UV} \lesssim -2.0$; §11.1.5) indicate young metal-poor stars and low dust extinction ($A_V < 0.1$). However, the most massive systems tend to have redder UV slopes ($\beta_{UV} \approx -1.8$), larger dust extinction (up to $A_V \approx 0.5$) and higher metallicities ($Z \approx 0.2$–$0.4 Z_\odot$). The rest-frame UV morphologies range from compact to multi-component systems. The sizes of these galaxies are small (0.2 kpc $< R_e <$ 1 kpc) and sometimes comparable to those of giant molecular gas regions in SFGs at $z \approx 0$. The average galaxy size decreases with increasing redshift and, for a fixed redshift, increases with luminosity. The typical SFRs, derived from the UV continuum and/or Ly$\alpha$ luminosities, are a few $\mathcal{M}_\odot$ yr$^{-1}$. The stellar masses are in the range $\mathcal{M}_\star \sim 10^{8-9}$ $\mathcal{M}_\odot$, and the sSFRs are around a few $10^{-9}$ yr$^{-1}$. Some SFGs at $z \gtrsim 6$ have high Ly$\alpha$ luminosities up to $\sim 10^{44}$ erg s$^{-1}$. In these cases, the SFRs reach several tens of $\mathcal{M}_\odot$ yr$^{-1}$. A very important quantity is the escape fraction ($f_{\rm esc}$) which indicates the percentage of hydrogen-ionising photons ($\lambda_{\rm rest} < 912$ Å) that can actually escape from the galaxies and reionise the IGM (§9.8 and §9.9.1). This quantity can vary significantly from galaxy to galaxy depending on the SFR, the geometrical distribution of the ISM and the amount of dust extinction. The estimates as a function of redshift are still uncertain and compatible with a wide range of values ($f_{\rm esc} \approx 5$–20% at $1 < z < 6$).

In addition to the dominant population of galaxies selected in the optical and NIR, IR–mm surveys unveiled dusty starbursts out to $z \gtrsim 6$. These systems would never have been found as LBGs and Ly$\alpha$ emitters because of the heavy dust obscuration which makes them very red or undetected at optical/NIR wavelengths. These extreme systems are characterised by high IR luminosities (up to $L_{\rm IR} > 10^{13}$ $L_\odot$), SFRs up to $\gtrsim 1000$ $\mathcal{M}_\odot$ yr$^{-1}$, large amounts of molecular gas ($M_{\rm H_2} \sim 10^{11}$ $\mathcal{M}_\odot$) concentrated in small regions of a few kpc, and dust masses $\sim 10^9$ $\mathcal{M}_\odot$.

*QSOs*. Besides SFGs, QSOs at $z > 6$ have also been identified. Their extreme luminosities ($L > 10^{47}$ erg s$^{-1}$) imply SMBHs with masses up to $\mathcal{M}_\bullet \sim 10^9$–$10^{10}$ $\mathcal{M}_\odot$. Explaining the existence of such SMBHs when the Universe was less than $\approx 1$ Gyr old is a major challenge for theoretical models (§9.7). The gas metallicities derived from the emission line spectra of the BLR and NLR (§3.6) are solar or even supersolar. These metal abundances require extremely fast SFHs to explain such a rapid enrichment of the ISM. The SFRs derived from the IR luminosities can be higher than 1000 $\mathcal{M}_\odot$ yr$^{-1}$. The dust masses are also very high ($\mathcal{M}_{\rm dust} \sim 10^8$ $\mathcal{M}_\odot$), and the direct observation of molecular (e.g. high-excitation CO lines) and atomic (e.g. [C I]$\lambda$369 $\mu$m and [C II]$\lambda$158 $\mu$m) gas shows the presence of reservoirs of gas with masses up to $\mathcal{M}_{\rm H_2} \gtrsim 10^{10}$ $\mathcal{M}_\odot$ extended over a few kpc. This gas can be rapidly exhausted by the extreme SFRs. However, strong outflows with velocities of a few hundred km s$^{-1}$ are also observed, suggesting that a large fraction of the gas could be expelled and therefore not used for star formation. In some cases, it has been possible to derive kinematic information on the gas motion and to detect the presence of ordered rotation indicating dynamical masses of $\mathcal{M}_{\rm dyn} \sim 10^{10}$–$10^{11}$ $\mathcal{M}_\odot$. Some high-$z$ QSOs have been found in overdense environments.

*Gamma-ray bursts*. As a final remark on the objects known at $z > 6$, it is important to include **gamma-ray bursts** (GRBs) as tracers of galaxies at cosmological distances. GRBs are unpredictable and rapid flashes of gamma-rays with durations between ~0.01 s and ~1000 s. They are extragalactic sources randomly distributed on the sky. GRBs emit most of their energy at 0.1–1 MeV, and in this range they have huge luminosities ($10^{50-54}$ erg s$^{-1}$). This makes it possible to detect them up to very high redshifts. There are two possible physical origins for the GRBs. The so-called **short GRBs** occur in quiescent galaxies and SFGs, and they are originated by the merging of two compact stellar remnants (neutron stars and/or black holes). The **long GRBs** are thought to be produced by the death of massive stars characterised by rapid rotation and are therefore located only in galaxies where star formation is ongoing. Each GRB is followed by the so-called **afterglow**, that is, the emission of radiation at longer wavelengths from the X-rays to the radio. The duration of the afterglow is longer than that of the GRB, and this helps to identify the GRB host galaxy once the available telescopes are rapidly alerted. Galaxies hosting GRBs have been identified up to $z > 8$.

### 11.2.15 Lessons Learned

The main lesson learned is that multi-wavelength observations are required for the identification of all galaxy types and AGNs at different redshifts. Any method based on colours is biased against galaxies with SEDs different from those adopted to define the colour selection itself. This means that galaxies with properties dissimilar from the targeted ones are inevitably missed unless the survey is extremely deep. Let us take the example of EROs (§11.2.6), which include distant quiescent galaxies (red because of the old stellar populations and the strong K correction; §11.2.3) and SFGs heavily reddened by dust extinction. EROs are bright enough to be easily selected in moderately deep NIR surveys (e.g. $K < 23$). However, they are so faint at shorter wavelengths ($R - K > 3.5$–$5$; Fig. 11.18) that very deep data ($R > 26.5$–$28$) are required not to miss them in optical samples. For this reason, EROs emerged only when NIR surveys became feasible.

The main differences between optical and NIR selections can be summarised as follows. On the one hand, optically selected (rest-frame UV; §11.2.5) samples of distant SFGs tend to be biased towards galaxies with low dust extinction, moderate SFRs, intermediate stellar masses and subsolar metallicities. On the other hand, SFGs selected in the NIR (rest-frame optical; §11.2.6) tend to have, on average, higher masses, SFRs and metal abundances, and more dust extinction. Moreover, NIR surveys turned out to be essential to unveil the unexpected population of massive, old and quiescent galaxies at $z > 1$. In retrospect, the differences between optically and NIR-selected galaxies were partly due to the rather bright limiting magnitudes of the first samples for which it was possible to obtain spectroscopic identifications ($R \lesssim 25.5$ and $K \lesssim 22$ for optical and NIR surveys, respectively). However, it remains inevitable that the optical selection is more sensitive to UV-luminous galaxies, whereas NIR surveys tend to include redder (i.e. older or more dust reddened) systems (Fig. 11.17).

Narrow-band imaging and integral field spectroscopy turned out to be a fundamental complement to identify galaxies (especially Ly$\alpha$ emitters) that are too faint for the typical broad-band surveys in the optical and NIR. Besides optical and NIR surveys, submm and IR observations are also essential to identify IR-luminous starburst galaxies so heavily obscured that they would never have been found at shorter wavelengths. Last but not least, X-ray and radio surveys are highly complementary to identify AGNs and SFGs at cosmological distances. Tab. 11.3 summarises the main selection methods for distant galaxies, AGNs and galaxy clusters.

**Table 11.3** Main methods to select distant galaxies

| Method | Observations | Redshift | Object type | Section |
|---|---|---|---|---|
| Lyman break | Opt–NIR ima | $z>3$ | SFG | §11.2.5, §11.2.14 |
| Red colours | Opt–NIR ima | $z>1$ | QG, SFG | §11.2.6 |
| sBzK colours | Opt–NIR ima | $1.4<z<2.5$ | SFG | §11.2.6 |
| pBzK colours | Opt–NIR ima | $1.4<z<2.5$ | QG | §11.2.6 |
| QSO colours | Opt–NIR ima | All | AGN, QSOs | §11.2.12, §11.2.14 |
| QSO absorptions | Opt–NIR spe | All | Gaseous systems | §9.8.1 |
| Narrow-band selection | Opt–NIR ima | All | SFG, AGN | §11.2.7, §11.2.14 |
| Slitless spectroscopy | Opt–NIR spe | All | All | §11.2.8 |
| Integral field spectroscopy | Opt–NIR spe | All | All | §11.2.8 |
| Pure flux-limited selection | All spe/ima | All | All | §11.2 |
| IR–mm selection | IR–mm ima | All | Dusty SFG | §11.2.9, §11.2.14 |
| Sunyaev–Zeldovich effect | mm ima | All | GC | §11.2.13 |
| Radio selection | Radio ima | All | SFG, AGN | §11.2.10 |
| X-ray selection | X-ray ima | All | SFG, AGN, GC | §11.2.11 |
| Gamma-ray bursts | Opt–NIR ima | All | All | §11.2.14 |

Opt, optical; ima, imaging; spe, spectroscopy; QG, quiescent galaxy; GC, galaxy cluster; pure flux-limited, selection with no criteria other than the limiting flux.

## 11.3 The Observation of Galaxy Evolution

The third part of this chapter describes how multi-wavelength observations can place empirical constraints on galaxy evolution. These studies are based on four main steps. (1) Selection of a sample of galaxies within a redshift range with the methods summarised in Tab. 11.3. (2) Distribution of the sample galaxies into redshift bins. (3) Assessment of the incompleteness effects and biases. (4) Study of galaxy properties as a function of the redshift bins, and comparison with galaxies at $z \approx 0$. As already pointed out at the beginning of this chapter, this is a relatively young and rapidly evolving research field, and only the most robust results are presented here.

### 11.3.1 The Evolution of Galaxy Morphologies and Sizes

One of the main questions of galaxy evolution is when the present-day Hubble morphological sequence (Fig. 1.1; §1.2; §3.1) originated. Addressing this question with observations requires two main issues to be faced. The first is the need for high angular resolution (§11.1.6). For this reason, space-based imaging with *HST* has been so far the primary source of information for morphological studies at high redshifts (Fig. 11.30). The second problem is the morphological K correction, i.e. the dependence of morphology on wavelength (§3.1.1). This makes galaxies (especially SFGs) more irregular and clumpy

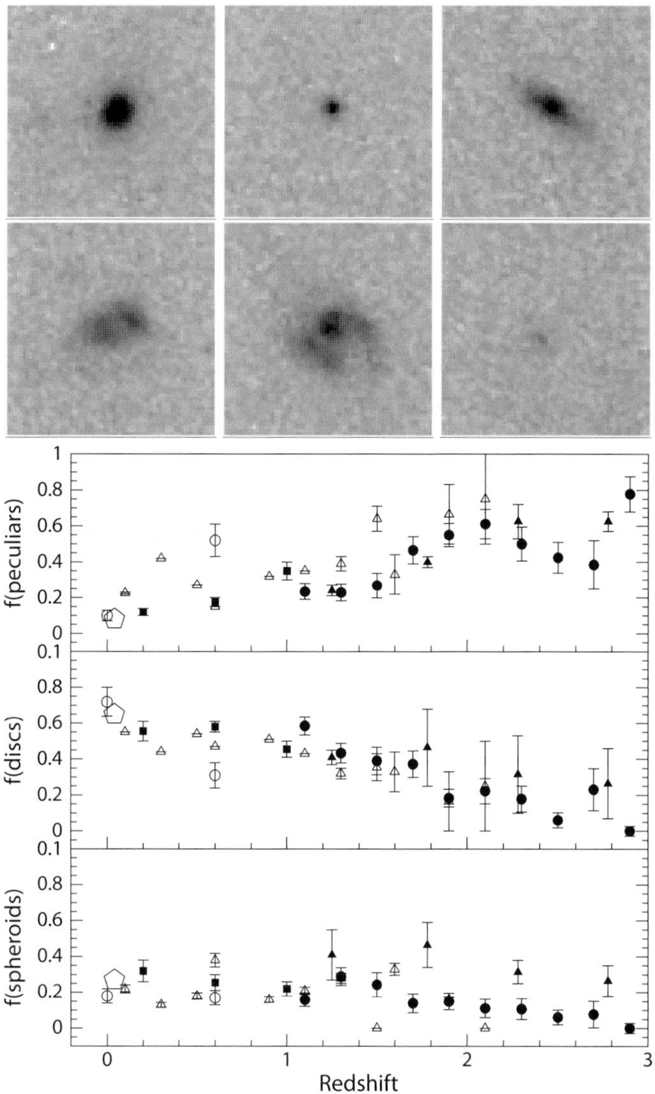


Fig. 11.30

*Top panel*. Examples of *HST* images of galaxies at $z > 1$ obtained with the WFC3 instrument and the F160W filter. Each snapshot is $4 \times 4$ arcsec, corresponding to $\approx 32.5 \times 32.5$ kpc and $\approx 31.4 \times 31.4$ kpc at $z = 1$ and $z = 3$, respectively. From *top left* to *bottom right*: elliptical ($z = 1.09$), compact ($z = 3.19$), disc ($z = 1.22$), peculiar/irregular ($z = 1.22$), peculiar/irregular ($z = 1.99$) and unclear classification ($z = 2.30$). This level of morphological detail is not achievable with ground-based observations, unless the observations are done with adaptive optics (§11.1.6). From Talia et al. (2014). *Bottom panel*. The evolution of the morphological fractions based on the results of Kelvin et al. (2014b) (pentagons), Cassata et al. (2005) (open triangles), Tasca et al. (2009) (filled squares), Delgado-Serrano et al. (2010) (open circles), Mortlock et al. (2013) (filled triangles) and Talia et al. (2014) (filled circles). The scatter is mainly due to the different criteria of sample selection and methods of morphological classification.

when observed at UV/blue wavelengths. Since the historical Hubble classification at $z \approx 0$ was defined in the optical, it is essential to observe high-$z$ galaxies at the same rest-frame optical wavelengths in order to allow a meaningful comparison with $z \approx 0$. For instance, $J$-, $H$- and $K$-band imaging is needed to observe the rest-frame $V$-band morphologies at $z \approx 1$, $z \approx 2$ and $z \approx 3$, respectively.

Once these effects have been properly taken into account, the observational results show a strong evolution of the morphological fractions. In particular, if galaxies are divided into the three broad classes of spheroids, discs and peculiars/irregulars, the major change is the rapid increase of the percentage of peculiars/irregulars with increasing redshift, from $\approx 5$–$10\%$ at $z \approx 0$ to $\approx 30\%$ at $z \approx 0.6$, and up to $\approx 60$–$70\%$ at $2.5 < z < 3$. In parallel, the fraction of spheroids and discs decreases rapidly from $z \approx 1$ to $z \approx 3$ (Fig. 11.30). This suggests that there is a continuous morphological transformation with increasing cosmic time. The combined fraction of spheroids + discs seems to equal that of peculiar galaxies at $z \approx 2$, suggesting that the Hubble sequence began to emerge about 10 Gyr ago. Interestingly, these evolutionary trends are mass-dependent because high-mass galaxies ($\mathcal{M}_\star > 10^{10.5}$–$10^{11}\ \mathcal{M}_\odot$) seem to display spheroid and disc morphologies earlier than low-mass systems. This is one of the several manifestations of the so-called **downsizing**, which is an evolutionary scenario in which massive galaxies evolved earlier and faster than low-mass ones.

The cosmic evolution of morphologies at $0 < z < 3$ is accompanied by a general increase of galaxy sizes with decreasing redshift. For $\mathcal{M}_\star > 5 \times 10^{10}\ \mathcal{M}_\odot$, the evolutionary trend of the average effective radius $R_e$ at fixed $\mathcal{M}_\star$ derived from the observations at $0 < z < 3$ is $R_e \propto (1+z)^\alpha$, where $\alpha \approx -0.7$ for discs and $\alpha \approx -1.5$ for spheroids (Fig. 11.31). This means that the size growth of spheroids was significantly faster than for discs. This increase of size (and mass) is ascribed to major and minor merging (§8.9.6 and §10.3.2), inside-out star formation (§10.7.4) and accretion of gas (§10.7.2). However, the details and the relative importance of these processes as a function of cosmic time are still unclear. An intriguing case is represented by massive quiescent spheroids at $z > 1.5$ (Fig. 11.23). The effective radii of these systems can be as small as $< 1$ kpc, and this implies very high stellar mass densities compared to E/S0 galaxies with the same mass at $z \approx 0$. This is also indicated by the higher velocity dispersions of these systems compared to present-day ETGs. A possibility is that some of these systems are the stellar relics of the dusty starbursts at $z > 2$ observed in the IR–mm where vigorous star formation takes place in compact regions of dense molecular gas (§11.2.9). The subsequent mass and size growth of these compact spheroids could have started from the dense core formed at early times, followed by a more gradual accretion of stars in the outer parts through minor mergers (§10.8).

### 11.3.2 The Evolution of Galaxy Mergers

In the ΛCDM cosmological framework, structures grow hierarchically through the merging of dark matter halos (§7.4.4). Thus, the observation of galaxy merging as a function of cosmic time is particularly important because of its relation with the evolution of the dark matter halos. Moreover, galaxy mergers are also important in several processes such as the triggering of starburst and AGN activities, the morphological transformation from discs to spheroids and the stellar mass growth. The identification of galaxy mergers at cosmological distances requires high-resolution images in order to detect pairs of galaxies or multiple systems. Typically, a projected separation of < 5–30 kpc and a line-of-sight velocity difference < 200 km $s^{-1}$ are the criteria adopted to isolate systems which will likely merge. Thus, the availability of spectroscopic redshifts is essential to ensure that pairs are indeed made of galaxies at the same redshift and are not the result of projection effects. Alternatively, photometric redshifts can also be used provided that the statistical contamination of spurious pairs is carefully estimated. Galaxy mergers can also be identified by selecting individual galaxies with morphological disturbances typical of merging systems based on the $CAS$, Gini and $M20$ parameters (§11.1.6). Once a sample of galaxy mergers has been selected, two main quantities are usually derived to trace their evolution. The first is the merger fraction ($f_{\rm merg}$; §6.1), which is the ratio between the number of mergers and the total number of galaxies at a given redshift. At $0 < z < 3$, the merger fraction increases rapidly with redshift as

$$f_{\rm merg} = f_{\rm merg,0}(1 + z)^{\alpha}, \tag{11.18}$$

where $f_{\rm merg,0}$ is the merger fraction at $z = 0$ and $\alpha \approx 1$ for galaxies with $\mathcal{M}_\star \gtrsim 10^{10}\ \mathcal{M}_\odot$. However, larger values of $\alpha$ have been also obtained depending on the sample selection and on the criteria adopted to identify merger candidates. Irrespectively of the precise value of $\alpha$, the results indicate unambiguously that mergers were more frequent in the past. Fig. 11.31 shows the redshift evolution of $f_{\rm merg}$ at $0 < z < 3.2$ for massive galaxies with $\mathcal{M}_\star > 10^{11}\ \mathcal{M}_\odot$.

Another important quantity is the merger rate ($\Gamma_{\rm merg} = n_{\rm merg}/\tau_{\rm merg}$; eq. 6.1), which measures the number of mergers per unit of comoving volume and time. However, the merger rate $\Gamma_{\rm merg}$ can be estimated only if an average timescale $\tau_{\rm merg}$ is assumed, usually based on numerical simulations ($\tau_{\rm merg} \approx 0.5$ Gyr). If $\tau_{\rm merg}(z)$ is assumed to be constant with redshift, $\Gamma_{\rm merg}(z)$ is nearly flat at $z < 1.5$ irrespective of galaxy mass, with typical values $\Gamma_{\rm merg} \approx 10^{-4}$ Mpc$^{-3}$ Gyr$^{-1}$ and $\approx 10^{-5}$ Mpc$^{-3}$ Gyr$^{-1}$ for galaxies with $\mathcal{M}_\star > 10^{10}\ \mathcal{M}_\odot$ and $\mathcal{M}_\star > 10^{11}\ \mathcal{M}_\odot$, respectively. At $1.5 < z < 3.2$, $\Gamma_{\rm merg}(z)$ gradually declines, as shown in Fig. 11.31. However, $\Gamma_{\rm merg}$ may be a monotonically increasing function of $z$ if $\tau_{\rm merg}(z)$ is assumed to decrease for increasing $z$, as expected in the standard cosmological model. The overall results suggest that, at $z < 1$, major mergers and star formation are comparable sources of galaxy stellar mass growth. However, at higher redshifts, major mergers play a minor role (with a contribution $\approx 10 - 100$ times smaller) compared to star formation in the build-up of galaxy stellar masses.

### 11.3.3 The Evolution of the Star Formation Rate

Tracing the evolution of star formation, stellar mass and metallicity is one of the main goals to understand the processes which drive the so-called baryon cycle (§1.6). The statistical distribution of the SFR for a sample of galaxies can be derived from their LF, provided that the luminosity is an SFR indicator such as in the UV, H$\alpha$, IR or radio (§11.1.2). The

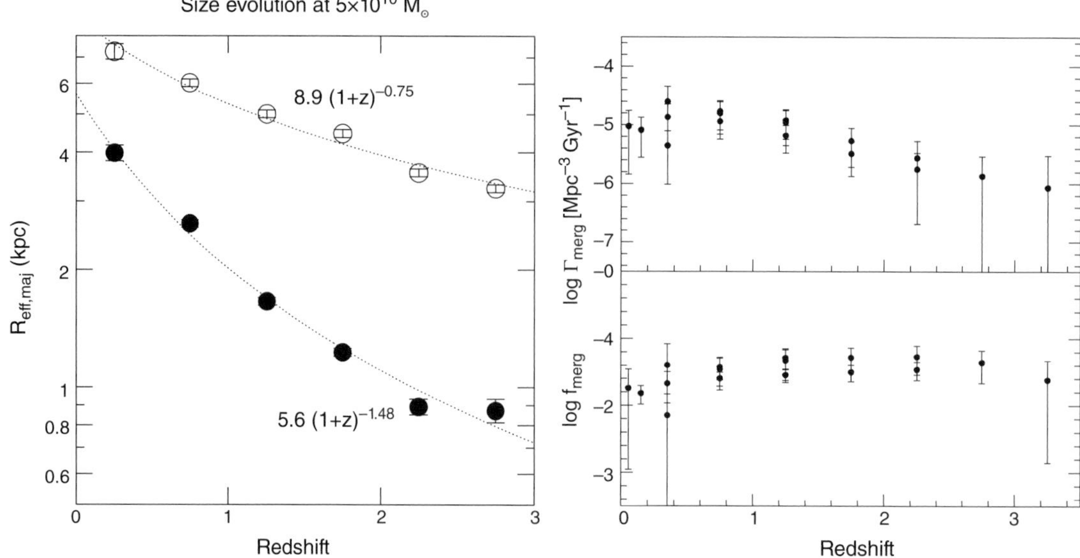


Fig. 11.31 *Left panel.* The evolution of the effective radius (measured along the major axis) for galaxies at fixed mass $\mathcal{M}_\star = 5 \times 10^{10}\ \mathcal{M}_\odot$. Filled and open circles are relative to spheroid and disc galaxies, respectively. The faster evolution of spheroids is evident. From van der Wel et al. (2014). © AAS, reproduced with permission. *Right panel.* The evolution of the major merger fraction ($f_{\rm merg}$) and the merger rate ($\Gamma_{\rm merg}$) for galaxies with stellar masses $\mathcal{M}_\star > 10^{11}\ \mathcal{M}_\odot$. Major mergers are defined here by a stellar mass ratio $\mathcal{M}_{\star,1}/\mathcal{M}_{\star,2} > 1/4$, where subscripts 1 and 2 indicate the two galaxies belonging to the merging system (assuming $\mathcal{M}_{\star,1} \leq \mathcal{M}_{\star,2}$). The results of $\Gamma_{\rm merg}(z)$ are derived assuming $\tau_{\rm merg}(z) =$ const, with $\tau_{\rm merg} = 0.60$ Gyr and $\tau_{\rm merg} = 0.32$ Gyr for galaxy-pair separations of 5–30 kpc and 5–20 kpc, respectively. For a given redshift, the different points indicate the results based on galaxy samples extracted from different sky fields, but analysed with the same method. Data from Mundy et al. (2017).

SFR-dependent LFs are found to evolve strongly with redshift. For instance, Fig. 11.32 shows the case of the IR LF where a remarkable evolution is visible. This is apparent from the progressive shift of the LF knee with increasing redshift. In particular, the characteristic luminosity $L^*_{\rm IR}$ shows a strong increase by more than one order of magnitude from $z = 0$ to $z \approx 2$–3, whereas the number density displays a weaker evolution. As $L_{\rm IR}$ of SFGs is proportional to the SFR (eq. 11.5), this implies that the galaxy SFRs were on average much higher in the past. The advantage of $L_{\rm IR}$ as an estimator of the SFR is that it is not affected by dust extinction (§11.1.2). However, due to the limited angular resolution of the IR surveys (especially at long wavelengths), the PSF (§11.1.6) can be so large that the uncertainties on the fluxes due to the blending of different galaxies must be carefully evaluated.

## 11.3.4 The Cosmic Star Formation Rate Density

Once a $\Phi(L)$ has been derived at a given redshift, it is possible to estimate the **cosmic star formation rate density** ($\rho_{\rm SFR}$). This is a measurement of the average SFR of the Universe

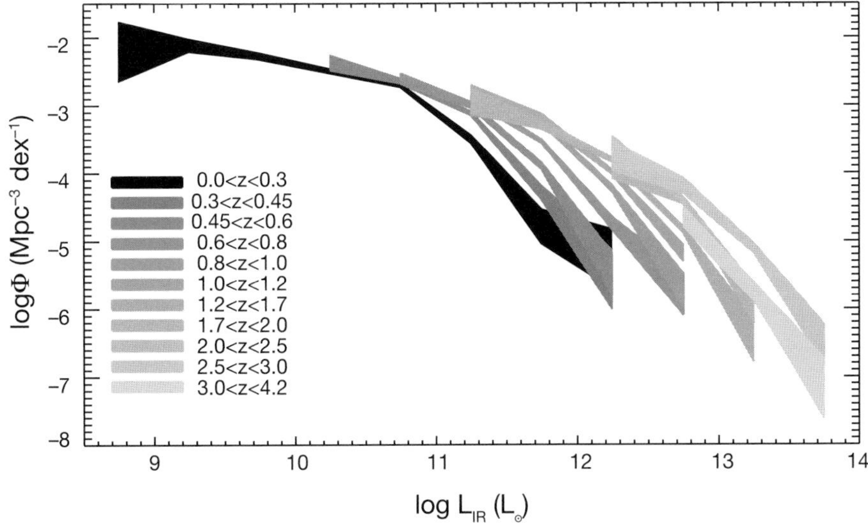


Fig. 11.32 The evolution of the IR luminosity function of SFGs as derived by the *Herschel* surveys. From Gruppioni et al. (2013).

per unit comoving volume at a given cosmic time. To estimate $\rho_{SFR}$, the first step is to integrate the $\Phi(L)$ and obtain the total **luminosity density** $\rho_L$ (eq. 3.10),

$$\rho_L(z) = \int_0^\infty L\Phi(L, z)\mathrm{d}L. \tag{11.19}$$

The units of $\rho_L(z)$ are erg s$^{-1}$ Mpc$^{-3}$ if based on a bolometric luminosity, or erg s$^{-1}$ Hz$^{-1}$ Mpc$^{-3}$ (or erg s$^{-1}$ Å$^{-1}$ Mpc$^{-3}$) if monochromatic. The second step is to convert $\rho_L$ into $\rho_{SFR}$ ($\mathcal{M}_\odot$ yr$^{-1}$ Mpc$^{-3}$) according to the relation between SFR and luminosity density $\rho_{SFR} = C\rho_L$ (see eq. 11.1). Fig. 11.33 shows that $\rho_{SFR}(z)$ increases rapidly from $z = 0$ to $z \approx 1$, reaches a maximum around $z \approx 2$ (the so-called **cosmic noon**), and then tends to decrease at higher redshifts. This demonstrates that the climax of cosmic star formation occurred around 10 billion years ago, when also the Hubble morphological sequence began to emerge (§11.3.1). At $z > 4$, the results are less clear because IR surveys are not sensitive enough and, as a consequence, $\rho_{SFR}(z)$ is estimated mostly from the rest-frame UV data (eq. 11.4) where dust extinction corrections introduce significant uncertainties. For more details, we refer the reader to the review of Madau and Dickinson (2014).

### 11.3.5 The Evolution of the Star Formation Main Sequence

At $z \approx 0$, the correlation between the SFR and stellar mass (the SFMS; §4.4.4) is one of the key properties of SFGs. The observation of distant galaxies shows that the SFMS persists up to high redshifts with a typical slope $\alpha \approx 0.8 \pm 0.1$ (SFR $\propto \mathcal{M}_\star^\alpha$). The slope and the scatter (about ±0.3 dex) of the SFMS seem to be weakly dependent on redshift, but the normalisation of the SFMS increases with redshift (Fig. 11.34). The SFR of galaxies

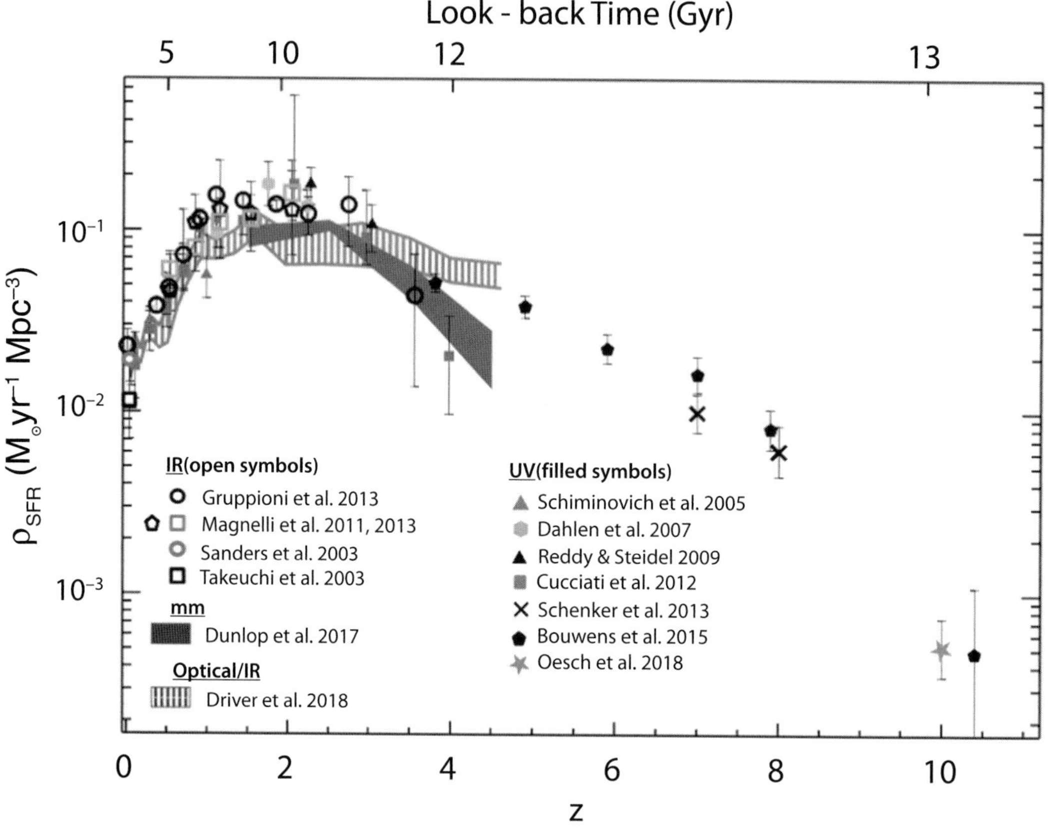


Fig. 11.33 The evolution of the cosmic SFR density based on a collection of literature data from multi-wavelength surveys (Sanders et al., 2003; Takeuchi et al., 2003; Schiminovich et al., 2005; Dahlen et al., 2007; Reddy and Steidel, 2009; Magnelli et al., 2011; Cucciati et al., 2012; Gruppioni et al., 2013; Magnelli et al., 2013; Schenker et al., 2013; Bouwens et al., 2015; Dunlop et al., 2017; Driver et al., 2018; Oesch et al., 2018). The data are homogeneised to a Salpeter IMF and corrected for dust extinction whenever appropriate. Courtesy of C. Gruppioni.

located on the SFMS depends on stellar mass and redshift according to the following relation, which provides a reasonable fit to the observations in the range $0 < z < 3$:

$$\mathrm{SFR_{MS}}(\mathcal{M}_\star, z) = 10^{(a \log \mathcal{M}_\star - b)} (1 + z)^{(c \log \mathcal{M}_\star + d)}, \tag{11.20}$$

where $a \approx 0.6$, $b \approx 5.8$, $c \approx 0.2$ and $d \approx 0.6$, and $\mathcal{M}_\star$ is in units of $\mathcal{M}_\odot$. The persistence of the SFMS as a function of redshift suggests that there is a dominant process which drives the formation of stars across cosmic time, and that this process evolves smoothly. According to this scenario, the sustenance of galaxies on the SFMS is due to the continuous accretion of cold gas from the filaments of the surrounding cosmic web over timescales of several Gyr (§8.2.4). As long as this process is at work, galaxies lie on the SFMS. The fact that the SFMS slope is essentially independent of redshift is taken as an indirect support of this scenario. When a given galaxy experiences a significant interaction with

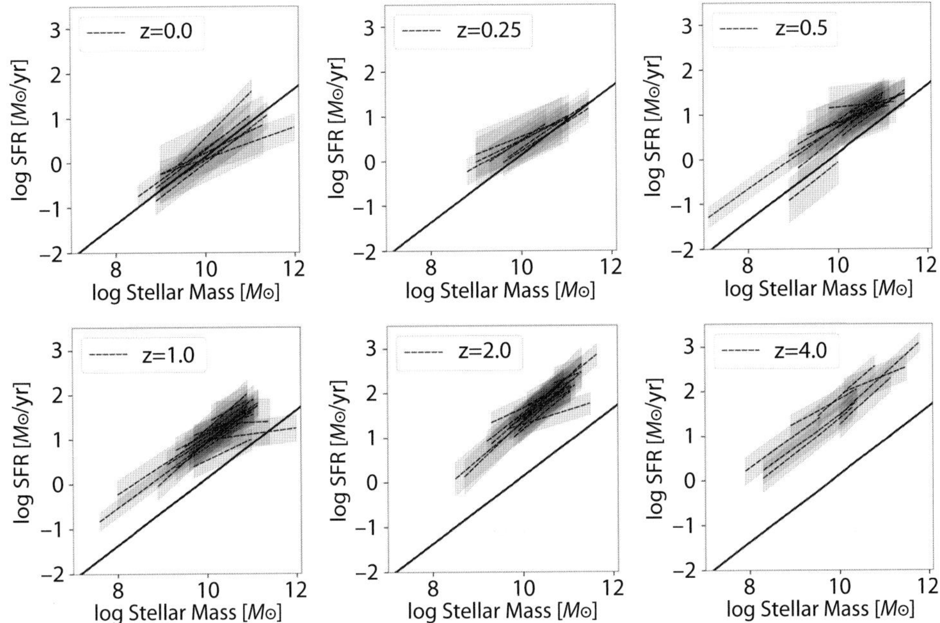


Fig. 11.34 The evolution of the SFMS at $0<z<4$ based on a collection of literature data. Each panel shows the best-fitting relations (dashed lines) to the observed SFMS relative to galaxy samples at the average redshift indicated by the label on the top. The shaded regions show the scatter around each best-fitting relation. The thick solid line repeated in each panel is the average SFMS at $z=0$. The gradual increase of the SFMS normalisation with increasing redshift is evident with respect to $z=0$. Data from Speagle et al. (2014). Courtesy of J.S. Speagle.

other gaseous systems (e.g. through major merging), the SFR is sporadically boosted and the galaxy leaves the SFMS, moving upwards into the starburst region of the SFMS plane. However, the contribution of the starburst phase to the global $\rho_{\rm SFR}(z)$ seems to be small. At $1.5<z<2.5$ and for $10<\log(\mathcal{M}_\star/\mathcal{M}_\odot)<11$, the number density of starbursts is only $\sim 10^{-5}$ Mpc$^{-3}$, and they represent only a few per cent of the total population of SFGs selected based on the stellar mass. In comparison, SFGs located on the SFMS are much more numerous ($\sim 10^{-4}$ Mpc$^{-3}$) and contribute to $\approx 90\%$ of the total $\rho_{\rm SFR}$ at $z\approx 2$, whereas starbursts account for the remaining $\approx 10\%$.

## 11.3.6 The Evolution of Specific Star Formation Rate

Other important constraints on galaxy evolution come from the sSFR (§4.4 and §11.1.2) because this quantity allows us to assess the relevance of the ongoing star formation in building up the stellar mass of a given galaxy (§10.10.3). The observations show that the sSFR increases dramatically with redshift (Fig. 11.35). This indicates that the SFR per unit stellar mass was much more important in the past than in the present-day Universe (§10.10.3). The sSFR allows us also to estimate the timescales on which galaxies assemble their stellar mass through star formation. For example, SFGs on the SFMS at $z\approx 2$

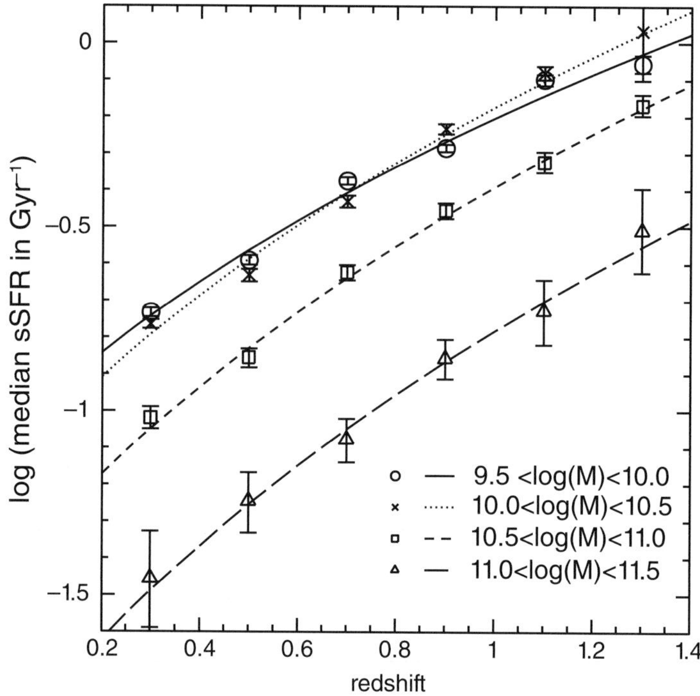


Fig. 11.35 The median sSFR as a function of redshift and in four bins of stellar mass (in solar units). Irrespectively of stellar mass, the sSFR increases rapidly with redshift. At each redshift, low-mass galaxies have systematically higher sSFRs. From Ilbert et al. (2015).

double their stellar mass on typical timescales of $\approx 0.5-1$ Gyr. The sSFR evolution is dependent on stellar mass and follows the downsizing trend: at a given redshift, massive galaxies with $\mathcal{M}_\star > 10^{11}\ \mathcal{M}_\odot$ have sSFRs systematically lower than those of lower-mass systems. This implies that massive galaxies formed the bulk of their stellar mass earlier and faster than the less massive ones. The dependence on stellar mass can be expressed as $\text{sSFR} \propto \mathcal{M}_\star^\alpha$ (where $-0.4 < \alpha < 0$), whereas the normalisation increases with redshift as $\text{sSFR}(z) \propto (1+z)^\beta$, with $\beta \approx 3$ from $z \approx 0$ to $z \approx 2$ and $\beta \approx 1.5$ at $z > 2$. However, uncertainties on the shape of the sSFR($z$) are still present at higher redshifts. The region below the SFMS is populated by systems with lower sSFR than SFMS galaxies. SFGs move to that region when they start to terminate their star formation and migrate downwards into the population of quiescent galaxies. The conclusion of this path occurs when no stars are formed any more, and galaxies become passive. This crucial phase is called quenching of star formation (§10.6). The observations suggest that the mass growth of SFGs terminates when they reach a critical stellar mass around $\mathcal{M}_\star \approx 10^{10.5}-10^{11}\ \mathcal{M}_\odot$. However, despite the several mechanisms proposed to explain the star formation quenching (§10.6), it is still unclear how galaxies cease to form stars. Different processes are in fact likely to act on different spatial and timescales, also depending on the environment where a given galaxy is located. For instance, in the case of massive galaxies at $z \approx 2$, some results based on integral field spectroscopy suggest that the star formation is quenched from the inside out. In this scenario, quenching develops from the inner region where the 'protobulge' experiences one or more processes (e.g. dissipative collapse, violent disc instability; §10.3.3) which cause a rapid exhaustion of the gas on rather short timescales ($\lesssim 0.5-1$ Gyr), leading to the formation of a compact and dense stellar system. Then, the quenching proceeds towards the outer regions of the galaxy where star formation is still fed by the accretion of external gas. Here, the termination of star formation seems to occur on longer timescales (a few Gyr), as a consequence of either mass or environmental quenching (§10.6). At $z \approx 2$, the frequency of AGN activity increases with galaxy mass, and this suggests that AGN feedback may indeed play an important role in quenching the star formation in massive systems.

### 11.3.7 The Evolution of Galaxy Stellar Mass

The strong evolution of the cosmic SFR density (§11.33) and merger rate (§11.3.1) imply that galaxies increase their stellar masses with cosmic time. Thus, the evolution of the stellar mass is another key probe to understand the physics of galaxy formation and the relationship between baryonic matter and dark matter halos. In this regard, fundamental constraints can be derived from the evolution of the galaxy SMF (§3.5.2; Fig. 3.11). As in the present-day Universe, the total SMF (i.e. the SMF which includes all galaxy types) is better reproduced by a double Schechter function (eq. 3.12) up to $z \approx 2.5-3$. At $0 < z < 3$, the SMF shows a clear evolution as a function of redshift (Fig. 11.36). In particular, the characteristic densities $\Phi_1^*(z)$ and $\Phi_2^*(z)$ increase with decreasing redshifts, whereas the characteristic mass $\mathcal{M}^*(z)$ does not strongly change. At $z > 3$, the SMF can be successfully fitted with a single Schechter function. The need for a double Schechter function to reproduce the total SMF at $z < 3$ is ascribed to the emergence of the quiescent galaxy population at $z \approx 2.5$. In Fig. 11.37 we see that the SMFs of star-forming and quiescent galaxies display clear differences. Quiescent galaxies build up their stellar mass gradually since $z \approx 3$, and they dominate the high-mass tail of the SMF at lower redshifts. The most significant increase of the number density of these galaxies occurs from $z \approx 2.5$ to $z \approx 1$. At fixed redshift and for all redshifts, the fraction of quiescent galaxies increases with mass. This is in agreement with the downsizing scenario where massive systems formed earlier and quenched faster than those with lower masses.

In the case of SFGs, a clear evolution of their SMF is also visible (Fig. 11.37, left). In particular, if these galaxies are further subdivided into classes according to sSFR, morphology or bulge dominance, a single Schechter function provides an accurate description of each SMF. Late spirals and irregulars dominate the low-mass end of the SMF and are more common at high redshifts, whereas Sa–Sc galaxies are more important at higher masses and lower redshifts. This evolution is qualitatively consistent with the inside-out quenching scenario of the bulge growth (§11.3.3).

### 11.3.8 The Influence of the Environment on Galaxy Evolution

In the present-day Universe, galaxy properties depend on the environment (§6.5), and a variety of environmental processes can contribute to the suppression of star formation

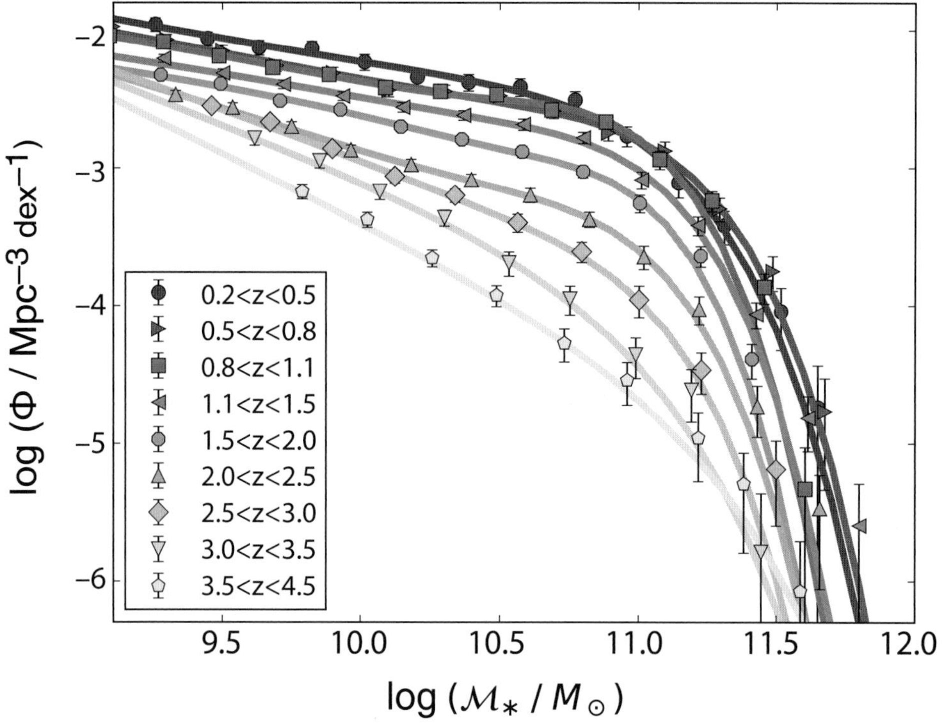


**Fig. 11.36** The evolution of the total SMF (i.e. including all galaxy types). The curves indicate the best fits to the data obtained with Schechter functions. A double Schechter function is fitted in each redshift bin up to $2.5 < z < 3$. At $z > 3$, the fits are based on a single Schechter function. A clear evolution is apparent: for a fixed stellar mass, the galaxy number density increases smoothly with $z$, and the high-mass tail becomes progressively populated by massive galaxies with decreasing $z$. Adapted from Davidzon et al. (2017).

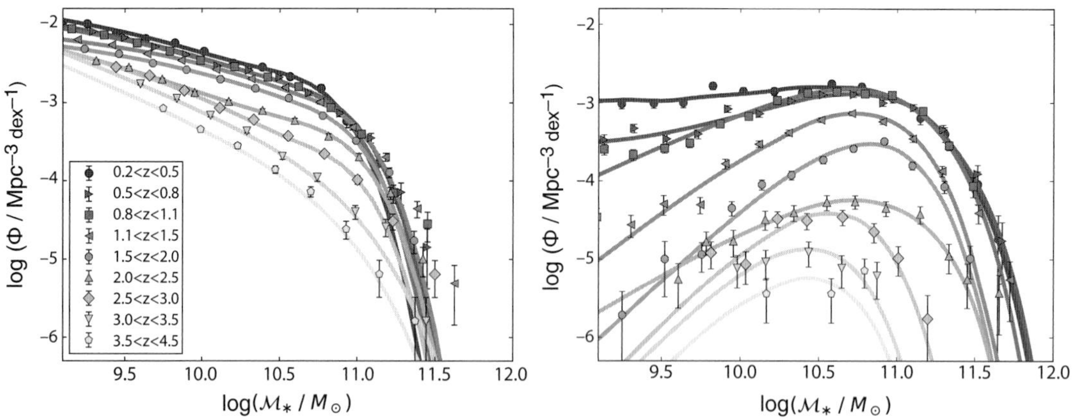


**Fig. 11.37** The evolution of the SMF of star-forming (*left*) and quiescent (*right*) galaxies. The curves indicate the best fits of the data obtained with Schechter functions as in Fig. 11.36. Adapted from Davidzon et al. (2017).

(§10.6). At higher redshifts, several results confirm that the role of the environment is important in galaxy evolution. For example, the fraction of SFGs is higher in distant clusters than in clusters at $z \approx 0$ (**Butcher–Oemler effect**). Moreover, the percentage of S0 galaxies in clusters increases with cosmic time at the expense of disc galaxies, whereas galaxies which terminated their star formation $\approx 0.5\text{–}1$ Gyr prior to the observation (**post-starburst galaxies**) were more numerous in distant clusters. These results indicate that the environment certainly plays a role in the suppression of star formation, and that the morphology–density relation at $z \approx 0$ (§6.5) is the end-point of this environment-dependent evolution. The influence of the environment can also be investigated by studying how the SMF evolution depends on the environmental density. In the redshift range $0 < z < 3$, the SFMS depends weakly on the environment, and mass quenching is dominant at $z > 1$, whereas environmental quenching is more likely to occur at lower redshifts. The fraction of quiescent galaxies depends mostly on the environment at $z < 1$, whereas the dependence on stellar mass is observed in a broader redshift range ($0 < z < 3$). Some studies have shown that at $z \gtrsim 1$ the SFR–environmental density relation reverses, so that the average SFR per galaxy is higher in denser environments than in the field. At $z < 2$, the SMF shows a marked dependence on the environment, whereas the influence of the density field seems to be reduced at higher redshifts. Quiescent galaxies are preferentially found in the densest environments out to $z \approx 2$, where the high-mass end of the SMF ($\mathcal{M}_\star > 10^{11}$ $\mathcal{M}_\odot$) is strongly enhanced. At $z < 1.5$, SFGs with $\mathcal{M}_\star < 10^{11}$ $\mathcal{M}_\odot$ appear to be more frequently located in low-density environments. At higher redshifts, protoclusters contain higher fractions of galaxies with high SFRs and starbursts than at $z \approx 0$. This suggests that these SFGs could be the precursors of the quiescent systems that are predominantly located in massive clusters at lower redshifts, such as the BCGs (§6.4.1).

### 11.3.9 The Cosmic Stellar Mass Density

The gradual increase of the galaxy stellar mass with cosmic time is also shown by the evolution of the **cosmic stellar mass density** $\rho_\star(z)$ (usually expressed in $\mathcal{M}_\odot$ Mpc$^{-3}$). Following the same approach as in eq. (11.19), $\rho_\star(z)$ can be derived by integrating the observed SMF as

$$\rho_\star(z) = \int_0^\infty \mathcal{M}_\star \Phi(\mathcal{M}_\star, z) \mathrm{d}\mathcal{M}_\star. \tag{11.21}$$

Fig. 11.38 shows the progressive increase of $\rho_\star(z)$ with decreasing redshift. As a reference, the stellar mass density of the Universe was $\approx 50\%$, 10%, and 1% of its present-day value at $z \approx 1.3$, $z \approx 2.9$ and $z \approx 5.7$, respectively. A consistency check is provided by the comparison of the observed $\rho_\star(z)$ with that expected from the integration of $\rho_{\rm SFR}(z)$ (eq. 11.19). Using $\mathrm{d}t = -\mathrm{d}z/[H(z)(1+z)]$ (eq. 2.38), we get

$$\rho_{\star,\rm exp}(z) = (1 - \mathcal{R}) \int_z^\infty \rho_{\rm SFR} \frac{\mathrm{d}z'}{H(z')(1+z')}, \tag{11.22}$$

where $\rho_{\star,\rm exp}(z)$ is the total mass density of stars and stellar remnants accumulated at the same redshift from previous star formation activity, and $\mathcal{R}$ is the fraction of stellar mass

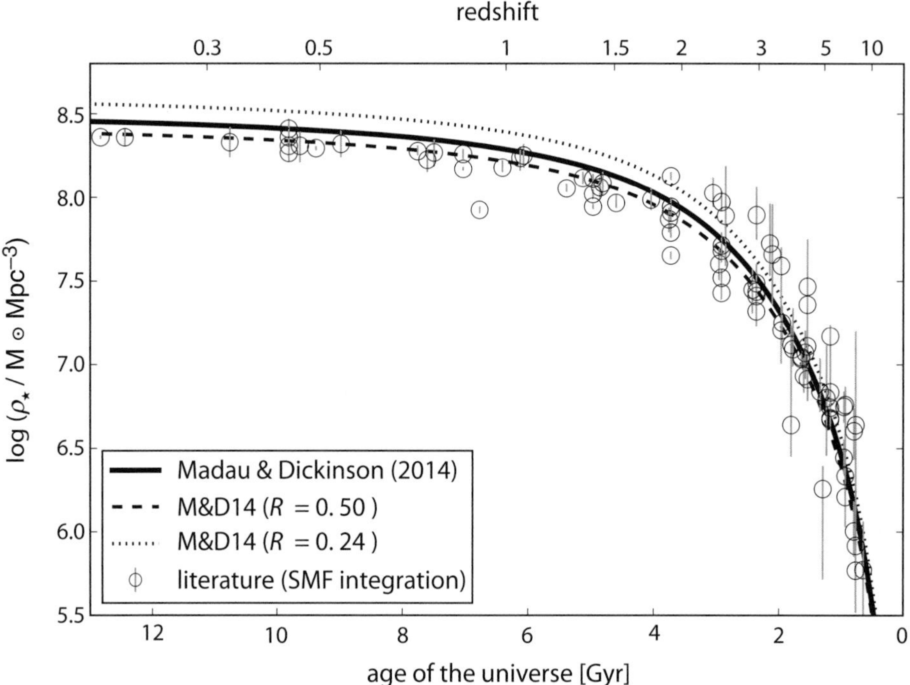


Fig. 11.38 The evolution of the stellar mass density derived from a collection of literature data and integrating directly the observed SMFs as functions of redshift (open circles). The three curves have been derived with eq. (11.22) using a best-fitting formula (eq. 15 of Madau and Dickinson, 2014) which reproduces the observed evolution of the cosmic SFR density $\rho_{\rm SFR}(z)$. The dashed, solid and dotted curves are relative to return fractions (§8.5.3) $\mathcal{R} = 0.24$, $\mathcal{R} = 0.27$ (the value adopted by Madau and Dickinson, 2014) and $\mathcal{R} = 0.50$, respectively. Adapted from Davidzon et al. (2017).

returned to the ISM during the evolution of the stellar populations due to winds, planetary nebulae, novae and SNe (§8.5.3). The shape and the normalisation of $\rho_{\star,\rm exp}(z)$ are in good agreement with the observed $\rho_\star(z)$ for $\mathcal{R} \approx 0.3$–$0.4$ (Fig. 11.38). Another check is given by the comparison of the observed rate of core-collapse (i.e. Type II) SNe with the one expected from the observed $\rho_{\rm SFR}(z)$. Since core-collapse SNe are produced by high-mass ($\mathcal{M} > 8\ \mathcal{M}_\odot$) short-lived stars, the higher the cosmic SFR density, the higher should be the rate of these SNe. The results up to $z \approx 1.4$ indicate a good agreement between the observed rate of Type II SNe and that predicted by $\rho_{\rm SFR}(z)$.

An interesting application of the SMF is the estimate of the total number of galaxies in the Universe. This can be done first by integrating the SMF between the minimum and maximum galaxy masses in order to obtain the total comoving number density of galaxies at a given redshift:

$$\Phi_{\rm tot}(z) = \int_{\mathcal{M}_{\star,\rm min}}^{\mathcal{M}_{\star,\rm max}} \Phi(\mathcal{M}'_\star, z)\mathrm{d}\mathcal{M}'_\star. \tag{11.23}$$

The second step is to derive the total number of galaxies $N_{tot}$ through the integral of $\Phi_{tot}(z)$ over the volume of the Universe defined by the entire sky and the redshift range where galaxies have been identified (see eq. 2.18 for the calculation of comoving volumes). For $0 < z < 8$, $\mathcal{M}_{\star,min} = 10^6\ \mathcal{M}_\odot$ and $\mathcal{M}_{\star,max} = \infty$, the total number of galaxies in the observable Universe turns out to be $N_{tot} \approx 2 \times 10^{12}$.

### 11.3.10 Linking Archaeological and Look-Back Results

As introduced in §1.5, two main observational approaches are possible to investigate galaxy evolution. The first is the direct observation of galaxies at different redshifts in order to trace the change of their properties with cosmic time (the **look-back approach**). The second is to treat the present-day galaxies as 'fossils' at $z \approx 0$ and to infer their evolution from the reconstruction of their past SFH (the **archaeological approach**). Galaxies in the Local Group (§6.3) are ideal archaeological probes thanks to the possibility to observe their individual stars and reconstruct the past SFHs. However, the look-back and archaeological approaches can be applied also beyond the Local Group through the inference of galaxy properties with SPS models (§8.6.2). The case of ETGs (§5.1.3) is particularly relevant to test the two approaches, and to verify if the interpretation of the high-$z$ and low-$z$ data is coherent. ETGs are crucial probes of galaxy formation because they are the most massive galaxies in the present-day Universe and contain a large fraction of the total stellar mass at $z = 0$ (see Renzini, 2006, for a review). Moreover, the absent or weak star formation, and the predominantly old (and often passively evolving) stellar populations make ETGs the simplest systems in which to apply SPS models (§8.6.2) and Lick indices (§11.1.3) to reconstruct their past SFH.

Let us start with the archaeological results. The results of spectral analysis obtained for ETGs at $z \approx 0$ are nicely summarised in Fig. 11.39, where the reconstructed sSFR is displayed as a function of redshift and for different dynamical masses $\mathcal{M}_{dyn}$. It is instructive to recall how these quantities have been derived from the spectra of ETGs. $\mathcal{M}_{dyn}$ (eq. 5.17) is estimated from the stellar velocity dispersion and the effective radius, whereas the fitting of the spectra and photometric SEDs with SPS models and/or the analysis of the Lick indices provide the stellar ages (and hence formation redshifts), the abundances of iron and $\alpha$-elements, and the stellar masses (§11.1.3). The average duration $\Delta t$ of the star formation can then be derived from the [$\alpha$/Fe] ratio (eq. 11.8). This allows the estimate of the SFR $\approx \mathcal{M}_\star/\Delta t$ and sSFR $\approx$ SFR/$\mathcal{M}_\star$. The downsizing scenario (§11.3.1) emerges clearly also from these archaeological studies. Fig. 11.39 shows that the most massive ETGs formed earlier ($z_{form} > 2$) and faster ($\Delta t \approx 0.1$–$0.3$ Gyr), whereas ETGs with lower masses formed the bulk of their stars more recently and during a more prolonged time. These archaeological results make crucial predictions that can be tested with the look-back approach. If the most massive ETGs terminated their star formation at $z > 2$–$3$ (as shown in Fig. 11.39), then a population of massive, quiescent ETGs should exist at $z > 1$, and these galaxies should have stellar ages consistent with the formation redshift predicted by the archaeological SFHs. As discussed in §11.2.6, these galaxies do indeed exist. Another archaeological prediction is that the number density of massive ($\mathcal{M}_\star > 10^{11}\ \mathcal{M}_\odot$) ETGs at

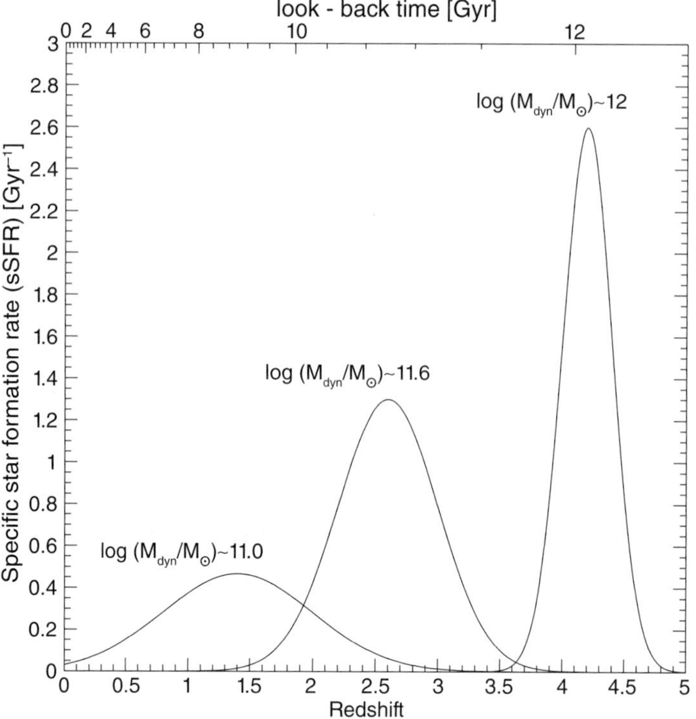


**Fig. 11.39** Sketch of the average SFHs of ETGs derived from the archaeological analysis of their spectra. The SFHs are shown as Gaussian curves because the true shapes remain uncertain. The average dynamical masses at $z = 0$ are indicated on the top of each SFH. Massive ETGs are characterised by narrower SFHs than ETGs with lower masses, i.e. they have star formation activity with a shorter duration. Moreover, the peak of the SFHs of massive ETGs occurs at higher redshifts than for lower-mass ETGs, and therefore they are older in the present-day Universe. This evolutionary trend is a manifestation of downsizing (§11.3.1). Adapted from Thomas et al. (2010).

$z \approx 0.5$ should be similar to that at $z \approx 0$ because, according to the downsizing scenario of Fig. 11.39, these galaxies should have assembled most of their mass at $z > 1$. This prediction is confirmed by the observed evolution of the SMF of ETGs which shows that the number density of massive ETGs with $\mathcal{M}_\star > 10^{11}\ \mathcal{M}_\odot$ is indeed nearly constant at $0 \lesssim z \lesssim 0.7$.

Finally, if massive ETGs formed most of their stellar mass at high redshifts in a short period of time, powerful starbursts with extremely high SFRs should exist at $z > 2$. As presented in §11.2.9 and §11.2.14, starburst galaxies with such properties have been identified out to $z > 6$. All these examples indicate a reasonable agreement between the properties expected from the archaeological results and the actual existence of high-redshift galaxies which have the characteristics required for being the progenitors of massive ETGs. However, this does not mean that all ETGs were formed from these progenitors because starbursts made only a limited fraction of the stellar mass (§11.3.3), whereas ETGs at $z \approx 0$ contain a substantial fraction of the whole stellar mass of the Universe.

### 11.3.11 The Evolution of the Interstellar Medium

As cosmic time goes on, the overall comoving **cosmic gas density** of the Universe (usually expressed in units of $\mathcal{M}_\odot$ Mpc$^{-3}$) decreases due to its progressive consumption by star formation. As stars are believed to form from molecular gas, it is crucial to trace the evolution of the cosmic density of the molecular hydrogen $\rho_{H_2}(z)$. This can be inferred from the conversion of the CO luminosity function following an approach similar to that described in §11.3.4 and §11.3.9. The cosmic $\rho_{H_2}(z)$ decreases by a factor of 3–10 from $z \approx 2$ to $z \approx 0$. When no data are available for the cold gas, $\rho_{H_2}(z)$ can be indirectly estimated via the evolution of the dust mass function adopting a $\mathcal{M}_{H_2}/\mathcal{M}_{\rm dust}$ ratio (§11.1.5), where the dust mass is derived with eq. (11.11). The evolution of the cosmic $\rho_{H_2}(z)$ depends on how the gas is exhausted as a function of cosmic time. The observations show that the gas fraction ($f_{H_2}$; eq. 11.9) of SFGs decreases with decreasing redshift, as qualitatively expected by the gradual gas consumption due to the evolution of $\rho_{\rm SFR}$. For instance, while at $z < 0.1$ SFGs with $\mathcal{M}_\star \sim 10^{10}$–$10^{11}$ $\mathcal{M}_\odot$ have typically $f_{H_2} < 0.1$, SFMS galaxies with the same mass have $f_{H_2} \approx 0.5$ at $z \approx 1$–$2$ and up to $f_{H_2} > 0.8$ at $z > 2$. An important result is that the gas fraction at a given $z$ is systematically lower in massive galaxies at all redshifts and is correlated with the sSFR. This is another manifestation of downsizing (§11.3.1) where massive galaxies exhausted their cold gas reservoirs more rapidly and earlier than lower-mass systems. Other important constraints come from the gas depletion time (eq. 4.20). The typical timescale for galaxies on the SFMS at $0 < z < 3$ is around 1 Gyr. This suggests that the star formation activity of low- and high-redshift galaxies located on the SFMS is characterised by similar physical processes across a wide range of cosmic times. Moreover, the depletion timescales are significantly shorter than the Hubble time ($t_{\rm dep}(z)/t_H \approx 0.1$–$0.2$) in all SFGs on the SFMS at $0 < z < 2$. This suggests that additional cold gas is nearly continuously accreted to maintain SFMS galaxies gas-rich over timescales of several billion years. This provides additional support to the accretion of intergalactic and circumgalactic cold gas (§8.2.4). Empirical scaling relations have been derived based on the observation of SFGs at different redshifts. Although these results are still in their infancy, we provide an example as a reference. At $0 < z < 3$ and for SFGs with $\mathcal{M}_\star > 10^{10}$ $\mathcal{M}_\odot$, the cold gas mass ($\mathcal{M}_{H_2}$) correlates with sSFR, stellar mass and redshift as

$$\mathcal{M}_{H_2} = A(1+z)^\alpha \zeta_{\rm MS}^\beta \left( \frac{\mathcal{M}_\star}{10^{10}\,\mathcal{M}_\odot} \right)^\gamma, \tag{11.24}$$

where $\zeta_{\rm MS} = {\rm sSFR}/{\rm sSFR}_{\rm MS}$ is the sSFR relative to the average value on the SFMS at redshift $z$, whereas the values of the constant $A$ and the exponents $\alpha$, $\beta$ and $\gamma$ depend on the details of the sample selection and the methods adopted to estimate $\mathcal{M}_{H_2}$. As a reference, $A \approx 7\times10^9 \mathcal{M}_\odot$, $\alpha \approx 1.8$, $\beta \approx 0.3$ and $\gamma \approx 0.3$. Also the gas fraction $f_{H_2}$ and the depletion time $t_{\rm dep}$ follow relations with the same functional form, although the values and signs of the exponents are different. The existence of such scaling relations can be explained within the scenario where SFGs spend most of their lifetime on the SFMS as long as the gas accretion continues from the surrounding circumgalactic and intergalactic media.

The observation of the atomic gas phase is also important because it provides complementary information with respect to the molecular gas. For instance, the intense star formation of high-$z$ galaxies may generate stronger fluxes of energetic cosmic rays than in SFGs at $z \approx 0$. These energetic particles may destroy CO and make this molecule more difficult to observe. Hence, observing also other tracers of gas in distant SFGs is essential to derive an unbiased view of their ISM. While the H I gas can be studied through the observation of absorption lines (§9.8), ALMA opened additional possibilities to observe the atomic gas in distant galaxies. An example is given by the observation of [C II] emission ($\lambda_{rest} \simeq 158\ \mu$m) at high redshifts. This transition is particularly important because it produces the strongest FIR emission line in SFGs. The [C II] luminosity correlates with the IR luminosity and accounts for $\approx 0.1\%-1\%$ of the galaxy bolometric luminosity. Moreover, it is a promising estimator of the SFR not affected by dust extinction, although it depends on the metallicity. Another important atomic transition is [C I] ($\lambda_{rest} \simeq 609\ \mu$m) because this line is associated with low-excitation CO emission in a wide range of star-forming environments and redshifts, and therefore represents a complementary tracer of the cold gas phase. This line has been detected with ALMA in high-$z$ SFGs. Also the [N II] at $\lambda_{rest} \simeq 205\ \mu$m is important to constrain the ionisation properties of the ISM at high redshifts. Regarding neutral hydrogen, SKA has been designed to observe directly the H I 21 cm line in emission also in high-redshift objects.

### 11.3.12 The Evolution of Metallicity of Gas and Stars

The evolution of the cosmic star formation and the parallel build-up of the stellar mass has the inevitable consequence that the global metallicity of the Universe must increase with time. The observations trace this evolution through the study of metallicity not only within galaxies (SFGs and ETGs), but also in the circumgalactic and intergalactic media (§9.8.4) and in the intracluster gas (§6.4.2).

*Star-forming galaxies*. The metallicity estimates in SFGs are mostly based on emission line ratio estimators (§11.1.4), and thus they measure the metal abundance of the ionised gas in H II regions. Fig. 11.40 shows that SFGs follow a mass–metallicity relation (§4.4.3) up to high redshifts. At all redshifts, the most massive galaxies are the most metal-rich. This is expected because massive galaxies have deeper gravitational potential wells, and therefore their gas is more easily retained and enriched during the star formation activity. However, Fig. 11.40 also shows that the average metallicity decreases with increasing redshift, as expected in the general scenario where the galaxy metallicity increases with cosmic time (§8.5). Some results suggest that the evolution of the mass–metallicity relation depends on the stellar mass, consistently with the downsizing scenario where massive galaxies completed their metal enrichment earlier and faster than the lower-mass ones. An anticorrelation is present between the sSFR and the metallicity at all redshifts. This indicates that galaxies with the lowest sSFRs (i.e. where the star formation activity is terminating) are reaching their highest metal abundances. Stellar mass, metallicity of the ionised gas and SFR are related with each other through a relation called the **fundamental metallicity relation** that can be described as a surface in 3D space ($\mathcal{M}_\star$–$Z$–SFR). This

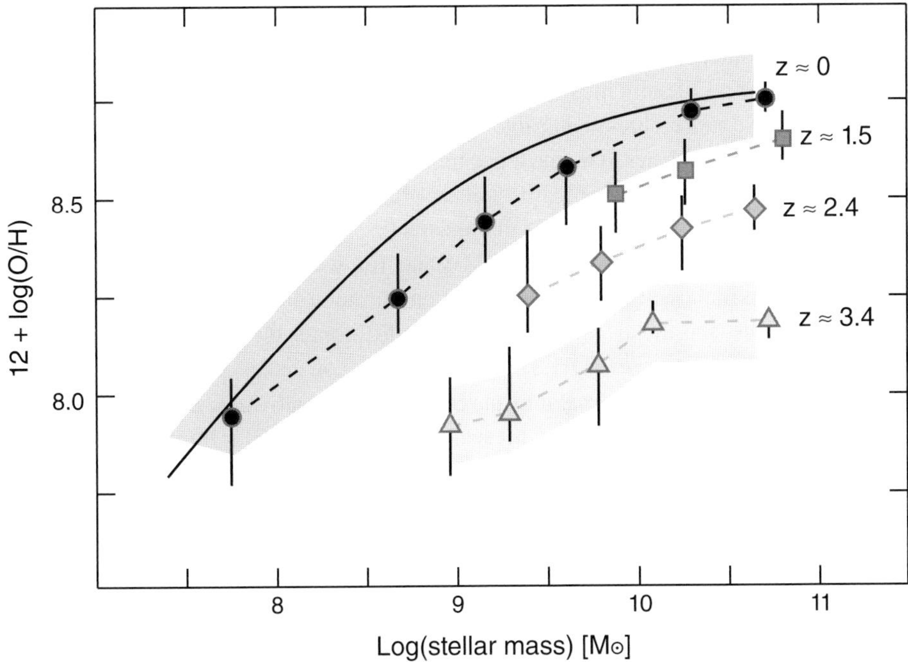


Fig. 11.40 The evolution of the mass–metallicity relation for SFGs based on the estimate of their ionised gas abundances. Here the metallicity is expressed as $12 + \log(O/H)$ (§C.6.1). The solid line indicates the reference mass–metallicity relation at $z \approx 0$. The dashed lines indicate the best fits to the relations obtained with samples of distant galaxies up to $z \approx 3.4$. Figure adapted from Hunt et al. (2016) with additional data from Sanders et al. (2015), Wuyts et al. (2016), Kashino et al. (2017) and Sanders et al. (2018). Courtesy of L. Hunt.

relation can be explained with models where the evolution of galaxies is regulated mainly by the interplay between the cold accretion (inflows) of low-metallicity gas, the outflows of material enriched by the evolution of the star formation, and the dependence of these processes on galaxy mass. For more details, we refer the reader to the review of Maiolino and Mannucci (2018).

Additional constraints can be derived from the study of metallicity gradients. Present-day galaxies always show the negative gradients expected in the inside-out evolution (§4.1.4 and §10.7.5). At high redshifts, metallicity gradients have been obtained up to $z \approx 3$ with integral field spectroscopy through the radial dependence of emission line ratios such as [N II]$\lambda$6583/H$\alpha$, [O III]$\lambda$5007/H$\beta$ and [O III]$\lambda$5007/[O II]$\lambda$3727 (§11.1.4). Due to the small angular sizes of high-redshift galaxies, these measurements are particularly challenging and prone to systematic effects such as beam smearing (§4.3.3). The results suggest that the metallicity profiles are usually flatter than in low-$z$ galaxies. However, there are also notable exceptions with negative or positive gradients. Flat profiles could be originated by outflows of more metal-rich gas from the inner regions of galaxies due to star formation and/or AGN feedback. These outflows may play a role through the redistribution of this enriched gas in the outer regions of the galaxies. Instead, positive (also called inverted) gradients are ascribed to the inflows of metal-poor gas from the circum/intergalactic medium, infalling into the inner regions of the galaxies and diluting their central metallicities. Regarding the stellar metallicity of distant SFGs, the measurements are more challenging due to the high $S/N$ ratio required to measure the weak stellar absorption lines. The most accurate results show that the stellar metallicities of SFGs at $z \approx 2-4$ are subsolar, with typical values around $\approx 0.1-0.2\ Z_{\odot}$, and that the metallicity increases with the galaxy stellar mass.

*Early-type galaxies*. In the case of ETGs, no (or weak) star formation is present, and therefore the metallicity estimators based on emission line ratios cannot be applied. As a consequence, the metal abundance is derived through the SPS modelling of their spectra or with Lick indices (§11.1.3). The metallicity of distant ($z > 1$) massive ETGs (§11.2.6) ranges from solar to supersolar. This indicates that these galaxies have metal abundances comparable with that of present-day ETGs (§5.1.3). Other important clues come from the supersolar abundances of $\alpha$-elements in these galaxies ([Mg/Fe] $\approx 0.3-0.6$). This brings another piece of evidence that these systems formed very rapidly with timescales of $\approx 0.1-0.5$ Gyr and that the metal enrichment was mostly caused by core-collapse SNe (§5.1.3). The large masses of these galaxies ($\mathcal{M}_{\star} > 10^{11}\ \mathcal{M}_{\odot}$) imply, again, that their progenitors were powerful starbursts at $z > 2$ characterised by vigorous star formation.

### 11.3.13 The Evolution of Scaling Relations Involving Kinematics

The evolution of the baryon cycle (§1.6) and the merging history change the main properties of galaxies across cosmic time. Thus, scaling relations such as the TFR (§4.4.1) and the fundamental plane (§5.4.1) are expected to be affected and to evolve too. The TFR of present-day disc galaxies consists of a tight relation between the rotation velocity ($v_{\rm rot}$) and the luminosity or the mass. In the case of stellar mass, it is expressed (§4.4.1; eq. 4.37) as

$$\log \mathcal{M}_{\star} = a \log v_{\rm rot} + b, \tag{11.25}$$

where $a$ and $b$ are the slope (typically $a \approx 4$) and zero-point of the correlation, respectively. The evolution of the TFR depends on how the gas fraction, star formation, stellar mass and their relative interplay with the dark matter halos evolve with redshift. Thus, understanding how this relation changes with redshift can place stringent constraints on the evolution of disc galaxies. However, tracing the evolution of the TFR with observations is challenging because of the faintness and small sizes of galaxies at high redshifts. The relation exists at least up to $z \approx 3$ (Fig. 11.41). Most results suggest that the slope $a$ does not change significantly. Instead, the evolution of the zero-point $b$ is more uncertain and the measured offsets with respect to $z = 0$ are in the wide range $-0.5 < \Delta b \leq 0$ depending on how the sample is selected, on the redshift range and on the specific form of the TFR adopted at $z = 0$.

The other main scaling relation is the fundamental plane (§5.4.1). This relation involves the size, velocity dispersion and surface brightness of ETGs according to eq. (5.54). The observations show that the fundamental plane is present out to the redshifts at which ETGs have been unambiguously identified ($z \approx 2-3$), and that the zero-point $\gamma$ displays a

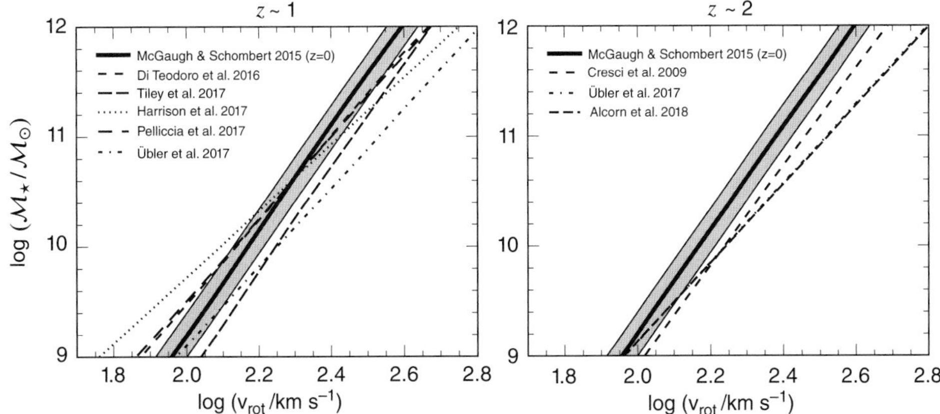


**Fig. 11.41** The stellar mass Tully–Fisher relation at $z \approx 1$ (*left panel*) and $z \approx 2$ (*right panel*). The thick solid line indicates the relation and its scatter (grey shaded band) at $z \approx 0$ (McGaugh and Schombert, 2015). The other lines indicate the best fits to the relations as derived in the different works (Cresci et al., 2009; Di Teodoro et al., 2016; Tiley et al., 2016; Harrison et al., 2017; Pelliccia et al., 2017; Übler et al., 2017; Alcorn et al., 2018). Courtesy of E. Di Teodoro.

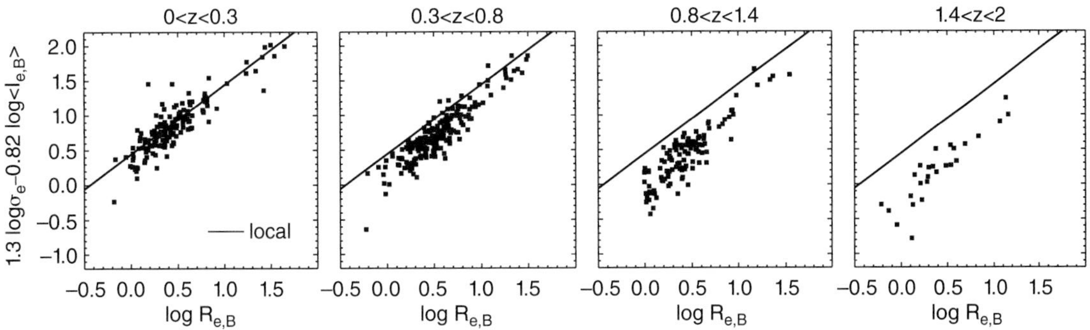


**Fig. 11.42** The evolution of the fundamental plane of the ETGs located in clusters of galaxies. The line indicates the fundamental plane relation at $z \approx 0$. A progressive evolution of the fundamental plane is clearly visible as a function of redshift with respect to $z \approx 0$. Adapted from Beifiori et al. (2017).

remarkable evolution with redshift compared to the fundamental plane at $z \approx 0$ (Fig. 11.42). This redshift dependence is at least partly due to the change of the luminosity $L(z)$ caused by the evolution (ageing) of the stellar populations, although size evolution may also play a role. Thus, under the assumption that $\alpha$ and $\beta$ of eq. (5.54) do not depend on redshift, the zero-point $\gamma(z)$ can be used to trace the evolution of the dynamical mass-to-light ratio $\mathcal{M}_{\rm dyn}/L$ according to the equation

$$\Delta \log \left(\frac{\mathcal{M}_{\rm dyn}}{L}\right)_z = \log \left(\frac{\mathcal{M}_{\rm dyn}}{L}\right)_z - \log \left(\frac{\mathcal{M}_{\rm dyn}}{L}\right)_{z=0}. \tag{11.26}$$

In a simplified model in which ETGs are assumed to evolve at constant $R_e$ and $\sigma_0$, this implies that $M_{dyn}$ (eq. 5.17) is also constant, and then $L$ is the only quantity which varies as a function of redshift due to the evolution of the stellar populations. Under these assumptions, it is possible to relate $\mathcal{M}_{dyn}/L$ to the SFH through SPS models. If the evolution of $\Delta \log(\mathcal{M}_{dyn}/L)_z$ is fitted with passively evolving SSPs (§8.6.2) where $L(z)$ is the only evolving quantity, it is possible to estimate the evolution of $\mathcal{M}_{dyn}/L$ and the redshift of the last major episode of star formation that occurred at $z = z_{form}$. Consistently with other results (§11.3.10), these results show that ETGs have old ages, high formation redshifts ($z_{form} > 2$) and evolve passively (Fig. 11.43). These results apply to ETGs both in clusters and in the field. However, field ETGs seem to be slightly younger by $\approx 1$ Gyr at most. Complementary constraints come from the stellar mass fundamental plane (§5.4.3; eq. 5.60). This plane does not depend on the variation of $\mathcal{M}_{dyn}/L$ due to the stellar population ageing because the surface brightness $\langle I_e \rangle$ (which depends on the luminosity $L$ of the stellar population) is not involved. While the traditional fundamental plane shows a remarkable evolution as a function of redshift (Fig. 11.42), the mass fundamental plane does not evolve out to $z \approx 2$. The absence of evolution of the mass fundamental plane supports the above interpretation of the evolution of the fundamental plane mostly driven by $L(z)$. Overall, the results on the fundamental plane are broadly consistent with the other constraints on ETG evolution previously illustrated based on the SMF evolution (§11.3.7) and the spectral properties of stellar populations (§11.3.10).

### 11.3.14 The Evolution of Massive Black Hole Accretion

Present-day galaxies show a remarkable correlation between the mass of the central SMBH and the stellar mass (or velocity dispersion) of the spheroidal component of the host galaxy (§5.4.4). If SMBHs at $z \approx 0$ are the fossils of previous accretion of matter onto the very central regions of galaxies, this relation can be interpreted as the end-point of the past evolution of the galaxy mass assembly and the **black hole accretion rate** (BHAR; in units of $\mathcal{M}_\odot$ yr$^{-1}$). Thus, in order to understand the physical origin of this **galaxy–SMBH coevolution**, it is fundamental to investigate the evolution of SMBHs and the interactions with their host galaxies as a function of redshift (see Alexander and Hickox, 2012, for a review). Based on eqs. (3.13) and (3.14), the BHAR can be written as

$$\mathrm{BHAR} \equiv \frac{\mathrm{d}\mathcal{M}_\bullet}{\mathrm{d}t} \approx \frac{L}{\epsilon_{rad} c^2}, \tag{11.27}$$

where $\mathrm{d}\mathcal{M}_\bullet/\mathrm{d}t$ is the rate at which the black hole accretes mass, $L$ is the emitted bolometric luminosity and $\epsilon_{rad}$ is the efficiency with which the accretion of matter is converted into radiation. According to eq. (11.27) and assuming a reasonable value of $\epsilon_{rad}$, the BHAR of a given AGN can be estimated provided that its bolometric luminosity is known from the multi-wavelength SED.

A strong correlation between BHAR and stellar mass exists for galaxies on the SFMS out to $z \approx 2.5$. Interestingly, starbursts seem to have an enhanced BHAR compared to normal SFGs, therefore suggesting a causal link between gas-rich galaxy mergers and

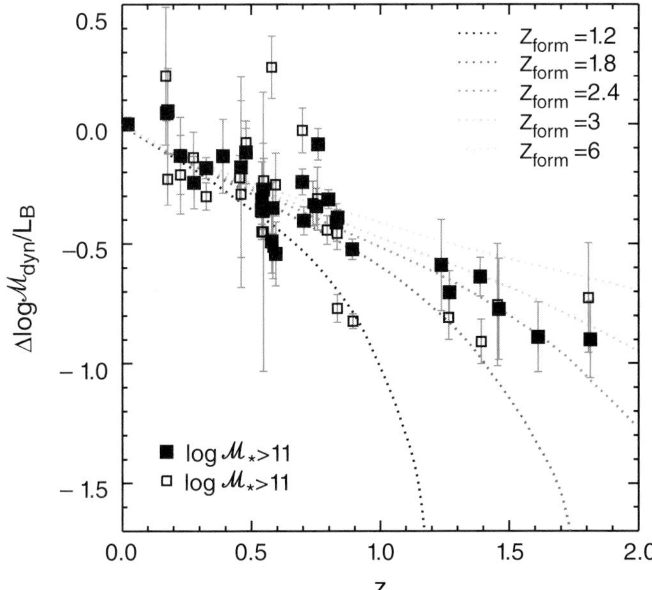


Fig. 11.43 The evolution of the dynamical mass-to-light ratio of ETGs as derived from the fundamental plane. The dotted curves indicate the values expected in the case of passive evolution and different redshifts of the last major episode of star formation ($z_{form}$). Adapted from Beifiori et al. (2017). © AAS, reproduced with permission.

the temporary increase of SMBH feeding in these systems. Instead, quiescent galaxies show a deficit of BHAR, suggesting that these systems are observed after the climax of star formation and BHAR. Once the BHAR has been estimated for galaxy samples at different redshifts, it is also possible to derive the **cosmic black hole accretion rate density**, $\rho_{BHAR}(z)$, and compare it with the evolutionary trend of the cosmic SFR density $\rho_{SFR}(z)$. It has been found that $\rho_{BHAR}(z)$ has a shape similar to $\rho_{SFR}(z)$ (Fig. 11.44). This strengthens the scenario of galaxy and AGN coevolution. Other important constraints have been obtained from the comparison of the evolution of the sSFR with that of the **specific black hole accretion rate**,

$$s\dot{\mathcal{M}}_\bullet = \frac{d\mathcal{M}_\bullet}{dt}\frac{1}{\mathcal{M}_\bullet}. \tag{11.28}$$

This quantity is equivalent to the sSFR as it measures the SMBH accretion rate per unit mass. Fig. 11.45 shows that the sSFR and $s\dot{\mathcal{M}}_\bullet$ coevolve out to $z \approx 3$, suggesting that SMBHs grow in step with the increase of stellar mass of their host galaxies. The similar trends of sSFR and $s\dot{\mathcal{M}}_\bullet$ with redshift are suggestive of a profound relationship between galaxy and AGN coevolution which probably gave origin to the correlation between $\mathcal{M}_\bullet$ and $\mathcal{M}_\star$ that we observe at $z \approx 0$ (§5.4.4). The deep-rooted connection between AGN activity and galaxy evolution suggests that AGN feedback indeed plays a significant role in influencing the evolution of the host galaxies (§8.8). For instance, when AGN activity

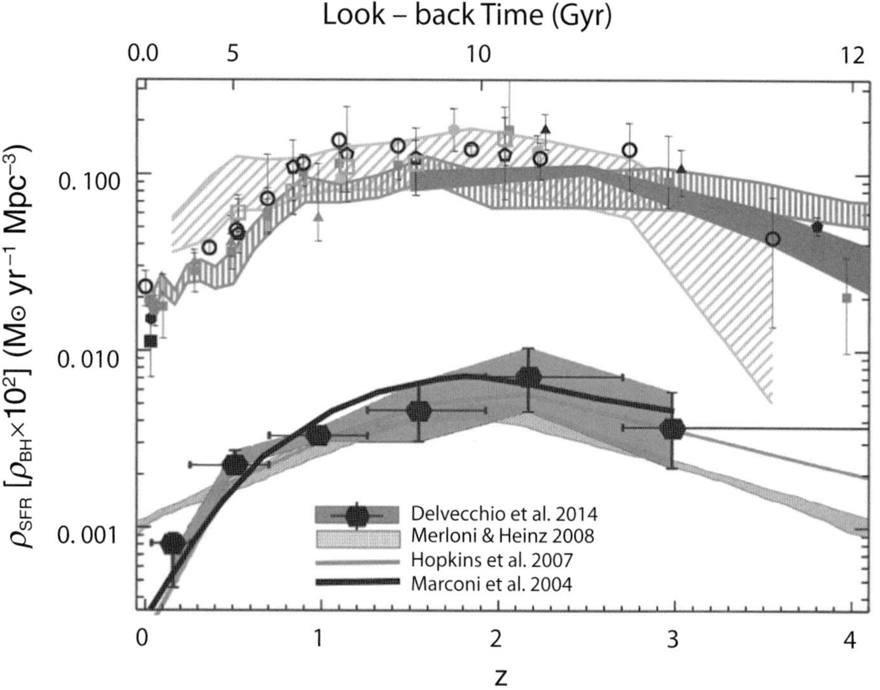


Fig. 11.44 The evolution of the cosmic BHAR density (bottom of the diagram) based on a collection of literature data (Marconi et al., 2004; Hopkins et al., 2007; Merloni and Heinz, 2008; Delvecchio et al., 2014). The SFR density (top of the diagram) is also shown to illustrate that its trend with redshift is similar to that of the BHAR density. Note that the BHAR density is multiplied by 100 to make it visible together with $\rho_{SFR}$. Courtesy of C. Gruppioni.

is present, SFGs at $z \approx 1-3$ often show kpc-scale outflows characterised by speeds and kinetic energies high enough to heat and/or unbind part of the gas, and therefore to lead to a reduction or the complete quenching of the star formation.

### 11.3.15 The Cosmic Multi-Wavelength Background

Once the foreground contribution of the Milky Way has been subtracted, the photons emitted by all objects of the Universe at all redshifts produce a background of diffuse radiation called the **cosmic background**. The definition of cosmic background radiation depends on the angular resolution and depth of the observations. For instance, the optical background has been fully resolved into its individual sources (i.e. galaxies and stars) by the deepest *HST* observations. Instead, a fraction of the X-ray background is still unresolved by current observations. Fig. 11.46 shows the cosmic background across the entire electromagnetic spectrum, including the CMB (§2.4). This can be seen as the global **spectrum of the Universe** originated by the sum of all radiative processes (continuum and

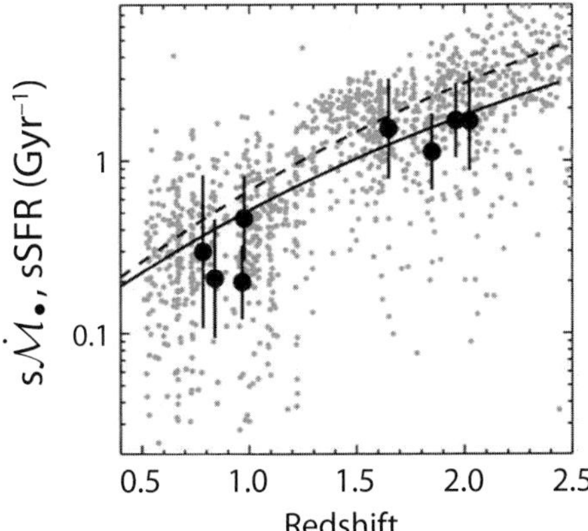


Fig. 11.45 The coevolution of the specific black hole accretion rate s$\dot{\mathcal{M}}_\bullet$ and galaxy sSFR. The small points indicate the sSFR. The black filled circles show the average values of the s$\dot{\mathcal{M}}_\bullet$ in the redshift bins where this quantity has been derived. The solid and dashed lines indicate the sSFR–$z$ relationships derived by Elbaz et al. (2011) and Pannella et al. (2009), respectively. From Mullaney et al. (2012). © AAS, reproduced with permission.

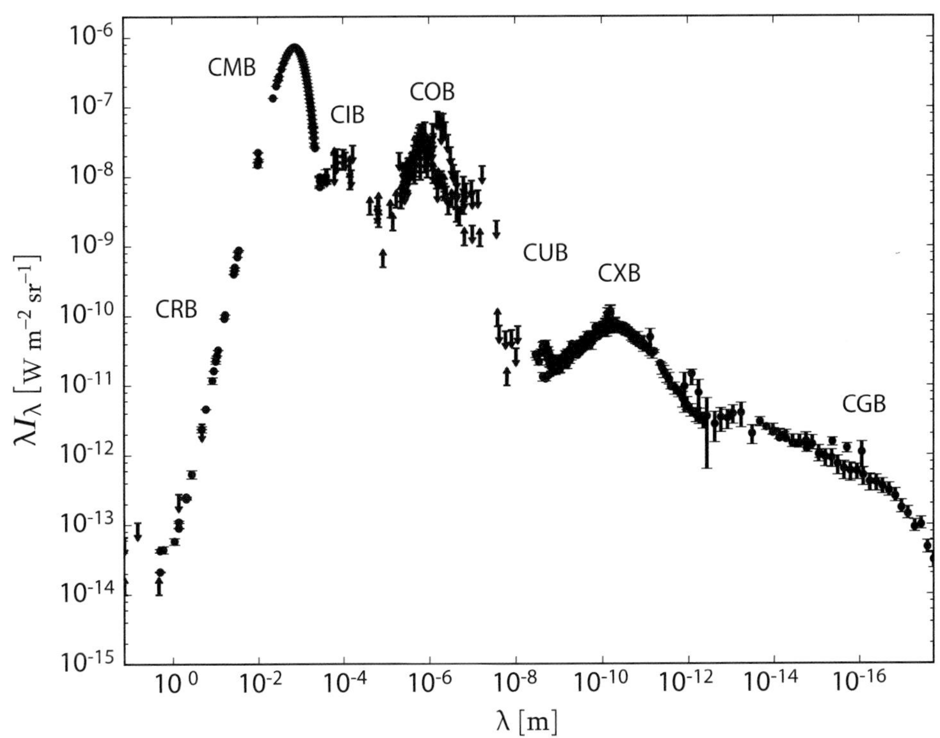


Fig. 11.46 The intensity of the cosmic extragalactic background as a function of observed wavelength. CRB, CMB, CIB, COB, CUB, CXB and CGB indicate the cosmic radio, microwave, IR, optical, UV, X-ray and gamma-ray backgrounds, respectively. Adapted from Hill et al. (2018).

line emissions) occurring in all extragalactic sources at all redshifts. For this reason, the shape and the absolute flux of the cosmic background can provide stringent constraints on galaxy formation and evolution. It is instructive to recall what the main contributors to this background are:

1. The main sources of the radio background are the Rayleigh–Jeans tail of the CMB, the bremsstrahlung continuum of SFGs and the synchrotron radiation emitted by AGNs and SNRs.
2. The FIR to MIR background is due to dust emission from all galaxy types.
3. The stellar light emitted by galaxies and Pop III objects at very high redshifts is thought to produce the NIR background.
4. The UV/optical background is mainly due to galaxy stellar light from sources at lower redshifts.
5. The X-ray and gamma-ray backgrounds are dominated by high-energy radiation emitted by the processes occurring in AGNs.

## 11.4 Summary

In this last section, we provide a summary of the main results on galaxy evolution based on the multi-wavelength observations of distant galaxies and AGNs.

- *Properties of star-forming galaxies.* Distant SFGs have been identified mostly with optical/NIR, FIR, submm/mm and radio surveys. They range from low-mass systems with weak SFRs and low dust extinction to dust-obscured starbursts with SFRs up to $\gtrsim 1000\ \mathcal{M}_\odot\ \mathrm{yr}^{-1}$, or high-mass SFGs with $\mathcal{M}_\star \gtrsim 10^{11}\ \mathcal{M}_\odot$. On average, the metallicities are lower than in present-day SFGs, whereas the SFRs, the sSFRs and the gas fractions are higher. Discs are present at least out to $z \sim 3$–4. The SFG sizes are smaller than at $z \sim 0$. The clustering depends on stellar mass, and the correlation length at $z \sim 2$ is $r_0 \approx 10$–$11\ h^{-1}$ Mpc for $\mathcal{M}_\star \approx 10^{11.4}\ \mathcal{M}_\odot$.
- *Properties of quiescent galaxies.* A substantial population of galaxies with weak or absent star formation (sSFR $< 10^{-11}\ \mathrm{yr}^{-1}$) have been identified with NIR surveys. These galaxies coexist with SFGs out to $z \approx 3$–4, and their number density drops rapidly at higher redshifts. They have spheroidal morphologies, but some of them show significant rotation. Their mature stellar populations (a few Gyr old), supersolar abundances of $\alpha$-elements, and high masses (up to $\mathcal{M}_\star > 10^{11}\ \mathcal{M}_\odot$) imply that the bulk of stellar mass was formed rapidly ($\lesssim 0.5$ Gyr) at $z > 2$–4. The sizes are much smaller and the velocity dispersions higher than in ETGs at $z \approx 0$. The clustering is similar to that of massive SFGs at the same redshift.
- *Morphology and size evolution.* The fraction of peculiars/irregulars increases with redshift from $\approx 5$–10% at $z \approx 0$ to $\approx 60$–70% at $z \approx 2$–3. The fraction of spheroids and discs increases rapidly from $z \approx 3$ to $z \approx 1$, and the Hubble sequence emerges at $z \approx 2$. ETGs increase their size with cosmic time faster than SFGs.

- *Evolution of mergers.* The merger fraction increases with redshift, indicating that mergers were more frequent in the past. If the merging timescale is assumed to be constant with redshift, the merger rate per unit time and cosmic volume is nearly independent of redshift at $z < 1.5$ irrespective of galaxy mass, whereas it gradually declines with $z$ at $z > 1.5$. At $z < 1$ major mergers are a source of stellar mass growth comparable with star formation, whereas at $z > 1$ they play a lesser role.
- *Evolution of star formation.* The cosmic SFR density of the Universe increases rapidly from $z = 0$ to $z \approx 1$, reaches a maximum around $z \approx 2$, and seems to gradually decrease at higher redshifts. The SFMS persists with a similar slope at least out to $z \approx 4$, whereas its normalisation increases with redshift.
- *Evolution of stellar mass.* At $0 < z < 3$, the total SMF shows a remarkable evolution where $\Phi^*(z)$ increases steadily with decreasing redshifts, whereas $\mathcal{M}^*(z)$ and the faint-end slope $\alpha$ do not strongly change with $z$. The major increase of the ETG number density occurs from $z \approx 2.5$ to $z \approx 1$. The cosmic stellar mass density of the Universe increases with decreasing $z$, and it is $\approx 50\%$, $10\%$ and $1\%$ of its present-day value at $z \approx 1.3$, $z \approx 2.9$ and $z \approx 5.7$, respectively.
- *Evolution of interstellar medium.* The molecular gas fraction of SFGs increases with redshift. At all $z$, it is systematically lower in massive galaxies and correlates with the sSFR. The depletion timescales of the order of 1 Gyr at $0 < z < 3$ imply that cold gas is nearly continuously accreted to maintain SFGs in the SFMS over timescales of several Gyr. The cold gas mass, its fraction and the depletion timescale correlate with the stellar mass, sSFR and redshift through simple scaling relations that can be explained if gas accretion is continuously taking place.
- *Evolution of metallicity.* At all redshifts, the metallicity of galaxies increases with galaxy mass because massive galaxies retain and enrich the gas more efficiently during star formation activity. The average metallicity of SFGs increases with decreasing redshift, as expected by the progressive enrichment of the ISM with cosmic time, and anticorrelates with sSFR at all redshifts. Stellar mass, metallicity of the ionised gas and SFR are related with each other. These relations are ascribed to the interplay between inflows of low-metallicity gas, outflows of enriched gas and galaxy mass. Instead, the metallicity of ETGs at $z > 1$ is typically solar, implying that these galaxies completed the bulk of their metal enrichment at earlier epochs.
- *Evolution of scaling relations involving kinematics.* The TFR is present out to $z \gtrsim 3$. Its slope does not change significantly with $z$, but the evolution of the zero-point is less constrained. The fundamental plane of ETGs persists up to $z \sim 2$, and its zero-point evolves with redshift. Assuming a negligible evolution of the ETG size and velocity dispersion, the inferred dynamical mass-to-light ratios indicate the stellar populations of massive ETGs formed at $z > 2$ and evolved passively thereafter.
- *Active galactic nuclei.* Luminous QSOs have been identified out to $z > 7$. A strong correlation between the BHAR and the stellar mass is present for main-sequence SFGs out to $z \approx 2$–$3$. ETGs show a deficit of BHAR, suggesting that these systems are observed in a later evolutionary phase. The similar evolution of the sSFR and the specific BHAR out to $z \sim 3$ suggests that the coevolution of galaxies and AGNs gave origin to the

$\mathcal{M}_\bullet$–$\mathcal{M}_\star$ tight relation observed at $z \approx 0$. The redshift evolution of the cosmic BHAR density has a shape similar to that of the cosmic SFR density, thus supporting the scenario where galaxies and AGNs coevolve. The strong influence of AGN feedback on galaxy evolution is supported by the observation of kpc-scale outflows able to heat and/or unbind part of the gas and thus to reduce or suppress star formation.

- *Mass- and environment-driven quenching.* Mass is the key driver of galaxy evolution. The archaeological and look-back time results consistently support the downsizing scenario where massive galaxies formed, matured and quenched their star formation earlier and faster than low-mass ones. However, also the environment is responsible for several processes which influence galaxy evolution. In particular, massive (central) galaxies tend to quench their star formation independently of the environmental density (mass quenching). Instead, satellite galaxies in dense environments are more likely to quench the star formation through environmental quenching independently of their mass.

# Appendix A Acronyms

**Table A.1** List of acronyms

| Acronym | Meaning |
|---|---|
| AGB | asymptotic giant branch |
| AGN | active galactic nucleus |
| ALMA | Atacama Large Millimetre Array |
| AMD | angular momentum distribution |
| AMR | adaptive mesh refinement |
| BAO | baryon acoustic oscillation |
| BCG | brightest cluster galaxy |
| BHAR | black hole accretion rate |
| BLR | broad-line region |
| BPT | Baldwin–Phillips–Terlevich |
| CDM | cold dark matter |
| CGM | circumgalatic medium |
| CMB | cosmic microwave background |
| CMD | colour–magnitude diagram |
| CNM | cold neutral medium |
| CSP | composite stellar population |
| DLA | damped Lyman-$\alpha$ absorber |
| DSS | Digital Sky Survey |
| EdS | Einstein–de Sitter |
| ELT | Extremely Large Telescope |
| EoR | epoch of reionisation |
| ERO | extremely red object |
| ETG | early-type galaxy |
| FIR | far-infrared |
| FR | Fanaroff–Riley |
| FSRQ | flat-spectrum radio-loud quasar |
| FWHM | full width at half-maximum |
| *GALEX* | *Galaxy Evolution Explorer* |
| GMC | giant molecular cloud |
| GRB | gamma-ray burst |
| HDM | hot dark matter |
| HSB | high surface brightness |
| *HST* | *Hubble Space Telescope* |

The list continues on the next page.

**Table A.1** List of acronyms (cont.)

| Acronym | Meaning |
|---|---|
| ICM | intracluster medium |
| IFU | integral field unit |
| IGM | intergalactic medium |
| IMF | initial mass function |
| IR | infrared |
| IRA | instantaneous recycling approximation |
| IRAM | Institut de Radioastronomie Millimétrique |
| ISM | interstellar medium |
| JCMT | James Clerk Maxwell Telescope |
| *JWST* | *James Webb Space Telescope* |
| LBG | Lyman-break galaxy |
| LF | luminosity function |
| LLS | Lyman-limit system |
| LMC | Large Magellanic Cloud |
| LOFAR | Low-Frequency Array |
| LOSVD | line-of-sight velocity distribution |
| LSB | low surface brightness |
| LSR | local standard of rest |
| LSS | large-scale structure |
| LTG | late-type galaxy |
| MIR | mid-infrared |
| MOND | modified Newtonian dynamics |
| NFW | Navarro–Frenk–White |
| NIR | near-infrared |
| NLR | narrow-line region |
| PAH | polycyclic aromatic hydrocarbon |
| PISN | pair instability supernova |
| PSF | point spread function |
| QSO | quasi-stellar object |
| RGB | red giant branch |
| SAM | semi-analytic model |
| SDSS | Sloan Digital Sky Survey |
| SED | spectral energy distribution |
| SFG | star-forming galaxy |
| SFH | star formation history |
| SFMS | star formation main sequence |
| SFR | star formation rate |
| SHMR | stellar-to-halo mass relation |
| SKA | Square Kilometre Array |
| SLED | spectral line energy distribution |
| SMBH | supermassive black hole |
| SMC | Small Magellanic Cloud |

The list continues on the next page.

**Table A.1** List of acronyms (cont.)

| Acronym | Meaning |
|---|---|
| SMF | stellar mass function |
| SMG | submillimetre galaxy |
| SN | supernova |
| SNR | supernova remnant |
| SPH | smoothed particle hydrodynamics |
| SPS | stellar population synthesis |
| sSFR | specific star formation rate |
| SSP | simple stellar population |
| SZE | Sunyaev–Zeldovich effect |
| TFR | Tully–Fisher relation |
| TP-AGB | thermally pulsating asymptotic giant branch |
| UFD | ultra-faint dwarf |
| ULIRG | ultra-luminous infrared galaxy |
| UV | ultraviolet |
| UVB | ultraviolet background |
| VLA | Very Large Array |
| VLT | Very Large Telescope |
| WDM | warm dark matter |
| *WFIRST* | *Wide Field Infrared Survey Telescope* |
| WHIM | warm–hot intergalactic medium |
| WIM | warm ionised medium |
| WIMP | weakly interacting massive particle |
| WNM | warm neutral medium |

# Appendix B Constants and Units

In this book, we adopt the centimetre–gram–second (cgs) system of units as commonly in use in the astrophysical community. Here we report the main physical constants (from Mohr et al., 2016) used throughout the book both in the cgs system (Tab. B.1) and in the metre–kilogram–second (mks) International System of Units (Tab. B.2), which is abbreviated SI from the French *Système International*. Tab. B.3 gives astronomical constants (from Cox, 2000) and Tab. B.4 useful conversions. When applicable, standard errors of the last digits are given in parentheses.

**Table B.1** Physical constants in the cgs system

| Constants | Symbol | Value | Units |
|---|---|---|---|
| Speed of light | $c$ | $2.99792458 \times 10^{10}$ | cm s$^{-1}$ |
| Gravitational constant | $G$ | $6.67408(31) \times 10^{-8}$ | cm$^3$ g$^{-1}$ s$^{-2}$ |
| Boltzmann constant | $k_{\rm B}$ | $1.380649 \times 10^{-16}$ | erg K$^{-1}$ |
| Planck constant | $h$ | $6.62607015 \times 10^{-27}$ | erg s |
| Electron mass | $m_{\rm e}$ | $9.10938356(11) \times 10^{-28}$ | g |
| Proton mass | $m_{\rm p}$ | $1.672621898(21) \times 10^{-24}$ | g |
| Elementary charge | $e$ | $4.80320471257 \times 10^{-10}$ | e.s.u. |
| Thomson cross section | $\sigma_{\rm T}$ | $6.6524587158(91) \times 10^{-25}$ | cm$^2$ |

**Table B.2** Physical constants in the mks system (SI)

| Constants | Symbol | Value | Units |
|---|---|---|---|
| Speed of light | $c$ | $2.99792458 \times 10^{8}$ | m s$^{-1}$ |
| Gravitational constant | $G$ | $6.67408(31) \times 10^{-11}$ | m$^3$ kg$^{-1}$ s$^{-2}$ |
| Boltzmann constant | $k_{\rm B}$ | $1.380649 \times 10^{-23}$ | J K$^{-1}$ |
| Planck constant | $h$ | $6.62607015 \times 10^{-34}$ | J s |
| Electron mass | $m_{\rm e}$ | $9.10938356(11) \times 10^{-31}$ | kg |
| Proton mass | $m_{\rm p}$ | $1.672621898(21) \times 10^{-27}$ | kg |
| Elementary charge | $e$ | $1.602176634 \times 10^{-19}$ | C |
| Thomson cross section | $\sigma_{\rm T}$ | $6.6524587158(91) \times 10^{-29}$ | m$^2$ |

**Table B.3** Astronomical constants and units

| Constants | Symbol | Value | cgs | SI |
|---|---|---|---|---|
| Parsec | pc | 3.0856776 | $\times 10^{18}$ cm | $\times 10^{16}$ m |
| Solar mass | $\mathcal{M}_\odot$ | 1.9891 | $\times 10^{33}$ g | $\times 10^{30}$ kg |
| Solar radius | $r_\odot$ | 6.95508 | $\times 10^{10}$ cm | $\times 10^{8}$ m |
| Solar bolometric luminosity | $L_\odot$ | 3.845(8) | $\times 10^{33}$ erg s$^{-1}$ | $\times 10^{26}$ W |
| Sidereal year | yr | 3.155815 | $\times 10^{7}$ s | $\times 10^{7}$ s |
| Hubble constant[a] | $H_0$ | $70\left(\frac{h}{0.7}\right)$ km s$^{-1}$ Mpc$^{-1}$ | | |
| Critical density[b] | $\rho_{\mathrm{crit},0}$ | $9.20\left(\frac{h}{0.7}\right)^2$ | $\times 10^{-30}$ g cm$^{-3}$ | $\times 10^{-27}$ kg m$^{-3}$ |
| Age of the Universe | $t_0$ | 13.8 Gyr | | |

[a] Here and below $h \equiv H_0/(100\,\mathrm{km\,s^{-1}\,Mpc^{-1}})$ is the dimensionless Hubble constant, not to be confused with the Planck constant (Tabs. B.1 and B.2).

[b] Of the Universe at the present time.

**Table B.4** Useful conversions

| | |
|---|---|
| Electron volt (eV) | $1.6021766208(98) \times 10^{-12}$ erg<br>$1.6021766208(98) \times 10^{-19}$ J |
| Jansky (Jy) | $10^{-23}$ erg s$^{-1}$ cm$^{-2}$ Hz$^{-1}$<br>$10^{-26}$ W m$^{-2}$ Hz$^{-1}$ |
| 1 km s$^{-1}$ | 1.0227 kpc Gyr$^{-1}$ |
| 1 radian | 206264.80625 arcsec<br>57.2957795° |
| [a]1 $\mathcal{M}_\odot$ pc$^{-2}$ | $1.2490 \times 10^{20} \mu^{-1}$particles cm$^{-2}$ |

[a] For a gas with mean molecular weight $\mu$.

# Appendix C Astronomical Compendium

## C.1 Spectral Regions

Astronomical observations are carried out across the whole electromagnetic spectrum. Tab. C.1 gives the indicative ranges in wavelengths and frequencies of the most important spectral regions. Some of these (e.g. optical, microwave and radio) are accessible from the ground, for others (e.g. X-ray, UV and far-IR) one needs to resort to telescopes in space. The last column of Tab. C.1 gives the most commonly used units in the various specific spectral regions (note that going from the gamma-rays to the radio band, units change from energy to wavelength and eventually frequency).

**Table C.1** Spectral regions

| Name | Wavelength (cm) | Frequency (Hz) | Typical units |
|---|---|---|---|
| Gamma-rays | $< 1 \times 10^{-9}$ | $> 2 \times 10^{19}$ | $> 100$ keV |
| Hard X-rays | $1 \times 10^{-9}$–$6 \times 10^{-8}$ | $5 \times 10^{17}$–$2 \times 10^{19}$ | 2–100 keV |
| Soft X-rays | $6 \times 10^{-8}$–$1 \times 10^{-6}$ | $2 \times 10^{16}$–$5 \times 10^{17}$ | 0.1–2 keV |
| Extreme ultraviolet | $1 \times 10^{-6}$–$1 \times 10^{-5}$ | $3 \times 10^{15}$–$3 \times 10^{16}$ | 10–100 nm |
| Far-ultraviolet[a] | $1$–$2 \times 10^{-5}$ | $1.7$–$2.3 \times 10^{15}$ | 1300–1800 Å |
| Near-ultraviolet | $2$–$3 \times 10^{-5}$ | $1$–$2 \times 10^{15}$ | 1800–3000 Å |
| Optical/visible | $4$–$9 \times 10^{-5}$ | $3$–$8 \times 10^{14}$ | 3900–9000 Å |
| Near-infrared (NIR) | $1$–$5 \times 10^{-4}$ | $6 \times 10^{13}$–$3 \times 10^{14}$ | 1–5 $\mu$m |
| Mid-infrared (MIR) | $5 \times 10^{-4}$–$0.004$ | $7 \times 10^{12}$–$6 \times 10^{13}$ | 5–40 $\mu$m |
| Far-infrared (FIR) | 0.004–0.03 | $1$–$7 \times 10^{12}$ | 40–300 $\mu$m |
| Submillimetre | 0.03–0.1 | $3 \times 10^{11}$–$1 \times 10^{12}$ | 300–1000 $\mu$m |
| Millimetre/microwaves[b] | 0.1–3 | $1 \times 10^{10}$–$3 \times 10^{11}$ | 10–300 GHz |
| Radio | $3$–$3 \times 10^{3}$ | $1 \times 10^{7}$–$1 \times 10^{10}$ | 10 MHz–10 GHz |

[a] For far- and near-UV we have approximately used the bands of *GALEX*.

[b] The distinction between the microwave and radio regions is somewhat loosely defined.

## C.2 Flux, Luminosity and Surface Brightness

In this section, we summarise the properties of electromagnetic radiation and the most useful photometric quantities. The reader can find more details in Karttunen et al. (2017). We define the **specific intensity** $\mathcal{I}_\nu$ as the radiation energy $\mathrm{d}E$ between frequencies $\nu$ and $\nu + \mathrm{d}\nu$, per unit time $\mathrm{d}t$ that passes through a surface $\mathrm{d}A$, at an angle $\theta$ from the normal to this surface, within a solid angle $\mathrm{d}\Omega$:

$$\mathcal{I}_\nu = \frac{\mathrm{d}E}{\mathrm{d}\nu\,\mathrm{d}t\,\mathrm{d}A\cos\theta\,\mathrm{d}\Omega}. \tag{C.1}$$

The specific intensity, also called simply **intensity**, has cgs units of $\mathrm{erg\,s^{-1}\,cm^{-2}\,sr^{-1}\,Hz^{-1}}$. By integrating over all frequencies, we obtain the **total intensity**.

When we observe an astronomical source we measure its energy over the collecting area of the detector for a certain period of time. The **flux density** at a frequency $\nu$ from a source is the intensity integrated over a solid angle,

$$F_\nu = \int_\Omega \mathcal{I}_\nu \cos\theta\,\mathrm{d}\Omega', \tag{C.2}$$

where $\Omega$ is typically the solid angle covering the astronomical source (for extended sources) or the angular resolution of the telescope (for point sources). It is common practice in the astronomical community to call the flux density simply **flux**. We adopt this choice in this book although be aware that, in other texts, the term 'flux' is used for other purposes, in particular energy flux (i.e. luminosity, see below). The flux $F_\nu$ has typically units of $\mathrm{erg\,s^{-1}\,cm^{-2}\,Hz^{-1}}$. In radio astronomy, this is sometimes indicated with $S$ and has units of Jy (Tab. B.4). In several astronomical applications, the source angular size is small ($\ll 1$ rad) and the instrument/telescope points directly at the source, thus $\cos\theta = 1$.

An equivalent expression to eq. (C.2) can be written using wavelengths instead of frequencies and thus $F_\lambda$. Given the equivalence $F_\nu(\nu)\mathrm{d}\nu = F_\lambda(\lambda)\mathrm{d}\lambda$ we can write

$$F_\lambda = F_\nu \left|\frac{\mathrm{d}\nu}{\mathrm{d}\lambda}\right| = \frac{c}{\lambda^2}F_\nu = \frac{\nu^2}{c}F_\nu, \tag{C.3}$$

which shows that the *shape* of the spectrum of a source is different if one uses $F_\nu$ or $F_\lambda$. It is therefore, sometimes, convenient to plot $\nu F_\nu$ or $\lambda F_\lambda$ because, from eq. (C.3), they have the same spectral shape. If we integrate the flux over the entire spectrum we obtain the **total flux** or **bolometric flux** of a source

$$F = \int_0^\infty F_\nu(\nu)\mathrm{d}\nu = \int_0^\infty F_\lambda(\lambda)\mathrm{d}\lambda. \tag{C.4}$$

In practical situations, it is common to have the flux in a spectral band $X$ (§C.4) indicated as $F_X$.

Given a source with a certain measured flux $F_\nu$, if we know its distance $d$, we can derive its monochromatic **luminosity** as

$$L_\nu = 4\pi d^2 F_\nu, \tag{C.5}$$

where one implicitly assumes that the source radiates isotropically. Analogously, we can do the same using $F_\lambda$ and obtain $L_\lambda = 4\pi d^2 F_\lambda$. Note that for sources at cosmological distances we need to use the luminosity distance $d_\mathrm{L}$ (eq. 2.11). As for the flux, the luminosity is usually given in a certain band $X$ and indicated as $L_X$. If we integrate eq. (C.5) over all frequencies we obtain the **total** or **bolometric luminosity** of the source

$$L = \int_0^\infty L_\nu(\nu)\mathrm{d}\nu, \tag{C.6}$$

with cgs units of $\mathrm{erg\,s^{-1}}$. This operation requires a knowledge of the shape of the whole electromagnetic spectrum of the source, which is often difficult to achieve. However, in some cases, the spectrum of a source is known (e.g. a star of a certain spectral type; §C.5) and one can obtain $L$ from observations in one or more bands.

For *extended* astronomical sources, one typically uses the surface brightness, for which there are two different definitions (see also Binney and Tremaine, 2008). Consider a source with monochromatic luminosity density or emissivity $j_\nu$ (§D.1.2). We can obtain its surface brightness by integrating the emissivity along the path length in the direction of the observer,

$$I_\nu = \int j_\nu(\boldsymbol{x})\mathrm{d}x_\mathrm{los} = \frac{\mathrm{d}L_\nu}{\mathrm{d}S}, \tag{C.7}$$

which is the source's luminosity per unit area, where the area $S$ is intended at the source. The monochromatic surface brightness $I_\nu$ in eq. (C.7) has an equivalent expression $I_\lambda$ using wavelengths. In practice, the surface brightness is typically given in a certain spectral range or band $X$ and indicated as $I_X$. It has units of $L_\odot\,\mathrm{pc^{-2}}$ or $\mathrm{mag\,arcsec^{-2}}$. When expressed in magnitudes over the angular area of 1 $\mathrm{arcsec^2}$, it is typically indicated with $\mu$ or $\mu_X$. The conversion between the two units reads

$$\frac{\mu_X}{\mathrm{mag\,arcsec^{-2}}} = -2.5\log\left(\frac{I_X}{L_\odot\,\mathrm{pc^{-2}}}\right) + M_{\odot,X} + 21.572, \tag{C.8}$$

where $M_{\odot,X}$ is the absolute magnitude of the Sun (§C.3.1) in the $X$ band.

The second definition of surface brightness is the flux received from the source per unit solid angle $\mathrm{d}\Omega$,

$$\hat{I}_\nu = \frac{\mathrm{d}F_\nu}{\mathrm{d}\Omega} = \frac{1}{4\pi}I_\nu, \tag{C.9}$$

or an equivalent expression with wavelengths. In this definition, the observer is located at the *apex* of the solid angle. Note that $\mathrm{d}\Omega = \mathrm{d}S/d^2$, and $F_\nu$ and $L_\nu$ are linked as in eq. (C.5). $\hat{I}_\nu$ has units of $\mathrm{erg\,s^{-1}\,cm^{-2}\,sr^{-1}\,Hz^{-1}}$, the same as intensity (eq. C.1); however, note that the intensity is the energy per solid angle *radiated* by the source, while the surface brightness is the energy *detected* per solid angle from an extended source.

If a galaxy has the same luminosity as another, but it is located, for instance, at twice the distance ($d$), its observed flux would be four times lower, scaling as $d^{-2}$. However, the surface brightness does not depend on distance, because also the angular area scales as $d^{-2}$. This principle is valid for galaxies not at cosmological distances (§2.1.4). Note also that, because the distance cancels out, one *does not* need to know the distance of the source to determine its surface brightness.

In radio astronomy, it is common to use the **brightness temperature** $T_b$ defined as the temperature that a black body (§D.1.3) should have to equal the source intensity: $B_\nu(T_b) = I_\nu$. At radio frequencies, typically $h\nu \ll k_B T_b$ and thus we can use the Rayleigh–Jeans approximation (eq. D.10) to write

$$T_b = \frac{\lambda^2}{2k_B} I_\nu = \frac{c^2}{2k_B \nu^2} I_\nu, \tag{C.10}$$

where $\lambda$ and $\nu$ are the wavelength and frequency of the observation.

In practical applications, for extended sources, the brightness temperature is a measure of the surface brightness with the following conversions:

$$T_b \simeq 1.36 \left(\frac{\lambda}{\text{cm}}\right)^2 \left(\frac{\theta_{\text{maj}}\theta_{\text{min}}}{\text{arcsec}^2}\right)^{-1} \left(\frac{\hat{I}}{\text{Jy beam}^{-1}}\right) \text{K} \tag{C.11}$$

$$\simeq 1.22 \times 10^3 \left(\frac{\nu}{\text{GHz}}\right)^{-2} \left(\frac{\theta_{\text{maj}}\theta_{\text{min}}}{\text{arcsec}^2}\right)^{-1} \left(\frac{\hat{I}}{\text{Jy beam}^{-1}}\right) \text{K}, \tag{C.12}$$

where $\theta_{\text{maj}}$ and $\theta_{\text{min}}$ are the angular major and minor axes of the (Gaussian) *beam* (angular resolution) of the observation and $\hat{I}$ is given in typical radio astronomical units of a flux (Jy) over a solid angle (beam).

### C.2.1 From 21 cm Surface Brightness to H I Column Density

We give a brief derivation of the relation between the 21 cm line surface brightness and the H I column density. We start from the radiative transfer equation (eq. D.4) written in the Rayleigh–Jeans regime ($h\nu \ll k_B T_b$). In this regime the brightness temperature (eq. C.10) can be written (see eq. D.4)

$$T_b(\tau_\nu) = T_b(0) e^{-\tau_\nu} + T_S(1 - e^{-\tau_\nu}), \tag{C.13}$$

where $T_b(0)$ is the brightness temperature of the background (the far side of the region of interest), $T_S$ is the temperature of the source and $\tau_\nu$ is the optical depth (eq. D.2). In the case of H I emission $T_S = T_{\text{spin}}$ is the spin temperature (excitation temperature; eq. 9.52).

If we observe an H I source (e.g. a galaxy), the background radiation (first term on the r.h.s. of eq. C.13) is a radio continuum flux that can usually be subtracted (unless there is a strong radio source behind our galaxy). Assuming an optically thin regime (§D.1.2), we expand the second term on the r.h.s. of eq. (C.13) for small $\tau_\nu$ to obtain

$$T_b(\tau_\nu) \simeq T_{\text{spin}} \tau_\nu, \tag{C.14}$$

where $\tau_\nu$ can be written as

$$\tau_\nu(\nu) = \frac{3}{32\pi} A_{10} \frac{hc^2}{k_B T_{\text{spin}} \nu_{\text{HI}}} N_{\text{HI}} \phi(\nu), \tag{C.15}$$

where $A_{10} \simeq 2.88 \times 10^{-15}\,\text{s}^{-1}$ is the Einstein spontaneous emission coefficient, $\nu_{\text{HI}} = 1420.4$ MHz, $N_{\text{HI}}$ is the H I column density that we aim to determine and $\phi(\nu)\text{d}\nu$ is the probability that the H I transition occurs in the frequency range between $\nu$ and $\nu + \text{d}\nu$. If we now make a substitution from frequency to line-of-sight velocity $dv = -(c/\nu)d\nu$ and integrate eq. (C.14) after the substitution of eq. (C.15), we obtain

$$N_{\rm H\,I} \simeq 1.82 \times 10^{18} \int \left(\frac{T_{\rm b}(v)}{\rm K}\right)\left(\frac{dv}{\rm km\,s^{-1}}\right) \quad {\rm cm}^{-2}, \tag{C.16}$$

where the integration in $v$ is made over the range of velocities covered by the source. We can also rewrite eq. (C.16) using eq. (C.11) as

$$N_{\rm H\,I} \simeq 1.10 \times 10^{21} \left(\frac{\rm arcsec^2}{\theta_{\rm max}\theta_{\rm min}}\right) \int \left(\frac{\hat{I}_{\rm H\,I}(v)}{\rm mJy/beam}\right)\left(\frac{dv}{\rm km\,s^{-1}}\right) \quad {\rm cm}^{-2}, \tag{C.17}$$

where $\hat{I}_{\rm H\,I}$ is the H I surface brightness (eq. C.9) and $\theta_{\rm max}$ and $\theta_{\rm min}$ are the angular size of the Gaussian beam of the observations.

## C.3 Magnitudes and Colours

Given two astronomical sources with fluxes $F_1$ and $F_2$ we can write

$$m_1 - m_2 = -2.5 \log\left(\frac{F_1}{F_2}\right), \tag{C.18}$$

where $m_1$ and $m_2$ are the **apparent magnitudes** of sources 1 and 2, respectively. The conversion back to fluxes is

$$\frac{F_1}{F_2} = 10^{-0.4(m_1 - m_2)}. \tag{C.19}$$

Thus, the brighter the source, the *lower* its apparent magnitude. In other words, the apparent magnitude is a measure of the 'faintness' of the source. For instance, if source 1 is 10 times brighter than source 2, its apparent magnitude is $m_2 - 2.5$; if $F_1 = 2F_2$ then $m_1 \simeq m_2 - 0.75$; and so forth.

As for fluxes, apparent magnitudes are defined in certain spectral bands that we describe in §C.4. Note that the apparent magnitude is often indicated with the letter of the band only, so for instance $B$ is used in place of $m_B$. The differences between two magnitudes of the same source in two different bands (ratio between their fluxes) are called **colours**. For instance, having observations in two bands $X_1$ and $X_2$ we can construct the colour

$$X_1 - X_2 = m_{X_1} - m_{X_2} = -2.5 \log\left(\frac{F_{X_1}}{F_{X_2}}\right) + {\rm const}, \tag{C.20}$$

where $F_{X_1}$ and $F_{X_2}$ are the fluxes in the respective bands, and the constant needs to be determined as we explain below.

Apparent magnitudes of astronomical objects are given in specific **magnitude systems**. These are standardised definitions of zero-points, which are fluxes $F_0$ at which the apparent magnitude ($m_0$) is typically fixed to zero, such that

$$m = -2.5 \log\left(\frac{F}{F_0}\right). \tag{C.21}$$

**Table C.2** Conversion between Vega and AB

| $U$ | $B$ | $V$ | $R$ | $I$ | $J$ | $H$ | $K_s$ |
|---|---|---|---|---|---|---|---|
| 0.72 | −0.13 | −0.01 | 0.18 | 0.41 | 0.87 | 1.34 | 1.81 |

Values of $m_X^*$ in eq. (C.23).

In some systems, these *calibrations* are given using so-called **standard stars**, which are non-variable stars with well known fluxes across the spectrum. In §C.4, we see that in the definition of the Johnson–Cousins system all the colours of Vega (type A0 star; Tab. C.4) are null, which means that its magnitude is the same in every band. Magnitudes given with this definition are called **Vega magnitudes**.

A second common definition of magnitudes, often used in extragalactic astronomy, is the **AB system**. In this system, there is no reference spectral type but the calibration is in absolute units. By definition, a source has zero magnitude in every filter if its flux is $F_\nu = 3.63 \times 10^{-20}\,\mathrm{erg\,s^{-1}\,cm^{-2}\,Hz^{-1}}$, which we can write

$$m^{\mathrm{AB}} = -2.5 \log \left( \frac{F_\nu}{\mathrm{erg\,s^{-1}\,cm^{-2}\,Hz^{-1}}} \right) - 48.6. \tag{C.22}$$

The conversion from Vega to **AB magnitudes** is

$$m_X^{\mathrm{AB}} = m_X^{\mathrm{Vega}} + m_X^*, \tag{C.23}$$

where $m_X^*$ depends on the filter. For instance, Tab. C.2 gives the values of $m_X^*$ for the bands of the extended Johnson–Cousins system (§C.4).

## C.3.1 Absolute Magnitude

The **absolute magnitude** of a source in a certain band $X$ ($M_X$) is defined as the apparent magnitude that this source would have if it were at a distance of 10 pc, so

$$m_X - M_X = 5 \log \left( \frac{d}{\mathrm{pc}} \right) - 5, \tag{C.24}$$

where $d$ is the distance of the source in parsecs and $m_X$ is its apparent magnitude. The quantity $m_X - M_X$, often indicated with $\mu$, is called the **distance modulus** and it is used as a measure of the distance.

The absolute magnitude of a source is a measure of its luminosity and thus it is an intrinsic property. Writing eq. (C.18) with $m_1$ as $m_X$ and $m_2$ as $m_{\odot,X}$, the apparent magnitude of the Sun, and then substituting $m_X$ and $m_{\odot,X}$ using eq. (C.24), we obtain the relation with the luminosity $L_X$ in the generic $X$ band:

$$M_X = M_{\odot,X} - 2.5 \log \left( \frac{L_X}{L_{\odot,X}} \right), \tag{C.25}$$

**Table C.3** Solar absolute magnitudes

| System | $U$ | $B$ | $V$ | $R$ | $I$ | $J$ | $H$ | $K_s$ | $u$ | $g$ | $r$ | $i$ | $z$ |
|---|---|---|---|---|---|---|---|---|---|---|---|---|---|
| Vega | 5.61 | 5.44 | 4.81 | 4.43 | 4.10 | 3.67 | 3.32 | 3.27 | 5.49 | 5.23 | 4.53 | 4.19 | 4.01 |
| AB | 6.33 | 5.31 | 4.80 | 4.60 | 4.51 | 4.54 | 4.66 | 5.08 | 6.39 | 5.11 | 4.65 | 4.53 | 4.50 |

Values from Willmer (2018).

where $M_{\odot,X}$ and $L_{\odot,X}$ are, respectively, the solar absolute magnitude and luminosity in the $X$ band. The luminosity as a function of absolute magnitude is

$$L_X = 10^{-0.4(M_X - M_{\odot,X})} L_{\odot,X}. \tag{C.26}$$

Values of the solar absolute magnitudes in different bands are reported in Tab. C.3. In analogy to the definition of the bolometric luminosity one can define an **absolute bolometric magnitude** $M$, given by eq. (C.25) without the subscript $X$. In practice, if we have a star of known spectral type, we can define the **bolometric correction** BC as the value that we have to subtract from a measured magnitude of the star in a band to obtain the bolometric magnitude $M_{\rm bol}$. Usually, one takes the $V$ band as reference and thus $M_{\rm bol} = M_V - \mathrm{BC}$.[1] For example, the Sun has $\mathrm{BC} \simeq 0.07$ and $M_{\rm bol} \simeq 4.74$, main-sequence stars between G and A types have very small positive bolometric corrections ($\mathrm{BC} \leq 0.3$), while stars of B and O types have BCs in the range 0.8–4.0.

## C.4 Filters

The detector of an instrument is designed to collect photons from astronomical sources in a specific range of wavelengths or frequencies. Typically, the electromagnetic radiation is selected using **filters**, each characterised by its own **transmission curve**, which describes the efficiency with which it lets photons through as a function of wavelength or frequency. We can write these functions generically as $T_X(\lambda)$ or $T_X(\nu)$, where $X$ is the filter bandpass. The shapes of these functions for some of the most used filters in the UV/optical and NIR are shown in Fig. C.1 (see also Fig. 3.6, bottom panel).

In the optical region of the electromagnetic spectrum, there are several standard definitions of filters and corresponding bands. The Johnson–Cousins system is defined such that Vega, used as a prototype of stars of spectral type A0 (§C.5), has $V = 0.03$ and all colours null, which implies that the magnitude of Vega is 0.03 in all bands. The $u'g'r'i'z'$ system (SDSS) is defined using more than 100 standard stars. In this system, the star BD +17°4708, a sub-dwarf F6 star with $B-V = 0.43$, has null colours. The *HST* system (ST magnitudes) is defined such that an object with flux $F_\lambda = 3.63 \times 10^{-9}$ erg s$^{-1}$ cm$^{-2}$ Å$^{-1}$

[1] Note that sometimes it is defined as $M_{\rm bol} = M_V + \mathrm{BC}$.

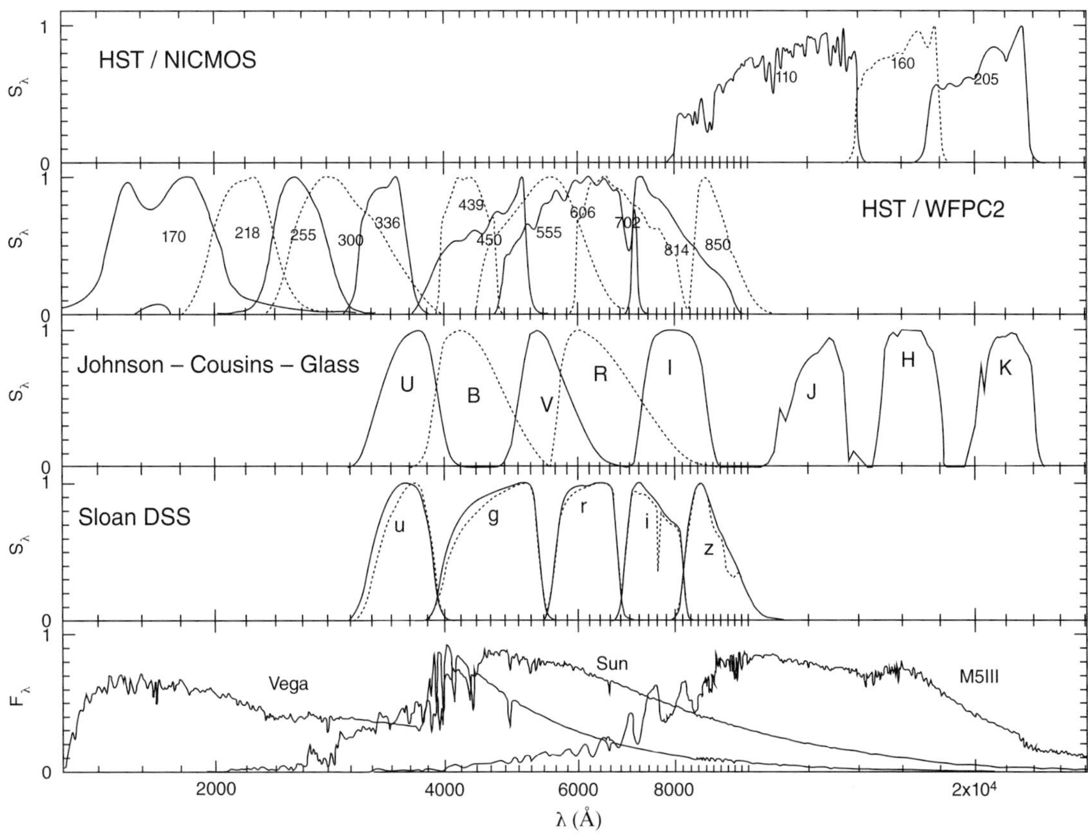


**Fig. C.1** Transmission curves for some of the most used astronomical filters from UV to NIR. *First* (*top*) and *second rows*. *HST* NIR and optical filters. *Third row*. Optical and NIR filters of the extended Johnson–Cousins system. *Fourth row*. SDSS optical filters. *Bottom row*. Spectra of Vega (A0-type star), the Sun (G2 type) and an M5-type star. Adapted from Girardi et al. (2002) and Girardi et al. (2004).

has zero magnitude in all filters. This definition is similar to that of the AB system (§C.3), but the flux is expressed per unit wavelength instead of frequency. In the NIR region, the system $JHK$ is an extension of the Johnson–Cousins system. The $K'$ and $K_s$ filters have a shorter cut-off wavelength than $K$ in order to mitigate the terrestrial thermal background radiation that begins to be relevant in this spectral region. At longer wavelengths, it is more common to use the central wavelength to identify the filter. For instance, the *Spitzer* bands are called 3.6 $\mu$m, 4.5 $\mu$m and so forth. For a review on standard photometric systems, see Bessell (2005).

All the above filters are **broad-band filters**. If we are interested in observing a specific emission line, for instance the H$\alpha$ line at 6563 Å, we should instead use the so-called **narrow-band filters**. The *HST* filters (see also Fig. 11.13) have intuitive names, for instance in F160W, 160 stands for 1.6 $\mu$m and W for *wide* filter (as opposed to M for medium or N for narrow). The same type of nomenclature is employed for the filters of the *JWST*.

## C.5 Stellar Spectral Types and Hertzsprung–Russell Diagram

Stars are classified in **spectral types** depending on the absorption lines present in their spectra. These types are indicated with the letters O, B, A, F, G, K and M, from the hottest to the coolest stars. Stars of types O and B are blue and have relatively few absorption lines. Typical elements are He (He II in O stars and He I in B stars) and ionised metals (e.g. C III). Hydrogen lines are also present. Stars of types A and F have strong H lines (A stars have the strongest Balmer lines) and lines of singly ionised metals. In G stars, the dominant features are Ca II lines, together with other ionised and neutral metals (e.g. Na I). K and M stars have mostly neutral metals (e.g. Ca I) and molecular bands, in particular CH and TiO bands. In Tab. C.4 we list the main properties of *main-sequence* stars of different spectral types. Each of these types is further divided into subtypes that are indicated with a number between 0 (most luminous) to 9 (dimmest).[2] For instance, the Vega star is of type A0 and the Sun is of type G2.

If we construct a diagram with spectral types versus absolute magnitudes, stars position themselves in specific regions in the shape of sequences and clumps. This is called a **Hertzsprung–Russell diagram** (Hertzsprung, 1911; Russell, 1914). An analogous diagram, which is very useful in practice for stars at the same distance (e.g. they belong to the same cluster or galaxy), is the colour–magnitude diagram (CMD), where one typically uses apparent magnitudes. A CMD has only photometric properties and it is more straightforward to build than the original Hertzsprung–Russell diagram although it contains the same type of information, as one can appreciate by looking at the correlations between the various quantities in Tab. C.4. A third version (more closely linked to theory) of the same type of diagram relates effective temperatures and luminosities. Fig. C.2 displays the CMD obtained using four million stars in the Milky Way observed with *Gaia*. Note that, in this case, the $y$-axis has absolute magnitudes, instead of apparent ones, because these stars are located at very different distances. These latter are however precisely measured

**Table C.4** Properties of main-sequence stars

| Type | $M_V$ (mag) | $B-V$ (mag) | $T_{\mathrm{eff}}$ (K) | Mass ($M_\odot$) | Lifetime (Myr) |
|---|---|---|---|---|---|
| O | $<-4.0$ | −0.32 | $3–6\times10^4$ | 10–100 | $\lesssim 10$ |
| B | −4.0 to 0 | −0.30 to −0.1 | $1–3\times10^4$ | few–10 | 10–few hundreds |
| A | 0.6–2 | −0.02 to 0.2 | 7300–9800 | $\approx 2$ | $\approx 10^3$ |
| F | 2.7–4 | 0.30–0.5 | 6000–7300 | 1–2 | a few $\times 10^3$ |
| G | 4.4–5 | 0.58–0.8 | 5300–5940 | $\approx 1$ | $\sim 10^4$ |
| K | 5.9–8 | 0.81–1.3 | 3900–5150 | 0.5–1 | $10^4–10^5$ |
| M | $\geq 8.8$ | $\geq 1.40$ | $\leq 3840$ | $<0.5$ | $\gtrsim 10^5$ |

Values from Cox (2000) and Pagel (2009).

[2] Note that the O0 type does not exist.

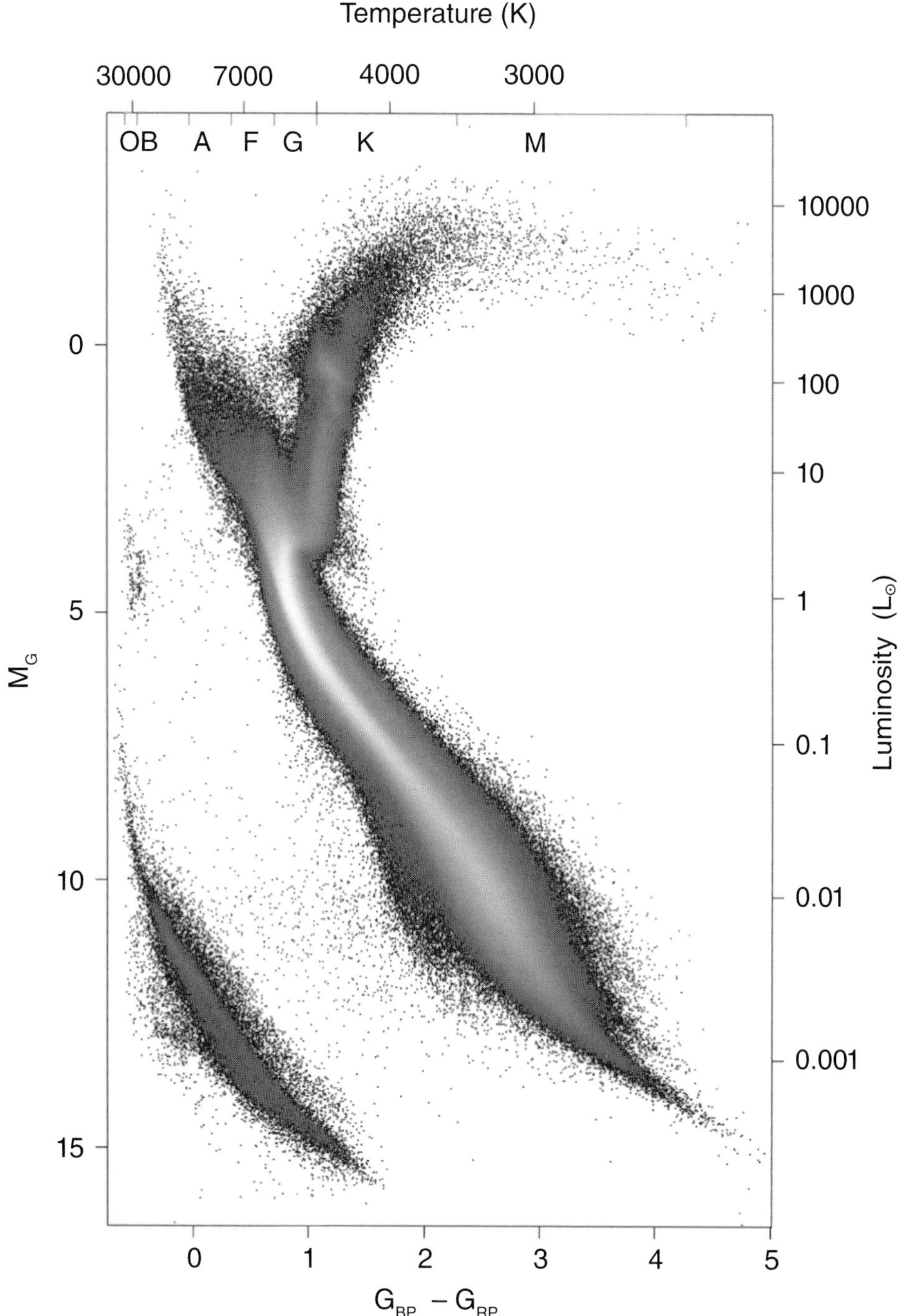


Fig. C.2

Diagram of colours versus absolute magnitudes of four million Galactic stars observed with *Gaia*. The stars have been selected to have low extinction, $E(B-V) < 0.015$ mag, and their distances are very precisely known from parallax measurements. The greyscale represents the square root of the density of stars: black is low and white is high density. Adapted from Gaia Collaboration (2018a). Figure courtesy of C. Babusiaux.

using parallaxes. The *Gaia G* passband is very broad and encompasses the whole optical region, while $G_{BP}$ and $G_{RP}$ are respectively its blue and red parts. Very roughly, $G_{BP}$ can be seen as a combination of Johnson $B$ and $V$, while $G_{RP}$ is similar to the sum of $R$ and $I$ (Fig. C.1).

The location of stars in the Hertzsprung–Russell diagram can be interpreted in detail using the theory of **stellar evolution**, for which we refer to specialised books like Salaris and Cassisi (2005). Fig. C.3 shows the evolutionary tracks of stars of different masses. After the evolutionary stages as a protostar described in §8.3.8, a star lands onto the **main sequence**, where the core thermonuclear reaction is the conversion of hydrogen into

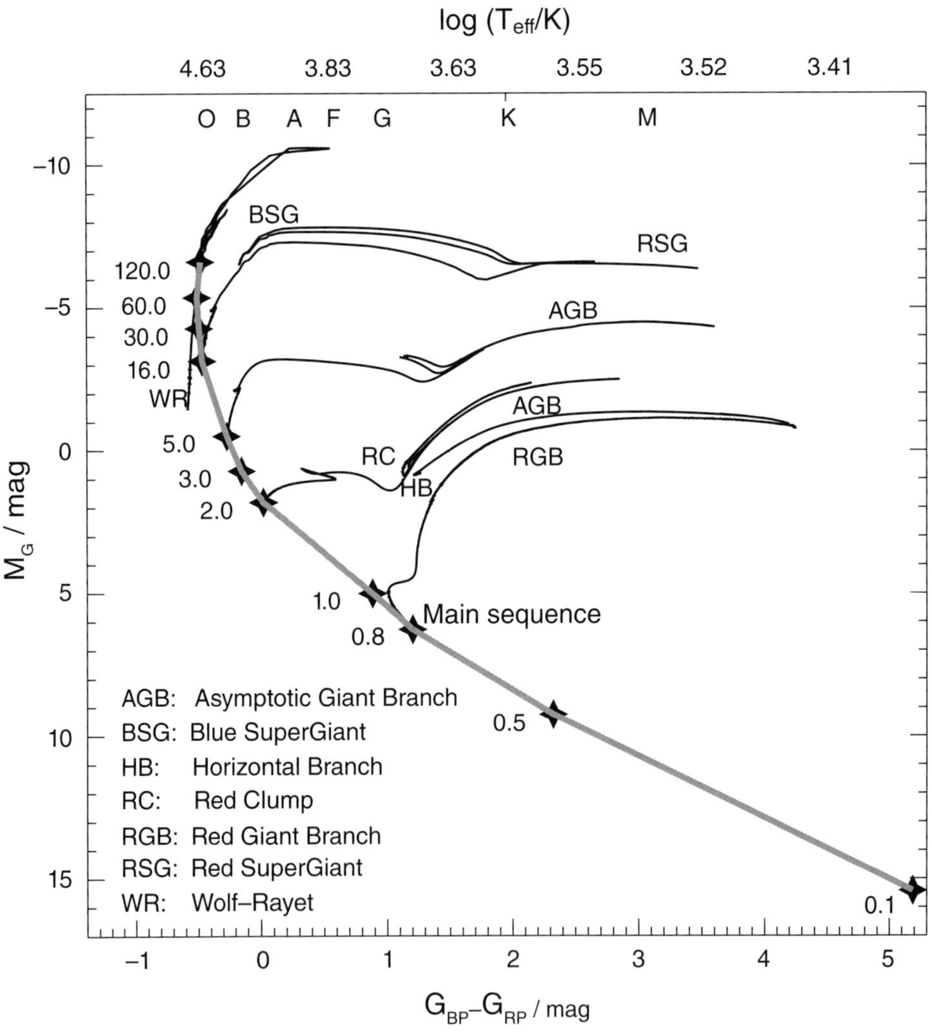


**Fig. C.3** Theoretical CMD with the zero-age main sequence and the evolution of stars with 0.8, 2.0, 5.0, 16.0 and 60 $\mathcal{M}_\odot$. The evolutionary tracks are computed with stellar evolution models from PARSEC (Bressan et al., 2012) with solar metallicity and in the *Gaia* bands. The main evolutionary phases are indicated. Courtesy of X. Fu.

helium. A star remains in the main sequence for most of its life (lifetimes in Tab. C.4). Once the hydrogen in the core has been converted into helium, this nuclear reaction proceeds in a shell while the inert helium core contracts. At this point, stars leave the main sequence and move to the **red giant branch** (RGB) as their radii expand by one or two orders of magnitude. This phase is also characterised by mass losses. When the temperature in the core is high enough, the burning of helium into carbon begins. This occurs suddenly (helium flash) for stars of $\mathcal{M} < 2.2\,\mathcal{M}_\odot$ and more slowly for higher masses. The stars move, at this point, to the **horizontal branch**. The subsequent evolution is determined by the mass of the star. High-mass ($\mathcal{M} > 8\,\mathcal{M}_\odot$) stars ignite nuclear fusions in the core of heavier and heavier elements. This fusion proceeds until the formation of iron, beyond which the reactions become endothermic and the nucleus collapses, producing a neutron star or a black hole, while the rest of the star explodes as a Type II SN (also called a **core-collapse supernova**). In stars with $\mathcal{M} < 8\,\mathcal{M}_\odot$, as the helium in the core is exhausted, carbon is typically not ignited while helium burns in a shell internal with respect to the hydrogen shell. Stars in this phase are located in the so-called **asymptotic giant branch** (AGB) that runs almost along the RGB. In the last AGB phase, the stars have periods of cessation and reignition of helium burning that produce pulsations (TP-AGB; §8.6.2). The core keeps contracting, eventually turning into a **white dwarf**, while the outer envelopes are ejected. Thus, stars of these masses suffer significant mass losses in this phase and release chemical elements into the ISM. They also produce the spectacular **planetary nebulae**. White dwarfs are much dimmer and less massive ($\mathcal{M} \approx 0.5{-}1\,\mathcal{M}_\odot$) than their original stars; they lie in the bottom left corner of Fig. C.2, parallel to the main sequence, on the so-called **cooling sequence**.

## C.6 Metallicity

The chemical elements heavier than helium are, in astrophysics, rather imprecisely referred to as **metals**.[3] Here below we see the various definitions that are used to describe the abundance of these elements.

### C.6.1 Definitions of Metallicity

Consider a gaseous system containing a mass of metals $\mathcal{M}_{\rm met}$ and a gas mass $\mathcal{M}_{\rm gas}$, then the **metallicity** is defined as

$$Z \equiv \frac{\mathcal{M}_{\rm met}}{\mathcal{M}_{\rm gas}}. \tag{C.27}$$

More generally, the metallicity of a star, a gas cloud or an entire galaxy is the fractional mass in metals with respect to the total mass in all chemical elements. The hydrogen and helium abundances are indicated with $X$ and $Y$, respectively. Clearly, the three quantities

[3] Note that, in other texts, metals are defined as elements from carbon upwards. The difference is only subtle given the typically negligible abundances of Li, Be and B.

must obey the relation $X + Y + Z = 1$. Our Sun has $Z_\odot \approx 0.013$–$0.014$: less than 2% of its mass is in metals (§C.6.2).

The second, most used, definition of metallicity is given as a ratio with respect to the solar abundance. Several elements maintain a relatively constant abundance ratio with respect to iron and thus the metallicity is defined as the logarithm of the ratio between the *number densities* of iron ($n_{\rm Fe}$) and hydrogen ($n_{\rm H}$) with respect to the solar value, and it is indicated with square brackets:

$$\left[\frac{\rm Fe}{\rm H}\right] = \log\left(\frac{\rm Fe}{\rm H}\right) - \log\left(\frac{\rm Fe}{\rm H}\right)_\odot, \tag{C.28}$$

where, on the r.h.s., Fe stands for $n_{\rm Fe}$ and H for $n_{\rm H}$. So, for instance, [Fe/H] = 0 corresponds to solar metallicity and [Fe/H] = −1 to 10% solar. If the isotopic composition of the Sun and the object under investigation are the same, then using number densities is equivalent to mass densities. Under the further assumption that the ratio between iron and other metals in the Sun and in the object are the same, this definition of metallicity becomes similar to that in eq. (C.27). However, eq. (C.28) refers to the abundance of hydrogen while eq. (C.27) refers to the total gas, including helium and metals. So, the two definitions become the closest if we also assume that the helium abundance in the Sun and in the object are the same. Given that the metals do not contribute substantially to the mass (§D.2.2), we can then say that, for instance, [Fe/H] = −1 corresponds to $Z \approx 0.1 Z_\odot$. In most situations, these assumptions are implicitly made and the two definitions are used interchangeably. By extension of the above definition, in this book we also employ the notation [M/H], intended as the logarithm of the ratio between the number densities of *all* metals and hydrogen with respect to the solar value.

A similar definition as eq. (C.28) can be applied to quantify the abundance of a given element relative to iron. A particularly relevant case is that of $\alpha$-elements (O, Ne, Mg, Si, S, Ar, Ca and Ti) that are released mostly by Type II SNe. The abundance of all $\alpha$-elements (or of a given one) relative to iron (always referred to the solar value) is written as

$$\left[\frac{\alpha}{\rm Fe}\right] = \log\left(\frac{\alpha}{\rm Fe}\right) - \log\left(\frac{\alpha}{\rm Fe}\right)_\odot, \tag{C.29}$$

where again the ratios are in number densities, e.g. $n_\alpha / n_{\rm Fe}$.

The abundance of single stars or entire galaxies can be derived from metal absorption lines in the stellar spectra, ideally Fe lines, although other lines (e.g. the Ca II triplet at $\lambda \approx 8500$ Å) have also been calibrated to give [Fe/H]. The direct measure of the metal abundance from stellar spectra involves the determination of the equivalent widths (eq. 3.6) of the lines and a careful modelling of the stellar photospheres. Metal abundances can also be estimated in the ISM, using absorption and emission lines. In the case of ionised gas, a common method is to use the oxygen forbidden emission lines (§11.1.4) to determine the abundance of oxygen with respect to hydrogen. Oxygen abundance is often expressed as

$$12 + \log\left(\frac{\rm O}{\rm H}\right). \tag{C.30}$$

The solar oxygen abundance is $12 + \log({\rm O/H}) \simeq 8.7$, which translates into an occurrence of oxygen versus hydrogen atoms in the Sun of ${\rm O/H} = 10^{(8.7-12)} \simeq 5 \times 10^{-4}$. This corresponds to an abundance in mass with respect to hydrogen $\mathcal{M}({\rm O})/\mathcal{M}({\rm H}) \simeq 16({\rm O/H}) \simeq 8 \times 10^{-3}$.

As the metallicity is strictly defined as the *abundance of iron*, it is not straightforwardly comparable with the oxygen abundance. If however we assume that the object under investigation has the same relative abundance between Fe and O as the Sun, we can convert between the two notations using

$$\left[\frac{\mathrm{Fe}}{\mathrm{H}}\right] \simeq 12 + \log\left(\frac{\mathrm{O}}{\mathrm{H}}\right) - 8.7. \quad \text{(C.31)}$$

## C.6.2 Abundance of the Sun

The chemical abundance of the Sun is the zero-point to define the abundance of any other astronomical object. It is important to specify that when we talk about the abundance of the Sun we are referring to the solar photosphere. Tab. C.5 reports the values for the most abundant chemical elements obtained by Asplund et al. (2009). Note that these recent determinations of $Z$ differ substantially from older classical values of $Z \simeq 0.02$.

**Table C.5** Abundance of chemical elements in the solar photosphere

| Element | Abundance[a] | Number density[b] |
|---|---|---|
| H | 12 | 1 |
| He | 10.93 ± 0.01 | $8.5 \times 10^{-2}$ |
| C | 8.43 ± 0.05 | $2.7 \times 10^{-4}$ |
| N | 7.83 ± 0.05 | $6.8 \times 10^{-5}$ |
| O | 8.69 ± 0.05 | $4.9 \times 10^{-4}$ |
| Ne | 7.93 ± 0.10 | $8.5 \times 10^{-5}$ |
| Na | 6.24 ± 0.04 | $1.7 \times 10^{-6}$ |
| Mg | 7.60 ± 0.04 | $4.0 \times 10^{-5}$ |
| Al | 6.45 ± 0.03 | $2.8 \times 10^{-6}$ |
| Si | 7.51 ± 0.03 | $3.2 \times 10^{-5}$ |
| S | 7.12 ± 0.03 | $1.3 \times 10^{-5}$ |
| Ar | 6.40 ± 0.13 | $2.5 \times 10^{-6}$ |
| Ca | 6.34 ± 0.04 | $2.2 \times 10^{-6}$ |
| Fe | 7.50 ± 0.04 | $3.2 \times 10^{-5}$ |
| Ni | 6.22 ± 0.04 | $1.7 \times 10^{-6}$ |
| Mass ratios | | |
| Hydrogen/total | $X$ | 0.7381 |
| Helium/total | $Y$ | 0.2485 |
| Metals/total | $Z$ | 0.0134 |
| Metals/hydrogen | $Z/X$ | 0.0181 |

[a] Given as log (X/H) + 12 where X and H are the number densities of element X and hydrogen, respectively.

[b] Normalised with respect to hydrogen. The abundance in mass can be calculated by multiplying these values by the average atomic mass of the various elements.

## C.7 Conversion Between Frequency/Wavelength and Velocity

The conversion between frequencies (or wavelengths) of a spectral feature (for instance, an emission line) and line-of-sight velocities is important to determine the relative velocity of a galaxy with respect to us and to study its internal kinematics. Unfortunately, the 'radio' and 'optical' communities tend to use two different conventions for this conversion. Note that, in both cases, these are approximations that only hold for velocities $v \ll c$. The optical definition is

$$v_{\rm opt} \equiv c\left(\frac{\nu_{\rm rest}-\nu_{\rm obs}}{\nu_{\rm obs}}\right) = c\left(\frac{\lambda_{\rm obs}-\lambda_{\rm rest}}{\lambda_{\rm rest}}\right), \tag{C.32}$$

where $\nu_{\rm rest}$ and $\lambda_{\rm rest}$ are the frequency and wavelength at rest, while $\nu_{\rm obs}$ and $\lambda_{\rm obs}$ are the observed frequency and wavelength, respectively. Note that the velocity in eq. (C.32) is meant *with respect to us observers* but it has this meaning only for very nearby galaxies (not in the Hubble flow; see §C.7.1) or other local objects. Radio astronomers, especially for the 21 cm H I emission line, tend to use

$$v_{\rm radio} \equiv c\left(\frac{\nu_{\rm rest}-\nu_{\rm obs}}{\nu_{\rm rest}}\right) = c\left(\frac{\lambda_{\rm obs}-\lambda_{\rm rest}}{\lambda_{\rm obs}}\right). \tag{C.33}$$

This notation is preferred for practical reasons, because equal increments in frequency (observed quantity) translate into equal increments in velocity as $\Delta v_{\rm radio} = -c\,\Delta\nu/\nu_{\rm rest}$, and this is handy to be used in observations (datacubes).

### C.7.1 Relative Velocities of Distant Sources

Consider that we aim to study the internal kinematics of a galaxy at relatively high redshift. It is important to realise that the recession velocity obtained with eq. (C.32) *does not* have the meaning of a relative velocity with respect to us, but is due to the Hubble flow (§2.1.2). One can therefore set the systemic velocity (§4.3.3) of the galaxy to zero and work with relative velocities. Suppose that we know the precise redshift corresponding to the systemic velocity of the source ($z_{\rm sys}$) due to the Hubble flow. Then we can define $\lambda_{\rm sys} \equiv (1+z_{\rm sys})\lambda_{\rm rest}$, with $\lambda_{\rm rest}$ the rest-frame wavelength of the (emission) line that we are detecting.

In different locations across the galaxy we measure $\lambda_{\rm obs}$, which may be slightly different from $\lambda_{\rm sys}$ because of the internal kinematics of the source (for instance, if we are observing an emission line from the ISM, this difference could be due to rotation or to an outflowing galactic wind). The difference $\Delta\lambda = \lambda_{\rm obs} - \lambda_{\rm sys}$ is related to the velocity difference by

$$\Delta v = c\frac{\Delta\lambda}{\lambda_{\rm sys}} = c\frac{\Delta\lambda}{(1+z_{\rm sys})\lambda_{\rm rest}} = -c\frac{\Delta\nu}{\nu_{\rm obs}}, \tag{C.34}$$

where $\Delta\nu = \nu_{\rm obs} - \nu_{\rm sys}$. A practical way to do this is to 'de-redshift' the spectrum (i.e. correct for the cosmological redshift by dividing all wavelengths by $1 + z_{\rm sys}$) and then use eq. (C.32) to calculate $\Delta v$ (using eq. C.33 would then make little difference provided that $\Delta v$ is small).

A similar situation occurs if we want to quantify the peculiar motion of a galaxy with respect to another, both being located at the same cosmological redshift. Let us call this redshift, in analogy to the above, $z_{\rm sys}$, although in this case it refers to the systemic redshift of the *galaxy pair*. The situation can be generalised to more than two galaxies if, for instance, we are observing a cluster of galaxies ($z_{\rm sys}$ being the redshift of the cluster as a whole). We call $z_{\rm pec} \simeq \Delta v/c$ the small redshift variation due to the peculiar motions of the galaxies, $\Delta v$ being their relative velocity. An equation analogous to eq. (C.34) using redshifts is

$$\Delta v = c\frac{\Delta z}{1 + z_{\rm sys}}, \tag{C.35}$$

where $\Delta z = z_{\rm obs} - z_{\rm sys}$. The observed redshift $z_{\rm obs} = \lambda_{\rm obs}/\lambda_{\rm rest} - 1$ is such that

$$1 + z_{\rm obs} = (1 + z_{\rm sys})(1 + z_{\rm pec}). \tag{C.36}$$

## C.8 Basic Properties of Collisionless Stellar Systems

Any gravitationally bound assembly of point masses (for instance stars or dark matter particles) is called a **stellar system**. A stellar system is collisionless if the two-body relaxation time, that is the characteristic timescale for energy exchange via two-body interactions, is much longer than the age of the system. The two-body relaxation times of galaxies typically exceed the Hubble time by orders of magnitude, so the stellar and dark matter components of a galaxy can be considered collisionless. The function that describes the distribution of particles (stars or dark matter) in the **phase space** $(\boldsymbol{x}, \boldsymbol{v})$, where $\boldsymbol{x}$ is the position and $\boldsymbol{v}$ is the velocity, is the **phase-space density** or **distribution function** $f(\boldsymbol{x}, \boldsymbol{v}, t)$, here defined so that $f\,{\rm d}^3\boldsymbol{x}\,{\rm d}^3\boldsymbol{v}$ is the mass of particles in the phase-space volume element ${\rm d}^3\boldsymbol{x}\,{\rm d}^3\boldsymbol{v}$ around $(\boldsymbol{x}, \boldsymbol{v})$ at time $t$. For a collisionless stellar system in a gravitational potential $\Phi$, $f$ obeys the **collisionless Boltzmann equation**

$$\frac{\partial f}{\partial t} + \frac{\partial f}{\partial \boldsymbol{x}} \cdot \boldsymbol{v} - \frac{\partial f}{\partial \boldsymbol{v}} \cdot \frac{\partial \Phi}{\partial \boldsymbol{x}} = 0, \tag{C.37}$$

which expresses the conservation of the phase-space density.

Under the assumption of equilibrium, each collisionless component of a galaxy (spheroid, disc or dark matter halo) can be described as a stationary stellar system, with time-independent distribution function $f(\boldsymbol{x}, \boldsymbol{v})$. We report here the definitions of a few quantities useful to describe stationary stellar systems.[4] The mass density distribution is

$$\rho(\boldsymbol{x}) = \int f(\boldsymbol{x}, \boldsymbol{v}){\rm d}^3\boldsymbol{v}. \tag{C.38}$$

[4] For in-depth descriptions of stellar systems, we refer the reader to Binney and Tremaine (2008) and Ciotti (in prep.).

The mean velocity at $\boldsymbol{x}$ is

$$\overline{\boldsymbol{v}}(\boldsymbol{x}) \equiv \frac{1}{\rho(\boldsymbol{x})} \int \boldsymbol{v} f(\boldsymbol{x}, \boldsymbol{v}) \mathrm{d}^3 \boldsymbol{v}. \tag{C.39}$$

The **velocity dispersion tensor** at $\boldsymbol{x}$ is

$$\sigma_{ij}^2(\boldsymbol{x}) \equiv \frac{1}{\rho(\boldsymbol{x})} \int (v_i - \overline{v}_i)(v_j - \overline{v}_j) f(\boldsymbol{x}, \boldsymbol{v}) \mathrm{d}^3 \boldsymbol{v}, \tag{C.40}$$

for $i = 1, 2, 3$ and $j = 1, 2, 3$, where $v_1$, $v_2$ and $v_3$ are the components of $\boldsymbol{v}$. The **random motion kinetic energy tensor** is

$$\Pi_{ij} = \int \rho(\boldsymbol{x}) \sigma_{ij}^2(\boldsymbol{x}) \mathrm{d}^3 \boldsymbol{x}. \tag{C.41}$$

The line-of-sight projected mass surface density is

$$\Sigma(\boldsymbol{R}) = \int \rho(\boldsymbol{x}) \mathrm{d}x_{\mathrm{los}}, \tag{C.42}$$

where $x_{\mathrm{los}} = \hat{\boldsymbol{n}} \cdot \boldsymbol{x}$ is the line-of-sight component of the position vector, $\boldsymbol{R} = \boldsymbol{x} - x_{\mathrm{los}} \hat{\boldsymbol{n}}$ is the projection of the position vector in the plane of the sky, and $\hat{\boldsymbol{n}}$ is a unit vector with direction along the line of sight. The **line-of-sight velocity distribution** (LOSVD) is

$$F(v_{\mathrm{los}}, \boldsymbol{R}) = \frac{1}{\Sigma(\boldsymbol{R})} \int f(\boldsymbol{x}, \boldsymbol{v}) \mathrm{d}x_{\mathrm{los}} \, \mathrm{d}^2 \boldsymbol{v}_{\boldsymbol{R}}, \tag{C.43}$$

where $v_{\mathrm{los}} = \hat{\boldsymbol{n}} \cdot \boldsymbol{v}$ and $\boldsymbol{v}_{\boldsymbol{R}} = \boldsymbol{v} - v_{\mathrm{los}} \hat{\boldsymbol{n}}$. The fundamental quantities characterising the LOSVD are its moments: the **mean line-of-sight velocity**

$$\overline{v}_{\mathrm{los}}(\boldsymbol{R}) = \int v_{\mathrm{los}} F(v_{\mathrm{los}}, \boldsymbol{R}) \mathrm{d}v_{\mathrm{los}} \tag{C.44}$$

and the **line-of-sight velocity dispersion** $\sigma_{\mathrm{los}}(\boldsymbol{R})$, defined by

$$\sigma_{\mathrm{los}}^2(\boldsymbol{R}) = \int (v_{\mathrm{los}} - \overline{v}_{\mathrm{los}})^2 F(v_{\mathrm{los}}, \boldsymbol{R}) \mathrm{d}v_{\mathrm{los}}. \tag{C.45}$$

In the particular case of a system with null streaming velocity ($\overline{\boldsymbol{v}} = 0$), the line-of-sight velocity dispersion is given by

$$\sigma_{\mathrm{los}}^2(\boldsymbol{R}) = \frac{1}{\Sigma(\boldsymbol{R})} \int \sum_{i,j=1}^{3} \hat{n}_i \sigma_{ij}^2(\boldsymbol{x}) \hat{n}_j \rho(\boldsymbol{x}) \mathrm{d}x_{\mathrm{los}}, \tag{C.46}$$

where $\hat{n}_1$, $\hat{n}_2$ and $\hat{n}_3$ are the components of $\hat{\boldsymbol{n}}$.

## C.9 Mapping Between Source and Images in Gravitational Lensing

We report here the equations describing, in the thin lens approximation, the mapping between source and images for a gravitational lensing system in the geometric configuration illustrated in the right panel of Fig. 5.11 (see Wambsganss, 1998, and Mollerach and Roulet, 2002, for the derivation of these equations). Let us consider a lens with mass surface density distribution $\Sigma(\boldsymbol{R})$ ($\boldsymbol{R}$ is a 2D vector in the lens plane). In the general case in which the lens is not circularly symmetric, the trajectory of the photon does not lie in a plane and the lens equation analogous to eq. (5.30) reads

$$\boldsymbol{\beta} = \boldsymbol{\theta} - \boldsymbol{\alpha}, \tag{C.47}$$

where $\boldsymbol{\beta}$ (source position angle), $\boldsymbol{\theta}$ (image position angle), $\boldsymbol{\alpha} = \hat{\boldsymbol{\alpha}} D_{\rm LS}/D_{\rm OS}$ (reduced deflection angle) and $\hat{\boldsymbol{\alpha}}$ (deflection angle) are now 2D vectors. For instance, the position of an image in the lens plane can be defined by $\boldsymbol{\theta} = (\theta_1, \theta_2)$ with $\theta_1 = x/D_{\rm OL}$ and $\theta_2 = y/D_{\rm OL}$, where $(x, y)$ is a Cartesian system of coordinates in the lens plane with origin in the lens centre (the angular diameter distances $D_{\rm OL}$, $D_{\rm LS}$ and $D_{\rm OS}$ are defined in Fig. 5.11, right). It can be shown that the deflection angle at $\boldsymbol{R} = D_{\rm OL}\boldsymbol{\theta}$ in the lens plane is

$$\hat{\boldsymbol{\alpha}}(\boldsymbol{R}) = \frac{2}{c^2} \nabla_{\boldsymbol{R}} \Phi_{\rm proj}(\boldsymbol{R}), \tag{C.48}$$

where $\nabla_{\boldsymbol{R}}$ is the spatial gradient in the lens plane and $\Phi_{\rm proj}$ is the projected gravitational potential (eq. 5.43). The lens equation (eq. C.47) can be rewritten in terms of the normalised projected gravitational potential (eq. 5.47) as

$$\boldsymbol{\beta} = \boldsymbol{\theta} - \nabla_{\boldsymbol{\theta}} \psi(\boldsymbol{\theta}), \tag{C.49}$$

where $\nabla_{\boldsymbol{\theta}}$ is the angular gradient in the lens plane.

From eq. (C.49) it follows that the mapping between the source and the images is given by

$$A \equiv \frac{\partial \boldsymbol{\beta}}{\partial \boldsymbol{\theta}} = \delta_{ij} - \psi_{ij} = \begin{pmatrix} 1 - \psi_{11} & -\psi_{12} \\ -\psi_{12} & 1 - \psi_{22} \end{pmatrix}, \tag{C.50}$$

where $\psi_{ij}$ are the second derivatives of $\psi$ with respect to $\boldsymbol{\theta}$ (eq. 5.46). The matrix $A$ (eq. C.50) can be rewritten as

$$A = \begin{pmatrix} 1 - \kappa - \gamma_1 & -\gamma_2 \\ -\gamma_2 & 1 - \kappa + \gamma_1 \end{pmatrix}, \tag{C.51}$$

in terms of the convergence $\kappa(\boldsymbol{\theta})$ and the shear $\gamma(\boldsymbol{\theta})$ defined in §5.3.4 (eqs. 5.45, 5.50 and 5.51).

# Appendix D Physics Compendium

## D.1 Radiative Processes

We give here a very brief account of the most relevant concepts concerning radiative processes that are used in this book. The interested reader is referred to more specialised texts (e.g. Rybicki and Lightman, 1986; Draine, 2011) for a more detailed treatment.

### D.1.1 Radiative Transfer and Optical Depth

Consider radiation that propagates through a medium before reaching the observer. The **radiative transfer equation** describes how the specific intensity $\mathcal{I}_\nu$ (eq. C.1) of the radiation evolves along the line of sight:

$$\frac{\mathrm{d}\mathcal{I}_\nu}{\mathrm{d}x_{\mathrm{los}}} = -\kappa_\nu \mathcal{I}_\nu + \mathcal{J}_\nu, \tag{D.1}$$

where $\mathrm{d}x_{\mathrm{los}}$ is an infinitesimal length along the path to the observer, while $\kappa_\nu$ and $\mathcal{J}_\nu$ are called the **absorption coefficient** (or **attenuation coefficient**) and **emission coefficient** of the medium, respectively. These two coefficients depend on the properties of the intervening medium and vary with frequency. If there is neither absorption nor emission, the r.h.s. of eq. (D.1) is null: the intensity does not change along the path. Note that eq. (D.1) neglects radiation scattered along the line of sight from other directions.

We use the absorption coefficient to define the **optical depth** along the line of sight,

$$\tau_\nu \equiv \int_{x_0}^{x_1} \kappa_\nu(x_{\mathrm{los}})\mathrm{d}x_{\mathrm{los}}, \tag{D.2}$$

where $x_0$ and $x_1$ are the extremities of the medium. If the medium covers the entire path length from us to a background object (e.g. the radiation comes from a star in the disc of our Galaxy and the medium is the ISM), then the integral is from the star to us. If we now divide eq. (D.1) by $\kappa_\nu$ and use $\mathrm{d}\tau_\nu = \kappa_\nu\,\mathrm{d}x_{\mathrm{los}}$, we obtain

$$\frac{\mathrm{d}\mathcal{I}_\nu}{\mathrm{d}\tau_\nu} + \mathcal{I}_\nu = S_\nu, \tag{D.3}$$

where $S_\nu \equiv \mathcal{J}_\nu/\kappa_\nu$ is called the **source function**. Eq. (D.3) is another form of the radiative transfer equation that can be solved to obtain

$$\mathcal{I}_\nu(\tau_\nu) = \mathcal{I}_\nu(0)e^{-\tau_\nu} + \int_0^{\tau_\nu} e^{\tau'-\tau_\nu} S_\nu(\tau')\mathrm{d}\tau', \tag{D.4}$$

where $\mathcal{I}_\nu(0)$ is the background (intrinsic) intensity that one would have in the absence of the intervening medium. The first term on the r.h.s. is thus this background radiation attenuated by a factor $e^{-\tau_\nu}$ (absorption of the medium), while the second term describes the emission and self-absorption from the medium itself. It is, indeed, the integral of the emission $S_\nu\, d\tau'$ attenuated by a factor $e^{\tau'-\tau_\nu}$, i.e. the attenuation from the point of emission over the path to the observer. A medium is called **optically thin** (little/no absorption) if $\tau_\nu \ll 1$ and **optically thick** (little/no radiation allowed through) if $\tau_\nu \gg 1$.

## D.1.2 Emission and Absorption Coefficients

The emission coefficient $\mathcal{J}_\nu$ is defined as the energy radiated per unit second, volume, frequency and solid angle. For isotropic emission we can integrate over the entire solid angle and obtain the (monochromatic) **emissivity**

$$j_\nu = 4\pi \mathcal{J}_\nu, \tag{D.5}$$

with units of erg $s^{-1}$ $Hz^{-1}$ $cm^{-3}$. If we further integrate over frequencies we obtain the **total emissivity** $j = \int j_\nu \, d\nu$.

The intensity of an object (eq. D.4) is determined by the emission $\mathcal{J}_\nu$ and absorption $\kappa_\nu$ coefficients. Here we see how these depend on physical properties like the particle density of the system. Consider a transition between upper ($k$) and lower ($i$) energy states that produces radiation at frequency $\nu_{ki}$ and has a probability $A_{ki}$ (Einstein coefficient for spontaneous emission). Its emission coefficient can be written

$$\mathcal{J}_\nu = \frac{1}{4\pi} n_k A_{ki} h\nu \phi_\nu, \tag{D.6}$$

where $n_k$ is the number density of atoms/molecules in the upper state and $\phi_\nu\, d\nu$ is the probability that the emitted photon has frequency between $\nu$ and $\nu + d\nu$ ($\int \phi_\nu \, d\nu = 1$).

Consider then the radiation going through a medium where atoms/molecules can absorb photons to jump from energy level $i$ to level $k$ with a cross section $\sigma_{ik}(\nu)$. The absorption coefficient is proportional to the *net* absorption and can be written as

$$\kappa_\nu = n_i \sigma_{ik}(\nu) - n_k \sigma_{ki}(\nu), \tag{D.7}$$

where $n_i$ and $n_k$ are, respectively, the number densities of atoms/molecules in the two levels, and the second term on the r.h.s. is the **stimulated emission** that can take place as a consequence of the incoming radiation.

## D.1.3 Black-Body Radiation

A medium in which the gas particles are in thermal equilibrium with the photons emits radiation that approaches that of a **black body**. The intensity of a black body is a function of its temperature $T$ called the **Planck function** and it is written

$$B_\nu(T) = \frac{2h\nu^3}{c^2} \frac{1}{\exp\left(\dfrac{h\nu}{k_B T}\right) - 1}, \tag{D.8}$$

where $h$ is the Planck constant. The Planck function can also be written using wavelengths as

$$B_\lambda(T) = \frac{2hc^2}{\lambda^5} \frac{1}{\exp\left(\frac{hc}{\lambda k_B T}\right) - 1}. \tag{D.9}$$

In the regime of low frequencies ($h\nu \ll k_B T$), typical of radio (long) wavelengths, one can use the **Rayleigh–Jeans approximation**:

$$B_\nu(T) = \frac{2k_B T \nu^2}{c^2} \quad \text{or} \quad B_\lambda(T) = \frac{2ck_B T}{\lambda^4}. \tag{D.10}$$

Eqs. (D.8) and (D.9) peak at

$$\nu_{\text{peak}} \simeq 5.879 \times 10^{10} \left(\frac{T}{\text{K}}\right) \text{Hz} \quad \text{and} \quad \lambda_{\text{peak}} \simeq 0.290 \left(\frac{T}{K}\right)^{-1} \text{cm}, \tag{D.11}$$

which are two forms of the so-called **Wien law**.

In realistic situations, the emission of a pure black body is attenuated by scattering and absorption; the system is thus called a **grey body**. The spectrum emitted by a grey body is $Q_\lambda B_\lambda(T)$, where $Q_\lambda$ is the efficiency factor for emission, i.e. the fraction of energy emitted at wavelength $\lambda$ relative to that emitted, at the same wavelength, by a black body (which has $Q_\lambda = 1$). For interstellar dust grains, $Q_\lambda \propto \lambda^{-\beta}$, where $\beta$ depends on the grain composition ($\beta \approx 1$ for amorphous material and $\beta \approx 2$ for metals and crystals).

### D.1.4 Boltzmann Law and Saha Equation

In a collisional system (ideal gas), due to encounters (elastic collisions), the speeds of the particles approach a **Maxwellian distribution**, which gives the probability that a particle has speed between $v$ and $v + \mathrm{d}v$ as

$$f(v)\mathrm{d}v = 4\pi \left(\frac{m}{2\pi k_B T}\right)^{3/2} v^2 \exp\left(-\frac{mv^2}{2k_B T}\right) \mathrm{d}v, \tag{D.12}$$

where $m$ is the mass of the particle and $T$ is the temperature of the gas. This distribution peaks at $\sqrt{2k_B T/m}$ and has a long tail at high speeds.

Consider now a gas in which collisions between particles also produce transitions to upper (excitation) and lower (de-excitation) energy levels. If the density of the gas is such that collisional transitions dominate over spontaneous emission or absorption ($n \gtrsim n_{\text{crit}}$, where $n_{\text{crit}}$ is the critical density; §4.2.3), then the relative populations of the energy levels approach the values that they would have in **thermodynamic equilibrium**. Under these conditions, the **Boltzmann law** is satisfied or nearly so. For an element X (atom or molecule) ionised $l$ times, the relative populations of two energy levels can be written as

$$\frac{n_k(\mathrm{X}^l)}{n_i(\mathrm{X}^l)} = \frac{g_k^l}{g_i^l} \exp\left(-\frac{\Delta E}{k_B T}\right), \tag{D.13}$$

where $g_k^l$ and $g_i^l$ are the statistical weights of the $k$ (upper) and $i$ (lower) levels of the element X ionised $l$ times and $\Delta E$ is the energy difference between the levels. At thermodynamic equilibrium, the intensity of radiation emitted by the gas is given by the Planck function (eq. D.8), i.e. $S_\nu = B_\nu(T)$.

In realistic situations, in the ISM of galaxies, the condition given by the Boltzmann law would typically hold only for *some* bound levels. The equilibrium is established in localised regions and at the *local* temperature of the medium. This condition is termed **local thermodynamic equilibrium**.

Consider then the balance between recombination and ionisation in thermodynamic equilibrium. The relative populations of an element X ionised $l$ times with respect to the same element ionised $l+1$ times are given by the **Saha equation**:

$$\frac{n(\mathrm{X}^{l+1})n_\mathrm{e}}{n(\mathrm{X}^l)} = \left(\frac{2\pi m_\mathrm{e} k_\mathrm{B} T}{h^2}\right)^{3/2} \frac{2g_1^{l+1}}{g_1^l} \exp\left(-\frac{\chi_l}{k_\mathrm{B}T}\right), \tag{D.14}$$

where $n_\mathrm{e}$ is the electron density, $g_1^{l+1}$ and $g_1^l$ are the statistical weights of atoms ionised $l+1$ and $l$ times, respectively, in their ground state, and $\chi_l$ is the energy needed to ionise the atom/ion $X^l$ from its ground state.

## D.1.5 Recombination Lines

Recombination lines are produced by an electron cascading down to lower energy levels from either a free state, if it has just been 'captured' by the atom (free–bound), or from an upper level (bound–bound). The wavelength of the transition between two energy levels ($k$ and $i$) for the hydrogen atom is given by the **Rydberg formula** that reads

$$\frac{1}{\lambda_{ki}} = R_\mathrm{H}\left(\frac{1}{n_i^2} - \frac{1}{n_k^2}\right), \tag{D.15}$$

where $R_\mathrm{H} \simeq 109677.58\ \mathrm{cm}^{-1}$ is the Rydberg constant, and $n_i$ and $n_k$ are two integer numbers with $n_k > n_i \geq 1$. Tab. D.1 gives the main hydrogen series.

If we consider hydrogen and assume that most atoms are ionised, the total emissivity (§D.1.2) from level $k$ to level $i$ is such that

$$j_\mathrm{rec} = 4\pi \int_0^\infty \mathcal{J}_{\nu,\mathrm{rec}}\, \mathrm{d}\nu = h\nu_{ki} n_\mathrm{e} n_\mathrm{p} \alpha_{ki}, \tag{D.16}$$

**Table D.1** Main hydrogen series

| Name | Transitions | Wavelengths |
|---|---|---|
| Lyman | $2\to1, 3\to1, ..., \infty\to1$ | Ly$\alpha \simeq 121.6$ nm, Ly$\beta \simeq 102.6$ nm, ..., lim $\simeq 91.2$ nm |
| Balmer | $3\to2, 4\to2, ..., \infty\to2$ | H$\alpha \simeq 656.3$ nm, H$\beta \simeq 486.1$ nm, ..., lim $\simeq 364.6$ nm |
| Paschen | $4\to3, 5\to3, ..., \infty\to3$ | Pa$\alpha \simeq 1.88\ \mu$m, Pa$\beta \simeq 1.28\ \mu$m, ..., lim $\simeq 820.4$ nm |
| Brackett | $5\to4, 6\to4, ..., \infty\to4$ | Br$\alpha \simeq 4.05\ \mu$m, Br$\beta \simeq 2.63\ \mu$m, ..., lim $\simeq 1.46\ \mu$m |

where

$$\alpha_{ki} \simeq 4.14 \times 10^{-16} b_k k^2 \left(\frac{A_{ki}}{\mathrm{s}^{-1}}\right)\left(\frac{T}{\mathrm{K}}\right)^{-3/2} \exp\left(-\frac{T_{\mathrm{ion}}}{k^2 T}\right) \mathrm{cm}^3\,\mathrm{s}^{-1} \tag{D.17}$$

is sometimes called the effective recombination coefficient, $n_\mathrm{e}$ and $n_\mathrm{p}$ are respectively the electron and proton densities, $A_{ki}$ is the Einstein coefficient, $T_\mathrm{ion} \simeq 1.58 \times 10^5$ K is the equivalent temperature of the hydrogen ionisation and $b_k$ is a correction factor of order unity that takes into account departures from thermodynamic equilibrium.

### D.1.6 Thermal Bremsstrahlung (Free–Free) Radiation

The continuous scattering between electrons and ions in a highly ionised plasma produces the deviation of the former and a consequent radiation due to their deceleration (Larmor radiation). This radiation is often referred to with the German word bremsstrahlung, which means braking radiation, or as free–free radiation, as the electrons remain free (they are not captured by the ions). The emission coefficient for thermal bremsstrahlung, i.e. from a thermal distribution of electrons and ions of charge $Z_\mathrm{i}e$, can be written as

$$\mathcal{J}_{\nu,\mathrm{ff}} \simeq 5.44 \times 10^{-41} g_\mathrm{ff} Z_\mathrm{i}^2 \left(\frac{n_\mathrm{e} n_\mathrm{i}}{\mathrm{cm}^{-6}}\right)\left(\frac{T}{10^4\,\mathrm{K}}\right)^{-1/2} \exp\left(-\frac{h\nu}{k_\mathrm{B} T}\right) \tag{D.18}$$
$$\mathrm{erg}\,\mathrm{s}^{-1}\,\mathrm{cm}^{-3}\mathrm{sr}^{-1}\mathrm{Hz}^{-1},$$

where $n_\mathrm{e}$ and $n_\mathrm{i}$ are the densities of electrons and ions, respectively, and $g_\mathrm{ff}(\nu, T, Z)$ is the dimensionless Gaunt factor for the free–free emission, which is of order unity and slowly varying with frequency and temperature. In particular, at radio frequencies, $g_\mathrm{ff} \propto \nu^{-0.1} T^{0.15}$. As a consequence, $\mathcal{J}_{\nu,\mathrm{ff}}$ depends very mildly on frequency: it decreases very slowly as a consequence of the Gaunt factor down to the exponential cut-off due to the shape of the Maxwellian distribution of electron speeds (§D.1.4).

We can integrate eq. (D.18) in frequency to obtain the total emissivity by bremsstrahlung in an optically thin thermal plasma,

$$4\pi \int_0^\infty \mathcal{J}_{\nu,\mathrm{ff}}\,\mathrm{d}\nu \simeq 1.42 \times 10^{-25} \overline{g_\mathrm{ff}} Z_\mathrm{i}^2 \left(\frac{n_\mathrm{e} n_\mathrm{i}}{\mathrm{cm}^{-6}}\right)\left(\frac{T}{10^4\,\mathrm{K}}\right)^{1/2} \mathrm{erg}\,\mathrm{s}^{-1}\,\mathrm{cm}^{-3}, \tag{D.19}$$

where the value of the average Gaunt factor $\overline{g_\mathrm{ff}}$ is about $1.34[(T/10^4\,\mathrm{K})/(Z_\mathrm{i}^2)]^{0.05}$ in the vicinity of $10^4$ K, rising to $\approx 1.3$–$1.4$ at $T = 10^6$ K and declining to $\approx 1.1$ at $T = 10^8$ K (for $Z_\mathrm{i} = 1$–$2$).

### D.1.7 Synchrotron Emission

Relativistic charged particles (electrons in particular) spiralling around magnetic field lines produce an emission called **synchrotron radiation**. The emission is due to the acceleration of the electrons and it is collimated (beamed) in a narrow cone along the direction of motion. Given the Lorentz factor of a relativistic electron $\gamma = (1 - v^2/c^2)^{-1/2}$, a population of electrons with $\gamma$ distributed in a power law $N(\gamma) \propto \gamma^{-\delta}$ (where $N(\gamma)\mathrm{d}\gamma$ is the number of particles with Lorentz factors between $\gamma$ and $\gamma + \mathrm{d}\gamma$) has an emission coefficient for synchrotron radiation

$$\mathcal{J}_{\nu,\mathrm{sync}} \propto B^{(\delta+1)/2}\nu^{-(\delta-1)/2}, \tag{D.20}$$

where $B$ is the strength of the magnetic field, assumed uniform. In an optically thin medium, this leads to an intensity with the same proportionalities $\mathcal{I}_{\nu,\mathrm{sync}} \propto B^{(\delta+1)/2}\nu^{-(\delta-1)/2}$. At relatively high radio frequencies, the typical synchrotron flux in our Galaxy is $F_\nu \propto \nu^{-\alpha}$, with $\alpha \approx 0.7$. This corresponds to an electron spectrum with a power-law index $\delta \approx 2.4$, roughly the energy spectrum of the cosmic-ray electrons measured from Earth.

At low frequencies, the synchrotron radiation is often self-absorbed: a synchrotron photon emitted by one electron is absorbed by another. The intensity in these optically thick conditions becomes $\mathcal{I}_{\nu,\mathrm{sync}} \propto B^{-1/2}\nu^{5/2}$, leading to a flux $F_\nu \propto \nu^{5/2}$, independent of $\delta$. In the middle range, the spectrum has a relatively flat slope. For ageing sources, the synchrotron spectrum becomes steeper ($\alpha > 0.7$ for a Milky Way-like index) at the highest frequencies due to the more rapid ageing of the most energetic electrons. Given eq. (D.20), the synchrotron flux is also proportional to the strength of the magnetic field of the source with a power 1.7 for $\delta = 2.4$.

## D.2 Hydrodynamics

In this section, we report a few very basic equations and concepts of astrophysical hydrodynamics and magnetohydrodynamics. The interested reader is referred to specialised books such as Shu (1992) and Landau and Lifshitz (1959) for details.

### D.2.1 Continuity and Euler Equations

Consider a portion of fluid of density $\rho$ contained in a volume $V$. If there are no sources or sinks within $V$, the variation of the mass in the volume must be equal to the flow of fluid through the surface $S$ that surrounds $V$:

$$\frac{\mathrm{d}}{\mathrm{d}t}\int_V \rho\,\mathrm{d}V = -\int_S (\rho\boldsymbol{u})\cdot \mathrm{d}\boldsymbol{S}, \tag{D.21}$$

where $\boldsymbol{u}$ is the fluid velocity and $\mathrm{d}\boldsymbol{S}$ is a surface element vector normal to $S$. On the l.h.s. of eq. (D.21) we can move the derivative inside the integral given that the volume does not change with time. On the r.h.s. we can make use of the divergence theorem, thus obtaining two volume integrals. This identity is obviously valid for every volume $V$; thus eliminating the volume integrations we obtain

$$\frac{\partial \rho}{\partial t} + \boldsymbol{\nabla}\cdot(\rho\boldsymbol{u}) = 0, \tag{D.22}$$

which is the **mass conservation equation**, also called the **continuity equation** of the fluid.

The second fundamental conservation equation is the **momentum conservation equation**, referred to also as the **Euler equation** or **force equation**. Here, we do not give the derivation, but the equation in Lagrangian form[1] is rather intuitive:

$$\rho\frac{\mathrm{D}\boldsymbol{u}}{\mathrm{D}t}=-\boldsymbol{\nabla}P-\rho\boldsymbol{\nabla}\Phi, \tag{D.24}$$

where $P$ is the pressure and $\Phi$ is the gravitational potential. Eq. (D.24) can be seen simply as Newton's second law of dynamics for the fluid. If we multiply this equation by the infinitesimal volume $\delta V$ of a fluid element, on the l.h.s. we obtain its mass $\rho\,\delta V$ times its acceleration, while the r.h.s. gives the forces acting on it. In Eulerian form, the Euler equation is

$$\frac{\partial\boldsymbol{u}}{\partial t}+\boldsymbol{u}\cdot\nabla\boldsymbol{u}=-\frac{1}{\rho}\boldsymbol{\nabla}P-\boldsymbol{\nabla}\Phi. \tag{D.25}$$

### D.2.2 Equation of State

The fact that a gaseous system is collisional allows us to use the **equation of state of ideal gases**, which we write in the form

$$\frac{P}{\rho}=\frac{k_{\mathrm{B}}T}{\mu m_{\mathrm{p}}}, \tag{D.26}$$

where $\mu$ is the **mean atomic** (or **molecular**) **weight** and $m_{\mathrm{p}}$ is the mass of the proton. The mean atomic weight is the average mass of the particles in the fluid in units of the proton mass. In a neutral gas of hydrogen, helium ($\approx 9\%$ of particles) and metals (small contribution) one has $\mu\simeq 1.3$, in a fully ionised gas $\mu\simeq 0.6$, and in a molecular cloud (mostly made of $H_2$) $\mu\approx 2.2$–$2.33$, varying only slightly for different chemical compositions. In general, for ideal gases, the pressure depends on two thermodynamic variables, so for instance

$$P=P(\rho,T) \qquad \text{or} \qquad P=P(\rho,s), \tag{D.27}$$

where $s$ is the specific entropy. In some situations, one assumes that the fluid state is such that the pressure can be considered a function of only density,

$$P=P(\rho), \tag{D.28}$$

in which case the fluid is called **barotropic**. Peculiar cases of barotropic systems are the **isothermal fluids** ($P\propto\rho$) and the **adiabatic fluids** ($P\propto\rho^{\gamma}$), with $\gamma=C_{\mathrm{P}}/C_{\mathrm{V}}$, where $C_{\mathrm{P}}$ and $C_{\mathrm{V}}$ are the specific heats at constant pressure and constant volume, respectively; $\gamma=5/3$ for a monoatomic gas.

[1] The **Lagrangian derivative** ($\mathrm{D}/\mathrm{D}t$), also called the total derivative, is the derivative (of any physical quantity) taken with respect to a fluid element as it follows the motion of the fluid, as opposed to the **Eulerian derivative** ($\partial/\partial t$), which is instead taken at a fixed position in space. The relation between the two can be written as

$$\frac{\mathrm{D}}{\mathrm{D}t}\equiv\frac{\partial}{\partial t}+\boldsymbol{u}\cdot\boldsymbol{\nabla}, \tag{D.23}$$

where the second term on the r.h.s. is called the **advection operator**.

### D.2.3 Energy Equation

In addition to the continuity (mass conservation) equation and the Euler equation (momentum conservation) one can write an equation of energy conservation for the fluid. This equation partially derives from the first law of thermodynamics and it can be written in different forms. One form is

$$\frac{\partial \mathcal{E}_{\rm tot}}{\partial t} + \boldsymbol{\nabla} \cdot [(\mathcal{E}_{\rm tot} + P)\boldsymbol{u}] = -\rho(C - \mathcal{H}) + \rho\frac{\partial \Phi}{\partial t} - \boldsymbol{\nabla} \cdot \boldsymbol{F}_{\rm cond}, \tag{D.29}$$

where $\mathcal{E}_{\rm tot} = \rho u^2/2 + \rho\Phi + (3/2)nk_{\rm B}T$ is the total energy density, $C$ and $\mathcal{H}$ are radiative cooling and heating rates (energy loss/gain per unit time and mass) and $\boldsymbol{F}_{\rm cond}$ is the flux of thermal conduction. Note that eq. (D.29) is only valid if $\Phi$ is an external gravitational potential and the self-gravity of the gas is neglected.

Eq. (D.29) can be interpreted as follows. A volume of fluid can lose/gain energy with time (first term of the l.h.s.) for a number of reasons. These are: (1) a flow of energy from and to the volume as a consequence of the motion of the fluid plus the work done per unit time on the fluid by the pressure forces (second term on the l.h.s.); (2) emission of radiation or being heated by incident radiation or energetic particles (first term on the r.h.s.); (3) time variation of the potential (second term on the r.h.s); and (4) thermal conduction with a fluid at a different temperature outside the volume (third term on the r.h.s).

### D.2.4 Shock Waves

A *small* pressure perturbation propagates in a gas at the **sound speed** ($c_{\rm s}$) that for isothermal ($T = $const) perturbations is

$$c_{\rm s} \equiv \left[\left(\frac{\partial P}{\partial \rho}\right)_T\right]^{1/2} = \sqrt{\frac{k_{\rm B}T}{\mu m_{\rm p}}}, \tag{D.30}$$

while for adiabatic ($s = $const) perturbations it is

$$c_{\rm s} \equiv \left[\left(\frac{\partial P}{\partial \rho}\right)_s\right]^{1/2} = \sqrt{\gamma\frac{k_{\rm B}T}{\mu m_{\rm p}}}, \tag{D.31}$$

where $\gamma$ is the adiabatic index. The sound speed can also be seen as the characteristic speed within the medium, and, for a fluid with particles of nearly the same mass, it is of the same order as their thermal speed (eq. 4.9).

We define the **Mach number** as

$$\mathcal{M} \equiv \frac{u}{c_{\rm s}}, \tag{D.32}$$

where $u$ is the speed of a portion of a fluid. Motions with $\mathcal{M} < 1$ are called **subsonic**, while for $\mathcal{M} > 1$ we have **supersonic** motions. A supersonic motion in a fluid creates a shock wave, which is a discontinuity in the thermodynamic variables ($\rho$, $T$, $P$) that propagates at a velocity ($v_{\rm sh}$) exceeding the sound speed.

A shock wave is an *irreversible transformation* of the fluid that produces a perturbed medium with properties fundamentally different from those of its unperturbed state. To treat the problem, it is convenient to work in the rest frame of the shock. On one side of the shock, there is the unperturbed medium characterised by density, temperature, pressure and velocity $\rho_0$, $T_0$, $P_0$ and $u_0$, respectively, and, on the other side, there is the medium perturbed by the passage of the shock with $\rho_1$, $T_1$, $P_1$ and $u_1$. Note that $u_0$ and $u_1$ are velocity components orthogonal to the shock, while other velocity components are not affected by the shock and can thus taken to be null without loss of generality. Given that we are in the shock rest frame, both $u_0$ and $u_1$ are intended *with respect to* the shock. Using the conservation equations in §D.2.1 and §D.2.3, assuming stationarity ($\partial/\partial t = 0$) and that the two media behave adiabatically (null terms on the r.h.s. of eq. D.29), one can derive the **Rankine–Hugoniot jump conditions** that regulate the passage from the unperturbed to the perturbed state:

$$\begin{cases} \rho_0 u_0 & = \rho_1 u_1, \\ \rho_0 u_0^2 + P_0 & = \rho_1 u_1^2 + P_1, \\ \frac{1}{2}u_0^2 + \frac{\gamma}{\gamma - 1}\frac{P_0}{\rho_0} & = \frac{1}{2}u_1^2 + \frac{\gamma}{\gamma - 1}\frac{P_1}{\rho_1}. \end{cases} \tag{D.33}$$

These equations represent, respectively from top to bottom, the conservations of mass, momentum and energy fluxes across the shock.

The Rankine–Hugoniot equations can be easily solved for the relevant quantities, leading to the following relations:

$$\rho_1 = 4\rho_0, \tag{D.34}$$

$$u_1' = \frac{3}{4}v_{\rm sh}, \tag{D.35}$$

$$P_1 = \frac{3}{4}\rho_0 v_{\rm sh}^2, \tag{D.36}$$

$$T_1 = \frac{3}{16}\frac{\mu m_{\rm p}}{k_{\rm B}}v_{\rm sh}^2, \tag{D.37}$$

where we have taken $\gamma = 5/3$ and the velocity in eq. (D.35) is now given in the external rest frame, where the shock moves at $v_{\rm sh}$. We have also assumed that the shock is strong ($\mathcal{M}_0 = |u_0/c_{\rm s,0}| \gg 1$, with $c_{\rm s,0}$ the sound speed of the unperturbed medium), which implies that $u_0 \approx -v_{\rm sh}$ and $|u_0'/v_{\rm sh}| \ll 1$. Eq. (D.37) is particularly important because it shows how the gas temperature can jump to very high values as it is proportional to the square of the shock speed.

All the above refers to shocks in which the perturbed medium does not radiate appreciably. If instead the conditions are such that the cooling time (eq. 8.3) of the medium is short (with respect to the dynamical time or the age of the system) the jump conditions modify greatly. A common approach is to consider the extreme case of a medium that radiates very efficiently the energy that it has acquired from the shock to the point that its temperature goes back to the original temperature of the unperturbed medium. This approximation is called an **isothermal shock** and has jump conditions similar to those in eq. (D.33), except for the last condition that simply reads $T_1 = T_0$. The solutions in this case are

$$\rho_1 = \mathcal{M}_0^2 \rho_0, \tag{D.38}$$
$$u_1' = v_{\rm sh}, \tag{D.39}$$
$$P_1 = \rho_0 v_{\rm sh}^2, \tag{D.40}$$

where again we have assumed $\mathcal{M}_0 \gg 1$ and $u_1'$ is in the external rest frame. Note that, unlike in the adiabatic case where it is limited to a factor of 4, the density of the perturbed medium can now reach very high values (eq. D.38).

## D.2.5 Viscous Fluids

The fluids considered in the previous sections are called **ideal fluids** where interactions between particles only contribute to produce a Maxwellian distribution of their speeds. However, in real fluids, interactions between particles can produce stresses (forces per unit area) that deform the flow and dissipate kinetic energy, a process akin to internal friction. In these circumstances, the motion of the fluid is no longer described by the Euler equation, but the so-called **viscous terms** need to be added. The analogue of the Euler equation can be written as

$$\frac{\partial \boldsymbol{u}}{\partial t} + \boldsymbol{u} \cdot \boldsymbol{\nabla} \boldsymbol{u} = -\frac{1}{\rho} \boldsymbol{\nabla} P - \boldsymbol{\nabla} \Phi + \frac{\eta}{\rho} \left[ \nabla^2 \boldsymbol{u} + \frac{1}{3} \boldsymbol{\nabla}(\boldsymbol{\nabla} \cdot \boldsymbol{u}) \right] + \frac{\zeta}{\rho} \boldsymbol{\nabla}(\boldsymbol{\nabla} \cdot \boldsymbol{u}), \tag{D.41}$$

where $\eta$ and $\zeta$ are the coefficients of shear viscosity and **bulk viscosity**, respectively. Eq. (D.41), called the **Navier–Stokes equation**, describes the motion of viscous (non-ideal) fluids. Note that the coefficient $\nu \equiv \eta/\rho$, called the **kinematic viscosity**, is often employed. For incompressible fluids ($\mathrm{D}\rho/\mathrm{D}t = 0$; §8.3.7), the two $\boldsymbol{\nabla} \cdot \boldsymbol{u}$ terms on the r.h.s. are null and the equation simplifies substantially.

## D.2.6 Magnetised Fluid

If a magnetic field ($\boldsymbol{B}$) is present in a fluid, one should use the equations of **magnetohydrodynamics**. Their derivation relies on the Maxwell equations and a series of assumptions that allow us to add magnetic terms to the above hydrodynamic equations. One key assumption is that we can treat the magnetised gas as a *one-component* fluid and we do not need to write separate equations for charged particles (that feel the magnetic field through the Lorentz force) and neutral particles (that do not). This is justified by the fact that, even in a largely neutral medium, there is always a residual fraction of ions and electrons. These charged particles will then operate through the drag force on the neutrals, causing the whole gas to be *locked* to the field lines. For typical ionisation fractions of the neutral ISM, $n_{\rm e} \sim 10^{-3}$–$10^{-1}$ and the drag force is efficient enough to assume that the Lorentz force acts on the gas as a whole and not only on the charged particles. This allows us to write the Euler equation for a magnetised (but inviscid) gas as

$$\rho \frac{\partial \boldsymbol{u}}{\partial t} + \rho \boldsymbol{u} \cdot \boldsymbol{\nabla} \boldsymbol{u} = -\boldsymbol{\nabla} P - \rho \boldsymbol{\nabla} \Phi - \frac{1}{8\pi} \boldsymbol{\nabla} B^2 + \frac{1}{4\pi} \boldsymbol{B} \cdot \boldsymbol{\nabla} \boldsymbol{B}, \tag{D.42}$$

where $P$, $\rho$ and $\boldsymbol{u}$ are the pressure, density and velocity of the *whole* fluid. The Lorentz force gives the last two terms on the r.h.s., which are, respectively, the gradient of the magnetic pressure ($B^2/(8\pi)$) and the **magnetic tension**. As for the thermal pressure, the magnetic pressure counteracts compressions, while the magnetic tension opposes deformation of the magnetic field lines.

A fundamental equation of magnetohydrodynamics is the following:

$$\frac{\partial \boldsymbol{B}}{\partial t} + \boldsymbol{\nabla} \times (\boldsymbol{B} \times \boldsymbol{u}) = \frac{c^2}{4\pi\sigma_e}\nabla^2 \boldsymbol{B}, \tag{D.43}$$

where $\sigma_e$ is the **electrical conductivity**. We do not give more details here but just mention that the conductivity enters Ohm's law written for a density of charged particles as

$$\boldsymbol{J} = \sigma_e \boldsymbol{E}, \tag{D.44}$$

where $\boldsymbol{J}$ is the current density and $\boldsymbol{E}$ is the electric field. Thus high conductivity means little resistance to the motion of charged particles. In general, the ISM is a nearly perfect conductor and thus the r.h.s. of eq. (D.43) can be considered null. It can be shown (e.g. Shu, 1992) that eq. (D.43) with null r.h.s. is equivalent to the conservation of the magnetic flux through a generic surface (eq. 8.77). This means that the magnetic field follows the motion of the fluid, e.g. it gets compressed and rarefied, conserving its flux: we say that it is 'frozen' in the fluid and the phenomenon is called flux freezing.

### D.2.7 Kolmogorov Turbulence

We give here a very brief account of the Kolmogorov theory of fully developed turbulence. Consider a turbulent ($\mathcal{R}e \gg \mathcal{R}e_{crit}$; eq. 8.80) and incompressible fluid. The Kolmogorov theory assumes that the turbulent entities can be envisioned as eddies (of very different sizes) and that there is a continuous transfer of energy between these eddies. This transfer occurs, as an **energy cascade**, from the largest eddies of size $L$ to the progressively smaller eddies of size $\lambda$ (so $L = \lambda_{max}$: the largest scale). This cascade is interrupted and turbulent energy is dissipated (turned into heat) when we reach eddies of size $\lambda_0$, called the **dissipation scale**. This dissipation is due to viscous forces (§D.2.5). Clearly, for the process to be stationary, the system must continuously feed fresh kinetic energy into the largest eddies at the same rate at which energy is dissipated on the $\lambda_0$ scale.

In order to translate the above into formulae, it is useful to consider a unit mass of fluid. Given that the medium is highly turbulent, energy is essentially only in the form of (disordered) kinetic energy. The *specific* kinetic energy ($\epsilon$) for the largest eddies can then be simply written as

$$\epsilon \approx V^2, \tag{D.45}$$

where $V$ is the turbulent speed on scales $L$. The typical timescale at these large scales is $\tau \sim L/V$ and the energy transfer per unit mass and time is

$$\dot{\epsilon} \approx \frac{V^3}{L}. \tag{D.46}$$

By assumption, the energy in large eddies must be transferred to smaller eddies and it can be dissipated only at the scale $\lambda_0$ (there is no dissipation at larger scales). Thus, the energy transfer rate per unit mass must be a constant and, at intermediate scales $\lambda$, with $\lambda_0 < \lambda < L$, we can also write

$$\dot{\epsilon} \approx \frac{v_\lambda^3}{\lambda}, \tag{D.47}$$

where we have indicated with $v_\lambda$ the typical turbulent speed over a spatial scale $\lambda$. Substituting the $\dot{\epsilon}$ in eq. (D.47) with eq. (D.46) we obtain

$$v_\lambda = V\left(\frac{\lambda}{L}\right)^{1/3}, \tag{D.48}$$

which is the **Kolmogorov–Obukhov law** (Kolmogorov, 1941; Obukhov, 1962) that states that the typical turbulent speed scales with the sizes of the eddies elevated to the 1/3 power.

It is customary to rewrite the expression in eq. (D.48) as a relation between energy and wavenumber $k = 2\pi/\lambda$:

$$\epsilon \approx v_\lambda^2 \propto \lambda^{2/3} \propto k^{-2/3}. \tag{D.49}$$

Note that, given the proportionality with $\lambda^{2/3}$, the large eddies contain most of the energy. Thus eq. (D.49) represents nearly all the energy stored in eddies with sizes equal to and smaller than $\lambda$, or equivalently, with wavenumbers $\geq k$. We can therefore write

$$\epsilon \equiv \int_\infty^k \mathsf{E}(k')\mathrm{d}k' \propto k^{-2/3}, \tag{D.50}$$

where $\mathsf{E}(k)$ is the energy power spectrum that, to satisfy eq. (D.50), must be $\mathsf{E}(k) \propto k^{-5/3}$ (see eq. 8.83).

# References


Abell, G. O. 1958. The distribution of rich clusters of galaxies. *ApJS*, **3**, 211.

Adibekyan, V. Z., Sousa, S. G., Santos, N. C., et al. 2012. Chemical abundances of 1111 FGK stars from the HARPS GTO planet search program. Galactic stellar populations and planets. *A&A*, **545**, A32.

Agertz, O., Teyssier, R., and Moore, B. 2009. Disc formation and the origin of clumpy galaxies at high redshift. *MNRAS*, **397**, L64–L68.

Alcorn, L. Y., Tran, K.-V., Glazebrook, K., et al. 2018. ZFIRE: 3D modeling of rotation, dispersion, and angular momentum of star-forming galaxies at $z \sim 2$. *ApJ*, **858**, 47.

Alexander, D. M., and Hickox, R. C. 2012. What drives the growth of black holes? *NewAR*, **56**, 93–121.

Allgood, B., Flores, R. A., Primack, J. R., et al. 2006. The shape of dark matter haloes: dependence on mass, redshift, radius and formation. *MNRAS*, **367**, 1781–1796.

Alves, J., Lombardi, M., and Lada, C. J. 2007. The mass function of dense molecular cores and the origin of the IMF. *A&A*, **462**, L17–L21.

Angrick, C., and Bartelmann, M. 2010. Triaxial collapse and virialisation of dark-matter haloes. *A&A*, **518**, A38.

Armillotta, L., Fraternali, F., Werk, J. K., et al. 2017. The survival of gas clouds in the circumgalactic medium of Milky Way-like galaxies. *MNRAS*, **470**, 114–125.

Asplund, M., Grevesse, N., Sauval, A. J., et al. 2009. The chemical composition of the Sun. *ARA&A*, **47**, 481–522.

Auger, M. W., Treu, T., Bolton, A. S., et al. 2010. The Sloan Lens ACS Survey. X. Stellar, dynamical, and total mass correlations of massive early-type galaxies. *ApJ*, **724**, 511–525.

Aumer, M., Binney, J., and Schönrich, R. 2016. The quiescent phase of galactic disc growth. *MNRAS*, **459**, 3326–3348.

Balbus, S. A. 1995. Thermal instability. Page 328 of: Ferrara, A., McKee, C. F., Heiles, C., et al. (eds), *The Physics of the Interstellar Medium and Intergalactic Medium*. Astronomical Society of the Pacific Conference Series, vol. 80.

Baldwin, J. A., Phillips, M. M., and Terlevich, R. 1981. Classification parameters for the emission-line spectra of extragalactic objects. *PASP*, **93**, 5–19.

Balogh, M. L., Baldry, I. K., Nichol, R., et al. 2004. The bimodal galaxy color distribution: dependence on luminosity and environment. *ApJ*, **615**, L101–L104.

Bardeen, J. M., Bond, J. R., Kaiser, N., et al. 1986. The statistics of peaks of Gaussian random fields. *ApJ*, **304**, 15–61.

Barkana, R., and Loeb, A. 2001. In the beginning: the first sources of light and the reionization of the universe. *Phys. Rep.*, **349**, 125–238.

Battaglia, G., Sollima, A., and Nipoti, C. 2015. The effect of tides on the Fornax dwarf spheroidal galaxy. *MNRAS*, **454**, 2401–2415.

Bauer, M., Pietsch, W., Trinchieri, G., et al. 2008. XMM-Newton observations of the diffuse X-ray emission in the starburst galaxy NGC 253. *A&A*, **489**, 1029–1046.

Becker, R. H., Fan, X., White, R. L., et al. 2001. Evidence for reionization at $z \sim 6$: detection of a Gunn–Peterson trough in a $z = 6.28$ quasar. *AJ*, **122**, 2850–2857.

Behroozi, P. S., Wechsler, R. H., and Conroy, C. 2013. The average star formation histories of galaxies in dark matter halos from $z = 0 - 8$. *ApJ*, **770**, 57.

Beifiori, A., Mendel, J. T., Chan, J. C. C., et al. 2017. The KMOS Cluster Survey (KCS). I. The fundamental plane and the formation ages of cluster galaxies at redshift $1.4 < z < 1.6$. *ApJ*, **846**, 120.

Belli, S., Newman, A. B., and Ellis, R. S. 2015. Stellar populations from spectroscopy of a large sample of quiescent galaxies at $z > 1$: measuring the contribution of progenitor bias to early size growth. *ApJ*, **799**, 206.

Bennett, C. L., Larson, D., Weiland, J. L., et al. 2013. Nine-year Wilkinson Microwave Anisotropy Probe (WMAP) observations: final maps and results. *ApJS*, **208**, 20.

Bensby, T., Feltzing, S., and Oey, M. S. 2014. Exploring the Milky Way stellar disk. A detailed elemental abundance study of 714 F and G dwarf stars in the solar neighbourhood. *A&A*, **562**, A71.

Bertin, G. 2014. *Dynamics of Galaxies*, 2nd Edition. Cambridge University Press.

Bessell, M. S. 2005. Standard photometric systems. *ARA&A*, **43**, 293–336.

Binney, J. 1977. The physics of dissipational galaxy formation. *ApJ*, **215**, 483–491.

Binney, J., and Merrifield, M. 1998. *Galactic Astronomy*. Princeton University Press.

Binney, J., and Tremaine, S. 2008. *Galactic Dynamics*, 2nd Edition. Princeton University Press.

Birkinshaw, M. 1999. The Sunyaev–Zel'dovich effect. *Phys. Rep.*, **310**, 97–195.

Birnboim, Y., and Dekel, A. 2003. Virial shocks in galactic haloes? *MNRAS*, **345**, 349–364.

Bland-Hawthorn, J., and Gerhard, O. 2016. The galaxy in context: structural, kinematic, and integrated properties. *ARA&A*, **54**, 529–596.

Blanton, M. R., Hogg, D. W., Bahcall, N. A., et al. 2003. The galaxy luminosity function and luminosity density at redshift $z = 0.1$. *ApJ*, **592**, 819–838.

Blanton, M. R., Bershady, M. A., Abolfathi, B., et al. 2017. Sloan Digital Sky Survey IV: Mapping the Milky Way, nearby galaxies, and the distant universe. *AJ*, **154**, 28.

Blumenthal, G. R., Faber, S. M., Flores, R., et al. 1986. Contraction of dark matter galactic halos due to baryonic infall. *ApJ*, **301**, 27–34.

Bode, P., Ostriker, J. P., and Turok, N. 2001. Halo formation in warm dark matter models. *ApJ*, **556**, 93–107.

Bond, J. R., Cole, S., Efstathiou, G., et al. 1991. Excursion set mass functions for hierarchical Gaussian fluctuations. *ApJ*, **379**, 440–460.

Bondi, H. 1952. On spherically symmetrical accretion. *MNRAS*, **112**, 195.

Bonnarel, F., Fernique, P., Bienaymé, O., et al. 2000. The ALADIN interactive sky atlas. A reference tool for identification of astronomical sources. *A&AS*, **143**, 33–40.

Bonnor, W. B. 1956. Boyle's law and gravitational instability. *MNRAS*, **116**, 351.

Boomsma, R., Oosterloo, T. A., Fraternali, F., et al. 2008. HI holes and high-velocity clouds in the spiral galaxy NGC 6946. *A&A*, **490**, 555–570.

Bouwens, R. J., Illingworth, G. D., Oesch, P. A., et al. 2015. UV luminosity functions at redshifts $z \sim 4$ to $z \sim 10$: 10,000 galaxies from HST legacy fields. *ApJ*, **803**, 34.

Bregman, J. N. 2007. The search for the missing baryons at low redshift. *ARA&A*, **45**, 221–259.

Bressan, A., Marigo, P., Girardi, L., et al. 2012. PARSEC: stellar tracks and isochrones with the PAdova and TRieste Stellar Evolution Code. *MNRAS*, **427**, 127–145.

Bromm, V. 2013. The first stars and galaxies – basic principles. Page 3 of: De Rossi, M. E., Pedrosa, S., and Pellizza, L. J. (eds), *From the First Structures to the Universe Today*. Asociacion Argentina de Astronomia Book Series, vol. 4.

Bromm, V., and Yoshida, N. 2011. The first galaxies. *ARA&A*, **49**, 373–407.

Bruzual, G., and Charlot, S. 2003. Stellar population synthesis at the resolution of 2003. *MNRAS*, **344**, 1000–1028.

Bryan, G. L., and Norman, M. L. 1998. Statistical properties of X-ray clusters: analytic and numerical comparisons. *ApJ*, **495**, 80–99.

Bullock, J. S., and Boylan-Kolchin, M. 2017. Small-scale challenges to the ΛCDM paradigm. *ARA&A*, **55**, 343–387.

Bullock, J. S., Dekel, A., Kolatt, T. S., et al. 2001. A universal angular momentum profile for galactic halos. *ApJ*, **555**, 240–257.

Calzetti, D. 2001. The dust opacity of star-forming galaxies. *PASP*, **113**, 1449–1485.

Calzetti, D. 2013. Star formation rate indicators. Page 419 of: Falcón-Barroso, J., and Knapen, J. H. (eds), *Secular Evolution of Galaxies*. Cambridge University Press.

Calzetti, D., Armus, L., Bohlin, R. C., et al. 2000. The dust content and opacity of actively star-forming galaxies. *ApJ*, **533**, 682–695.

Cappellari, M. 2016. Structure and kinematics of early-type galaxies from integral field spectroscopy. *ARA&A*, **54**, 597–665.

Cappellari, M., Emsellem, E., Krajnović, D., et al. 2011. The ATLAS$^{3D}$ project – I. A volume-limited sample of 260 nearby early-type galaxies: science goals and selection criteria. *MNRAS*, **413**, 813–836.

Carilli, C. L., Holdaway, M. A., Ho, P. T. P., et al. 1992. Discovery of a synchrotron-emitting halo around NGC 253. *ApJ*, **399**, L59–L62.

Carroll, S. M., Press, W. H., and Turner, E. L. 1992. The cosmological constant. *ARA&A*, **30**, 499–542.

Casey, C. M., Narayanan, D., and Cooray, A. 2014. Dusty star-forming galaxies at high redshift. *Phys. Rep.*, **541**, 45–161.

Cassata, P., Cimatti, A., Franceschini, A., et al. 2005. The evolution of the galaxy *B*-band rest-frame morphology to $z \sim 2$: new clues from the K20/GOODS sample. *MNRAS*, **357**, 903–917.

Catinella, B., Saintonge, A., Janowiecki, S., et al. 2018. xGASS: total cold gas scaling relations and molecular-to-atomic gas ratios of galaxies in the local Universe. *MNRAS*, **476**, 875–895.

Cavagnolo, K. W., Donahue, M., Voit, G. M., et al. 2009. Intracluster medium entropy profiles for a Chandra archival sample of galaxy clusters. *ApJS*, **182**, 12–32.

Cayrel, R., Depagne, E., Spite, M., et al. 2004. First stars V – Abundance patterns from C to Zn and supernova yields in the early Galaxy. *A&A*, **416**, 1117–1138.

Cebrián, M., and Trujillo, I. 2014. The effect of the environment on the stellar mass–size relationship for present-day galaxies. *MNRAS*, **444**, 682–699.

Chabrier, G. 2003. Galactic stellar and substellar initial mass function. *PASP*, **115**, 763–795.

Chabrier, G. 2005. The initial mass function: from Salpeter 1955 to 2005. Page 41 of: Corbelli, E., Palla, F., and Zinnecker, H. (eds), *The Initial Mass Function 50 Years Later*, Astrophysics and Space Science Library, vol. 327. Springer.

Chandrasekhar, S. 1943. Dynamical friction. I. General considerations: the coefficient of dynamical friction. *ApJ*, **97**, 255.

Churazov, E., Sunyaev, R., Forman, W., et al. 2002. Cooling flows as a calorimeter of active galactic nucleus mechanical power. *MNRAS*, **332**, 729–734.

Cimatti, A., Cassata, P., Pozzetti, L., et al. 2008. GMASS ultradeep spectroscopy of galaxies at $z \sim 2$. II. Superdense passive galaxies: how did they form and evolve? *A&A*, **482**, 21–42.

Ciotti, L. in prep. *Introduction to Stellar Dynamics*. Cambridge University Press.

Ciotti, L., and Bertin, G. 1999. Analytical properties of the $R^{1/m}$ law. *A&A*, **352**, 447–451.

Coil, A. L. 2013. The large-scale structure of the Universe. Page 387 of: Oswalt, T. D., and Keel, W. C. (eds), *Planets, Stars and Stellar Systems*, Vol. 6, *Extragalactic Astronomy and Cosmology*. Springer.

Cole, A. A., Skillman, E. D., Tolstoy, E., et al. 2007. Leo A: a late-blooming survivor of the epoch of reionization in the Local Group. *ApJ*, **659**, L17–L20.

Coles, P., and Lucchin, F. 2002. *Cosmology: The Origin and Evolution of Cosmic Structure,* 2nd Edition. Wiley-VCH.

Conroy, C. 2013. Modeling the panchromatic spectral energy distributions of galaxies. *ARA&A*, **51**, 393–455.

Conroy, C., Graves, G. J., and van Dokkum, P. G. 2014. Early-type galaxy archeology: ages, abundance ratios, and effective temperatures from full-spectrum fitting. *ApJ*, **780**, 33.

Conselice, C. J. 2003. The relationship between stellar light distributions of galaxies and their formation histories. *ApJS*, **147**, 1–28.

Cooper, A. P., D'Souza, R., Kauffmann, G., et al. 2013. Galactic accretion and the outer structure of galaxies in the CDM model. *MNRAS*, **434**, 3348–3367.

Correa, C. A., Wyithe, J. S. B., Schaye, J., et al. 2015. The accretion history of dark matter haloes – III. A physical model for the concentration–mass relation. *MNRAS*, **452**, 1217–1232.

Courteau, S., Dutton, A. A., van den Bosch, F. C., et al. 2007. Scaling relations of spiral galaxies. *ApJ*, **671**, 203–225.

Cox, A. N. 2000. *Allen's Astrophysical Quantities*. Springer.

Cox, T. J., and Loeb, A. 2008. The collision between the Milky Way and Andromeda. *MNRAS*, **386**, 461–474.

Crain, R. A., Schaye, J., Bower, R. G., et al. 2015. The EAGLE simulations of galaxy formation: calibration of subgrid physics and model variations. *MNRAS*, **450**, 1937–1961.

Cresci, G., Hicks, E. K. S., Genzel, R., et al. 2009. The SINS survey: modeling the dynamics of $z \sim 2$ galaxies and the high-$z$ Tully–Fisher relation. *ApJ*, **697**, 115–132.

Cucciati, O., Tresse, L., Ilbert, O., et al. 2012. The star formation rate density and dust attenuation evolution over 12 Gyr with the VVDS surveys. *A&A*, **539**, A31.

da Cunha, E., Charlot, S., and Elbaz, D. 2008. A simple model to interpret the ultraviolet, optical and infrared emission from galaxies. *MNRAS*, **388**, 1595–1617.

Daddi, E., Cimatti, A., Renzini, A., et al. 2004. A new photometric technique for the joint selection of star-forming and passive galaxies at $1.4 < z < 2.5$. *ApJ*, **617**, 746–764.

Daflon, S., and Cunha, K. 2004. Galactic metallicity gradients derived from a sample of OB stars. *ApJ*, **617**, 1115–1126.

Dahlen, T., Mobasher, B., Dickinson, M., et al. 2007. Evolution of the luminosity function, star formation rate, morphology, and size of star-forming galaxies selected at rest-frame 1500 and 2800 Å. *ApJ*, **654**, 172–185.

Davidzon, I., Ilbert, O., Laigle, C., et al. 2017. The COSMOS2015 galaxy stellar mass function. Thirteen billion years of stellar mass assembly in ten snapshots. *A&A*, **605**, A70.

Dawson, K. S., Schlegel, D. J., Ahn, C. P., et al. 2013. The baryon oscillation spectroscopic survey of SDSS-III. *AJ*, **145**, 10.

Deharveng, L., Schuller, F., Anderson, L. D., et al. 2010. A gallery of bubbles. The nature of the bubbles observed by Spitzer and what ATLASGAL tells us about the surrounding neutral material. *A&A*, **523**, A6.

Dekel, A., Zolotov, A., Tweed, D., et al. 2013. Toy models for galaxy formation versus simulations. *MNRAS*, **435**, 999–1019.

Delgado-Serrano, R., Hammer, F., Yang, Y. B., et al. 2010. How was the Hubble sequence 6 Gyr ago? *A&A*, **509**, A78.

Delvecchio, I., Gruppioni, C., Pozzi, F., et al. 2014. Tracing the cosmic growth of supermassive black holes to $z \sim 3$ with Herschel. *MNRAS*, **439**, 2736–2754.

de Blok, W. J. G., Walter, Fabian, Ferguson, Annette M. N., et al. 2018. A High-resolution Mosaic of the Neutral Hydrogen in the M81 Triplet. *ApJ*, **865**(1), 26.

de Vaucouleurs, G. 1948. Recherches sur les nebuleuses extragalactiques. *AnAp*, **11**, 247.

de Vaucouleurs, G. 1959. Classification and morphology of external galaxies. *Handb Physik*, **53**, 275.

Di Teodoro, E. M., Fraternali, F., and Miller, S. H. 2016. Flat rotation curves and low velocity dispersions in KMOS star-forming galaxies at $z \sim 1$. *A&A*, **594**, A77.

Djorgovski, S., and Davis, M. 1987. Fundamental properties of elliptical galaxies. *ApJ*, **313**, 59–68.

D'Onghia, E., Vogelsberger, M., and Hernquist, L. 2013. Self-perpetuating spiral arms in disk galaxies. *ApJ*, **766**, 34.

Draine, B. T. 2003. Interstellar dust grains. *ARA&A*, **41**, 241–289.

Draine, B. T. 2011. *Physics of the Interstellar and Intergalactic Medium*. Princeton University Press.

Dressler, A. 1980. Galaxy morphology in rich clusters – Implications for the formation and evolution of galaxies. *ApJ*, **236**, 351–365.

Dressler, A., Lynden-Bell, D., Burstein, D., et al. 1987. Spectroscopy and photometry of elliptical galaxies. I – A new distance estimator. *ApJ*, **313**, 42–58.

Driver, S. P., Wright, A. H., Andrews, S. K., et al. 2016. Galaxy And Mass Assembly (GAMA): panchromatic data release (far-UV–far-IR) and the low-$z$ energy budget. *MNRAS*, **455**, 3911–3942.

Driver, S. P., Andrews, S. K., da Cunha, E., et al. 2018. GAMA/G10-COSMOS/3D-HST: the $0 < z < 5$ cosmic star formation history, stellar-mass, and dust-mass densities. *MNRAS*, **475**, 2891–2935.

Dunlop, J. S., McLure, R. J., Biggs, A. D., et al. 2017. A deep ALMA image of the Hubble Ultra Deep Field. *MNRAS*, **466**, 861–883.

Dutton, A. A., and Macciò, A. V. 2014. Cold dark matter haloes in the Planck era: evolution of structural parameters for Einasto and NFW profiles. *MNRAS*, **441**, 3359–3374.

Dyson, J. E., and Williams, D. A. 1997. *The Physics of the Interstellar Medium*. Institute of Physics Publishing.

Ebert, R. 1955. Über die Verdichtung von H I-Gebieten. Mit 5 Textabbildungen. *ZAp*, **37**, 217.

Eddington, A. S. 1913. On a formula for correcting statistics for the effects of a known error of observation. *MNRAS*, **73**, 359–360.

Eddington, A. S. 1921. Das Strahlungsgleichgewicht der Sterne. *Z. Physik*, **7**, 351–397.

Eggen, O. J., Lynden-Bell, D., and Sandage, A. R. 1962. Evidence from the motions of old stars that the Galaxy collapsed. *ApJ*, **136**, 748.

Einasto, J. 1965. On the construction of a composite model for the galaxy and on the determination of the system of galactic parameters. *Trudy Astrofiz. Inst. Alma-Ata*, **5**, 87–100.

Einstein, A. 1916. Die Grundlage der allgemeinen Relativitätstheorie. *Ann. Physik*, **354**, 769–822.

Einstein, A., and de Sitter, W. 1932. On the relation between the expansion and the mean density of the Universe. *PNAS*, **18**, 213–214.

Elbaz, D., Daddi, E., Le Borgne, D., et al. 2007. The reversal of the star formation-density relation in the distant universe. *A&A*, **468**, 33–48.

Elbaz, D., Dickinson, M., Hwang, H. S., et al. 2011. GOODS-Herschel: an infrared main sequence for star-forming galaxies. *A&A*, **533**, A119.

Elmegreen, B. G., and Scalo, J. 2004. Interstellar turbulence I: Observations and processes. *ARA&A*, **42**, 211–273.

Emsellem, E., Cappellari, M., Krajnović, D., et al. 2011. The ATLAS$^{3D}$ project – III. A census of the stellar angular momentum within the effective radius of early-type galaxies: unveiling the distribution of fast and slow rotators. *MNRAS*, **414**, 888–912.

Faber, S. M., and Jackson, R. E. 1976. Velocity dispersions and mass-to-light ratios for elliptical galaxies. *ApJ*, **204**, 668–683.

Faber, S. M., Willmer, C. N. A., Wolf, C., et al. 2007. Galaxy luminosity functions to $z \sim 1$ from DEEP2 and COMBO-17: implications for red galaxy formation. *ApJ*, **665**, 265–294.

Fakhouri, O., Ma, C.-P., and Boylan-Kolchin, M. 2010. The merger rates and mass assembly histories of dark matter haloes in the two Millennium simulations. *MNRAS*, **406**, 2267–2278.

Fall, S. M. 1983. Galaxy formation – some comparisons between theory and observation. Pages 391–398 of: Athanassoula, E. (ed.), *Internal Kinematics and Dynamics of Galaxies*. IAU Symposium, vol. 100.

Fall, S. M., and Efstathiou, G. 1980. Formation and rotation of disc galaxies with haloes. *MNRAS*, **193**, 189–206.

Fan, X., Strauss, M. A., Becker, R. H., et al. 2006. Constraining the evolution of the ionizing background and the epoch of reionization with $z \sim 6$ quasars. II. A sample of 19 quasars. *AJ*, **132**, 117–136.

Fanaroff, B. L., and Riley, J. M. 1974. The morphology of extragalactic radio sources of high and low luminosity. *MNRAS*, **167**, 31P–36P.

Ferland, G. J., Porter, R. L., van Hoof, P. A. M., et al. 2013. The 2013 release of Cloudy. *RMxAA*, **49**, 137–163.

Field, G. B. 1958. Excitation of the hydrogen 21-cm line. *Proc. IRE*, **46**, 240–250.

Field, G. B. 1965. Thermal instability. *ApJ*, **142**, 531.

Field, G. B., Goldsmith, D. W., and Habing, H. J. 1969. Cosmic-ray heating of the interstellar gas. *ApJ*, **155**, L149.

Finkelstein, S. L. 2016. Observational searches for star-forming galaxies at $z > 6$. *PASA*, **33**, e037.

Fitzpatrick, E. L., and Massa, D. 2007. An analysis of the shapes of interstellar extinction curves. V. The IR- through-UV curve morphology. *ApJ*, **663**, 320–341.

Fox, A., and Davé, R. (eds). 2017. *Gas Accretion onto Galaxies*. Astrophysics and Space Science Library, vol. 430. Springer.

Francis, P. J., Hewett, P. C., Foltz, C. B., et al. 1991. A high signal-to-noise ratio composite quasar spectrum. *ApJ*, **373**, 465–470.

Fraternali, F., van Moorsel, G., Sancisi, R., et al. 2002. Deep H I survey of the spiral galaxy NGC 2403. *AJ*, **123**, 3124–3140.

Fraternali, F., Oosterloo, T., and Sancisi, R. 2004. Kinematics of the ionised gas in the spiral galaxy NGC 2403. *A&A*, **424**, 485–495.

Freeman, K. C. 1970. On the disks of spiral and S0 galaxies. *ApJ*, **160**, 811.

Fridman, A. M., Polyachenko, V. L., Aries, A. B., et al. 1984. *Physics of Gravitating Systems. I. Equilibrium and Stability.* Springer.

Friedmann, A. 1922. Über die Krümmung des Raumes. *Z. Physik*, **10**, 377–386.

Frinchaboy, P. M., Thompson, B., Jackson, K. M., et al. 2013. The Open Cluster Chemical Analysis and Mapping Survey: local galactic metallicity gradient with APOGEE using SDSS DR10. *ApJ*, **777**, L1.

Fukugita, M., and Peebles, P. J. E. 2004. The cosmic energy inventory. *ApJ*, **616**, 643–668.

Furlanetto, S. R., Lidz, A., Loeb, A., et al. 2009. Cosmology from the highly-redshifted 21 cm line. In: *Astro2010: The Astronomy and Astrophysics Decadal Survey*. ArXiv e-prints, arXiv:0902.3259.

Gaia Collaboration (Babusiaux, C., van Leeuwen, F., et al.). 2018a. Gaia data release 2: observational Hertzsprung–Russell diagrams. *A&A*, **616**, A10.

Gaia Collaboration (Brown, A. G. A., Vallenari, A., et al.). 2018b. Gaia data release 2. Summary of the contents and survey properties. *A&A*, **616**, A1.

Gallazzi, A., Charlot, S., Brinchmann, J., et al. 2006. Ages and metallicities of early-type galaxies in the Sloan Digital Sky Survey: new insight into the physical origin of the colour–magnitude and the $Mg_2$–$\sigma_V$ relations. *MNRAS*, **370**, 1106–1124.

Galli, D., and Palla, F. 2013. The dawn of chemistry. *ARA&A*, **51**, 163–206.

Gavazzi, R., Treu, T., Rhodes, J. D., et al. 2007. The Sloan Lens ACS Survey. IV. The mass density profile of early-type galaxies out to 100 effective radii. *ApJ*, **667**, 176–190.

Genovali, K., Lemasle, B., Bono, G., et al. 2014. On the fine structure of the Cepheid metallicity gradient in the Galactic thin disk. *A&A*, **566**, A37.

Gentile, G., Józsa, G. I. G., Serra, P., et al. 2013. HALOGAS: extraplanar gas in NGC 3198. *A&A*, **554**, A125.

Gerhard, O. E. 1993. Line-of-sight velocity profiles in spherical galaxies: breaking the degeneracy between anisotropy and mass. *MNRAS*, **265**, 213.

Girardi, L., Bertelli, G., Bressan, A., et al. 2002. Theoretical isochrones in several photometric systems. I. Johnson–Cousins–Glass, HST/WFPC2, HST/NICMOS, Washington, and ESO Imaging Survey filter sets. *A&A*, **391**, 195–212.

Girardi, L., Grebel, E. K., Odenkirchen, M., et al. 2004. Theoretical isochrones in several photometric systems. II. The Sloan Digital Sky Survey *ugriz* system. *A&A*, **422**, 205–215.

Gnedin, N. Y. 2000. Effect of reionization on structure formation in the universe. *ApJ*, **542**, 535–541.

Gobat, R., Daddi, E., Onodera, M., et al. 2011. A mature cluster with X-ray emission at $z = 2.07$. *A&A*, **526**, A133.

Gordon, K. D., Clayton, G. C., Misselt, K. A., et al. 2003. A quantitative comparison of the Small Magellanic Cloud, Large Magellanic Cloud, and Milky Way ultraviolet to near-infrared extinction curves. *ApJ*, **594**, 279–293.

Graham, A. W., Erwin, P., Trujillo, I., et al. 2003. A new empirical model for the structural analysis of early-type galaxies, and a critical review of the Nuker model. *AJ*, **125**, 2951–2963.

Grcevich, J., and Putman, M. E. 2009. H I in Local Group dwarf galaxies and stripping by the galactic halo. *ApJ*, **696**, 385–395.

Gruppioni, C., Pozzi, F., Rodighiero, G., et al. 2013. The Herschel PEP/HerMES luminosity function – I. Probing the evolution of PACS selected galaxies to $z \simeq 4$. *MNRAS*, **432**, 23–52.

Gunn, J. E., and Peterson, B. A. 1965. On the density of neutral hydrogen in intergalactic space. *ApJ*, **142**, 1633–1641.

Haardt, F., and Madau, P. 2012. Radiative transfer in a clumpy universe. IV. New synthesis models of the cosmic UV/X-ray background. *ApJ*, **746**, 125.

Harrison, C. M., Johnson, H. L., Swinbank, A. M., et al. 2017. The KMOS Redshift One Spectroscopic Survey (KROSS): rotational velocities and angular momentum of $z \approx 0.9$ galaxies. *MNRAS*, **467**, 1965–1983.

Harrison, E. R. 1970. Fluctuations at the threshold of classical cosmology. *Phys. Rev. D*, **1**, 2726–2730.

Hashimoto, T., Laporte, N., Mawatari, K., et al. 2018. The onset of star formation 250 million years after the Big Bang. *Nature*, **557**, 392–395.

Hayashi, C. 1961. Stellar evolution in early phases of gravitational contraction. *PASJ*, **13**, 450–452.

Heiles, C., and Troland, T. H. 2004. The Millennium Arecibo 21 centimeter absorption-line survey. III. Techniques for spectral polarization and results for Stokes V. *ApJS*, **151**, 271–297.

Heiter, U., Soubiran, C., Netopil, M., et al. 2014. On the metallicity of open clusters. II. Spectroscopy. *A&A*, **561**, A93.

Hendel, D., and Johnston, K. V. 2015. Tidal debris morphology and the orbits of satellite galaxies. *MNRAS*, **454**, 2472–2485.

Herpich, J., Tremaine, S., and Rix, H.-W. 2017. Galactic disc profiles and a universal angular momentum distribution from statistical physics. *MNRAS*, **467**, 5022–5032.

Hertzsprung, E. 1911. Über die Verwendung photographischer effecktiver Wellenlängen zur Bestimmung von Farbenäquivalenten. *Publikationen des Astrophysikalischen Observatoriums zu Potsdam*, **22**, A1–A40.1.

Hickson, P. 1982. Systematic properties of compact groups of galaxies. *ApJ*, **255**, 382–391.

Hill, R., Masui, K. W., and Scott, D. 2018. The spectrum of the Universe. *Appl. Spectrosc.*, **72**, 663–688.

Hirschmann, M., De Lucia, G., and Fontanot, F. 2016. Galaxy assembly, stellar feedback and metal enrichment: the view from the GAEA model. *MNRAS*, **461**, 1760–1785.

Hogg, D. W., Baldry, I. K., Blanton, M. R., et al. 2002. The K correction. ArXiv e-prints, arXiv:astro-ph/0210394.

Hoopes, C. G., Walterbos, R. A. M., and Greenwalt, B. E. 1996. Diffuse ionized gas in three Sculptor Group galaxies. *AJ*, **112**, 1429.

Hopkins, P. F., Bundy, K., Hernquist, L., et al. 2007. Observational evidence for the coevolution of galaxy mergers, quasars, and the blue/red galaxy transition. *ApJ*, **659**, 976–996.

Hopkins, P. F., Kereš, D., Oñorbe, J., et al. 2014. Galaxies on FIRE (Feedback In Realistic Environments): stellar feedback explains cosmologically inefficient star formation. *MNRAS*, **445**, 581–603.

Hu, E. M., Cowie, L. L., McMahon, R. G., et al. 2002. A redshift $z = 6.56$ galaxy behind the cluster Abell 370. *ApJ*, **568**, L75–L79.

Hubble, E. P. 1926. Extragalactic nebulae. *ApJ*, **64**, 321–369.

Hubble, E. 1929. A relation between distance and radial velocity among extra-galactic nebulae. *PNAS*, **15**, 168–173.

Hunt, L., Dayal, P., Magrini, L., et al. 2016. Coevolution of metallicity and star formation in galaxies to $z \simeq 3.7$ – I. A fundamental plane. *MNRAS*, **463**, 2002–2019.

Hyde, J. B., and Bernardi, M. 2009. Curvature in the scaling relations of early-type galaxies. *MNRAS*, **394**, 1978–1990.

Ilbert, O., Arnouts, S., Le Floc'h, E., et al. 2015. Evolution of the specific star formation rate function at $z < 1.4$: dissecting the mass–SFR plane in COSMOS and GOODS. *A&A*, **579**, A2.

Jablonka, P., North, P., Mashonkina, L., et al. 2015. The early days of the Sculptor dwarf spheroidal galaxy. *A&A*, **583**, A67.

Jeans, J. H. 1902. The stability of a spherical nebula. *Phil. Trans. R. Soc. A*, **199**, 1–53.

Jeans, J. H. 1915. On the theory of star-streaming and the structure of the universe. *MNRAS*, **76**, 70–84.

Jiang, F., and van den Bosch, F. C. 2016. Statistics of dark matter substructure – I. Model and universal fitting functions. *MNRAS*, **458**, 2848–2869.

Johansson, J., Thomas, D., and Maraston, C. 2012. Chemical element ratios of Sloan Digital Sky Survey early-type galaxies. *MNRAS*, **421**, 1908–1926.

Kamphuis, J., Sancisi, R., and van der Hulst, T. 1991. An H I superbubble in the spiral galaxy M 101. *A&A*, **244**, L29–L32.

Karakas, A. I. 2010. Updated stellar yields from asymptotic giant branch models. *MNRAS*, **403**, 1413–1425.

Karttunen, H., Kröger, P., Oja, H., et al. 2017. *Fundamental Astronomy*. Springer.

Kashino, D., More, S., Silverman, J. D., et al. 2017. The FMOS-COSMOS survey of star-forming galaxies at $z \sim 1.6$. V: Properties of dark matter halos containing H$\alpha$ emitting galaxies. *ApJ*, **843**, 138.

Kauffmann, G., Heckman, T. M., Tremonti, C., et al. 2003. The host galaxies of active galactic nuclei. *MNRAS*, **346**, 1055–1077.

Kelvin, L. S., Driver, S. P., Robotham, A. S. G., et al. 2014a. Galaxy And Mass Assembly (GAMA): stellar mass functions by Hubble type. *MNRAS*, **444**, 1647–1659.

Kelvin, L. S., Driver, S. P., Robotham, A. S. G., et al. 2014b. Galaxy And Mass Assembly (GAMA): *ugrizYJHK* Sérsic luminosity functions and the cosmic spectral energy distribution by Hubble type. *MNRAS*, **439**, 1245–1269.

Kenney, J. D. P., van Gorkom, J. H., and Vollmer, B. 2004. VLA H I Observations of Gas Stripping in the Virgo Cluster Spiral NGC 4522. *AJ*, **127**, 3361–3374.

Kennicutt, R. C. 1998. Star formation in galaxies along the Hubble sequence. *ARA&A*, **36**, 189–232.

Kennicutt, R. C., and Evans, N. J. 2012. Star formation in the Milky Way and nearby galaxies. *ARA&A*, **50**, 531–608.

Kewley, L. J., Dopita, M. A., Sutherland, R. S., et al. 2001. Theoretical modeling of starburst galaxies. *ApJ*, **556**, 121–140.

Kewley, L. J., Groves, B., Kauffmann, G., et al. 2006. The host galaxies and classification of active galactic nuclei. *MNRAS*, **372**, 961–976.

King, I. R. 1972. Density data and emission measure for a model of the Coma Cluster. *ApJ*, **174**, L123.

Kinney, A. L., Calzetti, D., Bohlin, R. C., et al. 1996. Template ultraviolet to near-infrared spectra of star-forming galaxies and their application to K-corrections. *ApJ*, **467**, 38.

Kolmogorov, A. 1941. The local structure of turbulence in incompressible viscous fluid for very large Reynolds' numbers. *Akad. Nauk SSSR Dokl.*, **30**, 301–305.

Koppelman, H., Helmi, A., and Veljanoski, J. 2018. One large blob and many streams frosting the nearby stellar halo in *Gaia* DR2. *ApJ*, **860**, L11.

Kormendy, J. 1977. Brightness distributions in compact and normal galaxies. II – Structure parameters of the spheroidal component. *ApJ*, **218**, 333–346.

Kormendy, J., and Bender, R. 2012. A revised parallel-sequence morphological classification of galaxies: structure and formation of S0 and spheroidal galaxies. *ApJS*, **198**, 2.

Kormendy, J., Fisher, D. B., Cornell, M. E., et al. 2009. Structure and formation of elliptical and spheroidal galaxies. *ApJS*, **182**, 216–309.

Kovetz, E. D., Viero, M. P., Lidz, A., et al. 2017. Line-intensity mapping: 2017 status report. ArXiv e-prints, arXiv:1709.09066.

Krajnović, D. 2011. Viewing galaxies in 3D. *Phys. World*, **24**(11), 26–30.

Kroupa, P. 2002. The initial mass function of stars: evidence for uniformity in variable systems. *Science*, **295**, 82–91.

Krumholz, M. R. 2014. The big problems in star formation: the star formation rate, stellar clustering, and the initial mass function. *Phys. Rep.*, **539**, 49–134.

Kurk, J., Cimatti, A., Daddi, E., et al. 2013. GMASS ultradeep spectroscopy of galaxies at $z \sim 2$. VII. Sample selection and spectroscopy. *A&A*, **549**, A63.

Lacey, C. G., Baugh, C. M., Frenk, C. S., et al. 2016. A unified multiwavelength model of galaxy formation. *MNRAS*, **462**, 3854–3911.

Landau, L. D., and Lifshitz, E. M. 1959. *Fluid Mechanics*. Pergamon Press.

Landoni, M., Falomo, R., Treves, A., et al. 2013. ESO Very Large Telescope optical spectroscopy of BL Lacertae objects. IV. New spectra and properties of the full sample. *AJ*, **145**, 114.

Larson, R. B. 1981. Turbulence and star formation in molecular clouds. *MNRAS*, **194**, 809–826.

Larson, R. B. 1998. Early star formation and the evolution of the stellar initial mass function in galaxies. *MNRAS*, **301**, 569–581.

Lauer, T. R., Postman, M., Strauss, M. A., et al. 2014. Brightest cluster galaxies at the present epoch. *ApJ*, **797**, 82.

Le Borgne, J.-F., Bruzual, G., Pelló, R., et al. 2003. STELIB: a library of stellar spectra at $R \sim 2000$. *A&A*, **402**, 433–442.

Leauthaud, A., Tinker, J., Bundy, K., et al. 2012. New constraints on the evolution of the stellar-to-dark matter connection: a combined analysis of galaxy–galaxy lensing, clustering, and stellar mass functions from $z = 0.2$ to $z = 1$. *ApJ*, **744**, 159.

Lehner, N., O'Meara, J. M., Howk, J. C., et al. 2016. The cosmic evolution of the metallicity distribution of ionized gas traced by Lyman limit systems. *ApJ*, **833**, 283.

Leitherer, C., Tremonti, C. A., Heckman, T. M., et al. 2011. An ultraviolet spectroscopic atlas of local starbursts and star-forming galaxies: the legacy of FOS and GHRS. *AJ*, **141**, 37.

Lelli, F., McGaugh, S. S., and Schombert, J. M. 2016a. SPARC: mass models for 175 disk galaxies with Spitzer photometry and accurate rotation curves. *AJ*, **152**, 157.

Lelli, F., McGaugh, S. S., and Schombert, J. M. 2016b. The small scatter of the baryonic Tully–Fisher relation. *ApJ*, **816**, L14.

Lemaître, G. 1927. Un Univers homogène de masse constante et de rayon croissant rendant compte de la vitesse radiale des nébuleuses extra-galactiques. *Ann. Soc. Sci. Bruxelles*, **47**, 49–59.

Limber, D. N. 1953. The analysis of counts of the extragalactic nebulae in terms of a fluctuating density field. *ApJ*, **117**, 134.

Lin, C. C., and Shu, F. H. 1964. On the spiral structure of disk galaxies. *ApJ*, **140**, 646.

Lin, C. C., and Shu, F. H. 1966. On the spiral structure of disk galaxies, II. Outline of a theory of density waves. *PNAS*, **55**, 229–234.

Lin, D. N. C., and Pringle, J. E. 1987. The formation of the exponential disk in spiral galaxies. *ApJ*, **320**, L87.

Lutz, D. 2014. Far-infrared surveys of galaxy evolution. *ARA&A*, **52**, 373–414.

Lynden-Bell, D. 1967. Statistical mechanics of violent relaxation in stellar systems. *MNRAS*, **136**, 101.

Mac Low, M.-M., and Klessen, R. S. 2004. Control of star formation by supersonic turbulence. *Rev. Mod. Phys.*, **76**, 125–194.

Mac Low, M.-M., McCray, R., and Norman, M. L. 1989. Superbubble blowout dynamics. *ApJ*, **337**, 141–154.

Madau, P., and Dickinson, M. 2014. Cosmic star-formation history. *ARA&A*, **52**, 415–486.

Magdis, G. E., Daddi, E., Béthermin, M., et al. 2012. The evolving interstellar medium of star-forming galaxies since $z = 2$ as probed by their infrared spectral energy distributions. *ApJ*, **760**, 6.

Magnelli, B., Elbaz, D., Chary, R. R., et al. 2011. Evolution of the dusty infrared luminosity function from $z = 0$ to $z = 2.3$ using observations from Spitzer. *A&A*, **528**, A35.

Magnelli, B., Popesso, P., Berta, S., et al. 2013. The deepest Herschel-PACS far-infrared survey: number counts and infrared luminosity functions from combined PEP/GOODS-H observations. *A&A*, **553**, A132.

Magorrian, J., Tremaine, S., Richstone, D., et al. 1998. The demography of massive dark objects in galaxy centers. *AJ*, **115**, 2285–2305.

Magrini, L., Randich, S., Zoccali, M., et al. 2010. Open clusters towards the Galactic centre: chemistry and dynamics. A VLT spectroscopic study of NGC 6192, NGC 6404, NGC 6583. *A&A*, **523**, A11.

Maiolino, R., and Mannucci, F. 2018. De Re Metallica: the cosmic chemical evolution of galaxies. *A&AR*, **27**, 187.

Majewski, S. R., Schiavon, R. P., Frinchaboy, P. M., et al. 2017. The Apache Point Observatory Galactic Evolution Experiment (APOGEE). *AJ*, **154**, 94.

Malmquist, K. G. 1922. On some relations in stellar statistics. *Medd. Lunds Astron. Observ. Serie I*, **100**, 1–52.

Marasco, A., Fraternali, F., van der Hulst, J. M., et al. 2017. Distribution and kinematics of atomic and molecular gas inside the solar circle. *A&A*, **607**, A106.

Maraston, C. 2005. Evolutionary population synthesis: models, analysis of the ingredients and application to high-$z$ galaxies. *MNRAS*, **362**, 799–825.

Marconi, A., Risaliti, G., Gilli, R., et al. 2004. Local supermassive black holes, relics of active galactic nuclei and the X-ray background. *MNRAS*, **351**, 169–185.

Martin, N. F., Ibata, R. A., McConnachie, A. W., et al. 2013. The PAndAS view of the Andromeda satellite system. I. A Bayesian search for dwarf galaxies using spatial and color–magnitude information. *ApJ*, **776**, 80.

Martínez-Delgado, D., Peñarrubia, J., Gabany, R. J., et al. 2008. The ghost of a dwarf galaxy: fossils of the hierarchical formation of the nearby spiral galaxy NGC 5907. *ApJ*, **689**, 184–193.

Martinez-Valpuesta, I., Shlosman, I., and Heller, C. 2006. Evolution of stellar bars in live axisymmetric halos: recurrent buckling and secular growth. *ApJ*, **637**, 214–226.

Mashian, N., Sturm, E., Sternberg, A., et al. 2015. High-*J* CO sleds in nearby infrared bright galaxies observed by Herschel/PACS. *ApJ*, **802**, 81.

Matteucci, F. 2012. *Chemical Evolution of Galaxies*. Springer.

McCarthy, P. J. 1993. High redshift radio galaxies. *ARA&A*, **31**, 639–688.

McCarthy, P. J. 2004. EROs and faint red galaxies. *ARA&A*, **42**, 477–515.

McClure-Griffiths, N. M., Dickey, J. M., Gaensler, B. M., et al. 2003. Loops, drips, and walls in the galactic chimney GSH 277+00+36. *ApJ*, **594**, 833–843.

McConnachie, A. W. 2012. The observed properties of dwarf galaxies in and around the Local Group. *AJ*, **144**, 4.

McCracken, H. J., Capak, P., Salvato, M., et al. 2010. The COSMOS-WIRCam near-infrared imaging survey. I. *BzK*-selected passive and star-forming galaxy candidates at $z < 1.4$. *ApJ*, **708**, 202–217.

McGaugh, S. S., and Schombert, J. M. 2015. Weighing galaxy disks with the baryonic Tully–Fisher relation. *ApJ*, **802**, 18.

McGaugh, S. S., Schombert, J. M., Bothun, G. D., et al. 2000. The baryonic Tully–Fisher relation. *ApJ*, **533**, L99–L102.

McKee, C. F., and Ostriker, J. P. 1977. A theory of the interstellar medium – three components regulated by supernova explosions in an inhomogeneous substrate. *ApJ*, **218**, 148–169.

McNamara, B. R., and Nulsen, P. E. J. 2007. Heating hot atmospheres with active galactic nuclei. *ARA&A*, **45**, 117–175.

Meiksin, A. A. 2009. The physics of the intergalactic medium. *Rev. Mod. Phys.*, **81**, 1405–1469.

Merloni, A., and Heinz, S. 2008. A synthesis model for AGN evolution: supermassive black holes growth and feedback modes. *MNRAS*, **388**, 1011–1030.

Mesinger, A., Greig, B., and Sobacchi, E. 2016. The evolution of 21 cm structure (EOS): public, large-scale simulations of Cosmic Dawn and reionization. *MNRAS*, **459**, 2342–2353.

Mie, G. 1908. Beiträge zur Optik trüber Medien, speziell kolloidaler Metallösungen. *Ann. Physik*, **330**, 377–445.

Milgrom, M. 1983. A modification of the Newtonian dynamics as a possible alternative to the hidden mass hypothesis. *ApJ*, **270**, 365–370.

Mo, H., van den Bosch, F. C., and White, S. 2010. *Galaxy Formation and Evolution*. Cambridge University Press.

Mohr, P. J., Newell, D. B., and Taylor, B. N. 2016. CODATA recommended values of the fundamental physical constants: 2014. *Rev. Mod. Phys.*, **88**(3), 035009.

Mollerach, S., and Roulet, E. 2002. *Gravitational Lensing and Microlensing*. World Scientific.

Monelli, M., Hidalgo, S. L., Stetson, P. B., et al. 2010. The ACS LCID Project. III. The star formation history of the Cetus dSph galaxy: a post-reionization fossil. *ApJ*, **720**, 1225–1245.

Morandi, A., and Limousin, M. 2012. Triaxiality, principal axis orientation and non-thermal pressure in Abell 383. *MNRAS*, **421**, 3147–3158.

Mortlock, A., Conselice, C. J., Hartley, W. G., et al. 2013. The redshift and mass dependence on the formation of the Hubble sequence at $z > 1$ from CANDELS/UDS. *MNRAS*, **433**, 1185–1201.

Mullaney, J. R., Daddi, E., Béthermin, M., et al. 2012. The hidden "AGN main sequence": evidence for a universal black hole accretion to star formation rate ratio since $z \sim 2$ producing an $M_{\rm BH}$–$M_*$ relation. *ApJ*, **753**, L30.

Mundy, C. J., Conselice, C. J., Duncan, K. J., et al. 2017. A consistent measure of the merger histories of massive galaxies using close-pair statistics – I. Major mergers at $z < 3.5$. *MNRAS*, **470**, 3507–3531.

Murray, N., Quataert, E., and Thompson, T. A. 2005. On the maximum luminosity of galaxies and their central black holes: feedback from momentum-driven winds. *ApJ*, **618**, 569–585.

Naab, T., and Ostriker, J. P. 2017. Theoretical challenges in galaxy formation. *ARA&A*, **55**, 59–109.

Navarro, J. F., Frenk, C. S., and White, S. D. M. 1996. The structure of cold dark matter halos. *ApJ*, **462**, 563.

Nelson, D., Genel, S., Pillepich, A., et al. 2016. Zooming in on accretion – I. The structure of halo gas. *MNRAS*, **460**, 2881–2904.

Newman, A. B., Treu, T., Ellis, R. S., et al. 2011. The dark matter distribution in A383: evidence for a shallow density cusp from improved lensing, stellar kinematic, and X-ray data. *ApJ*, **728**, L39.

Newman, A. B., Treu, T., Ellis, R. S., et al. 2013. The density profiles of massive, relaxed galaxy clusters. I. The total density over three decades in radius. *ApJ*, **765**, 24.

Nicastro, F., Kaastra, J., Krongold, Y., et al. 2018. Observations of the missing baryons in the warm–hot intergalactic medium. *Nature*, **558**, 406–409.

Nipoti, C., Londrillo, P., and Ciotti, L. 2006. Dissipationless collapse, weak homology and central cores of elliptical galaxies. *MNRAS*, **370**, 681–690.

Novak, G. S., Ostriker, J. P., and Ciotti, L. 2011. Feedback from central black holes in elliptical galaxies: two-dimensional models compared to one-dimensional models. *ApJ*, **737**, 26.

Nussbaumer, H., and Schmutz, W. 1984. The hydrogenic 2s–1s two-photon emission. *A&A*, **138**, 495.

Obukhov, A. M. 1962. Some specific features of atmospheric turbulence. *J. Geophys. Res.*, **67**, 3011.

Oesch, P. A., Bouwens, R. J., Illingworth, G. D., et al. 2018. The dearth of $z \sim 10$ galaxies in all HST legacy fields – the rapid evolution of the galaxy population in the first 500 Myr. *ApJ*, **855**, 105.

O'Meara, J. M., Tytler, D., Kirkman, D., et al. 2001. The deuterium to hydrogen abundance ratio toward a fourth QSO: HS 0105+1619. *ApJ*, **552**, 718–730.

Omukai, K. 2012. Do environmental conditions affect the dust-induced fragmentation in low-metallicity clouds? Effect of pre-ionization and far-ultraviolet/cosmic-ray fields. *PASJ*, **64**, 114.

Onodera, M., Carollo, C. M., Renzini, A., et al. 2015. The ages, metallicities, and element abundance ratios of massive quenched galaxies at $z \geq 1.6$. *ApJ*, **808**, 161.

Oser, L., Ostriker, J. P., Naab, T., et al. 2010. The two phases of galaxy formation. *ApJ*, **725**, 2312–2323.

Ostriker, J. P., and Peebles, P. J. E. 1973. A numerical study of the stability of flattened galaxies: or, can cold galaxies survive? *ApJ*, **186**, 467–480.

Ostriker, J. P., Choi, E., Ciotti, L., et al. 2010. Momentum driving: Which physical processes dominate active galactic nucleus feedback? *ApJ*, **722**, 642–652.

Overzier, R. A. 2016. The realm of the galaxy protoclusters. A review. *A&AR*, **24**, 14.

Padoan, P., and Nordlund, Å. 2002. The stellar initial mass function from turbulent fragmentation. *ApJ*, **576**, 870–879.

Padovani, P., Alexander, D. M., Assef, R. J., et al. 2017. Active galactic nuclei: What's in a name? *A&AR*, **25**, 2.

Pagel, B. E. J. 2009. *Nucleosynthesis and Chemical Evolution of Galaxies*. Cambridge University Press.

Pagel, B. E. J., and Patchett, B. E. 1975. Metal abundances in nearby stars and the chemical history of the solar neighborhood. *MNRAS*, **172**, 13–40.

Pannella, M., Carilli, C. L., Daddi, E., et al. 2009. Star formation and dust obscuration at $z \approx 2$: galaxies at the dawn of downsizing. *ApJ*, **698**, L116–L120.

Pascale, R., Posti, L., Nipoti, C., et al. 2018. Action-based dynamical models of dwarf spheroidal galaxies: application to Fornax. *MNRAS*, **480**, 927–946.

Peebles, P. J. E. 1968. Recombination of the primeval plasma. *ApJ*, **153**, 1.

Peebles, P. J. E. 1969. Origin of the angular momentum of galaxies. *ApJ*, **155**, 393.

Peebles, P. J. E. 2001. The galaxy and mass $N$-point correlation functions: a blast from the past. Page 201 of: Martínez, V. J., Trimble, V., and Pons-Bordería, M. J. (eds), *Historical Development of Modern Cosmology*. Astronomical Society of the Pacific Conference Series, vol. 252.

Pelliccia, D., Tresse, L., Epinat, B., et al. 2017. HR-COSMOS: kinematics of star-forming galaxies at $z \sim 0.9$. *A&A*, **599**, A25.

Penton, S. V., Stocke, J. T., and Shull, J. M. 2004. The local Ly$\alpha$ forest. IV. Space telescope imaging spectrograph G140M spectra and results on the distribution and baryon content of H I absorbers. *ApJS*, **152**, 29–62.

Pezzulli, G., and Fraternali, F. 2016. Accretion, radial flows and abundance gradients in spiral galaxies. *MNRAS*, **455**, 2308–2322.

Pillepich, A., Springel, V., Nelson, D., et al. 2018. Simulating galaxy formation with the IllustrisTNG model. *MNRAS*, **473**, 4077–4106.

Planck Collaboration (Aghanim, N., Akrami, Y., et al.) 2018. Planck 2018 results. VI. Cosmological parameters. ArXiv e-prints, arXiv:1807.06209.

Poggianti, B. M. 1997. K and evolutionary corrections from UV to IR. *A&AS*, **122**, 399–407.

Polletta, M., Tajer, M., Maraschi, L., et al. 2007. Spectral energy distributions of hard X-ray selected active galactic nuclei in the XMM-Newton medium deep survey. *ApJ*, **663**, 81–102.

Posti, L., Nipoti, C., Stiavelli, M., et al. 2014. The imprint of dark matter haloes on the size and velocity dispersion evolution of early-type galaxies. *MNRAS*, **440**, 610–623.

Posti, L., Fraternali, F., di Teodoro, E., et al. 2018. The angular momentum-mass relation: a fundamental law from dwarf irregulars to massive spirals. *A&A*, **612**, L6.

Power, C. 2013. Seeking observable imprints of small-scale structure on the properties of dark matter haloes. *PASA*, **30**, e053.

Press, W. H., and Schechter, P. 1974. Formation of galaxies and clusters of galaxies by self-similar gravitational condensation. *ApJ*, **187**, 425–438.

Prieto, M. A., Reunanen, J., Tristram, K. R. W., et al. 2010. The spectral energy distribution of the central parsecs of the nearest AGN. *MNRAS*, **402**, 724–744.

Pringle, J. E., and Rees, M. J. 1972. Accretion disc models for compact X-ray sources. *A&A*, **21**, 1.

Pritchard, J. R., and Loeb, A. 2012. 21 cm cosmology in the 21st century. *Rep. Prog. Phys.*, **75**(8), 086901.

Putman, M. E., Peek, J. E. G., and Joung, M. R. 2012. Gaseous galaxy halos. *ARA&A*, **50**, 491–529.

Reddy, N. A., and Steidel, C. C. 2009. A steep faint-end slope of the UV luminosity function at $z \sim 2$–3: implications for the global stellar mass density and star formation in low-mass halos. *ApJ*, **692**, 778–803.

Reed, S. L., McMahon, R. G., Martini, P., et al. 2017. Eight new luminous $z \geq 6$ quasars discovered via SED model fitting of VISTA, WISE and Dark Energy Survey Year 1 observations. *MNRAS*, **468**, 4702–4718.

Rees, M. J., and Ostriker, J. P. 1977. Cooling, dynamics and fragmentation of massive gas clouds – clues to the masses and radii of galaxies and clusters. *MNRAS*, **179**, 541–559.

Renzini, A. 2006. Stellar population diagnostics of elliptical galaxy formation. *ARA&A*, **44**, 141–192.

Riess, A. G., Macri, L. M., Hoffmann, S. L., et al. 2016. A 2.4% determination of the local value of the Hubble constant. *ApJ*, **826**, 56.

Rodríguez-Puebla, A., Behroozi, P., Primack, J., et al. 2016. Halo and subhalo demographics with Planck cosmological parameters: Bolshoi-Planck and MultiDark-Planck simulations. *MNRAS*, **462**, 893–916.

Romeo, A. B., and Falstad, N. 2013. A simple and accurate approximation for the Q stability parameter in multicomponent and realistically thick discs. *MNRAS*, **433**, 1389–1397.

Rood, R. T., Quireza, C., Bania, T. M., et al. 2007. The abundance gradient in galactic H II regions. Page 169 of: Vallenari, A., Tantalo, R., Portinari, L., et al. (eds), *From Stars to Galaxies: Building the Pieces to Build Up the Universe*. Astronomical Society of the Pacific Conference Series, vol. 374.

Russell, H. N. 1914. Relations between the spectra and other characteristics of the stars. II. Brightness and spectral class. *Nature*, **93**, 252–258.

Rybicki, G. B., and Lightman, A. P. 1986. *Radiative Processes in Astrophysics*. Wiley-VCH.

Ryden, B. 2017. *Introduction to Cosmology*, 2nd Edition. Cambridge University Press.

Saglia, R. P., Opitsch, M., Erwin, P., et al. 2016. The SINFONI black hole survey: the black hole fundamental plane revisited and the paths of (co)evolution of supermassive black holes and bulges. *ApJ*, **818**, 47.

Salaris, M., and Cassisi, S. 2005. *Evolution of Stars and Stellar Populations*. Wiley-VCH.

Salpeter, E. E. 1955. The luminosity function and stellar evolution. *ApJ*, **121**, 161.

Salpeter, E. E. 1964. Accretion of interstellar matter by massive objects. *ApJ*, **140**, 796–800.

Sancisi, R., Fraternali, F., Oosterloo, T., et al. 2008. Cold gas accretion in galaxies. *A&AR*, **15**, 189–223.

Sanders, D. B., Mazzarella, J. M., Kim, D.-C., et al. 2003. The IRAS revised bright galaxy sample. *AJ*, **126**, 1607–1664.

Sanders, J. S., and Fabian, A. C. 2006. Enrichment in the Centaurus cluster of galaxies. *MNRAS*, **371**, 1483–1496.

Sanders, R. L., Shapley, A. E., Kriek, M., et al. 2015. The MOSDEF survey: mass, metallicity, and star-formation rate at $z \sim 2.3$. *ApJ*, **799**, 138.

Sanders, R. L., Shapley, A. E., Kriek, M., et al. 2018. The MOSDEF survey: a stellar mass–SFR–metallicity relation exists at $z \sim 2.3$. *ApJ*, **858**, 99.

Scalo, J. M. 1986. The stellar initial mass function. *Fund. Cosmic Phys.*, **11**, 1–278.

Schaye, J., Crain, R. A., Bower, R. G., et al. 2015. The EAGLE project: simulating the evolution and assembly of galaxies and their environments. *MNRAS*, **446**, 521–554.

Schechter, P. 1976. An analytic expression for the luminosity function for galaxies. *ApJ*, **203**, 297–306.

Schenker, M. A., Robertson, B. E., Ellis, R. S., et al. 2013. The UV luminosity function of star-forming galaxies via dropout selection at redshifts $z \sim 7$ and 8 from the 2012 Ultra Deep Field campaign. *ApJ*, **768**, 196.

Schiminovich, D., Ilbert, O., Arnouts, S., et al. 2005. The GALEX-VVDS measurement of the evolution of the far-ultraviolet luminosity density and the cosmic star formation rate. *ApJ*, **619**, L47–L50.

Schmidt, M. 1959. The rate of star formation. *ApJ*, **129**, 243.

Schmidt, M. 1963. 3C 273: a star-like object with large red-shift. *Nature*, **197**, 1040.

Schneider, N., André, P., Könyves, V., et al. 2013. What determines the density structure of molecular clouds? A case study of Orion B with Herschel. *ApJ*, **766**, L17.

Schneider, P. 2015. *Extragalactic Astronomy and Cosmology: An Introduction*. Springer.

Schönrich, R., and McMillan, P. J. 2017. Understanding inverse metallicity gradients in galactic discs as a consequence of inside-out formation. *MNRAS*, **467**, 1154–1174.

Schwarzschild, K. 1916. Über das Gravitationsfeld eines Massenpunktes nach der Einsteinschen Theorie. *Sitzungsberichte der Königlich Preußischen Akademie der Wissenschaften (Berlin)*. Pages 189–196.

Schwarzschild, M. 1979. A numerical model for a triaxial stellar system in dynamical equilibrium. *ApJ*, **232**, 236–247.

Sedov, L. I. 1959. *Similarity and Dimensional Methods in Mechanics*. Academic Press.

Sellwood, J. A. 2013. Dynamics of disks and warps. Page 923 of: Oswalt, T. D., and Gilmore, G. (eds), *Stellar Systems and Galactic Structure*. Planets, Stars and Stellar Systems, vol. 5. Springer.

Sérsic, J. L. 1968. *Atlas de Galaxias Australes*. Observatorio Astronomico de Cordoba.

Seyfert, C. K. 1943. Nuclear emission in spiral nebulae. *ApJ*, **97**, 28.

Shakura, N. I., and Sunyaev, R. A. 1976. A theory of the instability of disk accretion on to black holes and the variability of binary X-ray sources, galactic nuclei and quasars. *MNRAS*, **175**, 613–632.

Shapley, A. E. 2011. Physical properties of galaxies from $z = 2$–4. *ARA&A*, **49**, 525–580.

Shen, S., Mo, H. J., White, S. D. M., et al. 2003. The size distribution of galaxies in the Sloan Digital Sky Survey. *MNRAS*, **343**, 978–994.

Sheth, R. K., and Tormen, G. 1999. Large-scale bias and the peak background split. *MNRAS*, **308**, 119–126.

Shu, F. H. 1992. *The Physics of Astrophysics*, Vol. II: *Gas Dynamics*. University Science Books.

Silk, J. 1977. On the fragmentation of cosmic gas clouds. I – The formation of galaxies and the first generation of stars. *ApJ*, **211**, 638–648.

Smith, B., Sigurdsson, S., and Abel, T. 2008. Metal cooling in simulations of cosmic structure formation. *MNRAS*, **385**, 1443–1454.

Soltan, A. 1982. Masses of quasars. *MNRAS*, **200**, 115–122.

Somerville, R. S., and Davé, R. 2015. Physical models of galaxy formation in a cosmological framework. *ARA&A*, **53**, 51–113.

Sonnenfeld, A., Treu, T., Gavazzi, R., et al. 2012. Evidence for dark matter contraction and a Salpeter initial mass function in a massive early-type galaxy. *ApJ*, **752**, 163.

Sorce, J. G., Courtois, H. M., Tully, R. B., et al. 2013. Calibration of the mid-infrared Tully–Fisher relation. *ApJ*, **765**, 94.

Sparke, L. S., and Gallagher, III, J. S. 2006. *Galaxies in the Universe*, 2nd Edition. Cambridge University Press.

Speagle, J. S., Steinhardt, C. L., Capak, P. L., et al. 2014. A highly consistent framework for the evolution of the star-forming "main sequence" from $z \sim 0$–6. *ApJS*, **214**, 15.

Spitzer, L. 1978. *Physical Processes in the Interstellar Medium*. Wiley-Interscience.

Springel, V., Wang, J., Vogelsberger, M., et al. 2008. The Aquarius Project: the subhaloes of galactic haloes. *MNRAS*, **391**, 1685–1711.

Springel, V., Pakmor, R., Pillepich, A., et al. 2018. First results from the IllustrisTNG simulations: matter and galaxy clustering. *MNRAS*, **475**, 676–698.

Stahler, S. W., and Palla, F. 2005. *The Formation of Stars*. Wiley-VCH.

Stark, D. P. 2016. Galaxies in the first billion years after the Big Bang. *ARA&A*, **54**, 761–803.

Starkenburg, E., Hill, V., Tolstoy, E., et al. 2013. The extremely low-metallicity tail of the Sculptor dwarf spheroidal galaxy. *A&A*, **549**, A88.

Stiavelli, M. 2009. *From First Light to Reionization: The End of the Dark Ages*. Wiley.

Strömgren, B. 1939. The physical state of interstellar hydrogen. *ApJ*, **89**, 526.

Sunyaev, R. A., and Chluba, J. 2009. Signals from the epoch of cosmological recombination (Karl Schwarzschild Award Lecture 2008). *AN*, **330**, 657.

Sunyaev, R. A., and Zeldovich, Y. B. 1970. The spectrum of primordial radiation, its distortions and their significance. *Comm. Astrophys. Space Phys.*, **2**, 66.

Sutherland, R. S., and Dopita, M. A. 1993. Cooling functions for low-density astrophysical plasmas. *ApJS*, **88**, 253–327.

Takeuchi, T. T., Yoshikawa, K., and Ishii, T. T. 2003. The luminosity function of IRAS point source catalog redshift survey galaxies. *ApJ*, **587**, L89–L92.

Talia, M., Cimatti, A., Mignoli, M., et al. 2014. Listening to galaxies tuning at $z \sim 2.5$–3.0: the first strikes of the Hubble fork. *A&A*, **562**, A113.

Tasca, L. A. M., Kneib, J.-P., Iovino, A., et al. 2009. The zCOSMOS redshift survey: the role of environment and stellar mass in shaping the rise of the morphology–density relation from $z \sim 1$. *A&A*, **503**, 379–398.

Taylor, G. 1950. The formation of a blast wave by a very intense explosion. I. Theoretical discussion. *Proc. R. Soc. A*, **201**, 159–174.

Telfer, R. C., Zheng, W., Kriss, G. A., et al. 2002. The rest-frame extreme-ultraviolet spectral properties of quasi-stellar objects. *ApJ*, **565**, 773–785.

Thomas, D., Maraston, C., Schawinski, K., et al. 2010. Environment and self-regulation in galaxy formation. *MNRAS*, **404**, 1775–1789.

Thompson, A. R., Moran, J. M., and Swenson, G. W., Jr. 2017. *Interferometry and Synthesis in Radio Astronomy,* 3rd Edition. Springer.

Tielens, A. G. G. M. 2005. *The Physics and Chemistry of the Interstellar Medium*. Springer.

Tiley, A. L., Stott, J. P., Swinbank, A. M., et al. 2016. The KMOS Redshift One Spectroscopic Survey (KROSS): the Tully–Fisher relation at $z \sim 1$. *MNRAS*, **460**, 103–129.

Tinsley, B. M. 1980. Evolution of the stars and gas in galaxies. *Fund. Cosmic Phys.*, **5**, 287–388.

Tolstoy, E., Hill, V., and Tosi, M. 2009. Star-formation histories, abundances, and kinematics of dwarf galaxies in the Local Group. *ARA&A*, **47**, 371–425.

Toomre, A. 1964. On the gravitational stability of a disk of stars. *ApJ*, **139**, 1217–1238.

Toomre, A. 1977. Mergers and some consequences. Page 401 of: Tinsley, B. M., Larson, R. B., and Gehret, D. C. (eds), *The Evolution of Galaxies and Stellar Populations*. Yale University Observatory.

Toomre, A. 1981. What amplifies the spirals. Pages 111–136 of: Fall, S. M., and Lynden-Bell, D. (eds), *The Structure and Evolution of Normal Galaxies*. Cambridge University Press.

Toomre, A., and Toomre, J. 1972. Galactic bridges and tails. *ApJ*, **178**, 623–666.

Treu, T. 2010. Strong lensing by galaxies. *ARA&A*, **48**, 87–125.

Tully, R. B., and Fisher, J. R. 1977. A new method of determining distances to galaxies. *A&A*, **54**, 661–673.

Tumlinson, J., Peeples, M. S., and Werk, J. K. 2017. The circumgalactic medium. *ARA&A*, **55**, 389–432.

Übler, H., Förster Schreiber, N. M., Genzel, R., et al. 2017. The evolution of the Tully–Fisher relation between $z \sim 2.3$ and $z \sim 0.9$ with KMOS$^{3D}$. *ApJ*, **842**, 121.

van Albada, T. S. 1982. Dissipationless galaxy formation and the $R$ to the 1/4-power law. *MNRAS*, **201**, 939–955.

van Albada, T. S., Bahcall, J. N., Begeman, K., et al. 1985. Distribution of dark matter in the spiral galaxy NGC 3198. *ApJ*, **295**, 305–313.

van den Hoek, L. B., and Groenewegen, M. A. T. 1997. New theoretical yields of intermediate mass stars. *A&AS*, **123**.

van der Hulst, J. M., van Albada, T. S., and Sancisi, R. 2001. The Westerbork HI Survey of Irregular and Spiral Galaxies, WHISP. Page 451 of: Hibbard, J. E., Rupen, M., and van Gorkom, J. H. (eds), *Gas and Galaxy Evolution*. Astronomical Society of the Pacific Conference Series, vol. 240.

van der Marel, R. P., and Franx, M. 1993. A new method for the identification of non-Gaussian line profiles in elliptical galaxies. *ApJ*, **407**, 525–539.

van der Wel, A., Franx, M., van Dokkum, P. G., et al. 2014. 3D-HST+CANDELS: the evolution of the galaxy size–mass distribution since $z=3$. *ApJ*, **788**, 28.

Vantyghem, A. N., McNamara, B. R., Russell, H. R., et al. 2014. Cycling of the powerful AGN in MS 0735.6+7421 and the duty cycle of radio AGN in clusters. *MNRAS*, **442**, 3192–3205.

Venn, K. A., Irwin, M., Shetrone, M. D., et al. 2004. Stellar chemical signatures and hierarchical galaxy formation. *AJ*, **128**, 1177–1195.

Villaescusa-Navarro, F., Genel, S., Castorina, E., et al. 2018. Ingredients for 21 cm intensity mapping. *ApJ*, **866**, 135.

Volonteri, M. 2012. The formation and evolution of massive black holes. *Science*, **337**, 544.

Wambsganss, J. 1998. Gravitational lensing in astronomy. *Living Rev. Relativ.*, **1**, 12.

Weaver, R., McCray, R., Castor, J., et al. 1977. Interstellar bubbles. II – Structure and evolution. *ApJ*, **218**, 377–395.

Wechsler, R. H., and Tinker, J. L. 2018. The connection between galaxies and their dark matter halos. *ARA&A*, **56**, 435–487.

Wegg, C., Gerhard, O., and Portail, M. 2015. The structure of the Milky Way's bar outside the bulge. *MNRAS*, **450**, 4050–4069.

Weinzirl, T., Jogee, S., Khochfar, S., et al. 2009. Bulge $n$ and $B/T$ in high-mass galaxies: constraints on the origin of bulges in hierarchical models. *ApJ*, **696**, 411–447.

White, S. D. M., and Rees, M. J. 1978. Core condensation in heavy halos – a two-stage theory for galaxy formation and clustering. *MNRAS*, **183**, 341–358.

Whittet, D. C. B. 1992. *Dust in the Galactic Environment*. Institute of Physics Publishing.

Willmer, C. N. A. 2018. The absolute magnitude of the Sun in several filters. *ApJS*, **236**, 47.

Woosley, S. E., and Weaver, T. A. 1995. The evolution and explosion of massive stars. II. Explosive hydrodynamics and nucleosynthesis. *ApJS*, **101**, 181.

Wouthuysen, S. A. 1952. On the excitation mechanism of the 21-cm (radio-frequency) interstellar hydrogen emission line. *AJ*, **57**, 31–32.

Wuyts, E., Wisnioski, E., Fossati, M., et al. 2016. The evolution of metallicity and metallicity gradients from $z=2.7$ to 0.6 with KMOS$^{3D}$. *ApJ*, **827**, 74.

Xue, Y.-J., and Wu, X.-P. 2000. The $L_X$–$T$, $L_X$–$\sigma$, and $\sigma$–T relations for groups and clusters of galaxies. *ApJ*, **538**, 65–71.

Yun, M. S., and Carilli, C. L. 2002. Radio-to-far-infrared spectral energy distribution and photometric redshifts for dusty starburst galaxies. *ApJ*, **568**, 88–98.

Zackrisson, E., Rydberg, C.-E., Schaerer, D., et al. 2011. The spectral evolution of the first galaxies. I. James Webb Space Telescope detection limits and color criteria for Population III galaxies. *ApJ*, **740**, 13.

Zahid, H. J., Kudritzki, R.-P., Conroy, C., et al. 2017. Stellar absorption line analysis of local star-forming galaxies: the relation between stellar mass, metallicity, dust attenuation, and star formation rate. *ApJ*, **847**, 18.

Zeldovich, Y. B. 1972. A hypothesis, unifying the structure and the entropy of the Universe. *MNRAS*, **160**, 1P.

Zeldovich, Y. B., and Novikov, I. D. 1964. The radiation of gravity waves by bodies moving in the field of a collapsing star. *Sov. Phys. Dokl.*, **9**, 246.

Zeldovich, Y. B., Kurt, V. G., and Sunyaev, R. A. 1968. Recombination of hydrogen in the hot model of the universe. *Zh. Eksp. Teor. Fiz.*, **55**, 278–286.

Zwicky, F. 1933. Die Rotverschiebung von extragalaktischen Nebeln. *Helv. Phys. Acta*, **6**, 110–127.

# Index